ELEMENTARY
DIFFERENTIAL EQUATIONS
WITH APPLICATIONS

2nd Edition

ELEMENTARY DIFFERENTIAL EQUATIONS WITH APPLICATIONS

C. H. Edwards, Jr.

David E. Penney

University of Georgia

PRENTICE HALL,
Englewood Cliffs, New Jersey 07632

Library of Congress Cataloging-in-Publication Data

Edwards, C. H. (Charles Henry), (date)
  Elementary differential equations with applications.

  Bibliography: p.
  Includes index.
  1. Differential equations.   I. Penney, David E.
II. Title.
QA371.E3 1989   515.3′5   88-23177
ISBN 0-13-254046-0

Editorial/production supervision: Maria McColligan
Interior and cover design: Meryl Poweski
Manufacturing buyer: Paula Massenaro
Cover photo: Courtesy Steuben Glass

© 1989, 1985 by Prentice-Hall, Inc.
A Division of Simon & Schuster
Englewood Cliffs, New Jersey 07632

Printed in the United States of America
10 9 8 7 6 5 4 3 2

ISBN 0-13-254046-0

Prentice-Hall International (UK) Limited, *London*
Prentice-Hall of Australia Pty. Limited, *Sydney*
Prentice-Hall Canada Inc., *Toronto*
Prentice-Hall Hispanoamericana, S.A., *Mexico*
Prentice-Hall of India Private Limited, *New Delhi*
Prentice-Hall of Japan, Inc., *Tokyo*
Simon & Schuster Asia Pte. Ltd., *Singapore*
Editora Prentice-Hall do Brasil, Ltda., *Rio de Janeiro*

To Alice and Carol

SUMMARY CONTENTS

Preface   xiii
1   Introduction and First Order Differential Equations   1
2   Linear Equations of Higher Order   106
3   Power Series Solutions of Linear Equations   211
4   The Laplace Transform   284
5   Linear Systems of Differential Equations   345
6   Numerical Methods   426
7   Qualitative Properties and Existence of Solutions   465
    References for Further Study   534
    Answers   536
    Index   559

CONTENTS

Preface   xiii

Chapter 1 Introduction and First Order Differential
           Equations   1

  1.1   Introduction   *2*
  1.2   Solution by Direct Integration   *10*
  1.3   Existence and Uniqueness of Solutions   *20*
  1.4   Separable Equations and Applications   *39*
  1.5   Linear First Order Equations   *52*
  1.6   Substitution Methods   *63*
  1.7   Exact Equations and Integrating Factors   *72*
 *1.8   Population Models   *81*
 *1.9   Motion with Variable Acceleration   *88*
        Chapter 1 Summary and a Look Ahead   *102*
        Review Problems   *105*

Chapter 2 Linear Equations of Higher Order   106

  2.1   Introduction   *107*
  2.2   General Solutions of Linear Equations   *116*
  2.3   Homogeneous Equations with Constant Coefficients   *126*
  2.4   Mechanical Vibrations   *137*
  2.5   Nonhomogeneous Equations and the Method of Undetermined
        Coefficients   *149*
  2.6   Reduction of Order and Euler-Cauchy Equations   *158*
  2.7   Variation of Parameters   *173*
 *2.8   Forced Oscillations and Resonance   *181*

*2.9   Electrical Circuits   *193*
*2.10 Endpoint Problems and Eigenvalues   *201*

Chapter 3   Power Series Solutions of Linear Equations   211

3.1   Introduction and Review of Power Series   *212*
3.2   Series Solutions Near Ordinary Points   *225*
3.3   Regular Singular Points   *234*
*3.4   Method of Frobenius—The Exceptional Cases   *248*
3.5   Bessel's Equation   *261*
*3.6   Applications of Bessel Functions   *271*
*3.7   Appendix on Infinite Series and the Atom   *277*

Chapter 4   The Laplace Transform   284

4.1   Laplace Transforms and Inverse Transforms   *285*
4.2   Transformation of Initial Value Problems   *296*
4.3   Translation and Partial Fractions   *305*
4.4   Derivatives, Integrals, and Products of Transforms   *314*
*4.5   Periodic and Piecewise Continuous Forcing Functions   *322*
*4.6   Impulses and Delta Functions   *333*
Table of Laplace Transforms   *344*

Chapter 5   Linear Systems of Differential Equations   345

5.1   Introduction to Systems   *346*
5.2   The Method of Elimination   *357*
5.3   Linear Systems and Matrices   *366*
*5.4   Mechanical Applications of Linear Systems   *384*
5.5   The Eigenvalue Method for Homogeneous Systems   *393*
5.6   Nonhomogeneous Linear Systems   *408*
*5.7   Matrix Exponentials and Linear Systems   *418*

Chapter 6   Numerical Methods   426

6.1   Introduction: Euler's Method   *427*
6.2   A Closer Look at the Euler Method, and Improvements   *435*
6.3   The Runge-Kutta Method   *445*
6.4   Systems of Differential Equations   *453*

Chapter 7   Qualitative Properties and Existence of Solutions   465

7.1   Introduction to Stability   *466*
7.2   Stability and the Phase Plane   *472*

CONTENTS

7.3 Linear and Almost Linear Systems *481*
7.4 Nonlinear Mechanical Systems *494*
7.5 Ecological Applications—Predators and Competitors *506*
7.6 Existence and Uniqueness of Solutions *518*

References for Further Study

534

Answers

536

Index

559

PREFACE

We wrote this book to provide a concrete and readable text for the traditional course in elementary differential equations that science, engineering, and mathematics students take following calculus. It includes enough material appropriately arranged for different courses lasting one or two quarters (or a single semester). Our treatment is shaped throughout by the goal of an exposition that students will find accessible, attractive, and interesting. We hope that we have anticipated and addressed most of the questions and difficulties that students typically encounter when they study differential equations for the first time.

The book begins and ends with discussions and examples of the mathematical modeling of real-world situations. The fact that differential equations have diverse and important applications is too familiar for extensive comment here. But these applications have played a singular role in the historical development of this subject. Whole areas of the subject exist mainly because of their applications. So in teaching it, we want our students to learn first to solve those differential equations that enjoy the most frequent application.

We therefore make consistent use of appealing applications for both motivation and illustration of the standard elementary techniques of solution of differential equations. A number of the more substantial applications are placed in optional sections, each marked with an asterisk (in the table of contents and in the text). These sections can be omitted without loss of continuity, but their availability can provide instructors with flexibility for variations in emphasis.

While according real-world applications their due, we also think the first course in differential equations should be a window on the world of mathematics. Matters of definition, classification, and logical structure deserve (and receive here) careful attention—for the first time in the mathematical experience of many of the students (and perhaps for the last time in some cases). While it is neither feasible nor desirable to include proofs of the fundamental existence and uniqueness theorems along the way in an elementary course, students need to see precise and clearcut statements of these theorems, and to understand their role in the subject.

We do include some existence and uniqueness proofs in Section 7.6, and occasionally refer to them in the main body of the text.

The list of introductory topics in differential equations is quite standard, and a glance at our chapter titles will reveal no major surprises, though in the fine structure we have attempted to add a bit of zest here and there. A number of different permutations in the order of topics are possible, and the table that follows this preface exhibits the logical dependence between chapters. In most chapters the principal ideas of the topic are introduced in the first few sections of the chapter, and the remaining sections are devoted to extensions and applications. Hence the instructor has a wide range of choice regarding breadth and depth of coverage.

At various points our approach reflects the widespread use of computer programs for the numerical solution of differential equations. Nevertheless, we continue to believe that the traditional elementary analytical methods of solution are important for students to learn. One reason is that effective and reliable use of numerical methods often requires preliminary analysis using standard elementary techniques; the construction of a realistic numerical model often is based on the study of a simpler analytical model.

SECOND EDITION FEATURES

In preparing this revision we have taken advantage of many valuable comments and suggestions from users of the first edition. In addition to the specific changes mentioned below, we have rewritten many discussions for greater clarity and have added new remarks, applications, examples, problems, and computational details throughout the book. We hope that the additional computer-generated artwork that we have included will help students to visualize better the geometric aspects of differential equations.

In this text we restrict our attention to ordinary differential equations and their applications. Chapters 1 through 6 correspond in order to the first six chapters of the first edition, while Chapter 7 treats qualitative properties of solutions. An expanded version of this book, *Elementary Differential Equations with Boundary Value Problems*, includes additional chapters on Fourier series and boundary value problems (corresponding to Chapters 7 and 8 of the first edition).

Chapter 1 naturally treats first order equations, with separable equations (Section 1.4), linear equations (Section 1.5), substitution methods (Section 1.6), and exact equations (Section 1.7) comprising the core of the chapter. Chapter 2 is devoted to linear equations of higher order. In order to make the concepts of linear independence and general solutions more concrete and tangible, we discuss only second order equations in Section 2.1, and follow with the nth order case in Section 2.2.

Chapter 3 begins with a review of the basic facts about power series that will be needed. The first three sections of the chapter treat the standard power series techniques for the solution of linear equations with variable coefficients. We devote more attention than usual to certain matters—such as shifting indices of summation—that are mathematically routine but nevertheless troublesome for many students. In Section 3.4 (optional) we include for reference more detail on the method of Frobenius than ordinarily will be covered in the classroom. Similarly, we go slightly further than is customary in Section 3.6 (optional) with applications of Bessel functions. Chapter 4 on Laplace transforms is rather standard, though our discussion in Section 4.6 (optional) of impulses and Dirac delta functions may have some merit.

There is much variation in the treatment of linear systems in introductory courses, depending on the background in linear algebra that is assumed. The first two sections of Chapter 5 can stand alone as an introduction to linear systems without the use of linear algebra and matrices. The last five sections of Chapter 5 employ the notation and terminology (though not so much of the theory) of elementary linear algebra. For ready reference, we have included in Section 5.3 a complete and self-contained account of the needed notation and terminology of determinants, matrices, and vectors. For this edition we have added Section 5.7, a brief introduction to matrix exponentials and their applications to linear systems.

Many instructors will choose to proceed directly from Chapter 5 to the study of qualitative properties of linear and nonlinear systems in Chapter 7. This chapter is a considerable expansion of the corresponding chapter in the first edition. We believe the importance of qualitative analysis of differential equations for elementary students is increasing, and therefore have made a special effort to make this material accessible to these students. We now devote two initial sections to introducing stability and phase plane concepts. Section 7.4 on applications of stability to nonlinear mechanical systems is new. Section 7.5 contains applications to competition, survival, and extinction of species. Section 7.6 closes the chapter with discussion and proofs of the basic existence and uniqueness theorems that are stated and applied earlier at appropriate places in the book.

NUMERICAL METHODS AND COMPUTING

Chapter 6 on numerical methods has been completely rewritten—with new examples, discussions, applications, and problems throughout—and requires some special comment. Personal computers are now everywhere, and they affect the perspective in which we view our subject. With ready accessibility to substantial computing power, students can now envision the numerical approximation of solutions as a routine and commonplace matter. Our viewpoint in Chapter 6 is that understanding and appreciation of numerical algorithms is enhanced (and rendered more concrete to students) by discussion of their computer implementations. We have included illustrative BASIC programs because no flowchart has the convincing tangibility of a program that actually runs (and produces the results claimed). BASIC is the *lingua franca* of personal computing, and in BASIC we could include simple programs that without extensive discussion would be intelligible and informative to students with little or no programming experience.

While we feel that BASIC is best for elementary textbook exposition in mathematics, serious scientific programming is done on most campuses in either FORTRAN or Pascal. Consequently a computer diskette (with instructions) that is available for use with this book contains BASIC, FORTRAN, Pascal, and APL versions of the programs included in the text. It can be used for classroom demonstrations, but is intended mainly for student use, either on an individual basis or in a computer laboratory setting. This diskette (which is not copy protected) is provided to instructors for classroom and coursework use, and copies can be obtained by writing to the College Textbook Division of Prentice Hall.

In another vein, it is pointed out in the Chapter 1 summary that much of the numerical work in Chapter 6 can be covered at any point in the course subsequent to Chapter 1. In particular, instructors who are experimenting with the use of computers in teaching differential equations may wish to cover numerical methods earlier than has been the custom in the past.

PROBLEMS AND SOLUTIONS

Probably in no other mathematics course beyond calculus are the exercises and problem sets so crucial to student learning as in the introductory differential equations course. We therefore devoted great effort to the development and selection of the approximately 1500 problems in this book. Each section contains more computational problems ("solve the following equations," and so on) than any class will ordinarily use, plus an ample number of applied problems. We were, however, very sparing in our inclusion of purely theoretical problems. The answer section includes the answers to all odd-numbered problems and to some of the even-numbered ones.

In addition there is a Solutions Manual that accompanies this book. It includes answers to all of the problems, together with either complete or partial solutions of most of the problems in the text that are not so routine that an answer alone suffices.

ACKNOWLEDGMENTS

In preparing this revision we profited greatly from the advice of the following very able reviewers: Gertrude Ehrlich, University of Maryland; Robert Glassey, Indiana University; Terry Herdman, Virginia Polytechnic Institute and State University; S. F. Neustadter, San Francisco State University; Thomas Rousseau, Siena College; and Juan A. Gatica, University of Iowa. In addition we would like to acknowledge again the assistance of the following reviewers of the first edition: W. Dan Curtis, Kansas State University; Bruce Conrad, Temple University; James W. Cushing, University of Arizona; James L. Heitsch, University of Illinois at Chicago; Erich Zauderer, Polytechnic Institute of New York; Anthony Peressini, University of Illinois; William Rundell, Texas A & M University; and especially George Feissner, State University of New York at Cortland. We thank also our editor, David Ostrow, for his enthusiastic and efficient coordination of the entire process. Once again, we are unable to express adequately our debts to Alice F. Edwards and Carol W. Penney for their continued assistance, encouragement, support, and patience.

C. H. E., Jr.
D. E. P.

DEPENDENCE OF CHAPTERS

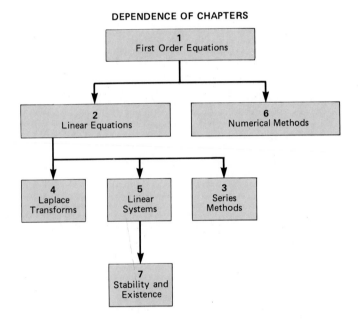

ELEMENTARY
DIFFERENTIAL EQUATIONS
WITH APPLICATIONS

CHAPTER *1*

INTRODUCTION AND FIRST ORDER DIFFERENTIAL EQUATIONS

1.1 INTRODUCTION

1.2 SOLUTION BY DIRECT INTEGRATION

1.3 EXISTENCE AND UNIQUENESS OF SOLUTIONS

1.4 SEPARABLE EQUATIONS AND APPLICATIONS

1.5 LINEAR FIRST ORDER EQUATIONS

1.6 SUBSTITUTION METHODS

1.7 EXACT EQUATIONS AND INTEGRATING FACTORS

***1.8** POPULATION MODELS

***1.9** MOTION WITH VARIABLE ACCELERATION

CHAPTER 1 SUMMARY AND A LOOK AHEAD

REVIEW PROBLEMS

Introduction

The laws of the universe are written largely in the language of mathematics. Algebra is sufficient to solve many static problems, but the most interesting natural phenomena involve change and are best described by equations that relate changing quantities.

Because the derivative $dy/dt = f'(t)$ of the function f may be regarded as the rate at which the quantity $y = f(t)$ changes with respect to the independent variable t, it is natural that equations involving derivatives are those that describe the changing universe. An equation involving an unknown function and one or more of its derivatives is called a **differential equation**, and the study of differential equations has two principal goals:

1. To discover the differential equation that describes a specified physical situation;

2. To find the appropriate solution of that equation.

Unlike algebra, in which we seek the unknown numbers that satisfy an equation such as $x^3 + 7x^2 - 11x + 41 = 0$, in solving a differential equation we are challenged to find the unknown functions $y = g(x)$ for which an identity such as $g'(x) - 2xg(x) = 0$, that is,

$$\frac{dy}{dx} - 2xy = 0,$$

holds on some interval of real numbers. Ordinarily, we will want to find *all* solutions of the differential equation if possible.

EXAMPLE 1   If C is a constant and

$$y(x) = Ce^{x^2} \tag{1}$$

then

$$\frac{dy}{dx} = C(2xe^{x^2}) = (2x)(Ce^{x^2}) = 2xy.$$

Thus every function $y(x)$ of the form in (1) is a solution of the differential equation

$$\frac{dy}{dx} = 2xy \tag{2}$$

for all x. In particular, Eq. (1) defines an *infinite* family of different solutions of this differential equation, one for each choice of the "arbitrary constant" C. By the method of separation of variables (Section 1.4) it can be shown that *every* solution of the differential equation in (2) is of the form in Eq. (1).

DIFFERENTIAL EQUATIONS AND MATHEMATICAL MODELS

The following three examples illustrate the process of translating scientific laws and principles into differential equations, by interpreting rates of change as derivatives. In each of these examples the independent variable is time t, but we will see numerous applications in which some quantity other than time is the independent variable.

EXAMPLE 2 Newton's law of cooling may be stated in the following form: The *time rate of change* (the rate of change with respect to time t) of the temperature $T(t)$ of a body is proportional to the difference between T and the temperature A of the surrounding medium. That is,

$$\frac{dT}{dt} = k(A - T) \tag{3}$$

where k is a positive constant. Observe that if $T > A$, then $dT/dt < 0$, so the temperature $T(t)$ is a decreasing function of t and the body is cooling. On the other hand, if $T < A$, then $dT/dt > 0$, so T is increasing.

Thus the physical law is translated into a differential equation. If we are given the values of k and A, we hope to find an explicit formula for $T(t)$, and then—with the aid of this formula—we can predict the future temperature of the body.

EXAMPLE 3 The *time rate of change* of a population $P(t)$ with constant birth and death rates is, in many simple cases, proportional to the size of the population. That is,

$$\frac{dP}{dt} = kP \tag{4}$$

where k is the constant of proportionality.

EXAMPLE 4 Torricelli's law implies that the *time rate of change* of the volume V of water in a draining tank is proportional to the square root of the depth y of the water in the tank:

$$\frac{dV}{dt} = -ky^{1/2} \tag{5}$$

where k is constant. If the tank is a cylinder with cross-sectional area A, then $V = Ay$, and so $dV/dt = A(dy/dt)$. In this case Eq. (5) takes the form

$$\frac{dy}{dt} = -hy^{1/2} \tag{6}$$

where $h = k/A$.

Let us discuss Example 3 further. Note first that each function of the form

$$P(t) = Ce^{kt} \tag{7}$$

is a solution of the differential equation

$$\frac{dP}{dt} = kP$$

in (4). We verify this assertion as follows:

$$P'(t) = Cke^{kt} = k(Ce^{kt}) = kP(t)$$

for all real numbers t. Because substitution of each function of the form given in (7) into Eq. (4) produces an identity, all these functions are solutions of Eq. (4).

Thus, even if the value of the constant k is known, the differential equation $dP/dt = kP$ has *infinitely* many different solutions of the form $P(t) = Ce^{kt}$—one for each choice of the "arbitrary" constant C. This is typical of differential equations in general. It is also fortunate, because it allows us to use additional information to select from all the solutions a particular one that fits the situation under study.

EXAMPLE 5 Suppose that $P(t) = Ce^{kt}$ is the population of a bacterial colony at time t, that the population at time $t = 0$ (hours, h) was 1000, and that the population doubled after 1 h. This additional information about the function $P(t)$ yields the following equations:

$$1000 = P(0) = Ce^0 = C,$$

$$2000 = P(1) = Ce^k.$$

It follows that $C = 1000$ and that $e^k = 2$, so $k = \ln 2 \approx 0.69315$. With this value of k the differential equation in (4) is

$$\frac{dP}{dt} = (\ln 2)P \approx (0.69315)P,$$

and the value $C = 1000$ yields the particular solution

$$P(t) = 1000e^{t \ln 2} \approx 1000e^{(0.69315)t}$$

that satisfies the given conditions. Therefore, we can predict the population at any future time; for example, the population at time $t = 90$ minutes (min) (1.5 h) will be $P(1.5) = 1000e^{(1.5) \ln 2}$, approximately 2828 bacteria.

The condition $P(0) = 1000$ is called an **initial condition** because we frequently write differential equations for which $t = 0$ is the "starting time." Figure 1.1 shows a number of graphs of the form $P(t) = Ce^{kt}$ for which $k = \ln 2$. The graphs of all the solutions of $dP/dt = (\ln 2)P$ in fact fill up the entire two-dimensional plane, and no two intersect. Moreover, the selection of any point on the P-axis amounts to a determination of the value $P(0)$. Because exactly one solution passes through each such point, we see in this case that an initial condition $P(0) = P_0$ may determine a unique solution agreeing with the given data.

Nevertheless, is is possible that none of these solutions fits *all* the known information. In such a case we must suspect that the differential equation—a "mathematical model" of the physical phenomenon in question—may not adequately describe the real world. The solutions of Eq. (4) are of the form $P(t) = Ce^{kt}$ where C is a positive constant, but for *no* choice of the constants k and C does $P(t)$ accurately describe the actual growth of the human population of the world over the past hundred years. We must therefore write a more complicated differential equation, one that takes into account the effects of population pressure on the birth rate, the declining food supply, and other factors. This should not be regarded as a failure of the model of Example 3, but as an insight into what additional factors must be considered in studying the growth of populations. Indeed, Eq. (4) is quite accurate under certain circumstances—for example, the growth of a bacterial population under conditions of unlimited food and space.

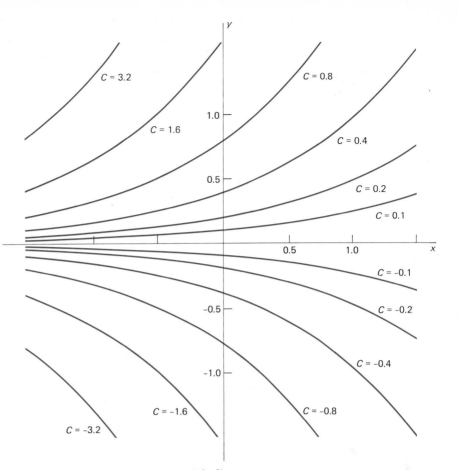

FIGURE 1.1   Graphs of $P(t) = C \exp(t \ln 2)$.

This brief discussion of population growth illustrates the crucial process of *mathematical modeling* (see Fig. 1.2), which involves:

1. The formulation of a real-world problem in mathematical terms—that is, the construction of a mathematical model;

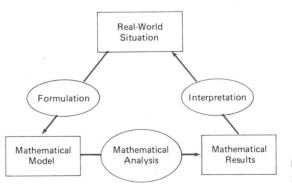

FIGURE 1.2   The process of mathematical modeling.

2. The analysis or solution of the resulting mathematical problem;

3. The interpretation of the mathematical results in the context of the original real-world situation—for example, answering the question originally posed.

In the population example, the real-world problem is that of determining the population at some future time. A **mathematical model** consists of a list of variables (P and t) that describe the given situation, together with one or more equations relating these variables ($dP/dt = kP$, $P(0) = P_0$) that are known or are assumed to hold. The mathematical analysis consists of solving these equations (here, for P as a function of t). Finally, we apply these mathematical results to answer the original real-world question.

But in our population example we ignored the effects of such factors as varying birth and death rates. This made the mathematical analysis quite simple, perhaps unrealistically so. A satisfactory mathematical model is subject to two contradictory requirements: It must be sufficiently detailed to represent the real-world situation with relative accuracy, yet it must be sufficiently simple to make the mathematical analysis practical. If the model is so detailed that it fully represents the physical situation, the mathematical analysis may be too difficult to carry out. If the model is too simple, the results may be so inaccurate as to be useless. Thus there is an inevitable trade-off between what is physically realistic and what is mathematically possible. The construction of a model that adequately bridges this gap between realism and feasibility is therefore the most crucial and delicate step in the process. Ways must be found to simplify the model mathematically without sacrificing essential features of the real-world situation.

Mathematical models are discussed throughout this book. The remainder of this introductory section is devoted to simple examples and to standard terminology used in discussing differential equations and their solutions.

EXAMPLES AND TERMINOLOGY

EXAMPLE 6   If C is a constant and $y = 1/(C - x)$, then

$$\frac{dy}{dx} = \frac{1}{(C - x)^2} = y^2$$

if $x \neq C$. Thus

$$y(x) = \frac{1}{C - x} \tag{8}$$

defines a solution of the differential equation

$$y' = y^2 \tag{9}$$

on any interval of real numbers not containing the point $x = C$. Actually, Eq. (8) defines a *one-parameter family* of solutions of $y' = y^2$, one for each value of the arbitrary constant or "parameter" C. With $C = 1$ we get the particular solution $y = 1/(1 - x)$ on the interval $(-\infty, 1)$ that satisfies the initial condition $y(0) = 1$.

EXAMPLE 7 Verify that the function $y = 2x^{1/2} - x^{1/2} \ln x$ satisfies the differential equation

$$4x^2 y'' + y = 0 \tag{10}$$

for all $x > 0$.

SOLUTION First we compute the derivatives

$$y' = -\tfrac{1}{2}x^{-1/2} \ln x$$

and

$$y'' = \tfrac{1}{4}x^{-3/2} \ln x - \tfrac{1}{2}x^{-3/2}.$$

Then substitution into Eq. (10) yields

$$4x^2 y'' + y = 4x^2(\tfrac{1}{4}x^{-3/2} \ln x - \tfrac{1}{2}x^{-3/2}) + 2x^{1/2} - x^{1/2} \ln x = 0$$

if x is positive, and so the differential equation is satisfied for all $x > 0$.

The fact that we can write a differential equation is not enough to guarantee that it has a solution. For example, it is clear that the differential equation

$$(y')^2 + y^2 = -1 \tag{11}$$

has *no* (real-valued) solution because the sum of nonnegative numbers cannot be negative. For another variation on this theme, note that the equation

$$(y')^2 + y^2 = 0 \tag{12}$$

obviously has only the (real-valued) solution $y(x) \equiv 0$. In our previous examples any differential equation having at least one solution indeed had infinitely many.

The **order** of a differential equation is the order of the highest derivative that appears in it. The differential equation of Example 7 is of second order, our previous examples are first order equations, and

$$y^{(4)} + x^2 y^{(3)} + x^5 y = \sin x$$

is a fourth order equation. The most general form of an **nth order** differential equation with independent variable x and unknown function or dependent variable $y = y(x)$ is

$$F(x, y, y', y'', \ldots, y^{(n)}) = 0 \tag{13}$$

where F is a specific real-valued function of $n + 2$ variables.

Our usage of the word *solution* has until now been somewhat informal. More precisely, we say that the function $y = u(x)$ is a **solution** of the differential equation in (13) **on the interval** I provided that the derivatives $u', u'', \ldots, u^{(n)}$ exist and that

$$F(x, u, u', u'', \ldots, u^{(n)}) = 0$$

for all x in I. When brevity is needed, we say that $y = u(x)$ **satisfies** the differential equation in (13) on I.

EXAMPLE 8 If A and B are constants and

$$y(x) = A \cos 3x + B \sin 3x, \tag{14}$$

then two successive differentiations yield

$$y'(x) = -3A \sin 3x + 3B \cos 3x,$$

$$y''(x) = -9A \cos 3x - 9B \sin 3x = -9y(x)$$

for all x. Consequently, Eq. (14) defines what it is natural to call a *two-parameter family* of solutions of the second order differential equation

$$y'' + 9y = 0 \tag{15}$$

on the whole real line.

Although the differential equations in (11) and (12) are exceptions to the general rule, we will see that an nth order equation ordinarily has an n-parameter family of solutions—one involving n different arbitrary constants or parameters.

In both Eqs. (11) and (12), the appearance of y' as an implicitly defined function causes complications. For this reason, we will ordinarily assume that any differential equation under study can be solved explicitly for the highest derivative that appears; that is, that the equation may be written in the form

$$y^{(n)} = G(x, y, y', y'', \ldots, y^{(n-1)}) \tag{16}$$

where G is a real-valued function of $n + 1$ variables. In addition, we will always seek only real-valued solutions unless we warn the reader to the contrary.

All the differential equations we have mentioned so far are **ordinary** differential equations, meaning that the unknown function (dependent variable) depends on only a *single* independent variable. For this reason only ordinary derivatives appear in the equation. If the dependent variable is a function of two or more independent variables, then partial derivatives are likely to be involved; if so, the equation is called a **partial** differential equation. For example, the temperature $u = u(x, t)$ of a long thin rod at the point x at time t satisfies (under appropriate simple conditions) the partial differential equation

$$\frac{\partial u}{\partial t} = k \frac{\partial^2 u}{\partial x^2},$$

where k is a constant (called the *thermal diffusivity* of the rod). In Chapters 1 through 7 we will be concerned only with *ordinary* differential equations, and will refer to them simply as differential equations.

In this chapter we concentrate our attention on first order differential equations of the general form

$$\frac{dy}{dx} = f(x, y). \tag{17}$$

We also will sample the wide range of applications of such equations. A typical mathematical model of an applied situation will be an **initial value problem**, consisting of a differential equation of the form in (17) together with an **initial condition**

$y(x_0) = y_0$. Note that we call $y(x_0) = y_0$ an initial condition whether or not $x_0 = 0$. To solve the initial value problem

$$\frac{dy}{dx} = f(x, y), \qquad y(x_0) = y_0 \qquad (18)$$

means to find a differentiable function $y(x)$ that satisfies both conditions in Eq. (18).

EXAMPLE 9   Given the solution $y = 1/(C - x)$ of the differential equation $y' = y^2$ discussed in Example 6, solve the initial value problem

$$y' = y^2, \qquad y(1) = 2.$$

SOLUTION   We need only find a value of C so that the solution $y(x) = 1/(C - x)$ satisfies the initial condition $y(1) = 2$. Substitution of the values $x = 1$ and $y = 2$ in the solution yields

$$2 = y(1) = \frac{1}{C - 1},$$

so $2C - 2 = 1$, and hence $C = \frac{3}{2}$. With this value of C we get the desired solution

$$y(x) = \frac{1}{\frac{3}{2} - x} = \frac{2}{3 - 2x}.$$

The central question of greatest interest to us is this: If we are given a differential equation known to have a solution satisfying a given initial condition, how do we actually *find* or *compute* that solution? And, once found, what can we *do* with it? We will see that a relatively few simple techniques—separation of variables (Section 1.4), solution of linear equations (Section 1.5), substitution methods (Section 1.6), multiplication by integrating factors (Section 1.7)—are enough to enable us to solve a diversity of first order equations having impressive applications.

1.1   PROBLEMS

In each of Problems 1–12, verify by substitution that each given function is a solution of the given differential equation.

1. $y' = 3x^2$; $y = x^3 + 7$

2. $y' + 2y = 0$; $y = 3e^{-2x}$

3. $y'' + 4y = 0$; $y_1 = \cos 2x, y_2 = \sin 2x$

4. $y'' = 9y$; $y_1 = e^{3x}, y_2 = e^{-3x}$

5. $y' = y + 2e^{-x}$; $y = e^x - e^{-x}$

6. $y'' + 4y' + 4y = 0$; $y_1 = e^{-2x}, y_2 = xe^{-2x}$

7. $y'' - 2y' + 2y = 0$; $y_1 = e^x \cos x, y_2 = e^x \sin x$

8. $y'' + y = 3 \cos 2x$; $y_1 = \cos x - \cos 2x$, $y_2 = \sin x - \cos 2x$

9. $y' + 2xy^2 = 0$; $y = \dfrac{1}{1 + x^2}$

10. $x^2y'' + xy' - y = \ln x$; $y_1 = x - \ln x, y_2 = \dfrac{1}{x} - \ln x$

11. $x^2y'' + 5xy' + 4y = 0$; $y_1 = \dfrac{1}{x^2}, y_2 = \dfrac{\ln x}{x^2}$

12. $x^2y'' - xy' + 2y = 0$; $y_1 = x \cos (\ln x), y_2 = x \sin (\ln x)$

In each of Problems 13–16, substitute $y = e^{rx}$ into the given differential equation to determine all values of r for which $y = e^{rx}$ is a solution of the equation.

13. $3y' = 2y$

14. $4y'' = y$

15. $y'' + y' - 2y = 0$ **16.** $3y'' + 3y' - 4y = 0$

In each of Problems 17–26, first verify that $y(x)$ satisfies the given differential equation. Then determine a value of the constant C so that $y(x)$ satisfies the given initial condition.

17. $y' + y = 0;$ $y(x) = Ce^{-x}$, $y(0) = 2$

18. $y' = 2y;$ $y(x) = Ce^{2x}$, $y(0) = 3$

19. $y' = y + 1;$ $y(x) = Ce^{x} - 1$, $y(0) = 5$

20. $y' = x - y;$ $y(x) = Ce^{-x} + x - 1$, $y(0) = 10$

21. $y' + 3x^2y = 0;$ $y(x) = Ce^{-x^3}$, $y(0) = 7$

22. $e^y y' = 1;$ $y = \ln(x + C)$, $y(0) = 0$

23. $x \dfrac{dy}{dx} + 3y = 2x^5;$ $y(x) = \frac{1}{4}x^5 + Cx^{-3}$, $y(2) = 1$

24. $xy' - 3y = x^3;$ $y(x) = x^3(C + \ln x)$, $y(1) = 17$

25. $y' = 3x^2(y^2 + 1);$ $y(x) = \tan(x^3 + C)$, $y(0) = 1$

26. $y' + y \tan x = \cos x;$ $y(x) = (x + C)\cos x$, $y(\pi) = 0$

In each of Problems 27–31, a function $y = g(x)$ is described by some geometric property of its graph. Write a differential equation of the form $y' = f(x, y)$ having the function $g(x)$ as its solution (or as one of its solutions).

27. The slope of the graph of g at the point (x, y) is the sum of x and y.

28. The tangent line to the graph of g at the point (x, y) intersects the x-axis at the point $(x/2, 0)$.

29. Every straight line normal to the graph of g passes through the point $(0, 1)$.

30. The graph of g is normal to every curve of the form $y = k + x^2$ (k is a constant) where they meet.

31. The line tangent to the graph of g at (x, y) passes through the point $(-y, x)$.

In each of Problems 32–36, write—in the manner of Eqs. (3)–(6) of this section—a differential equation that is a mathematical model of the situation described.

32. The time rate of change of a population P is proportional to the square root of P.

33. The time rate of change of the velocity v of a coasting motorboat is proportional to the square of v.

34. The acceleration dv/dt of a certain sports car is proportional to the difference between 250 km/h and the velocity of the car.

35. In a city having a fixed population of P persons, the time rate of change of the number N of those persons who have heard a certain rumor is proportional to the number of those who have not yet heard the rumor.

36. In a city with a fixed population of P persons, the time rate of change of the number N of persons infected with a certain disease is proportional to the product of the number who have the disease and the number who do not.

In each of Problems 37–42, determine by inspection at least one solution of the given differential equation. That is, use your knowledge of derivatives to make an intelligent guess, then test your hypothesis.

37. $y'' = 0$ **38.** $y' = y$

39. $xy' + y = 3x^2$ **40.** $(y')^2 + y^2 = 1$

41. $y' + y = e^x$ **42.** $y'' + y = 0$

43. In Example 6 we saw that $y(x) = 1/(C - x)$ defines a one-parameter family of solutions of the differential equation $y' = y^2$. (a) Determine a value of C so that $y(10) = 10$. (b) Is there a value of C so that $y(0) = 0$? Can you nevertheless find by inspection a solution of $y' = y^2$ such that $y(0) = 0$?

44. (a) Show that $y(x) = Cx^4$ defines a one-parameter family of solutions of the differential equation $xy' = 4y$. (b) Show that

$$y(x) = \begin{cases} -x^4 & \text{if } x \le 0, \\ +x^4 & \text{if } x \ge 0 \end{cases}$$

defines a solution of $xy' = 4y$ for all x but which is not of the form $y = Cx^4$.

1.2

Solution by Direct Integration

The first order differential equation $y' = f(x, y)$ takes an especially simple form if the function f is independent of the dependent variable y:

$$\frac{dy}{dx} = f(x). \tag{1}$$

In this special case we need only integrate both sides of Eq. (1) to obtain

$$y = \int f(x)\, dx + C. \tag{2}$$

This is a **general solution** of Eq. (1), meaning that it involves an arbitrary constant C, and for every choice of C it is a solution of the differential equation. If $G(x)$ is a particular antiderivative of $f(x)$—that is, if $G'(x) \equiv f(x)$—then

$$y(x) = G(x) + C. \tag{3}$$

To satisfy an initial condition $y(x_0) = y_0$, we need only substitute $x = x_0$ and $y = y_0$ into Eq. (3) to obtain $y_0 = G(x_0) + C$, so that $C = y_0 - G(x_0)$. With this choice of C, we obtain the **particular solution** of (1) satisfying the initial value problem

$$y' = f(x), \qquad y(x_0) = y_0.$$

We will see that this is the typical pattern for solutions of first order differential equations. Ordinarily, we will first find a *general solution* involving an arbitrary constant C. We then can attempt to obtain, by appropriately choosing C, a *particular solution* satisfying a given initial condition $y(x_0) = y_0$.

Remark: As the term is used above, a *general solution* of a first order differential equation is simply a one-parameter family of solutions. A natural question is whether a given general solution contains *every* particular solution of the differential equation. When this is known to be so, we call it **the** general solution of the differential equation. For instance, because any two antiderivatives of the same function $f(x)$ can differ only by a constant, it follows that every solution of Eq. (1) is of the form in (2). Thus Eq. (2) serves to define *the* general solution of (1).

EXAMPLE 1   Solve the initial value problem

$$\frac{dy}{dx} = \frac{x}{(x^2 + 9)^{1/2}}, \qquad y(4) = 2.$$

SOLUTION   Integration immediately yields the general solution

$$y = \int \frac{x}{(x^2 + 9)^{1/2}}\, dx = (x^2 + 9)^{1/2} + C.$$

The substitution $x = 4$, $y = 2$ gives $2 = 5 + C$ and hence $C = -3$, so the desired particular solution is

$$y(x) = (x^2 + 9)^{1/2} - 3.$$

The observation that the special first order equation $dy/dx = f(x)$ is readily solvable (provided the function $f(x)$ can be integrated) extends to second order differential equations of the special form

$$\frac{d^2 y}{dx^2} = g(x), \tag{4}$$

in which the given function on the right-hand side involves neither the independent variable y nor its derivative y'. We simply integrate once to obtain

$$\frac{dy}{dx} = G(x) + C_1,$$

where $G(x)$ is an antiderivative of $g(x)$ and C_1 is an arbitrary constant. Then another integration yields

$$y = \int G(x)\,dx + C_1 x + C_2$$

where C_2 is a second arbitrary constant. In effect, the second order differential equation in (4) is one that can be solved by solving successively the *first order* differential equations

$$\frac{dv}{dx} = g(x) \quad \text{and} \quad \frac{dy}{dx} = v(x).$$

VELOCITY AND ACCELERATION

Direct integration is sufficient to allow us to solve a number of important problems concerning the motion of a particle (or *mass point*) in terms of the forces acting upon it. The motion of a particle along a straight line (the x-axis) is described by its **position function**

$$x = f(t) \tag{5}$$

giving its x-coordinate at time t. The **velocity** $v(t)$ of the particle is defined to be

$$v(t) = f'(t); \quad \text{that is,} \quad v = \frac{dx}{dt}. \tag{6}$$

Its **acceleration** $a(t)$ is $a(t) = v'(t) = x''(t)$; in alternative notation,

$$a = \frac{dv}{dt} = \frac{d^2 x}{dt^2}. \tag{7}$$

Newton's *second law of motion* implies that if a force $F(t)$ acts on the particle and is directed along its line of motion, then

$$ma(t) = F(t); \quad \text{that is,} \quad F = ma, \tag{8}$$

where m is the mass of the particle. If the force $F(t)$ is known, then the equation $x''(t) = F(t)/m$ can be integrated twice to obtain the position function $x(t)$ in terms of two constants of integration. These two arbitrary constants are frequently determined by the **initial position** $x(0) = x_0$ and the **initial velocity** $v(0) = v_0$ of the particle.

For instance, suppose that the force F, and therefore the acceleration $a = F/m$, are constant. Then we begin with the equation

$$\frac{dv}{dt} = a \qquad (a \text{ is a constant}) \tag{9}$$

and integrate both sides to obtain $v = at + C_1$. We know that $v = v_0$ when $t = 0$, and substitution of this information into the last equation yields the fact that

$C_1 = v_0$. So

$$v = \frac{dx}{dt} = at + v_0. \tag{10}$$

A second integration gives $x = \frac{1}{2}at^2 + v_0 t + C_2$, and the substitution $t = 0$, $x = x_0$ gives $C_2 = x_0$; therefore,

$$x = \frac{1}{2}at^2 + v_0 t + x_0. \tag{11}$$

Thus with Eq. (10) we can find the velocity, and with (11) the position, of the particle at any time t in terms of its *constant* acceleration a, its initial velocity v_0, and its initial position x_0.

EXAMPLE 2 A lunar lander is falling freely toward the surface of the moon at a speed of 1000 miles per hour (mi/h). Its retrorockets, when fired, provide a deceleration of 20,000 miles per hour per hour (mi/h^2) (the gravitational acceleration produced by the moon is assumed to be included in the given deceleration). At what height above the lunar surface should the retrorockets be activated to ensure a "soft touchdown" ($v = 0$ at impact)?

SOLUTION We denote by $x(t)$ the height of the lunar lander above the surface, as indicated in Fig. 1.3. Then $v_0 = -1000$ (mi/h—negative because the height is decreasing), and $a = +20,000$ because an upward thrust increases the velocity

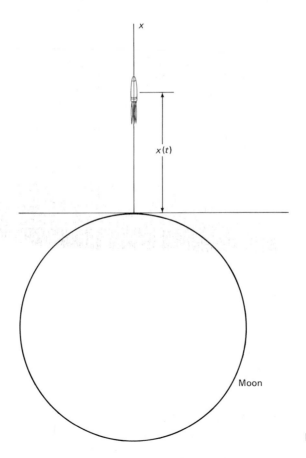

FIGURE 1.3 The lunar lander.

v (although it decreases the *speed* $|v|$). Then Eqs. (10) and (11) become

$$v = 20{,}000t - 1000 \qquad (12)$$

and

$$x = 10{,}000t^2 - 1000t + x_0, \qquad (13)$$

where x_0 is the height of the lander above the lunar surface at the time $t = 0$ at which the retrorockets should be activated.

From Eq. (12) we see that $v = 0$ (soft touchdown) occurs when $t = \frac{1}{20}$ (h; thus 3 min); then substitution of $t = \frac{1}{20}$, $x = 0$ into (13) yields

$$x_0 = -10{,}000(\tfrac{1}{20})^2 + 1000(\tfrac{1}{20}) = 25$$

(mi). Thus the retrorockets should be activated when the lunar lander is 25 mi above the surface, and it will reach the surface 3 min later.

GRAVITATIONAL ACCELERATION AND PHYSICAL UNITS

A common application of Eqs. (10) and (11) involves vertical motion near the surface of the earth. Recall that the **weight** W of a body is the force exerted by gravity upon the body at the surface of the earth. Hence Newton's second law implies that

$$W = mg \qquad (8')$$

where m is the mass of the body and g is the gravitational acceleration at the surface of the earth. (The value of g depends upon the precise location of its measurement, but it is nearly constant; we assume it constant in the remainder of this section.) Although it was convenient to use units of miles and hours in Example 2, we will ordinarily employ one of the three systems of units summarized in the following table.

	fps system	cgs system	mks system
Force	pound (lb)	dyne (dyn)	newton (N)
Mass	slug	gram (g)	kilogram (kg)
Distance	foot (ft)	centimeter (cm)	meter (m)
Time	second (s)	second (s)	second (s)
g	32 ft/s^2	980 cm/s^2	9.8 m/s^2

The values listed for g are approximations; a more accurate value for g at most locations near sea level is 32.2 ft/s^2. All three unit systems are compatible with Newton's law (8′). For example, the weight of a mass of 1 slug is

$$W = (1 \text{ slug})(32 \text{ ft/s}^2) = 32 \text{ lb}.$$

Similarly, a mass of 1 g has a weight of 980 dyn, whereas a mass of 1 kg has a weight of 9.8 N. It follows that $9.8 \text{ N} = 1000 \times 980$ dyn, so

$$1 \text{ newton} = 10^5 \text{ dynes}.$$

CHAPTER 1: Introduction and First Order Differential Equations

For conversions between fps and metric units it helps to remember that

$$1 \text{ ft} = 12 \text{ in.} \times 2.54 \frac{\text{cm}}{\text{in.}} = 30.48 \text{ cm}$$

and that

$$1 \text{ lb} = 454 \text{ g} \quad \text{(approximately)},$$

meaning that a *mass* of 454 g has a *weight* of 1 lb (and hence a mass of $\frac{1}{32}$ slug). Consequently

$$1 \text{ lb} = 454 \text{ g} \times 980 \frac{\text{dyn}}{\text{g}} \approx 4.45 \times 10^5 \text{ dyn} = 4.45 \text{ N}.$$

Because we intend to deal here with vertical motion, it is natural to choose the y-axis as the coordinate system for position. If we choose the upward direction as the positive direction, then the effect of gravity on the body is to *decrease* its height and also to *decrease* its velocity $v = dy/dt$, so we see that if air resistance is ignored, then the acceleration of the body is

$$a = \frac{dv}{dt} = -g = -32 \text{ (ft/s}^2).$$

Equations (10) and (11) then take the forms

$$v = -32t + v_0 \tag{10'}$$

and

$$y = -16t^2 + v_0 t + y_0. \tag{11'}$$

Here y_0 is the initial height of the body in feet and v_0 its initial velocity in feet per second. If the initial ($t = 0$) values v_0 and y_0 are given, then the subsequent velocity and height of the body are given by Eqs. (10') and (11').

EXAMPLE 3   For instance, suppose that a ball is thrown straight upward from the ground ($y_0 = 0$) with initial velocity $v_0 = 96$ ft/s. Then it reaches its maximum height when its velocity is zero,

$$v = -32t + 96 = 0,$$

and thus when $t = 3$ s. Hence the maximum height the ball attains is

$$y(3) = -(16)(3)^2 + (96)(3) = 144 \text{ ft}.$$

*THE DEFLECTION OF A UNIFORM BEAM

We include now an example of the use of a relatively simple differential equation to explain a complicated physical phenomenon—the shape of a horizontal beam upon which a vertical force is acting.

Consider the horizontal beam shown in Fig. 1.4, uniform both in cross section and in material. If it is supported only at its ends, then the force of its own weight

FIGURE 1.4   Distortion of a horizontal beam.

_____
* Asterisks denote material that can be omitted without loss of continuity.

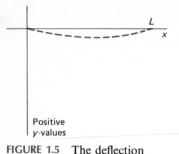

Positive
y-values

FIGURE 1.5 The deflection curve.

distorts its longitudinal axis of symmetry into the curve shown as a dashed line in the figure. We want to investigate the shape $y = y(x)$ of this curve, the **deflection curve** of the beam. We will use the coordinate system indicated in Fig. 1.5, with the positive y-axis directed downward.

A consequence of the theory of elasticity is that, for relatively small deflections of such a beam (so small that $(y')^2$ is negligible in comparison with unity), an adequate mathematical model of the deflection curve is the fourth order differential equation

$$EIy^{(4)} = F(x), \tag{14}$$

where

E denotes the Young's modulus of the material of the beam,

I denotes the moment of inertia of the cross section of the beam about a horizontal line through the centroid of the cross section, and

$F(x)$ denotes the density of *downward* force acting vertically on the beam at the point x.

Density of force? Yes; this means that the force acting downward on a very short segment $[x, x + \Delta x]$ of the beam is approximately $F(x) \Delta x$. The units of $F(x)$ are units of force per unit length, such as pounds per foot. We will consider here the case in which the only force distributed along the beam is its own weight, w pounds per foot, so that $F(x) \equiv w$. Then Eq. (14) takes the form

$$EIy^{(4)} = w \tag{15}$$

where E, I, and w are all constant.

Note: We assume no previous familiarity with elasticity or with Eqs. (14) and (15) here. It is important to be able to begin with a differential equation that arises in a specific applied discipline and then analyze its implications; thus we develop an understanding of the equation by examining its solutions. Observe that, in essence, Eq. (15) implies that the fourth derivative $y^{(4)}$ is proportional to the weight density w. This proportionality involves, however, *two* constants: E, which depends only upon the material in the beam, and I, which depends only upon the shape of the cross section of the beam. Values of the Young's modulus E of various materials can be found in handbooks of physical constants; $I = \frac{1}{4}\pi a^4$ for a circular cross section of radius a.

Though (15) is a fourth order differential equation, its solution involves only the solution of simple first order equations by successive simple integrations. One integration of (15) yields

$$EIy''' = wx + C_1;$$

a second yields

$$EIy'' = \tfrac{1}{2}wx^2 + C_1x + C_2;$$

another yields

$$EIy' = \tfrac{1}{6}wx^3 + \tfrac{1}{2}C_1x^2 + C_2x + C_3;$$

CHAPTER 1: Introduction and First Order Differential Equations

x = 0   x = L

Simply supported or hinged

x = 0   x = L

Built-in

FIGURE 1.6 Two ways of supporting a beam.

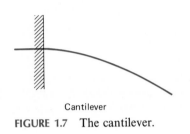

Cantilever

FIGURE 1.7 The cantilever.

a final integration gives

$$EIy = \tfrac{1}{24}wx^4 + \tfrac{1}{6}C_1x^3 + \tfrac{1}{2}C_2x^2 + C_3x + C_4,$$

where C_1, C_2, C_3, and C_4 are arbitrary constants. Thus we obtain a solution of Eq. (15) of the form

$$y(x) = \frac{w}{24EI}x^4 + Ax^3 + Bx^2 + Cx + D, \tag{16}$$

where A, B, C, and D are constants resulting from the four integrations.

These last four constants are determined by the way in which the beam is supported at its ends, where $x = 0$ and $x = L$. Figure 1.6 shows two common types of support. A beam might also be supported one way at one end but another way at the other end. For instance, Fig. 1.7 shows a **cantilever**—a beam firmly fastened at $x = 0$ but *free* (no support whatsoever) at $x = L$. The following table shows the **boundary**, or **endpoint**, **conditions** corresponding to the three most common cases. We will see that these conditions are applied readily in beam problems, though a discussion here of their origin would take us too far afield.

Support	Endpoint Conditions
Simply supported	$y = y'' = 0$
Built-in or fixed end	$y = y' = 0$
Free end	$y'' = y^{(3)} = 0$

For example, the deflection curve of the cantilever in Fig. 1.7 would be given by Eq. (16), with the coefficients A, B, C, and D determined by the conditions

$$y(0) = y'(0) = 0 \quad \text{and} \quad y''(L) = y^{(3)}(L) = 0 \tag{17}$$

corresponding to the fixed end at $x = 0$ and the free end at $x = L$. The conditions in (17) together with the differential equation in (16) constitute an **endpoint value problem** (or boundary value problem). An endpoint value problem calls for a solution of a differential equation on a closed interval, satisfying conditions imposed at *both* endpoints of the interval. By contrast, in an initial value problem conditions on the solution are imposed only at a single point.

EXAMPLE 4 Determine the shape of the deflection curve of a uniform horizontal beam of length L and weight w per unit length and simply supported at each end.

SOLUTION We have the endpoint conditions $y(0) = y''(0) = 0 = y(L) = y''(L)$. Rather than imposing them directly on Eq. (16), let us begin with the differential equation $EIy^{(4)} = w$ and determine the constants as we proceed with the four successive integrations. The first two integrations yield

$$EIy^{(3)} = wx + A;$$

$$EIy'' = \tfrac{1}{2}wx^2 + Ax + B.$$

Hence $y''(0) = 0$ implies that $B = 0$, and then $y''(L) = 0$ gives

$$0 = \tfrac{1}{2}wL^2 + AL.$$

It follows that $A = -wL/2$ and thus that

$$EIy'' = \tfrac{1}{2}wx^2 - \tfrac{1}{2}wLx.$$

Then two more integrations give

$$EIy' = \tfrac{1}{6}wx^3 - \tfrac{1}{4}wLx^2 + C,$$

and finally,

$$EIy(x) = \tfrac{1}{24}wx^4 - \tfrac{1}{12}wLx^3 + Cx + D. \tag{18}$$

Now $y(0) = 0$ implies that $D = 0$; then, because $y(L) = 0$,

$$0 = \tfrac{1}{24}wL^4 - \tfrac{1}{12}wL^4 + CL.$$

It follows that $C = wL^3/24$. Hence from Eq. (18) we obtain

$$y(x) = \frac{w}{24EI}(x^4 - 2Lx^3 + L^3x) \tag{19}$$

as the shape of the supported beam. It is apparent from symmetry (see also Problem 35) that the *maximum deflection* $y_{\max}$ of the beam occurs at its midpoint $x = L/2$, and thus has the value

$$y_{\max} = y\left(\frac{L}{2}\right) = \frac{w}{24EI}\left(\frac{1}{16}L^4 - \frac{2}{8}L^4 + \frac{1}{2}L^4\right);$$

that is,

$$y_{\max} = \frac{5wL^4}{384EI}. \tag{20}$$

For instance, suppose that we want to calculate the maximum deflection of a simply supported steel rod 20 ft long with a circular cross section 1 inch (in.) in diameter. From a handbook we find that typical steel has density $\delta = 7.75$ g/cm^2, and that its Young's modulus is $E = 2 \times 10^{12}$ g/cm·s^2, so it will be more convenient to work in cgs units. Thus our rod has

$$\text{Length:} \quad L = (20 \text{ ft})\left(30.48\ \frac{\text{cm}}{\text{ft}}\right) = 609.60 \text{ cm}$$

and

$$\text{Radius:} \quad a = \left(\frac{1}{2}\text{ in.}\right)\left(2.54\ \frac{\text{cm}}{\text{in.}}\right) = 1.27 \text{ cm}.$$

Its *linear* mass density (that is, its mass per unit length) is

$$\rho = \pi a^2 \delta = \pi(1.27)^2(7.75) \approx 39.27\ \frac{\text{g}}{\text{cm}},$$

so

$$w = \rho g = \left(39.27 \frac{\text{g}}{\text{cm}}\right)\left(980 \frac{\text{cm}}{\text{s}^2}\right)$$

$$\approx 38{,}484.6 \frac{\text{dyn}}{\text{cm}}.$$

The area moment of inertia of a circular disk of radius a about a diameter is $I = \pi a^4/4$, so

$$I = \tfrac{1}{4}\pi(1.27)^4 \approx 2.04 \text{ cm}^4.$$

Therefore, Eq. (20) yields

$$y_{\max} \approx \frac{(5)(38484.6)(609.60)^4}{(384)(2 \times 10^{12})(2.04)} \approx 16.96 \text{ cm},$$

or about 6.68 in. as the maximum deflection of the rod at its midpoint. It is interesting to note that $y_{\max}$ is proportional to L^4, so if our rod were 10 ft long, its maximum deflection would be only one-sixteenth as much—only about 0.42 in. Because $I = \pi a^4/4$, we see from Eq. (20) that the same reduction in maximum deflection could be achieved by doubling the radius a of the rod.

1.2 PROBLEMS

In each of Problems 1–10, find a function $y = f(x)$ satisfying the given differential equation and the prescribed initial condition.

1. $\dfrac{dy}{dx} = 2x + 1;\quad y(0) = 3$

2. $\dfrac{dy}{dx} = (x - 2)^3;\quad y(2) = 1$

3. $\dfrac{dy}{dx} = x^{1/2};\quad y(4) = 0$

4. $\dfrac{dy}{dx} = \dfrac{1}{x^2};\quad y(1) = 5$

5. $\dfrac{dy}{dx} = (x + 2)^{-1/2};\quad y(2) = -1$

6. $\dfrac{dy}{dx} = x(x^2 + 9)^{1/2};\quad y(-4) = 0$

7. $\dfrac{dy}{dx} = \dfrac{10}{x^2 + 1};\quad y(0) = 0$

8. $\dfrac{dy}{dx} = \cos 2x;\quad y(0) = 1$

9. $\dfrac{dy}{dx} = (1 - x^2)^{-1/2};\quad y(0) = 0$

10. $\dfrac{dy}{dx} = xe^{-x};\quad y(0) = 1$

In Problems 11–17, find the position function $x(t)$ of a moving particle with the given acceleration $a(t)$, initial position $x_0 = x(0)$, and initial velocity $v_0 = v(0)$.

11. $a(t) = 50,\ v_0 = 10,\ x_0 = 20$

12. $a(t) = -20,\ v_0 = -15,\ x_0 = 5$

13. $a(t) = 3t,\ v_0 = 5,\ x_0 = 0$

14. $a(t) = 2t + 1,\ v_0 = -7,\ x_0 = 4$

15. $a(t) = 4(t + 3)^2,\ v_0 = -1,\ x_0 = 1$

16. $a(t) = \dfrac{3}{(t + 4)^{1/2}},\ v_0 = -1,\ x_0 = 1$

17. $a(t) = \dfrac{1}{(t + 1)^3},\ v_0 = 0,\ x_0 = 0$

18. A ball is dropped from the top of a building that is 400 ft high. How long does it take to reach the ground? With what speed does the ball strike the ground?

19. The brakes of a car are applied when it is moving at 100 km/h and provide a constant deceleration of 10 meters per second per second (m/s²). How far does the car travel before coming to a stop?

20. A ball is thrown straight upward from ground level with an initial speed of 160 ft/s. What is the maximum height that the ball attains? How long does it remain aloft?

21. A ball is thrown straight downward from the top of a tall building. The initial speed of the ball is 10 m/s. It strikes the ground with a speed of 60 m/s. How tall is the building?

22. A baseball is thrown straight downward with an initial speed of 40 ft/s from the top of the Washington Monument (555 ft high). How long does it take to reach the ground, and with what speed does the baseball strike the ground?

23. A bomb is dropped from a balloon hovering at an altitude of 800 ft. A gun emplacement is located on the ground directly below the balloon. The gun fires a projectile straight upward toward the bomb exactly 2 s after the bomb is released. With what initial speed should the projectile be fired in order to intercept the bomb at an altitude of exactly 400 ft?

24. A car traveling at 60 mi/h skids 176 ft after its brakes are suddenly applied. Under the assumption that the braking system provides constant deceleration, what is that deceleration? How many seconds does the skid continue?

25. The skid marks made by an automobile indicate that its brakes were fully applied for a distance of 225 ft before it came to a stop. The car in question is known to have a constant deceleration of 50 feet per second per second (ft/s²) under the conditions we describe. How fast was the car traveling when the brakes were first applied?

26. Suppose that a car skids 15 m if it is moving at 50 km/h when the brakes are applied. Assuming that the car has the same constant deceleration, how far will it skid if it is moving at 100 km/h when the brakes are applied?

27. On the planet Gzyx, a ball dropped from a height of 20 ft hits the ground in 2 s. If a ball is dropped from the top of a 200-ft-tall building on Gzyx, how long will it take to hit the ground? With what speed will it hit?

28. A person can throw a ball straight upward from the surface of the earth to a maximum height of 144 ft. How high could the person throw the ball on the planet Gzyx of Problem 27?

29. A stone is dropped from rest at initial height h above the surface of the earth. Show that the speed at which it strikes the ground is $v = (2gh)^{1/2}$.

30. If a woman has enough "spring" in her legs to jump vertically to a height of 2.25 ft on the earth, how high could she jump on the moon, where the surface gravitational acceleration is (approximately) 5.3 ft/s²?

31. At noon a car starts from rest at point A and proceeds with constant acceleration along a straight road toward point

B. If the car reaches B at 12:50 P.M. with a velocity of 60 mi/h, what is the distance from A to B?

32. At noon a car starts from rest at point A and proceeds with constant acceleration along a straight road toward point C, 35 mi distant. If the constantly accelerated car arrives at C with a velocity of 60 mi/h, at what time does it arrive?

33. (a) A cantilever beam is fixed at $x = 0$ and free at $x = L$ (its other end). Show that its shape is given by

$$y = \frac{w}{24EI}(x^4 - 4Lx^3 + 6L^2x^2).$$

(b) Show that $y'(x) = 0$ only at $x = 0$, and thus that it follows (why?) that the maximum deflection of the cantilever is $y_{max} = y(L) = wL^4/8EI$.

34. (a) Suppose that a beam is fixed at its ends $x = 0$ and $x = L$. Show that its shape is given by

$$y = \frac{w}{24EI}(x^4 - 2Lx^3 + L^2x^2).$$

(b) Show that the roots of $y'(x) = 0$ are $x = 0$, $x = L$, and $x = L/2$, and so it follows (why?) that the maximum deflection of the beam is $y_{max} = y(L/2) = wL^4/(384EI)$, one-fifth that of a beam with simply supported ends.

35. For the simply supported beam whose deflection curve is given by Eq. (19), show that the only root of $y'(x) = 0$ in $[0, L]$ is $x = L/2$, so it follows (why?) that the maximum deflection is indeed given by Eq. (20).

36. (a) A beam is fixed at its left end $x = 0$ but is simply supported at the other end $x = L$. Show that its deflection curve is

$$y = \frac{w}{48EI}(2x^4 - 5Lx^3 + 3L^2x^2).$$

(b) Show that its maximum deflection occurs where $x = (15 - \sqrt{33})L/16$, and is about 41.6% of the maximum deflection that would occur if it were simply supported at each end.

Existence and Uniqueness of Solutions*

In the case of a general first order differential equation of the form

$$y' = f(x, y), \tag{1}$$

---

* This section may be deferred to any point later in the chapter.

we cannot simply integrate each side as in Section 1.2, because now the right-hand side involves the unknown function $y(x)$. Before one spends much time trying to solve a differential equation, it is best to know that solutions actually *exist*. We may also want to know whether there is only one solution of the equation satisfying a given initial condition—that is, whether solutions are *unique*. For instance, such a simple-looking initial value problem as

$$\frac{dy}{dx} = 2\sqrt{y}, \qquad y(0) = 0 \tag{2}$$

has the two different solutions $y_1(x) = x^2$ and $y_2(x) \equiv 0$.

The questions of existence and uniqueness also bear on the process of mathematical modeling. Suppose we are studying a physical system whose behavior is completely determined by certain initial conditions, but that our proposed mathematical model involves a differential equation *not* having unique solutions. Then this raises an immediate question as to whether the mathematical model adequately represents the physical system.

SLOPE FIELDS AND SOLUTION CURVES

In order to investigate the possible behavior of solutions of a differential equation of the form $y' = f(x, y)$, we may think of it in a very geometric way: At various points (x, y) of the two-dimensional plane, the value of $f(x, y)$ determines a slope y'. A *solution* of the above differential equation is a differentiable function with graph having slope $y' = f(x, y)$ at each point (x, y). The *graph* of a solution of a differential equation is sometimes called a **solution curve** of the equation. From a geometric viewpoint, a solution curve of the differential equation $y' = f(x, y)$ is then a curve in the plane whose tangent line at each point (x, y) has slope $m = f(x, y)$.

This idea of a solution curve suggests the following *graphical method* for constructing approximate solutions of the differential equation $y' = f(x, y)$. Through each of a representative collection of points (x, y) we draw a short line segment having slope $m = f(x, y)$. The set of all these line segments is called a **slope field** (or a *direction field*) for the equation $y' = f(x, y)$. We can attempt to sketch a solution curve that threads its way through the slope field in such a way that it is tangent to each of the short line segments that it intersects.

EXAMPLE 1   Figure 1.8 shows the slope field and some typical solution curves for the differential equation

$$y' = -y. \tag{3}$$

Note that the slope field yields important qualitative information about the set of all solutions of (3). For instance, it seems apparent from Fig. 1.8 that every solution $y(x)$ of (3) approaches zero as $x \to +\infty$. In this particular example it is also clear that $y = Ce^{-x}$ is a general solution of $y' = -y$, so the solution curves shown in Fig. 1.8 do indeed have the correct shape.

It is easy to instruct a computer to draw a slope field for a given differential equation $y' = f(x, y)$, and programs for doing this are readily available. There is

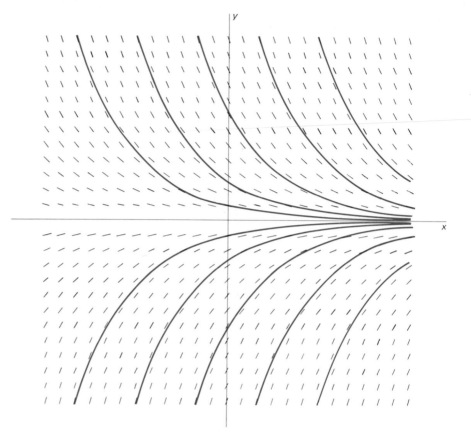

FIGURE 1.8 Slope field and solution curves of $y' = -y$.

also an older manual method still useful on occasion. An **isocline** of the differential equation $y' = f(x, y)$ is a curve of the form

$$f(x, y) = c \qquad (c \text{ is a constant}) \tag{4}$$

on which the slope y' is *constant*. If these isoclines are simple and familiar curves, we first sketch several of them, then draw short line segments with the same slope c at representative points of each isocline $f(x, y) = c$.

For example, $f(x, y) = -y$ for the equation $y' = -y$ of Example 1, so the isoclines are horizontal straight lines of the form $y = -c$. Several of these isoclines, each embellished with short line segments of slope c, are shown in Fig. 1.9. Observe that the resulting slope field is consistent with the more detailed one shown in Fig. 1.8.

EXAMPLE 2 The typical isocline of the differential equation

$$\frac{dy}{dx} = x^2 + y^2 \tag{5}$$

has the equation $x^2 + y^2 = c > 0$, and thus is a circle centered at the origin with radius $r = \sqrt{c}$. Several of these circles are shown in Fig. 1.10. Figure 1.11 shows

CHAPTER 1: Introduction and First Order Differential Equations

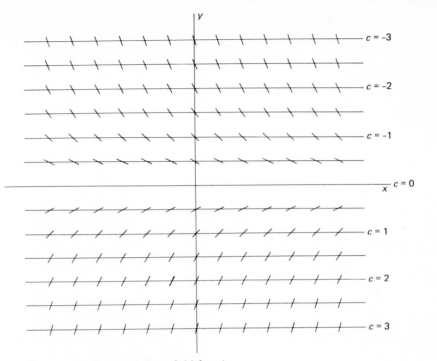

FIGURE 1.9 Isoclines and slope field for $y' = -y$.

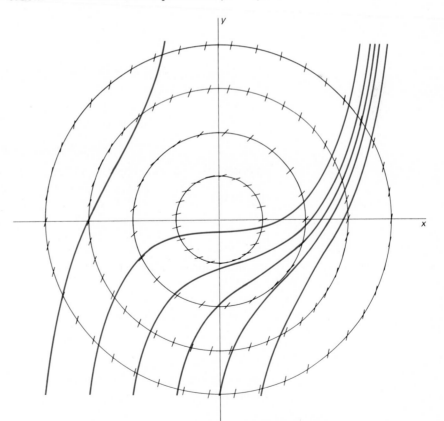

FIGURE 1.10 Isoclines ($c = 0.25, 1, 2.25, 4$) and slope field for $y' = x^2 + y^2$.

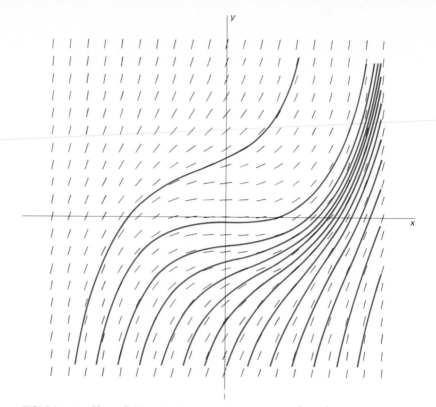

FIGURE 1.11 Slope field and solution curves of $y' = x^2 + y^2$.

the corresponding slope field and some typical solution curves of the equation $y' = x^2 + y^2$. It seems apparent that on each solution curve, $y(x) \to +\infty$ as x increases. Indeed, it can be shown—see Problem 16 of Section 3.6—that the solution of the initial value problem

$$y' = x^2 + y^2, \qquad y(0) = 0$$

approaches $+\infty$ as x approaches the approximate value 2.003147 from the left.

EXAMPLE 3 The isoclines of the differential equation

$$\frac{dy}{dx} = \sin(x - y) \qquad (6)$$

are of the form

$$\sin(x - y) = c;$$

that is, $x - y = \sin^{-1} c$, and thus of the form

$$y = x - \sin^{-1} c.$$

Therefore, the isoclines of Eq. (6) are all straight lines of slope 1 in the xy-plane. The slope field and some typical solution curves are shown in Fig. 1.12. Most of the solution curves one sees in this figure appear to exhibit the oscillatory behavior

CHAPTER 1: Introduction and First Order Differential Equations

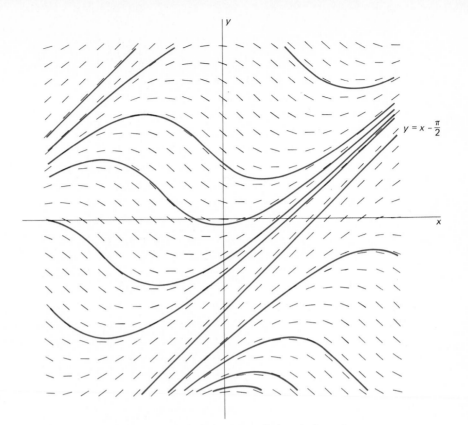

$$y = x - \frac{\pi}{2}$$

FIGURE 1.12 Slope field and solution curves of $y' = \sin(x - y)$.

suggested by the appearance of the sine function in (6). Yet there is one solution curve that appears to be a straight line! This observation prompts us to inspect Eq. (6) more closely. When we do, we spot the particular solution

$$y = x - \frac{\pi}{2}$$

for which $y' \equiv 1$ and $x - y = \pi/2$, so that $\sin(x - y) \equiv 1$ as well.

EXAMPLE 4 Consider the differential equation

$$y^2 + x^2 y' = 0. \tag{7}$$

We show a slope field for this equation in Fig. 1.13. Indeed, because $y' = -(y/x)^2$ we see that each differentiable solution must be decreasing except on the coordinate axes (where x or y is zero). Much additional qualitative information can be read from the slope field of Eq. (7).

In Section 1.4 we will solve equations like this one—known as **separable** differential equations—by separating the variables; we first write

$$\frac{dy}{dx} = -\frac{y^2}{x^2},$$

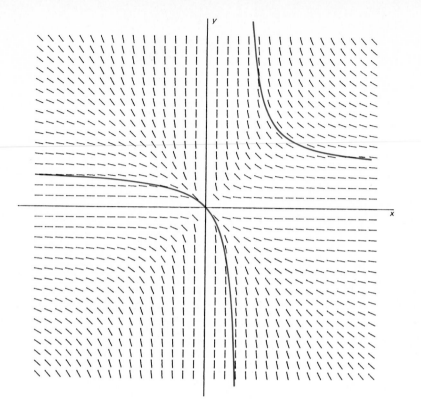

FIGURE 1.13 Slope field for $y^2 + x^2y' = 0$ and the graph of one solution of the equation.

then

$$-\frac{dy}{y^2} = \frac{dx}{x^2};$$

next, we antidifferentiate to obtain

$$\frac{1}{y} = -\frac{1}{x} + C.$$

Finally, we solve for y to obtain

$$y = y(x) = \frac{x}{Cx - 1}. \tag{8}$$

We have drawn a number of solution curves for various values of C; these are shown in Fig. 1.14. Each solution curve of the above form has asymptotes $y = 1/C$ and $x = 1/C$ provided that $C \neq 0$, and in fact each such solution curve consists of the two branches of a rectangular hyperbola, one branch of which passes through the origin. For $C = 0$ we obtain the "different" solution $y = -x$. Moreover, the above solution process excludes the function $y \equiv 0$, which (by substitution in Eq. (7)) is yet another solution.

CHAPTER 1:   Introduction and First Order Differential Equations

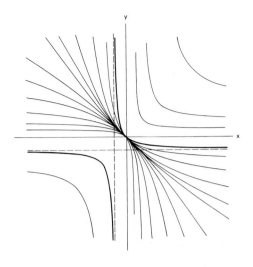

FIGURE 1.14 Graphs of some solutions of Eq. (7).

The remarkable feature of the differential equation in (7) and its solutions is this: First, there are *infinitely many* different solutions satisfying the initial condition $y(0) = 0$. But it follows from Eq. (7) that $y(0) = 0$, so if $b \neq 0$, then there is *no* solution satisfying the initial condition $y(0) = b$. Finally, if $a \neq 0$ and b is arbitrary, there is exactly one solution with $y(a) = b$. All these observations are evident from inspection of Fig. 1.14.

EXISTENCE AND UNIQUENESS

Example 4 shows that an initial value problem may have either no solution, a unique solution, or many—even infinitely many—solutions. The following theorem provides sufficient conditions to ensure existence and uniqueness of solutions, so that neither extreme case (no solution or nonunique solutions) can occur. Methods of proving existence and uniqueness theorems are discussed in Section 7.6.

Theorem *Existence and Uniqueness of Solutions*

Suppose that the real-valued function $f(x, y)$ is continuous on some rectangle in the plane containing the point (a, b) in its interior. Then the initial value problem

$$\frac{dy}{dx} = f(x, y); \qquad y(a) = b \tag{9}$$

has at least one solution on some open interval J containing the point $x = a$. If, in addition, the partial derivative $\partial f / \partial y$ is continuous on that rectangle, then the solution is unique on some (perhaps smaller) open interval J_0 containing the point $x = a$.

In the case of the differential equation $y' = -y$ of Example 1, both the function $f(x, y) = -y$ and the partial derivative $\partial f/\partial y = -1$ are continuous everywhere, so the theorem implies the existence of a unique solution for any initial data (a, b). Although the theorem ensures existence only on some open interval about $x = a$, each solution $y(x) = Ce^{-x}$ actually is defined for all x.

In the case of the differential equation $y' = 2\sqrt{y}$ in Eq. (2), the function $f(x, y) = 2\sqrt{y}$ is continuous wherever $y > 0$, but the partial derivative $\partial f/\partial y = 1/\sqrt{y}$ is discontinuous when $y = 0$, and hence at the point $(0, 0)$. This is why it is possible for there to exist two different solutions $y_1(x) = x^2$ and $y_2(x) \equiv 0$, each of which satisfies the initial condition $y(0) = 0$.

In Example 4 we analyzed the differential equation $y^2 + x^2 y' = 0$ and found that there was no solution passing through $(0, 1)$. If we take $f(x, y) = -(y/x)^2$, we see that the above theorem cannot guarantee existence of a solution through $(0, 1)$ because f is not continuous there. (Note that f is also not continuous at $(0, 0)$, but some solutions *do* pass through this point. Thus continuity of f is a sufficient condition, but not a necessary condition, for the existence of solutions.)

Finally, in Example 6 of Section 1.1, we examined the especially simple differential equation $y' = y^2$. Here we have $f(x, y) = y^2$ and $\partial f/\partial y = 2y$. Each of these functions is continuous everywhere in the plane, and in particular on the rectangle $-2 < x < 2, 0 < y < 2$. Because the point $(0, 1)$ lies in the interior of this rectangle, the existence and uniqueness theorem guarantees a unique solution—necessarily a continuous function—of the initial value problem

$$y' = y^2, \qquad y(0) = 1$$

on *some* open interval containing $x_0 = 0$. This is the solution

$$y(x) = \frac{1}{1 - x}$$

that we discussed in the example mentioned above. But $1/(1 - x)$ is discontinuous at $x = 1$, so we do *not* have *existence* of a solution on the entire interval $-2 < x < 2$. This means that the interval J of the theorem may not be as wide as the rectangle; even though the hypotheses of the theorem are satisfied for all x in an interval I (and for all y in an appropriate interval), the solution may, as in this example, exist only on a smaller interval J. Similarly, the interval J_0 of uniqueness may be even smaller than J.

Nevertheless, there is one important case in which *global* existence and uniqueness are assured. The *linear* first order differential equation

$$\frac{dy}{dx} = a(x)y + b(x) \tag{10}$$

is particularly important in applications. It is most common for $a(x)$ and $b(x)$ to be continuous functions on some open interval, and we will see in Section 1.5 that this is enough to guarantee existence and uniqueness of the solution of any initial value problem involving Eq. (10) on that (entire) interval. The equation in (10) is called **linear** because only the first powers of the dependent variable and its derivative y' appear (we do not require that $a(x)$ or $b(x)$ be linear functions of x). The differential equation $y' = 2x^3 y + \cos x$ is linear, with $a(x) = 2x^3$ and $b(x) = \cos x$ each continuous on the entire real line, whereas the equations in Examples 2, 3, and 4 above are nonlinear.

EXAMPLE 5 The first order differential equation

$$xy' = 2y \tag{11}$$

is linear with $a(x) = 2/x$ and $b(x) \equiv 0$, but $a(x)$ is not continuous at points of the y-axis where $x = 0$. Applying the existence-uniqueness theorem with $f(x, y) = 2y/x$, we see that Eq. (11) has a unique solution near any point where $x \neq 0$. Indeed, we see readily that

$$y(x) = Cx^2 \tag{12}$$

satisfies Eq. (11) for any value of the constant C and for all values of x.

With these preliminary observations, let us consider the initial value problem

$$xy' = 2y, \qquad y(-1) = 1. \tag{13}$$

The initial condition $y(-1) = 1$ implies that $C = 1$ in (12), so on some open interval about $x = -1$ we have the unique solution $y = x^2$ which passes through the origin $(0, 0)$. But to the right of the origin we may choose any value we please of C in (12). That is, for any fixed C, a solution of the initial value problem in (13) is defined by

$$y(x) = \begin{cases} x^2 & \text{if } x \leq 0, \\ Cx^2 & \text{if } x > 0. \end{cases}$$

Thus this initial value problem has infinitely many different solutions, despite the fact that (in accordance with the theorem) there is a unique solution in *some* open interval about the point $x = -1$. Figure 1.15 shows three of these different solutions of (13).

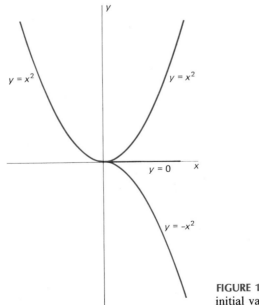

$y = x^2$

$y = x^2$

$y = 0$

x

$y = -x^2$

FIGURE 1.15 Three solutions of the initial value problem in (13).

The point is that the theorem guarantees uniqueness *near* the point (a, b), but the solution curve may branch elsewhere, and uniqueness will be lost. Similarly, the theorem can guarantee existence *near* the point (a, b), but the differential equation may have no solution for some other values of x.

1.3   PROBLEMS

In each of Problems 1–16, we have provided the slope field of a differential equation (also given). Sketch three or four typical solutions for each. (One method: Photocopy the slope field, and draw your solutions in a second color. Another method: Use tracing paper.)

1. $y' = y$

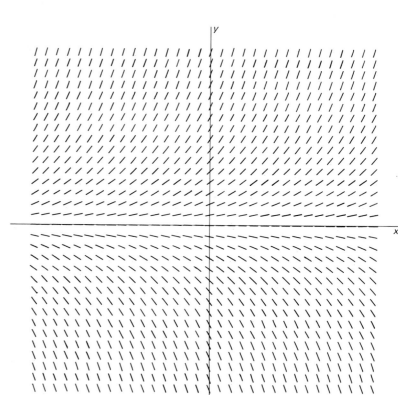

FIGURE 1.16   Slope field for Problem 1.

2. $y' = \dfrac{-y}{2}$

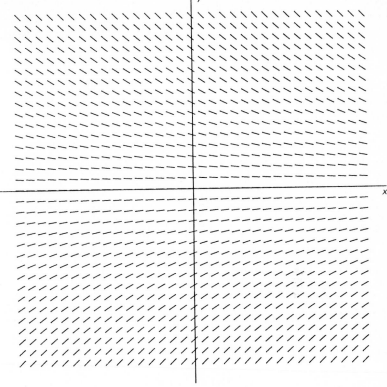

FIGURE 1.17 Slope field for Problem 2.

3. $y' = y^2 + 1$

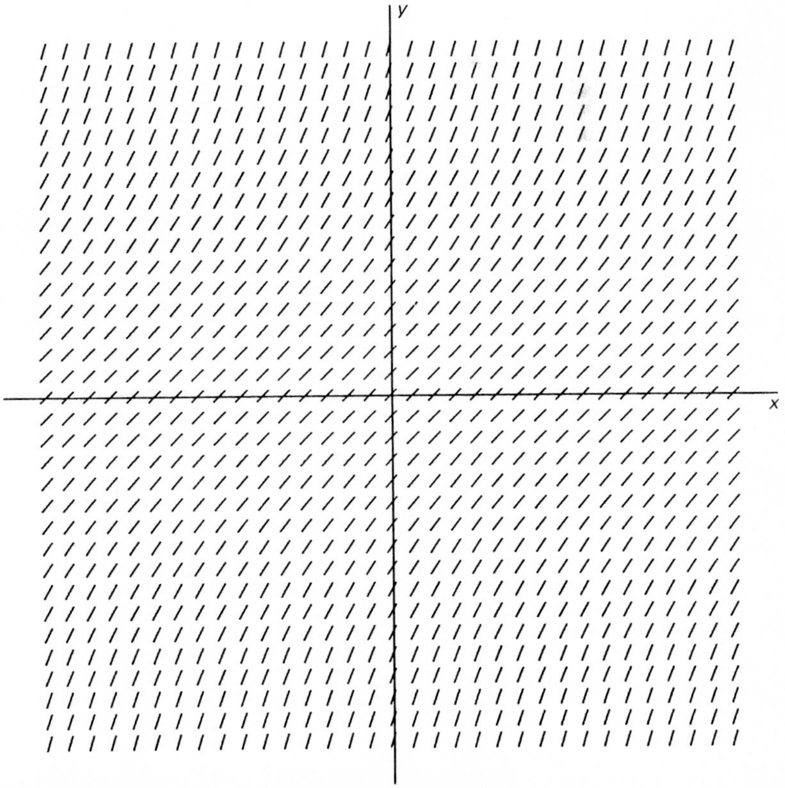

FIGURE 1.18 Slope field for Problem 3.

4. $y' = x^2 - y^2$

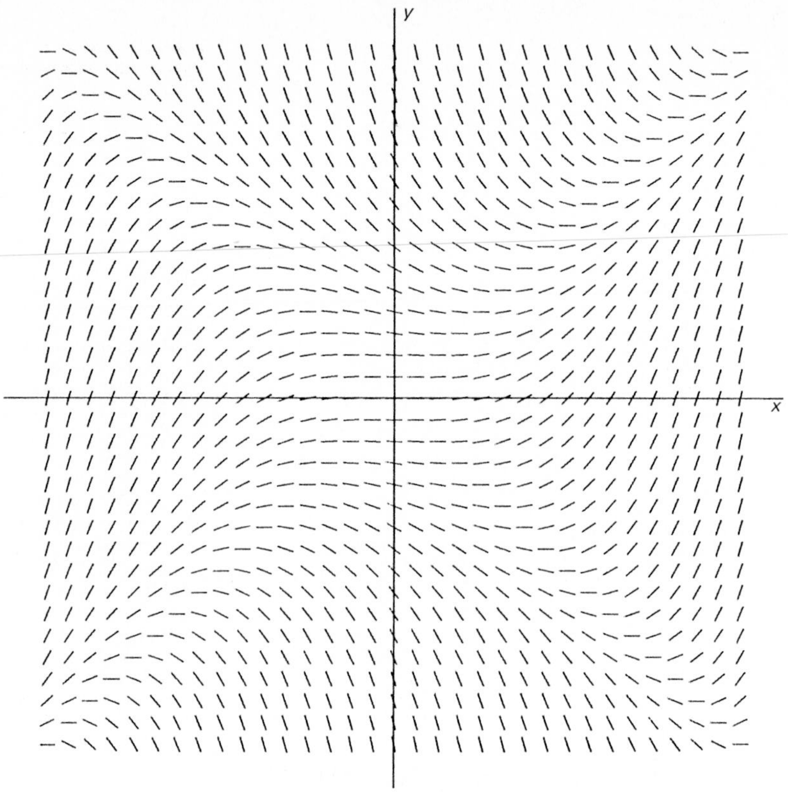

FIGURE 1.19 Slope field for Problem 4.

5. $y' = xy + 1$

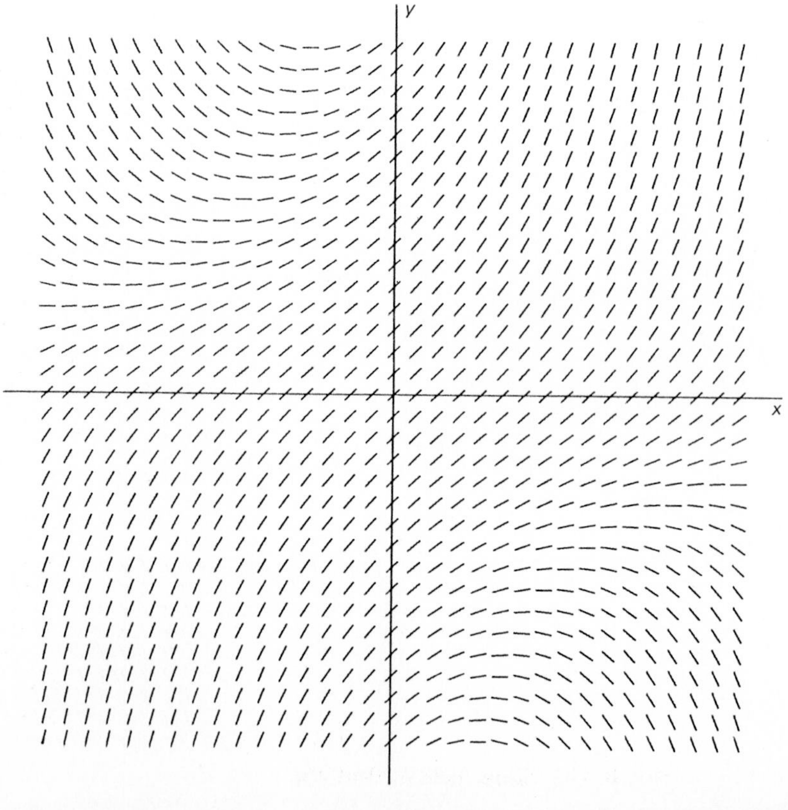

FIGURE 1.20 Slope field for Problem 5.

6. $y' = x + y$

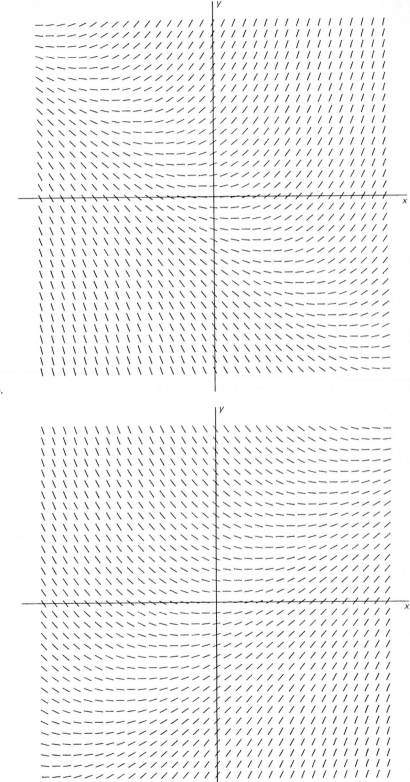

FIGURE 1.21 Slope field for Problem 6.

7. $y' = x - y$

FIGURE 1.22 Slope field for Problem 7.

8. $y' = xy$

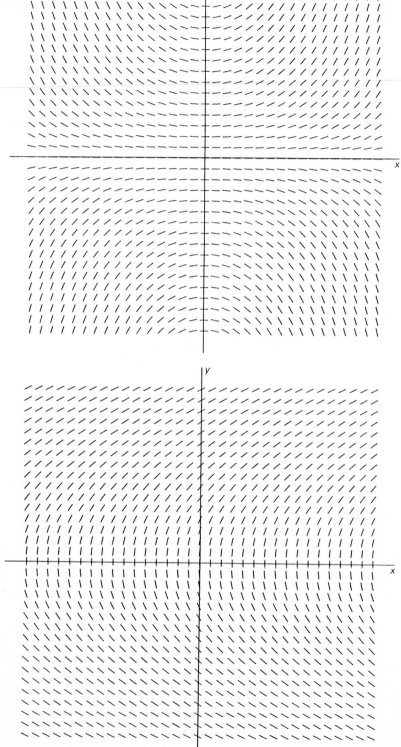

FIGURE 1.23   Slope field for Problem 8.

9. $y' = \dfrac{1}{y}$

FIGURE 1.24   Slope field for Problem 9.

10. $y' = \dfrac{x}{y}$

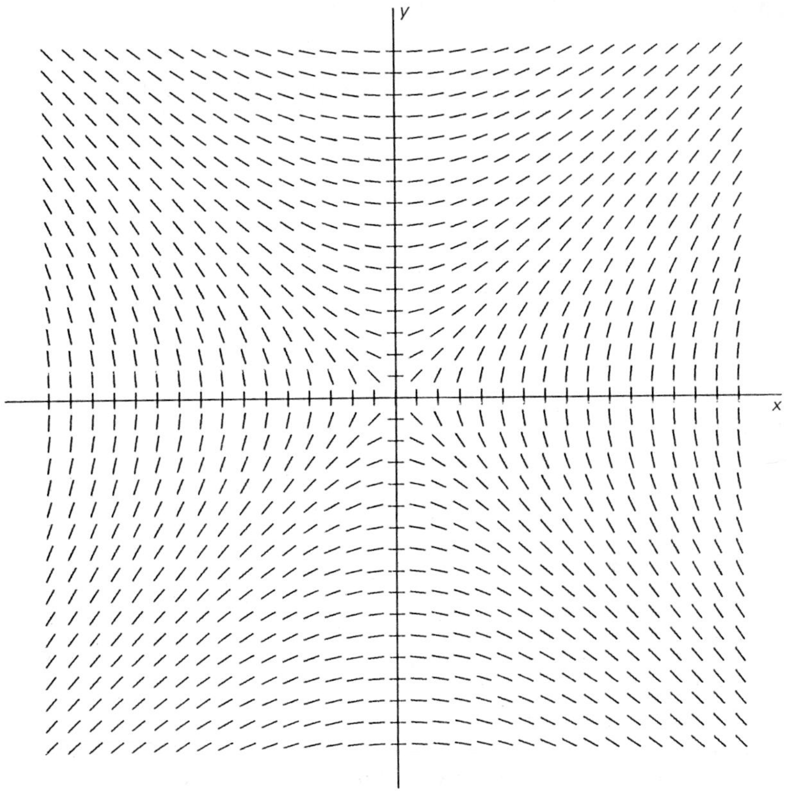

FIGURE 1.25 Slope field for Problem 10.

11. $y' = y^3$

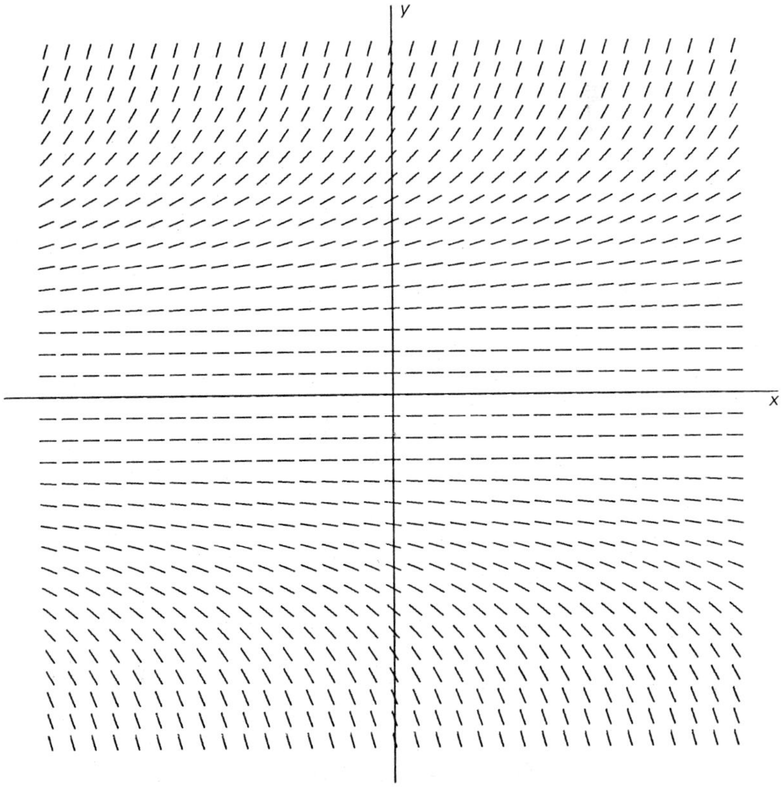

FIGURE 1.26 Slope field for Problem 11.

12. $y' = y^{2/3}$

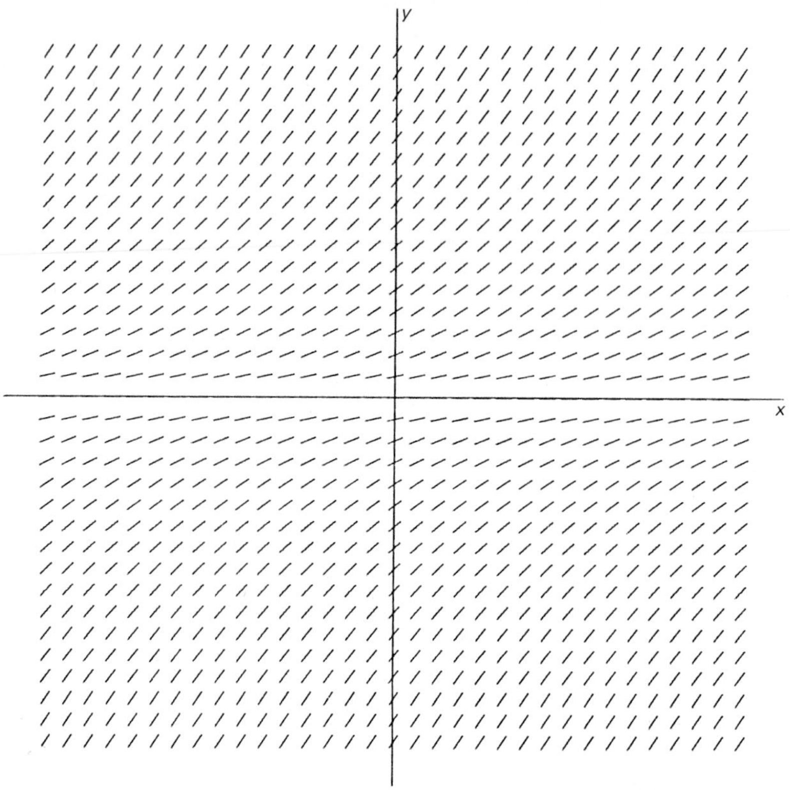

FIGURE 1.27 Slope field for Problem 12.

13. $y' = \sin y$

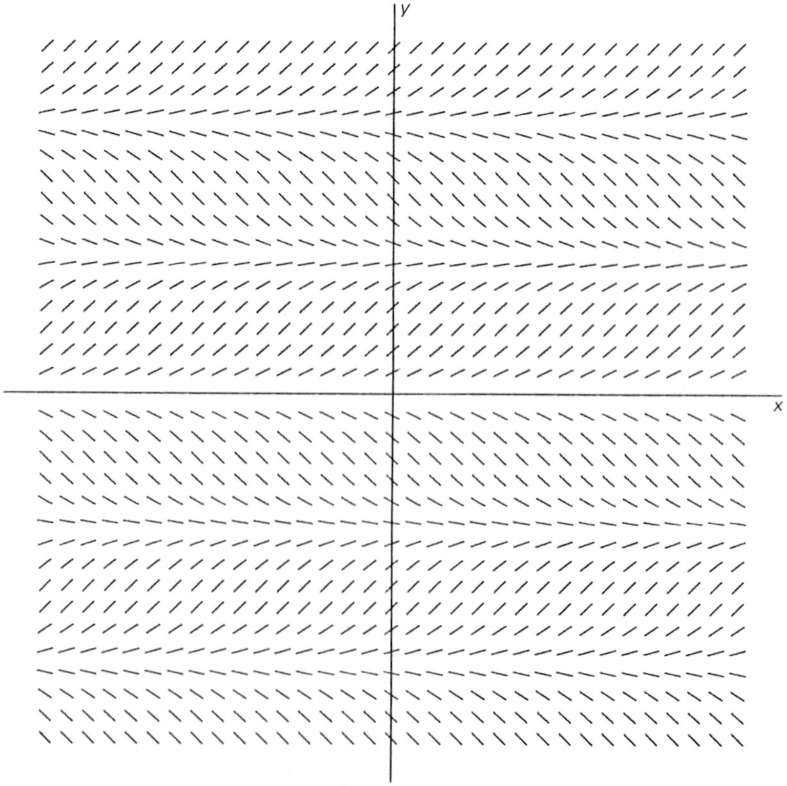

FIGURE 1.28 Slope field for Problem 13.

14. $y' = \sin xy$

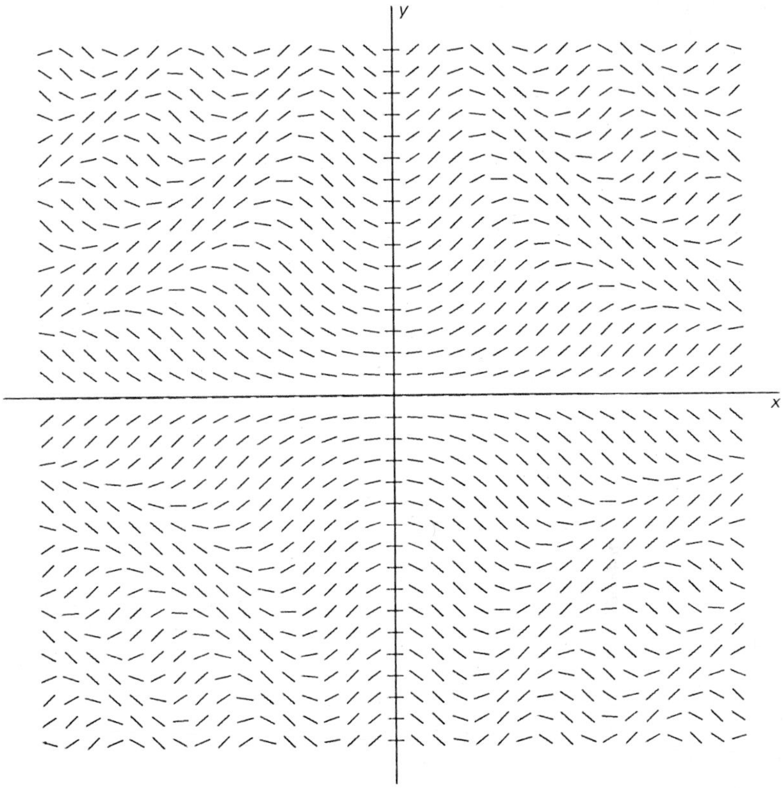

FIGURE 1.29 Slope field for Problem 14.

15. $y' = xe^{-y}$

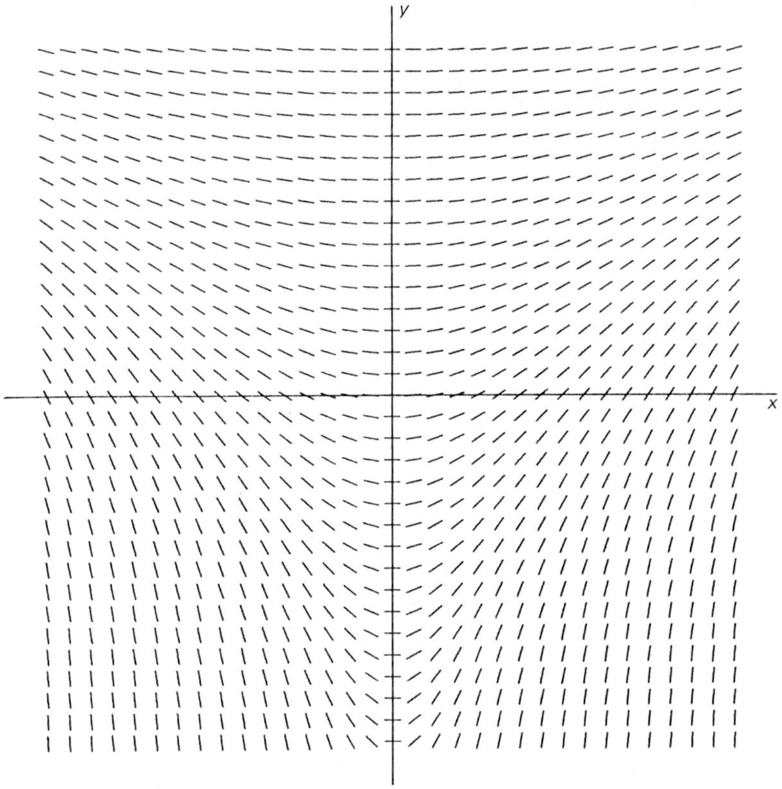

FIGURE 1.30 Slope field for Problem 15.

16. $y' = \ln(1 + y^2)$

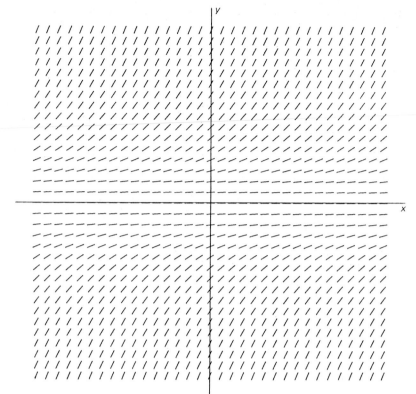

FIGURE 1.31 Slope field for Problem 16.

In each of Problems 17–26, identify the isoclines of the given differential equation. Draw a sketch showing several of these isoclines, each marked with short line segments having the appropriate slope.

17. $y' = x - 1$

18. $y' = x + y$

19. $y' = y^2$

20. $y' = y^{1/3}$

21. $y' = \dfrac{y}{x}$

22. $y' = x^2 - y^2$

23. $y' = xy$

24. $y' = x - y^2$

25. $y' = y - x^2$

26. $y' = ye^{-x}$

In each of Problems 27–36, determine whether the existence and uniqueness theorem of this section does or does not guarantee existence of a solution of the given initial value problem. If existence is guaranteed, determine whether the theorem does or does not guarantee uniqueness of that solution.

27. $y' = 2x^2y^2$; $y(1) = -1$

28. $y' = x \ln y$; $y(1) = 1$

29. $y' = y^{1/3}$; $y(0) = 1$

30. $y' = y^{1/3}$; $y(0) = 0$

31. $y' = (x - y)^{1/2}$; $y(2) = 2$

32. $y' = (x - y)^{1/2}$; $y(2) = 1$

33. $y' = \dfrac{x}{y}$; $y(0) = 1$

34. $y' = \dfrac{x}{y}$; $y(1) = 0$

35. $y' = \ln(1 + y^2)$; $y(0) = 0$

36. $y' = x^2 - y^2$; $y(0) = 1$

The next six problems illustrate that if the hypotheses of the existence and uniqueness theorem fail at a point (a, b), then there may be no solutions, finitely many solutions, or infinitely many solutions passing through (a, b).

37. Show that, on the interval $0 \leq x \leq \pi$, the functions $y_1(x) \equiv 1$ and $y_2(x) = \cos x$ each satisfy the initial value problem

$$y' + (1 - y^2)^{1/2} = 0, \qquad y(0) = 1.$$

Why does this fact not contradict the existence and uniqueness theorem stated in this section? Explain your answer carefully.

38. Find by inspection two different solutions of the initial value problem

$$y' = 3y^{2/3}, \qquad y(0) = 0.$$

Why does this not contradict the theorem stated in this section?

CHAPTER 1: Introduction and First Order Differential Equations

39. Use Fig. 1.32 as a suggestion for showing that the initial value problem

$$y' = 3y^{2/3}, \qquad y(-1) = -1$$

has infinitely many solutions. Why does this not contradict the theorem stated in this section?

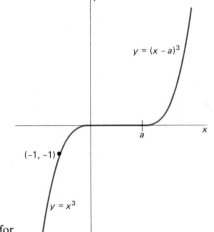

FIGURE 1.32 Figure for Problem 39.

40. Verify that if k is a constant, then $y = kx$ satisfies the differential equation $xy' = y$. Hence conclude that the initial value problem

$$xy' = y, \qquad y(0) = 0$$

has infinitely many solutions on any open interval containing $x = 0$.

41. Use isoclines to construct a slope field for the differential equation $xy' = y$. Explain why this slope field suggests that the initial value problem

$$xy' = y, \qquad y(a) = b$$

has (i) a unique solution if $a \neq 0$; (ii) no solution if $a = 0$ and $b \neq 0$; (iii) infinitely many solutions if $a = b = 0$. Are these results consistent with the existence-uniqueness theorem of this section?

42. Consider the differential equation $y' = 4x\sqrt{y}$ for $y \geq 0$. Apply the existence and uniqueness theorem of this section to find those points (a, b) such that a solution of the differential equation *must* exist on an open interval J containing a. Find those points (a, b) such that a unique solution exists on an open interval J_0 containing a. Then find how many solutions pass through $(0, 0)$. (*Suggestion:* Note that $y_1(x) \equiv 0$ is a solution, as is $y_2(x) = (x^2 + C)^2$ if $x^2 + C \geq 0$.)

1.4

Separable Equations and Applications

The first order differential equation

$$\frac{dy}{dx} = H(x, y) \tag{1}$$

is called *separable* provided that $H(x, y)$ can be written as the product of a function of x and a function of y or, equivalently, as a quotient $H(x, y) = g(x)/f(y)$. In this case the variables x and y can be *separated*—isolated on opposite sides of an equation—by writing informally the equation $f(y)\,dy = g(x)\,dx$, which we understand to be compact notation for the differential equation

$$f(y)\frac{dy}{dx} = g(x). \tag{2}$$

It is easy to solve this special type of differential equation simply by integrating both sides with respect to x:

$$\int f(y(x))\frac{dy}{dx}\,dx = \int g(x)\,dx + C;$$

more concisely,

$$\int f(y)\, dy = \int g(x)\, dx + C. \tag{3}$$

All that is required is that the antiderivatives $F(y) = \int f(y)\, dy$ and $G(x) = \int g(x)\, dx$ can be found. To see that (2) and (3) are equivalent, note the following consequence of the chain rule:

$$D_x F(y(x)) = F'(y(x))y'(x) = f(y)\frac{dy}{dx} = g(x) = D_x G(x),$$

which in turn is equivalent to

$$F(y(x)) = G(x) + C, \tag{4}$$

because two functions have the same derivative on an interval if and only if they differ by a constant on that interval.

EXAMPLE 1 Solve the initial value problem

$$\frac{dy}{dx} = -6xy, \qquad y(0) = 7.$$

SOLUTION Informally, we divide each side of the differential equation by y and multiply each side by dx to get

$$\frac{dy}{y} = -6x\, dx.$$

Hence

$$\int \frac{dy}{y} = \int (-6x)\, dx;$$

$$\ln |y| = -3x^2 + C.$$

We see from the initial condition $y(0) = 7$ that $y(x)$ is positive near $x = 0$, so we may delete the absolute value symbols:

$$\ln y = -3x^2 + C,$$

and hence

$$y(x) = e^{-3x^2 + C} = e^{-3x^2}e^C = Ae^{-3x^2},$$

where $A = e^C$. The condition $y(0) = 7$ yields $A = 7$, so the desired solution is

$$y(x) = 7e^{-3x^2}.$$

EXAMPLE 2 Solve the differential equation

$$x^2 \frac{dy}{dx} = \frac{x^2 + 1}{3y^2 + 1}.$$

SOLUTION   We multiply each side by the formal expression $(3y^2 + 1)\,dx/x^2$ to obtain

$$(3y^2 + 1)\,dy = \left(1 + \frac{1}{x^2}\right)dx.$$

Integration of both sides then gives

$$y^3 + y = x - \frac{1}{x} + C.$$

As Example 2 illustrates, it may not be possible or practical to solve Eq. (4) explicitly for y as a function of x. If not, we call (4) an *implicit solution* of the differential equation in (2). Because Eq. (4) contains the arbitrary constant C, we also call it a **general solution** of (2). Given an initial condition $y(x_0) = y_0$, the choice $C_0 = F(y_0) - G(x_0)$ of C yields the equation

$$F(y) = G(x) + C_0$$

that implicitly defines a **particular solution** (if any) of the initial value problem

$$f(y)\frac{dy}{dx} = g(x), \qquad y(x_0) = y_0.$$

For instance, to solve the initial value problem

$$x^2\frac{dy}{dx} = \frac{x^2 + 1}{3y^2 + 1}, \qquad y(1) = 2,$$

we substitute $x = 1$ and $y = 2$ in the general solution found in Example 2. We find that $C = 10$, so the desired particular solution is defined implicitly through the equation

$$y^3 + y = x - \frac{1}{x} + 10.$$

Although it would not be convenient to solve this equation explicitly for y as a function of x, it can be solved numerically for y if a specific value of x is substituted. Thus if $x = 2$, we find that the resulting equation $y^3 + y = 11.5$ has the unique real solution $y \approx 2.10973$. It follows that $y(2) \approx 2.10973$ for the solution $y(x)$ of the initial value problem above.

In general, the equation $K(x, y) = 0$ is called an **implicit solution** of a differential equation if it is satisfied (on some interval) by some solution $y = y(x)$ of the differential equation. But note that a particular solution $y = y(x)$ of $K(x, y) = 0$ may or may not satisfy a given initial condition. For example, differentiation of $x^2 + y^2 = 4$ yields

$$x + y\frac{dy}{dx} = 0,$$

so $x^2 + y^2 = 4$ is an implicit solution of the differential equation $x + yy' = 0$. But only the first of the two explicit solutions $y = +(4 - x^2)^{1/2}$ and $y = -(4 - x^2)^{1/2}$ satisfies the initial condition $y(0) = 2$.

The argument preceding Example 1 shows that every particular solution of (2) satisfies (4) for some choice of C; *this* is why it is appropriate to call (4) a general solution of (2).

Warning: Suppose, however, that we begin with the differential equation

$$\frac{dy}{dx} = g(x)h(y) \tag{5}$$

and divide by $h(y)$ to obtain the separated equation

$$\frac{1}{h(y)}\frac{dy}{dx} = g(x). \tag{6}$$

If y_0 is a root of the equation $h(y) = 0$—that is, if $h(y_0) = 0$—then the constant function $y(x) \equiv y_0$ is clearly a solution of (5), but may *not* be contained in the general solution of (6). Thus solutions of a differential equation may be lost upon division by a vanishing factor. (Indeed, false solutions may be gained upon multiplication by a vanishing factor. This phenomenon is similar to the introduction of extraneous roots in solving algebraic equations.)

In Section 1.5 we shall see that every particular solution of a *linear* first order differential equation is contained in its general solution. By contrast, it is common for a nonlinear first order differential equation to have both a general solution involving an arbitrary constant C and one or several particular solutions that cannot be obtained by selecting a value for C. These exceptional solutions are frequently called **singular solutions**. In Example 4 of Section 1.3, we found that the solution $y(x) \equiv 0$ was a singular solution of the equation $y^2 + x^2 y' = 0$; this solution cannot be obtained from the general solution $y(x) = x/(Cx - 1)$ by any choice of the constant C.

EXAMPLE 3   Find all solutions of the differential equation

$$\frac{dy}{dx} = 2x\sqrt{y - 1}.$$

SOLUTION   We note first the constant solution $y(x) \equiv 1$. If $y \neq 1$, we can divide each side by $\sqrt{y - 1}$ to obtain

$$(y - 1)^{-1/2}\frac{dy}{dx} = 2x.$$

Integration gives $2\sqrt{y - 1} = x^2 + C$; upon solving for y we get the general solution

$$y(x) = 1 + \tfrac{1}{4}(x^2 + C)^2.$$

Note that no value of C gives the particular solution $y \equiv 1$. It was lost when we divided by $\sqrt{y - 1}$.

EXAMPLE 4   Solve the initial value problem

$$\frac{dy}{dx} = xy + x - 2y - 2, \qquad y(0) = 2.$$

SOLUTION   Sometimes a factorization is not obvious at first glance; here we have

$$\frac{dy}{dx} = (x - 2)(y + 1).$$

The constant function $y \equiv -1$ satisfies the differential equation but does not satisfy the initial condition, so we cannot lose the solution of the initial value problem by dividing by $y + 1$. We therefore perform that division and integrate:

$$\int \frac{dy}{y + 1} = \int (x - 2)\, dx;$$

$$\ln |y + 1| = \frac{1}{2} x^2 - 2x + C;$$

$$\ln (y + 1) = \frac{1}{2} x^2 - 2x + C.$$

In the final step we use the fact that $y + 1 > 0$ near the initial value $y = 2$. Now apply the exponential function to each side of the last equation. With the observation that $e^{\ln z} = z$, we get

$$y + 1 = \exp\left(\tfrac{1}{2} x^2 - 2x + C\right),$$

so that

$$y = A \exp\left(\tfrac{1}{2} x^2 - 2x\right) - 1$$

where $A = e^C$. Note in passing that we did not lose the constant solution $y \equiv -1$; it corresponds to $A = 0$. But to conclude the example, the initial condition $y(0) = 2$ implies that $A = 3$, so the desired solution is

$$y(x) = 3 \exp\left(\tfrac{1}{2} x^2 - 2x\right) - 1.$$

NATURAL GROWTH AND DECAY

The differential equation

$$\frac{dx}{dt} = kx \qquad (k \text{ a constant}) \tag{7}$$

serves as a mathematical model for a remarkably wide range of natural phenomena—any involving a quantity whose time rate of change is proportional to its current value. Here are some examples.

POPULATION GROWTH   Suppose that $P(t)$ is the number of individuals in a population (of humans, or insects, or bacteria) having *constant* birth and death rates β and δ (in births or deaths per individual per unit of time). Then, during a short time interval Δt, approximately $\beta P(t)\, \Delta t$ births and $\delta P(t)\, \Delta t$ deaths occur, so the change in $P(t)$ is given approximately by

$$\Delta P = (\beta - \delta)P(t)\, \Delta t,$$

and therefore

$$\frac{dP}{dt} = \lim_{\Delta t \to 0} \frac{\Delta P}{\Delta t} = kP \tag{8}$$

where $k = \beta - \delta$.

COMPOUND INTEREST   Let $A(t)$ be the number of dollars in a savings account at time t, and suppose that the interest is *compounded continuously* at an annual interest rate r. (Note that 10% annual interest means that $r = 0.10$.) Continuous compounding means that, during a short time interval Δt, the amount of interest added to the account is approximately $\Delta A = rA(t) \Delta t$, so that

$$\frac{dA}{dt} = \lim_{\Delta t \to 0} \frac{\Delta A}{\Delta t} = rA. \tag{9}$$

RADIOACTIVE DECAY   Consider a sample of material that contains $N(t)$ atoms of a certain radioactive isotope at time t. It has been observed that a constant fraction of these radioactive atoms will spontaneously decay (into atoms of another element or into another isotope of the same element) during each unit of time. Consequently, the sample behaves exactly like a population with a constant death rate, but with no births occurring. To write a model for $N(t)$, we use Eq. (8) with N in place of P, with $k > 0$ in place of δ, and with $\beta = 0$. We thus get the differential equation

$$\frac{dN}{dt} = -kN. \tag{10}$$

The value of k depends upon the particular radioactive isotope.

The key to the method of *radiocarbon dating* is that a constant proportion of the carbon atoms in any living creature is made up of the radioactive isotope C^{14} of carbon. This proportion remains constant because the fraction of C^{14} in the atmosphere remains almost constant, and living matter is continuously taking up carbon from the air or is consuming other living matter containing the same constant ratio of C^{14} atoms to ordinary carbon atoms. The same ratio permeates all life, because organic processes seem to make no distinction between the two isotopes.

The ratio of C^{14} to normal carbon remains constant in the atmosphere because, though C^{14} is radioactive and slowly decays, the amount is continuously replenished through the conversion of nitrogen to C^{14} by cosmic rays in the upper atmosphere. Over the long history of the planet, this decay and replenishment process has come into nearly steady state.

Of course, when a living organism dies, it ceases its metabolism of carbon, and the process of radioactive decay begins to deplete its C^{14} content. There is no replenishment of C^{14}, and consequently the ratio of C^{14} to normal carbon begins to drop. By measuring this ratio, the amount of time elapsed since the death of the organism can be estimated. For such purposes, it is necessary to measure the decay constant; for C^{14}, it's known that k is approximately 0.0001216.

(Matters are not so simple as we have made them appear. In applying the technique of radiocarbon dating, extreme care must be taken to avoid contaminating the sample with organic matter, or even with ordinary fresh air. In addition, the cosmic ray levels apparently have not been constant, so the ratio of radioactive carbon in the atmosphere has varied over the past centuries. By using independent methods of dating samples, researchers in this area have compiled tables of correction factors to enhance the accuracy of the process.)

DRUG ELIMINATION   In many cases the amount $A(t)$ of a certain drug in the bloodstream, measured by the excess over the natural level of the drug, will decline at a rate proportional to the current excess amount. That is,

$$\frac{dA}{dt} = -\lambda A \tag{11}$$

where $\lambda > 0$. The parameter λ is called the **elimination constant** of the drug.

The prototype differential equation $x' = kx$ with $x(t) > 0$ and k a constant (either positive or negative) is readily solved by separating the variables and integrating:

$$\int \frac{1}{x} \, dx = \int k \, dt;$$

$$\ln x = kt + C.$$

Then we solve for x:

$$e^{\ln x} = e^{kt+C};$$

$$x = e^C e^{kt} = A e^{kt}.$$

Because C is a constant, so is $A = e^C$. It is also clear that $A = x(0) = x_0$, so the particular solution of Eq. (7) with the initial condition $x(0) = x_0$ is simply

$$x(t) = x_0 e^{kt}. \tag{12}$$

Because of the presence of the natural exponential function in its solution, the differential equation

$$x' = kx \tag{13}$$

is often called the **exponential** or **natural growth equation**. Figure 1.33 shows a typical graph of $x(t)$ in the case $k > 0$; the case $k < 0$ is illustrated in Fig. 1.34.

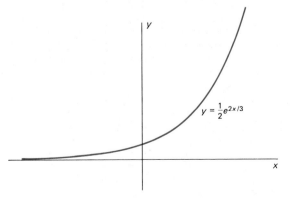

$$y = \frac{1}{2} e^{2x/3}$$

FIGURE 1.33   Natural growth.

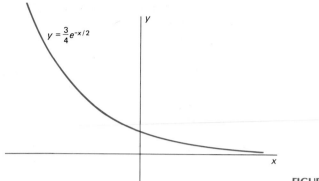

$$y = \frac{3}{4}e^{-x/2}$$

FIGURE 1.34   Natural decay.

EXAMPLE 5   In mid-1982, the world population was 4.5 billion and it was then increasing at the rate of a quarter million persons each day. Assuming constant birth and death rates, when should a world population of 10 billion be expected?

SOLUTION   From Eq. (12) with P in place of x, we know that $P(t) = P_0 e^{kt}$. We measure the world population $P(t)$ in billions and measure time t in years. We shall take $t = 0$ to correspond to 1982, so that $P_0 = 4.5$. The fact that P is increasing by a quarter million, or 0.00025 billion, persons per day at time $t = 0$ means that

$$P'(0) = (0.00025)(365.25) \approx 0.0913$$

billion per year. From Eq. (8) we now obtain

$$k = \frac{P'(0)}{P(0)} \approx \frac{0.0913}{4.5} \approx 0.0203.$$

Thus the population is growing at the rate of about 2.03% per year.

To find when the population will be 10 billion, we need only solve the equation

$$10 = P(T) = (4.5)e^{(0.0203)T}$$

for

$$T = \frac{\ln(10/4.5)}{0.0203} \approx 39 \text{ years,}$$

which corresponds to the year 2021.

The decay constant of a radioactive isotope is often specified in terms of another empirical constant, the *half-life* of the isotope, because this parameter is more convenient. The **half-life** τ of a radioactive isotope is the time required for *half* of it to decay. To find the relationship between k and τ, we set $t = \tau$ and $N = \frac{1}{2}N_0$ in the equation $N(t) = N_0 e^{-kt}$, so that $\frac{1}{2}N_0 = N_0 e^{-k\tau}$. When we solve for τ, we find that

$$\tau = \frac{\ln 2}{k}. \tag{14}$$

For example, the half-life of C^{14} is $\tau = (\ln 2)/(0.0001216)$, approximately 5700 years.

CHAPTER 1:   Introduction and First Order Differential Equations

EXAMPLE 6 A specimen of charcoal found at Stonehenge turns out to contain 63% as much C^{14} as a sample of present-day charcoal of equal mass. What is the age of the sample?

SOLUTION We take $t = 0$ as the time of the death of the tree from which the Stonehenge charcoal was made and N_0 as the number of C^{14} atoms it contained then. We are given that $N = (0.63)N_0$ now, so we solve the equation $(0.63)N_0 = N_0 e^{-kt}$ with the value $k = 0.0001216$. Thus we find that

$$t = -\frac{\ln (0.63)}{0.0001216} \approx 3800$$

years. The sample is about 3800 years old, and if it has any connection with the builders of Stonehenge, our computations suggest that this observatory, monument, or temple—whichever it may be—dates from 1800 B.C. or earlier.

COOLING AND HEATING

According to Newton's law of cooling (Eq. (3) of Section 1.1), the time rate of change of the temperature $T(t)$ of a body immersed in a medium of constant temperature A is proportional to the difference $A - T$. That is,

$$\frac{dT}{dt} = k(A - T) \tag{15}$$

where k is a positive constant. This is an instance of the linear first order differential equation with constant coefficients:

$$\frac{dx}{dt} = ax + b. \tag{16}$$

It includes the exponential equation as a special case ($b = 0$) and is also easy to solve by separation of variables.

EXAMPLE 7 A 5-lb roast, initially at 50°F, is put into a 375°F oven at 5:00 P.M.; it is found that the temperature $T(t)$ of the roast is 125°F after 75 min. When will the roast be 150°F (medium rare)?

SOLUTION We take time t in minutes, with $t = 0$ corresponding to 5:00 P.M. We also assume (not altogether realistically) that at any instant the temperature $T(t)$ of the roast is uniform throughout. We have $T(t) < A = 375$, $T(0) = 50$, and $T(75) = 125$. Hence

$$\frac{dT}{dt} = k(375 - T);$$

$$\int \frac{1}{375 - T} \, dT = \int k \, dt;$$

$$-\ln (375 - T) = kt + C;$$

$$375 - T = Be^{-kt}.$$

Now $T(0) = 50$ implies that $B = 325$, so $T(t) = 375 - 325e^{-kt}$. We also know that $T = 125$ when $t = 75$. Substitution of these values in the last equation yields

$$k = -\tfrac{1}{75} \ln \left(\tfrac{250}{325}\right) \approx 0.0035.$$

Hence we finally solve the equation

$$150 = 375 - 325e^{(-0.0035)t}$$

for $t = -[\ln (225/325)]/(0.0035) \approx 105$ min, the total cooking time required. Because the roast was put in the oven at 5:00 P.M., it should be removed at about 6:45 P.M.

TORRICELLI'S LAW

Suppose that a water tank has a hole with area a at its bottom, from which water is leaking. Denote by $y(t)$ the depth of the water in the tank at time t, and by $V(t)$ the volume of water in the tank then. It is plausible—as well as true under ideal conditions—that the velocity of water exiting through the hole is

$$v = \sqrt{2gy}, \tag{17}$$

which is the velocity a drop of water would acquire in falling freely from the surface of the water to the hole (see Problem 29 of Section 1.2). Under real conditions, taking into account the constriction of a water jet from an orifice, $v = c\sqrt{2gy}$ where c is an empirical constant between 0 and 1 (usually about 0.6 for a small continuous stream of water). For simplicity we take $c = 1$ in the following discussion.

As a consequence of Eq. (17) we have

$$\frac{dV}{dt} = -av = -a\sqrt{2gy}; \tag{18}$$

this is a statement of Torricelli's law for a draining tank. If $A(y)$ denotes the horizontal cross-sectional area of the tank at height y, then the method of volume by cross sections gives

$$V = \int_0^y A(y)\, dy,$$

so the fundamental theorem of calculus implies that $dV/dy = A(y)$ and therefore that

$$\frac{dV}{dt} = \frac{dV}{dy}\frac{dy}{dt} = A(y)\frac{dy}{dt}. \tag{19}$$

From Eqs. (18) and (19) we finally obtain

$$A(y)\frac{dy}{dt} = -a\sqrt{2gy}, \tag{20}$$

an alternative form of Torricelli's law.

EXAMPLE 8   A hemispherical tank has top radius 4 ft and at time $t = 0$ is full of water. At that moment a circular hole of diameter 1 in. is opened in the bottom of the tank. How long will it take for all the water to drain from the tank?

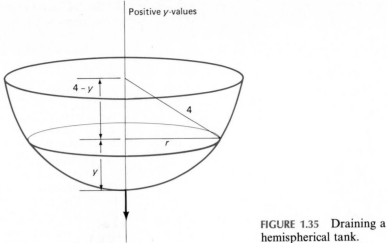

Positive *y*-values

4 − *y*

4

r

y

FIGURE 1.35 Draining a hemispherical tank.

SOLUTION   From the right triangle in Fig. 1.35, we see that

$$A(y) = \pi r^2 = \pi[16 - (4-y)^2] = \pi(8y - y^2).$$

With $g = 32$ ft/s², Eq. (20) becomes

$$(8y - y^2)\frac{dy}{dt} = -\left(\frac{1}{24}\right)^2 \sqrt{64y};$$

$$\int (8y^{1/2} - y^{3/2})\, dy = -\int \frac{dt}{72};$$

$$\frac{16}{3} y^{3/2} - \frac{2}{5} y^{5/2} = -\frac{1}{72} t + C.$$

Now $y(0) = 4$, so $C = \frac{16}{3}(4)^{3/2} - \frac{2}{5}(4)^{5/2} = \frac{448}{15}$. The tank is empty when $y = 0$, thus when

$$t = (72)(\tfrac{448}{15}) \approx 2150 \text{ s};$$

that is, about 35 min 50 s. So it takes slightly less than 36 min for the tank to drain.

1.4   PROBLEMS

Find general solutions (implicit if necessary, explicit if convenient) of the differential equations in Problems 1–18. Primes denote derivatives with respect to x.

1. $y' + 2xy = 0$

2. $y' + 2xy^2 = 0$

3. $y' = y \sin x$

4. $(1 + x)y' = 4y$

5. $2\sqrt{x}\,\dfrac{dy}{dx} = \sqrt{1 - y^2}$

6. $\dfrac{dy}{dx} = 3\sqrt{xy}$

7. $\dfrac{dy}{dx} = (64xy)^{1/3}$

8. $\dfrac{dy}{dx} = 2x \sec y$

9. $(1 - x^2)\dfrac{dy}{dx} = 2y$

10. $(1 + x)^2 \dfrac{dy}{dx} = (1 + y)^2$

11. $\dfrac{dy}{dx} = xy^3$

12. $y\dfrac{dy}{dx} = x(y^2 + 1)$

13. $y^3 \dfrac{dy}{dx} = (y^4 + 1)\cos x$

14. $\dfrac{dy}{dx} = \dfrac{1 + \sqrt{x}}{1 + \sqrt{y}}$

15. $\dfrac{dy}{dx} = \dfrac{(x-1)y^5}{x^2(2y^3 - y)}$

16. $(x^2 + 1)\dfrac{dy}{dx}\tan y = x$

17. $\dfrac{dy}{dx} = 1 + x + y + xy$

18. $x^2 y' = 1 - x^2 + y^2 - x^2 y^2$

Find explicit particular solutions of the initial value problems in Problems 19–26.

19. $\dfrac{dy}{dx} = y e^x;\quad y(0) = 2e$

20. $\dfrac{dy}{dx} = 3x^2(y^2 + 1);\quad y(0) = 1$

21. $2y\dfrac{dy}{dx} = x(x^2 - 16)^{-1/2};\quad y(5) = 2$

22. $\dfrac{dy}{dx} = 4x^3 y - y;\quad y(1) = -3$

23. $\dfrac{dy}{dx} + 1 = 2y;\quad y(1) = 1$

24. $y'\tan x = y;\quad y\left(\dfrac{\pi}{2}\right) = \dfrac{\pi}{2}$

25. $x\dfrac{dy}{dx} - y = 2x^2 y;\quad y(1) = 1$

26. $\dfrac{dy}{dx} = 2xy^2 + 3x^2 y^2;\quad y(1) = -1$

27. (Population growth) A certain city had a population of 25,000 in 1960 and a population of 30,000 in 1970. Assume that its population will continue to grow exponentially at a constant rate. What population can its city planners expect in the year 2000?

28. (Population growth) In a certain culture of bacteria, the number of bacteria increased sixfold in 10 h. How long did it take for the population to double its initial number?

29. (Radiocarbon dating) Carbon extracted from an ancient skull contained only one-sixth as much radioactive C^{14} as carbon extracted from present-day bone. How old is the skull?

30. (Radiocarbon dating) Carbon taken from a purported relic of the time of Christ contained 4.6×10^{10} atoms of C^{14} per gram. Carbon extracted from a present-day specimen of the same substance contained 5.0×10^{10} atoms of C^{14} per gram. Compute the approximate age of the relic. What is your opinion as to its authenticity?

31. (Continuously compounded interest) Upon the birth of their first child, a couple deposited $5000 in a savings account that draws 8% interest compounded continuously. The interest payments are allowed to accumulate. How much will the account contain on the child's eighteenth birthday?

32. (Continuously compounded interest) Suppose that you discover in your attic an overdue library book on which your great-great-grandfather owed a fine of 30 cents 100 years ago.

If an overdue fine grows exponentially at a 5% annual rate compounded continuously, how much would you have to pay if you returned the book today?

33. (Drug elimination)   Suppose that sodium pentobarbitol is used to anesthetize a dog: The dog is anesthetized when its bloodstream concentration contains at least 45 milligrams (mg) of sodium pentobarbitol per kilogram of the dog's body weight. Suppose also that sodium pentobarbitol is eliminated exponentially from the dog's bloodstream, with a half-life of 5 h. What single dose should be administered in order to anesthetize a 50-kg dog for 1 h?

34. The half-life of radioactive cobalt is 5.27 years. Suppose that a nuclear accident has left the level of cobalt radiation in a certain region at 100 times the level acceptable for human habitation. How long will it be before the region is again habitable? (Ignore the probable presence of other radioactive elements.)

35. Suppose that a mineral body formed in an ancient cataclysm—perhaps the formation of the earth itself—originally contained the uranium isotope U^{238} (which has a half-life of 4.51×10^9 years) but no lead, the end product of the radioactive decay of U^{238}. If today the ratio of U^{238} atoms to lead atoms in the mineral body is 0.9, when did the cataclysm occur?

36. A certain moon rock was found to contain equal numbers of potassium and argon atoms. Assume that all the argon is the result of radioactive decay of potassium (its half-half is about 1.28×10^9 years) and that one of every nine potassium atom disintegrations yields an argon atom. What is the age of the rock, measured from the time it contained only potassium?

37. A pitcher of buttermilk initially at 25°C is to be cooled by setting it on the front porch, where the temperature is 0°C. Suppose that the temperature of the buttermilk has dropped to 15°C after 20 min. When will it be at 5°C?

38. When sugar is dissolved in water, the amount A that remains undissolved after t minutes satisfies the differential equation $dA/dt = -kA \ (k > 0)$. If 25% of the sugar dissolves after 1 min, how long does it take for half the sugar to dissolve?

39. The intensity I of light at a depth x meters below the surface of a lake satisfies the differential equation $dI/dx = (-1.4)I$. (a) At what depth is the intensity half the intensity I_0 at the surface (where $x = 0$)? (b) What is the intensity at a depth of 10 m (as a fraction of I_0)? (c) At what depth will the intensity be $\frac{1}{100}$ of that at the surface?

40. The barometric pressure p (in inches of mercury) at an altitude x miles above sea level satisfies the initial value problem $dp/dx = (-0.2)p,\ p(0) = 29.92$. (a) Calculate the barometric pressure at 10,000 ft and again at 30,000 ft. (b) Without prior conditioning, few people can survive when the pressure drops to less than 15 in. of mercury. How high is that?

41. Consider a savings account that contains A_0 dollars initially and earns interest at the annual rate r compounded continuously. Suppose that deposits are added to this account at the rate of Q dollars per year. To simplify the mathematical model, assume that these deposits are made continuously rather than (for instance) monthly. (a) Derive the differential equation for the amount $A(t)$ in the account at time t years. (b) Suppose that you wish to arrange, at the time of her birth, for your daughter to have $40,000 available for college expenses at her eighteenth birthday. You plan to do so by making frequent small—essentially continuous—deposits in a savings account, at the rate of Q thousand dollars each year. This account accumulates interest at 11% annual interest compounded continuously. What should Q be so that you may achieve your goal?

42. According to one cosmological theory, there were equal amounts of the two uranium isotopes U^{235} and U^{238} at the creation of the universe in the "big bang." At present there are 137.7 U^{238} atoms for each atom of U^{235}. Using the half-lives 4.51 billion years for U^{238} and 0.71 billion years for U^{235}, calculate the age of the universe.

43. A cake is removed from an oven at 210°F and left to cool at room temperature, which is 70°F. After 30 min the temperature of the cake is 140°F. When will it be 100°F?

44. (a) Payments are made on a mortgage (original loan) of P_0 dollars continuously at the constant rate of c dollars per month. Let $P(t)$ denote the principal (amount still owed) after t months, and let r denote the monthly interest rate paid by the borrower (for instance, $r = 0.12/12 = 0.01$ if the annual interest rate is 12%). Derive the differential equation

$$\frac{dP}{dt} = rP - c, \qquad P(0) = P_0.$$

(b) An automobile loan of $10,800 is to be paid off continuously over a period of 60 months. Determine the monthly payment required if the annual interest rate is (i) 12%; (ii) 18%.

45. A certain piece of dubious information about phenylthiourea in the drinking water began to spread one day in a city with a population of 100,000. Within a week, 10,000 people had heard this rumor. Assume that the rate of increase of the number who have heard the rumor is proportional to the number who have not yet heard it. How long will it be before half the population of the city has heard this piece of information?

46. A tank is shaped like a vertical cylinder; it initially contains water to a depth of 9 ft, and a bottom plug is pulled at time $t = 0$ (hours). After 1 h the depth has dropped to 4 ft. How long does it take for all the water to run out of this tank?

47. Suppose that the tank of Problem 46 has a radius of 3 ft and that its bottom hole is circular with radius 1 in. How long will it take the water (initially 9 ft deep) to drain completely?

48. A cylindrical tank with length 5 ft and radius 3 ft is situated with its axis horizontal. If a circular bottom hole with a radius of 1 in. is opened, and the tank is initially half full of xylene, how long will it take for the liquid to drain completely?

49. A spherical tank with radius 4 ft is full of gasoline when a circular bottom hole with radius 1 in. is opened. How long will be required for all the gasoline to drain from the tank?

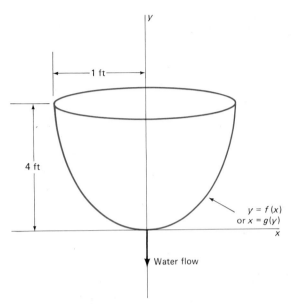

FIGURE 1.36 The clepsydra.

50. (The clepsydra, or water clock) A 12-h water clock is to be designed with the dimensions shown in Fig. 1.36, shaped like the surface obtained by revolving the curve $y = f(x)$ around the y-axis. What should be this curve, *and* what should be the radius of the circular bottom hole, in order that the water level will fall at the *constant* rate of 4 inches per hour (in./h)?

51. Suppose that a cylindrical tank initially containing V_0 gallons of water drains (through a bottom hole) in T minutes. Use Torricelli's law to show that the volume of water in the tank after $t \leq T$ minutes is $V = V_0[1 - (t/T)]^2$.

52. At time $t = 0$ the bottom plug (at the vertex) of a full conical water tank 16 ft high is removed. After 1 h the water in the tank is 9 ft deep. When will the tank be empty?

53. Just before midday the body of an apparent homicide victim is found in a room that is kept at a constant temperature of 70°F. At 12 noon the temperature of the body is 80°F and at 1 P.M. it is 75°F. Assume that the temperature of the body at the time of death was 98.6°F and that it has cooled in accord with Newton's law. What was the time of death?

54. Early one morning it began to snow at a constant rate. At 7:00 A.M. a snowplow set off to clear a road. By 8:00 A.M. it had traveled 2 miles, but it took two more hours (until 10:00 A.M.) for the snowplow to go an additional 2 miles. (a) Let $t = 0$ when it began to snow and let x denote the distance traveled by the snowplow at time t. Assuming that the snowplow clears snow from the road at a constant rate (in cubic feet per hour, say) show that

$$k\frac{dx}{dt} = \frac{1}{t}$$

where k is a constant. (b) What time did it start snowing? (*Answer:* 6:00 A.M.)

55. A snowplow sets off at 7:00 A.M. as in Problem 54. Suppose now that by 8:00 A.M. it had traveled 4 miles and that by 9:00 A.M. it had moved an additional 3 miles. What time did it start snowing? This is a more difficult snowplow problem because now a transcendental equation must be solved numerically in order to find the value of k. (*Answer:* 4:27 A.M.)

1.5

Linear First Order Equations

In Section 1.4 we saw how to solve a separable differential equation by integrating *after* multiplying each side by an appropriate factor. For instance, to solve the equation

$$\frac{dy}{dx} = 2xy \qquad (y > 0), \tag{1}$$

we multiply each side by the factor $1/y$ to get

$$\frac{1}{y}\frac{dy}{dx} = 2x; \quad \text{that is,} \quad D_x[\ln y] = D_x[x^2]. \tag{2}$$

Because each side of the equation in (2) is recognizable as a *derivative* (with respect to the independent variable x), all that remains is a simple integration, which yields $\ln y = x^2 + C$. For this reason, the function $\rho = 1/y$ is called an *integrating factor* for the original equation in (1). An **integrating factor** for a differential equation is a function $\rho(x, y)$ such that multiplication of each side of the differential equation by $\rho(x, y)$ yields an equation in which each side is recognizable as a derivative.

With the aid of the appropriate integrating factor, there is a standard technique for solving the ***linear* first order equation**

$$\frac{dy}{dx} + P(x)y = Q(x) \tag{3}$$

on an interval where the coefficient functions $P(x)$ and $Q(x)$ are continuous. We multiply each side in Eq. (3) by the integrating factor

$$\rho = \rho(x) = e^{\int P(x)\,dx}. \tag{4}$$

The result is

$$e^{\int P(x)\,dx}\frac{dy}{dx} + P(x)e^{\int P(x)\,dx}y = Q(x)e^{\int P(x)\,dx}. \tag{5}$$

Because $D_x[\int P(x)\,dx] = P(x)$, the left-hand side is the derivative of the *product* $y \cdot e^{\int P(x)\,dx}$, so (5) is equivalent to

$$D_x[y(x)e^{\int P(x)\,dx}] = Q(x)e^{\int P(x)\,dx}.$$

Integration of both sides of this equation gives

$$y(x)e^{\int P(x)\,dx} = \int \left(Q(x)e^{\int P(x)\,dx}\right) dx + C.$$

Finally solving for y, we obtain the general solution of the linear first order equation in (3):

$$y(x) = e^{-\int P(x)\,dx}\left[\int \left(Q(x)e^{\int P(x)\,dx}\right) dx + C\right]. \tag{6}$$

The formula in (6) should not be memorized. In a specific problem it is generally simpler to use the *method* by which we developed this formula. That is, in order to solve Eq. (3), carry out the following steps:

1. Begin by calculating the integrating factor

   $$\rho(x) = e^{\int P(x)\,dx}.$$

2. Then multiply each side of the differential equation by $\rho(x)$.
3. Next, recognize the left-hand side of the resulting equation as the derivative of a product:

   $$D_x[\rho(x)y(x)] = \rho(x)Q(x).$$

4. Finally, integrate this equation,

   $$\rho(x)y(x) = \int \rho(x)Q(x)\,dx + C,$$

   then solve for y to obtain the general solution of the original differential equation.

Moreover, given an initial condition $y(x_0) = y_0$, we can substitute $x = x_0$ and $y = y_0$ in (6) to solve for the value of C yielding the particular solution of (3) that satisfies this initial condition.

The integrating factor $\rho(x)$ is determined only to within a multiplicative constant. If we replace $\int P(x)\,dx$ by $\int P(x)\,dx + c$ in (4), the result is

$$\rho(x) = e^{(\int P(x)\,dx)+c} = e^c e^{\int P(x)\,dx}.$$

But the constant factor e^c does not affect the result of multiplying both sides of the differential equation in (3) by $\rho(x)$. Hence we may choose for $\int P(x)\,dx$ any convenient antiderivative of $P(x)$.

EXAMPLE 1 Solve the initial value problem

$$\frac{dy}{dx} - 3y = e^{2x}, \qquad y(0) = 3.$$

SOLUTION Here we have $P(x) = -3$ and $Q(x) = e^{2x}$, so the integrating factor is

$$\rho = e^{\int(-3)\,dx} = e^{-3x}.$$

Multiplication of each side of the given equation by e^{-3x} yields

$$e^{-3x}\frac{dy}{dx} - 3e^{-3x}y = e^{-x},$$

which we recognize as

$$\frac{d}{dx}(e^{-3x}y) = e^{-x}.$$

Hence integration with respect to x gives

$$e^{-3x}y = \int e^{-x}\, dx = -e^{-x} + C,$$

so multiplication by e^{3x} yields the general solution

$$y(x) = Ce^{3x} - e^{2x}.$$

Substitution of $x = 0$ and $y = 3$ now gives $C = 4$. Thus the desired particular solution is

$$y(x) = 4e^{3x} - e^{2x}.$$

EXAMPLE 2 Find a general solution of

$$(x^2 + 1)\frac{dy}{dx} + 3xy = 6x.$$

SOLUTION After division of each side of the equation by $x^2 + 1$, we recognize the result

$$\frac{dy}{dx} + \frac{3x}{x^2 + 1}y = \frac{6x}{x^2 + 1}$$

as a first order linear equation with $P(x) = 3x/(x^2 + 1)$ and $Q(x) = 6x/(x^2 + 1)$. Multiplication by

$$\rho = \exp\left(\int \frac{3x}{x^2 + 1}\, dx\right)$$

$$= \exp\left(\frac{3}{2}\ln(x^2 + 1)\right) = (x^2 + 1)^{3/2}$$

yields

$$(x^2 + 1)^{3/2}\frac{dy}{dx} + 3x(x^2 + 1)^{1/2}y = 6x(x^2 + 1)^{1/2},$$

and thus

$$D_x[(x^2 + 1)^{3/2}y] = 6x(x^2 + 1)^{1/2}.$$

Integration then yields

$$(x^2 + 1)^{3/2}y = \int 6x(x^2 + 1)^{1/2}\, dx = 2(x^2 + 1)^{3/2} + C.$$

Multiplication of both sides by $(x^2 + 1)^{-3/2}$ gives the general solution

$$y(x) = 2 + C(x^2 + 1)^{-3/2}$$

Frequently the integrations are not so simple as those in Examples 1 and 2. We may apply the fundamental theorem of calculus, however, to write an antiderivative of an arbitrary continuous function $f(x)$ in the form

$$\int_a^x f(t)\, dt.$$

EXAMPLE 3   Solve the initial value problem

$$x^2 \frac{dy}{dx} + xy = \sin x, \qquad y(1) = 2.$$

SOLUTION   Division by x^2 gives

$$\frac{dy}{dx} + \frac{1}{x} y = \frac{\sin x}{x^2},$$

so the integrating factor is $\rho = e^{\int (1/x)\, dx} = e^{\ln x} = x$. Hence

$$x \frac{dy}{dx} + y = \frac{\sin x}{x};$$

$$D_x(xy) = \frac{\sin x}{x};$$

$$xy = \int_0^x \frac{\sin t}{t}\, dt + C. \tag{7}$$

The lower limit of integration $a = 0$ is permissible because $(\sin t)/t \to 1$ as $t \to 0$. Consequently, we can define the integrand in (7) to have the value 1 when $t = 0$. Then the integral is not improper. Substitution of the initial condition $x = 1, y = 2$ in (7) yields

$$C = 2 - \int_0^1 \frac{\sin t}{t}\, dt,$$

so

$$y(x) = \frac{1}{x} \int_0^x \frac{\sin t}{t}\, dt + \frac{C}{x}$$

$$= \frac{1}{x} \int_0^x \frac{\sin t}{t}\, dt + \frac{2}{x} - \frac{1}{x} \int_0^1 \frac{\sin t}{t}\, dt$$

$$= \frac{2}{x} + \frac{1}{x} \int_1^x \frac{\sin t}{t}\, dt.$$

For example,

$$y(2) = 1 + \frac{1}{2} \int_1^2 \frac{\sin t}{t}\, dt.$$

In general, an integral such as this one would have to be approximated numerically (for example, by using Simpson's rule). In this case, however, we have the **sine**

integral function

$$\text{Si}(x) = \int_0^x \frac{\sin t}{t} \, dt,$$

which appears with sufficient frequency in applications that its values have been tabulated. A good set of tables of special functions is Abramowitz and Stegun, *Handbook of Mathematical Functions* (New York: Dover, 1965). From Table 5.1 of this reference, we find that

$$y(2) = 1 + \tfrac{1}{2}[\text{Si}(2) - \text{Si}(1)] \cdot$$
$$\approx 1 + \tfrac{1}{2}(1.60541 - 0.94608) \approx 1.32967.$$

In the sequel we will see that it is the rule, rather than the exception, that a solution to a differential equation cannot be expressed in terms of elementary functions. We will study various devices for obtaining good approximations to the values of the nonelementary functions we encounter; in Chapter 6 we will discuss numerical integration of differential equations in some detail.

The derivation above of the solution in (6) of the linear first order equation in (3) bears a closer examination. Suppose that the functions $P(x)$ and $Q(x)$ are continuous on the (possibly unbounded) open interval I. Then the antiderivatives

$$\int P(x) \, dx \quad \text{and} \quad \int (Q(x) e^{\int P(x) \, dx}) \, dx$$

exist on I. Our derivation of Eq. (6) shows that *if $y = y(x)$ is a solution of Eq. (3) on I, then* $y(x)$ is given by the formula in (6) for some choice of the constant C. Conversely, you may verify by direct substitution (Problem 31) that the function $y(x)$ given in Equation (6) satisfies Eq. (3). Finally, given a point x_0 of I and any number y_0, there is (as previously noted) a unique value of C such that $y(x_0) = y_0$. Consequently, we have proved the following existence-uniqueness theorem.

Theorem   *The Linear First Order Equation*

If the functions $P(x)$ and $Q(x)$ are continuous on the open interval I containing the point x_0, then the initial value problem

$$\frac{dy}{dx} + P(x)y = Q(x), \qquad y(x_0) = y_0$$

has a unique solution $y(x)$ on I, given by the formula in Eq. (6) with an appropriate value of the constant C.

Remark 1:   This theorem gives a solution on the *whole* interval I for a *linear* differential equation, in contrast with the basic existence-uniqueness theorem stated in Section 1.3, which guarantees only a solution on a possibly smaller interval J.

Remark 2:   The theorem above tells us that *every* solution of Eq. (3) is included in the general solution given in (6). Thus a *linear* first order differential equation has *no* singular solutions.

EXAMPLE 4 If we write the equation

$$x\frac{dy}{dx} + y = 1$$

in the standard form shown in (3), we see that it has coefficient functions $P(x) = Q(x) = 1/x$ that are continuous on the intervals $x < 0$ and $x > 0$, but discontinuous at $x = 0$. Because the equation as written is equivalent to $D_x(xy) = 1$, integration gives $xy = x + C$; thus

$$y(x) = 1 + \frac{C}{x}.$$

Some typical **solution curves**—graphs of solutions—are shown in Fig. 1.37. The solution with $y(1) = 2$ is $y = 1 + (1/x)$ for $x > 0$, while the solution with $y(-1) = 2$ is $y = 1 - (1/x)$ for $x < 0$. The only solution that is continuous on the whole real line is the constant solution $y(x) \equiv 1$. We see also that, if $y_0 \neq 1$, there is *no* solution with $y(0) = y_0$. Note that, in accord with the theorem, all solutions are continuous except possibly at the discontinuity $x = 0$ of $P(x)$ and $Q(x)$, and that most solutions are singular at the singularity $x = 0$ of the differential equation.

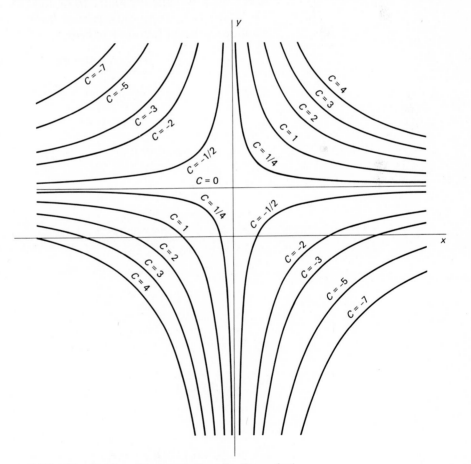

FIGURE 1.37 Graphs of some solutions for Example 4.

MIXTURE PROBLEMS

As a first application of linear first order equations, we consider a tank containing a solution—a mixture of solute and solvent—such as salt dissolved in water. There is both inflow and outflow, and we want to compute the *amount* $x(t)$ of solute in the tank at time t, given the amount $x(0) = x_0$ at time $t = 0$. Suppose that solution with a concentration of c_i grams of solute per liter of solution flows into the tank at the constant rate of r_i liters per second, and that the solution in the tank—kept thoroughly mixed by stirring—flows out at the constant rate of r_o liters per second.

To set up a differential equation for $x(t)$, we estimate the change Δx in x during the brief time interval $[t, t + \Delta t]$. The amount of solute that flows into the tank during Δt seconds is $r_i c_i \, \Delta t$ grams. To check this, note how the cancellation of dimensions checks our computation:

$$\left(r_i \frac{\text{liters}}{\text{second}} \right) \left(c_i \frac{\text{grams}}{\text{liter}} \right) (\Delta t \text{ seconds})$$

yields a quantity measured in grams.

The amount of solute that flows out of the tank during the same time interval depends on the concentration $c_o(t)$ in the tank at time t. But, as noted in Fig. 1.38, $c_o(t) = x(t)/V(t)$, where $V(t)$ denotes the volume (not constant unless $r_i = r_o$) of solution in the tank at time t. Then

$$\Delta x = \{\text{grams input}\} - \{\text{grams output}\}$$
$$\approx r_i c_i \, \Delta t - r_o c_o \, \Delta t$$

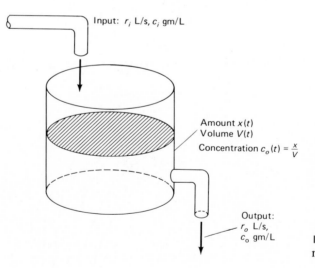

Input: r_i L/s, c_i gm/L

Amount $x(t)$
Volume $V(t)$
Concentration $c_o(t) = \frac{x}{V}$

Output:
r_o L/s,
c_o gm/L

FIGURE 1.38 The single-tank mixture problem.

We now divide by Δt:

$$\frac{\Delta x}{\Delta t} \approx r_i c_i - r_o c_o.$$

Finally, we take the limit as $\Delta t \to 0$; if all the functions involved are continuous and $x(t)$ is differentiable, then the error in the approximation also approaches zero, and we obtain the differential equation

$$\frac{dx}{dt} = r_i c_i - r_o c_o, \tag{8}$$

in which r_i, c_i, and r_o are constants, but c_o denotes the variable concentration

$$c_o(t) = \frac{x(t)}{V(t)} \tag{9}$$

of solute in the tank at time t. Thus the *amount* $x(t)$ of solute in the tank satisfies the differential equation

$$\frac{dx}{dt} = r_i c_i - \frac{r_o}{V} x. \tag{10}$$

If $V_0 = V(0)$, then $V(t) = V_0 + (r_i - r_o)t$, so Eq. (10) is a linear first order differential equation for the amount $x(t)$ of solute in the tank.

Important: Equation (10) is *not* one you should commit to memory. It is the process we used to obtain that equation—examination of the behavior of the system over a short time interval $[t, t + \Delta t]$—that you should strive to understand, because it is a very useful tool for obtaining all sorts of differential equations.

EXAMPLE 5 Lake Erie has a volume of 458 km³, and its rate of inflow and outflow are both 175 km³/year. Suppose that at the time $t = 0$ (years), its pollutant concentration is 0.05%, and that thereafter the concentration of pollutants in the inflowing water is 0.01%. Assuming that the outflow is perfectly mixed lake water, how long will it take to reduce the pollution concentration in the lake to 0.02%?

SOLUTION Here we have $V = 458$ (km³) and $r_i = r_o = 175$ (km³/year). Let $x(t)$ denote the volume of pollutants in the lake at time t. We are given

$$x(0) = (0.0005)(458) = 0.2290 \text{ (km}^3\text{)},$$

and we want to find when

$$x(t) = (0.0002)(458) = 0.0916 \text{ (km}^3\text{)}.$$

The change Δx in Δt years is

$$\Delta x \approx (0.0001)(175)\,\Delta t - \left(\frac{x}{458}\right)(175)\,\Delta t$$

$$= (0.0175 - 0.3821x)\,\Delta t,$$

so our differential equation is

$$\frac{dx}{dt} + (0.3821)x = 0.0175.$$

The integrating factor is $e^{(0.3821)t}$; it yields

$$D_t[e^{(0.3821)t}x] = (0.0175)e^{(0.3821)t};$$

$$[e^{(0.3821)t}]x = (0.0458)e^{(0.3821)t} + C.$$

Substitution of $x(0) = 0.2290$ into the last equation gives

$$C = 0.2290 - 0.0458 = 0.1832,$$

so the solution is

$$x(t) = 0.0458 + (0.1832)e^{-(0.3821)t}.$$

Finally, we solve the equation

$$0.0916 = 0.0458 + (0.1832)e^{-(0.3821)t}$$

for

$$t = \frac{-1}{0.3821} \ln \frac{0.0916 - 0.0458}{0.1832} \approx 3.63.$$

The answer to the problem, then, is after about 3.63 years.

EXAMPLE 6   A 120-gallon (gal) tank initially contains 90 lb of salt dissolved in 90 gal of water. Brine containing 2 lb/gal of salt flows into the tank at the rate of 4 gal/min, and the mixture flows out of the tank at the rate of 3 gal/min. How much salt does the tank contain when it is full?

SOLUTION   The interesting feature of this example is that, due to the differing rates in inflow and outflow, the volume of brine in the tank increases steadily with $V(t) = 90 + t$ (gallons). The change Δx in the amount x of salt in the tank from time t to time $t + \Delta t$ (minutes) is given by

$$\Delta x \approx (4)(2) \, \Delta t - 3 \left(\frac{x}{90 + t} \right) \Delta t,$$

so our differential equation is

$$\frac{dx}{dt} + \frac{3}{90 + t} x = 8.$$

The integrating factor is

$$\exp \left(\int \frac{3 \, dt}{90 + t} \right) = e^{3 \ln (90 + t)} = (90 + t)^3,$$

which gives

$$D_t[(90 + t)^3 x] = 8(90 + t)^3;$$
$$(90 + t)^3 x = 2(90 + t)^4 + C.$$

Substitution of $x(0) = 90$ gives $C = -(90)^4$, so the amount of salt in the tank at time t is

$$x(t) = 2(90 + t) - \frac{(90)^4}{(90 + t)^3}.$$

The tank is full after 30 min, and when $t = 30$ we have

$$x(30) = 2(90 + 30) - \frac{(90)^4}{(120)^3} \approx 202 \text{ (lb)}$$

of salt.

1.5 PROBLEMS

Find general solutions of the differential equations in Problems 1–25. If an initial condition is given, find the corresponding particular solution. Throughout, primes denote derivatives with respect to x.

1. $y' + y = 2; \quad y(0) = 0$
2. $y' - 2y = 3e^{2x}; \quad y(0) = 0$
3. $y' + 3y = 2xe^{-3x}$
4. $y' - 2xy = e^{x^2}$
5. $xy' + 2y = 3x; \quad y(1) = 5$
6. $xy' + 5y = 7x^2; \quad y(2) = 5$
7. $2xy' + y = 10\sqrt{x}$
8. $3xy' + y = 12x$
9. $xy' - y = x; \quad y(1) = 7$
10. $2xy' - 3y = 9x^3$
11. $xy' + y = 3xy; \quad y(1) = 0$
12. $xy' + 3y = 2x^5; \quad y(2) = 1$
13. $y' + y = e^x; \quad y(0) = 1$
14. $xy' - 3y = x^3; \quad y(1) = 10$
15. $y' + 2xy = x; \quad y(0) = -2$
16. $y' = (1 - y)\cos x; \quad y(\pi) = 2$
17. $(1 + x)y' + y = \cos x; \quad y(0) = 1$
18. $xy' = 2y + x^3 \cos x$
19. $y' + y \cot x = \cos x$
20. $y' = 1 + x + y + xy; \quad y(0) = 0$
21. $xy' = 3y + x^4 \cos x; \quad y(2\pi) = 0$
22. $y' = 2xy + 3x^2 \exp(x^2); \quad y(0) = 5$
23. $xy' + (2x - 3)y = 4x^4$
24. $(x^2 + 4)y' + 3xy = x; \quad y(0) = 1$
25. $(x^2 + 1)y' + 3x^3y = 6x \exp\left(-\frac{3x^2}{2}\right); \quad y(0) = 1$

Solve the differential equations in Problems 26–28 by regarding y as the independent variable rather than x.

26. $(1 - 4xy^2)y' = y^3$
27. $(x + ye^y)y' = 1$
28. $(1 + 2xy)y' = 1 + y^2$

29. Express the general solution of $y' = 1 + 2xy$ in terms of the **error function**

$$\text{erf}(x) = \frac{2}{\sqrt{\pi}} \int_0^x e^{-t^2} \, dt.$$

30. Express the solution of the initial value problem

$$2xy' = y + 2x \cos x, \quad y(1) = 0$$

as an integral as in Example 3.

31. (a) Show that $y_c(x) = Ce^{-\int P(x)\,dx}$ is a general solution of $y' + P(x)y = 0$. (b) Show that

$$y_p(x) = e^{-\int P(x)\,dx}\left[\int \left(Q(x)e^{\int P(x)\,dx}\right) dx\right]$$

is a particular solution of $y' + P(x)y = Q(x)$. (c) If $y_c(x)$ is any general solution of $y' + P(x)y = 0$ and $y_p(x)$ is any particular solution of $y' + P(x)y = Q(x)$, then show that $y(x) = y_c(x) + y_p(x)$ is a general solution of $y' + P(x)y = Q(x)$.

32. (a) Find constants A and B such that $y_p(x) = A \sin x + B \cos x$ is a solution of $y' + y = 2 \sin x$. (b) Use the result of part (a) and the method of Problem 31 to find the general solution of $y' + y = 2 \sin x$. (c) Solve the initial value problem $y' + y = 2 \sin x, y(0) = 1$.

33. A tank contains 1000 liters (L) of a solution consisting of 100 kg of salt dissolved in water. Pure water is pumped into the tank at the rate of 5 L/s, and the mixture—kept uniform by stirring—is pumped out at the same rate. How long will it be until only 10 kg of salt remain in the tank?

34. Consider a reservoir with a volume of 8 billion cubic feet (ft^3) and an initial pollutant concentration of 0.25%. There is a daily inflow of 500 million ft^3 of water with a pollutant concentration of 0.05% and an equal daily outflow of the well-mixed water of the reservoir. How long will it take to reduce the pollutant concentration in the reservoir to 0.10%?

35. Rework Example 5 for the case of Lake Ontario; the only differences are that this lake has a volume of 1636 km^3 and an inflow-outflow rate of 209 km^3/year.

36. A tank initially contains 60 gal of pure water. Brine containing 1 lb of salt per gallon enters the tank at 2 gal/min, and the (perfectly mixed) solution leaves the tank at 3 gal/min; the tank is empty after exactly 1 h. (a) Find the amount of salt in the tank after t minutes. (b) What is the maximum amount of salt ever in the tank?

37. A 400-gal tank initially contains 100 gal of brine containing 50 lb of salt. Brine containing 1 lb of salt per gallon enters the tank at the rate of 5 gal/s, and the mixed brine in the tank flows out at the rate of 3 gal/s. How much salt will the tank contain when it is full of brine?

38. Consider the *cascade* of two tanks shown in Fig. 1.39, with $V_1 = 100$ (gal) and $V_2 = 200$ (gal) the volumes of brine in the two tanks. Each tank initially contains 50 lb of salt. The three flow rates are each 5 gal/s, with pure water flowing into Tank 1. (a) Find the amount $x(t)$ of salt in Tank 1 at time t. (b) Suppose that $y(t)$ is the amount of salt in Tank 2 at time t. Show first that

$$\frac{dy}{dt} = \frac{5x}{100} - \frac{5y}{200},$$

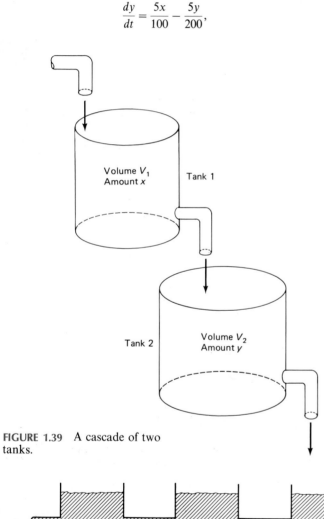

FIGURE 1.39 A cascade of two tanks.

and then solve for $y(t)$, using the value of $x(t)$ found in part (a). (c) Finally, find the maximum amount of salt ever in Tank 2.

39. Suppose that in the cascade shown in Fig. 1.39, Tank 1 initially contains 100 gal of pure ethyl alcohol and Tank 2 initially contains 100 gal of pure water. Pure water flows into Tank 1 at 10 gal/min, and the other two flow rates are also 10 gal/min. (a) Find the amounts $x(t)$ and $y(t)$ of alcohol in the two tanks. (b) Find the maximum amount of alcohol ever in Tank 2.

40. A multiple cascade is shown in Fig. 1.40. At time $t = 0$, Tank 0 contains 1 gal of alcohol and 1 gal of water, while each of the other tanks contains 2 gal of pure water. Pure water is pumped into Tank 0 at 1 gal/min, and the varying mixture in each tank is pumped into the one to its right at the same rate. Assume, as usual, that the mixtures are kept perfectly uniform by stirring. Let $x_n(t)$ denote the amount of alcohol in Tank n at time t. (a) Show that $x_0(t) = e^{-t/2}$. (b) Show by induction on n that

$$x_n(t) = \frac{t^n e^{-t/2}}{n!2^n} \quad \text{for } n > 0.$$

(c) Show that the maximum value of $x_n(t)$ for $n > 0$ is $M_n = n^n e^{-n}/n!$. (d) Conclude from **Stirling's approximation** $n! \approx (2\pi n)^{1/2} n^n e^{-n}$ that $M_n \approx (2\pi n)^{-1/2}$.

41. A 30-year-old woman accepts an engineering position with a starting salary of \$30,000 per year. Her salary $S(t)$ increases exponentially, with $S(t) = 30e^{t/20}$ thousand dollars after t years. Meanwhile, 12% of her salary is deposited continuously in a retirement account, which accumulates interest at a continuous annual rate of 6%. (a) Estimate ΔA in terms of Δt to derive the differential equation for the amount $A(t)$ in her retirement account after t years. (b) Compute $A(40)$, the amount available for her retirement at age 70.

42. Suppose that a falling hailstone with density $\delta = 1$ starts from rest with negligible radius $r = 0$. Thereafter its radius is $r = kt$ (k is constant) as it grows by accretion during its fall. Set up and solve the initial value problem

$$\frac{d}{dt}(mv) = mg, \qquad v(0) = 0$$

where m is the variable mass of the hailstone, $v = dy/dt$ is its velocity, and the positive y-axis points downward. Then show that $dv/dt = g/4$. Thus the hailstone falls as though it were under *one-fourth* the influence of gravity.

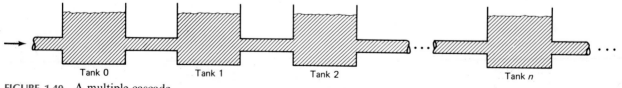

Tank 0   Tank 1   Tank 2   Tank n

FIGURE 1.40 A multiple cascade.

CHAPTER 1: Introduction and First Order Differential Equations

Substitution Methods

The first order differential equations we have solved in the previous sections have all been either separable or linear. But many applications involve differential equations that are neither separable nor linear. In this section we discuss substitution methods that sometimes can be used to transform a given differential equation into one that we already know how to solve.

Given a differential equation

$$\frac{dy}{dx} = f(x, y), \tag{1}$$

with dependent variable y and independent variable x, we can transform it into a new differential equation with the new dependent variable v by making the substitution

$$y = \phi(x, v). \tag{2}$$

Such a substitution may be suggested by the particular form of the differential equation, as in Example 1 below. At any rate, note that if the function ϕ has partial derivatives ϕ_x and ϕ_v with respect to x and v, then application of the chain rule—regarding v as an (unknown) function of x—yields

$$\frac{dy}{dx} = \phi_x(x, v) + \phi_v(x, v)\frac{dv}{dx}. \tag{3}$$

Substitution of (2) and (3) in (1) gives

$$\phi_x(x, v) + \phi_v(x, v)\frac{dv}{dx} = f(x, \phi(x, v)). \tag{4}$$

This is our *new* differential equation of the form

$$\frac{dv}{dx} = g(x, v) \tag{5}$$

that we get by solving Eq. (4) for dv/dx, and with *new* dependent variable v. If this new equation is either separable or linear, then we can apply the methods of preceding sections to solve it.

If $v = v(x)$ is a solution of Eq. (5), then $y = \phi(x, v(x))$ is a solution of the original Eq. (1). The trick is to select a substitution $y = \phi(x, v)$ such that the transformed Eq. (5) is one we can solve. Even when possible, this is not always easy; it may require a fair amount of ingenuity or trial and error.

Sometimes it is more convenient to begin with a relation $v = \psi(x, y)$ that can be solved for $y = \phi(x, v)$. Occasionally, as in the following example, the presence of a conspicuous combination of x and y in the differential equation will suggest a promising substitution.

EXAMPLE 1 Solve the differential equation

$$\frac{dy}{dx} = (x + y + 3)^2.$$

SOLUTION It seems reasonable to try the substitution

$$v = x + y + 3; \quad \text{that is,} \quad y = v - x - 3.$$

Then

$$\frac{dy}{dx} = \frac{dv}{dx} - 1,$$

so our transformed equation is

$$\frac{dv}{dx} = 1 + v^2.$$

This is a separable equation, and we have no difficulty in obtaining its solution

$$x = \int \frac{dv}{1 + v^2} = \tan^{-1} v + C.$$

So $v = \tan(x - C)$. Because $v = x + y + 3$, the general solution of the original equation $y' = (x + y + 3)^2$ is $x + y + 3 = \tan(x - C)$; that is,

$$y(x) = \tan(x - C) - x - 3.$$

Example 1 illustrates the fact that any differential equation of the form

$$\frac{dy}{dx} = F(ax + by + c) \tag{6}$$

can be transformed into a separable equation by use of the substitution $v = ax + by + c$ (see Problem 31). The remainder of this section deals with other classes of first order equations for which there are standard substitutions that are known to succeed.

HOMOGENEOUS EQUATIONS

A **homogeneous** first order differential equation is one that can be written in the form

$$\frac{dy}{dx} = F\left(\frac{y}{x}\right). \tag{7}$$

If we make the substitutions

$$v = \frac{y}{x}, \qquad y = vx, \qquad \frac{dy}{dx} = v + x\frac{dv}{dx}, \tag{8}$$

then Eq. (7) is transformed into the *separable* equation

$$x\frac{dv}{dx} = F(v) - v.$$

Thus any homogeneous first order differential equation can be reduced to an integration problem by means of the substitutions in (8).

EXAMPLE 2   Solve the differential equation

$$2xy \frac{dy}{dx} = 4x^2 + 3y^2.$$

SOLUTION   This equation is neither separable nor linear, but we recognize it as a homogeneous equation by writing it in the form

$$\frac{dy}{dx} = \frac{4x^2 + 3y^2}{2xy} = 2\left(\frac{x}{y}\right) + \frac{3}{2}\left(\frac{y}{x}\right).$$

The substitutions in (8) then take the form

$$y = vx, \qquad \frac{dy}{dx} = v + x \frac{dv}{dx},$$

$$v = \frac{y}{x}, \quad \text{and} \quad \frac{1}{v} = \frac{x}{y}.$$

These yield

$$v + x \frac{dv}{dx} = \frac{2}{v} + \frac{3}{2} v,$$

and hence

$$x \frac{dv}{dx} = \frac{2}{v} + \frac{v}{2} = \frac{v^2 + 4}{2v};$$

$$\int \frac{2v}{v^2 + 4} \, dv = \int \frac{1}{x} \, dx;$$

$$\ln (v^2 + 4) = \ln |x| + \ln C.$$

We apply the exponential function to each side of the last equation to obtain

$$v^2 + 4 = C|x|;$$

$$\frac{y^2}{x^2} + 4 = C|x|;$$

$$y^2 + 4x^2 = Cx^3,$$

because the sign of x can be absorbed by the arbitrary constant C.

EXAMPLE 3   Solve the initial value problem

$$x \frac{dy}{dx} = y + (x^2 - y^2)^{1/2}, \qquad y(1) = 0.$$

SOLUTION   We divide both sides by x and find that

$$\frac{dy}{dx} = \frac{y}{x} + \left[1 - \left(\frac{y}{x}\right)^2\right]^{1/2},$$

so we make the substitutions in (8); we get

$$v + x\frac{dv}{dx} = v + [1 - v^2]^{1/2};$$

$$\int \frac{1}{(1 - v^2)^{1/2}}\, dv = \int \frac{1}{x}\, dx;$$

$$\sin^{-1} v = \ln x + C.$$

There is no need to use $\ln |x|$ because $x > 0$ near $x = 1$ (part of the given initial condition). Next we note that $v(1) = y(1)/1 = 0$, so

$$C = \sin^{-1} 0 - \ln 1 = 0.$$

Hence

$$v = \frac{y}{x} = \sin (\ln x),$$

and therefore $y = x \sin (\ln x)$ is the desired solution.

BERNOULLI EQUATIONS

A first order differential equation of the form

$$\frac{dy}{dx} + P(x)y = Q(x)y^n \tag{9}$$

is called a **Bernoulli equation**. If either $n = 0$ or $n = 1$, then (9) is linear. Otherwise, as we ask you to show in Problem 32, the substitution

$$v = y^{1-n} \tag{10}$$

transforms (9) into the linear equation

$$\frac{dv}{dx} + (1 - n)P(x)v = (1 - n)Q(x).$$

Rather than memorizing the form of this transformed equation, it is more efficient to make the substitution in (10) explicitly, as in the following examples.

EXAMPLE 4 If we rewrite the homogeneous equation $2xyy' = 4x^2 + 3y^2$ of Example 2 in the form

$$\frac{dy}{dx} - \frac{3}{2x}y = \frac{2x}{y},$$

we see that it is also a Bernoulli equation with $P(x) = -3/(2x)$, $Q(x) = 2x$, $n = -1$, and $1 - n = 2$. Hence we substitute

$$v = y^2, \quad y = v^{1/2}, \quad \text{and} \quad \frac{dy}{dx} = \frac{dy}{dv}\frac{dv}{dx} = \frac{1}{2}v^{-1/2}\frac{dv}{dx}.$$

This gives

$$\frac{1}{2}v^{-1/2}\frac{dv}{dx} - \frac{3}{2x}v^{1/2} = 2xv^{-1/2}.$$

CHAPTER 1: Introduction and First Order Differential Equations

Multiplication by $2v^{1/2}$ produces the linear equation

$$\frac{dv}{dx} - \frac{3}{x}v = 4x$$

with integrating factor $\rho = e^{\int (-3/x)\,dx} = x^{-3}$. So we obtain

$$D_x(x^{-3}v) = \frac{4}{x^2};$$

$$x^{-3}v = -\frac{4}{x} + C;$$

$$x^{-3}y^2 = -\frac{4}{x} + C;$$

$$y^2 = -4x^2 + Cx^3.$$

EXAMPLE 5   The equation

$$x\frac{dy}{dx} + 6y = 3xy^{4/3}$$

is neither separable nor linear nor homogeneous, but it is a Bernoulli equation with $n = \frac{4}{3}$, $1 - n = -\frac{1}{3}$. The substitutions

$$v = y^{-1/3}, \qquad y = v^{-3}, \quad \text{and} \quad \frac{dy}{dx} = \frac{dy}{dv}\frac{dv}{dx} = -3v^{-4}\frac{dv}{dx}$$

transform it into

$$-3xv^{-4}\frac{dv}{dx} + 6v^{-3} = 3xv^{-4}.$$

Division by $-3xv^{-4}$ yields the linear equation

$$\frac{dv}{dx} - \frac{2}{x}v = -1$$

with integrating factor $\rho = e^{\int (-2/x)\,dx} = x^{-2}$. This gives

$$D_x(x^{-2}v) = -\frac{1}{x^2};$$

$$x^{-2}v = \frac{1}{x} + C;$$

$$v = x + Cx^2;$$

and finally,

$$y(x) = \frac{1}{(x + Cx^2)^3}.$$

EXAMPLE 6   The equation

$$2xe^{2y}\frac{dy}{dx} = 3x^4 + e^{2y} \tag{11}$$

is neither separable, linear, nor homogeneous, nor is it a Bernoulli equation. But we observe that y appears only in the combinations e^{2y} and $(e^{2y})' = 2e^{2y}y'$. This prompts the substitution

$$v = e^{2y}, \qquad \frac{dv}{dx} = 2e^{2y}\frac{dy}{dx}$$

that transforms Eq. (11) into the linear equation $xv' = 3x^4 + v$; that is,

$$\frac{dv}{dx} - \frac{1}{x}v = 3x^3.$$

After multiplying by the integrating factor $\rho = x^{-1}$, we find that

$$vx^{-1} = \int 3x^2\,dx = x^3 + C,$$

so

$$e^{2y} = v = x^4 + Cx,$$

and hence

$$y(x) = \tfrac{1}{2}\ln\left|x^4 + Cx\right|.$$

FLIGHT TRAJECTORIES

Suppose that an airplane starts at the point $(a, 0)$ located due east of its intended destination—an airport located at the origin $(0, 0)$. The plane travels with constant speed v_0 relative to the wind, which is blowing due north with constant speed w. As indicated in Fig. 1.41, we assume that the plane's pilot maintains its heading directly toward the origin.

As indicated in Fig. 1.42, the plane's velocity components relative to the ground are

$$\frac{dx}{dt} = -v_0\cos\theta = -\frac{v_0 x}{\sqrt{x^2 + y^2}},$$

$$\frac{dy}{dt} = -v_0\sin\theta + w = -\frac{v_0 y}{\sqrt{x^2 + y^2}} + w.$$

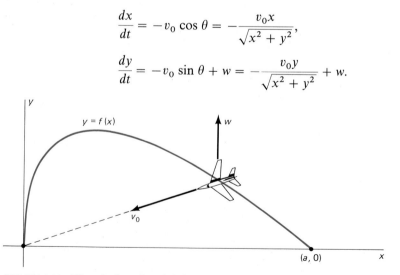

FIGURE 1.41 The airplane headed for the origin.

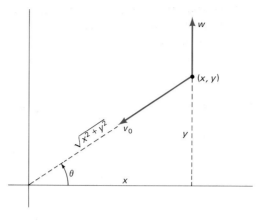

FIGURE 1.42 The plane's velocity components.

Hence the trajectory $y = f(x)$ of the plane satisfies the differential equation

$$\frac{dy}{dx} = \frac{dy/dt}{dx/dt} = \frac{1}{v_0 x}(v_0 y - w\sqrt{x^2 + y^2}). \tag{12}$$

If we set

$$k = \frac{w}{v_0}, \tag{13}$$

the ratio of the windspeed to the plane's airspeed, then Eq. (12) takes the homogeneous form

$$\frac{dy}{dx} = \frac{y}{x} - k\left[1 + \left(\frac{y}{x}\right)^2\right]^{1/2}. \tag{14}$$

The substitution $y = vx$, $y' = v + xv'$ then leads routinely to

$$\int \frac{dv}{\sqrt{1 + v^2}} = -\int \frac{k}{x}\,dx. \tag{15}$$

By trigonometric substitution, or by consulting a table for the integral on the left, we find that

$$\ln\left(v + \sqrt{1 + v^2}\right) = -k \ln x + C, \tag{16}$$

and the initial condition $v(a) = y(a)/a = 0$ yields

$$C = -k \ln a. \tag{17}$$

According to Problem 44, the result of substituting (17) in (16) and then solving for v is

$$v = \frac{1}{2}\left[\left(\frac{x}{a}\right)^{-k} - \left(\frac{x}{a}\right)^{k}\right]. \tag{18}$$

Because $y = vx$, we finally obtain

$$y = \frac{a}{2}\left[\left(\frac{x}{a}\right)^{1-k} - \left(\frac{x}{a}\right)^{1+k}\right] \tag{19}$$

for the equation of the plane's trajectory.

Note that only in the case $k < 1$ (that is, $w < v_0$) does the curve in (19) pass through the origin, so that the plane reaches its destination. If $w = v_0$ (so that $k = 1$), then Eq. (19) takes the form $y = \frac{1}{2}a(1 - x^2/a^2)$, so the plane's trajectory passes through $(0, a/2)$ rather than $(0, 0)$; in this case the wind blows the plane permanently off course. The situation is even worse if $w > v_0$ (so $k > 1$)—in this case it follows from (19) that $y \to +\infty$ as $x \to 0$.

EXAMPLE 7 If $a = 200$ mi, $v_0 = 500$ mi/h, and $w = 100$ mi/h, then $k = w/v_0 = \frac{1}{5}$, so the plane will succeed in reaching the airport at $(0, 0)$. With these values, Eq. (19) yields

$$y(x) = 100\left[\left(\frac{x}{200}\right)^{4/5} - \left(\frac{x}{200}\right)^{6/5}\right]. \tag{20}$$

Now suppose that we want to find the maximum amount by which the plane is blown off course during its trip. That is, what is the maximum value of $y(x)$ for $0 \leq x \leq 200$?

SOLUTION Differentiation of (20) yields

$$\frac{dy}{dx} = \frac{1}{2}\left[\frac{4}{5}\left(\frac{x}{200}\right)^{-1/5} - \frac{6}{5}\left(\frac{x}{200}\right)^{1/5}\right],$$

and we readily solve the equation $y'(x) = 0$ for $(x/200)^{2/5} = \frac{2}{3}$. Hence

$$y_{\max} = 100\left[\left(\frac{2}{3}\right)^2 - \left(\frac{2}{3}\right)^3\right] = \frac{400}{27} \approx 14.81.$$

Thus the plane is blown almost 15 mi north at one point during its westward trip. (The graph of the function in Eq. (20) is the one used to construct Fig. 1.41. The vertical scale there is exaggerated by a factor of 4.)

1.6 PROBLEMS

Find general solutions of the differential equations in Problems 1–30. Primes denote derivatives with respect to x throughout.

1. $(x + y)y' = x - y$
2. $2xyy' = x^2 + 2y^2$
3. $xy' = y + 2(xy)^{1/2}$
4. $(x - y)y' = x + y$
5. $x(x + y)y' = y(x - y)$
6. $(x + 2y)y' = y$
7. $xy^2y' = x^3 + y^3$
8. $x^2y' = xy + x^2e^{y/x}$
9. $x^2y' = xy + y^2$
10. $xyy' = x^2 + 3y^2$
11. $(x^2 - y^2)y' = 2xy$
12. $xyy' = y^2 + x(4x^2 + y^2)^{1/2}$
13. $xy' = y + (x^2 + y^2)^{1/2}$
14. $yy' + x = (x^2 + y^2)^{1/2}$
15. $x(x + y)y' + y(3x + y) = 0$
16. $y' = (x + y + 1)^{1/2}$
17. $y' = (4x + y)^2$
18. $(x + y)y' = 1$
19. $x^2y' + 2xy = 5y^3$
20. $y^2y' + 2xy^3 = 6x$
21. $y' = y + y^3$

22. $x^2y' + 2xy = 5y^4$
23. $xy' + 6y = 3xy^{4/3}$
24. $2xy' + y^3e^{-2x} = 2xy$
25. $y^2(xy' + y)(1 + x^4)^{1/2} = x$
26. $3y^2y' + y^3 = e^{-x}$
27. $3xy^2y' = 3x^4 + y^3$
28. $xe^yy' = 2(e^y + x^3e^{2x})$
29. $(2x \sin y \cos y)y' = 4x^2 + 3 \sin^2 y$
30. $(x + e^y)y' = xe^{-y} - 1$

31. Show that the substitution $v = ax + by + c$ transforms the differential equation $y' = F(ax + by + c)$ into a separable equation.

32. Suppose that $n \neq 0$ and $n \neq 1$. Show that the substitution $v = y^{1-n}$ transforms the Bernoulli equation $y' + P(x)y = Q(x)y^n$ into the linear equation

$$v' + (1 - n)P(x)v = (1 - n)Q(x).$$

CHAPTER 1: Introduction and First Order Differential Equations

33. Show that the substitution $v = \ln y$ transforms the differential equation $y' + P(x)y = Q(x)y \ln y$ into the linear equation $v' + P = Qv$.

34. Use the method of Problem 33 to solve the equation $xy' - 4x^2 y + 2y \ln y = 0$.

35. Solve the equation

$$\frac{dy}{dx} = \frac{x - y - 1}{x + y + 3}$$

by finding h and k so that the substitutions $x = u + h$, $y = v + k$ transform it into the homogeneous equation

$$\frac{dv}{du} = \frac{u - v}{u + v}.$$

36. Use the method of Problem 35 to solve the equation

$$\frac{dy}{dx} = \frac{2y - x + 7}{4x - 3y - 18}.$$

37. Make an appropriate substitution to find a general solution of the equation $y' = \sin(x - y)$. Does this general solution contain the solution $y = x - \pi/2$ that we noted in Example 3 of Section 1.3?

38. Show that the solution curves of the homogeneous equation

$$\frac{dy}{dx} = -\frac{y(2x^3 - y^3)}{x(2y^3 - x^3)}$$

are of the form $x^3 + y^3 = 3Cxy$.

39. The equation $dy/dx = A(x)y^2 + B(x)y + C(x)$ is called a **Riccati equation**. Suppose that one particular solution $y_1(x)$ of

this equation is known. Show that the substitution

$$y = y_1 + \frac{1}{v}$$

transforms it into the *linear* equation

$$\frac{dv}{dx} + (B + 2Ay_1)v = -A.$$

Use the method of Problem 39 to solve the equations in Problems 40 and 41, given that $y_1 = x$ is a solution of each.

40. $y' + y^2 = 1 + x^2$ **41.** $y' + 2xy = 1 + x^2 + y^2$

42. An equation of the form

$$y = xy' + g(y') \tag{21}$$

is called a **Clairaut equation**. Show that the one-parameter family of straight lines described by

$$y = Cx + g(C) \tag{22}$$

is a general solution of Eq. (21).

43. Consider the Clairaut equation

$$y = xy' - \tfrac{1}{4}(y')^2$$

with $g(y') = -\tfrac{1}{4}(y')^2$ in Eq. (21). Show that the line

$$y = Cx - \tfrac{1}{4}C^2$$

is tangent to the parabola $y = x^2$ at the point $(C/2, C^2/4)$. Explain why this implies that $y = x^2$ is a singular solution of the given Clairaut equation. This singular solution and the one-parameter family of straight line solutions are illustrated in Fig. 1.43.

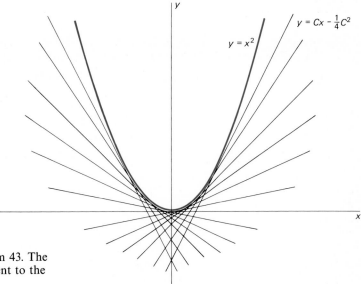

FIGURE 1.43 Solutions of the Clairaut equation of Problem 43. The "typical" straight line with equation $y = Cx - \tfrac{1}{4}C^2$ is tangent to the parabola at the point $(\tfrac{1}{2}C, \tfrac{1}{4}C^2)$.

44. Derive Eq. (18) in the text from Eqs. (16) and (17).

45. In the situation of Example 7, suppose that $a = 100$ mi, $v_0 = 400$ mi/h, and $w = 40$ mi/h. Now how far northward does the wind blow the airplane?

46. As in the text discussion, suppose that an airplane maintains a heading toward an airport at the origin. If $v_0 = 500$ mi/h and $w = 50$ mi/h (with the wind blowing north), and the plane begins at the point (200, 150), show that its trajectory is defined by

$$y + \sqrt{x^2 + y^2} = 2(200x^9)^{1/10}.$$

47. A river 100 ft wide is flowing north at w ft/s. A dog starts at (100, 0) on the east bank and swims at $v_0 = 4$ ft/s, always heading toward a tree at (0, 0) on the west bank directly across from the dog's starting point. (a) If $w = 2$ ft/s, show that the dog reaches the tree. (b) If $w = 4$ ft/s, show that the dog reaches instead the point on the west bank 50 ft north of the tree. (c) If $w = 6$ ft/s, show that the dog never reaches the west bank.

1.7

Exact Equations and Integrating Factors

We have seen that a general solution $y(x)$ of a first order differential equation is often defined implicitly by an equation of the form

$$F(x, y(x)) = C \tag{1}$$

where C is a constant. On the other hand, given the identity in (1), we can recover the original differential equation by differentiating each side with respect to x. Provided that Eq. (1) implicitly defines y as a differentiable function of x, this gives the differential equation in the form

$$\frac{\partial F}{\partial x} + \frac{\partial F}{\partial y}\frac{dy}{dx} = 0;$$

that is,

$$M(x, y) + N(x, y)\frac{dy}{dx} = 0, \tag{2}$$

where $M(x, y) = F_x(x, y)$ and $N(x, y) = F_y(x, y)$.

It is sometimes convenient to rewrite Eq. (2) in the form

$$M(x, y)\, dx + N(x, y)\, dy = 0, \tag{3}$$

called its **differential** form. The general first order differential equation $y' = f(x, y)$ can be written in this form with $M = f(x, y)$ and $N = -1$. The discussion above shows that, *if* there exists a function $F(x, y)$ such that

$$\frac{\partial F}{\partial x} = M \quad \text{and} \quad \frac{\partial F}{\partial y} = N,$$

then the equation

$$F(x, y) = C$$

implicitly defines a general solution of Eq. (3). In this case, Eq. (3) is called an

exact differential equation—the differential

$$dF = F_x \, dx + F_y \, dy$$

of $F(x, y)$ is exactly $M \, dx + N \, dy$.

A natural question is this: How can we determine whether the differential equation in (3) is exact? And if it is exact, how can we find the function $F(x, y)$ such that $F_x = M$ and $F_y = N$? To answer the first question, let us recall that if the mixed second order partial derivatives F_{xy} and F_{yx} are continuous on an open set in the xy-plane, then they are equal: $F_{xy} \equiv F_{yx}$. If Eq. (3) is exact and M and N have continuous partial derivatives, it then follows that

$$\frac{\partial M}{\partial y} = F_{xy} = F_{yx} = \frac{\partial N}{\partial x}.$$

Thus the equation

$$\frac{\partial M}{\partial y} = \frac{\partial N}{\partial x} \tag{4}$$

is a *necessary condition* that the differential equation $M \, dx + N \, dy = 0$ be exact. That is, if $M_y \neq N_x$, then the differential equation in question is not exact, so we need not attempt to find a function $F(x, y)$ such that $F_x = M$ and $F_y = N$: There is no such function.

EXAMPLE 1 The differential equation

$$y^3 \, dx + 3xy^2 \, dy = 0 \tag{5}$$

is exact because we can immediately see that the function $F(x, y) = xy^3$ has the property that $F_x(x, y) = y^3$ and $F_y(x, y) = 3xy^2$. Thus a general solution of (5) is

$$xy^3 = C;$$

if you prefer, $y = kx^{-1/3}$.

Suppose, however, that we divide each term of the differential equation in Example 1 by y^2 to obtain

$$y \, dx + 3x \, dy = 0. \tag{6}$$

This equation is not exact because $M = y$ and $N = 3x$, so that

$$\frac{\partial M}{\partial y} = 1 \neq 3 = \frac{\partial N}{\partial x}.$$

Hence the necessary condition in (4) is not satisfied.

We are confronted with a curious situation here: The differential equations in (5) and (6) are essentially equivalent—and they have the same solutions—yet one is exact and the other is not. This observation opens the way for an exciting possibility: that we may be able to convert an equation that is not exact into one that is, simply by multiplying each term by an appropriate factor. In brief, whether a given differential equation is exact or not has to do with the precise form $M \, dx + N \, dy = 0$ in which it is written. The following theorem tells us that, subject to differentiability conditions that usually are satisfied in practice, the necessary condition in (4) is also a *sufficient* condition for exactness. That is, if $M_y = N_x$, then the differential equation $M\,dx + N\,dy = 0$ is exact.

Theorem *Criterion for Exactness*

Suppose that the functions $M(x, y)$ and $N(x, y)$ are continuous and have continuous first order partial derivatives in the open rectangle $R: a < x < b$, $c < y < d$. Then the differential equation

$$M(x, y)\,dx + N(x, y)\,dy = 0 \tag{3}$$

is exact in R if and only if

$$\frac{\partial M}{\partial y} = \frac{\partial N}{\partial x} \tag{4}$$

at each point of R. That is, there exists a function $F(x, y)$ defined on R with $\partial F/\partial x = M$ and $\partial F/\partial y = N$ if and only if Eq. (4) holds on R.

Proof We have seen already that it is necessary for Eq. (4) to hold if Eq. (3) is to be exact. To prove the converse, we must show that if (4) holds, then we can construct a function $F(x, y)$ such that $\partial F/\partial x = M$ and $\partial F/\partial y = N$. Note first that, for *any* function $g(y)$, the function

$$F(x, y) = \int M(x, y)\,dx + g(y) \tag{7}$$

satisfies the condition $\partial F/\partial x = M$. (In (7), the notation $\int M(x, y)\,dx$ denotes an antiderivative of $M(x, y)$ with respect to x.) We want to choose $g(y)$ so that

$$N = \frac{\partial F}{\partial y} = \frac{\partial}{\partial y} \int M(x, y)\,dx + g'(y)$$

as well; that is, so that

$$g'(y) = N - \frac{\partial}{\partial y} \int M(x, y)\,dx. \tag{8}$$

To see that there *is* such a function of y, it suffices to show that the right-hand side in (8) is a function of y alone. We can then find $g(y)$ by integrating with respect to y. Because the right-hand side in (8) is defined on a rectangle, and hence on an interval as a function of x, it suffices to show that its derivative with respect to x is identically zero. But

$$\frac{\partial}{\partial x}\left(N - \frac{\partial}{\partial y} \int M(x, y)\,dx \right) = \frac{\partial N}{\partial x} - \frac{\partial}{\partial x}\frac{\partial}{\partial y} \int M(x, y)\,dx$$

$$= \frac{\partial N}{\partial x} - \frac{\partial}{\partial y}\frac{\partial}{\partial x} \int M(x, y)\,dx$$

$$= \frac{\partial N}{\partial x} - \frac{\partial M}{\partial y} = 0$$

CHAPTER 1: Introduction and First Order Differential Equations

by hypothesis. So we can, indeed, find the desired function $g(y)$ by integrating Eq. (8). We substitute this result in Eq. (7) to obtain

$$F(x, y) = \int M(x, y)\, dx + \int \left[N(x, y) - \frac{\partial}{\partial y} \int M(x, y)\, dx \right] dy \qquad (9)$$

as the desired function with $F_x = M$ and $F_y = N$. ∎

Instead of remembering Eq. (9), it is usually better to solve an exact equation $M\, dx + N\, dy = 0$ by carrying out the process indicated by Eqs. (7) and (8). First we integrate $M(x, y)$ with respect to x and write

$$F(x, y) = \int M(x, y)\, dx + g(y),$$

thinking of the function $g(y)$ as an "arbitrary constant of integration" as far as the variable x is concerned. Then we determine $g(y)$ by imposing the condition that $\partial F/\partial y = N(x, y)$. This yields a general solution in the implicit form $F(x, y) = C$.

EXAMPLE 2 Solve the differential equation

$$(6xy - y^3)\, dx + (4y + 3x^2 - 3xy^2)\, dy = 0.$$

SOLUTION This equation is exact because

$$\frac{\partial M}{\partial y} = 6x - 3y^2 = \frac{\partial N}{\partial x}.$$

Integrating $\partial F/\partial x = M$ with respect to x, we get

$$F(x, y) = \int (6xy - y^3)\, dx = 3x^2 y - xy^3 + g(y).$$

Then we differentiate with respect to y and set $\partial F/\partial y$ equal to N. This yields

$$\frac{\partial F}{\partial y} = 3x^2 - 3xy^2 + g'(y) = 4y + 3x^2 - 3xy^2,$$

and it follows that $g'(y) = 4y$. Hence

$$g(y) = \int 4y\, dy = 2y^2 + C_1,$$

and thus

$$F(x, y) = 3x^2 y - xy^3 + 2y^2 + C_1.$$

Therefore, a general solution of the given differential equation is defined implicitly by the equation

$$3x^2 y - xy^3 + 2y^2 = C$$

(we have absorbed the constant C_1 into the constant C).

INTEGRATING FACTORS

We sometimes see nonexact differential equations written in the form $M\, dx + N\, dy = 0$. As suggested in the discussion following Example 1, it may be possible to convert such an equation into an exact equation by multiplying its terms by a suitable integrating factor. For example, the equation $y\,dx - x\,dy = 0$ is not exact as it stands, but multiplication by $1/y^2$ gives

$$\frac{y\,dx - x\,dy}{y^2} = 0.$$

The left-hand side is the differential of $F(x, y) = x/y$, and thus we see that the general solution of $y\,dx - x\,dy = 0$ is $x/y = C$; if you prefer, $y = kx$.

The function $1/y^2$ used above is an example of an integrating factor. An **integrating factor** for the differential equation in the differential form

$$M(x, y)\,dx + N(x, y)\,dy = 0 \tag{10}$$

is a function $\rho(x, y)$ such that the equation

$$\rho(x, y)M(x, y)\,dx + \rho(x, y)N(x, y)\,dy = 0$$

is exact; that is,

$$\frac{\partial}{\partial y}(\rho M) = \frac{\partial}{\partial x}(\rho N). \tag{11}$$

EXAMPLE 3   The separable equation

$$y\,dx + \sec x\,dy = 0$$

is not exact. We actually separate the variables by multiplying the equation by the integrating factor $\rho(x, y) = 1/(y \sec x)$ to obtain the exact equation

$$\cos x\,dx + \frac{1}{y}\,dy = 0$$

with general solution $\sin x + \ln y = C$. This illustrates the fact that whenever we separate the variables in a separable equation, we are using an integrating factor.

$F(x, y)$	dF
xy	$y\,dx + x\,dy$
$\dfrac{x}{y}$	$\dfrac{y\,dx - x\,dy}{y^2}$
$\dfrac{1}{2}\ln(x^2 + y^2)$	$\dfrac{x\,dx + y\,dy}{x^2 + y^2}$
$\tan^{-1}\left(\dfrac{y}{x}\right)$	$\dfrac{-y\,dx + x\,dy}{x^2 + y^2}$

FIGURE 1.44   Important forms for the method of grouping.

Unfortunately, even when a differential equation of the form in (10) is known to have solutions, no useful general method for finding an explicit integrating factor is known. Sometimes we can spot an integrating factor by recognizing the presence of propitious combinations such as those in the table shown in Fig. 1.44. This is the basis for the *method of grouping* illustrated next in Example 4.

EXAMPLE 4 Solve the equation

$$(x^2 + y^2 - y)\, dx + x\, dy = 0.$$

SOLUTION This equation is not exact, nor is it susceptible to any of the previous methods of this chapter. But if we multiply each term by the integrating factor $\rho(x, y) = 1/(x^2 + y^2)$, we get the equation

$$dx + \frac{-y\, dx + x\, dy}{x^2 + y^2} = 0.$$

This equation can be integrated immediately, and its general solution is

$$x + \tan^{-1}\left(\frac{y}{x}\right) = C;$$

that is, $y = x \tan (C - x)$.

The integrating factor of Example 4 was obtained by a judicious guess. But *if* the equation $M\, dx + N\, dy = 0$ has an integrating factor ρ that is either a function of x alone or a function of y alone, then ρ can be found systematically. This situation is summarized in the table in Fig. 1.45.

Case	Integrating factor
$\dfrac{1}{N} (M_y - N_x) = f(x)$	$\rho(x) = e^{\int f(x)\,dx}$
$\dfrac{1}{M} (N_x - M_y) = g(y)$	$\rho(y) = e^{\int g(y)\,dy}$

FIGURE 1.45 Special integrating factors.

The meaning of the first line in Fig. 1.45 is that the equation $M\, dx + N\, dy = 0$ has an integrating factor that is a function of x alone if and only if $(M_y - N_x)/N$ is a function $f(x)$ of x alone, in which case $\rho(x) = \exp\left(\int f(x)\, dx\right)$ is an integrating factor of that equation. To see why, suppose that

$$\frac{1}{N} (M_y - N_x) = f(x),$$

and define $\rho(x)$ to be $\exp\left(\int f(x)\,dx\right)$. Then

$$\frac{\partial}{\partial x}(\rho N) = \frac{\partial \rho}{\partial x} N + \rho \frac{\partial N}{\partial x}$$

$$= \frac{\rho}{N}(M_y - N_x)N + \rho \frac{\partial N}{\partial x} \qquad \text{(because } \rho'(x) = \rho(x)f(x)\text{)}$$

$$= \rho \frac{\partial M}{\partial y}$$

$$= \frac{\partial}{\partial y}(\rho M)$$

(because ρ is a function of x alone). Thus the equation $\rho M\,dx + \rho N\,dy = 0$ is exact, and we have verified that $\rho(x)$ is an integrating factor of $M\,dx + N\,dy = 0$. In Problem 33 we ask you to show that the condition $(M_y - N_x)/N = f(x)$ is also a necessary condition that $M\,dx + N\,dy = 0$ have an integrating factor that is a function of x alone.

EXAMPLE 5 Solve the differential equation

$$y^2 \cos x\,dx + (4 + 5y \sin x)\,dy = 0.$$

SOLUTION With $M = y^2 \cos x$ and $N = 4 + 5y \sin x$, we find that

$$\frac{M_y - N_x}{N} = \frac{2y \cos x - 5y \cos x}{4 + 5y \sin x} = -\frac{3y \cos x}{4 + 5y \sin x}$$

is not a function of x alone, so the given equation has no integrating factor of the form $\rho(x)$. But

$$\frac{N_x - M_y}{M} = \frac{3y \cos x}{y^2 \cos x} = \frac{3}{y},$$

so the equation has the integrating factor

$$\rho(y) = e^{\int (3/y)\,dy} = e^{3 \ln y} = y^3.$$

After we multiply each term by y^3, we get the equation

$$y^5 \cos x\,dx + (4y^3 + 5y^4 \sin x)\,dy = 0,$$

which we easily verify to be exact. Integrating the coefficient of dx, we write

$$F(x, y) = \int y^5 \cos x\,dx = y^5 \sin x + g(y).$$

Then we set $\partial F/\partial y$ equal to the coefficient of dy to obtain

$$5y^4 \sin x + g'(y) = 4y^3 + 5y^4 \sin x.$$

Thus $g'(y) = 4y^3$, so we may choose $g(y) = y^4$. Therefore, a general solution of the original differential equation is

$$y^5 \sin x + y^4 = C.$$

EXAMPLE 6* Suppose that a flexible 4-ft rope starts with 3 ft of its length arranged in a heap right at the edge of a high horizontal table, with the remaining foot hanging (at rest) off the table. At time $t = 0$ the heap begins to unwind and the rope begins gradually to fall off the table, under the force of gravity pulling on the overhanging part. Under the assumption that frictional forces of all sorts are negligible, how long will it take for all the rope to fall off the table?

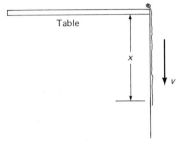

Table

FIGURE 1.46 The heap of rope unwinding off a table.

SOLUTION We assume that some unspecified device prevents the entire heap from simply falling off the edge all at once. Let $x(t)$ denote the length of overhanging rope at time t, as shown in Fig. 1.46, and let $v(t)$ denote its velocity then, with the positive direction downward. Then

$$x(0) = 1 \quad \text{and} \quad v(0) = 0. \tag{12}$$

If the linear density of the rope is w (slugs/foot), then the mass of the over-hanging rope is $m = wx$ and the gravitational force acting upon it is $F = mg = wgx$. Then Newton's second law in the form

$$F = \frac{d}{dt}(mv)$$

yields

$$wgx = \frac{d}{dt}(wxv) = w\left(x\frac{dv}{dt} + v\frac{dx}{dt}\right).$$

Because $dx/dt = v$, we need to solve the differential equation

$$gx = x\frac{dv}{dt} + v^2 = x\frac{dv}{dx}\frac{dx}{dt} + v^2 = xv\frac{dv}{dx} + v^2;$$

that is,

$$\left(\frac{v^2}{x} - g\right)dx + v\,dv = 0. \tag{13}$$

With $M(x, v) = v^2/x - g$ and $N(x, v) = v$, we have $M_v = 2v/x$ but $N_x = 0$, so Eq. (13) is not exact. But

$$\frac{M_v - N_x}{N} = \frac{(2v/x) - 0}{v} = \frac{2}{x},$$

so it has the integrating factor

$$\rho(x) = e^{\int (2/x)\,dx} = e^{2\ln x} = x^2.$$

Multiplication of the terms in (13) by x^2 gives the exact equation

$$(xv^2 - gx^2)\,dx + x^2v\,dv = 0.$$

---

* We are indebted to Dr. Carol W. Penney for this interesting variant of a standard example (see Problem 36).

Its general solution $\frac{1}{2}x^2v^2 - \frac{1}{3}gx^3 = C$ is readily found. The initial conditions in (12) imply that $C = -g/3$. With this value of C we solve for

$$v = \frac{dx}{dt} = \left(\frac{2g}{3}\right)^{1/2} \frac{(x^3 - 1)^{1/2}}{x},$$

taking the positive square root because $v \geq 0$. Hence

$$t = \left(\frac{3}{2g}\right)^{1/2} \int \frac{x}{(x^3 - 1)^{1/2}}\, dx.$$

Because $x = 1$ when $t = 0$, the desired value of t when $x = 4$ is

$$T = \left(\frac{3}{2g}\right)^{1/2} \int_1^4 \frac{x}{(x^3 - 1)^{1/2}}\, dx.$$

This is a nonelementary improper integral, but the substitution $x^3 = \sec^2 u$ converts it into the proper integral

$$T = \left(\frac{2}{3g}\right)^{1/2} \int_0^{\cos^{-1}(1/8)} (\sec u)^{4/3}\, du.$$

Simpson's rule with $n = 100$ subintervals finally gives $T \approx 0.541$ s for the time required for all the rope to fall off the table.

1.7   PROBLEMS

In each of Problems 1–12, verify that the given differential equation is exact and then solve it.

1. $(2x + 3y)\, dx + (3x + 2y)\, dy = 0$

2. $(4x - y)\, dx + (6y - x)\, dy = 0$

3. $(3x^2 + 2y^2)\, dx + (4xy + 6y^2)\, dy = 0$

4. $(2xy^2 + 3x^2)\, dx + (2x^2y + 4y^3)\, dy = 0$

5. $\left(x^3 + \dfrac{y}{x}\right) dx + (y^2 + \ln x)\, dy = 0$

6. $(1 + ye^{xy})\, dx + (2y + xe^{xy})\, dy = 0$

7. $(\cos x + \ln y)\, dx + \left(\dfrac{x}{y} + e^y\right) dy = 0$

8. $(x + \tan^{-1} y)\, dx + \dfrac{x + y}{1 + y^2}\, dy = 0$

9. $(3x^2y^3 + y^4)\, dx + (3x^3y^2 + y^4 + 4xy^3)\, dy = 0$

10. $(e^x \sin y + \tan y)\, dx + (e^x \cos y + x \sec^2 y)\, dy = 0$

11. $\left(\dfrac{2x}{y} - \dfrac{3y^2}{x^4}\right) dx + \left(\dfrac{2y}{x^3} - \dfrac{x^2}{y^2} + \dfrac{1}{y^{1/2}}\right) dy = 0$

12. $\dfrac{2x^{5/2} - 3y^{5/3}}{2x^{5/2}y^{2/3}}\, dx + \dfrac{3y^{5/3} - 2x^{5/2}}{3x^{3/2}y^{5/3}}\, dy = 0$

Use the method of grouping to solve the differential equations in Problems 13–20.

13. $y\, dx - x\, dy = y^3\, dy$

14. $x\, dy - y\, dx = x^2e^x\, dx$

15. $(y - 1)\, dx + (x - 2)\, dy = 0$

16. $(xy^2 - x)\, dx + (x^2y - y)\, dy = 0$

17. $3x^2y\, dx + (x^3 + e^y)\, dy = 0$

18. $(2x + x^2 + y^2)\, dx + (2y - x^2 - y^2)\, dy = 0$

19. $(x^2 + y^2 - y)\, dx + x\, dy = 0$

20. $(y^2 + e^x \sin y)\, dx + (2xy + e^x \cos y)\, dy = 0$

Solve each of the differential equations in Problems 21–28 by finding an integrating factor that is a function of a single variable.

21. $4y\, dx + x\, dy = 0$

22. $(4x + 3y^3)\, dx + 3xy^2\, dy = 0$

23. $2xy\, dx + (y^2 - x^2)\, dy = 0$

24. $(4x^2 + 3 \cos y)\, dx - x \sin y\, dy = 0$

25. $(y \ln y + ye^x)\, dx + (x + y \cos y)\, dy = 0$

26. $(4xy^2 + y)\, dx + (6y^3 - x)\, dy = 0$

27. $2x \, dx + x^2 \cot y \, dy = 0$

28. $\left(1 + \dfrac{1}{x}\right) \tan y \, dx + \sec^2 y \, dy = 0$

29. Solve $(7x^4 y - 3y^8) \, dx + (2x^5 - 9xy^7) \, dy = 0$, given that there exists an integrating factor of the form $x^m y^n$.

30. Use the technique suggested in Problem 29 to solve the equation $(3y^2 + 10xy) \, dx + (5xy + 12x^2) \, dy = 0$.

Use the observation that

$$D_x\left[\tan^{-1}\frac{x}{y}\right] = \frac{y \, dx - x \, dy}{x^2 + y^2}$$

to solve the differential equations in Problems 31 and 32.

31. $\left[y + x(x^2 + y^2)^{1/2}\right] dx + \left[y(x^2 + y^2)^{1/2} - x\right] dy = 0$

32. $\left(y + \dfrac{x}{x^2 + y^2}\right) dx + \left(\dfrac{y}{x^2 + y^2} - x\right) dy = 0$

33. Suppose that $\rho(x)$ is a function of x alone and is an integrating factor of $M(x, y) \, dx + N(x, y) \, dy = 0$. Show that $(M_y - N_x)/N$ is a function of x alone.

34. Suppose that $(N_x - M_y)/M = g(y)$ is a function of y alone.

Prove that $\rho(y) = \exp\left(\int g(y) \, dy\right)$ is an integrating factor of $M \, dx + N \, dy = 0$.

35. To solve the linear equation $y' + P(x)y = Q(x)$, rewrite it in the form $M \, dx + N \, dy = 0$ with $M = Py - Q$ and $N = 1$. Then show that this last differential equation has an integrating factor of the form $\rho(x)$, and use this fact to derive the solution

$$y(x) = e^{-\int P(x) \, dx}\left[C + \int\left(Q(x)e^{\int P(x) \, dx}\right) dx\right].$$

36. In Example 6 suppose that the 3 ft of rope on the table, rather than being heaped in a pile at the edge of the table, are initially stretched out on the table perpendicular to its edge. At time $t = 0$, when 1 ft of rope is hanging from the table, the rope begins to slide off the table. (a) Using the notation in the text, derive the differential equation

$$Lx'' = gx; \quad \text{that is,} \quad x'' = 8x,$$

because $L = 4$ (ft) and $g = 32$ (ft/s²). (b) Note that the independent variable t is missing from the differential equation derived in part (a), and that the initial conditions are $x(0) = 1$ and $x'(0) = 0$. Derive the solution $x(t) = \cosh(t\sqrt{8})$. (c) How long will it take until the last bit of rope leaves the table? (*Answer:* About 0.730 s.) Hint: Substitute $x' = v$, $x'' = dv/dt = v(dv/dx)$.

---

*1.8

Population Models

In Section 1.4 we introduced the exponential differential equation $dP/dt = kP$, with solution $P(t) = P_0 e^{kt}$, as a mathematical model for natural population growth that occurs as a result of constant birth and death rates. Here we present a more general population model that accommodates birth and death rates that are not necessarily constant. As before, however, our population function $P(t)$ will be a *continuous* approximation to the actual population, which of course grows by integral increments.

Suppose that the population changes only by the occurrence of births and deaths—there is no immigration or emigration. Let $B(t)$ and $D(t)$ denote, respectively, the numbers of births and deaths that have occurred (since $t = 0$) by time t. Then the *birth rate* $\beta(t)$ and *death rate* $\delta(t)$, in births or deaths per individual per unit of time, are defined as follows:

$$\left.\begin{aligned}
\beta(t) &= \lim_{h \to 0} \frac{B(t + h) - B(t)}{hP(t)} = \frac{1}{P}\frac{dB}{dt}, \\[2mm]
\delta(t) &= \lim_{h \to 0} \frac{D(t + h) - D(t)}{hP(t)} = \frac{1}{P}\frac{dD}{dt}.
\end{aligned}\right\} \tag{1}$$

Thus the birth rate $\beta(t)$ is the number of births per unit time per unit of population, while the death rate $\delta(t)$ is the number of deaths per unit time per unit of population.

Then

$$P'(t) = \lim_{h \to 0} \frac{P(t + h) - P(t)}{h}$$

$$= \lim_{h \to 0} \frac{[B(t + h) - B(t)] - [D(t + h) - D(t)]}{h}$$

$$= B'(t) - D'(t).$$

Thus

$$P'(t) = [\beta(t) - \delta(t)]P(t). \tag{2}$$

We will also use the briefer notation

$$\frac{dP}{dt} = (\beta - \delta)P.$$

Equation (2) is the **general population equation**. If β and δ are constant, it reduces to the natural growth equation with $k = \beta - \delta$. But it also includes the possibility that β and δ are variable functions of t. The birth and death rates need not be known in advance; they may well depend upon the unknown function $P(t)$.

LIMITED POPULATIONS

In situations as diverse as the human population of a nation and a fruit fly population in a closed container, it is often observed that the birth rate decreases as the population itself increases. The reasons may range from increased scientific or cultural sophistication to a limited food supply. Suppose, for example, that the birth rate β is a *linear* decreasing function of the population P, so that $\beta = \beta_0 - \beta_1 P$, where β_0 and β_1 are positive constants. If the death rate $\delta = \delta_0$ remains constant, then Eq. (2) takes the form

$$\frac{dP}{dt} = (\beta_0 - \beta_1 P - \delta_0)P;$$

that is,

$$\frac{dP}{dt} = kP(M - P) \tag{3}$$

where $k = \beta_1$ and $M = (\beta_0 - \delta_0)/\beta_1$. We assume $\beta_0 > \delta_0$ so that $M > 0$.

Equation (3) is called the **logistic equation**. If we assume that $P < M$, it may be solved by separation of variables as follows.

$$\int \frac{dP}{P(M - P)} = \int k \, dt;$$

$$\frac{1}{M} \int \left(\frac{1}{P} + \frac{1}{M - P} \right) dP = \int k \, dt;$$

$$\ln \left(\frac{P}{M - P} \right) = kMt + C.$$

Exponentiation gives

$$\frac{P}{M - P} = Ae^{kMt}$$

where $A = e^C$. We substitute $t = 0$ into each side of this last equation to find that $A = P_0/(M - P_0)$. So

$$\frac{P}{M - P} = \frac{P_0 e^{kMt}}{M - P_0}.$$

This equation is easy to solve for

$$P(t) = \frac{MP_0}{P_0 + (M - P_0)e^{-kMt}}. \tag{4}$$

While we made the assumption that $P < M$ in order to derive Eq. (4), this restriction is unnecessary, because we can verify by direct substitution into Eq. (3) that $P(t)$ as given in (4) satisfies the logistic equation whether $P < M$ or $P \geqq M$.

If the initial population satisfies $P_0 < M$, then (4) shows that $P(t) < M$ for all $t \geqq 0$, and also that

$$\lim_{t \to \infty} P(t) = M. \tag{5}$$

Thus a population that satisfies the logistic equation is *not* like a naturally growing population; it does not grow without bound, but instead approaches the finite **limiting population** M as $t \to +\infty$. But because $dP/dt = kP(M - P) > 0$ in this case, we see that the population is steadily increasing. Moreover, differentiation with respect to t gives

$$\frac{d^2P}{dt^2} = \left[\frac{d}{dP}\left(\frac{dP}{dt} \right) \right]\left(\frac{dP}{dt} \right) = (kM - 2kP)[kP(M - P)].$$

Hence the graph of $P(t)$ has an inflection point where $P = M/2$. Therefore, the graph of P has the shape of one of the lower curves in Fig. 1.47. The population increases at an increasing rate until $P = M/2$, and thereafter increases at a decreasing rate. If $P_0 > M$, then a similar analysis (see Problem 8) shows that $P(t)$ is a steadily decreasing function with a graph resembling one of the upper curves in Fig. 1.47.

In 1845 the Belgian demographer Verhulst used the 1790–1840 U.S. population data to predict the U.S. population through the year 1930, under his assumption that it would continue to satisfy the logistic equation. With $P_0 = 3.9$ (in millions), $M = 197.3$ (in millions), and $k = 0.0001589$, Eq. (4) gives the remarkable results shown in the table in Fig. 1.48. Of course, the assumed limiting population of $M = 197.3$ millions has now been exceeded, so the population of the United States has not continued to satisfy the logistic equation in the decades following 1930.

EXAMPLE 1   Suppose that in 1885 the population of a certain country was 50 million and was growing at the rate of 750,000 people per year at that time. Suppose also that in 1940 its population was 100 million and was then growing at the rate of 1 million per year. Assume that this population satisfies the logistic

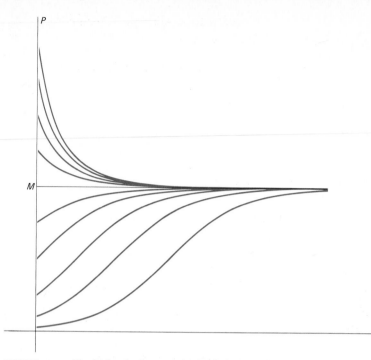

FIGURE 1.47 Typical solutions of the logistic equation.

Year	Actual United States population (millions)	Calculated population
1790	3.9	3.9
1800	5.3	5.3
1810	7.2	7.2
1820	9.6	9.7
1830	12.9	13.0
1840	17.1	17.4
1850	23.2	23.1
1860	31.4	30.2
1870	38.6	39.2
1880	50.2	49.9
1890	62.9	62.5
1900	76.0	76.6
1910	92.0	91.7
1920	106.5	107.1
1930	123.2	122.1

FIGURE 1.48 U.S. population data.

CHAPTER 1: Introduction and First Order Differential Equations

equation. Determine both the limiting population M and the predicted population for the year 2000.

SOLUTION We substitute the two given pairs of data in Eq. (3) and find that

$$0.75 = 50k(M - 50),$$

$$1.00 = 100k(M - 100).$$

We solve simultaneously for $M = 200$ and $k = 0.0001$. Thus the limiting population of the country in question is 200 million. With these values of M and k and with $t = 0$ corresponding to the year 1940 (in which $P_0 = 100$), we find that, as a consequence of Eq. (4), the population in the year 2000 will be

$$P(60) = \frac{(100)(200)}{100 + (200 - 100)e^{-(0.0001)(200)(60)}},$$

or about 153.7 million people.

We next describe some situations that illustrate the varied circumstances in which the logistic equation is a satisfactory mathematical model.

1. *Limited environment situation.* A certain environment can support a population of at most M individuals. It's then reasonable to expect the growth rate $\beta - \delta$ (the combined birth and death rates) to be proportional to $M - P$, because we may think of $M - P$ as the potential for further expansion. Then $\beta - \delta = k(M - P)$, so that

$$\frac{dP}{dt} = (\beta - \delta)P = kP(M - P).$$

   The classic example of a limited environment situation is a fruit fly population in a closed container.

2. *Competition situation.* If the birth rate β is constant but the death rate δ is proportional to P, so that $\delta = \alpha P$, then

$$\frac{dP}{dt} = (\beta - \alpha P)P = kP(M - P).$$

   This might be a reasonable working hypothesis in a study of a cannibalistic population, in which all deaths result from chance encounters between individuals. Of course, competition between individuals is not usually so deadly, nor its effects so immediate and decisive.

3. *Joint proportion situation.* Let $P(t)$ denote the number of individuals in a constant susceptible population M who are infected with a certain contagious and incurable disease. The disease in question is spread by chance encounters. Then $P'(t)$ should be proportional to the product of the number P of the individuals having the disease and the number $M - P$ of those not having it, so that $dP/dt = kP(M - P)$. Again we discover that the mathematical model is the logistic equation. The mathematical description of the spread of a rumor in a population of M individuals is identical.

EXAMPLE 2   Suppose that at time $t = 0$, half of a population of 100,000 persons have heard a certain rumor, and that the number of those who have heard it is then increasing at the rate of 1000 persons per day. How long will it take for this rumor to spread to 80% of the population?

SOLUTION   Let us work in units of thousands of persons. Then we take $M = 100$ for the total fixed population. We substitute $M = 100$, $P'(0) = 1$, and $P_0 = 50$ into the logistic equation, and thereby obtain

$$1 = k(50)(100 - 50), \quad \text{so that} \quad k = 0.0004.$$

If t denotes the number of days until 80 thousand people have heard the rumor, then Eq. (4) gives

$$80 = \frac{(50)(100)}{50 + (100 - 50)e^{-(0.04)t}},$$

so that t is approximately 34.66. Thus the rumor will have spread to 80% of the population in a little less than 35 days.

DOOMSDAY VERSUS EXTINCTION

Consider a population $P(t)$ of unsophisticated animals that rely solely on chance encounters to meet mates for reproductive purposes. It is reasonable to expect such encounters to occur at a rate that is proportional to the product of the number $P/2$ of males and the number $P/2$ of females, hence at a rate proportional to P^2. Hence we assume that births occur at the rate $dB/dt = kP^2$ (k constant), so that $\beta = kP$ by Eq. (1). If the death rate δ is constant, then the general population equation in (2) yields the differential equation

$$\frac{dP}{dt} = kP^2 - \delta P = kP(P - a), \tag{6}$$

where $a = \delta/k > 0$, as a mathematical model of the population. The solution of Eq. (6) depends upon whether the initial population $P_0 = P(0)$ is greater than or less than a.

Case 1: $P_0 > a$. From Eq. (6) we see that $P'(0) = kP_0(P_0 - a) > 0$, so $P(t)$ starts out increasing. Hence $P'(t)$ remains positive, so $P(t)$ continues to increase, and therefore $P(t) > a$ for all $t > 0$. We note that

$$\frac{1}{P(P - a)} = -\frac{1}{a}\left(\frac{1}{P} - \frac{1}{P - a}\right).$$

We separate variables in Eq. (6) and integrate as follows:

$$\int \frac{dP}{P(P - a)} = \int k \, dt;$$

$$\int \left(\frac{1}{P} - \frac{1}{P - a}\right) dP = -\int ka \, dt;$$

$$\ln \frac{P}{P - a} = -kat + C_1.$$

Substitution of P_0 for P and 0 for t then gives

$$C_1 = \ln \frac{P_0}{P_0 - a} = \ln C$$

where $C = P_0/(P_0 - a) > 1$. Exponentiation then yields $P/(P - a) = Ce^{-kat}$, which we solve for

$$P(t) = \frac{Cae^{-kat}}{Ce^{-kat} - 1}. \tag{7}$$

Note that the denominator in (7) approaches zero as

$$t \to T = \frac{\ln C}{ka} = \frac{1}{ka} \ln \frac{P_0}{P_0 - a} > 0.$$

Thus $\lim\limits_{t \to T} P(t) = +\infty$. This is a *doomsday* situation.

Case 2: $P_0 < a$. In this case $P'(0) < 0$, and it follows that $P(t) < a$ for all $t > 0$. A similar separation of variables (Problem 14) now leads to

$$P(t) = \frac{Cae^{-kat}}{Ce^{-kat} + 1} \tag{8}$$

where $C = P_0/(a - P_0) > 0$. The difference is that the denominator in Eq. (8) remains greater than 1, and so $\lim\limits_{t \to \infty} P(t) = 0$. This is an *extinction* situation.

Thus the population either explodes or is an endangered species threatened with extinction, depending upon its initial size. An approximation to this phenomenon is sometimes observed with actual animal populations, such as the alligator populations in certain areas of the southern United States.

1.8 PROBLEMS

1. Suppose that the fish population $P(t)$ in a lake is attacked by disease at time $t = 0$, with the results that the fish cease to reproduce (so that the birth rate is $\beta = 0$) and the death rate δ (deaths per week per fish) is thereafter proportional to $\sqrt{1/P}$. If there were initially 900 fish in the lake and 441 were left after 6 weeks, how long did it take all the fish in the lake to die?

2. Suppose that when a certain lake is stocked with fish, the birth and death rates β and δ are both inversely proportional to $\sqrt{P}$. (a) Show that $P(t) = (\frac{1}{2}kt + \sqrt{P_0})^2$ where k is a constant. (b) If $P_0 = 100$ and after 6 months there are 169 fish in the lake, how many will there be after 1 year?

3. Consider a prolific breed of rabbits whose birth and death rates, β and δ, are each proportional to the rabbit population $P = P(t)$, with $\beta > \delta$. (a) Show that

$$P(t) = \frac{P_0}{1 - kP_0 t}, \qquad k \text{ constant.}$$

Note that $P(t) \to +\infty$ as $t \to 1/(kP_0)$. This is doomsday. (b) Suppose that $P_0 = 2$ and that there are four rabbits after 3 months. When does doomsday occur?

4. Repeat Problem 3(a) in the case $\beta < \delta$. What now happens to the rabbit population in the long run?

5. Consider a human population $P(t)$ with constant birth and death rates β and δ, but also with I persons per year entering the country (immigration). Estimate the change ΔP during the short time interval Δt to derive the differential equation

$$\frac{dP}{dt} = kP + I \qquad (k = \beta - \delta).$$

Use the result of Problem 5 in Problems 6 and 7.

6. A certain city had a population of 1.5 million in 1980. Assume that it grows continuously at a 4% annual rate (this implies that $\beta - \delta = 0.04$) and also absorbs 50,000 newcomers per year. What will be its population in the year 2000?

7. Consider the U.S. population with $P_0 = 222$ million in 1980. (a) Compute the population in the year 2000, assuming a natural growth rate of 1% annually. (b) Rework part (a) with the additional assumption that immigration will remain at a constant 500,000 people per year.

8. Derive the solution in (4) of the logistic equation in (3) under the assumption that $P > M$, and show in this case that the graph of $P(t)$ resembles the upper curves in Fig. 1.47.

9. Suppose that as a certain salt dissolves in a solvent, the number $x(t)$ of grams of the salt in solution after t seconds satisfies the logistic equation $dx/dt = (0.8)x - (0.004)x^2$. (a) What is the maximum amount of the salt that will dissolve in this solvent? (b) If $x = 50$ when $t = 0$, how long will it take for an additional 50 g of the salt to dissolve?

10. Suppose that a community contains 15,000 people who are susceptible to a spreading contagious disease. At time $t = 0$ the number $N(t)$ of people who have the disease is 5000 and is increasing by 500 per day. How long will it take for another 5000 people to contract the disease? Assume that $N'(t)$ is proportional to the product of the numbers of those who have the disease and those who do not.

11. The data in the table in Fig. 1.49 are given for a certain population $P(t)$ that satisfies the logistic equation in (3). (a) What is

Year	P (millions)
1909	24.63
1925	25.00
1926	25.38
.	.
.	.
.	.
1974	47.04
1975	47.54
1976	48.04

FIGURE 1.49 Population data for Problem 11.

the limiting population M? (*Suggestion:* Use the approximation

$$P'(t) \approx \frac{P(t+h) - P(t-h)}{2h}$$

with $h = 1$ to estimate the values of dP/dt when $P = 25.00$ and when $P = 47.54$. Then substitute these values in the logistic equation and solve for k and M.) (b) Use the values of k and M found in part (a) to determine when $P = 75$. (*Suggestion:* Take $t = 0$ to correspond to the year 1925.)

12. A population $P(t)$ of small rodents has birth rate $\beta = (0.001)P$ (births per month per rodent) and *constant* death rate δ. If $P(0) = 100$ and $P'(0) = 8$, how long (in months) will it take this population to double to 200 rodents? (*Suggestion:* First find the value of δ.)

13. Consider an animal population $P(t)$ with constant death rate $\delta = 0.01$ and with birth rate β proportional to P. Suppose that $P(0) = 200$ and $P'(0) = 2$. (a) When is $P = 1000$? (b) When does doomsday occur?

14. Derive the solution in (8) of Eq. (7) in the case $P(0) < a$.

15. A tumor may be regarded as a population of multiplying cells. It is found empirically that the "birth rate" of the cells in a tumor decreases exponentially with time, so that $\beta(t) = \beta_0 e^{-\alpha t}$ (where α and β_0 are positive constants), and hence

$$\frac{dP}{dt} = \beta_0 e^{-\alpha t} P, \qquad P(0) = P_0.$$

Solve this initial value problem for

$$P(t) = P_0 \exp\left[\frac{\beta_0}{\alpha}\left(1 - e^{-\alpha t}\right)\right].$$

Observe that $P(t)$ approaches the finite limiting population $P_0 \exp(\beta_0/\alpha)$ as $t \to +\infty$.

16. For the tumor of Problem 15, suppose that at time $t = 0$ there are $P_0 = 10^6$ cells and that $P(t)$ is then increasing at the rate of 3×10^5 cells per month. After 6 months the tumor has doubled (in size and in number of cells). Solve numerically for α, and then find the limiting population of the tumor.

Motion with Variable Acceleration

In Section 1.2 we discussed vertical motion near the surface of the earth under the influence of constant gravitational acceleration, $g \approx 32$ ft/s^2. Beginning with the acceleration $dv/dt = -g$, we integrated twice to obtain the equations

$$\frac{dy}{dt} = v = -gt + v_0$$

and

$$y = -\tfrac{1}{2}gt^2 + v_0 t + y_0$$

for the velocity v and height y at time t. We saw that these equations can be used to answer specific questions.

EXAMPLE 1   Suppose that we want to predict the maximum height that will be attained by a bolt shot from a crossbow aimed directly upward from ground level ($y_0 = 0$). A modern crossbow can easily impart an initial velocity of 288 ft/s to a lightweight bolt, so let us take $v_0 = 288$. We take $g = 32$ (exactly) for simplicity, and the equations of motion of the bolt take the form

$$v = -32t + 288;$$

$$y = -16t^2 + 288t.$$

Of course, here we assume that the crossbow is fired at time $t = 0$ s. Then $v = 0$ when $t = 9$, so the maximum height attained by the bolt will be

$$y_{\max} = y(9) = -(16)(9^2) + (288)(9) = 1296 \text{ (ft)}.$$

In the computation of Example 1 we have, however, ignored the effect of air resistance. This makes the mathematical analysis quite simple, but perhaps unrealistically so. Next we want to construct a more complete mathematical model, one that takes air resistance into account.

Empirical studies indicate that the force F_R of air resistance on a moving body with velocity v is approximately of the form $F_R = kv^p$, where $1 \leq p \leq 2$ and the value of k depends upon the size and shape of the body, as well as the density and viscosity of the air. Generally speaking, $p = 1$ for relatively low speeds and $p = 2$ for high speeds, while $1 < p < 2$ for intermediate speeds. But how slow "low speed" and how fast "high speed" are depends upon the same factors that determine the value of the coefficient k.

Thus air resistance is a complicated physical phenomenon. But the simplifying assumption that F_R is *exactly* of the form given above, with either $p = 1$ or $p = 2$, yields a tractable mathematical model that exhibits the most important qualitative features of motion with resistance.

RESISTANCE PROPORTIONAL TO VELOCITY

So let us consider the vertical motion of a body with mass m near the surface of the earth, subject to two forces: a downward gravitational force F_G and a force F_R of air resistance that is proportional to velocity (so that $p = 1$) and of course directed opposite the direction of motion of the body. If we set up a coordinate system with the positive y-direction upward and $y = 0$ at ground level, then $F_G = -mg$ and

$$F_R = -kv, \tag{1}$$

where k is a positive constant and $v = dy/dt$ is the velocity of the body. Note that the minus sign in Eq. (1) makes F_R positive (an upward force) if the body is falling

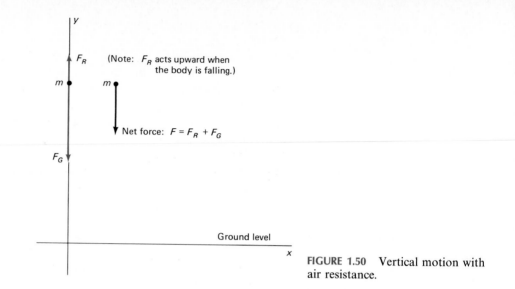

(Note: F_R acts upward when the body is falling.)

Net force: $F = F_R + F_G$

Ground level

FIGURE 1.50 Vertical motion with air resistance.

(v is negative), while it makes F_R negative (a downward force) if the body is rising (v is positive). As indicated in Fig. 1.50, the net force acting on the body is then

$$F = F_R + F_G = -kv - mg,$$

and Newton's law of motion $F = m(dv/dt)$ yields the equation

$$m \frac{dv}{dt} = -kv - mg.$$

Thus

$$\frac{dv}{dt} = -\rho v - g \tag{2}$$

where $\rho = k/m > 0$. The reader should verify that, if the positive y-axis were directed downward, then Eq. (2) would take the form $v' = -\rho v + g$.

Equation (2) is a separable first order differential equation, and its solution is

$$v(t) = \left(v_0 + \frac{g}{\rho} \right) e^{-\rho t} - \frac{g}{\rho}. \tag{3}$$

Here, $v_0 = v(0)$ is the initial velocity of the body. Note that

$$\lim_{t \to \infty} v(t) = -\frac{g}{\rho}. \tag{4}$$

Thus the speed of a body falling with air resistance does *not* increase indefinitely; instead, it approaches a *finite* limiting speed, or **terminal speed**,

$$v_\tau = \frac{g}{\rho} = \frac{mg}{k}. \tag{5}$$

This fact is what makes the parachute a practical invention; it even helps explain the occasional survival of people who fall without parachutes from high-flying airplanes.

CHAPTER 1: Introduction and First Order Differential Equations

We now rewrite Eq. (3) in the form

$$\frac{dy}{dt} = (v_0 + v_\tau)e^{-\rho t} - v_\tau. \tag{6}$$

Integration gives

$$y(t) = -\frac{1}{\rho}(v_0 + v_\tau)e^{-\rho t} - v_\tau t + C.$$

We substitute 0 for t and let $y_0 = y(0)$ denote the initial height of the body. Thus we find that $C = y_0 + (v_0 + v_\tau)/\rho$, and so

$$y(t) = y_0 - v_\tau t + \frac{1}{\rho}(v_0 + v_\tau)(1 - e^{-\rho t}). \tag{7}$$

Equations (6) and (7) give the velocity v and height y of a body moving vertically under the influence of gravity and air resistance. The formulas depend upon the initial height y_0 of the body, its initial velocity v_0, and what might be called the *drag coefficient* ρ, the constant such that the acceleration due to air resistance is $a_R = -\rho v$. The two equations also involve the terminal speed v_τ.

EXAMPLE 2   We consider again a bolt shot straight upward from a crossbow at ground level with initial velocity $v_0 = 288$ ft/s. But now let us also take into account air resistance, with $\rho = 0.04$. What will be the maximum height reached by the bolt, and when will that maximum be attained? When and with what speed will the bolt strike the ground?

SOLUTION   We substitute $y_0 = 0$, $v_0 = 288$, and $v_\tau = g/\rho = 800$ into Eqs. (3) and (7). We obtain

$$v = 1088e^{-t/25} - 800$$

and

$$y = 27200(1 - e^{-t/25}) - 800t.$$

To find the time required for the bolt to reach its maximum height (when $v = 0$), we solve the equation

$$1088e^{-t/25} - 800 = 0$$

for $t = 25 \ln (34/25) \approx 7.687$ s. Its maximum height is then

$$y_{\max} \approx 27200(1 - e^{-0.3075}) - (800)(7.687) \approx 1050 \text{ (ft)}.$$

To find when the bolt strikes the ground ($y = 0$), we must solve the equation $(27200)(1 - e^{-t/25}) - 800t = 0$, which can be written in the simpler form

$$(34)(1 - e^{-t/25}) - t = 0.$$

To solve the latter equation, we use the iterative formula of Newton's method:

$$t_{n+1} = t_n - \frac{f(t_n)}{f'(t_n)}. \tag{8}$$

Recall that this iteration successively improves an initial estimate t_0 of a solution of the equation $f(t) = 0$. (See, for example, Section 3.10 of Edwards and Penney, *Calculus and Analytic Geometry*, 2nd ed. (Englewood Cliffs, N.J.: Prentice-Hall, 1986).) With

$$f(t) = (34)(1 - e^{-t/25}) - t,$$

Equation (8) takes the form

$$t_{n+1} = t_n - \frac{34[1 - \exp(-t_n/25)] - t_n}{(34/25)\exp(-t_n/25) - 1}. \tag{9}$$

If we begin with $t_0 = 18$ s (the total time aloft in the case of no air resistance), the iteration in (9) yields $t_1 = 16.374$ s and $t_2 = t_3 = 16.252$ s. Thus the bolt is in the air for about 16.25 s and hits the ground with velocity

$$v_{\text{imp}} = 1088 \exp(-0.65) - 800 \approx -232 \text{ (ft/s)}.$$

The effect of air resistance is to decrease the maximum height, the total time spent aloft, and the final impact speed. Note also that when the effect of air resistance is included, the bolt spends more time in descent (about 8.565 s) than in ascent (about 7.687 s).

For a person descending with the aid of a parachute, a typical value of ρ is 1.5, which corresponds to a terminal speed of $v_\tau \approx 21.3$ ft/s, or about 14.5 mi/h. With an unbuttoned overcoat flapping in the wind in place of a parachute, an unlucky skydiver might increase ρ to perhaps as much as 0.5, which gives a terminal speed of $v_\tau \approx 64$ ft/s, or about 44 mi/h. See Problems 10 and 11 for some parachute-jump computations.

RESISTANCE PROPORTIONAL TO SQUARE OF VELOCITY

Now let us assume that the force of air resistance is proportional to the *square* of the velocity:

$$F_R = \pm kv^2 \tag{10}$$

where k is a positive constant. Then Newton's second law gives

$$m \frac{dv}{dt} = F_G + F_R = \pm mg \pm kv^2;$$

that is,

$$\frac{dv}{dt} = \pm g \pm \rho v^2 \tag{11}$$

where $\rho = k/m > 0$. The choice of signs in Eq. (11) depends upon the direction of motion, as well as the choice of the direction of the positive y-axis. Consequently, we must discuss the two cases—upward and downward motion—separately.

DOWNWARD MOTION   Suppose that the body is dropped from a given height with initial velocity $v(0) = 0$. In this case it is more convenient to choose the y-axis pointing *downward*, with $y_0 = y(0) = 0$, so $y(t)$ will denote the distance the body has fallen at time t. Then F_G is a downward force, thus positive, while F_R is an upward (negative) force. So we choose the signs in Eq. (11) to obtain

$$\frac{dv}{dt} = +g - \rho v^2 = g\left(1 - \frac{\rho}{g}v^2\right). \tag{12}$$

Recall that the hyperbolic tangent function may be defined as

$$\tanh u = \frac{\sinh u}{\cosh u} = \frac{e^u - e^{-u}}{e^u + e^{-u}}.$$

In Problem 13 we ask you to apply the integral

$$\int \frac{du}{1 - u^2} = \tanh^{-1} u + C$$

to solve Eq. (12) with $v(0) = 0$ for

$$v(t) = \sqrt{\frac{g}{\rho}} \tanh(t\sqrt{\rho g}). \tag{13}$$

Because $v = dy/dt$, another integration (see Problem 14) yields

$$y(t) = \frac{1}{\rho} \ln \cosh(t\sqrt{\rho g}) \tag{14}$$

for the distance the body has fallen at time t.

In the case of downward motion with resistance proportional to the velocity, $F_R = -kv$, we saw earlier that a body acquires a terminal speed $v_\tau = g/\rho$ (with $\rho = k/m$). Because

$$\lim_{t \to \infty} \tanh at = \lim_{t \to \infty} \frac{e^{at} - e^{-at}}{e^{at} + e^{-at}} = 1$$

if $a > 0$, it follows from Eq. (13) that, in the present case of downward motion with resistance $F_R = kv^2$, the body approaches a terminal speed

$$v_\tau = \sqrt{\frac{g}{\rho}} = \sqrt{\frac{mg}{k}}. \tag{15}$$

UPWARD MOTION   Now let us consider a body that is projected straight upward from ground level $y_0 = 0$ with initial velocity $v(0) = v_0$ and with the positive y-axis now directed upward. Our differential equation now takes the form

$$\frac{dv}{dt} = -g - \rho v^2 \tag{16}$$

because F_G and F_R are each directed downward.

In Problem 16 we ask you to deduce that the velocity and height of the body at time t are given by

$$v = \sqrt{\frac{g}{\rho}} \tan(C - t\sqrt{\rho g}) \tag{17}$$

and

$$y = \frac{1}{\rho} \ln \frac{\cos (C - t\sqrt{\rho g})}{\cos C} \qquad (18)$$

where

$$C = \tan^{-1}\left(v_0\sqrt{\frac{\rho}{g}}\right). \qquad (19)$$

The motion of the body continues to be described by Eqs. (17) and (18) as long as the velocity given in (17) is still positive; that is, as long as the body is still moving upward. It reaches its maximum height when $t\sqrt{\rho g} = C$ (so that $v = 0$); hence Eq. (18) gives

$$y_{max} = -\frac{1}{\rho} \ln \cos C. \qquad (20)$$

Thereafter the body falls back toward the ground, with its (downward) velocity and distance fallen (t seconds after beginning to fall) given by Eqs. (13) and (14).

Let us now reconsider the crossbow bolt of Example 2, which we assumed shot straight upward from the ground with an initial velocity $v_0 = 288$ ft/s. Instead of air resistance proportional to velocity v, we assume air resistance proportional to v^2, with $\rho = 0.0002$ in Eqs. (12) and (16). In Problems 18 and 19 we ask you to verify the entries in the last line of the following table.

Air Resistance	Maximum Height (ft)	Ascent Time (s)	Descent Time (s)	Impact Speed (ft/s)
0	1296	9.000	9.000	288
$0.04v$	1050	7.687	8.565	232
$0.0002v^2$	1044	7.800	8.362	234

Comparison of the last two lines of data here suggests a qualitative similarity between the cases $F_R = kv$ and $F_R = kv^2$.

VARIABLE GRAVITATIONAL ACCELERATION

Unless a projectile in vertical motion remains in the immediate vicinity of the earth's surface, the gravitational acceleration acting upon it is not constant. According to Newton's law of gravitation, the gravitational force of attraction between two point masses M and m located at a distance r apart is given by

$$F = \frac{GMm}{r^2} \qquad (21)$$

where G is a certain empirical constant. The formula is also valid if either or both of the two masses are homogeneous spheres; in this case, the distance r is measured between the centers of the spheres.

Let M denote the mass of the earth and R its radius. We obtain the gravitational acceleration $a = g$ of a particle of mass m at the earth's surface by combining Eq. (21) above with the equation $F = ma$:

$$ma = mg = \frac{GMm}{r^2} = \frac{GMm}{R^2},$$

so that

$$g = \frac{GM}{R^2} \quad \text{(about 32 ft/s}^2\text{).} \tag{22}$$

We will use Eq. (22) from time to time to remove the necessity for the actual determination of G in our problems and examples. For instance, the mass M_1 of the moon is approximately $(0.0123)M$ and its radius R_1 is approximately $(0.2725)R$, so we can use Eq. (22) to compute the lunar gravitational acceleration g_1 at the surface of the moon:

$$g_1 = \frac{GM_1}{R_1^2} = G\frac{(0.0123)M}{(0.2725R)^2} \approx (0.1656)\frac{GM}{R^2}$$

$$= (0.1656)g \approx (0.1656)(32) \approx 5.3 \text{ ft/s}^2.$$

We now calculate the *escape velocity* from the earth—the minimum initial velocity with which an object must be launched straight upward from the earth's surface so that it will continue forever to move away from the earth. The condition that the upward motion will continue forever becomes the relation $v = dy/dt > 0$ for all $t \geq 0$. We will use $y = y(t)$ to denote the distance from the object to the earth's *center* (because of the way in which Newton's law of gravitation is phrased). Our launch site, though, will be on the earth's *surface*, where we take $y_0 = y(0) = 6370$ km; that is, 6,370,000 m. The solution to the escape velocity problem will be the initial velocity $v_0 = v(0)$ that is just sufficient to insure that $v = v(t)$ is never zero or negative.

Now imagine the launched object as it is shown in Fig. 1.51, with mass m, at distance $y = y(t)$ from the center of the earth at time $t > 0$, and with velocity $v = v(t)$ then. The only force acting on the mass m as it ascends is the pull of gravity of the earth, given by Eq. (21), Newton's law of gravitation, as $F = -GMm/y^2$. The resulting acceleration of m is then given, with the *same* values of F and m, by Newton's second law of motion $F = ma$. We eliminate F by equating the other sides of each equation:

$$ma = m\frac{dv}{dt} = -\frac{GMm}{y^2},$$

which we simplify to

$$\frac{dv}{dt} = -\frac{GM}{y^2}.$$

To solve this differential equation, we use the chain rule:

$$\frac{dv}{dt} = \frac{dv}{dy}\frac{dy}{dt} = v\frac{dv}{dy}.$$

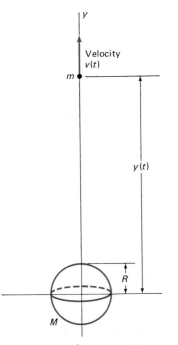

FIGURE 1.51 A mass m at a great distance from the earth.

Thus

$$v \frac{dv}{dy} = -\frac{GM}{y^2}.$$

That is,

$$D_y\left(\frac{1}{2} v^2\right) = D_y\left(\frac{GM}{y}\right).$$

Note that we think of y as the independent variable, rather than the more natural t. It follows immediately that

$$\frac{1}{2} v^2 = \frac{GM}{y} + C. \tag{23}$$

To evaluate the constant C, we use the fact that when $t = 0$, we have $v = v_0$ and $y = R$. Thus

$$\frac{1}{2} v_0^2 = \frac{GM}{R} + C;$$

we substitute the resulting value of C in Eq. (23) to obtain

$$v^2 = v_0^2 + 2GM\left(\frac{1}{y} - \frac{1}{R}\right).$$

In particular,

$$v^2 > v_0^2 - \frac{2GM}{R}.$$

Therefore v will remain positive provided that $v_0^2 \geq 2GM/R$. With the aid of Eq. (22), we can now write a formula for the escape velocity for the earth:

$$v_0 = \sqrt{\frac{2GM}{R}} = \sqrt{2Rg}. \tag{24}$$

With $g = 9.8$ m/s^2 and $R = 6.37 \times 10^6$ m, this gives $v_0 \approx 11{,}174$ m/s, about 36,000 ft/s, about 24,995 mi/h, or about 6.94 mi/s.

The following example is a refinement of Example 2 of Section 1.2; we now take lunar gravity into account.

EXAMPLE 3   A lunar lander is free-falling toward the moon's surface at a speed of 1000 mi/h. Its retrorockets, when fired in free space, provide a deceleration of 33,000 mi/h^2. At what height above the lunar surface should the retrorockets be activated to insure a "soft" touchdown ($v = 0$ at impact)?

SOLUTION   We use units of kilomiles and hours and denote by $y(t)$ the lunar lander's distance from the center of the moon at time t. The moon's radius is about 1.08 kilomiles, so we want $v = 0$ when $y = 1.08$. The moon's *surface* gravitational acceleration is, as computed above, 5.3 ft/s^2; this is approximately 13 kilomiles/h^2. Because gravitational acceleration is inversely proportional to the square of the distance, the lunar gravitational acceleration at distance y (from its center) is

$$\left(\frac{1.08}{y}\right)^2 (13) = \frac{15.16}{y^2}$$

kilomiles/h^2. We subtract this from the retrorocket acceleration of 33 kilomiles/h^2 to obtain

$$\frac{dv}{dt} = 33 - \frac{15.16}{y^2}$$

where $v = dy/dt$. We change the independent variable to y by using the chain rule to write

$$\frac{dv}{dt} = \frac{dv}{dy}\frac{dy}{dt} = v\frac{dv}{dy} = 33 - \frac{15.16}{y^2}.$$

Integration with respect to y gives

$$\frac{1}{2}v^2 = 33y + \frac{15.16}{y} + C.$$

The desired condition $v = 0$ when $y = 1.08$ implies that $C = -49.68$, so

$$\frac{1}{2}v^2 = 33y + \frac{15.16}{y} - 49.68.$$

Finally, we want to know the value of $y = y_0$ when $v_0 = -1$ (the initial velocity of the rocket, in kilomiles/h.) We substitute $v = -1$ into the last equation and simplify to obtain the quadratic equation

$$33y^2 - (50.18)y + 15.16 = 0.$$

The only roots of this equation are $y = 1.105$ and $y = 0.416$ (approximately). The second corresponds to a height *below* the lunar surface, so the first of the two is the solution we seek. And $1.105 - 1.080 = 0.025$ kilomiles, so the retrorockets should be activated at a distance of 25 miles above the lunar surface.

ROCKET PROPULSION

Suppose that the rocket of Fig. 1.52 blasts off straight upward from the surface of the earth at time $t = 0$. We want to compute its height y and velocity $v = dy/dt$ at time t. The rocket is propelled by exhaust gases that exit with constant speed c (relative to the rocket). Because of the combustion of its fuel, the mass $m = m(t)$ of the rocket is variable.

To derive the equation of motion of the rocket, we need Newton's second law in the form

$$\frac{dP}{dt} = F \tag{25}$$

where P is momentum (the product of mass and velocity) and F is the net external force (gravity, air resistance, and so on). If m is constant, then Eq. (25) takes the standard form $F = m(dv/dt)$.

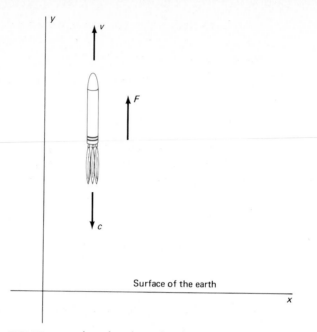

FIGURE 1.52 A rocket departing from the earth.

Suppose that m changes to $m + \Delta m$ and v changes to $v + \Delta v$ during the short time interval from t to $t + \Delta t$. The change in the momentum of the *rocket itself* is

$$(m + \Delta m)(v + \Delta v) - mv = m\,\Delta v + v\,\Delta m + \Delta m\,\Delta v.$$

But the system also includes the exhaust gases expelled during this time interval, with mass $-\Delta m$ and velocity $v - c$. Hence the total change in momentum during the time interval Δt is

$$\Delta P \approx (m\,\Delta v + v\,\Delta m + \Delta m\,\Delta v) + (-\Delta m)(v - c)$$
$$= m\,\Delta v + c\,\Delta m + \Delta m\,\Delta v.$$

Dividing by Δt and taking the limit as $\Delta t \to 0$, we get

$$\frac{dP}{dt} = m\frac{dv}{dt} + c\frac{dm}{dt}$$

because Δm and Δv approach zero as $\Delta t \to 0$. We substitute this expression for dP/dt into Newton's law (25) and thereby obtain the **rocket propulsion equation**

$$m\frac{dv}{dt} + c\frac{dm}{dt} = F. \tag{26}$$

We shall assume that $F = F_G + F_R$, where $F_G = -mg$ is a constant force of gravity and $F_R = -kv$ is a force of resistance proportional to velocity. This gives

$$m\frac{dv}{dt} + c\frac{dm}{dt} = -mg - kv. \tag{27}$$

Now suppose that the fuel of the rocket is consumed at a constant rate β (burn rate) during the time interval $[0, \tau]$, during which time the mass of the rocket decreases from m_0 to m_1. Thus

$$m(0) = m_0, \qquad m(\tau) = m_1,$$

$$m(t) = m_0 - \beta t, \qquad \frac{dm}{dt} = -\beta \quad \text{for } t \leqq \tau, \qquad (28)$$

with burnout occurring at time $t = \tau$.

We proceed to compute the velocity at time $t \leqq \tau$. If we substitute the expressions in (28) into Eq. (27), the result is

$$(m_0 - \beta t) \frac{dv}{dt} - \beta c = -(m_0 - \beta t)g - kv,$$

which we can rewrite in the form

$$\frac{dv}{dt} + \frac{k}{m_0 - \beta t} v = -g + \frac{\beta c}{m_0 - \beta t}. \qquad (29)$$

This is a linear first order equation for v as a function of t, with integrating factor

$$\rho = \exp\left(\int \frac{k \, dt}{m_0 - \beta t} \right) = (m_0 - \beta t)^{-k/\beta}.$$

The details are somewhat tedious (Problem 31), but the result of integrating Eq. (29) is

$$v(t) = v_0 M^{k/\beta} + \frac{\beta c}{k} (1 - M^{k/\beta}) + \frac{g m_0}{\beta - k} (M - M^{k/\beta}), \qquad (30)$$

where $v_0 = v(0)$ and

$$M = \frac{m_0 - \beta t}{m_0} = \frac{m(t)}{m_0}$$

is the **fractional mass** of the rocket at time t.

NO RESISTANCE

Let us simplify the situation by ignoring the force of resistance to the motion of the rocket. Then $k = 0$. In Problem 29 we ask you to take the limit of the right-hand side of Eq. (30) as $k \to 0$, using l'Hôpital's rule, to obtain

$$v(t) = v_0 - gt + c \ln \frac{m_0}{m_0 - \beta t} \qquad (31)$$

for the case of no resistance. You can get the same result by setting $k = 0$ in the differential equation in (29) and then integrating (Problem 30). Because $m_0 - \beta \tau = m_1$, the velocity of the rocket at burnout ($t = \tau$) is

$$v(\tau) = v_0 - g\tau + c \ln \left(\frac{m_0}{m_1} \right). \qquad (32)$$

We can compute the height of the rocket at time $t \leq \tau$ by integrating Eq. (31):

$$y(t) = (v_0 + c \ln m_0)t - \frac{1}{2} gt^2 - c \int_0^t \ln (m_0 - \beta t) \, dt,$$

assuming that $y_0 = y(0) = 0$. Using the integration formula $\int \ln u \, du = u \ln u - u + C$, we find after some simplifications that

$$y(t) = (v_0 + c)t - \frac{1}{2} gt^2 - \frac{c}{\beta} (m_0 - \beta t) \ln \frac{m_0}{m_0 - \beta t}. \qquad (33)$$

When we substitute $t = \tau$, we find the height at burnout to be

$$y(\tau) = (v_0 + c)\tau - \frac{1}{2} g\tau^2 - \frac{cm_1}{\beta} \ln \frac{m_0}{m_1}. \qquad (34)$$

FREE SPACE

Suppose finally that the rocket is accelerating in free space, where there is neither gravity nor resistance, so that $g = k = 0$. With $g = 0$ in Eq. (32), we see that as the mass of the rocket decreases from m_0 to m_1, its increase in velocity is

$$\Delta v = v(\tau) - v_0 = c \ln \frac{m_0}{m_1}. \qquad (35)$$

Note that Δv depends only upon the exhaust gas speed c and the initial-to-final mass ratio m_0/m_1, but does *not* depend upon the burn rate β. For example, if $c = 2$ mi/s and $m_0/m_1 = 10$, then $\Delta v = 2 \ln 10 \approx 4.61$ mi/s. Thus if a rocket initially consists predominately of fuel, then it can attain speeds significantly greater than the (relative) speed of its exhaust gases.

1.9 PROBLEMS

1. The acceleration of a certain sports car is proportional to the difference between 250 km/h and the velocity of the sports car. If this machine can accelerate from rest to 100 km/h in 10 s, how long will it take for the car to accelerate from rest to 200 km/h?

2. Suppose that a body moves through a resisting medium with resistance proportional to its velocity v, so that $dv/dt = -kv$. (a) Show that its velocity and position at time t are given by

$$v = v_0 e^{-kt} \quad \text{and} \quad x = x_0 + \left(\frac{v_0}{k}\right)(1 - e^{-kt}).$$

(b) Conclude that the body travels only a *finite* distance, and find that distance.

3. Suppose that a motorboat is moving at 40 ft/s when its motor suddenly quits, and that 10 s later the boat has slowed to 20 ft/s. Assume, as in Problem 2, that the resistance it en-

counters while coasting is proportional to its velocity. How far will the boat coast in all?

4. Consider a body that moves horizontally through a medium whose resistance is proportional to the *square* of the velocity, so that $dv/dt = -kv^2$. Show that

$$v(t) = \frac{v_0}{1 + v_0 kt}$$

and that

$$x(t) = x_0 + \frac{1}{k} \ln (1 + v_0 kt).$$

Note that, in contrast to the result of Problem 2, $x(t) \to \infty$ as $t \to +\infty$.

5. Assuming resistance proportional to the square of the velocity as in Problem 4, how far does the motorboat of Problem 3 coast in the first minute after its motor quits?

CHAPTER 1: Introduction and First Order Differential Equations

6. Assume that a body moving with velocity v encounters resistance of the form $dv/dt = -kv^{3/2}$. Show that

$$v(t) = \frac{4v_0}{(kt\sqrt{v_0} + 2)^2}$$

and

$$x(t) = x_0 + \frac{2}{k}\sqrt{v_0}\left(1 - \frac{2}{kt\sqrt{v_0} + 2}\right).$$

Conclude that under a $\frac{3}{2}$-power resistance a body travels only a finite distance before coasting to a stop.

7. Calculate $v(t)$ and $x(t)$ under the assumption that

$$\frac{dv}{dt} = -kv^{1+r} \quad \text{where} \quad 0 < r < 1.$$

Also compute the coasting distance.

8. A car starts from rest ($x_0 = v_0 = 0$) and travels along a straight road. Its engine provides a constant acceleration of 18 ft/s^2. The combination of air and road resistance provides a deceleration of 0.12 ft/s^2 for every foot per second of the car's velocity v. (a) Set up and solve a first order differential equation for $v(t)$. (b) Find v when $t = 10$ s and also the limiting velocity as $t \to +\infty$, both in miles per hour.

9. A motorboat weighs 32,000 lb, and its motor provides a thrust of 5000 lb. Assume that the water resistance is 100 pounds for each foot per second of the speed v of the boat. Then

$$1000\frac{dv}{dt} = 5000 - 100v.$$

If the boat starts from rest, what is the maximum velocity that it can attain?

10. A woman bails out of an airplane at an altitude of 10,000 ft, falls freely for 20 s, then opens her parachute. How long will it take her to reach the ground? Assume that $\rho = 0.15$ without the parachute and that $\rho = 1.5$ with the parachute. (*Suggestion:* First determine her height and velocity when the parachute opens.)

11. According to a newspaper account, a paratrooper survived a training jump from 1200 ft when his parachute failed to open but provided some resistance by flapping unopened in the wind. Allegedly he hit the ground at 100 mi/h after falling for 8 s. Test the accuracy of this account. (*Suggestion:* Find ρ (in Eq. (2)) by assuming a terminal velocity of 100 mi/h. Then calculate the time required to fall 1200 ft.)

12. It is proposed to dispose of nuclear wastes—in drums with weight $W = 640$ lb and volume 8 ft^3—by dropping them into the ocean ($v_0 = 0$). The force equation for a drum falling through water is

$$m\frac{dv}{dt} = -W + B + F_R,$$

where the buoyant force B is equal to the weight of the volume of water displaced by the drum (Archimedes' principle) and F_R is the force of water resistance, found empirically to be 1 lb per foot per second of the velocity of a drum. If the drums are likely to burst upon an impact of more than 75 ft/s, what is the maximum depth to which they can be dropped in the ocean without likelihood of bursting?

13. Derive Eq. (13).

14. Derive Eq. (14); note that

$$\int \tanh u \, du = \int \frac{\sinh u}{\cosh u} \, du = \ln \cosh u + C.$$

15. A motorboat starts from rest (initial velocity $v(0) = 0$). Its motor provides a constant acceleration of 4 ft/s^2, but water resistance causes a deceleration of $v^2/400$ ft/s^2. Find v when $t = 10$ s, and also find the *limiting velocity* as $t \to +\infty$ (that is, the maximum possible speed of the boat).

16. Derive Eqs. (17)–(19) in the text.

17. If a ball is projected upward from the ground with initial velocity v_0, deduce from Eq. (20) that the maximum height it attains is

$$y_{\max} = \frac{1}{2\rho}\ln\left(1 + \frac{\rho v_0^2}{g}\right).$$

18. Suppose that a crossbow bolt is shot straight upward with initial velocity 288 ft/s. Assume that air resistance is proportional to v^2, the square of the velocity, with $\rho = 0.0002$ in Eqs. (12) and (16). Find the maximum height the bolt attains and its time of ascent.

19. In a continuation of Problem 18, find the descent time of the bolt and the speed with which it strikes the ground.

20. Suppose that $\rho = 0.075$ in Eq. (12) for a paratrooper falling with parachute open. If he jumps from an altitude of 10,000 ft and opens his parachute immediately, what will be his terminal speed? How long will it take him to reach the ground?

21. Suppose that the paratrooper of Problem 20 freefalls for 30 s with $\rho = 0.00075$ before opening his parachute. How long will it now take him to reach the ground?

22. The mass of the sun is 329,320 times that of the earth, and its radius is 109 times the radius of the earth. (a) To what radius (in meters) would the earth have to be compressed in order for it to become a *black hole*—the escape velocity from its surface equal to the velocity $c = 3 \times 10^8$ m/s of light? (b) Repeat part (a) with the sun in place of the earth.

23. (a) Show that if a projectile is launched upward from the surface of the earth with initial velocity v_0 less than escape velocity, then the maximum distance from the center of the earth that the projectile will attain is

$$y_{\max} = \frac{2gR^2}{2gR - v_0^2}.$$

where g and R are the surface gravity and radius of the earth. (b) With what initial velocity v_0 (in miles per hour) must such a projectile be launched to yield a maximum height of 100 mi above the surface of the earth? (c) Find the maximum distance from the center of the earth, expressed in terms of earth radii, attained by a projectile launched from the surface of the earth with 90% of escape velocity.

24. (a) Suppose that a body is dropped ($v_0 = 0$) from a distance y_0 from the earth's center, so that its gravitational acceleration is $dv/dt = -k/y^2$ with $k = GM = R^2g$. Show that it reaches the height y at time

$$t = \sqrt{\frac{y_0}{2R^2g}} \left(\sqrt{yy_0 - y^2} + y_0 \cos^{-1} \sqrt{\frac{y}{y_0}} \right).$$

(*Suggestion:* Substitute $y = y_0 \cos^2 \theta$ to evaluate the antiderivative of $\sqrt{y/(y_0 - y)}$.) (b) If a body is dropped from a height of 1000 mi above the earth's surface, how long does it fall and with what speed will it strike the earth's surface?

25. Suppose that a projectile is fired straight upward from the surface of the earth with initial velocity v_0. Then its height $x(t)$ at time t satisfies the initial value problem

$$\frac{d^2x}{dt^2} = -\frac{gR^2}{(x+R)^2}; \qquad x(0) = 0, \qquad x'(0) = v_0.$$

Let us use the values $g = 32.15$ (ft/s^2) or 0.006089 (mi/s^2) for the gravitational acceleration of the earth at its surface and $R = 3960$ (mi) as the radius of the earth. (a) Substitute $dv/dt = v(dv/dx)$ and then integrate to obtain

$$v^2 = \frac{v_0^2 R - (2gR - v_0^2)x}{R + x}$$

for the velocity of the projectile at altitude x. If $v_0 = 1$ (mi/s), what is the maximum height x_{max} attained by the projectile? (*Answer:* About 84 mi.) (b) Assume that $v_0^2 < 2gR$. Conclude

from part (a) that the time required for the projectile to ascend to its maximum height x_{max} is

$$t_{max} = \int_0^{x_{max}} \sqrt{\frac{R + x}{\alpha^2 - \beta^2 x}}\, dx,$$

where $\alpha^2 = v_0^2 R$ and $\beta^2 = 2gR - v_0^2$. Make the rationalizing substitution $u^2 = (R + x)/(\alpha^2 - \beta^2 x)$, and then integrate to obtain

$$t_{max} = \frac{2gR^2}{\beta^3} \left[\frac{\pi}{2} - \tan^{-1}\left(\frac{\beta}{v_0}\right) + \frac{\beta v_0}{2gR} \right].$$

In the case $v_0 = 1$ (mi/s), conclude that $t_{max} \approx 169$ s.

26. A rocket has an initial weight of 25 tons, of which 20 tons consists of fuel mixture that burns at the rate of 1 ton/s. Its exhaust gas speed is 1 mi/s. It blasts off at time $t = 0$ with $y_0 = v_0 = 0$. Find its height and velocity (in miles per hour) at burnout. Ignore air resistance and use $g = 32$ ft/s^2.

27. For the rocket of Problem 26, how large must its exhaust speed be in order for it to get off the ground?

28. For a rocket in free space, show that Eq. (26) can be written in the form $dv/dm = -c/m$. Integrate this equation to obtain the velocity of the rocket as given in Eq. (35).

29. Derive Eq. (31) by taking the limit as $k \to 0$ in Eq. (30).

30. Derive Eq. (31) by solving Eq. (29) with the value $k = 0$.

31. Derive Eq. (30) by solving Eq. (29) in the case $k > 0$.

32. The V-2 rocket of World War II had an initial weight of 28,300 lb (so $m_0 = 878.88$ slugs, using $g = 32.2$ ft/s^2); 68.5% of its mass was fuel. This fuel burned uniformly for 70 s with an exhaust velocity of 1.25 mi/s. Assume an air resistance of $v/10$ lb (with v in ft/s). Find the velocity and height of the V-2 at burnout under the assumption that it is fired vertically upward. Begin by solving Eq. (29) with the numerical parameters given here.

Chapter 1 Summary and a Look Ahead

In this chapter we have discussed applications of and solution methods for several important types of first order differential equations, including those that are separable (Section 1.4), linear (Section 1.5), or exact (Section 1.7). In Section 1.6 we discussed substitution techniques that can sometimes be used to transform a given first order differential equation into one that is either separable, linear, or exact.

Lest it appear that these methods constitute a "grab bag" of special and unrelated techniques, it is important to note that they are all versions of a single idea. Given a differential equation

$$f(x, y, y') = 0, \tag{1}$$

we attempt to write it in the form

$$\frac{d}{dx}[G(x, y)] = 0. \tag{2}$$

It is precisely to attain the form in (2) that we multiply the terms in Eq. (1) by an appropriate integrating factor (even if all we are doing is separating the variables). But once we have found a function $G(x, y)$ such that (1) and (2) are equivalent, a general solution is defined implicitly by means of the equation

$$G(x, y) = C \tag{3}$$

that one obtains by integrating (2).

Given a specific first order differential equation to be solved, we can attack it by means of the following steps:

- Is it *separable*? If so, separate the variables and integrate (Section 1.4).
- Is it *linear*? That is, can it be written in the form

$$\frac{dy}{dx} + P(x)y = Q(x)?$$

  If so, multiply by the integrating factor $\rho = \exp(\int P\, dx)$ of Section 1.5.
- Is it *exact*? That is, when the equation is written in the form $M\, dx + N\, dy = 0$, is $\partial M/\partial y = \partial N/\partial x$? Or can terms be grouped or an integrating factor found that produces an exact equation (Section 1.7)?
- If the equation as it stands is not separable, linear, or exact, is there a plausible substitution that will make it so? For instance, is it homogeneous (Section 1.6)?

Many first order differential equations succumb to the line of attack outlined above. Nevertheless, many more do not. Because of the wide availability of computers—even inexpensive pocket computers—numerical techniques are being used with increasing frequency to *approximate* the solutions of differential equations that cannot easily be solved explicitly by the methods of this chapter.

To illustrate very briefly a standard technique of numerical approximation, let us consider the initial value problem

$$\frac{dy}{dx} = y, \qquad y(0) = 1. \tag{4}$$

In order to *approximate* the exact solution $y(x) = e^x$ on the interval $0 \le x \le 1$, we begin with a subdivision of the interval into (say) 10 subintervals each of length $h = 0.1$, by means of the points

$$x_0 = 0.0, x_1 = 0.1, x_2 = 0.2, \ldots, x_{10} = 1.0.$$

We know that $y_0 = y(0) = 1$, and we want to find (for each $n = 1, 2, 3, \ldots, 10$) an approximation y_n to the actual value $y(x_n)$ of the solution at x_n. If we have found

```
10 LET X = 0
20 LET Y = 1
30 LET H = 0.1
40 FOR N = 1 TO 10
50    LET X = X + H
60    LET Y = (1 + H)*Y
70    PRINT N, X, Y
80 NEXT N
90 END
```

FIGURE 1.53 A BASIC program for Euler's method for $y' = y$, $y(0) = 1$.

y_n somehow, then the mean value theorem yields

$$y_{n+1} = y(x_n) + y'(\bar{x}_n)h$$
$$\approx y(x_n) + y'(x_n)h$$
$$= (1 + h)y(x_n),$$

because our differential equation is $y' = y$. Thus each value of y should be approximately $1 + h$ times the preceding value of y, so we choose

$$y_{n+1} = (1 + h)y_n. \tag{5}$$

Beginning with the initial value $y_0 = 1$, Eq. (5) is an iterative formula for computing successively the values $y_1, y_2, \ldots, y_{10}$. This will give us a table of *approximate* values of the solution of the initial value problem in (4).

Figure 1.53 is a listing of a simple BASIC program that carries out the iteration in (5). It can be run on any pocket computer that is programmable in BASIC and will display at each iteration the values of n, x_n, y_n, and $y(x_n) = \exp(x_n)$. Figure 1.54 shows a version of the same program, written for the IBM Personal Computer, fully documented, and formatted for ease of examination of what the program does. The REM (remark) statements are not executed by the computer; their purpose is explanation and spacing for the benefit of the reader.

```
100 REM--Euler Method Program
110 REM
120 REM--To approximate the solution on  [0,1]  of the
130 REM--differential equation  y' = y  with initial
140 REM--value  y(0) = 1  and with step size  h = 0.1
150 REM
160 REM--Initialization:
170 REM
180       X = 0  :  Y = 1              'Initial values
190       H = 0.1                      'Step size
200       K = 1/H                      'No. of subintervals
210 REM
220 REM--Loop to update X and Y:
230 REM
240       FOR N = 1 TO K               'Do it  k  times
250           X = X + H                'New  X  value
260           Y = (1 + H)*Y            'New  Y  value
270           PRINT N, X, Y, EXP(X)    'Display results
280       NEXT N
290 REM
300       END
```

FIGURE 1.54 The program of Figure 1.53 with comment lines and improved readability.

Figure 1.55 shows the output of this program as executed by the IBM Personal Computer. Note that $y_{10} = 2.593743$ is about 5% less than the actual value $y(1) = e \approx 2.718282$. A better approximation could be obtained by decreasing the value of h, and thereby increasing the number of subintervals. Such a table of approximate values can be used numerically or can be used to construct a curve that approximates the graph of the actual solution of the differential equation.

CHAPTER 1: Introduction and First Order Differential Equations

n	x_n	y_n	$y(x_n) = exp(x_n)$
1	.1	1.1	1.105171
2	.2	1.21	1.221403
3	.3	1.331	1.349859
4	.4	1.4641	1.491825
5	.5	1.61051	1.648721
6	.6	1.771561	1.822119
7	.7	1.948717	2.013753
8	.8	2.143589	2.225541
9	.9	2.357948	2.459603
10	1	2.593743	2.718282

FIGURE 1.55 Output of the program of Figure 1.54.

Chapter 6 is devoted to numerical solutions of differential equations, and the first two or three sections of Chapter 6 can well be studied at this time, before proceeding to higher order differential equations in Chapter 2.

PROBLEMS

For each of the following initial value problems, apply the iterative method described above to compute approximate values $y_1, y_2, \ldots, y_{10}$ of the solution on the interval $0 \leq x \leq 1$. In each case, apply the mean value theorem to derive first the indicated iterative formula for y_{n+1} in terms of x_n, y_n, and $h = 0.1$. Also compare your approximate values with the actual values of the solution.

1. $y' = -y$, $y(0) = 1$; $y_{n+1} = (1 - h)y_n$

2. $y' = 2y$, $y(0) = 1$; $y_{n+1} = (1 + 2h)y_n$

3. $y' = y^2$, $y(0) = 0.5$; $y_{n+1} = y_n + hy_n^2$

4. $y' = 2xy$, $y(0) = 1$; $y_{n+1} = (1 + 2hx_n)y_n$

5. $y' = y + e^x$, $y(0) = 2$; $y_{n+1} = (1 + h)y_n + he^{x_n}$

CHAPTER 1 REVIEW PROBLEMS

Find general solutions of the differential equations in Problems 1–30.

1. $x^3 + 3y - xy' = 0$

2. $xy^2 + 3y^2 - x^2y' = 0$

3. $xy + y^2 - x^2y' = 0$

4. $2xy^3 + e^x + (3x^2y^2 + \sin y)y' = 0$

5. $3y + x^4y' = 2xy$

6. $2xy^2 + x^2y' = y^2$

7. $2x^2y + x^3y' = 1$

8. $2xy + x^2y' = y^2$

9. $xy' + 2y = 6x^2\sqrt{y}$

10. $y' = 1 + x^2 + y^2 + x^2y^2$

11. $x^2y' = xy + 3y^2$

12. $6xy^3 + 2y^4 + (9x^2y^2 + 8xy^3)y' = 0$

13. $4xy^2 + y' = 5x^4y^2$

14. $x^3y' = x^2y - y^3$

15. $y' + 3y = 3x^2e^{-3x}$

16. $y' = x^2 - 2xy + y^2$

17. $e^x + ye^{xy} + (e^y + xe^{xy})y' = 0$

18. $2x^2y - x^3y' = y^3$

19. $3x^5y^2 + x^3y' = 2y^2$

20. $xy' + 3y = 3x^{-3/2}$

21. $(x^2 - 1)y' + (x - 1)y = 1$

22. $xy' = 6y + 12x^4y^{2/3}$

23. $(e^y + y \cos x) + (xe^y + \sin x)y' = 0$

24. $9x^2y^2 + x^{3/2}y' = y^2$

25. $2y + (x + 1)y' = 3x + 3$

26. $9x^{1/2}y^{4/3} - 12x^{1/5}y^{3/2} + (8x^{3/2}y^{1/3} - 15x^{6/5}y^{1/2})y' = 0$

27. $3y + x^3y^4 + 3xy' = 0$

28. $y + xy' = 2e^{2x}$

29. $(2x + 1)y' + y = (2x + 1)^{3/2}$

30. $y' = \sqrt{x + y}$

CHAPTER *2*

LINEAR EQUATIONS OF HIGHER ORDER

2.1 INTRODUCTION

2.2 GENERAL SOLUTIONS OF LINEAR EQUATIONS

2.3 HOMOGENEOUS EQUATIONS WITH CONSTANT COEFFICIENTS

2.4 MECHANICAL VIBRATIONS

2.5 NONHOMOGENEOUS EQUATIONS AND THE METHOD OF UNDETERMINED COEFFICIENTS

2.6 REDUCTION OF ORDER AND EULER-CAUCHY EQUATIONS

2.7 VARIATION OF PARAMETERS

*__2.8__ FORCED OSCILLATIONS AND RESONANCE

*__2.9__ ELECTRICAL CIRCUITS

*__2.10__ ENDPOINT PROBLEMS AND EIGENVALUES

Introduction

Only in very special cases can an nth order differential equation of the form $G(x, y, y', y'', \ldots, y^{(n)}) = 0$ be solved exactly and explicitly. In this chapter we restrict our attention to *linear* equations of order $n > 1$. The general **nth order linear equation** has the form

$$a_n(x)\frac{d^n y}{dx^n} + a_{n-1}(x)\frac{d^{n-1}y}{dx^{n-1}} + \cdots + a_1(x)\frac{dy}{dx} + a_0(x)y = F(x). \tag{1}$$

Unless otherwise noted, we will always assume that the coefficient functions $a_i(x)$ and $F(x)$ are continuous on some open interval I (perhaps unbounded) on which we wish to solve the differential equation, but they need not be linear functions. Thus the differential equation

$$e^x y'' + (\cos x)y' + (1 + \sqrt{x})y = \tan^{-1} x$$

is linear because the dependent variable y and its derivatives appear linearly, whereas the equations

$$y'' = yy' \quad \text{and} \quad y'' + 3(y')^2 + 4y^3 = 0$$

are not linear because products and powers of y or its derivatives appear.

If the function $F(x)$ on the right-hand side of (1) vanishes identically on I, then we call (1) a **homogeneous** linear equation; otherwise, it is **nonhomogeneous**. For example, the second order equation

$$x^2 y'' + 2xy' + 3y = \cos x$$

is nonhomogeneous; its *associated* homogeneous equation is

$$x^2 y'' + 2xy' + 3y = 0.$$

The homogeneous linear equation **associated** with Eq. (1) is

$$a_n(x)\frac{d^n y}{dx^n} + a_{n-1}(x)\frac{d^{n-1}y}{dx^{n-1}} + \cdots + a_1(x)\frac{dy}{dx} + a_0(x)y = 0. \tag{2}$$

In order to solve an nth order differential equation we must, at least in principle, integrate n times (to get from $y^{(n)}$ to y). It is therefore natural to expect Eq. (2) to have a general solution involving n arbitrary constants $c_1, c_2, \ldots, c_n$ (of integration). Indeed, we will see in Section 2.2 that if $a_n(x)$ is nonzero on the interval I, then (2) has a general solution of the especially pleasant form

$$y = c_1 y_1 + c_2 y_2 + \cdots + c_n y_n, \tag{3}$$

a *linear combination* of n particular solutions $y_1, y_2, \ldots, y_n$. The general theory of homogeneous linear equations parallels the case of second order linear equations (the case $n = 2$) which is discussed below in this section.

With regard to the crucial matter of actually finding a general solution like (3), there is a substantial difference between the case $n = 1$ and the higher order cases $n \geq 2$. In Section 1.5 we saw that there exists a systematic procedure by which a general solution of any *first order* linear differential equation can always be found explicitly. By contrast, there does *not* exist a formula for computing the general solution of an arbitrary higher order linear equation with variable coefficients. Fortunately, many important applications involve only homogeneous equations with *constant* coefficients, and we will see in Section 2.3 how to solve such equations in a routine fashion.

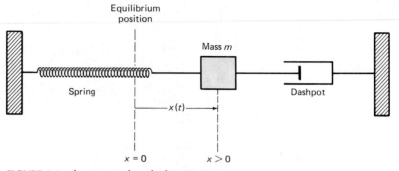

FIGURE 2.1 A mass-spring-dashpot system.

Linear differential equations with constant coefficients frequently appear as mathematical models of mechanical systems and electrical circuits. For example, suppose that a mass m is attached both to a spring that exerts on it a force F_S and to a dashpot (shock absorber) that exerts a force F_R on the mass (see Fig. 2.1). Assume that the restoring force F_S of the spring is proportional to the displacement x (positive to the right, negative to the left) of the mass from equilibrium, and that the dashpot force F_R is proportional to the velocity $v = dx/dt$ of the mass. With the aid of Fig. 2.2, we also get the appropriate directions of action of these two forces:

$$F_S = -kx \quad \text{and} \quad F_R = -cv \qquad (k, c > 0).$$

The minus signs are correct—F_S and F_R are negative when x and v are positive. Newton's law $F = ma$ now gives

$$mx'' = F_S + F_R; \tag{4}$$

that is,

$$m\frac{d^2x}{dt^2} + c\frac{dx}{dt} + kx = 0. \tag{5}$$

This homogeneous second order linear equation governs the *free vibrations* of the mass; we will return to this problem in detail in Section 2.4.

If, in addition to F_S and F_R, the mass m is acted on by an external force $F(t)$—which must then be added to the right-hand side in Eq. (4)—the resulting equation is

$$m\frac{d^2x}{dt^2} + c\frac{dx}{dt} + kx = F(t). \tag{6}$$

This nonhomogeneous linear differential equation governs the *forced vibrations* of the mass under the influence of the external force $F(t)$.

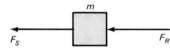

FIGURE 2.2 Directions of the forces acting on m.

CHAPTER 2: Linear Equations of Higher Order

As this example illustrates, an external force in a mechanical system typically corresponds to a nonhomogeneous term in the differential equation(s) describing the system. In Section 2.2 we will see that a general solution of a nonhomogeneous linear differential equation is the sum $y = y_c + y_p$ of (1) a general solution y_c of the associated homogeneous equation and (2) a single particular solution y_p of the given nonhomogeneous equation. We will take up the problem of finding y_p in Sections 2.5 and 2.7.

SECOND ORDER LINEAR EQUATIONS

Consider the general second order linear differential equation

$$A(x)y'' + B(x)y' + C(x)y = F(x) \tag{7}$$

where the coefficient functions A, B, C, and F are continuous on the open interval I. Here we will assume in addition that $A(x) \neq 0$ at each point of I, so we can divide each term in (7) by $A(x)$ and write it in the form

$$y'' + p(x)y' + q(x)y = f(x). \tag{8}$$

We will discuss first the associated homogeneous equation

$$y'' + p(x)y' + q(x)y = 0. \tag{9}$$

A particularly useful property of this *homogeneous* linear equation is the fact that the sum of any two solutions of (9) is again a solution, as is any constant multiple of a solution. This is the central idea of the following theorem.

Theorem 1 *Principle of Superposition*

Let y_1 and y_2 be two solutions of the homogeneous linear Eq. (9) on the interval I. If c_1 and c_2 are constants, then the linear combination

$$y = c_1 y_1 + c_2 y_2 \tag{10}$$

is also a solution of (9) on I.

Proof The conclusion follows almost immediately from the *linearity* of the operation of differentiation, which gives

$$y' = c_1 y_1' + c_2 y_2' \quad \text{and} \quad y'' = c_1 y_1'' + c_2 y_2''.$$

Then

$$\begin{aligned}
y'' + py' + qy &= (c_1 y_1 + c_2 y_2)'' + p(c_1 y_1 + c_2 y_2)' + q(c_1 y_1 + c_2 y_2) \\
&= (c_1 y_1'' + c_2 y_2'') + p(c_1 y_1' + c_2 y_2') + q(c_1 y_1 + c_2 y_2) \\
&= c_1(y_1'' + py_1' + qy_1) + c_2(y_2'' + py_2' + qy_2) \\
&= (c_1)(0) + (c_2)(0) = 0
\end{aligned}$$

because y_1 and y_2 are solutions. Thus $y = c_1 y_1 + c_2 y_2$ is also a solution. ∎

EXAMPLE 1 We can see by inspection that

$$y_1 = \cos x \quad \text{and} \quad y_2 = \sin x$$

are two solutions of the equation

$$y'' + y = 0.$$

Theorem 1 tells us that any linear combination of these solutions, such as

$$y = 3y_1 - 2y_2 = 3\cos x - 2\sin x$$

is also a solution. We will see later that, conversely, *every* solution of $y'' + y = 0$ is a linear combination of these two particular solutions y_1 and y_2. Thus a general solution of $y'' + y = 0$ is

$$y(x) = c_1 \cos x + c_2 \sin x.$$

We gave above the linear equation $mx'' + cx' + kx = F(t)$ as a mathematical model of the motion of the mass of Fig. 2.1. Physical considerations suggest that the motion of the mass ought to be determined by its initial position and initial velocity. Hence, given any preassigned values of $x(0)$ and $x'(0)$, Eq. (6) ought to have a *unique* solution satisfying these initial conditions. More generally, in order to be a "good" mathematical model of a deterministic physical situation, a differential equation must have unique solutions satisfying any appropriate initial conditions. The following existence and uniqueness theorem (proved in Section 7.6) gives us this assurance for the general second order linear equation.

Theorem 2   *Existence and Uniqueness*

Suppose that the functions p, q, and f are continuous on the open interval I containing the point a. Then, given two numbers b_0 and b_1, Eq. (8),

$$y'' + p(x)y' + q(x)y = f(x),$$

has a unique (that is, one and only one) solution on the entire interval I that satisfies the initial conditions

$$y(a) = b_0, \qquad y'(a) = b_1. \tag{11}$$

Equation (8) and the conditions in (11) constitute a second order linear **initial value problem**. Theorem 2 tells us that any such initial value problem has a unique solution on the *whole* interval I where the coefficient functions in (8) are continuous. Recall from Section 1.3 that a *nonlinear* differential equation generally has a unique solution on only a smaller interval.

EXAMPLE 1 CONTINUED   We saw above that $y = 3\cos x - 2\sin x$ is a solution (on the entire real line) of $y'' + y = 0$. It has the initial values $y(0) = 3$, $y'(0) = -2$. Theorem 2 tells us that this is the *only* solution with these initial values.

110   CHAPTER 2:   Linear Equations of Higher Order

More generally, the solution

$$y(x) = b_0 \cos x + b_1 \sin x$$

satisfies the *arbitrary* initial conditions $y(0) = b_0$, $y'(0) = b_1$; this illustrates the *existence* of such a solution, also as guaranteed by Theorem 2.

Example 1 suggests how, given a *homogeneous* second order linear equation, we might actually find the solution whose existence is assured by Theorem 2. First, we find two "essentially different" solutions y_1 and y_2; second, we attempt to impose on the general solution

$$y = c_1 y_1 + c_2 y_2 \tag{12}$$

the initial conditions $y(a) = b_0$, $y'(a) = b_1$. That is, we attempt to solve the simultaneous equations

$$\left. \begin{array}{l} c_1 y_1(a) + c_2 y_2(a) = b_0, \\ c_1 y'_1(a) + c_2 y'_2(a) = b_1 \end{array} \right\} \tag{13}$$

for the coefficients c_1 and c_2.

EXAMPLE 2 Verify that the functions

$$y_1 = e^x \quad \text{and} \quad y_2 = x e^x$$

are solutions of the differential equation

$$y'' - 2y' + y = 0,$$

and then find a solution satisfying the initial conditions $y(0) = 3$, $y'(0) = 1$.

SOLUTION The verification is routine; we omit it. We impose the given initial conditions on the general solution

$$y = c_1 e^x + c_2 x e^x,$$

for which

$$y' = (c_1 + c_2)e^x + c_2 x e^x,$$

to obtain the simultaneous equations

$$\begin{cases} y(0) = c_1 = 3, \\ y'(0) = c_1 + c_2 = 1. \end{cases}$$

The resulting solution is $c_1 = 3$, $c_2 = -2$. Hence the solution of the original initial value problem is

$$y(x) = 3e^x - 2x e^x.$$

In order for the procedure of Example 2 to succeed, the two solutions y_1 and y_2 must have the elusive property that the equations in (13) can always be solved for c_1 and c_2, no matter what the initial conditions b_0 and b_1 might be. The following definition tells precisely how different the two functions y_1 and y_2 must be.

> **Definition** *Linear Independence of Two Functions*
>
> Two functions defined on an open interval are called **linearly independent** provided that neither is a constant multiple of the other.

Two functions are said to be **linearly dependent** if they are not linearly independent; that is, one of them *is* a constant multiple of the other. We can always determine whether two given functions f and g are linearly dependent on an interval I by noting at a glance whether either of the two quotients f/g or g/f is a constant on I. Thus it is clear that the following pairs of functions are linearly independent on the entire real line:

$$\sin x \quad \text{and} \quad \cos x;$$
$$e^x \quad \text{and} \quad e^{-2x};$$
$$e^x \quad \text{and} \quad xe^x;$$
$$x + 1 \quad \text{and} \quad x^2;$$
$$x \quad \text{and} \quad |x|.$$

But the identically zero function and any other function g are linearly dependent on every interval because $0 = (0)(g(x))$. Also, the functions

$$f(x) = \sin 2x \quad \text{and} \quad g(x) = \sin x \cos x$$

are linearly dependent on any interval because $f(x) = 2g(x)$ for all x (a familiar trigonometric identity).

But does the homogeneous equation

$$y'' + py' + qy = 0$$

always have two linearly independent solutions? Theorem 2 says yes! We need only choose y_1 and y_2 so that

$$y_1(a) = 1, \ y_1'(a) = 0 \quad \text{and} \quad y_2(a) = 0, \ y_2'(a) = 1.$$

It is then impossible that either $y_1 = ky_2$ or $y_2 = ky_1$ because $(k)(0) \neq 1$ for any constant k. Theorem 2 tells us that two such linearly independent solutions *exist*; actually finding them is a crucial matter that we will defer until later sections (beginning in Section 2.3).

We want to show, finally, that given *any* two linearly independent solutions y_1 and y_2 of the homogeneous equation

$$y'' + p(x)y' + q(x)y = 0, \tag{9}$$

every solution y of (9) can be expressed as a linear combination

$$y = c_1 y_1 + c_2 y_2 \tag{12}$$

of y_1 and y_2. This means that the function in (12) is a *general solution* of Eq. (9).

As suggested by the equations in (13), the determination of the constants c_1 and c_2 in (12) depends upon a certain 2×2 determinant of values of y_1, y_2, and their derivatives. Given two functions f and g, the **Wronskian** of f and g is the determinant

$$W = \begin{vmatrix} f & g \\ f' & g' \end{vmatrix} = fg' - f'g.$$

We write either $W(f, g)$ or $W(x)$, depending upon whether we wish to emphasize the two functions or the point x at which their Wronskian is to be evaluated. For example,

$$W(\cos x, \sin x) = \begin{vmatrix} \cos x & \sin x \\ -\sin x & \cos x \end{vmatrix} = \cos^2 x + \sin^2 x = 1$$

while

$$W(e^x, xe^x) = \begin{vmatrix} e^x & xe^x \\ e^x & e^x + xe^x \end{vmatrix} = e^{2x}.$$

These are examples of linearly *independent* pairs of solutions of differential equations (see Examples 1 and 2). Note that in each case the Wronskian is everywhere *nonzero*.

On the other hand, if the functions f and g are linearly dependent, with $f = kg$ (for example), then

$$W(f, g) = \begin{vmatrix} kg & g \\ kg' & g' \end{vmatrix} = kgg' - kg'g \equiv 0.$$

Thus the Wronskian of two linearly *dependent* functions is identically zero. In Section 2.2 we will prove that, if the two functions y_1 and y_2 are solutions of a homogeneous second order linear equation, then the strong converse stated in part (b) of the following theorem holds.

Theorem 3   Wronskians of Solutions

Suppose that y_1 and y_2 are two solutions of the homogeneous second order linear equation (Eq. (9))

$$y'' + p(x)y' + q(x)y = 0$$

on an open interval I on which p and q are continuous.

(a) If y_1 and y_2 are linearly dependent, then $W(y_1, y_2) \equiv 0$ on I;
(b) If y_1 and y_2 are linearly independent, then $W(y_1, y_2) \neq 0$ at each point of I.

Thus, given two solutions of (9), there are just two possibilities: The Wronskian W is identically zero if the solutions are linearly dependent; the Wronskian is never zero if the solutions are linearly independent. The latter fact is what we need in order to show that $y = c_1y_1 + c_2y_2$ is a general solution of Eq. (9) if y_1 and y_2 are linearly independent solutions.

Theorem 4 *General Solutions*

Let y_1 and y_2 be two linearly independent solutions of the homogeneous equation (Eq. (9))

$$y'' + p(x)y' + q(x)y = 0$$

with p and q continuous on the open interval I. If Y is any solution whatsoever of Eq. (9), then there exist numbers c_1 and c_2 such that

$$Y(x) = c_1y_1(x) + c_2y_2(x)$$

for all x in I.

In essence, Theorem 4 tells us that when we have found *two* linearly independent solutions of the homogeneous equation in (9), then we have found *all* of its solutions.

Proof Choose a point a of I, and consider the simultaneous equations

$$\left. \begin{array}{l} c_1y_1(a) + c_2y_2(a) = Y(a), \\ c_1y_1'(a) + c_2y_2'(a) = Y'(a). \end{array} \right\} \tag{14}$$

The determinant of the coefficients in this system of linear equations in the unknowns c_1 and c_2 is simply the Wronskian $W(y_1, y_2)$ evaluated at a. By Theorem 3, this determinant is nonzero, so by elementary algebra it follows that the equations in (14) can be solved for c_1 and c_2. With these values of c_1 and c_2, we define the solution

$$G(x) = c_1y_1(x) + c_2y_2(x)$$

of Eq. (9); then

$$G(a) = c_1y_1(a) + c_2y_2(a) = Y(a)$$

and

$$G'(a) = c_1y_1'(a) + c_2y_2'(a) = Y'(a).$$

Thus the two solutions Y and G have the same initial values at a, as do Y' and G'. By the uniqueness of a solution determined by such initial values (Theorem 2), it follows that Y and G agree on I. Thus we see that

$$Y(x) \equiv G(x) = c_1y_1(x) + c_2y_2(x),$$

as desired. ∎

EXAMPLE 3 It is evident that

$$y_1 = e^{2x} \quad \text{and} \quad y_2 = e^{-2x}$$

are linearly independent solutions of

$$y'' - 4y = 0. \tag{15}$$

But $y_3 = \cosh 2x$ and $y_4 = \sinh 2x$ are also solutions of (15) because

$$(\cosh 2x)'' = (2 \sinh 2x)' = 4 \cosh 2x$$

and similarly, $(\sinh 2x)'' = 4 \sinh 2x$. It therefore follows from Theorem 4 that $\cosh 2x$ and $\sinh 2x$ can be expressed as linear combinations of $y_1 = e^{2x}$ and $y_2 = e^{-2x}$. Of course this is no surprise, because

$$\cosh 2x = \tfrac{1}{2}e^{2x} + \tfrac{1}{2}e^{-2x}$$

and

$$\sinh 2x = \tfrac{1}{2}e^{2x} - \tfrac{1}{2}e^{-2x}$$

by the definitions of the hyperbolic cosine and sine.

2.1 PROBLEMS

In each of Problems 1–16, a homogeneous second order linear differential equation, two functions y_1 and y_2, and a pair of initial conditions are given. First verify that y_1 and y_2 are solutions of the differential equation, and then find a particular solution of the form $y = c_1 y_1 + c_2 y_2$ that satisfies the given initial conditions.

1. $y'' - y = 0;$ $y_1 = e^x, y_2 = e^{-x};$ $y(0) = 0, y'(0) = 5$

2. $y'' - 9y = 0;$ $y_1 = e^{3x}, y_2 = e^{-3x};$ $y(0) = -1, y'(0) = 15$

3. $y'' + 4y = 0;$ $y_1 = \cos 2x, y_2 = \sin 2x;$ $y(0) = 3, y'(0) = 8$

4. $y'' + 25y = 0;$ $y_1 = \cos 5x, y_2 = \sin 5x;$ $y(0) = 10,$ $y'(0) = -10$

5. $y'' - 3y' + 2y = 0;$ $y_1 = e^x, y_2 = e^{2x};$ $y(0) = 1, y'(0) = 0$

6. $y'' + y' - 6y = 0;$ $y_1 = e^{2x}, y_2 = e^{-3x};$ $y(0) = 7,$ $y'(0) = -1$

7. $y'' + y' = 0;$ $y_1 = 1, y_2 = e^{-x};$ $y(0) = -2, y'(0) = 8$

8. $y'' - 3y' = 0;$ $y_1 = 1, y_2 = e^{3x};$ $y(0) = 4, y'(0) = -2$

9. $y'' + 2y' + y = 0;$ $y_1 = e^{-x}, y_2 = xe^{-x};$ $y(0) = 2,$ $y'(0) = -1$

10. $y'' - 10y' + 25y = 0;$ $y_1 = e^{5x}, y_2 = xe^{5x};$ $y(0) = 3,$ $y'(0) = 13$

11. $y'' - 2y' + 2y = 0;$ $y_1 = e^x \cos x, y_2 = e^x \sin x;$ $y(0) = 0, y'(0) = 5$

12. $y'' + 6y' + 13y = 0;$ $y_1 = e^{-3x} \cos 2x, y_2 = e^{-3x} \sin 2x;$ $y(0) = 2, y'(0) = 0$

13. $x^2 y'' - 2xy' + 2y = 0;$ $y_1 = x, y_2 = x^2;$ $y(1) = 3,$ $y'(1) = 1$

14. $x^2 y'' + 2xy' - 6y = 0;$ $y_1 = x^2, y_2 = \dfrac{1}{x^3};$ $y(2) = 10,$ $y'(2) = 15$

15. $x^2 y'' - xy' + y = 0;$ $y_1 = x, y_2 = x \ln x;$ $y(1) = 7,$ $y'(1) = 2$

16. $x^2 y'' + xy' + y = 0;$ $y_1 = \cos(\ln x), y_2 = \sin(\ln x);$ $y(1) = 2, y'(1) = 3$

The following three problems illustrate the fact that the superposition principle does not generally hold for nonlinear equations.

17. Show that $y = 1/x$ is a solution of $y' + y^2 = 0$, but that if $c \neq 0, 1$, then $y = c/x$ is not a solution.

18. Show that $y = x^3$ is a solution of $yy'' = 6x^4$, but that if $c^2 \neq 1$, then $y = cx^3$ is not a solution.

19. Show that $y_1 = 1$ and $y_2 = x^{1/2}$ are solutions of $yy'' + (y')^2 = 0$, but that their sum $y = y_1 + y_2$ is not a solution.

Determine whether the pairs of functions in Problems 20–26 are linearly independent or linearly dependent on the real line.

20. $f(x) = \pi, g(x) = \cos^2 x + \sin^2 x$

21. $f(x) = x^3, g(x) = x^2 |x|$

22. $f(x) = 1 + x, g(x) = 1 + |x|$

23. $f(x) = xe^x, g(x) = |x|e^x$

24. $f(x) = \sin^2 x, g(x) = 1 - \cos 2x$

25. $f(x) = e^x \sin x, g(x) = e^x \cos x$

26. $f(x) = 2 \cos x + 3 \sin x, g(x) = 3 \cos x - 2 \sin x$

27. Let y_p be a particular solution of the nonhomogeneous equation $y'' + py' + qy = f(x)$, and let y_c be a solution of its associated homogeneous equation. Show that $y = y_c + y_p$ is a solution of the given nonhomogeneous equation.

28. With $y_p \equiv 1$ and $y_c = c_1 \cos x + c_2 \sin x$ in the notation of Problem 27, find a solution of $y'' + y = 1$ satisfying the initial conditions $y(0) = -1 = y'(0)$.

29. Show that $y_1 = x^2$ and $y_2 = x^3$ are two different solutions of $x^2 y'' - 4xy' + 6y = 0$, both satisfying the initial conditions $y(0) = 0 = y'(0)$. Explain why these facts do not contradict Theorem 2 (with respect to the guaranteed uniqueness).

30. (a) Show that $y_1 = x^3$ and $y_2 = |x^3|$ are linearly independent solutions on the real line of the equation $x^2 y'' - 3xy' + 3y = 0$. (b) Verify that $W(y_1, y_2)$ is identically zero. Why do these facts not contradict Theorem 3?

31. Show that $y_1 = \sin x^2$ and $y_2 = \cos x^2$ are linearly independent functions, but that their Wronskian vanishes at $x = 0$. Why does this imply that there is *no* differential equation of the form $y'' + p(x)y' + q(x)y = 0$, with p and q both continuous everywhere, having both y_1 and y_2 as solutions?

32. Let y_1 and y_2 be two solutions of $A(x)y'' + B(x)y' + C(x)y = 0$ on an open interval I where A, B, and C are continuous and

$A(x)$ is never zero. (a) Let $W = W(y_1, y_2)$. Show that

$$A(x) \frac{dW}{dx} = y_1(Ay_2'') - y_2(Ay_1'').$$

Then substitute for Ay_2'' and Ay_1'' from the original differential equation to show that

$$A(x) \frac{dW}{dx} = -B(x)W(x).$$

(b) Solve this first order equation to deduce **Abel's formula**

$$W(x) = K \exp\left(-\int \frac{B(x)}{A(x)} dx\right)$$

where K is a constant. (c) Why does Abel's formula imply that the Wronskian $W(y_1, y_2)$ is either zero everywhere or nonzero everywhere (as stated in Theorem 3)?

33. (a) Take as given the fact that $D_x e^{ix} = ie^{ix}$, where i is the complex number $\sqrt{-1}$. Show that $u = e^{ix}$ is a solution of $y'' + y = 0$. Why does it follow that e^{ix} is a linear combination of $\cos x$ and $\sin x$? (b) **Euler's formula** is the identity $e^{ix} = \cos x + i \sin x$. Deduce Euler's formula from the facts that $u(0) = 1$ and $u'(0) = i$.

General Solutions of Linear Equations

We now show that our discussion in Section 2.1 of second order linear equations generalizes in a very natural way to the general nth order *linear* differential equation of the form

$$P_0(x)y^{(n)} + P_1(x)y^{(n-1)} + \cdots + P_{n-1}(x)y' + P_n(x)y = F(x). \tag{1}$$

Unless otherwise noted, we will always assume that the coefficient functions $P_i(x)$ and $F(x)$ are continuous on some open interval I (perhaps unbounded) where we wish to solve the equation. Under the additional assumption that $P_0(x) \neq 0$ at each point x of I, we can divide each term in (1) by $P_0(x)$ to obtain an equation with leading coefficient 1, of the form

$$y^{(n)} + p_1(x)y^{(n-1)} + \cdots + p_{n-1}(x)y' + p_n(x)y = f(x). \tag{2}$$

The *homogeneous* linear equation *associated* with (2) is

$$y^{(n)} + p_1(x)y^{(n-1)} + \cdots + p_{n-1}(x)y' + p_n(x)y = 0. \tag{3}$$

Just as in the second order case, a *homogeneous* nth order linear differential equation has the nice property that any superposition, or *linear combination*, of solutions of the equation is again a solution. The proof of the following theorem is essentially the same—a routine verification—as that of Theorem 1 of Section 2.1.

Theorem 1 *Principle of Superposition*

Let $y_1, y_2, \ldots, y_n$ be n solutions of the homogeneous linear equation in (3) on the interval I. If $c_1, c_2, \ldots, c_n$ are constants, then the linear combination

$$y = c_1 y_1 + c_2 y_2 + \cdots + c_n y_n \qquad (4)$$

is also a solution of (3) on I.

EXAMPLE 1   It is easy to verify that the three functions

$$y_1 = e^{3x}, \qquad y_2 = \cos 2x, \quad \text{and} \quad y_3 = \sin 2x$$

are all solutions of the homogeneous third order equation

$$y^{(3)} - 3y'' + 4y' - 12y = 0$$

on the entire real line. Theorem 1 tells us that any linear combination of these solutions, such as

$$y = 7y_1 + 3y_2 - 2y_3 = 7e^{3x} + 3\cos 2x - 2\sin 2x,$$

is also a solution on the entire real line. We will see that, conversely, every solution of the differential equation above is a linear combination of the three particular solutions y_1, y_2, and y_3. Thus its general solution has the form

$$y(x) = c_1 e^{3x} + c_2 \cos 2x + c_3 \sin 2x.$$

We saw in Section 2.1 that a particular solution of a *second order* linear differential equation is determined by *two* initial conditions. Similarly, a particular solution of an *n*th order linear differential equation is determined by *n* initial conditions. The following theorem, proved in Section 7.6, is the natural generalization of Theorem 2 of Section 2.1.

Theorem 2   *Existence and Uniqueness*

Suppose that the functions $p_1, p_2, \ldots, p_n$, and f are continuous on the open interval I containing the point a. Then, given n numbers $b_0, b_1, b_2, \ldots, b_{n-1}$, the *n*th order linear equation (Eq. (2))

$$y^{(n)} + p_1(x)y^{(n-1)} + \cdots + p_{n-1}(x)y' + p_n(x)y = f(x)$$

has a unique (that is, one and only one) solution on the entire interval I that satisfies the n initial conditions

$$y(a) = b_0, \; y'(a) = b_1, \ldots, y^{(n-1)}(a) = b_{n-1}. \qquad (5)$$

Equation (2) and the conditions in (5) constitute an *n*th order **initial value problem**. Theorem 2 tells us that any such initial value problem has a unique solution on the *whole* interval I where the coefficient functions in (2) are continuous.

It tells us nothing, however, about how to find this solution. In Section 2.3 we will see how to construct explicit solutions of initial value problems in the *constant coefficient* case that occurs often in applications.

EXAMPLE 1 CONTINUED We saw earlier that

$$y(x) = 7e^{3x} + 3\cos 2x - 2\sin 2x$$

is a solution of

$$y^{(3)} - 3y'' + 4y' - 12y = 0$$

on the real line. It has the initial values $y(0) = 10$, $y'(0) = 17$, and $y''(0) = 51$. Theorem 2 assures us that this is the *only* solution with these initial values.

Note that Theorem 2 implies that the *trivial* solution $y(x) \equiv 0$ is the only solution of the *homogeneous* equation

$$y^{(n)} + p_1(x)y^{(n-1)} + \cdots + p_{n-1}(x)y' + p_n(x)y = 0 \tag{3}$$

that satisfies the *trivial* initial conditions

$$y(a) = y'(a) = \cdots = y^{(n-1)}(a) = 0.$$

EXAMPLE 2 It is easy to verify that

$$y_1 = x^2 \quad \text{and} \quad y_2 = x^3$$

are two different solutions of

$$x^2 y'' - 4xy' + 6y = 0,$$

and that each satisfies the initial conditions $y(0) = y'(0) = 0$. Why does this not contradict the uniqueness part of Theorem 2? It is because the leading coefficient in this differential equation vanishes at $x = 0$, so the equation cannot be written in the form of Eq. (3) with coefficient functions *continuous* on an open interval containing the point $x = 0$.

On the basis of our knowledge of general solutions of second order linear equations, we anticipate that a general solution of the *homogeneous* nth order linear equation

$$y^{(n)} + p_1(x)y^{(n-1)} + \cdots + p_{n-1}(x)y' + p_n(x)y = 0 \tag{3}$$

will be a linear combination

$$y = c_1 y_1 + c_2 y_2 + \cdots + c_n y_n, \tag{4}$$

where $y_1, y_2, \ldots, y_n$ are particular solutions of Eq. (3). But these n particular solutions must be "sufficiently independent" that we can always choose the coefficients $c_1, c_2, \ldots, c_n$ in (4) to satisfy initial conditions of the form in (5). The question is this: What should be meant by *independence* of three or more functions?

Recall that *two* functions f_1 and f_2 are linearly *dependent* if one is a constant multiple of the other; that is, if either $f_1 = kf_2$ or $f_2 = kf_1$ for some constant k. If we rewrite these equations as

$$(1)f_1 + (-k)f_2 = 0 \quad \text{or} \quad (k)f_1 + (-1)f_2 = 0,$$

we see that the linear dependence of f_1 and f_2 implies that there exist two constants c_1 and c_2 *not both zero* such that

$$c_1 f_1 + c_2 f_2 = 0. \tag{6}$$

Conversely, if c_1 and c_2 are not both zero, then (6) certainly implies that f_1 and f_2 are linearly dependent.

By analogy with Eq. (6), we say that n functions $f_1, f_2, \ldots, f_n$ are *linearly dependent* provided that some *nontrivial* linear combination

$$c_1 f_1 + c_2 f_2 + \cdots + c_n f_n$$

of them vanishes identically; "nontrivial" means that *not all* of the coefficients $c_1, c_2, \ldots, c_n$ are zero (although some of them may be).

Definition *Linear Dependence of Functions*

The n functions $f_1, f_2, \ldots, f_n$ are said to be **linearly dependent** on the interval I provided that there exist constants $c_1, c_2, \ldots, c_n$ not all zero such that

$$c_1 f_1 + c_2 f_2 + \cdots + c_n f_n = 0 \tag{7}$$

on I; that is,

$$c_1 f_1(x) + c_2 f_2(x) + \cdots + c_n f_n(x) = 0$$

for all x in I.

If not all the coefficients in (7) are zero, then clearly we can solve for at least one of the functions as a linear combination of the others, and conversely. Thus the functions $f_1, f_2, \ldots, f_n$ are linearly dependent if and only if at least one of them is a linear combination of the others.

EXAMPLE 3 The functions

$$f_1(x) = \sin 2x, \quad f_2(x) = \sin x \cos x, \quad \text{and} \quad f_3(x) = e^x$$

are linearly dependent on the real line because

$$(1)f_1 + (-2)f_2 + (0)f_3 = 0$$

(by the familiar trigonometric identity $\sin 2x = 2 \sin x \cos x$).

The n functions $f_1, f_2, \ldots, f_n$ are called **linearly independent** on the interval I provided they are not linearly dependent there. Equivalently, they are linearly independent on I provided that the identity

$$c_1 f_1 + c_2 f_2 + \cdots + c_n f_n = 0 \tag{7}$$

holds on I only in the trivial case

$$c_1 = c_2 = \cdots = c_n = 0;$$

that is, *no* nontrivial linear combination of these functions vanishes on I. Put another way, the functions $f_1, f_2, \ldots, f_n$ are linearly independent if no one of them is a linear combination of the others. (Why?)

Sometimes one can show that n given functions are linearly dependent by finding, as in Example 3, nontrivial values of the coefficients so that Eq. (7) holds. But in order to show that n given functions are linearly independent, we must prove that nontrivial values of the coefficients *cannot* be found, and this is seldom easy to do in any direct or obvious manner.

Fortunately, in the case of n solutions of a homogeneous nth order linear equation, there is a tool that makes the determination of their linear dependence or independence a routine matter. This tool is the Wronskian determinant, which we introduced (for the case $n = 2$) in Section 2.1. Suppose that the n functions $f_1, f_2, \ldots, f_n$ are each $n - 1$ times differentiable. Then their **Wronskian** is the $n \times n$ determinant

$$W = \begin{vmatrix} f_1 & f_2 & \cdots & f_n \\ f_1' & f_2' & \cdots & f_n' \\ \vdots & \vdots & & \vdots \\ f_1^{(n-1)} & f_2^{(n-1)} & \cdots & f_n^{(n-1)} \end{vmatrix}. \tag{8}$$

We write $W(f_1, f_2, \ldots, f_n)$ or $W(x)$, depending upon whether we wish to emphasize the functions or the point x at which their Wronskian is to be evaluated. The Wronskian is named after the Polish mathematician J. M. H. Wronski (1778–1853), most of whose other mathematical work is now forgotten.

We saw in Section 2.1 that the Wronskian of two linearly dependent functions vanishes identically. More generally, *the Wronskian of n linearly dependent functions $f_1, f_2, \ldots, f_n$ is identically zero.* To prove this, assume that Eq. (7) holds on the interval I for some choice of the constants $c_1, c_2, \ldots, c_n$ not all zero. We then differentiate these equations $n - 1$ times in succession, obtaining the n equations

$$\left. \begin{aligned} c_1 f_1(x) + c_2 f_2(x) \quad + \cdots + \quad c_n f_n(x) &= 0, \\ c_1 f_1'(x) + c_2 f_2'(x) \quad + \cdots + \quad c_n f_n'(x) &= 0, \\ \vdots \\ c_1 f_1^{(n-1)}(x) + c_2 f_2^{(n-1)}(x) + \cdots + c_n f_n^{(n-1)}(x) &= 0, \end{aligned} \right\} \tag{9}$$

which hold for all x in I. We recall from linear algebra that a system of n linear *homogeneous* equations in n unknowns has a nontrivial solution if and only if its determinant of coefficients vanishes. In (9) the unknowns are the constants $c_1, c_2, \ldots, c_n$ and the determinant of coefficients is simply the Wronskian $W(f_1, f_2, \ldots, f_n)$ evaluated at the typical point x of I. Because we know that the c_i are not all zero, it follows that $W(x) \equiv 0$, as we wanted to prove.

Therefore, in order to show that the functions $f_1, f_2, \ldots, f_n$ are *linearly independent* on the interval I, it suffices to show that their Wronskian is nonzero at just one point of I.

EXAMPLE 4 Show that the functions $y_1 = e^{3x}$, $y_2 = \cos 2x$, and $y_3 = \sin 2x$ (of Example 1) are linearly independent.

SOLUTION Their Wronskian is

$$W = \begin{vmatrix} e^{3x} & \cos 2x & \sin 2x \\ 3e^{3x} & -2\sin 2x & 2\cos 2x \\ 9e^{3x} & -4\cos 2x & -4\sin 2x \end{vmatrix}$$

$$= e^{3x}\begin{vmatrix} -2\sin 2x & 2\cos 2x \\ -4\cos 2x & -4\sin 2x \end{vmatrix} - 3e^{3x}\begin{vmatrix} \cos 2x & \sin 2x \\ -4\cos 2x & -4\sin 2x \end{vmatrix}$$

$$+ 9e^{3x}\begin{vmatrix} \cos 2x & \sin 2x \\ -2\sin 2x & 2\cos 2x \end{vmatrix} = 26e^{3x} \neq 0.$$

Because $W \neq 0$ everywhere, it follows that y_1, y_2, and y_3 are linearly independent on any open interval (including the whole real line).

EXAMPLE 5 Show first that the three solutions

$$y_1 = x, \qquad y_2 = x \ln x, \quad \text{and} \quad y_3 = x^2$$

of the third order equation

$$x^3 y^{(3)} - 4x^3 y'' + 5xy' - 2y = 0 \tag{10}$$

are linearly independent on the open interval $x > 0$. Then find a particular solution of (10) that satisfies the initial conditions

$$y(1) = 3, \qquad y'(1) = 2, \qquad y''(1) = 1. \tag{11}$$

SOLUTION Note that for $x > 0$, we could divide each term in (10) by x^3 to obtain a homogeneous linear equation of the standard form in (3). When we compute the Wronskian of the three given solutions, we find that

$$W = \begin{vmatrix} x & x \ln x & x^2 \\ 1 & 1 + \ln x & 2x \\ 0 & \dfrac{1}{x} & 2 \end{vmatrix} = x.$$

Thus $W \neq 0$ for $x > 0$, so y_1, y_2, and y_3 are linearly independent on the interval $x > 0$. To find the desired particular solution, we impose the initial conditions in (11) on

$$y(x) = c_1 x + c_2 x \ln x \qquad + c_3 x^3,$$

$$y'(x) = c_1 \quad + c_2(1 + \ln x) + 2c_3 x,$$

$$y''(x) = 0 \quad + \frac{c_2}{x} \qquad + 2c_3.$$

This yields the simultaneous equations

$$y(1) = c_1 \qquad + c_3 = 3,$$
$$y'(1) = c_1 + c_2 + 2c_3 = 2,$$
$$y''(1) = \qquad c_2 + 2c_3 = 1;$$

we solve to find $c_1 = 1$, $c_2 = -3$, and $c_3 = 2$. Thus our particular solution is

$$y(x) = x - 3x \ln x + 2x^2.$$

Provided that $W(y_1, y_2, \ldots, y_n) \neq 0$, it turns out (Theorem 4) that we can always find values of the coefficients in the linear combination

$$y = c_1 y_1 + c_2 y_2 + \cdots + c_n y_n$$

in order to satisfy any given initial conditions of the form in (5). The following theorem provides the necessary nonvanishing of W in the case of linearly independent solutions.

Theorem 3 *Wronskians of Solutions*

Suppose that $y_1, y_2, \ldots, y_n$ are n solutions of the homogeneous nth order linear equation

$$y^{(n)} + p_1(x)y^{(n-1)} + \cdots + p_{n-1}(x)y' + p_n(x)y = 0 \qquad (3)$$

on an open interval I where each p_i is continuous. Let

$$W = W(y_1, y_2, \ldots, y_n).$$

(a) If $y_1, y_2, \ldots, y_n$ are linearly dependent, then $W \equiv 0$ on I.
(b) If $y_1, y_2, \ldots, y_n$ are linearly independent, then $W \neq 0$ at each point of I.

Thus there are just two possibilities: Either $W = 0$ everywhere on I, or $W \neq 0$ everywhere on I.

Proof We have already proven part (a). To prove part (b), it is sufficient to assume that $W(a) = 0$ for some point a of I, and show that this implies that the solutions $y_1, y_2, \ldots, y_n$ are linearly dependent. But $W(a)$ is simply the determinant of coefficients of the system of n homogeneous linear equations

$$\left.\begin{array}{l} c_1 y_1(a) + c_2 y_2(a) \quad + \cdots + \quad c_n y_n(a) = 0, \\ c_1 y_1'(a) + c_2 y_2'(a) \quad + \cdots + \quad c_n y_n'(a) = 0, \\ \qquad\qquad\qquad\qquad\qquad\vdots \\ c_1 y_1^{(n-1)}(a) + c_2 y_2^{(n-1)}(a) + \cdots + c_n y_n^{(n-1)}(a) = 0 \end{array}\right\} \qquad (12)$$

in the n unknowns $c_1, c_2, \ldots, c_n$. Because $W(a) = 0$, the basic fact from linear algebra quoted just following (9) above implies that the equations in (12) have a nontrivial solution. That is, the numbers $c_1, c_2, \ldots$, and c_n cannot all be zero.

CHAPTER 2: Linear Equations of Higher Order

We now use these values to define the particular solution

$$Y(x) = c_1 y_1(x) + c_2 y_2(x) + \cdots + c_n y_n(x) \qquad (13)$$

of Eq. (3). Equations (12) then imply that Y satisfies the trivial initial conditions

$$Y(a) = Y'(a) = \cdots = Y^{(n-1)}(a) = 0.$$

Theorem 2 (uniqueness) therefore implies that $Y(x) \equiv 0$ on I. In view of (13) and the fact that $c_1, c_2, \ldots, c_n$ are not all zero, this is the desired conclusion that the solutions $y_1, y_2, \ldots, y_n$ are linearly dependent. This completes the proof of Theorem 3.　∎

We can now show that every solution of a *homogeneous* nth order linear equation is a linear combination of n given linearly independent solutions. Using the fact from Theorem 3 that the Wronskian of n linearly independent solutions is nonzero, the proof of the following theorem is essentially the same as the proof of Theorem 4 of Section 2.1 (the case $n = 2$).

Theorem 4　General Solutions

Let $y_1, y_2, \ldots, y_n$ be n linearly independent solutions of the homogeneous equation

$$y^{(n)} + p_1(x)y^{(n-1)} + \cdots + p_{n-1}(x)y' + p_n(x)y = 0 \qquad (3)$$

on an open interval I where the p_i are continuous. If Y is any solution whatsoever of Eq. (3), then there exist numbers $c_1, c_2, \ldots, c_n$ such that

$$Y(x) = c_1 y_1(x) + c_2 y_2(x) + \cdots + c_n y_n(x)$$

for all x in I.

Thus *every* solution of a homogeneous nth order linear differential equation is a linear combination

$$y = c_1 y_1 + c_2 y_2 + \cdots + c_n y_n$$

of *any* n given linearly independent solutions. On this basis we call such a linear combination a **general solution** of the differential equation.

NONHOMOGENEOUS EQUATIONS

We now consider the *nonhomogeneous* nth order linear differential equation

$$Ly = y^{(n)} + p_1(x)y^{(n-1)} + \cdots + p_{n-1}(x)y' + p_n(x)y = f(x) \qquad (2)$$

with associated homogeneous equation

$$Ly = y^{(n)} + p_1(x)y^{(n-1)} + \cdots + p_{n-1}(x)y' + p_n(x)y = 0. \qquad (3)$$

Here we introduce the symbol L to represent an **operator**; given an n times differentiable function y, L operates on y (as suggested in Fig. 2.3) to produce the linear

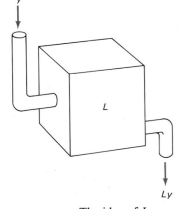

FIGURE 2.3　The idea of L "operating" on the function y.

combination

$$Ly = y^{(n)} + p_1 y^{(n-1)} + \cdots + p_{n-1} y' + p_n y \tag{14}$$

of y and its first n derivatives. The principle of superposition (Theorem 1) means simply that the operator L is *linear*; that is,

$$L(c_1 y_1 + c_2 y_2) = c_1 L y_1 + c_2 L y_2 \tag{15}$$

if c_1 and c_2 are constants.

Suppose that a single particular solution y_p of the nonhomogeneous equation in (2) is known and that Y is any other solution of (2). Then (15) implies that

$$L(Y - y_p) = LY - L y_p = f - f = 0.$$

Thus $y_c = Y - y_p$ is a solution of the associated homogeneous equation in (3). Then

$$Y = y_c + y_p, \tag{16}$$

and it follows from Theorem 4 that

$$y_c = c_1 y_1 + c_2 y_2 + \cdots + c_n y_n \tag{17}$$

where $y_1, y_2, \ldots, y_n$ are linearly independent solutions of the associated *homogeneous* equation. We call y_c a **complementary function** of the nonhomogeneous equation and have thus proved that a *general solution* of the nonhomogeneous equation in (2) is the sum of its complementary function y_c and a single particular solution y_p of (2).

Theorem 5 *Solutions of Nonhomogeneous Equations*

Let y_p be a particular solution of the nonhomogeneous equation in (2) on an open interval I where the functions p_i and f are continuous. Let $y_1, y_2, \ldots, y_n$ be linearly independent solutions of the associated homogeneous equation in (3). If Y is any solution whatsoever of Eq. (2) on I, then there exist numbers $c_1, c_2, \ldots, c_n$ such that

$$Y(x) = c_1 y_1(x) + c_2 y_2(x) + \cdots + c_n y_n(x) + y_p(x) \tag{18}$$

for all x in I.

EXAMPLE 6 It is evident that $y_p(x) = 3x$ is a particular solution of the equation

$$y'' + 4y = 12x, \tag{19}$$

and that $y_c = c_1 \cos 2x + c_2 \sin 2x$ is its complementary function. Find a solution of (19) that satisfies the initial conditions $y(0) = 5$, $y'(0) = 7$.

SOLUTION The general solution of Eq. (19) is

$$y = c_1 \cos 2x + c_2 \sin 2x + 3x.$$

Now

$$y' = -2c_1 \sin 2x + 2c_2 \cos 2x + 3.$$

Hence the initial conditions give

$$y(0) = c_1 \qquad = 5,$$
$$y'(0) = 2c_2 + 3 = 7.$$

We find that $c_1 = 5$ and $c_2 = 2$. Thus the desired solution is

$$y(x) = 5 \cos 2x + 2 \sin 2x + 3x.$$

2.2 PROBLEMS

In each of Problems 1–6, show directly that the given functions are linearly dependent on the real line. That is, find a nontrivial linear combination of the given functions that vanishes identically.

1. $f(x) = 2x$, $g(x) = 3x^2$, $h(x) = 5x - 8x^2$

2. $f(x) = 5$, $g(x) = 2 - 3x^2$, $h(x) = 10 + 15x^2$

3. $f(x) = 0$, $g(x) = \sin x$, $h(x) = e^x$

4. $f(x) = 17$, $g(x) = 2 \sin^2 x$, $h(x) = 3 \cos^2 x$

5. $f(x) = 17$, $g(x) = \cos^2 x$, $h(x) = \cos 2x$

6. $f(x) = e^x$, $g(x) = \cosh x$, $h(x) = \sinh x$

In each of Problems 7–12, use the Wronskian to prove that the given functions are linearly independent on the indicated interval.

7. $f(x) = 1$, $g(x) = x$, $h(x) = x^2$; the real line

8. $f(x) = e^x$, $g(x) = e^{2x}$, $h(x) = e^{3x}$; the real line

9. $f(x) = e^x$, $g(x) = \cos x$, $h(x) = \sin x$; the real line

10. $f(x) = e^x$, $g(x) = x^{-2}$, $h(x) = x^{-2} \ln x$; $x > 0$

11. $f(x) = x$, $g(x) = xe^x$, $h(x) = x^2 e^x$; the real line

12. $f(x) = x$, $g(x) = \cos(\ln x)$, $h(x) = \sin(\ln x)$; $x > 0$

In each of Problems 13–20, a third order homogeneous linear equation and three linearly independent solutions are given. Find a particular solution satisfying the given initial conditions.

13. $y^{(3)} + 2y'' - y' - 2y = 0$; $y(0) = 1$, $y'(0) = 2$, $y''(0) = 0$;
$y_1 = e^x$, $y_2 = e^{-x}$, $y_3 = e^{-2x}$

14. $y^{(3)} - 6y'' + 11y' - 6y = 0$; $y(0) = 0$, $y'(0) = 0$, $y''(0) = 3$;
$y_1 = e^x$, $y_2 = e^{2x}$, $y_3 = e^{3x}$

15. $y^{(3)} - 3y'' + 3y' - y = 0$; $y(0) = 2$, $y'(0) = 0$, $y''(0) = 0$;
$y_1 = e^x$, $y_2 = xe^x$, $y_3 = x^2 e^x$

16. $y^{(3)} - 5y'' + 8y' - 4y = 0$; $y(0) = 1$, $y'(0) = 4$, $y''(0) = 0$;
$y_1 = e^x$, $y_2 = e^{2x}$, $y_3 = xe^{2x}$

17. $y^{(3)} + 9y' = 0$; $y(0) = 3$, $y'(0) = -1$, $y''(0) = 2$; $y_1 = 1$,
$y_2 = \cos 3x$, $y_3 = \sin 3x$

18. $y^{(3)} - 3y'' + 4y' - 2y = 0$; $y(0) = 1$, $y'(0) = 0$, $y''(0) = 0$;
$y_1 = e^x$, $y_2 = e^x \cos x$, $y_3 = e^x \sin x$

19. $x^3 y^{(3)} - 3x^2 y'' + 6xy' - 6y = 0$; $y(1) = 6$, $y'(1) = 14$,
$y''(1) = 22$; $y_1 = x$, $y_2 = x^2$, $y_3 = x^3$

20. $x^3 y^{(3)} + 6x^2 y'' + 4xy' - 4y = 0$; $y(1) = 1$, $y'(1) = 5$,
$y''(1) = -11$; $y_1 = x$, $y_2 = x^{-2}$, $y_3 = x^{-2} \ln x$

In each of Problems 21–24, a nonhomogeneous differential equation, a complementary solution y_c, and a particular solution y_p are given. Find a solution satisfying the given initial conditions.

21. $y'' + y = 3x$; $y(0) = 2$, $y'(0) = -2$;
$y_c = c_1 \cos x + c_2 \sin x$; $y_p = 3x$

22. $y'' - 4y = 12$; $y(0) = 0$, $y'(0) = 10$; $y_c = c_1 e^{2x} + c_2 e^{-2x}$;
$y_p = -3$

23. $y'' - 2y' - 3y = 6$; $y(0) = 3$, $y'(0) = 11$;
$y_c = c_1 e^{-x} + c_2 e^{3x}$; $y_p = -2$

24. $y'' - 2y' + 2y = 2x$; $y(0) = 4$, $y'(0) = 8$,
$y_c = c_1 e^x \cos x + c_2 e^x \sin x$; $y_p = x + 1$

25. Let $Ly = y'' + py' + qy$. Suppose that y_1 and y_2 are two functions such that $Ly_1 = f(x)$ and $Ly_2 = g(x)$. Show that their sum $y = y_1 + y_2$ satisfies the nonhomogeneous equation $Ly = f(x) + g(x)$.

26. (a) Find by inspection particular solutions of the two nonhomogeneous equations $y'' + 2y = 4$ and $y'' + 2y = 6x$. (b) Use the method of Problem 25 to find a particular solution of the differential equation $y'' + 2y = 6x + 4$.

27. Prove directly that the functions $f_1(x) \equiv 1$, $f_2(x) = x$, and $f_3(x) = x^2$ are linearly independent on the whole real line. (*Suggestion*: Assume that $c_1 + c_2 x + c_3 x^2 = 0$. Differentiate this equation twice. You now have three equations that must be satisfied by any x, including $x = 0$. Conclude that $c_1 = c_2 = c_3 = 0$.)

28. Generalize the method of Problem 27 to prove directly that the functions $f_0(x) \equiv 1$, $f_1(x) = x$, $f_2(x) = x^2$, ..., $f_n(x) = x^n$ are linearly independent on the real line.

29. Use the result of Problem 28 and the definition of linear independence to prove directly that, for any constant r, the functions $f_0(x) = e^{rx}$, $f_1(x) = xe^{rx}, \ldots, f_n(x) = x^n e^{rx}$ are linearly independent on the whole real line.

30. Verify that $y_1 = x$ and $y_2 = x^2$ are linearly independent solutions on the entire real line of the equation $x^2 y'' - 2xy' + 2y = 0$, but that $W(x, x^2)$ vanishes at $x = 0$. Why do these observations not contradict part (b) of Theorem 3?

31. This problem indicates why we can impose *only* n initial conditions on a solution of an nth order linear differential equation. (a) Given the equation $y'' + py' + qy = 0$, explain why the value of $y''(a)$ is determined by the values of $y(a)$ and $y'(a)$. (b) Prove that the equation $y'' - 2y' - 5y = 0$ has a solution satisfying the conditions $y(0) = 1$, $y'(0) = 0$, $y''(0) = C$ if and only if $C = 5$.

32. Prove that an nth order homogeneous linear differential equation satisfying the hypotheses of Theorem 2 has n *linearly independent* solutions $y_1, y_2, y_3, \ldots, y_n$. (*Suggestion:* Let y_i be the unique solution such that $y_i^{(i-1)}(a) = 1$ and $y_i^{(k)}(a) = 0$ if $k \neq i - 1$.)

33. Suppose that the three numbers r_1, r_2, and r_3 are distinct. Show that the three functions $\exp(r_1 x)$, $\exp(r_2 x)$, and $\exp(r_3 x)$ are linearly independent by showing that their Wronskian

$$W = \exp([r_1 + r_2 + r_3]x) \cdot \begin{vmatrix} 1 & 1 & 1 \\ r_1 & r_2 & r_3 \\ r_1^2 & r_2^2 & r_3^2 \end{vmatrix}$$

is nonzero for all x.

34. Assume as known that the **Vandermonde determinant**

$$V = \begin{vmatrix} 1 & 1 & \cdots & 1 \\ r_1 & r_2 & \cdots & r_n \\ r_1^2 & r_2^2 & \cdots & r_n^2 \\ \vdots & \vdots & & \vdots \\ r_1^{n-1} & r_2^{n-1} & \cdots & r_n^{n-1} \end{vmatrix}$$

is nonzero if the numbers $r_1, r_2, \ldots, r_n$ are distinct. Prove by the method of Problem 33 that the functions

$$f_i(x) = \exp(r_i x), \qquad 1 \leq i \leq n$$

are linearly independent.

35. According to Problem 32 of Section 2.1, the Wronskian $W(y_1, y_2)$ of two solutions of the second order equation

$$y'' + p_1(x)y' + p_2(x)y = 0$$

is given by Abel's formula

$$W(x) = K \exp\left(-\int p_1(x)\,dx\right)$$

for some constant K. It can be shown that the Wronskian of n solutions $y_1, y_2, \ldots, y_n$ of the nth order equation

$$y^{(n)} + p_1(x)y^{(n-1)} + \cdots + p_{n-1}(x)y' + p_n(x)y = 0$$

satisfies the same identity. Prove this for the case $n = 3$ as follows: (a) The derivative of a determinant of functions is the sum of the determinants obtained by separately differentiating the rows of the original determinant. Conclude that

$$W' = \begin{vmatrix} y_1 & y_2 & y_3 \\ y_1' & y_2' & y_3' \\ y_1^{(3)} & y_2^{(3)} & y_3^{(3)} \end{vmatrix}.$$

(b) Substitute for $y_1^{(3)}$, $y_2^{(3)}$, and $y_3^{(3)}$ from the differential equation $y^{(3)} + p_1 y'' + p_2 y' + p_3 y = 0$, and then show that $W' = -p_1 W$. Integration now gives Abel's formula.

Homogeneous Equations with Constant Coefficients

In the first two sections of this chapter we saw that a general solution of an nth order homogeneous linear equation is a linear combination of n linearly independent particular solutions, but we said little about how to actually find even a single solution. The solution of a linear differential equation with *variable* coefficients ordinarily requires infinite series methods (Chapter 3) or numerical methods (Chapter 6). But we can now show how to find, explicitly and in a rather straightforward way, n linearly independent solutions of a given nth order homogeneous linear equation if it has constant coefficients. The general such equation may be written in the form

$$a_n y^{(n)} + a_{n-1} y^{(n-1)} + \cdots + a_2 y'' + a_1 y' + a_0 y = 0, \tag{1}$$

where the coefficients $a_0, a_1, a_2, \ldots, a_n$ are real constants with $a_n \neq 0$.

We first look for a *single* solution of Eq. (1), and begin with the observation that

$$\frac{d^k}{dx^k}(e^{rx}) = r^k e^{rx}, \tag{2}$$

so any derivative of e^{rx} is a constant multiple of e^{rx}. Hence, if we substituted $y = e^{rx}$ in Eq. (1), each term would be a constant multiple of e^{rx}, with the constant coefficients depending upon r and the coefficients a_i. This suggests that we try to find r so that all these multiples of e^{rx} will have sum zero, in which case $y = e^{rx}$ will be a solution of Eq. (1).

For example, if we substitute $y = e^{rx}$ in the equation

$$y'' - 5y' + 6y = 0,$$

we obtain

$$r^2 e^{rx} - 5r e^{rx} + 6 e^{rx} = 0;$$

division by e^{rx} and factoring then gives the equation

$$r^2 - 5r + 6 = (r - 2)(r - 3) = 0.$$

Hence $y = e^{rx}$ will be a solution if either $r = 2$ or $r = 3$, and not otherwise. So, in searching for a single solution, we actually have found *two* solutions: $y_1 = e^{2x}$ and $y_2 = e^{3x}$.

To carry out this technique in the general case, we substitute $y = e^{rx}$ in Eq. (1), and with the aid of (2) we find the result to be

$$a_n r^n e^{rx} + a_{n-1} r^{n-1} e^{rx} + \cdots + a_2 r^2 e^{rx} + a_1 r e^{rx} + a_0 e^{rx} = 0;$$

that is,

$$e^{rx}(a_n r^n + a_{n-1} r^{n-1} + \cdots + a_2 r^2 + a_1 r + a_0) = 0.$$

Because e^{rx} is never zero, we see that $y = e^{rx}$ will be a solution of Eq. (1) precisely when r is a root of the equation

$$a_n r^n + a_{n-1} r^{n-1} + \cdots + a_2 r^2 + a_1 r + a_0 = 0. \tag{3}$$

This equation is called the **characteristic equation**, or **auxiliary equation**, of the differential equation in (1). Our problem, then, is reduced to the solution of this purely algebraic equation.

According to the fundamental theorem of algebra, every nth degree polynomial—such as the one in (3)—has n zeros, though not necessarily distinct and not necessarily real. Finding the exact values of these zeros may be difficult or even impossible; the quadratic formula is sufficient for second degree equations, but for equations of high degree we may need to spot a fortuitous factorization, or resort to numerical methods (such as Newton's method for finding real zeros or Müller's method for finding complex zeros—see any numerical analysis text).

Whatever the method we use, let us suppose that we have solved the characteristic equation. Then we can always write a general solution of the differential equation. The situation is slightly more complicated in the case of repeated roots or complex roots of Eq. (3), so let us first examine the simplest case—in which the characteristic equation has n *distinct* (no two equal) *real* roots $r_1, r_2, \ldots, r_n$. Then the functions

$$e^{r_1 x}, e^{r_2 x}, \ldots, e^{r_n x}$$

are all solutions of Eq. (1), and (by Problem 34 of Section 2.2) these n solutions are linearly independent (on the entire real line). In summary, we have proved Theorem 1.

Theorem 1   *Distinct Real Roots*

If the n roots $r_1, r_2, \ldots, r_n$ of the characteristic equation in (3) are real and distinct, then

$$y(x) = c_1 e^{r_1 x} + c_2 e^{r_2 x} + \cdots + c_n e^{r_n x} \tag{4}$$

is a general solution of Eq. (1).

EXAMPLE 1   Solve the initial value problem

$$y'' + 2y' - 8y = 0; \qquad y(0) = 5, \qquad y'(0) = -2.$$

SOLUTION   According to Theorem 1 we need only solve the characteristic equation

$$r^2 + 2r - 8 = (r - 2)(r + 4) = 0$$

for $r = 2, -4$. Hence a general solution is

$$y(x) = c_1 e^{2x} + c_2 e^{-4x}.$$

Then

$$y'(x) = 2c_1 e^{2x} - 4c_2 e^{-4x},$$

so the given initial conditions yield the equations

$$y(0) = \ c_1 + \ c_2 = 5,$$
$$y'(0) = 2c_1 - 4c_2 = -2$$

that we readily solve for $c_1 = 3$, $c_2 = 2$. Thus the desired particular solution is

$$y(x) = 3e^{2x} + 2e^{-4x}.$$

EXAMPLE 2   Find a general solution of

$$y^{(3)} - y'' - 6y' = 0.$$

SOLUTION The characteristic equation of this differential equation is

$$r^3 - r^2 - 6r = 0,$$

which we solve by factoring:

$$r(r^2 - r - 6) = r(r - 3)(r + 2) = 0,$$

so the three roots are $r = 0$, $r = 3$, and $r = -2$. They are real and distinct, and therefore—because $e^0 = 1$—a general solution of the given differential equation is

$$y(x) = c_1 + c_2 e^{3x} + c_3 e^{-2x}.$$

REPEATED ROOTS

If the roots of the characteristic equation in (3) are *not* distinct—there are repeated roots—then we cannot produce n linearly independent solutions of Eq. (1) by the method of Theorem 1. For example, if the roots are 1, 2, 2, and 2, we obtain only the *two* functions e^x and e^{2x}. The problem, then, is to produce the missing linearly independent solutions. For this purpose it is convenient to adopt the operator notation introduced near the conclusion of Section 2.2. Equation (1) corresponds to the operator equation $Ly = 0$ where L is the operator

$$L = a_n \frac{d^n}{dx^n} + a_{n-1} \frac{d^{n-1}}{dx^{n-1}} + \cdots + a_2 \frac{d^2}{dx^2} + a_1 \frac{d}{dx} + a_0. \tag{5}$$

We also denote by $D = d/dx$ the operation of differentiation with respect to x, so that

$$Dy = y', \qquad D^2 y = y'', \qquad D^3 y = y^{(3)},$$

and so on. In terms of D, the operator L in (5) may be written

$$L = a_n D^n + a_{n-1} D^{n-1} + \cdots + a_2 D^2 + a_1 D + a_0, \tag{6}$$

and we will find it useful to think of the right-hand side in (6) as a (formal) nth degree polynomial in the "variable" D; it is a **polynomial operator**.

A first degree polynomial operator has the form $D - a$ where a is a real number. It operates on the function $y = y(x)$ to produce

$$(D - a)y = Dy - ay = y' - ay.$$

The important fact about such operators is that any two of them *commute*:

$$(D - a)(D - b)y = (D - b)(D - a)y \tag{7}$$

for any twice differentiable function $y = y(x)$. The proof of the formula in (7) is the following computation:

$$\begin{aligned}
(D - a)(D - b)y &= (D - a)(y' - by) \\
&= D(y' - by) - a(y' - by) = y'' - (b + a)y' + aby \\
&= y'' - (a + b)y' + bay = D(y' - ay) - b(y' - ay) \\
&= (D - b)(y' - ay) = (D - b)(D - a)y.
\end{aligned}$$

Let us now consider the possibility that the characteristic equation

$$a_n r^n + a_{n-1} r^{n-1} + \cdots + a_1 r + a_0 = 0 \qquad (3)$$

has *repeated* roots. For example, suppose that Eq. (3) has only two distinct roots, r_0 of multiplicity 1 and r_1 of multiplicity $k > 1$. Then (3) can be rewritten in the form

$$(r - r_1)^k (r - r_0) = (r - r_0)(r - r_1)^k = 0. \qquad (8)$$

Similarly, the operator L in (6) can be written as

$$L = (D - r_1)^k (D - r_0) = (D - r_0)(D - r_1)^k, \qquad (9)$$

the order of the factors making no difference because of the formula in (7).

Two solutions of the differential equation $Ly = 0$ are certainly $y_0 = e^{r_0 x}$ and $y_1 = e^{r_1 x}$. This is, however, not sufficient; we need $k + 1$ linearly independent solutions in order to construct a general solution, because the equation is of order $k + 1$. To find the missing $k - 1$ solutions, we note that

$$Ly = (D - r_0)[(D - r_1)^k y] = 0.$$

Consequently *every* solution of the kth order equation

$$(D - r_1)^k y = 0 \qquad (10)$$

will also be a solution of the original equation $Ly = 0$. Hence our problem is reduced to that of finding the general solution of the differential equation in (10).

The fact that $e^{r_1 x}$ is one solution of (10) suggests that we try the substitution

$$y(x) = u(x) e^{r_1 x}, \qquad (11)$$

where $u(x)$ is a function yet to be determined. Observe that

$$(D - r_1)[u e^{r_1 x}] = (Du) e^{r_1 x} + r_1 u e^{r_1 x} - r_1 u e^{r_1 x},$$

so

$$(D - r_1)[u e^{r_1 x}] = (Du) e^{r_1 x}. \qquad (12)$$

It therefore follows by induction on k that

$$(D - r_1)^k [u e^{r_1 x}] = (D^k u) e^{r_1 x} \qquad (13)$$

for any function $u(x)$. Hence $y = u e^{r_1 x}$ will be a solution of (10) if and only if $D^k u = u^{(k)} = 0$. But this is so if and only if

$$u(x) = c_1 + c_2 x + c_3 x^2 + \cdots + c_k x^{k-1},$$

a polynomial of degree at most $k - 1$. Hence our desired solution of (10) is

$$y(x) = u e^{r_1 x} = (c_1 + c_2 x + \cdots + c_k x^{k-1}) e^{r_1 x}.$$

In particular, we see here the additional solutions $x e^{r_1 x}, x^2 e^{r_1 x}, \ldots, x^{k-1} e^{r_1 x}$ of our original differential equation $Ly = 0$.

The analysis above can be carried out with the operator $D - r_0$ replaced with an arbitrary polynomial operator. When this is done, the result is a proof of the following theorem.

CHAPTER 2:   Linear Equations of Higher Order

> **Theorem 2** *Repeated Roots*
>
> If the characteristic equation in (3) has a repeated root r of multiplicity k, then the part of a general solution of the differential equation in (1) corresponding to r is of the form
>
> $$(c_1 + c_2 x + c_3 x^2 + \cdots + c_k x^{k-1})e^{rx}. \tag{14}$$

We may observe that, according to Problem 29 of Section 2.2, the k functions $e^{rx}, xe^{rx}, x^2 e^{rx}, \ldots,$ and $x^{k-1}e^{rx}$ involved in (14) are linearly independent on the real line. Thus a root of multiplicity k corresponds to k linearly independent solutions of the differential equation.

EXAMPLE 3 The differential equation

$$y'' + 4y' + 4y = 0$$

has the characteristic equation

$$r^2 + 4r + 4 = (r + 2)^2 = 0$$

with double root $r = -2, -2$. Hence with $k = 2$ Theorem 2 yields the general solution

$$y(x) = (c_1 + c_2 x)e^{-2x}.$$

EXAMPLE 4 Find a general solution of

$$y^{(4)} + 3y^{(3)} + 3y'' + y' = 0.$$

SOLUTION The characteristic equation of this differential equation is

$$r^4 + 3r^3 + 3r^2 + r = r(r + 1)^3 = 0.$$

It has the simple root $r_1 = 0$, which contributes $y_1 = c_1$ to the general solution. It has also the triple ($k = 3$) root $r_2 = -1$, which contributes $y_2 = (c_2 + c_3 x + c_4 x^2)e^{-x}$ to the solution. Hence a general solution of the differential equation is

$$y(x) = c_1 + (c_2 + c_3 x + c_4 x^2)e^{-x}.$$

COMPLEX ROOTS

Because we have assumed that the coefficients of the differential equation and its characteristic equation are real, any complex (nonreal) roots will occur in complex conjugate pairs $a \pm bi$ where a and b are real and $i = \sqrt{-1}$. This raises the question as to what might be meant by an exponential such as $\exp([a + bi]x)$.

To answer this question, we recall from elementary calculus the Taylor series for the exponential function:

$$e^t = \sum_{n=0}^{\infty} \frac{t^n}{n!} = 1 + t + \frac{t^2}{2!} + \frac{t^3}{3!} + \frac{t^4}{4!} + \cdots.$$

If we substitute $t = ix$ in this series, we get

$$e^{ix} = \sum_{n=0}^{\infty} \frac{(ix)^n}{n!}$$

$$= 1 + ix - \frac{x^2}{2!} - \frac{ix^3}{3!} + \frac{x^4}{4!} + \frac{ix^5}{5!} - \cdots$$

$$= \left(1 - \frac{x^2}{2!} + \frac{x^4}{4!} - \cdots\right) + i\left(x - \frac{x^3}{3!} + \frac{x^5}{5!} - \cdots\right).$$

Because the two real series in the last line are the Taylor series for $\cos x$ and $\sin x$, respectively, this implies that

$$e^{ix} = \cos x + i \sin x. \qquad (15)$$

This result is known as *Euler's formula*. Because of it we *define* the exponential function e^z, for $z = x + iy$ an arbitrary complex number, to be

$$e^z = e^{x+iy} = e^x e^{iy} = e^x(\cos y + i \sin y). \qquad (16)$$

Thus it appears that complex roots of the characteristic equation will lead to complex-valued solutions of the differential equation. A **complex-valued function** F of the real variable x associates with each real number x (in its domain of definition) the complex number

$$y = F(x) = f(x) + ig(x). \qquad (17)$$

The real-valued functions f and g are called the **real** and **imaginary** parts, respectively, of F. If they are differentiable, we define the **derivative** F' of F to be

$$F'(x) = f'(x) + ig'(x). \qquad (18)$$

Thus we simply differentiate the real and imaginary parts of F separately. Similarly, we say that the complex-valued function $y = F(x)$ **satisfies** the differential equation in (1) provided that its real and imaginary parts separately satisfy that differential equation.

The particular complex-valued functions of interest here are of the form $F(x) = e^{rx}$ where $r = a \pm bi$. We note from Euler's formula that

$$e^{(a+bi)x} = e^{ax}(\cos bx + i \sin bx) \qquad (19a)$$

and

$$e^{(a-bi)x} = e^{ax}(\cos bx - i \sin bx). \qquad (19b)$$

The most important property of e^{rx} is that

$$D_x(e^{rx}) = re^{rx} \qquad (20)$$

even if r should be a complex number. The proof of this assertion is a straightforward computation based on the definitions and formulas given above:

$$D_x(e^{rx}) = D_x(e^{ax} \cos bx) + iD_x(e^{ax} \sin bx)$$
$$= [ae^{ax} \cos bx - be^{ax} \sin bx] + i[ae^{ax} \sin bx + be^{ax} \cos bx]$$
$$= (a + bi)(e^{ax} \cos bx + ie^{ax} \sin bx) = re^{rx}.$$

It follows from (20) that, when r is complex (just as when r is real), e^{rx} will be a solution of the differential equation in (1) if and only if r is a root of its characteristic equation. If the complex conjugate pair of roots $r_1 = a + bi$ and $r_2 = a - bi$ are simple (nonrepeated), then the corresponding part of a general solution of (1) is

$$C_1 e^{(a+bi)x} + C_2 e^{(a-bi)x} = C_1 e^{ax}(\cos bx + i \sin bx)$$
$$+ C_2 e^{ax}(\cos bx - i \sin bx)$$
$$= e^{ax}(c_1 \cos bx + c_2 \sin bx)$$

where $c_1 = C_1 + C_2$ and $c_2 = (C_1 - C_2)i$. Thus the conjugate pair of roots $a \pm bi$ leads to the linearly independent *real-valued* solutions $e^{ax} \cos bx$ and $e^{ax} \sin bx$. This yields the following result.

Theorem 3 *Complex Roots*

If the characteristic equation in (3) has an unrepeated pair of complex conjugate roots $a \pm bi$ (with $b \neq 0$), then the corresponding part of a general solution of (1) is of the form

$$e^{ax}(c_1 \cos bx + c_2 \sin bx). \tag{21}$$

EXAMPLE 5   The characteristic equation of

$$y'' + b^2 y = 0 \qquad (b > 0)$$

is $r^2 + b^2 = 0$, with roots $\pm bi$. So Theorem 3 (with $a = 0$) gives the general solution

$$y(x) = c_1 \cos bx + c_2 \sin bx.$$

EXAMPLE 6   Find the particular solution of

$$y'' - 4y' + 5y = 0$$

for which $y(0) = 1$ and $y'(0) = 5$.

SOLUTION   The characteristic equation is

$$r^2 - 4r + 5 = (r - 2)^2 + 1 = 0,$$

with roots $2 + i$ and $2 - i$. Hence a general solution is

$$y(x) = e^{2x}(c_1 \cos x + c_2 \sin x).$$

Then

$$y' = 2e^{2x}(c_1 \cos x + c_2 \sin x) + e^{2x}(-c_1 \sin x + c_2 \cos x),$$

so the initial conditions give

$$y(0) = c_1 = 1 \quad \text{and} \quad y'(0) = 2c_1 + c_2 = 5.$$

It follows that $c_2 = 3$, and so the desired particular solution is

$$y(x) = e^{2x}(\cos x + 3 \sin x).$$

EXAMPLE 7 Find a general solution of $y^{(4)} + k^4 y = 0$ $(k > 0)$.

SOLUTION The characteristic equation is

$$r^4 + k^4 = (r^2 + k^2 i)(r^2 - k^2 i) = 0,$$

and its four roots are $\pm k \sqrt{\pm i}$. Now $i = e^{i\pi/2}$ and $-i = e^{3i\pi/2}$, so

$$\sqrt{i} = (e^{i\pi/2})^{1/2} = e^{i\pi/4} = \frac{1 + i}{\sqrt{2}}$$

and

$$\sqrt{-i} = (e^{3i\pi/2})^{1/2} = e^{3i\pi/4} = \frac{-1 + i}{\sqrt{2}}.$$

Thus the four (distinct) roots of the characteristic equation are $r = \pm(1 \pm i)a$ where $a = k/\sqrt{2}$. These two pairs of complex conjugate roots, $a \pm ai$ and $-a \pm ai$, give a general solution

$$y(x) = e^{ax}(c_1 \cos ax + c_2 \sin ax) + e^{-ax}(c_3 \cos ax + c_4 \sin ax)$$

of $y^{(4)} + k^4 y = 0$, where $a = k/\sqrt{2}$.

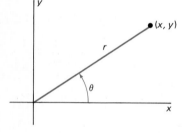

FIGURE 2.4 Modulus and argument of the complex number $x + iy$.

In Example 7 we employed the **polar form**

$$x + iy = re^{i\theta} \tag{22}$$

of a complex number. The relation between the real and imaginary parts x, y and the **modulus** r and **argument** θ is indicated in Fig. 2.4. One consequence of (22) is that the nonzero complex number $x + iy$ has the two square roots

$$\pm(x + iy)^{1/2} = \pm(re^{i\theta})^{1/2} = \pm r^{1/2} e^{i\theta/2}. \tag{23}$$

REPEATED COMPLEX ROOTS

Theorem 2 holds for repeated complex roots. If the conjugate pair $a \pm bi$ has multiplicity k, then the corresponding part of the general solution is of the form

$$(A_1 + A_2 x + \cdots + A_k x^{k-1})e^{(a+bi)x} + (B_1 + B_2 x + \cdots + B_k x^{k-1})e^{(a-bi)x}$$

$$= \sum_{p=0}^{k-1} x^p e^{ax}(c_i \cos bx + d_i \sin bx). \tag{24}$$

It can be shown that the $2k$ functions

$$x^p e^{ax} \cos bx, \qquad x^p e^{ax} \sin bx, \qquad 0 \le p \le k - 1$$

that appear in (24) are linearly independent.

EXAMPLE 8 Find a general solution of $(D^2 + 2D + 4)^2 y = 0$.

SOLUTION The characteristic equation $(r^2 + 2r + 4)^2 = 0$ has as its roots the conjugate pair $-1 \pm i\sqrt{3}$ of multiplicity 2. Hence (24) gives a general solution

$$y(x) = e^{-x}(c_1 \cos x\sqrt{3} + d_1 \sin x\sqrt{3})$$
$$+ xe^{-x}(c_2 \cos x\sqrt{3} + d_2 \sin x\sqrt{3}).$$

In applications we are seldom presented in advance with a factorization as convenient as the one in Example 8. Often the most difficult part of solving a homogeneous linear equation is finding the roots of its characteristic equation. Sometimes we can find one root r_1, either by guessing (if it is an integer, as is the root $r_1 = 1$ of the equation $r^3 + 4r - 5 = 0$), or by a numerical method as in the following example. We can then carry out a division to obtain a partial factorization.

EXAMPLE 9 Find a general solution of $y^{(3)} + y' + y = 0$.

SOLUTION The characteristic equation is $p(r) = r^3 + r + 1 = 0$. Because $p(-1) = -1$ and $p(0) = +1$, there is a root between -1 and 0. We begin with the initial estimate $r_0 = -0.5$ and use the iteration

$$r_{n+1} = r_n - \frac{p(r_n)}{p'(r_n)} = r_n - \frac{r_n^3 + r_n + 1}{3r_n^2 + 1}$$

of Newton's method. This technique yields the approximate root -0.68233. The simple BASIC program shown in Fig. 2.5 may be used to find this root to the degree of accuracy determined by the stopping criterion in line 250; if your computer is capable of handling numbers with more significant digits, then you may, of course, obtain greater accuracy by choosing E closer to zero.

```
100 REM--Program NEWTON
110 REM--Applies Newton's method to solve the
120 REM--equation f(x) = 0.   Lines 220 and 230
130 REM--must be edited to read FCTN = f(x) and
140 REM--DERIV = f'(x) respectively.
150 REM
160       INPUT "INITIAL GUESS"; X
170       N = 1  :   E = 0.000001
180 REM
190 REM--Newton's iteration:
200 REM
210       PRINT N, X
220       FCTN = X*X*X + X + 1
230       DERIV = 3*X*X + 1
240       XNEW = X - FCTN/DERIV
250       IF ABS(XNEW - X) < E*ABS(X) THEN
              PRINT "SOLUTION = ";XNEW : STOP
260       X = XNEW : N = N + 1
270       GOTO 210
280 REM
300       END                        FIGURE 2.5
```

The long division

$$(r + 0.68233)\overline{)r^3 + r + 1}$$

then produces as its quotient the approximate quadratic factor

$$r^2 - (0.68233)r + 1.46557.$$

With the aid of the quadratic formula, we find that the zeros of the latter are approximately $0.34116 \pm (1.16154)i$. These roots give an approximate general solution

$$y(x) = c_1 e^{-(0.68233)x} + e^{(0.34116)x}[c_2 \cos (1.16154)x + c_3 \sin (1.16154)x].$$

2.3 PROBLEMS

Find a general solution of each of the differential equations given in Problems 1–20.

1. $y'' - 4y = 0$

2. $2y'' - 3y' = 0$

3. $y'' + 3y' - 10y = 0$

4. $2y'' - 7y' + 3y = 0$

5. $y'' + 6y' + 9y = 0$

6. $y'' + 5y' + 5y = 0$

7. $4y'' - 12y' + 9y = 0$

8. $y'' - 6y' + 13y = 0$

9. $y'' + 8y' + 25y = 0$

10. $5y^{(4)} + 3y^{(3)} = 0$

11. $y^{(4)} - 8y^{(3)} + 16y'' = 0$

12. $y^{(4)} - 3y^{(3)} + 3y'' - y' = 0$

13. $9y^{(3)} + 12y'' + 4y' = 0$

14. $y^{(4)} + 3y'' - 4y = 0$

15. $y^{(4)} - 8y'' + 16y = 0$

16. $y^{(4)} + 18y'' + 81y = 0$

17. $6y^{(4)} + 11y'' + 4y = 0$

18. $y^{(4)} = 16y$

19. $y^{(3)} + y'' - y' - y = 0$

20. $y^{(4)} + 2y^{(3)} + 3y'' + 2y' + y = 0$ (*Suggestion:* Compute $(r^2 + r + 1)^2$.)

Solve each of the initial value problems given in Problems 21–26.

21. $y'' - 4y' + 3y = 0$; $y(0) = 7$, $y'(0) = 11$

22. $9y'' + 6y' + 4y = 0$; $y(0) = 3$, $y'(0) = 4$

23. $y'' - 6y' + 25y = 0$; $y(0) = 3$, $y'(0) = 1$

24. $2y^{(3)} - 3y'' - 2y' = 0$; $y(0) = 1$, $y'(0) = -1$, $y''(0) = 3$

25. $3y^{(3)} + 2y'' = 0$; $y(0) = -1$, $y'(0) = 0$, $y''(0) = 1$

26. $y^{(3)} + 10y'' + 25y' = 0$; $y(0) = 3$, $y'(0) = 4$, $y''(0) = 5$

Find a general solution of each of the equations in Problems 27–30. First find a small integral root of the characteristic equation by inspection; then factor by division.

27. $y^{(3)} + 3y'' - 4y = 0$

28. $2y^{(3)} - y'' - 5y' - 2y = 0$

29. $y^{(3)} + 27y = 0$

30. $y^{(4)} - y^{(3)} + y'' - 3y' - 6y = 0$

Find general solutions of the equations in Problems 31–33. First find numerically (as in Example 7) a root of the characteristic equation.

31. $y^{(3)} - 3y'' + y = 0$

32. $y^{(3)} + 3y' + 5y = 0$

33. $y^{(4)} + 4y' - 7y = 0$

In each of Problems 34–36, one solution of the differential equation is given. Find the general solution.

34. $3y^{(3)} - 2y'' + 12y' - 8y = 0$; $y = e^{2x/3}$

35. $6y^{(4)} + 5y^{(3)} + 25y'' + 20y' + 4y = 0$; $y = \cos 2x$

36. $9y^{(3)} + 11y'' + 4y' - 14y = 0$; $y = e^{-x} \sin x$

Problems 37–41 pertain to the solution of differential equations with complex coefficients.

37. (a) Use Euler's formula to show that every complex number can be written in the form $re^{i\theta}$, where $r \geq 0$ and $-\pi < \theta \leq \pi$. (b) Express the numbers 4, -2, $3i$, $1 + i$, and $-1 + i\sqrt{3}$ in the form $re^{i\theta}$. (c) The two square roots of $re^{i\theta}$ are $\pm\sqrt{r}e^{i\theta/2}$. Find the square roots of the numbers $2 - 2i\sqrt{3}$ and $-2 + 2i\sqrt{3}$.

38. Use the quadratic formula to solve the following equations. Note in each case that the roots are not complex conjugates. (a) $x^2 + ix + 2 = 0$. (b) $x^2 - 2ix + 3 = 0$.

39. Find a general solution of $y'' - 2iy' + 3y = 0$.

40. Find a general solution of $y'' - iy' + 6y = 0$.

41. Find a general solution of $y'' = (-2 + 2i\sqrt{3})y$.

42. Solve the initial value problem

$$y^{(3)} = y; \qquad y(0) = 1, \quad y'(0) = y''(0) = 0.$$

(*Suggestion:* Impose the given initial conditions on the general solution

$$y(x) = Ae^x + Be^{\alpha x} + Ce^{\beta x},$$

where α and β are the complex conjugate roots of $r^3 - 1 = 0$, to derive the solution

$$y(x) = \frac{1}{3}\left(e^x + 2e^{-x/2}\cos\frac{\sqrt{3}}{2}x.\right)$$

43. Solve the initial value problem

$$y^{(4)} = y^{(3)} + y'' + y' + 2y;$$

$$y(0) = y'(0) = y''(0) = 0, \qquad y^{(3)}(0) = 30.$$

44. The differential equation

$$y'' + (\text{sgn } x)y = 0 \qquad (25)$$

has the discontinuous coefficient function

$$\text{sgn } x = \begin{cases} +1 & \text{if } x > 0, \\ -1 & \text{if } x < 0. \end{cases}$$

Show that Eq. (25) nevertheless has two linearly independent

solutions $y_1(x)$ and $y_2(x)$ defined for all x such that:

- Each satisfies Eq. (25) at each point $x \neq 0$.
- Each has a continuous derivative at $x = 0$.
- $y_1(0) = y_2'(0) = 1$ and $y_2(0) = y_1'(0) = 0$.

(*Suggestion:* Each $y_i(x)$ will be defined by one formula for $x \leq 0$ and by another for $x \geq 0$.)

2.4

Mechanical Vibrations

The motion of a mass attached to a spring serves as a relatively simple exemplar of the vibrations that occur in more complex mechanical systems. For many such systems the analysis of these vibrations is a problem in the solution of linear differential equations with constant coefficients.

   We consider a body of mass m attached to one end of an ordinary spring that resists compression as well as stretching; the other end of the spring is attached to a fixed wall, as shown in Fig. 2.6. Assume that the body rests on a frictionless horizontal plane, so that it can move only back and forth as the spring stretches and compresses. Denote by x the distance of the body from its **equilibrium position**—its position when the spring is unstretched. We take $x > 0$ when the spring is stretched and $x < 0$ when it is compressed.

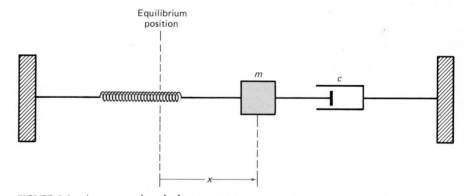

FIGURE 2.6   A mass-spring-dashpot system.

   According to Hooke's law, the restorative force F_S that the spring exerts on the mass is proportional to the distance that the spring has been stretched or compressed. Because this is the same as the displacement of the mass m from its equilibrium position, it follows that

$$F_S = -kx. \qquad (1)$$

The positive constant of proportionality k is called the **spring constant**. Note that F_S and x have opposite signs: $F_S < 0$ when $x > 0$, $F_S > 0$ when $x < 0$.

Figure 2.6 also shows the mass attached to a dashpot—a device, like a shock absorber, that provides a force directed opposite to the instantaneous direction of motion of the mass m. We assume the dashpot is designed so that this force F_R is proportional to the velocity $v = dx/dt$ of the mass, so that

$$F_R = -cv = -c\frac{dx}{dt}.\qquad(2)$$

The positive constant c is the **damping constant** of the dashpot. More generally, we may regard (2) as specifying frictional forces in our system (including air resistance to the motion of m).

If, in addition to the forces F_S and F_R, the mass is subjected to a given **external force** $F_E = F(t)$, then the total force acting on the mass is $F = F_S + F_R + F_E$. Using Newton's law

$$F = ma = m\frac{d^2x}{dt^2} = mx'',$$

we obtain the second order linear differential equation

$$mx'' + cx' + kx = F(t)\qquad(3)$$

that governs the motion of the mass.

If there is no dashpot (and we ignore all frictional forces), then we set $c = 0$ in (3) and call the motion **undamped**; it is **damped** motion if $c > 0$. If there is no external force we replace $F(t)$ by 0 in (3). We refer to the motion as **free** in this case and **forced** in the case $F(t) \neq 0$. Thus the homogeneous equation

$$mx'' + cx' + kx = 0\qquad(4)$$

describes free motion of a mass on a spring with dashpot but with no external forces applied. We will defer discussion of forced motion until Section 2.8.

For an alternative example, we might attach the mass to the lower end of a spring that is suspended vertically from a fixed support, as in Fig. 2.7. In this case the weight $W = mg$ of the mass would stretch the spring a distance s_0 determined by Eq. (1) with $F_S = -W$ and $x = s_0$. That is, $mg = ks_0$, so that $s_0 = mg/k$. This

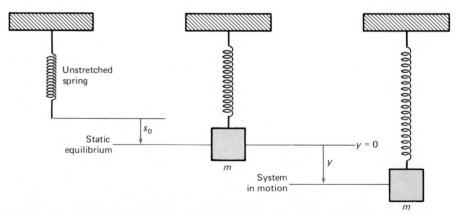

FIGURE 2.7 A mass suspended vertically from a spring.

CHAPTER 2: Linear Equations of Higher Order

gives the **static** equilibrium position of the mass. If y denotes the displacement of the mass in motion, measured downward from its static equilibrium position, then we ask you to show in Problem 9 that y satisfies Eq. (3); specifically, that

$$my'' + cy' + ky = F(t) \tag{5}$$

if we include damping and external forces.

The importance of the differential equation that appears in Eqs. (3) and (5) stems from the fact that it describes the motion of many other simple mechanical systems. For example, a **simple pendulum** consists of a mass m swinging back and forth in a vertical plane on the end of a string (or, better, a *massless rod*) of length L, as shown in Fig. 2.8. We may specify the position of the mass at time t by giving the counterclockwise angle $\theta = \theta(t)$ that the string or rod makes with the vertical at time t. To analyze the motion of the mass m, we will apply the law of the conservation of mechanical energy, according to which the sum of the kinetic energy and the potential energy of m remains constant.

The distance along the circular arc from 0 to m is $s = L\theta$, so the velocity of the mass is $v = ds/dt = L(d\theta/dt)$, and therefore its kinetic energy is

$$T = \frac{1}{2}mv^2 = \frac{1}{2}m\left(\frac{ds}{dt}\right)^2 = \frac{1}{2}mL^2\left(\frac{d\theta}{dt}\right)^2.$$

We next choose as reference point the lowest point O reached by the mass (see Fig. 2.8). Then its potential energy V is the product of its weight mg and its vertical height $h = L(1 - \cos\theta)$ above O, so that

$$V = mgL(1 - \cos\theta).$$

The fact that the sum of T and V is a constant C therefore gives

$$\frac{1}{2}mL^2\left(\frac{d\theta}{dt}\right)^2 + mgL(1 - \cos\theta) = C.$$

We differentiate each side of this identity with respect to t to obtain

$$mL^2\left(\frac{d\theta}{dt}\right)\left(\frac{d^2\theta}{dt^2}\right) + mgL(\sin\theta)\frac{d\theta}{dt} = 0,$$

and so

$$\frac{d^2\theta}{dt^2} + \frac{g}{L}\sin\theta = 0 \tag{6}$$

after removal of the common factor $mL^2(d\theta/dt)$.

Now recall that $\sin\theta \approx \theta$ when θ is small; in particular, θ and $\sin\theta$ agree to two decimal places when θ is at most $\pi/12$ (that is, $15°$). In a typical pendulum clock, for example, θ would certainly never exceed $15°$. It therefore seems reasonable to simplify our mathematical model of the simple pendulum by replacing $\sin\theta$ with θ in Eq. (6). If we also insert a term $c\theta'$ to account for the frictional resistance of the surrounding medium, the result is an equation of the form of Eq. (4):

$$\theta'' + c\theta' + k\theta = 0 \tag{7}$$

where $k = g/L$. Note that this equation is independent of the mass m on the end of the rod. We might, however, expect the effects of the discrepancy between θ and

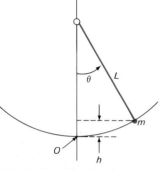

FIGURE 2.8 The simple pendulum.

sin θ to accumulate over a period of time, so that Eq. (7) might not describe accurately the actual motion of the pendulum over a long period of time.

In the remainder of this section, we first analyze free undamped motion and then free damped motion.

FREE UNDAMPED MOTION

If we have only a mass on a spring, with neither damping nor external force, then Eq. (3) takes the simpler form

$$mx'' + kx = 0. \tag{8}$$

It is convenient to define

$$\omega_0 = \sqrt{\frac{k}{m}}, \tag{9}$$

and rewrite Eq. (8) as

$$x'' + \omega_0^2 x = 0. \tag{8'}$$

The general solution of Eq. (8') is

$$x(t) = A \cos \omega_0 t + B \sin \omega_0 t. \tag{10}$$

To analyze the motion described by this solution, we choose constants C and α so that

$$C = (A^2 + B^2)^{1/2}, \quad \cos \alpha = \frac{A}{C}, \quad \text{and} \quad \sin \alpha = \frac{B}{C}, \tag{11}$$

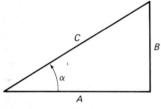

FIGURE 2.9 The angle α.

as indicated in Fig. 2.9. Note that, while tan $\alpha = B/A$, the angle α is *not* given by the principal branch of the inverse tangent function (which gives values only in the interval $(-\pi/2, \pi/2)$). Instead, α is the angle between 0 and 2π whose cosine and sine have the signs given in (11).

In any event, from (10) and (11) we get

$$x(t) = C\left(\frac{A}{C} \cos \omega_0 t + \frac{B}{C} \sin \omega_0 t\right)$$

$$= C(\cos \alpha \cos \omega_0 t + \sin \alpha \sin \omega_0 t);$$

with the aid of the cosine addition formula, we find that

$$x(t) = C \cos (\omega_0 t - \alpha). \tag{12}$$

Thus the mass oscillates to-and-fro about its equilibrium position with

Amplitude	C,
Circular frequency	ω_0,

and

Phase angle	α.

Such motion is called **simple harmonic motion**. A typical graph of $x(t)$ is shown in Fig. 2.10. If time t is measured in seconds, the circular frequency ω_0 has dimensions

CHAPTER 2: Linear Equations of Higher Order

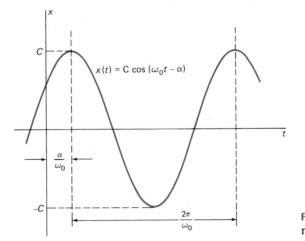

$x(t) = C \cos(\omega_0 t - \alpha)$

FIGURE 2.10   Simple harmonic motion.

of radians per second (rad/s). The **period** of the motion is the time required for the system to complete one full oscillation, so is given by

$$T = \frac{2\pi}{\omega_0} \tag{13}$$

seconds; its **frequency** is

$$\frac{1}{T} = \frac{\omega_0}{2\pi} \tag{14}$$

in hertz (Hz), which measures the number of complete cycles per second. Note that frequency is measured in cycles per second, while circular frequency has the dimensions of radians per second.

If the initial position $x(0) = x_0$ and initial velocity $x'(0) = v_0$ of the mass are given, we first determine the values of the coefficients A and B in (10), and then find the amplitude C and phase angle α by carrying out the transformation of $x(t)$ to the form in (12), as indicated above.

EXAMPLE 1   A body that weighs 16 lb is attached to the end of a spring which is stretched 2 ft by a force of 100 lb. It is set in motion with initial position $x_0 = 0.5$ (ft) and initial velocity $v_0 = -10$ (ft/s). (Note that these data indicate that the body is displaced to the right and moving to the left at time $t = 0$.) Find the position function of the body, as well as the amplitude, frequency, period of oscillation, and phase angle of its motion.

SOLUTION   We take $g = 32$ ft/s^2. The mass of the body is then $m = W/g = 0.5$ (slugs). The spring constant is $k = 100/2 = 50$ (lb/ft), so Eq. (8) yields $\frac{1}{2}x'' + 50x = 0$; that is, $x'' + 100x = 0$. Consequently, the circular frequency will be $\omega_0 = 10$ (rad/s). So the body will oscillate with

$$\text{Frequency:} \quad \frac{10}{2\pi} \approx 1.59 \text{ Hz}$$

and

$$\text{period: } \frac{2\pi}{10} \approx 0.63 \text{ s.}$$

We now impose the initial conditions $x(0) = 0.5$ and $x'(0) = -10$ in the general solution $x = A \cos 10t + B \sin 10t$, and it follows that $A = 0.5$ and $B = -1$. So the position function of the body is

$$x(t) = \tfrac{1}{2} \cos 10t - \sin 10t.$$

Hence its amplitude of motion is

$$C = [(\tfrac{1}{2})^2 + (1)^2]^{1/2} = \tfrac{1}{2}\sqrt{5} \approx 1.12 \text{ (ft).}$$

To find the phase angle we write

$$x = \frac{\sqrt{5}}{2}\left(\frac{1}{\sqrt{5}} \cos 10t - \frac{2}{\sqrt{5}} \sin 10t\right) = \frac{\sqrt{5}}{2} \cos (10t - \alpha).$$

Thus we require $\cos \alpha = 1/\sqrt{5} > 0$ and $\sin \alpha = -2\sqrt{5} < 0$. Hence α is the fourth-quadrant angle

$$\alpha = 2\pi - \tan^{-1}\left(\frac{2/\sqrt{5}}{1/\sqrt{5}}\right) \approx 5.1760 \text{ (rad).}$$

In the form in which the amplitude and phase angle are made explicit, the position function is

$$x(t) \approx \frac{\sqrt{5}}{2} \cos (10t - 5.1760).$$

FREE DAMPED MOTION

With damping but no external force, the differential equation we have been studying takes the form $mx'' + cx' + kx = 0$; alternatively,

$$x'' + 2px' + \omega_0^2 x = 0, \tag{15}$$

where $\omega_0 = \sqrt{k/m}$ is the corresponding *undamped* circular frequency and

$$p = \frac{c}{2m} > 0. \tag{16}$$

The characteristic equation $r^2 + 2pr + \omega_0^2 = 0$ of (15) has roots

$$r_1, r_2 = -p \pm (p^2 - \omega_0^2)^{1/2} \tag{17}$$

that depend upon the sign of

$$p^2 - \omega_0^2 = \frac{c^2}{4m^2} - \frac{k}{m} = \frac{c^2 - 4km}{4m^2}.$$

The **critical damping** c_{CR} is given by $c_{CR} = \sqrt{4km}$, and we distinguish three cases, according as $c > c_{CR}$, $c = c_{CR}$, or $c < c_{CR}$.

Overdamped Case: $c > c_{CR}$ $(c^2 > 4km)$

Because c is relatively large in this case, we are dealing with a strong resistance in comparison with a relatively weak spring or a small mass. Then (17) gives distinct real roots r_1 and r_2, both of which are negative. The position function has the form

$$x(t) = c_1 e^{r_1 t} + c_2 e^{r_2 t}. \tag{18}$$

It is easy to see that $x(t) \to 0$ as $t \to +\infty$ and that the body settles to its equilibrium position without any oscillations (Problem 27). Figure 2.11 shows some typical graphs of the position function for the overdamped case; we chose x_0 a fixed positive number and illustrated the effects of changing the initial velocity v_0. In every case the would-be oscillations are damped out.

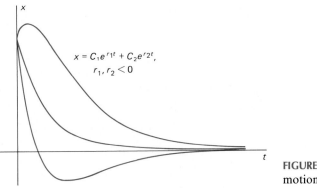

$x = C_1 e^{r_1 t} + C_2 e^{r_2 t},$
$r_1, r_2 < 0$

FIGURE 2.11 Overdamped motion.

Critically Damped Case: $c = c_{CR}$ $(c^2 = 4km)$

In this case (17) gives equal roots $r_1 = r_2 = -p$ of the characteristic equation, so the general solution is

$$x(t) = e^{-pt}(c_1 + c_2 t). \tag{19}$$

Because $e^{-pt} > 0$ and $c_1 + c_2 t$ has at most one positive zero, the body passes through its equilibrium position at most once, and it is clear that $x(t) \to 0$ as $t \to +\infty$. Some graphs of the motion in the critically damped case appear in Fig. 2.12, and they resemble those of the overdamped case (Fig. 2.11). In the critically damped case, the resistance of the dashpot is just large enough to damp

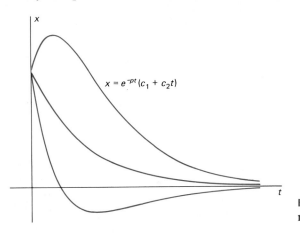

$x = e^{-pt}(c_1 + c_2 t)$

FIGURE 2.12 Critically damped motion.

out any oscillations, but even a slight decrease in resistance will bring us to the remaining case; the one that shows the most dramatic behavior.

Underdamped Case: $c < c_{CR}$ $(c^2 < 4km)$

The characteristic equation now has the two complex conjugate roots $-p \pm i(\omega_0^2 - p^2)^{1/2}$, and the general solution is

$$x(t) = e^{-pt}(c_1 \cos \omega_1 t + c_2 \sin \omega_1 t), \tag{20}$$

where

$$\omega_1 = (\omega_0^2 - p^2)^{1/2} = \frac{\sqrt{4km - c^2}}{2m}. \tag{21}$$

Using the cosine addition formula as in the derivation of Eq. (12), we may rewrite (20) as

$$x(t) = Ce^{-pt} \cos(\omega_1 t - \alpha), \tag{22}$$

where $C = (c_1^2 + c_2^2)^{1/2}$ and $\tan \alpha = c_2/c_1$.

The solution in (22) represents exponentially damped oscillations of the body about its equilibrium position. The graph of $x(t)$ lies between the curves $x = Ce^{-pt}$ and $x = -Ce^{-pt}$ and touches them when $\omega_1 t - \alpha$ is an integral multiple of π. The motion is not actually periodic, but it nevertheless is useful to call ω_1 its **circular frequency**, $T_1 = 2\pi/\omega_1$ its **pseudoperiod** of oscillation, and Ce^{-pt} its **time-varying amplitude**. Most of these quantities are shown in the typical graph of underdamped motion shown in Fig. 2.13. Note from (21) that in this case ω_1 is less than the undamped circular frequency ω_0, so T_1 is larger than the period T of oscillation of the same mass without damping on the same spring. Thus the damping of the dashpot has at least three effects:

1. It exponentially damps the oscillations, in accord with the time-varying amplitude.

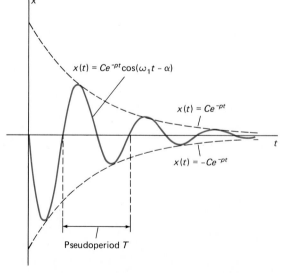

FIGURE 2.13 Underdamped motion.

CHAPTER 2: Linear Equations of Higher Order

2. It slows the motion; that is, the dashpot decreases the frequency of the motion.

3. It delays the motion—this is the effect of the phase angle in Eq. (22).

EXAMPLE 2   The mass and spring of Example 1 are now attached also to a dashpot that provides 6 lb of resistance for each foot per second of velocity. The mass is set in motion with the same initial position $x(0) = 0.5$ (ft) and the same initial velocity $x'(0) = -10$ (ft/s). Find the position function of the mass, its new frequency and pseudoperiod, its phase angle, and the amplitudes of its first four (local) maxima and minima.

SOLUTION   Rather than memorizing the various formulas given above, it is far better practice in a particular case to set up the differential equation and then solve it directly. Recall that $m = 1/2$ and $k = 50$; we are now given $c = 6$ in fps units. Hence Eq. (3) is $\frac{1}{2}x'' + 6x' + 50x = 0$; that is,

$$x'' + 12x' + 100x = 0.$$

The roots of the characteristic equation $r^2 + 12r + 100 = 0$ are

$$\frac{-12 \pm \sqrt{144 - 400}}{2} = -6 \pm 8i,$$

so the general solution is

$$x(t) = e^{-6t}(A \cos 8t + B \sin 8t). \tag{23}$$

The new circular frequency is $\omega_1 = 8$ (rad/s) and the pseudoperiod and new frequency are

$$T_1 = \frac{2\pi}{8} \approx 0.79 \text{ (s)}$$

and

$$\frac{1}{T_1} = \frac{8}{2\pi} \approx 1.27 \text{ (Hz)}$$

(at opposed to 0.63 s and 1.59 Hz, respectively, in the undamped case).
From (23) we compute

$$x'(t) = e^{-6t}(-8A \sin 8t + 8B \cos 8t) - 6e^{-6t}(A \cos 8t + B \sin 8t).$$

Our initial conditions therefore give the equations

$$x(0) = A = \tfrac{1}{2}$$

and

$$x'(0) = -6A + 8B = -10,$$

so $A = \tfrac{1}{2}$ and $B = -\tfrac{7}{8}$. Thus

$$x(t) = e^{-6t}(\tfrac{1}{2} \cos 8t - \tfrac{7}{8} \sin 8t),$$

so with $C = [(\frac{1}{2})^2 + (\frac{7}{8})^2]^{1/2} = \sqrt{65}/8$ we have

$$x(t) = \frac{\sqrt{65}}{8} e^{-6t} \left(\frac{4}{\sqrt{65}} \cos 8t - \frac{7}{\sqrt{65}} \sin 8t \right).$$

We require $\cos \alpha = 4/\sqrt{65} > 0$ and $\sin \alpha = -7/\sqrt{65} < 0$, so α is the fourth-quadrant angle

$$\alpha = 2\pi - \tan^{-1}(\tfrac{7}{4}) \approx 5.2315 \text{ (rad.)}$$

Finally,

$$x(t) \approx \frac{\sqrt{65}}{8} e^{-6t} \cos(8t - 5.2315). \tag{24}$$

The local maxima and minima of $x(t)$ occur when

$$0 = x'(t) \approx \frac{\sqrt{65}}{8} [-6e^{-6t} \cos(8t - 5.2315) - 8e^{-6t} \sin(8t - 5.2315)],$$

and thus when

$$\tan(8t - 5.2315) \approx -0.75.$$

Because $\tan^{-1}(0.75) \approx -0.6435$, we want the first four positive values of t such that $8t - 5.2315$ is equal to -0.6435 plus an integral multiple of π. These values of t and the corresponding values of x computed using (24) are as follows:

t (s)	0.1808	0.5735	0.9662	1.3589
x (ft)	−0.2725	0.0258	−0.0024	0.0002

We see that the oscillations are damped out very rapidly, with their amplitude decreasing by a factor of about 10 every half-cycle. See Problems 30 and 31 for a more general discussion of this phenomenon.

2.4 PROBLEMS

1. Determine the period and frequency of the simple harmonic motion of a 4-kg mass on the end of a spring with spring constant 16 N/m.

2. Determine the period and frequency of the simple harmonic motion of a body weighing 24 lb on the end of a spring with spring constant 48 lb/ft.

3. A mass of 3 kg is attached to the end of a spring that is stretched 20 cm by a force of 15 N. It is set in motion with initial position $x_0 = 0$ and initial velocity $v_0 = -10$ m/s. Find the amplitude, period, and frequency of the resulting motion.

4. A body weighing 8 lb is attached to the end of a spring that is stretched 1 in. by a force of 3 lb. At time $t = 0$ the body is pulled 1 ft to the right (spring stretched) and set in motion with an initial velocity of 5 ft/s to the left. (a) Find $x(t)$ in the form $C \cos(\omega_0 t + \alpha)$. (b) Find the amplitude and period of the motion of the body.

In Problems 5–8, assume that the differential equation of a simple pendulum of length L is $L\theta'' + g\theta = 0$, where $g = GM/R^2$ is the gravitational acceleration at the location of the pendulum (at distance R from the center of the earth; M denotes the mass of the earth).

5. Two pendulums are of lengths L_1 and L_2, and—when located at the respective distances R_1 and R_2 from the center of the earth—have periods p_1 and p_2. Show that

$$\frac{p_1}{p_2} = \frac{R_1\sqrt{L_1}}{R_2\sqrt{L_2}}.$$

6. A certain pendulum clock keeps perfect time in Paris, where the radius of the earth is $R = 3956$ (mi). But this clock loses 2 min 40 s per day at a location on the equator. Use the result of Problem 5 to find the amount of the equatorial bulge of the earth.

7. A pendulum of length 100.10 in., located at a point at sea level where the radius of the earth is $R = 3960$ (mi), has the same period as does a pendulum of length 100.00 in. atop a nearby mountain. Use the result of Problem 5 to find the height of the mountain.

8. Most grandfather clocks have pendulums with lengths that are adjustable. One such clock loses 10 min per day when the length of the pendulum is 30 in. With what length pendulum will this clock keep perfect time?

9. Derive Eq. (5) describing the motion of a mass attached to the bottom of a vertically suspended spring. (*Suggestion:* First denote by $x(t)$ the displacement of the mass below the unstretched position of the spring; set up the differential equation for x. Then substitute $y = x - s_0$ in this differential equation.)

10. Consider a floating cylindrical buoy with radius r, height h, and density $\rho \leq 0.5$ (recall that the density of water is 1 g/cm³). The buoy is initially suspended at rest with its bottom at the top surface of the water and is released at time $t = 0$. Thereafter it is acted upon by two forces: a downward gravitational force equal to its weight $mg = \rho\pi r^2 hg$ and an upward force of buoyancy equal to the weight $\pi r^2 xg$ of water displaced, where $x = x(t)$ is the depth of the bottom of the buoy beneath the surface at time t. Conclude that the buoy undergoes simple harmonic motion about the equilibrium position $x_e = \rho h$ with period $p = 2\pi\sqrt{\rho h/g}$. Compute p and the amplitude of the motion if $\rho = 0.5$ g/cm³, $h = 200$ cm, and $g = 980$ cm/s².

11. A cylindrical buoy weighing 100 lb floats in water with its axis vertical (as in Problem 10). When depressed slightly and released, it oscillates up and down four times every 10 s. Assume that friction is negligible. Find the radius of the buoy.

12. Assume that the earth is a solid sphere of uniform density, with mass M and radius $R = 3960$ (mi). For a particle of mass m *within* the earth at distance r from the center of the earth, the gravitational force attracting m toward the center is $F_r = -GM_r m/r^2$, where M_r is the mass of the part of the earth within a sphere of radius r. (a) Show that $F_r = -GMmr/R^3$. (b) Now suppose that a hole is drilled straight through the center of the earth, thus connecting two antipodal points on its surface. Let a particle of mass m be dropped at time $t = 0$ into this hole with initial speed zero, and let $r(t)$ be its distance from the center of the earth at time t. Conclude from Newton's second law and part (a) that $r''(t) = -k^2 r(t)$, where $k^2 = GM/R^3 = g/R$. (c) Take $g = 32.2$ ft/s², and conclude from part (b) that the particle undergoes simple harmonic motion back and forth between the ends of the hole, with a period of about 84 min. (d) Look up (or derive) the period of a satellite that just skims the surface of the earth; compare the value with the results in part (c). How do you explain the coincidence? Or *is* it a coincidence? (e) With what speed (in miles per hour) does the particle pass through the center of the earth? (f) Look up or derive the orbital velocity of a satellite that just skims the surface of the earth; compare the value with the result in part (e). How do you explain the coincidence? Or *is* it a coincidence?

The remaining problems in this section deal with free damped motion. In Problems 13–19, a mass m is attached to both a spring (with given spring constant k) and a dashpot (with given damping constant c). The mass is set in motion with initial position x_0 and initial velocity v_0. Find the position function $x(t)$ and determine whether the motion is overdamped, critically damped, or underdamped. If it is underdamped, write $x(t)$ in the form $Ce^{-pt}\cos(\omega_1 t - \alpha)$.

13. $m = \frac{1}{2}$, $c = 3$, $k = 4$; $x_0 = 2$, $v_0 = 0$

14. $m = 3$, $c = 30$, $k = 63$; $x_0 = 2$, $v_0 = 2$

15. $m = 1$, $c = 8$, $k = 16$; $x_0 = 5$, $v_0 = -10$

16. $m = 2$, $c = 12$, $k = 50$; $x_0 = 0$, $v_0 = -8$

17. $m = 4$, $c = 20$, $k = 169$; $x_0 = 4$, $v_0 = 16$

18. $m = 2$, $c = 16$, $k = 40$; $x_0 = 5$, $v_0 = 4$

19. $m = 1$, $c = 10$, $k = 125$; $x_0 = 6$, $v_0 = 50$

20. A 12-lb weight is attached both to a vertically suspended spring that it stretches 6 in. and to a dashpot that provides 3 lb of resistance for every foot per second of velocity. (a) If the weight is pulled down 1 ft below its static equilibrium position and then released from rest at time $t = 0$, find its position function $x(t)$. (b) Find the frequency, time-varying amplitude, and phase angle of the motion.

21. This problem deals with a highly simplified model of a car of weight 3200 lb. Assume that the suspension system of the car acts like a single spring and its shock absorbers like a single dashpot, so that its vertical vibrations satisfy Eq. (4) with appropriate values of the coefficients. (a) Find the stiffness coefficient k of the spring if the car undergoes free vibrations at 80 cycles per minute (cycles/min) when its shock absorbers are disconnected. (b) With the shock absorbers connected the car is set into vibration by driving it over a bump, and the resulting damped vibrations have a frequency of 78 cycles/min. After how long will the time-varying amplitude be 1% of its original value?

Each of Problems 22–32 deals with a mass-spring-dashpot system having position function $x(t)$ satisfying Eq. (4). We write $x_0 = x(0)$ and $v_0 = x'(0)$ and recall that $p = c/2m$, $\omega_0^2 = k/m$, and $\omega_1^2 = \omega_0^2 - p^2$. The system is critically damped, overdamped, or underdamped, as specified in each problem.

22. (Critically damped)   Show in this case that

$$x(t) = (x_0 + v_0 t + p x_0 t) e^{-pt}.$$

23. (Critically damped)   Deduce from Problem 22 that the mass passes through $x = 0$ at some instant $t > 0$ if and only if x_0 and $v_0 + p x_0$ have opposite signs.

24. (Critically damped)   Deduce from Problem 22 that $x(t)$ has a local maximum or minimum at some instant $t > 0$ if and only if v_0 and $v_0 + p x_0$ have the same sign.

25. (Overdamped)   Show in this case that

$$x(t) = \frac{1}{2\gamma} \left[(v_0 - r_2 x_0) \exp (r_1 t) - (v_0 - r_1 x_0) \exp (r_2 t) \right],$$

where $r_1, r_2 = -p \pm (p^2 - \omega_0^2)^{1/2}$ and $\gamma = (r_1 - r_2)/2 > 0$.

26. (Overdamped)   If $x_0 = 0$ deduce from Problem 25 that

$$x(t) = \frac{v_0}{\gamma} e^{-pt} \sinh \gamma t.$$

27. (Overdamped)   Prove that in this case the mass can pass through its equilibrium position $x = 0$ at most once.

28. (Underdamped)   Show that in this case

$$x(t) = e^{-pt} \left[x_0 \cos \omega_1 t + \frac{v_0 + p x_0}{\omega_1} \sin \omega_1 t \right].$$

29. (Underdamped)   If the damping constant c is small in comparison with $\sqrt{8mk}$, apply the binomial series to show that

$$\omega_1 \approx \omega_0 \left(1 - \frac{c^2}{8mk} \right).$$

30. (Underdamped)   Show that the local maxima and minima of $x(t) = Ce^{-pt} \cos (\omega_1 t - \alpha)$ occur where

$$\tan (\omega_1 t - \alpha) = \frac{-p}{\omega_1}.$$

Conclude that $t_2 - t_1 = 2\pi/\omega_1$ if two consecutive maxima occur at times t_1 and t_2.

31. (Underdamped)   Let x_1 and x_2 be two consecutive local maximum values of $x(t)$. Deduce from the result of Problem 30 that

$$\ln \frac{x_1}{x_2} = \frac{2\pi p}{\omega_1}.$$

The constant $\Delta = 2\pi p/\omega_1$ is called the **logarithmic decrement** of the oscillation. Note also that $c = m\omega_1 \, \Delta/\pi$ because $p = c/2m$.

Note: The result of Problem 31 provides an accurate method for measuring the *viscosity* of a fluid, which is an important parameter in fluid dynamics, but which is not easily measured directly. According to Stokes' drag law, a spherical body of radius a moving at a (relatively slow) speed v through a fluid of viscosity μ experiences a resistive force $F_R = 6\pi\mu av$. Thus if a spherical mass on a spring is immersed in the fluid and set in motion, this drag resistance damps its oscillations with damping constant $c = 6\pi a\mu$. The frequency ω_1 and logarithmic decrement Δ of the oscillations can be measured by direct observation. The final formula in Problem 31 then gives c and hence the viscosity of the fluid.

32. (Underdamped)   A body weighing 100 lb is oscillating attached to a spring and a dashpot. Its first two maximum displacements of 6.73 in. and 1.46 in. are observed to occur at times 0.34 s and 1.17 s, respectively. Compute the damping constant (in pound-seconds per foot) and spring constant (in pounds per foot).

DIFFERENTIAL EQUATIONS AND DETERMINISM
Given a mass m, a dashpot constant c, and a spring constant k, Theorem 2 of Section 2.1 implies that the equation

$$mx'' + cx' + kx = 0 \tag{25}$$

has a *unique* solution for $t \geq 0$ satisfying given initial conditions $x(0) = x_0$, $x'(0) = v_0$. Thus the future motion of an ideal mass-spring-dashpot system is completely determined by the differential equation and the initial conditions. Of course in a real physical system it is impossible to measure the parameters m, c, and k *precisely*. Problems 33–36 explore the resulting uncertainty in predicting the future behavior of a physical system.

33. Suppose that $m = 1$, $c = 2$, and $k = 1$ in Eq. (25). Show that the solution with $x(0) = 0$ and $x'(0) = 1$ is

$$x_1(t) = te^{-t}.$$

34. Suppose that $m = 1$ and $c = 2$ but $k = 1 - 10^{-2n}$. Show that the solution of (25) with $x(0) = 0$ and $x'(0) = 1$ is

$$x_2(t) = 10^n e^{-t} \sinh 10^{-n} t.$$

35. Suppose that $m = 1$ and $c = 2$ but $k = 1 + 10^{-2n}$. Show that the solution of (25) with $x(0) = 0$ and $x'(0) = 1$ is

$$x_3(t) = 10^n e^{-t} \sin 10^{-n} t.$$

36. Whereas the graphs of $x_1(t)$ and $x_2(t)$ resemble those shown in Figs. 2.11 and 2.12, the graph of $x_3(t)$ exhibits damped oscillations like those illustrated in Fig. 2.13, but with a very long period. Nevertheless, show that for each fixed t it is true that

$$\lim_{n \to \infty} x_2(t) = \lim_{n \to \infty} x_3(t) = x_1(t).$$

Conclude that *on a given finite time interval* the three solutions are in "practical" agreement if n is sufficiently large.

CHAPTER 2:   Linear Equations of Higher Order

Nonhomogeneous Equations and the Method of Undetermined Coefficients

We learned in Section 2.3 how to solve homogeneous linear equations with constant coefficients, but we saw in Section 2.4 that an external force in a simple mechanical system contributes a nonhomogeneous term to its differential equation. The general nonhomogeneous nth order linear equation with constant coefficients has the form

$$a_n y^{(n)} + a_{n-1} y^{(n-1)} + \cdots + a_1 y' + a_0 y = f(x). \tag{1}$$

By Theorem 5 of Section 2.2, a general solution of (1) is of the form

$$y = y_c + y_p \tag{2}$$

where the complementary function y_c is a general solution of the associated homogeneous equation

$$a_n y^{(n)} + a_{n-1} y^{(n-1)} + \cdots + a_1 y' + a_0 y = 0, \tag{3}$$

and y_p is a single particular solution of (1). Thus our remaining task is to find y_p.

The **method of undetermined coefficients** is a straightforward way of doing this when the given function $f(x)$ in (1) is sufficiently simple that we can make an intelligent guess as to the general form of y_p. For example, suppose that $f(x)$ is a polynomial of degree m. Then, because the derivatives of a polynomial are themselves polynomials of lower degree, it is reasonable to suspect a particular solution

$$y_p = A_m x^m + A_{m-1} x^{m-1} + \cdots + A_1 x + A_0$$

that is also a polynomial of degree m, but with as yet undetermined coefficients. We may, therefore, substitute this form of y_p into (1), and then—by equating coefficients of like powers of x on the two sides of the resulting equation—attempt to determine the coefficients $A_0, A_1, \ldots, A_m$ so that y_p will, indeed, be a particular solution.

Similarly, suppose that

$$f(x) = a \cos kx + b \sin kx.$$

Then it is reasonable to suspect a particular solution of the same form:

$$y_p = A \cos kx + B \sin kx,$$

a linear combination with undetermined coefficients A and B. The reason is that any derivative of such a linear combination of $\cos kx$ and $\sin kx$ has the same form. We may therefore substitute this form of y_p in (1), and then—by equating coefficients of $\cos kx$ and $\sin kx$ on the two sides of the resulting equation—attempt to determine the coefficients A and B so that y_p will, indeed, be a particular solution.

It turns out that this approach does succeed whenever all the derivatives of $f(x)$ have the same form as $f(x)$ itself. Before describing the method in full generality, we illustrate it with several preliminary examples.

EXAMPLE 1   Find a particular solution of $y'' + 3y' + 4y = 3x + 2$.

SOLUTION   Here $f(x) = 3x + 2$, a polynomial of degree 1, so our guess is

$$y_p = Ax + B.$$

Then $y_p' = A$ and $y_p'' = 0$, so y_p will satisfy the differential equation provided that

$$(0) + 3(A) + 4(Ax + B) = 3x + 2.$$

This is so if and only if $4A = 3$ and $3A + 4B = 2$. These two equations yield $A = \frac{3}{4}$ and $B = -\frac{1}{16}$, so we have found the particular solution

$$y_p(x) = \tfrac{3}{4}x - \tfrac{1}{16}.$$

EXAMPLE 2   Find a particular solution of $y'' - 4y = 2e^{3x}$.

SOLUTION   Any derivative of e^{3x} is a constant multiple of e^{3x}, so it is reasonable to try

$$y_p = Ae^{3x}.$$

Then $y_p'' = 9Ae^{3x}$, so the differential equation will be satisfied provided that

$$9Ae^{3x} - 4(Ae^{3x}) = 2e^{3x};$$

that is, $5A = 2$, so that $A = \frac{2}{5}$. Thus our particular solution is $y_p(x) = \frac{2}{5}e^{3x}$.

EXAMPLE 3   Find a particular solution of $3y'' + y' - 2y = 2\cos x$.

SOLUTION   Our first guess might be $A \cos x$, but the presence of y' on the left-hand side signals that we may need a term involving $\sin x$ as well. So we try

$$y_p = A \cos x + B \sin x,$$
$$y_p' = -A \sin x + B \cos x,$$
$$y_p'' = -A \cos x - B \sin x.$$

Then substitution of y_p into the differential equations yields

$$3(-A \cos x - B \sin x) + (-A \sin x + B \cos x) - 2(A \cos x + B \sin x) = 2 \cos x.$$

We equate the coefficient of $\cos x$ on the left-hand side with that of $\cos x$ on the right and do the same with $\sin x$ to get the equations

$$-5A + B = 2,$$
$$-A - 5B = 0,$$

with solution $A = -\frac{5}{13}$, $B = \frac{1}{13}$. Hence our particular solution is

$$y_p(x) = -\tfrac{5}{13} \cos x + \tfrac{1}{13} \sin x.$$

CHAPTER 2:   Linear Equations of Higher Order

The following example, which superficially resembles Example 2, indicates that the method of undetermined coefficients is not always quite so simple as it appears.

EXAMPLE 4 Find a particular solution of $y'' - 4y = 2e^{2x}$.

SOLUTION If we try $y_p = Ae^{2x}$, we find that

$$y_p'' - 4y_p = 4Ae^{2x} - 4Ae^{2x} = 0 \neq 2e^{2x};$$

so, no matter what A is, Ae^{2x} cannot satisfy the given nonhomogeneous equation. In fact, the above computation shows that Ae^{2x} satisfies instead the associated *homogeneous* equation. Therefore, we should begin with a trial function whose derivative involves both e^{2x} *and something else* that can cancel upon substitution into the differential equation to leave the e^{2x} term that we need. A reasonable guess is

$$y_p = Axe^{2x},$$

with

$$y_p' = Ae^{2x} + 2Axe^{2x} \quad \text{and} \quad y_p'' = 4Ae^{2x} + 4Axe^{2x}.$$

Substitution yields

$$(4Ae^{2x} + 4Axe^{2x}) - 4(Axe^{2x}) = 2e^{2x}.$$

The terms involving xe^{2x} obligingly cancel, leaving only $4Ae^{2x} = 2e^{2x}$, so that $A = \frac{1}{2}$. Thus our particular solution is $y_p(x) = \frac{1}{2}xe^{2x}$.

Our initial difficulty in Example 4 resulted from the fact that $f(x) = 2e^{2x}$ satisfies the associated homogeneous equation. Rule 1, given shortly, tells what to do when we do not have this difficulty, and Rule 2 tells what to do when we do have it.

The method of undetermined coefficients applies whenever the function $f(x)$ in Eq. (1) is a linear combination of (finite) products of functions of the following three types:

1. A polynomial in x;
2. An exponential function e^{rx};   (4)
3. $\cos kx$ or $\sin kx$.

Any such function, for example

$$f(x) = (3 - 4x^2)e^{5x} - 4x^3 \cos 10x,$$

has the crucial property that only *finitely* many linearly independent functions appear as terms (summands) in $f(x)$ and its derivatives of all orders. In Rules 1 and 2 below we assume that $Ly = f(x)$ is a nonhomogeneous linear equation with constant coefficients and that $f(x)$ is a function of this kind.

> **Rule 1** *Method of Undetermined Coefficients*
>
> Suppose that no term appearing either in $f(x)$ or in any of its derivatives satisfies the associated homogeneous equation $Ly = 0$. Then take as a trial solution for y_p a linear combination of all linearly independent such terms. Determine the coefficients by substitution of this trial solution into the non-homogeneous equation.

Note that this rule is not a theorem requiring proof; it is merely a procedure to be followed in searching for a particular solution y_p. If we succeed in finding y_p, then nothing more need be said. It can be proved, however, that this procedure will always succeed under the conditions specified.

In practice we check the supposition made in Rule 1 by first using the characteristic equation to find the complementary function y_c, and then write a list of all the terms appearing in $f(x)$ and its successive derivatives. If none of the terms in this list duplicates a term in y_c, then we're in business!

EXAMPLE 5 Find a particular solution of

$$y'' + 4y = 3x^3. \tag{5}$$

SOLUTION The (familiar) complementary function of this equation is

$$y_c = c_1 \cos 2x + c_2 \sin 2x.$$

The function $f(x) = 3x^3$ and its derivatives are constant multiples of the linearly independent functions x^3, x^2, x, and 1. Because none of these appears in y_c, we try

$$y_p = Ax^3 + Bx^2 + Cx + D,$$

$$y_p' = 3Ax^2 + 2Bx + C,$$

$$y_p'' = 6Ax + 2B.$$

Substitution in (5) gives

$$\begin{aligned}
y_p'' + 4y_p &= (6Ax + 2B) + 4(Ax^3 + Bx^2 + Cx + D) \\
&= 4Ax^3 + 4Bx^2 + (6A + 4C)x + (2B + D) \\
&= 3x^3.
\end{aligned}$$

We equate coefficients of like powers of x to get the equations

$$4A = 3, \qquad\qquad 4B = 0,$$

$$6A + 4C = 0, \qquad 2B + D = 0$$

with solution $A = \frac{3}{4}$, $B = 0$, $C = -\frac{9}{8}$, and $D = 0$. Hence a particular solution of (5) is

$$y_p(x) = \tfrac{3}{4}x^3 - \tfrac{9}{8}x.$$

EXAMPLE 6 Solve the initial value problem

$$y'' - 3y' + 2y = 3e^{-x} - 10 \cos 3x;$$

$$y(0) = 1, \qquad y'(0) = 2.$$

(6)

SOLUTION The characteristic equation $r^2 - 3r + 2 = 0$ has roots $r = 1$ and $r = 2$, so the complementary function is

$$y_c = c_1 e^x + c_2 e^{2x}.$$

The terms involved in $f(x) = 3e^{-x} - 10 \cos 3x$ and its derivatives are e^{-x}, $\cos 3x$, and $\sin 3x$. Because none of these appears in y_c, we try

$$y_p = Ae^{-x} + B \cos 3x + C \sin 3x,$$

$$y_p' = -Ae^{-x} - 3B \sin 3x + 3C \cos 3x,$$

$$y_p'' = Ae^{-x} - 9B \cos 3x - 9C \sin 3x.$$

After we substitute these expressions into (6) and collect coefficients, we get

$$y_p'' - 3y_p' + 2y_p = 6Ae^{-x} + (-7B - 9C) \cos 3x + (9B - 7C) \sin 3x$$
$$= 3e^{-x} - 10 \cos 3x.$$

We equate the coefficients of the terms involving e^{-x}, $\cos 3x$, and $\sin 3x$ to obtain

$$6A = 3,$$

$$-7B - 9C = -10,$$

$$9B - 7C = 0$$

with solution $A = \frac{1}{2}$, $B = \frac{7}{13}$, and $C = \frac{9}{13}$. This gives the particular solution

$$y_p(x) = \frac{1}{2}e^{-x} + \frac{7}{13} \cos 3x + \frac{9}{13} \sin 3x$$

which, however, does not have the required initial values.

To satisfy the initial conditions, we begin with the *general* solution

$$y = y_c + y_p$$
$$= c_1 e^x + c_2 e^{2x} + \frac{1}{2}e^{-x} + \frac{7}{13} \cos 3x + \frac{9}{13} \sin 3x$$

with derivative

$$y' = c_1 e^x + 2c_2 e^{2x} - \frac{1}{2}e^{-x} - \frac{21}{13} \sin 3x + \frac{27}{13} \cos 3x.$$

The initial conditions lead to the equations

$$y(0) = c_1 + c_2 + \frac{1}{2} + \frac{7}{13} = 1,$$

$$y'(0) = c_1 + 2c_2 - \frac{1}{2} + \frac{27}{13} = 2$$

with solution $c_1 = -\frac{1}{2}$, $c_2 = \frac{6}{13}$. The desired particular solution is therefore

$$y(x) = -\frac{1}{2}e^x + \frac{6}{13}e^{2x} + \frac{1}{2}e^{-x} + \frac{7}{13} \cos 3x + \frac{9}{13} \sin 3x.$$

EXAMPLE 7 Find the general form of a particular solution of

$$y^{(3)} + 9y' = x \sin x + x^2 e^{2x}.$$

(7)

SOLUTION The characteristic equation $r^3 + 9r = 0$ has roots $r = 0$, $-3i$, and $3i$. So the complementary function is

$$y_c = c_1 + c_2 \cos 3x + c_3 \sin 3x.$$

The derivatives of the right-hand side in (7) involve the terms

$$\cos x, \quad \sin x, \quad x \cos x, \quad x \sin x, \quad e^{2x}, \quad xe^{2x}, \quad \text{and} \quad x^2 e^{2x}.$$

Because there is no duplication with the terms of the complementary function, the trial function takes the form

$$y_p = A \cos x + B \sin x + Cx \cos x + Dx \sin x + Ee^{2x} + Fxe^{2x} + Gx^2 e^{2x}.$$

Upon substituting y_p in (7) and equating coefficients, we would get seven equations determining the seven coefficients A, B, C, D, E, F, and G.

Now we turn our attention to the situation in which Rule 1 does not apply: Some of the terms involved in $f(x)$ and its derivatives satisfy the associated homogeneous equation. For instance, suppose that we want to find a particular solution of the differential equation

$$(D - r)^3 y = (2x - 3)e^{rx}. \tag{8}$$

Proceeding as in Rule 1, our first guess would be

$$y_p = Ae^{rx} + Bxe^{rx}. \tag{9}$$

This form of y_p will not be adequate because the complementary function of (8) is

$$y_c = c_1 e^{rx} + c_2 xe^{rx} + c_3 x^2 e^{rx}, \tag{10}$$

so substitution of (9) in the left-hand side of (8) would yield zero rather than $(2x - 3)e^{rx}$.

To see how to amend our first guess, we observe that

$$(D - r)^2[(2x - 3)e^{rx}] = [D^2(2x - 3)]e^{rx} = 0$$

by Eq. (13) of Section 2.3. If y is *any* solution of Eq. (8) and we apply the operator $(D - r)^2$ to both sides, we therefore see that y is also a solution of the equation $(D - r)^5 y = 0$. The general solution of this *homogeneous* equation can be written as

$$y = \underbrace{c_1 e^{rx} + c_2 xe^{rx} + c_3 x^2 e^{rx}}_{y_c} + \underbrace{Ax^3 e^{rx} + Bx^4 e^{rx}}_{y_p}.$$

Thus *every* solution of our original equation in (8) is the sum of a complementary function and a *particular solution* of the form

$$y_p = Ax^3 e^{rx} + Bx^4 e^{rx}. \tag{11}$$

Note that (11) may be obtained by multiplying each term of our first guess in (9) by the least power of x (in this case, x^3) that suffices to eliminate duplication between the terms of the resulting trial solution y_p and the complementary function y_c given in (10). This procedure succeeds in the general case.

To simplify the general statement, we observe that, in order to find a particular solution of the nonhomogeneous linear differential equation

$$Ly = f_1(x) + f_2(x), \tag{12}$$

CHAPTER 2: Linear Equations of Higher Order

it suffices to find *separately* particular solutions Y_1 and Y_2 of the two equations

$$Ly = f_1(x) \quad \text{and} \quad Ly = f_2(x), \tag{13}$$

respectively. For linearity then gives

$$L[Y_1 + Y_2] = LY_1 + LY_2 = f_1(x) + f_2(x),$$

and therefore $y_p = Y_1 + Y_2$ is a particular solution of (2). This is the **principle of superposition** for nonhomogeneous linear equations.

Now our problem is to find a particular solution of the equation $Ly = f(x)$, where $f(x)$ is a linear combination of products of the elementary functions listed in (4). Thus $f(x)$ can be written as a sum of terms each of the form

$$P_m(x)e^{rx}\{\cos kx \quad \text{or} \quad \sin kx\} \tag{14}$$

where $P_m(x)$ is a polynomial in x. Note that any derivative of such a term is of the same form, but with *both* sines and cosines appearing. The procedure by which we arrived above at the particular solution in (11) of Eq. (8) can be generalized to show that the following procedure is always successful.

Rule 2 *Method of Undetermined Coefficients*

If the function $f(x)$ is of the form in (14), take as the trial solution

$$y_p = x^s[(A_0 + A_1x + \cdots + A_mx^m)e^{rx}\cos kx$$
$$+ (B_0 + B_1x + \cdots + B_mx^m)e^{rx}\sin kx], \tag{15}$$

where s is the smallest nonnegative integer such that no term in y_p duplicates a term in the complementary function y_c. Then determine the coefficients in (15) by substituting y_p into the nonhomogeneous equation.

In practice we seldom need to deal with a function $f(x)$ exhibiting the full generality in (14). The table in Fig. 2.14 lists the form of y_p in various common cases, corresponding to the possibilities $m = 0$, $r = 0$, and $k = 0$.

$f(x)$	y_p
$P_m(x) = b_0 + b_1x + b_2x^2 + \ldots + b_mx^m$	$x^s(A_0 + A_1x + A_2x^2 + \ldots + A_mx^m)$
$a\cos kx + b\sin kx$	$x^s(A\cos kx + B\sin kx)$
$e^{rx}(a\cos kx + b\sin kx)$	$x^s e^{rx}(A\cos kx + B\sin kx)$
$P_m(x)e^{rx}$	$x^s(A_0 + A_1x + A_2x^2 + \ldots + A_mx^m)e^{rx}$
$P_m(x)(a\cos kx + b\sin kx)$	$x^s[(A_0 + A_1x + \ldots + A_mx^m)\cos kx$ $+ (B_0 + B_1x + \ldots + B_mx^m)\sin kx]$

FIGURE 2.14 Substitutions in the method of undetermined coefficients.

On the other hand, it is not uncommon to have

$$f(x) = f_1(x) + f_2(x)$$

where $f_1(x)$ and $f_2(x)$ are different functions of the sort listed in the table of Fig. 2.14. In this event we take as y_p the sum of the indicated particular functions, choosing s *separately* for each part to eliminate duplication with the complementary function. This procedure is illustrated in the following examples.

EXAMPLE 8 Find a particular solution of

$$y^{(3)} + y'' = 3e^x + 4x^2. \qquad (16)$$

SOLUTION The characteristic equation $r^3 + r^2 = 0$ has roots $r_1 = r_2 = 0$ and $r_3 = -1$, so the complementary function is

$$y_c = c_1 + c_2x + c_3e^{-x}.$$

As a first step toward our particular solution, we form the sum

$$(Ae^x) + (B + Cx + Dx^2).$$

The part Ae^x corresponding to $3e^x$ does not duplicate any part of the complementary function, but the part $B + Cx + Dx^2$ must be multiplied by x^2 to eliminate duplication. Hence we take

$$y_p = Ae^x + Bx^2 + Cx^3 + Dx^4,$$

$$y_p' = Ae^x + 2Bx + 3Cx^2 + 4Dx^3,$$

$$y_p'' = Ae^x + 2B + 6Cx + 12Dx^2,$$

and

$$y_p^{(3)} = Ae^x + 6C + 24Dx.$$

Substitution of these derivatives in (16) yields

$$2Ae^x + (2B + 6C) + (6C + 24D)x + 12Dx^2 = 3e^x + 4x^2.$$

The system of equations

$$2A = 3, \qquad 2B + 6C = 0,$$

$$6C + 24D = 0, \qquad 12D = 4$$

has the solution $A = \frac{3}{2}$, $B = 4$, $C = -\frac{4}{3}$, and $D = \frac{1}{3}$. Hence the desired particular solution is

$$y_p(x) = \frac{3}{2}e^x + 4x^2 - \frac{4}{3}x^3 + \frac{1}{3}x^4.$$

EXAMPLE 9 Determine the appropriate form for a particular solution of $y'' + 6y' + 13y = e^{-3x} \cos 2x$.

SOLUTION The characteristic equation $r^2 + 6r + 13 = 0$ has roots $-3 \pm 2i$, so the complementary function is

$$y_c = e^{-3x}(c_1 \cos 2x + c_2 \sin 2x).$$

This is the same form as a first attempt $e^{-3x}(A \cos 2x + B \sin 2x)$ at a particular solution, so we must multiply by x to eliminate duplication. Hence we would take

$$y_p = e^{-3x}(Ax \cos 2x + Bx \sin 2x).$$

EXAMPLE 10 Determine the appropriate form for a particular solution of the fifth order equation

$$(D - 2)^3(D^2 + 9)y = x^2 e^{2x} + x \sin 3x.$$

SOLUTION The characteristic equation $(r - 2)^3(r^2 + 9) = 0$ has roots $r = 2$, 2, 2, 3i, and $-3i$ so the complementary function is

$$y_c = c_1 e^{2x} + c_2 x e^{2x} + c_3 x^2 e^{2x} + c_4 \cos 3x + c_5 \sin 3x.$$

As a first step toward the form of a particular solution, we examine the sum

$$[(A + Bx + Cx^2)e^{2x}] + [(D + Ex) \cos 3x + (F + Gx) \sin 3x].$$

To eliminate duplication with terms of y_c, the first part—corresponding to $x^2 e^{2x}$—must be multiplied by x^3, and the second part—corresponding to $x \sin 3x$—must be multiplied by x. Hence we would take

$$y_p = (Ax^3 + Bx^4 + Cx^5)e^{2x} + (Dx + Ex^2) \cos 3x + (Fx + Gx^2) \sin 3x.$$

Finally, let us point out the kind of situation in which the method of undetermined coefficients cannot be used. Consider, for instance, the equation

$$y'' + y = \tan x, \tag{17}$$

which at first glance may appear similar to those considered in the examples above. Not so; the function $f(x) = \tan x$ has *infinitely many* linearly independent derivatives

$$\sec^2 x, \ 2 \sec^2 x \tan x, \ 4 \sec^2 x \tan^2 x + 2 \sec^4 x, \dots.$$

Therefore we do not have available a *finite* linear combination to use as a trial solution. In Section 2.7 (Variation of Parameters) we will discuss an alternative method that can be applied to nonhomogeneous equations such as (17).

2.5 PROBLEMS

In each of Problems 1–20, find a particular solution y_p of the given equation.

1. $y'' + 16y = e^{3x}$

2. $y'' - y' - 2y = 3x + 4$

3. $y'' - y' - 6y = 2 \sin 3x$

4. $4y'' + 4y' + y = 3xe^x$

5. $y'' + y' + y = \sin^2 x$

6. $2y'' + 4y' + 7y = x^2$

7. $y'' - 4y = \sinh x$

8. $y'' - 4y = \cosh 2x$

9. $y'' + 2y' - 3y = 1 + xe^x$

10. $y'' + 9y = 2 \cos 3x + 3 \sin 3x$

11. $y^{(3)} + 4y' = 3x - 1$

12. $y^{(3)} + y' = 2 - \sin x$

13. $y'' + 2y' + 5y = e^x \sin x$

14. $y^{(4)} - 2y'' + y = xe^x$

15. $y^{(5)} + 5y^{(4)} - y = 17$

16. $y'' + 9y = 2x^2 e^{3x} + 5$

17. $y'' + y = \sin x + x \cos x$

18. $y^{(4)} - 5y'' + 4y = e^x - xe^{2x}$

19. $y^{(5)} + 2y^{(3)} + 2y'' = 3x^2 - 1$

20. $y^{(3)} - y = e^x + 7$

In each of Problems 21–30, set up the appropriate form of a particular solution y_p, but do not determine the values of the coefficients.

21. $y'' - 2y' + 2y = e^x \sin x$

22. $y^{(5)} - y^{(3)} = e^x + 2x^2 - 5$

23. $y'' + 4y = 3x \cos 2x$

24. $y^{(3)} - y'' - 12y' = x - 2xe^{-3x}$

25. $y'' + 3y' + 2y = x(e^{-x} - e^{-2x})$

26. $y'' - 6y' + 13y = xe^{3x} \sin 2x$

27. $y^{(4)} + 5y'' + 4y = \sin x + \cos 2x$

28. $y^{(4)} + 9y'' = (x^2 + 1) \sin 3x$

29. $(D - 1)^3(D^2 - 4)y = xe^x + e^{2x} + e^{-2x}$

30. $y^{(4)} - 2y'' + y = x^2 \cos x$

Solve the initial value problems in Problems 31–40.

31. $y'' + 4y = 2x;$ $y(0) = 1,$ $y'(0) = 2$

32. $y'' + 3y' + 2y = e^x;$ $y(0) = 0,$ $y'(0) = 3$

33. $y'' + 9y = \sin 2x;$ $y(0) = 1,$ $y'(0) = 0$

34. $y'' + y = \cos x;$ $y(0) = 1,$ $y'(0) = -1$

35. $y'' - 2y' + 2y = x + 1;$ $y(0) = 3,$ $y'(0) = 0$

36. $y^{(4)} - 4y'' = x^2;$ $y(0) = y'(0) = 1,$ $y''(0) = y^{(3)}(0) = -1$

37. $y^{(3)} - 2y'' + y' = 1 + xe^x;$ $y(0) = y'(0) = 0,$ $y''(0) = 1$

38. $y'' + 2y' + 2y = \sin 3x;$ $y(0) = 2,$ $y'(0) = 0$

39. $y^{(3)} + y'' = x + e^{-x};$ $y(0) = 1,$ $y'(0) = 0,$ $y''(0) = 1$

40. $y^{(4)} - y = 5;$ $y(0) = y'(0) = y''(0) = y^{(3)}(0) = 0$

41. Find a particular solution of the equation
$$y^{(4)} - y^{(3)} - y'' - y' - 2y = 8x^5.$$

42. Find the solution of the initial value problem consisting of the differential equation of Problem 41 and the initial conditions
$$y(0) = y'(0) = y''(0) = y^{(3)}(0) = 0.$$

43. (a) Write
$$\cos 3x + i \sin 3x = e^{3ix} = (\cos x + i \sin x)^3$$
by Euler's formula, expand, and equate real and imaginary parts to derive the identities
$$\cos^3 x = \tfrac{1}{4} \cos 3x + \tfrac{3}{4} \cos x,$$
$$\sin^3 x = \tfrac{3}{4} \sin x - \tfrac{1}{4} \sin 3x.$$
(b) Use the result of part (a) to find a general solution of
$$y'' + 4y = \cos^3 x.$$

Use trigonometric identities to find general solutions of the equations in Problems 44–46.

44. $y'' + y' + y = \sin x \sin 3x$

45. $y'' + 9y = \sin^4 x$

46. $y'' + y = x \cos^3 x$

Let $Ly = f(x)$ be a nonhomogeneous linear differential equation with constant coefficients. If M is a linear differential operator with constant coefficients such that $M[f(x)] \equiv 0$, then M is called an **annihilator** of $f(x)$. In this case every solution of $Ly = f(x)$ satisfies the homogeneous equation $MLy = 0$. If the general solution of $MLy = 0$ is written in the form $y = y_c + y_p$, where y_c is a general solution of $Ly = 0$, then y_p will be an appropriate form of a particular solution of the original nonhomogeneous equation $Ly = f(x)$. This method was used in the text, with $L = (D - r)^3$ and $M = (D - r)^2$, to discover the particular solution (11) of Eq. (8). Use this annihilator method in each of Problems 47–52 to discover the form of a particular solution of the given equation.

47. $y'' - 5y' = x^2;$ take $M = D^3$.

48. $y'' - 5y' = e^{5x};$ take $M = D - 5$.

49. $y'' - 5y' = xe^{5x};$ take $M = (D - 5)^2$.

50. $y'' + y = \cos x;$ take $M = D^2 + 1$.

51. $y^{(4)} + 8y'' + 16y = \sin 2x;$ take $M = D^2 + 4$.

52. $(D^2 - 2D + 2)^2 y = e^x \cos x;$ take $M = D^2 - 2D + 2$.

2.6

Reduction of Order and Euler-Cauchy Equations

In Section 2.3 we discussed fully the solution of homogeneous linear differential equations with constant coefficients and mentioned that, unfortunately, there exists no similar procedure for finding routinely a general solution of a higher order linear equation with variable coefficients. Sometimes, however, we can find *one* solution of such an equation by inspection. For example, consider the equation
$$x^2 y'' - 5xy' + 9y = 0. \tag{1}$$

Because the exponent matches the order of differentiation in each term, it is plausible that there is a solution of the form $y_1 = x^r$. Upon substitution of y_1 in (1), we obtain

$$x^2 \cdot r(r-1)x^{r-2} - 5x \cdot rx^{r-1} + 9x^r = 0,$$

so $y_1 = x^r$ will indeed be a solution of (1) provided that

$$r(r-1) - 5r + 9 = 0;$$

that is, if $(r-3)^2 = 0$. Thus $y_1 = x^3$ is one solution of the equation in (1), but this approach does not yield the second solution we need to form a general solution.

We discuss here the method of **reduction of order**, which enables us to use one known solution y_1 of a second order homogeneous linear differential equation to find a second linearly independent solution y_2. Consider the second order equation

$$y'' + p(x)y' + q(x)y = 0 \tag{2}$$

on an open interval I on which p and q are continuous functions. Suppose that we know one solution y_1 of (2). By Theorem 2 of Section 2.1, there exists a second linearly independent solution y_2; our problem is to find y_2. Equivalently, we would like to find the quotient

$$v(x) = \frac{y_2(x)}{y_1(x)}. \tag{3}$$

Once we know $v(x)$, y_2 will be given by

$$y_2(x) = v(x)y_1(x). \tag{4}$$

We begin by substituting the expression in (4) in Eq. (2), using the derivatives

$$y_2' = vy_1' + v'y_1,$$
$$y_2'' = vy_1'' + 2v'y_1' + v''y_1. \tag{5}$$

We get

$$[vy_1'' + 2v'y_1' + v''y_1] + p[vy_1' + v'y_1] + qvy_1 = 0,$$

and rearrangement gives

$$v[y_1'' + py_1' + qy_1] + v''y_1 + 2v'y_1' + pv'y_1 = 0.$$

But the bracketed terms in this last equation vanish because y_1 is a solution of (2). This leaves the equation

$$y_1v'' + (2y_1' + py_1)v' = 0. \tag{6}$$

The key to the success of this method is that Eq. (6) is *linear* in v'. Thus the substitution in (4) has reduced the second order linear equation in (2) to the first order (in v') linear equation in (6). If we write $u = v'$ and assume that y_1 never vanishes on I, then (6) yields

$$u' + \left(2\frac{y_1'}{y_1} + p(x)\right)u = 0. \tag{7}$$

An integrating factor for (7) is

$$\rho = \exp\left(\int\left(2\frac{y_1'}{y_1} + p(x)\right)dx\right)$$

$$= \exp\left(2\ln|y_1| + \int p(x)\,dx\right);$$

thus

$$\rho = y_1^2 e^{\int p(x)\,dx}.$$

We now integrate the equation in (7) to obtain

$$uy_1^2 e^{\int p(x)\,dx} = C,$$

so

$$v' = u = \frac{C}{y_1^2}e^{-\int p(x)\,dx}.$$

Another integration now gives

$$\frac{y_2}{y_1} = v = C\int\frac{e^{-\int p(x)\,dx}}{y_1^2}\,dx + K.$$

With the particular choices $C = 1$ and $K = 0$, we get

$$y_2 = y_1\int\frac{e^{-\int p(x)\,dx}}{y_1^2}\,dx. \tag{8}$$

This formula provides a second solution $y_2(x)$ of Eq. (2) on any interval where y_1 is never zero. Note that, because an exponential never vanishes, y_2 is a nonconstant multiple of y_1, so y_1 and y_2 are linearly independent solutions. We have therefore proved the following theorem.

Theorem *Reduction of Order*

If $y_1(x)$ is a solution of Eq. (2),

$$y'' + p(x)y' + q(x)y = 0,$$

on an interval I where p and q are continuous and y_1 is nonzero, then a second linearly independent solution of (2) on I is given by

$$y_2 = y_1\int\frac{e^{-\int p(x)\,dx}}{y_1^2}\,dx. \tag{8}$$

It can be shown that if y_1 is a known solution of an nth order homogeneous linear equation, then the substitution $y_2 = vy_1$ reduces this nth order equation to a homogeneous linear equation of order $n - 1$ in $u = v'$; hence the terminology *reduction of order*. Unfortunately, the resulting equation of order

$n - 1$ will generally have variable coefficients, so if $n > 2$, it may be as difficult to solve as the original equation of order n. Consequently the method of reduction of order is useful primarily for solving second order equations.

EXAMPLE 1   We noted above that $y_1 = x^3$ is one solution of the equation

$$x^2 y'' - 5xy' + 9y = 0. \tag{1}$$

To apply the formula in (8), we first rewrite Eq. (1) in the standard form

$$y'' - \frac{5}{x} y' + \frac{9}{x^2} y = 0 \qquad (x > 0)$$

with leading coefficient 1. Then $p(x) = -5/x$ and $e^{-\int p(x)\,dx} = e^{5 \ln x} = x^5$, so the formula in (8) gives a second solution

$$y_2 = x^3 \int \frac{x^5}{(x^3)^2} \, dx = x^3 \ln x,$$

and a general solution of (1) for $x > 0$ is

$$y(x) = c_1 x^3 + c_2 x^3 \ln x.$$

Sometimes it is just as convenient to substitute $y_2 = vy_1$ directly into the differential equation as to apply the formula in (8). If we substitute

$$y_2 = x^3 v, \qquad y_2' = 3x^2 v + x^3 v', \qquad y_2'' = 6xv + 6x^2 v' + x^3 v''$$

in (1), we get

$$(6x^3 v + 6x^4 v' + x^5 v'') - 5(3x^3 v + x^4 v') + 9x^3 v = 0;$$

$$x^4 v' + x^5 v'' = 0;$$

$$D_x[xv'] = 0;$$

$$v' = \frac{C}{x};$$

$$v = C \ln x.$$

Thus we obtain the same independent solution $y_2 = x^3 v = x^3 \ln x$ (if we choose $C = 1$).

EXAMPLE 2   Given the solution $y_1 = x$, find a general solution of

$$(x^2 - 1)y'' - 2xy' + 2y = 0 \qquad (x^2 < 1).$$

SOLUTION   First divide each term by $x^2 - 1$ to obtain

$$y'' - \frac{2x}{x^2 - 1} y' + \frac{2}{x^2 - 1} y = 0$$

with $p(x) = -2x/(x^2 - 1)$. Hence

$$e^{-\int p(x)\,dx} = \exp\left(\int \frac{2x}{x^2 - 1} \, dx \right) = \exp\left(\ln |x^2 - 1| \right) = 1 - x^2,$$

because $x^2 < 1$. Therefore, the formula in (8) yields

$$y_2 = x \int \frac{1 - x^2}{x^2} \, dx$$

$$= x \int (x^{-2} - 1) \, dx = -1 - x^2.$$

Thus a general solution is

$$y(x) = c_1 x + c_2(1 + x^2).$$

It is interesting to observe that we needed the condition $x^2 < 1$ both for the preliminary division and for the first integration; we also needed the condition $x \neq 0$ to apply the formula in (8). Yet you may verify without difficulty that the general solution found here is valid for all x.

EULER-CAUCHY EQUATIONS

Equation (1) above is an example of an Euler-Cauchy equation. The general nth order **Euler-Cauchy equation** is the equation

$$a_n x^n y^{(n)} + a_{n-1} x^{n-1} y^{(n-1)} + \cdots + a_2 x^2 y'' + a_1 x y' + a_0 y = 0 \tag{9}$$

where $a_0, a_1, \ldots, a_n$ are constants with $a_n \neq 0$. It is sometimes called an **equidimensional** equation because the exponent of each coefficient matches the order of the derivative; this implies that the substitution $y = x^r$ will yield terms all of the same degree. If we make this substitution in (9) and then divide by x^r, we get the polynomial equation

$$a_n r(r-1) \cdots (r-n+1) + \cdots + a_2 r(r-1) + a_1 r + a_0 = 0. \tag{10}$$

If the nth degree equation in (10) has n distinct roots $r_1, r_2, \ldots, r_n$, then we get n linearly independent solutions and a general solution

$$y(x) = c_1 x^{r_1} + c_2 x^{r_2} + \cdots + c_n x^{r_n} \tag{11}$$

of Eq. (9) for $x > 0$. For $x < 0$, x^r is not defined for all r. This is not a difficulty; it is easy to verify that the form of Eq. (9) is unchanged by the substitution $t = -x$, so we get the solution for $x < 0$ by replacing x by $-x = |x|$ in the solution for $x > 0$. Hence if the roots of Eq. (10) are distinct, we have a general solution:

$$y(x) = c_1 |x|^{r_1} + c_2 |x|^{r_2} + \cdots + c_n |x|^{r_n} \tag{12}$$

for all x.

To discuss the second order Euler-Cauchy equation in detail, let us divide each term by the leading coefficient to obtain

$$x^2 y'' + p_0 x y' + q_0 y = 0. \tag{13}$$

Equation (10) now takes the form

$$r(r-1) + p_0 r + q_0 = 0 \tag{14}$$

with roots

$$r_1, r_2 = \frac{-(p_0 - 1) \pm \sqrt{(p_0 - 1)^2 - 4q_0}}{2}$$ (15)

The usual three cases occur.

Distinct Real Roots: $r_1 \neq r_2$
As indicated above, a general solution of (13) for $x > 0$ is

$$y(x) = c_1 x^{r_1} + c_2 x^{r_2}.$$ (16)

Equal Real Roots: $r_1 = r_2$
In this case we begin with the single solution

$$y_1 = x^{r_1} \quad \text{where} \quad r_1 = -\tfrac{1}{2}(p_0 - 1).$$

Because

$$\exp\left(-\int \frac{p_0}{x}\, dx\right) = e^{-p_0 \ln x} = x^{-p_0},$$

the formula in (8) gives the second solution

$$y_2 = x^{r_1} \int \frac{x^{-p_0}}{x^{-(p_0 - 1)}}\, dx$$

$$= x^{r_1} \int \frac{1}{x}\, dx = x^{r_1} \ln x,$$

so a general solution for $x > 0$ is

$$y(x) = c_1 x^{r_1} + c_2 x^{r_1} \ln x.$$ (17)

Complex Conjugate Roots: $r_1 = a + bi$, $r_2 = a - bi$
Then

$$x^{(a \pm bi)} = e^{(a \pm bi) \ln x} = e^{a \ln x} e^{\pm i(b \ln x)}$$
$$= x^a [\cos (b \ln x) \pm i \sin (b \ln x)].$$

The real and imaginary parts are linearly independent solutions, so for $x > 0$ a general solution of (13) in this case is

$$y(x) = c_1 x^a \cos (b \ln x) + c_2 x^a \sin (b \ln x).$$ (18)

EXAMPLE 3 Find a general solution of $2x^2 y'' + xy' - 15y = 0$.

SOLUTION Equation (10) is

$$2r(r - 1) + r - 15 = (r - 3)(2r + 5) = 0$$

with distinct roots $r_1 = 3$ and $r_2 = -\tfrac{5}{2}$, so a general solution is

$$y(x) = c_1 x^3 + \frac{c_2}{x^{5/2}}.$$

EXAMPLE 4   Find a general solution of $x^2y'' + 7xy' + 13y = 0$.

SOLUTION   Equation (10) is

$$r(r-1) + 7r + 13 = r^2 + 6r + 13 = 0$$

with complex conjugate roots $-3 \pm 2i$, so the general solution given in (18) is

$$y(x) = c_1 \frac{\cos{(2 \ln x)}}{x^3} + c_2 \frac{\sin{(2 \ln x)}}{x^3}.$$

Finally, suppose that $n > 2$ and that the roots of Eq. (10) are not distinct. In this event we can discover the missing solutions by transforming Eq. (9) with the aid of the substitution $x = e^t$. This will always produce a homogeneous linear equation with constant coefficients, which can then be solved using the methods of Section 2.3. The following example illustrates this technique.

EXAMPLE 5   Find a general solution of

$$x^3 \frac{d^3y}{dx^3} + 6x^2 \frac{d^2y}{dx^2} + 7x \frac{dy}{dx} + y = 0.$$

SOLUTION   We must preface the substitution $x = e^t$ by some computations. First, $dx/dt = e^t = x$. This implies that

$$\frac{dy}{dx} = \frac{dy}{dt}\frac{dt}{dx} = \frac{1}{e^t}\frac{dy}{dt} = \frac{1}{x}\frac{dy}{dt},$$

so

$$x \frac{dy}{dx} = \frac{dy}{dt}. \tag{19}$$

Next,

$$\frac{d^2y}{dx^2} = \frac{1}{x}\frac{d}{dx}\left(\frac{dy}{dt}\right) + \frac{dy}{dt}\frac{d}{dx}\left(\frac{1}{x}\right)$$

$$= \frac{1}{x}\frac{d^2y}{dt^2}\frac{dt}{dx} - \frac{1}{x^2}\frac{dy}{dt}$$

$$= \frac{1}{x^2}\left(\frac{d^2y}{dt^2} - \frac{dy}{dt}\right),$$

so

$$x^2 \frac{d^2y}{dx^2} = \frac{d^2y}{dt^2} - \frac{dy}{dt}. \tag{20}$$

Another similar computation gives

$$x^3 \frac{d^3y}{dx^3} = \frac{d^3y}{dt^3} - 3\frac{d^2y}{dt^2} + 2\frac{dy}{dt}. \tag{21}$$

Now substitution of (19), (20), and (21) in the given equation yields

$$\left(\frac{d^3y}{dt^3} - 3\frac{d^2y}{dt^2} + 2\frac{dy}{dt}\right) + 6\left(\frac{d^2y}{dt^2} - \frac{dy}{dt}\right) + 7\frac{dy}{dt} + y = 0;$$

that is,

$$\frac{d^3y}{dt^3} + 3\frac{d^2y}{dt^2} + 3\frac{dy}{dt} + y = 0.$$

The characteristic equation

$$r^3 + 3r^2 + 3r + 1 = (r+1)^3 = 0$$

has the root -1 of multiplicity 3. Because $t = \ln x$, a general solution is therefore

$$y(x) = c_1 e^{-t} + c_2 t e^{-t} + c_3 t^2 e^{-t}$$

$$= \frac{c_1}{x} + \frac{c_2 \ln x}{x} + \frac{c_3 (\ln x)^2}{x}.$$

As Example 5 may suggest, it can be shown in general that a root r of multiplicity k of Eq. (10) corresponds to the k linearly independent solutions

$$x^r (\ln x)^m, \qquad m = 0, 1, 2, \ldots, k-1.$$

REDUCIBLE SECOND ORDER EQUATIONS

The general form of a second order differential equation is

$$F(x, y, y', y'') = 0. \tag{22}$$

If either the dependent variable y or the independent variable x is missing from a second order equation, then it is easily reduced to a first order equation by a change of variables.

If the dependent variable y is missing, so that the equation has the form

$$F(x, y', y'') = 0, \tag{23}$$

then we make the substitutions

$$p = y' = \frac{dy}{dx}, \qquad y'' = \frac{dp}{dx}. \tag{24}$$

These transform (23) into the first order equation

$$F\left(x, p, \frac{dp}{dx}\right) = 0$$

for p as a function of x. If we can find its general solution $p = p(x, C_1)$, then we get the general solution of the original second order equation in (23) by integrating:

$$y(x) = \int y'(x)\, dx = \int p\, dx = \int p(x, C_1)\, dx + C_2.$$

Note that the solution above contains *two* arbitrary constants. This is to be expected—solving a second order equation would normally require two integrations.

EXAMPLE 6   Solve the equation $xy'' + 2y' = 6x$.

SOLUTION   The substitutions in (24) give

$$x\frac{dp}{dx} + 2p = 6x; \quad \text{that is,} \quad \frac{dp}{dx} + \frac{2}{x}p = 6.$$

Upon multiplying both sides of the linear equation (on the right) by its integrating factor x^2, we get

$$D_x(x^2 p) = 6x^2;$$

$$x^2 p = 2x^3 + C_1;$$

$$\frac{dy}{dx} = p = 2x + \frac{C_1}{x^2}.$$

Another integration produces the solution:

$$y(x) = x^2 - \frac{C_1}{x} + C_2.$$

If the independent variable x is missing, so that the equation has the form

$$F(y, y', y'') = 0, \tag{25}$$

then we make the substitutions

$$p = y'(x), \quad y'' = \frac{dp}{dx} = \frac{dp}{dy}\frac{dy}{dx} = p\frac{dp}{dy}. \tag{26}$$

These transform (25) into the first order equation

$$F\left(y, p, p\frac{dp}{dy}\right) = 0$$

for p as a function of y. If we can find its general solution $p = p(y, C_1)$, then we get the general solution of the original second order equation by integrating:

$$x(y) = \int \frac{dx}{dy}\, dy = \int \frac{1}{p}\, dy = \int \frac{dy}{p(y, C_1)} + C_2.$$

For the same reason as in the first case, the final solution contains two arbitrary constants of integration. Note also that the solution expresses x as a function of y. Thus we obtain an implicit solution of (25) that is valid where $p = y'(x) \neq 0$.

EXAMPLE 7   Solve the equation

$$yy'' = (y')^2$$

under the assumption that y and y' are known to be positive.

SOLUTION The substitutions in (26) yield

$$yp\frac{dp}{dy} = p^2,$$

so

$$\int \frac{dp}{p} = \int \frac{dy}{y};$$

$$\ln p = \ln y + C \qquad \text{(because } y > 0 \text{ and } p > 0\text{)};$$

$$p = C_1 y$$

where $C_1 = e^C$. Thus

$$\frac{dx}{dy} = \frac{1}{p} = \frac{1}{C_1 y};$$

$$C_1 x = \int \frac{dy}{y} = \ln y + C_2.$$

Hence the general solution of $yy'' = (y')^2$ is

$$y(x) = \exp(C_1 x - C_2) = Ae^{Bx}$$

where $A = \exp(-C_2)$ and $B = C_1$.

THE HANGING CABLE

An an application, we now investigate the shape of a hanging cable made of a perfectly flexible yet inelastic material. We assume that the only *external* force acting on the cable (except for the supporting forces at its two ends) is the gravitational force of its own weight. We make the simple assumption that the weight of the cable is uniformly distributed along its length, and thus is *not* uniformly distributed horizontally.

Let w denote the density of the cable, measured in units such as pounds per foot of length. Figure 2.15 shows the cable with lowest point P and with a coordinate system set up so that the y-axis passes through the lowest point of the cable and the x-axis is somewhere below that point.

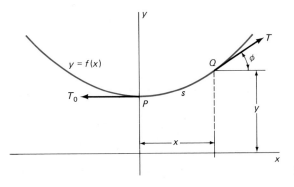

FIGURE 2.15 The hanging cable.

Now consider a section PQ of the cable of length s. We plan to obtain a differential equation for the shape $y = f(x)$ of the hanging cable by balancing horizontal and vertical components of the forces acting on PQ. These forces are:

T_0:   the horizontal tension pulling on the cable at P;

T: the tangential tension pulling on the cable at Q;

ws: the force of gravity pulling downward on the section PQ.

When we equate the horizontal and vertical components, we find that

$$T \cos \phi = T_0 \quad \text{and} \quad T \sin \phi = ws, \tag{27}$$

where, as indicated in Fig. 2.15, ϕ is the angle the tangent to the cable at Q makes with the horizontal.

We divide the second equation above by the first to find that

$$\frac{dy}{dx} = \tan \phi = \frac{T \sin \phi}{T \cos \phi} = \frac{ws}{T_0}. \tag{28}$$

Because s is a function of x, this differential equation is not as simple as we might hope. But after we differentiate both sides, we find that

$$\frac{d^2y}{dx^2} = \frac{w}{T_0} \frac{ds}{dx}.$$

But $ds/dx = [1 + (dy/dx)^2]^{1/2}$, and so

$$\frac{d^2y}{dx^2} = \frac{w}{T_0} \left[1 + \left(\frac{dy}{dx} \right)^2 \right]^{1/2}. \tag{29}$$

This is the differential equation we must solve to find the shape $y = y(x)$ of the hanging cable.

Because the dependent variable y is missing, we substitute

$$p = \frac{dy}{dx} \quad \text{and} \quad \frac{dp}{dx} = \frac{d^2y}{dx^2}$$

in Eq. (29). This yields the simpler equation

$$\frac{dp}{dx} = \frac{w}{T_0} (1 + p^2)^{1/2},$$

which we rewrite in the form

$$\frac{1}{(1 + p^2)^{1/2}} \frac{dp}{dx} = \frac{w}{T_0}.$$

Because $D_x \sinh^{-1} x = (1 + x^2)^{-1/2}$, integration of both sides of the equation above gives

$$\sinh^{-1} p = \frac{wx}{T_0} + C_1.$$

Now $p = 0$ when $x = 0$ because the cable has a horizontal tangent at its lowest point P. This tells us that $C_1 = \sinh^{-1} 0 = 0$, and thus that $\sinh^{-1} p =$

CHAPTER 2:   Linear Equations of Higher Order

wx/T_0. Therefore

$$\frac{dy}{dx} = p = \sinh\left(\frac{wx}{T_0}\right).$$

Finally, we integrate both sides of this last equation, and thus we find that

$$y = \frac{T_0}{w}\cosh\left(\frac{wx}{T_0}\right) + C_2.$$

If y_0 is the height of the point P above the x-axis, then $C_2 = y_0 - T_0/w$, and so the shape of the hanging cable is given by

$$y = \frac{T_0}{w}\cosh\left(\frac{wx}{T_0}\right) + y_0 - \frac{T_0}{w}. \tag{30}$$

We may choose the location of the x-axis so that $y_0 = T_0/w$, and then Eq. (30) takes the simple form

$$y = \frac{T_0}{w}\cosh\left(\frac{wx}{T_0}\right), \tag{31}$$

or, if we let $a = T_0/w$, the form $y = a \cosh(x/a)$. This curve is frequently called a *catenary*, from the Latin word *catena* (chain—a good physical approximation to a cable that is both perfectly flexible and inelastic).

Problems 56–60 deal with the relationship between the length S of the cable, the distance $2L$ between the two points from which it is suspended (at equal heights), and the dip or sag H of the cable at its middle. All these lengths are shown in Fig. 2.16.

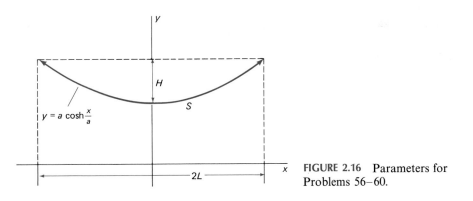

FIGURE 2.16 Parameters for Problems 56–60.

A PURSUIT PROBLEM

Suppose that a cat starts at the origin and runs with speed a straight and due north toward a tree located at the point $T(0, d)$ on the y-axis. At the same time a dog starts at the point $(c, 0)$ on the x-axis, running with speed b, and pursuing the cat by always running directly toward the cat. We wish to determine whether the cat can safely reach the tree before the dog catches it.

Let $C(0, at)$ and $D(x, y)$ denote the locations of the cat and the dog, respectively, at time t. As indicated in Fig. 2.17, the slope of the tangent line to the

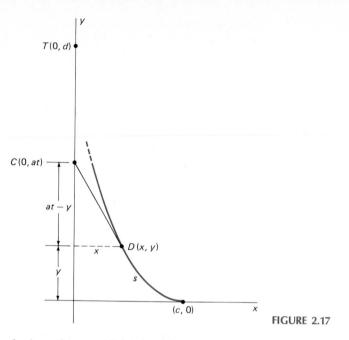

FIGURE 2.17

dog's path $y = y(x)$ is given by

$$\frac{dy}{dx} = -\frac{at - y}{x},$$

so that

$$xy' = y - at. \tag{32}$$

Regarding x as a decreasing differentiable function of t, so that t is a differentiable function of x, we differentiate each side in Eq. (32) with respect to x and obtain

$$xy'' = -a\frac{dt}{dx}. \tag{33}$$

If s measures arc length from $(c, 0)$ along the dog's path, then $b = ds/dt$ is the dog's speed, and

$$ds = \sqrt{(dx)^2 + (dy)^2} = -\sqrt{1 + (y')^2}\, dx.$$

(The minus sign results from the fact that s increases as x decreases.) Hence

$$\frac{dt}{dx} = \frac{dt}{ds}\frac{ds}{dx} = -\frac{1}{b}\sqrt{1 + (y')^2}. \tag{34}$$

Substitution of (34) in (33) finally yields the differential equation for the dog's path:

$$xy'' = k\sqrt{1 + (y')^2}, \tag{35}$$

where

$$k = \frac{a}{b} \tag{36}$$

denotes the ratio of the speed a of the cat to the speed b of the dog.

CHAPTER 2:   Linear Equations of Higher Order

Note that the dependent variable y is missing in Eq. (35). In Problem 61 we ask you to show that the substitution in (24) leads to

$$\frac{dy}{dx} = \frac{1}{2}\left[\left(\frac{x}{c}\right)^k - \left(\frac{c}{x}\right)^k\right].\tag{37}$$

If $a \geq b$, then obviously the dog can never catch the cat. Hence we assume that $a < b$, so that $0 < k < 1$. Then, as a consequence of Problem 62, integration of (37) yields the equation

$$y = \frac{c}{2}\left[\frac{1}{1+k}\left(\frac{x}{c}\right)^{1+k} - \frac{1}{1-k}\left(\frac{x}{c}\right)^{1-k}\right] + \frac{ck}{1-k^2}\tag{38}$$

that describes the dog's path.

EXAMPLE 8   Suppose that the cat's speed is $a = 20$ ft/s, the dog's speed is $b = 30$ ft/s, and that $c = 150$ ft. Then $k = \frac{2}{3}$, and Eq. (38) becomes

$$y = 75\left[\frac{3}{5}\left(\frac{x}{150}\right)^{5/3} - 3\left(\frac{x}{150}\right)^{1/3}\right] + 180.$$

Note that $y = 180$ when $x = 0$. This means that if the cat's initial distance d from the tree is greater than 180 ft, the dog catches it at the point $(0, 180)$. But if $d < 180$ ft, the cat reaches the tree safely.

2.6   PROBLEMS

In each of Problems 1–10, a differential equation and one solution y_1 are given. Apply the formula in (8) to find a second linearly independent solution y_2.

1. $y'' - 4y' + 4y = 0;\quad y_1 = e^{2x}$

2. $x^2y'' + 3xy' + y = 0;\quad y_1 = \dfrac{1}{x}$

3. $4x^2y'' + y = 0;\quad y_1 = \sqrt{x}$

4. $xy'' + (x - 1)y' - y = 0;\quad y_1 = e^{-x}$

5. $xy'' - 2(x + 1)y' + 4y = 0;\quad y_1 = e^{2x}$

6. $(x^2 + 1)y'' - 2xy' + 2y = 0;\quad y_1 = x$

7. $x^2y'' - 2xy' + (x^2 + 2)y = 0;\quad y_1 = x\cos x$

8. $xy'' - y' + 4x^3y = 0;\quad y_1 = \cos x^2$

9. $(x + x^2)y'' - (2x + 1)y' + 2y = 0;\quad y_1 = x^2$

10. $(x^4 - x^2)y'' - (3x^3 - x)y' + 8y = 0;\quad y_1 = x^4$

In each of Problems 11–17, a differential equation and one solution y_1 are given. Substitute $y_2 = vy_1$ in the equation to find a second linearly independent solution y_2.

11. $4y'' - 4y' + y = 0;\quad y_1 = e^{x/2}$

12. $xy'' - 3y' = 0;\quad y_1 = 1$

13. $x^2y'' + xy' - 9y = 0;\quad y_1 = x^3$

14. $(x + 1)y'' - (x + 2)y' + y = 0;\quad y_1 = e^x$

15. $x^2y'' - x(x + 2)y' + (x + 2)y = 0;\quad y_1 = x$

16. $(1 + x^3)y'' - 3x^2y' + 3xy = 0;\quad y_1 = x$

17. $x^2y'' - 4xy' + (x^2 + 6)y = 0;\quad y_1 = x^2\sin x$

18. Note that $y_1 = x$ is one solution of Legendre's equation of order 1, $(1 - x^2)y'' - 2xy' + 2y = 0$, and use the method of reduction of order to derive the second solution,

$$y_2 = 1 - \frac{x}{2}\ln\frac{1 + x}{1 - x}$$

for $-1 < x < 1$.

19. Verify that $y_1 = x^{-1/2}\cos x$ is one solution of Bessel's equation of order $\frac{1}{2}$, $x^2y'' + xy' + (x^2 - \frac{1}{4})y = 0$ $(x > 0)$, and derive the second solution $y_2 = x^{-1/2}\sin x$.

20. Show that $y_1 = 3x^2 - 1$ is one solution of Legendre's equation of order 2, $(1 - x^2)y'' - 2xy' + 6y = 0$. Then apply the formula in (8) to derive the second solution,

$$y_2 = y_1\ln\frac{1 + x}{1 - x} - 6x.$$

Note that the integral

$$\int \frac{dx}{(3x^2 - 1)^2(1 - x^2)}$$

can be evaluated by the method of partial fractions.

Find general solutions of the Euler-Cauchy equations in Problems 21–29.

21. $x^2y'' + xy' = 0$ **22.** $x^2y'' + xy' - y = 0$

23. $x^2y'' + 2xy' - 12y = 0$ **24.** $x^2y'' - 3xy' + 4y = 0$

25. $4x^2y'' + 8xy' - 3y = 0$ **26.** $9x^2y'' + 3xy' - 8y = 0$

27. $x^2y'' - xy' + 2y = 0$ **28.** $x^2y'' + 7xy' + 25y = 0$

29. $2x^2y'' + 5y = 0$

30. Verify Eq. (21) of the text.

Apply the method of Example 5 to find general solutions of the Euler-Cauchy equations in Problems 31–36.

31. $x^3y^{(3)} - x^2y'' + xy' = 0$

32. $x^3y^{(3)} + 6x^2y'' + 4xy' = 0$

33. $x^3y^{(3)} + 3x^2y'' + xy' = 0$

34. $x^3y^{(3)} - 2xy' + 2y = 0$

35. $x^3y^{(3)} + 3x^2y'' + xy' + y = 0$

36. $x^3y^{(3)} + 2x^2y'' - 4y = 0$

Find general solutions of the reducible second order equations in Problems 37–48.

37. $xy'' = y'$

38. $yy'' + (y')^2 = 0$ (Assume y and y' positive.)

39. $y'' + 4y = 0$

40. $xy'' + y' = 4x$

41. $y'' = (y')^2$ (Assume y and y' positive.)

42. $x^2y'' + 3xy' = 2$

43. $yy'' + (y')^2 = yy'$ (Assume y and y' positive.)

44. $y'' = (x + y')^2$ **45.** $y'' = 2y(y')^3$

46. $y^3y'' = 1$ **47.** $y'' = 2yy'$

48. $yy'' = 3(y')^2$

49. The substitution $y = e^{rx}$ yields the single solution $y_1 = e^{kx}$ of the linear equation

$$y'' - 2ky' + k^2y = 0.$$

Use reduction of order to find the second linearly independent solution $y_2 = xe^{kx}$.

50. Substitute $p = x'(t)$ to derive the general solution $x(t) = A \cos \omega t + B \sin \omega t$ of the differential equation $x'' + \omega^2x = 0$.

51. In the calculus of plane curves one learns that the curvature κ of the curve $y = y(x)$ at the point (x, y) is given by

$$\kappa = \frac{|y''(x)|}{(1 + [y'(x)]^2)^{3/2}},$$

and that the curvature of a circle of radius R is $\kappa = 1/R$. Conversely, substitute $p = y'$ to derive the general solution of the equation

$$Ry'' = [1 + (y')^2]^{3/2}$$

(with R constant) in the form

$$(x - a)^2 + (y - b)^2 = R^2.$$

Thus a circle of radius R (or a part thereof) is the *only* curve with constant curvature $1/R$.

52. Show that the substitution $y = (x^2 + 1)u$ transforms the differential equation

$$(x^2 + 1)y'' = 2y$$

into the equation

$$(x^2 + 1)u'' + 4xu' = 0.$$

The latter equation is readily solved for u'. Hence derive the general solution

$$y(x) = (x^2 + 1)\left[A + B\left(\frac{x}{x^2 + 1} + \tan^{-1} x\right)\right].$$

53. One solution of the equation

$$x(x - 1)(x + 1)^2y'' + 2x(x - 3)(x + 1)y' - 2(x - 1)y = 0$$

is $y_1 = x/(x + 1)^2$. Derive by reduction of order the second solution

$$y_2 = \frac{x^2 - 1 - 2x \ln x}{(x + 1)^2}.$$

54. (a) Show that the substitution $u = y'/y$ transforms the general homogeneous second order linear equation

$$y'' + p(x)y' + q(x)y = 0 \tag{39}$$

into the first order **associated Riccati equation**

$$\frac{du}{dx} + u^2 + p(x)u + q(x) = 0. \tag{40}$$

(b) Conclude that every solution of (39) can be written in the form

$$y(x) = C \exp\left[\int u(x)\, dx\right]$$

where $u(x)$ is a solution of Eq. (40).

55. (a) Note that the Riccati equation associated with the second order equation $y'' + \omega^2y = 0$ is

$$u' + u^2 + \omega^2 = 0.$$

Hence use the result of Problem 54 to derive the general solution $y = A\cos\omega x + B\sin\omega x$. (b) Note that the Riccati equation associated with the second order equation $y'' - 2ky' + k^2y = 0$ is

$$u' + (u - k)^2 = 0.$$

Hence use the result of Problem 54 to derive the general solution $y = Ae^{kx} + Bxe^{kx}$.

56. Consider a hanging cable, of the sort to which Eq. (31) applies. Show that the length of the section lying above the interval $[0, x]$ is $s = (T_0/w)\sinh(wx/T_0)$.

57. Deduce from Eq. (31) that the sag of the cable at its middle is

$$H = \frac{T_0}{w}\left[\cosh\left(\frac{wL}{T_0}\right) - 1\right].$$

58. Deduce from Eq. (27) that the tension in the cable satisfies the equation $T^2 = T_0^2 + w^2s^2$.

59. Use the results of Problems 56 and 58 to show that the tension in the cable at the point (x, y) is $T = T_0\cosh(wx/T_0) = wy$, the weight of a cable of the same density and of length y.

60. A high-voltage transmission line 205 ft long and weighing 1 lb/ft is strung between two towers 200 ft apart. (a) Use the result of Problem 56 to find the minimum tension T_0. (*Suggestion:* Use Newton's method to solve for $u = 100/T_0$.) (b) Use Problem 58 to find the maximum tension in the transmission line. (c) Use Problem 57 to find the sag H at its middle.

61. Derive Eq. (37) in the text by solving Eq. (35) with the initial condition $y'(c) = 0$.

62. Derive Eq. (38) by integrating Eq. (37) with the initial condition $y(c) = 0$.

63. Solve Eq. (35) with $k = 1$ to verify that if $a = b$ then the dog can never catch the cat.

2.7

Variation of Parameters

In Section 2.5 we presented the method of undetermined coefficients which, when it is applicable, is usually the simplest method of finding a particular solution of a nonhomogeneous linear differential equation with constant coefficients. We pointed out there, however, that this method cannot succeed with an equation such as $y'' + y = \tan x$, because the function $f(x) = \tan x$ has infinitely many linearly independent derivatives. In addition, the method of undetermined coefficients can be used only with equations having constant coefficients.

We discuss here the method of **variation of parameters** which—in principle (that is, if the integrals that appear can be evaluated)—can always be used to find a particular solution of the nonhomogeneous linear differential equation

$$y^{(n)} + p_{n-1}(x)y^{(n-1)} + \cdots + p_1(x)y' + p_0(x)y = f(x), \tag{1}$$

provided that we already know the general solution

$$y_c = c_1y_1 + c_2y_2 + \cdots + c_ny_n \tag{2}$$

of the associated homogeneous equation

$$y^{(n)} + p_{n-1}(x)y^{(n-1)} + \cdots + p_1(x)y' + p_0(x)y = 0. \tag{3}$$

Here, in brief, is the basic idea of the method of variation of parameters. Suppose that we replace the constants, or *parameters*, $c_1, c_2, \ldots, c_n$ in the complementary function in (2) by variables: functions $u_1, u_2, \ldots, u_n$ of x. We ask whether it is possible to choose these functions in such a way that the combination

$$y_p(x) = u_1(x)y_1(x) + u_2(x)y_2(x) + \cdots + u_n(x)y_n(x) \tag{4}$$

is a particular solution of the nonhomogeneous Eq. (1). It turns out that this *is* always possible.

The method is essentially the same for all orders $n \geq 2$. We will describe it in detail for the case $n = 2$, and then state the generalization. So we begin with the second order nonhomogeneous equation

$$L[y] = y'' + P(x)y' + Q(x)y = f(x) \tag{5}$$

with complementary function

$$y_c = c_1 y_1 + c_2 y_2 \tag{6}$$

on some open interval I where the functions P and Q are continuous. We want to find functions $u_1(x)$ and $u_2(x)$ such that

$$y_p = u_1 y_1 + u_2 y_2 \tag{7}$$

is a particular solution of (5).

One condition on the two functions u_1 and u_2 is that $L[y_p] = f(x)$. Because two conditions are required to determine two functions, we are free to impose an additional condition of our choice. We will do this in a way that simplifies the computations as much as possible. But first, to impose the condition $L[y_p] = f(x)$, we must compute the derivatives y_p' and y_p''. The product rule gives

$$y_p' = (u_1 y_1' + u_2 y_2') + (u_1' y_1 + u_2' y_2).$$

Our first imposed condition will be that the second sum here must vanish:

$$u_1' y_1 + u_2' y_2 = 0. \tag{8}$$

Then

$$y_p' = u_1 y_1' + u_2 y_2', \tag{9}$$

and the product rule gives

$$y_p'' = (u_1 y_1'' + u_2 y_2'') + (u_1' y_1' + u_2' y_2'). \tag{10}$$

But both y_1 and y_2 satisfy the homogeneous equation

$$y'' + Py' + Qy = 0$$

associated with the nonhomogeneous equation in (5), so

$$y_i'' = -Py_i' - Qy_i \tag{11}$$

for $i = 1, 2$. It therefore follows from (10) that

$$y_p'' = (u_1' y_1' + u_2' y_2') - P(u_1 y_1' + u_2 y_2') - Q(u_1 y_1 + u_2 y_2).$$

In view of Eqs. (7) and (9), this means that

$$y_p'' = (u_1' y_1' + u_2' y_2') - Py_p' - Qy_p;$$

that is, that

$$L[y_p] = u_1' y_1' + u_2' y_2'. \tag{12}$$

The requirement that y_p satisfy the nonhomogeneous equation in (5)—that is, that $L[y_p] = f(x)$—therefore implies that

$$u_1' y_1' + u_2' y_2' = f(x). \tag{13}$$

Finally, Eqs. (8) and (13) determine the functions u_1 and u_2 that we need. Collecting them together, we have a system

$$\left.\begin{aligned} u_1'y_1 + u_2'y_2 &= 0, \\ u_1'y_1' + u_2'y_2' &= f(x) \end{aligned}\right\} \tag{14}$$

of two linear equations in the two *derivatives* u_1' and u_2'. Note that the determinant of coefficients in (14) is simply the Wronskian $W(y_1, y_2)$. Once we have solved the equations in (14) for the derivatives u_1' and u_2', we integrate each to obtain the functions u_1 and u_2 such that

$$y_p = u_1 y_1 + u_2 y_2$$

is the desired particular solution of Eq. (5).

In the general case of the nth order nonhomogeneous equation in (1), a generalization of the procedure described above yields the system

$$\left.\begin{aligned} u_1'y_1 \quad &+ u_2'y_2 \quad + \cdots + u_n'y_n \quad = 0, \\ u_1'y_1' \quad &+ u_2'y_2' \quad + \cdots + u_n'y_n' \quad = 0, \\ u_1'y_1'' \quad &+ u_2'y_2'' \quad + \cdots + u_n'y_n'' \quad = 0, \\ &\qquad\qquad\qquad\qquad \vdots \\ u_1'y_1^{(n-1)} &+ u_2'y_2^{(n-1)} + \cdots + u_n'y_n^{(n-1)} = f(x) \end{aligned}\right\} \tag{15}$$

of n linear equations in the *derivatives* of the n functions $u_1, u_2, \ldots, u_n$. The nonzero determinant of coefficients in (15) is the Wronskian $W = W(x)$ of the independent solutions $y_1, y_2, \ldots, y_n$. If $W_i(x)$ denotes the determinant obtained from $W(x)$ upon replacing its ith column

$$\begin{bmatrix} y_i \\ y_i' \\ \vdots \\ y_i^{(n-1)} \end{bmatrix} \qquad \text{with the column} \qquad \begin{bmatrix} 0 \\ 0 \\ \vdots \\ 1 \end{bmatrix},$$

then Cramer's rule for the solution of a linear system of equations yields

$$u_i' = \frac{W_i(x)f(x)}{W(x)}; \tag{16}$$

finally, integration yields

$$u_i(x) = \int \frac{W_i(x)f(x)}{W(x)}\, dx \tag{17}$$

for $i = 1, 2, \ldots, n$. Because we are looking only for a single particular solution, the choice of constants of integration in (17) is not material.

On substituting (17) in the formula in (4), we get the general **variation of parameters formula**

$$y_p(x) = \sum_{i=1}^{n} y_i(x) \int \frac{W_i(x)f(x)}{W(x)}\, dx \tag{18}$$

for a particular solution of our original nth order nonhomogeneous linear differential equation in (1). In the case $n = 2$, this formula reduces to the formula

$$y_p(x) = -y_1(x) \int \frac{y_2(x)f(x)}{W(x)} \, dx + y_2(x) \int \frac{y_1(x)f(x)}{W(x)} \, dx \qquad (19)$$

for a particular solution in terms of two linearly independent solutions $y_1(x)$ and $y_2(x)$ of the associated homogeneous equation.

The formula in (19) leads to the concept of **Green's functions**, important in more advanced studies of differential equations (see Problems 34–37). In Problem 33 we ask you to complete the proof of the following theorem.

Theorem *Variation of Parameters*

Suppose that the functions P, Q, and f are continuous on the interval I containing the point x_0. Let u'_1 and u'_2 be the functions obtained by solving the system in (14). If

$$u_i(x) = \int_{x_0}^{x} u'_i(t) \, dt \qquad (20)$$

for $i = 1, 2$, then $y_p = u_1 y_1 + u_2 y_2$ is a particular solution of

$$y'' + P(x)y' + Q(x)y = f(x)$$

such that $y_p(x_0) = y'_p(x_0) = 0$.

Note that the specification of the reference point x_0 in (20) amounts to a specific choice of the constants of integration in (19).

Remark: In using the formulas just given, it is important to remember that we began with the nonhomogeneous equation in (1) written in *standard form* with leading coefficient 1. Hence we always begin by dividing each term of the given differential equation by its leading coefficient. Moreover, rather than memorizing (18) or (19), it is ordinarily better practice to set up the system of equations in (15) and then solve them explicitly for the derivatives $u'_1, u'_2, \ldots, u'_n$. The equations in (15) are easy to remember by virtue of the fact that their determinant of coefficients is the Wronskian $W = W(y_1, y_2, \ldots, y_n)$.

Thus the method of variation of parameters involves carrying out the following steps in order to find a particular solution y_p of the nonhomogeneous equation $L[y] = f(x)$.

1. Find the general solution

$$y_c = c_1 y_1 + c_2 y_2 + \cdots + c_n y_n$$

of the associated homogeneous equation $L[y] = 0$.

2. Calculate the derivatives $y_i^{(j)}$ in order to set up the linear system in (15).
3. Solve this system for the derivatives $u_1', u_2', \ldots, u_n'$.
4. Integrate those derivatives to obtain the functions $u_1, u_2, \ldots, u_n$.
5. Finally, form the desired particular solution

$$y_p = u_1 y_1 + u_2 y_2 + \cdots + u_n y_n.$$

EXAMPLE 1 Find a particular solution of $y'' + y = \tan x$.

SOLUTION The complementary function is $y_c = c_1 \cos x + c_2 \sin x$, so

$$y_1 = \cos x, \qquad y_2 = \sin x,$$
$$y_1' = -\sin x, \qquad y_2' = \cos x.$$

Hence the equations in (15) with $n = 2$ are

$$u_1'(\cos x) \quad + u_2'(\sin x) = 0,$$
$$u_1'(-\sin x) + u_2'(\cos x) = \tan x.$$

We easily solve these equations for

$$u_1' = -\sin x \tan x = \frac{-\sin^2 x}{\cos x} = \cos x - \sec x,$$

$$u_2' = \cos x \tan x = \sin x.$$

Hence we take

$$u_1 = \int (\cos x - \sec x)\, dx = \sin x - \ln |\sec x + \tan x|$$

and

$$u_2 = \int \sin x\, dx = -\cos x.$$

Thus our particular solution is

$$y_p = u_1 y_1 + u_2 y_2$$
$$= (\sin x - \ln |\sec x + \tan x|)(\cos x) + (-\cos x)(\sin x);$$

that is,

$$y_p(x) = -(\cos x) \ln |\sec x + \tan x|.$$

EXAMPLE 2 Find a particular solution of the nonhomogeneous equation

$$x^3 y^{(3)} + x^2 y'' - 6xy' + 6y = 30x,$$

given the fact that the associated homogeneous (Euler-Cauchy) equation has linearly independent solutions

$$y_1 = x, \qquad y_2 = x^3, \quad \text{and} \quad y_3 = x^{-2}.$$

SOLUTION We first write the equation in standard form:

$$y^{(3)} + \frac{1}{x}\, y'' - \frac{6}{x^2}\, y' + \frac{6}{x^3}\, y = 30x^{-2},$$

and thereby note that $f(x) = 30x^{-2}$. Inserting the derivatives of y_1, y_2, and y_3 into the equations in (15) with $n = 3$, we obtain the system

$$(x)u_1' + (x^3)u_2' + (x^{-2})u_3' = 0,$$
$$(1)u_1' + (3x^2)u_2' + (-2x^{-3})u_3' = 0,$$
$$(0)u_1' + (6x)u_2' + (6x^{-4})u_3' = 30x^{-2}.$$

The determinant of coefficients is the Wronskian

$$W = \begin{vmatrix} x & x^3 & x^{-2} \\ 1 & 3x^2 & -2x^{-3} \\ 0 & 6x & 6x^{-4} \end{vmatrix} = \frac{30}{x}.$$

Cramer's rule then gives

$$u_1' = \frac{x}{30} \begin{vmatrix} 0 & x^3 & x^{-2} \\ 0 & 3x^2 & -2x^{-3} \\ 30x^{-2} & 6x & 6x^{-4} \end{vmatrix} = -\frac{5}{x},$$

$$u_2' = \frac{x}{30} \begin{vmatrix} x & 0 & x^{-2} \\ 1 & 0 & -2x^{-3} \\ 0 & 30x^{-2} & 6x^{-4} \end{vmatrix} = \frac{3}{x^3},$$

and

$$u_3' = \frac{x}{30} \begin{vmatrix} x & x^3 & 0 \\ 1 & 3x^2 & 0 \\ 0 & 6x & 30x^{-2} \end{vmatrix} = 2x^2.$$

Integrating these derivatives, we get

$$u_1 = -5 \ln x, \qquad u_2 = -\frac{3}{2x^2}, \qquad u_3 = \frac{2}{3}\, x^3.$$

Thus our particular solution (for $x > 0$) is

$$y_p(x) = (-5 \ln x)(x) + \left(-\frac{3}{2x^2} \right)(x^3) + \left(\frac{2}{3}\, x^3 \right)(x^{-2}) = -\frac{5}{6}\, x - 5x \ln x.$$

The integral formulas in (17) are most useful precisely when the integrals are nonelementary, as in the following example.

EXAMPLE 3 Solve the initial value problem

$$y'' + y = \frac{1}{\sqrt{2\pi x}}, \qquad y(\pi) = y'(\pi) = 0.$$

SOLUTION   Here $y_1 = \cos x$ and $y_2 = \sin x$, so

$$W = \begin{vmatrix} \cos x & \sin x \\ -\sin x & \cos x \end{vmatrix} = \cos^2 x + \sin^2 x = 1;$$

$$W_1 = \begin{vmatrix} 0 & \sin x \\ 1 & \cos x \end{vmatrix} = -\sin x,$$

$$W_2 = \begin{vmatrix} \cos x & 0 \\ -\sin x & 1 \end{vmatrix} = \cos x.$$

Hence the formulas in (16) give

$$u_1' = -\frac{\sin x}{\sqrt{2\pi x}}, \qquad u_2' = \frac{\cos x}{\sqrt{2\pi x}}.$$

For u_1 and u_2 we take the specific antiderivatives

$$u_1(x) = -\frac{1}{\sqrt{2\pi}} \int_0^x \frac{\sin t}{\sqrt{t}} \, dt = -S_2(x),$$

$$u_2(x) = \frac{1}{\sqrt{2\pi}} \int_0^x \frac{\cos t}{\sqrt{t}} \, dt = C_2(x);$$

note that the improper integrals on the right converge. The *Fresnel integrals* $S_2(x)$ and $C_2(x)$ are nonelementary functions, but are tabulated in Table 7.7 of Abramowitz and Stegun, *Handbook of Mathematical Functions* (New York: Dover, 1965).

The general solution of $y'' + y = 1/\sqrt{2\pi x}$ is therefore

$$y = y_c + y_p = Ay_1 + By_2 + u_1 y_1 + u_2 y_2$$
$$= A \cos x + B \sin x - S_2(x) \cos x + C_2(x) \sin x,$$

and its derivative is

$$y' = -A \sin x + B \cos x - \frac{\sin x \cos x}{\sqrt{2\pi x}} + \frac{\cos x \sin x}{\sqrt{2\pi x}}$$

$$+ S_2(x) \sin x + C_2(x) \cos x;$$

that is,

$$y' = -A \sin x + B \cos x + S_2(x) \sin x + C_2(x) \cos x.$$

The initial conditions $y(\pi) = y'(\pi) = 0$ now give $A = S_2(\pi)$ and $B = -C_2(\pi)$. Thus the solution of the initial value problem is

$$y = [S_2(\pi) - S_2(x)] \cos x - [C_2(\pi) - C_2(x)] \sin x.$$

From the tables mentioned earlier, we find that $S_2(\pi) \approx 0.714$ and $C_2(\pi) \approx 0.529$. Because

$$S_2(\pi) - S_2(x) = \frac{1}{\sqrt{2\pi}} \int_x^\pi \frac{\sin t}{\sqrt{t}} \, dt$$

(and similarly for $C_2(\pi) - C_2(x)$), an interesting alternative form of our solution is

$$y = \frac{\cos x}{\sqrt{2\pi}} \int_x^\pi \frac{\sin t}{\sqrt{t}}\, dt - \frac{\sin x}{\sqrt{2\pi}} \int_x^\pi \frac{\cos t}{\sqrt{t}}\, dt;$$

thus

$$y(x) = \frac{1}{\sqrt{2\pi}} \int_x^\pi \frac{\sin (t - x)}{\sqrt{t}}\, dt,$$

with the aid of a familiar trigonometric identity. If a table of values of the Fresnel integrals were not available, we could use this formula to approximate the value $y(x)$ for any given x by numerical integration (for instance, by using Simpson's rule).

2.7 PROBLEMS

In each of Problems 1–18, use the method of variation of parameters to find a particular solution of the given differential equation.

1. $y'' + 3y' + 2y = 4e^x$

2. $y'' - 2y' - 8y = 3e^{-2x}$

3. $y'' - 4y' + 4y = 2e^{2x}$

4. $y'' - 4y = \sinh 2x$

5. $y'' + 4y = \cos 3x$

6. $y'' + 9y = \sin 3x$

7. $y'' + 9y = 2 \sec 3x$

8. $y'' + y = \csc^2 x$

9. $y'' + 4y = \sin^2 x$

10. $y'' - 4y = xe^x$

11. $y'' - 2y' + y = x^{-2}e^x$

12. $x^2 y'' - 4xy' + 6y = x^3$

13. $x^2 y'' - 3xy' + 4y = x^4$

14. $4x^2 y'' - 4xy' + 3y = 8x^{4/3}$

15. $x^2 y'' + xy' + y = \ln x$

16. $y^{(3)} - y'' - 2y' = x^2$

17. $y^{(3)} + 3y'' + 3y' + y = e^{-x}$

18. $y^{(3)} + 4y' = \cot 2x$

In Problems 19–24, find particular solutions involving nonelementary indefinite integrals (as in Example 3).

19. $y'' - y = x^{-2}e^x$

20. $y'' + y = x^{1/2}$

21. $y'' + 4y = \exp(-x^2)$

22. $y^{(3)} + y'' + y' + y = \dfrac{1}{x}$

23. $y^{(3)} - y'' = \ln x$

24. $y^{(4)} - y = \tanh x$

25. Find a particular solution of the equation

$$x^3 y^{(3)} + 5x^2 y'' + 2xy' - 2y = x^4,$$

given that its complementary function is $y_c = c_1 x + c_2 x^{-1} + c_3 x^{-2}$.

26. Find a particular solution of the equation

$$24x^3 y^{(3)} + 46x^2 y'' + 7xy' - y = 24x^3,$$

given that its complementary function is $y_c = c_1 x^{1/2} + c_2 x^{1/3} + c_3 x^{1/4}$.

27. Find a particular solution of the equation

$$(x^2 - 1)y'' - 2xy' + 2y = x^2 - 1;$$

recall from Example 2 of Section 2.6 that its complementary function is $y_c = c_1 x + c_2(1 + x^2)$.

28. Find a particular solution of the equation

$$x^2 y'' + xy' + (x^2 - \tfrac{1}{4})y = x^{3/2} \cos x;$$

recall from Problem 19 of Section 2.6 that its complementary function is $y_c = x^{-1/2}(c_1 \cos x + c_2 \sin x)$.

29. Express the solution of the initial value problem

$$y'' - y = \frac{1}{x}, \qquad y(1) = y'(1) = 1$$

in terms of integrals of the form

$$\int_1^x \frac{1}{t} e^{\pm t}\, dt.$$

30. Express the solution of the initial value problem

$$y'' - 3y' + 2y = \sin x^2; \qquad y(0) = 2, \qquad y'(0) = 3$$

in terms of integrals of the form

$$\int_0^x e^{-at} \sin t^2\, dt.$$

31. Generalize the method of Example 3 to derive the formula

$$y(x) = \int_a^x f(t) \sin (x - t)\, dt$$

for the solution of the initial value problem $y'' + y = f(x)$, $y(a) = 0 = y'(a)$.

CHAPTER 2:   Linear Equations of Higher Order

32. Use variation of parameters to derive the particular solution

$$y_p = \tfrac{1}{2} \int_0^x f(t)[e^{x-t} - e^{t-x}]\, dt$$

of the equation $y'' - y = f(x)$.

33. Suppose that $u_1(x)$ and $u_2(x)$ are defined as in Eq. (20). Verify that $y_p = u_1 y_1 + u_2 y_2$ satisfies the initial conditions $y_p(x_0) = y_p'(x_0) = 0$.

34. Show that the formula in (19) for a particular solution of the second order linear equation $y'' + Py' + Qy = f(x)$ can be written in the form

$$y_p(x) = \int_{x_0}^x G(x, t)f(t)\, dt \qquad (21)$$

where x_0 is a fixed reference point for evaluation of the integral and the function $G(x, t)$ is defined by

$$G(x, t) = \frac{1}{W(t)} \begin{vmatrix} y_1(t) & y_1(x) \\ y_2(t) & y_2(x) \end{vmatrix}, \qquad (22)$$

with $W(t)$ being the Wronskian of $y_1(t)$ and $y_2(t)$.

GREEN'S FUNCTIONS
The function $G(x, t)$ defined in (22) is a **Green's function** for the nonhomogeneous equation

$$y'' + P(x)y' + Q(x)y = f(x) \qquad (23)$$

with complementary function $y_c = c_1 y_1 + c_2 y_2$. Note that G depends only on the complementary solutions y_1 and y_2, *not* on f. Once $G(x, t)$ has been calculated, the formula in (21) expresses a particular solution of (23) in terms of the given "forcing function" $f(x)$. The following three problems describe Green's functions for linear second order equations of the form

$$y'' + by' + cy = f(x)$$

with constant coefficients.

35. Suppose that the characteristic equation $r^2 + br + c = 0$ has distinct real roots $r_1 \neq r_2$, so that $y_1(x) = e^{r_1 x}$ and $y_2(x) = e^{r_2 x}$. Show that $G(x, t) = g(x - t)$ where

$$g(z) = \frac{e^{r_1 z} - e^{r_2 z}}{r_1 - r_2}.$$

36. In the case in which $r^2 + br + c = 0$ has equal real roots $r_1 = r_2$, show that $G(x, t) = g(x - t)$ where

$$g(z) = z e^{r_1 z}.$$

37. Suppose that $r^2 + br + c = 0$ has complex conjugate roots $\alpha \pm \beta i$. Show that $G(x, t) = g(x - t)$ where

$$g(z) = \frac{1}{\beta} e^{\alpha z} \sin \beta z.$$

*2.8

Forced Oscillations and Resonance

In Section 2.4 we derived the differential equation

$$mx'' + cx' + kx = F(t) \qquad (1)$$

that governs the one-dimensional motion of a mass m that is attached to a spring (with constant k) and a dashpot (with constant c) and is also acted on by an external force $F(t)$. Machines with rotating components commonly involve mass-spring systems (or their equivalents) in which the external force is simple harmonic:

$$F(t) = F_0 \cos \omega t \quad \text{or} \quad F(t) = F_0 \sin \omega t, \qquad (2)$$

where the constant F_0 is the amplitude of the periodic force and ω is its circular frequency.

For an example of how a rotating machine component can provide a simple harmonic force, consider the cart with a rotating vertical flywheel shown in Fig. 2.18. The cart has mass $m - m_0$, not including the flywheel of mass m_0. The centroid of the flywheel is off-center at distance a from its center, and its angular speed is ω radians per second. The cart is attached to a spring (with constant k) as shown. Assume that the centroid of the cart itself is directly beneath the center

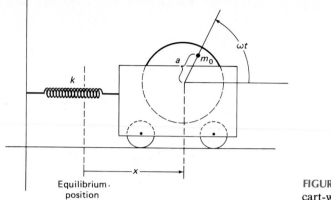

FIGURE 2.18 The cart-with-flywheel system.

of the flywheel, and denote by $x(t)$ its displacement from its equilibrium position (where the spring is unstretched). Figure 2.18 helps us to see that the displacement $\bar{x}$ of the centroid of the combined cart plus flywheel is given by

$$\bar{x} = \frac{(m - m_0)x + m_0(x + a \cos \omega t)}{m} = x + \frac{m_0 a}{m} \cos \omega t.$$

Let us ignore friction and apply Newton's second law $m\bar{x}'' = -kx$, because the force exerted by the spring is $-kx$. We substitute for $\bar{x}$ in the last equation to obtain

$$mx'' - m_0 a\omega^2 \cos \omega t = -kx;$$

that is,

$$mx'' + kx = m_0 a\omega^2 \cos \omega t. \tag{3}$$

Thus the cart with its rotating flywheel acts like a mass on a spring under the influence of a simple harmonic external force with amplitude $F_0 = m_0 a\omega^2$. Such a system is a reasonable model of a front-loading washing machine with the clothes being washed loaded off-center. This illustrates the practical importance of analyzing solutions of Eq. (1) with external forces as in (2).

UNDAMPED FORCED OSCILLATIONS

To study undamped oscillations under the influence of the external force $F(t) = F_0 \cos \omega t$, we set $c = 0$ in Eq. (1), and thereby begin with the equation

$$mx'' + kx = F_0 \cos \omega t \tag{4}$$

whose complementary function is $x_c = c_1 \cos \omega_0 t + c_2 \sin \omega_0 t$. Here,

$$\omega_0 = \sqrt{\frac{k}{m}}$$

is the (circular) **natural frequency** of the mass-spring system. Let us assume initially that the external and natural frequencies are *unequal*: $\omega \neq \omega_0$. We substitute $x_p = A \cos \omega t$ in (4) to find a particular solution. (No sine term is needed in x_p because

CHAPTER 2: Linear Equations of Higher Order

there is no term involving x' on the left-hand side in Eq. (4).) This gives

$$-m\omega^2 A \cos \omega t + kA \cos \omega t = F_0 \cos \omega t,$$

so

$$A = \frac{F_0}{k - m\omega^2} = \frac{F_0/m}{\omega_0^2 - \omega^2}, \tag{5}$$

and thus

$$x_p(t) = \frac{F_0/m}{\omega_0^2 - \omega^2} \cos \omega t. \tag{6}$$

Therefore the general solution $x = x_c + x_p$ is given by

$$x(t) = c_1 \cos \omega_0 t + c_2 \sin \omega_0 t + \frac{F_0/m}{\omega_0^2 - \omega^2} \cos \omega t, \tag{7}$$

where the constants c_1 and c_2 are determined by the initial values $x(0)$ and $x'(0)$. Equivalently, as in Eq. (12) of Section 2.4, we can rewrite (7) as

$$x(t) = C \cos (\omega_0 t - \alpha) + \frac{F_0/m}{\omega_0^2 - \omega^2} \cos \omega t, \tag{8}$$

so we see that the resulting motion is a superposition of two oscillations, one with natural circular frequency ω_0, the other with the external circular frequency ω.

EXAMPLE 1   Suppose that $m = 1$, $k = 9$, $F_0 = 80$, and $\omega = 5$, so the differential equation in (4) is

$$x'' + 9x = 80 \cos 5t.$$

Find $x(t)$ if $x(0) = x'(0) = 0$.

SOLUTION   Here the natural frequency $\omega_0 = 3$ and the external frequency $\omega = 5$ are unequal, as in the preceding discussion. First we substitute $x_p = A \cos 5t$ in the differential equation and find that $-25A + 9A = 80$, so that $A = -5$. Thus a particular solution is

$$x_p(t) = -5 \cos 5t.$$

The complementary function is $x_c = c_1 \cos 3t + c_2 \sin 3t$, so the general solution of the given nonhomogeneous equation is

$$x(t) = c_1 \cos 3t + c_2 \sin 3t - 5 \cos 5t,$$

with derivative

$$x'(t) = -3c_1 \sin 3t + 3c_2 \cos 3t + 25 \sin 5t.$$

The initial conditions $x(0) = 0$ and $x'(0) = 0$ now yield $c_1 = 5$ and $c_2 = 0$, so the desired solution is

$$x(t) = 5 \cos 3t - 5 \cos 5t.$$

If we impose the initial conditions $x(0) = x'(0) = 0$ on the solution in (7), we find that $c_1 = -F_0/m(\omega_0^2 - \omega^2)$ and $c_2 = 0$, so the particular solution is

$$x(t) = \frac{F_0/m}{\omega_0^2 - \omega^2}(\cos \omega t - \cos \omega_0 t). \tag{9}$$

The trigonometric identity $2 \sin A \sin B = \cos(A - B) - \cos(A + B)$, applied with $A = \frac{1}{2}(\omega_0 + \omega)t$ and $B = \frac{1}{2}(\omega_0 - \omega)t$, enables us to rewrite (9) as

$$x(t) = \frac{2F_0}{m(\omega_0^2 - \omega^2)} \sin \frac{1}{2}(\omega_0 - \omega)t \sin \frac{1}{2}(\omega_0 + \omega)t. \tag{10}$$

Suppose now that $\omega \approx \omega_0$, so that $\omega_0 + \omega$ is large in comparison with $|\omega_0 - \omega|$. Then $\sin \frac{1}{2}(\omega_0 + \omega)t$ is a *rapidly* varying function, while $\sin \frac{1}{2}(\omega_0 - \omega)t$ is a *slowly* varying function. We may therefore interpret (10) as a rapid oscillation with circular frequency $\frac{1}{2}(\omega_0 + \omega)$,

$$x(t) = A(t) \sin \frac{1}{2}(\omega_0 + \omega)t,$$

but with a slowly varying amplitude

$$A(t) = \frac{2F_0}{m(\omega_0^2 - \omega^2)} \sin \frac{1}{2}(\omega_0 - \omega)t.$$

The graph of $x(t)$ is shown in Fig. 2.19. An oscillation such as this, with a slowly varying periodic amplitude, exhibits the phenomenon of *beats*. For example, if two horns not exactly attuned to one another simultaneously play their middle C, one at $\omega_0/2\pi = 258$ Hz and the other at $\omega/2\pi = 254$ Hz, then one hears a beat—an audible variation in the *amplitude* of the combined sound—with a frequency of

$$\frac{(\omega_0 - \omega)/2}{2\pi} = \frac{1}{2}(258 - 254) = 2 \text{ Hz}.$$

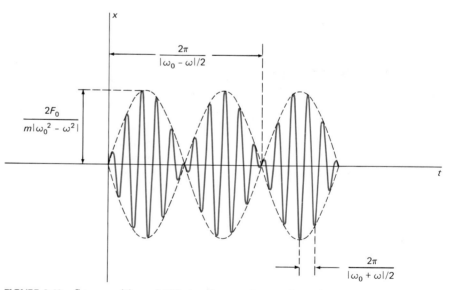

FIGURE 2.19 Superposition of differing frequencies produces beats.

CHAPTER 2: Linear Equations of Higher Order

Looking at Eq. (6), we see that the amplitude A of x_p is large when the natural and external frequencies ω_0 and ω are approximately equal. It is sometimes useful to rewrite (5) in the form

$$A = \frac{F_0}{k - m\omega^2} = \frac{F_0/k}{1 - (\omega/\omega_0)^2} = \pm \frac{\rho F_0}{k}, \tag{11}$$

where F_0/k is the **static displacement** of a spring with constant k due to a *constant* force F_0, and the **amplification factor** ρ is defined to be

$$\rho = \frac{1}{\left|1 - (\omega/\omega_0)^2\right|}. \tag{12}$$

It is clear that $\rho \to +\infty$ as $\omega \to \omega_0$. This is the phenomenon of **resonance**—the increase without bound (as $\omega \to \omega_0$) in the amplitude of the oscillations of an undamped system with natural frequency ω_0 in response to an external force with frequency ω.

We have been assuming that $\omega \neq \omega_0$. What sort of catastrophe should one expect if ω and ω_0 are precisely equal? Then Eq. (4), upon division of each term by m, becomes

$$x'' + \omega_0^2 x = \frac{F_0}{m} \cos \omega_0 t. \tag{13}$$

Because $\cos \omega_0 t$ is a term of the complementary function, the method of undetermined coefficients calls for us to try

$$x_p = t(A \cos \omega_0 t + B \sin \omega_0 t).$$

We substitute this in (13), and thereby find that $A = 0$ and $B = F_0/2m\omega_0$. Hence the particular solution is

$$x_p(t) = \frac{F_0}{2m\omega_0} t \sin \omega_0 t. \tag{14}$$

The graph of $x_p(t)$ in Fig. 2.20 shows vividly how the amplitude of the oscillation theoretically would increase without bound in this case of *pure resonance*, $\omega = \omega_0$. We may interpret the phenomenon as reinforcement of the natural vibrations of the system by externally impressed vibrations at the same frequency.

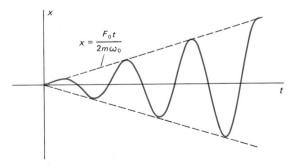

FIGURE 2.20   Pure resonance.

EXAMPLE 2 Suppose that $m = 4$ slugs and that $k = 400$ lb/ft in the cart with the flywheel of Fig. 2.18. Then the natural frequency is $\omega_0 = \sqrt{k/m} = 10$ rad/s; that is, $10/2\pi \approx 1.59$ Hz. We would therefore expect oscillations of very large amplitude to occur if the flywheel revolves at about $(1.59)(60) \approx 95$ revolutions per minute (rpm).

In practice, a mechanical system with very little damping can be destroyed by resonance vibrations. A spectacular example can occur when a column of soldiers marches in step over a bridge. Any complicated structure such as a bridge has many natural frequencies of vibration. If the frequency of the solders' cadence is approximately equal to one of the natural frequencies of the structure, then—just as in our simple example of a mass on a spring—resonance will occur. Indeed, the resulting resonance vibrations can be of such large amplitude that the bridge will collapse. This has actually happened—for example, the collapse of Broughton Bridge near Manchester, England, in 1831—and it is the reason for the now-standard practice of breaking cadence when crossing a bridge. Resonance may have been involved in the 1981 Kansas City disaster in which a hotel balcony (called a *skywalk*) collapsed with dancers on it. The collapse of a building in an earthquake is sometimes due to resonance vibrations caused by the ground oscillating at one of the natural frequencies of the structure; this happened to many buildings in the Mexico City earthquake of September 19, 1985. On occasion an airplane has crashed because of resonant wing oscillations caused by vibrations of the engines. It is reported that for some of the first commercial jet aircraft, the natural frequency of the vertical vibrations of the airplane during turbulence was almost exactly that of the mass-spring system consisting of the pilot's head (mass) and spine (spring). Resonance occurred, causing pilots to have difficulty in reading the instruments. The newer wide-bodied jets have different natural frequencies, so that this resonance problem no longer occurs.

The avoidance of destructive resonance vibrations is a constant factor in the design of mechanical structures and systems of all types. Often the most important step in determining the natural frequency of vibration of a system is the formulation of its differential equation. In addition to Newton's law $F = ma$, the principle of conservation of energy is sometimes useful for this purpose (as in the derivation of the pendulum equation in Section 2.4). The following kinetic and potential energy formulas are often useful:

1. *Kinetic energy:* $T = \frac{1}{2}mv^2$ for translation of a mass with velocity v;
2. *Kinetic energy:* $T = \frac{1}{2}I\omega^2$ for rotation of a body of moment of inertia I with angular velocity ω;
3. *Potential energy:* $V = \frac{1}{2}kx^2$ for a spring with constant k stretched or compressed a distance x;
4. *Potential energy:* $V = mgh$ for the gravitational potential energy of a mass m at height h above the reference level (the level at which $V = 0$).

EXAMPLE 3 Find the natural frequency of a mass m on a spring (with constant k) if, instead of sliding without friction, it is a uniform disk of radius a that rolls without slipping, as shown in Fig. 2.21.

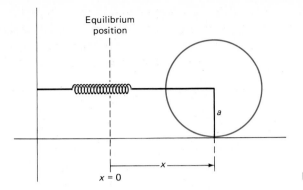

Equilibrium
position

$x = 0$

x

FIGURE 2.21 The rolling disk.

SOLUTION With the notation above, the principle of conservation of energy gives

$$\tfrac{1}{2}mv^2 + \tfrac{1}{2}I\omega^2 + \tfrac{1}{2}kx^2 = E$$

where E is a constant (the total mechanical energy of the system). We note that $v = a\omega$, and recall that $I = ma^2/2$ for a uniform circular disk. Thus we may simplify the above equation to

$$\tfrac{3}{4}mv^2 + \tfrac{1}{2}kx^2 = E.$$

Differentiation ($v = x'$, $v' = x''$) now gives

$$\tfrac{3}{2}mx'x'' + kxx' = 0.$$

We divide each term by $\tfrac{3}{2}mx'$ to obtain

$$x'' + \frac{2k}{3m}x = 0.$$

Thus the natural circular frequency is $\omega_0 = \sqrt{2k/3m}$, which is $\sqrt{\tfrac{2}{3}}$ times the frequency in the previous situation of sliding without friction.

EXAMPLE 4 Assume that a car weighing 1600 lb oscillates vertically as if it were a mass $m = 50$ slugs on a single spring (with constant $k = 4800$ lb/ft), attached to a single dashpot (with constant $c = 200$ lb-s/ft). Suppose that this car with the dashpot *disconnected* is driven along a washboard road surface with an amplitude of 2 in. and a wavelength of $L = 30$ ft (see Fig. 2.22). At what car speed (in miles per hour) will resonance vibrations occur?

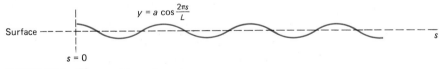

Surface

$y = a \cos \dfrac{2\pi s}{L}$

$s = 0$

s

FIGURE 2.22 The washboard road.

SOLUTION We think of the car as a unicycle, as pictured in Fig. 2.23. Let $x(t)$ denote the upward displacement of the mass m from its equilibrium position; we ignore the force of gravity, because it merely displaces the equilibrium position as

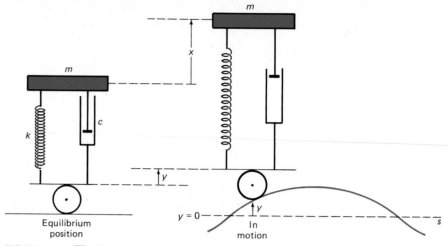

FIGURE 2.23 The "unicycle model" of a car.

in Problem 9 of Section 2.4. We write the equation of the road surface as

$$y = a \cos \frac{2\pi s}{L} \qquad \left(a = \frac{1}{6} \text{ ft}, L = 30 \text{ ft}\right).$$ (15)

When the car is in motion, the spring is stretched by the amount $x - y$, so Newton's second law, $F = ma$, gives

$$mx'' = -k(x - y);$$

that is,

$$mx'' + kx = ky.$$ (16)

If the velocity of the car is v, then $s = vt$ in (15), so (16) takes the form

$$mx'' + kx = ka \cos \frac{2\pi vt}{L}.$$ (16′)

This is the differential equation governing the vertical oscillations of the car. In comparing it with Eq. (4), we see that we have forced oscillations with circular frequency $\omega = 2\pi v/L$. Resonance will occur when $\omega = \omega_0 = \sqrt{k/m}$. We insert our numerical data to find the speed of the car at resonance:

$$v = \frac{L}{2\pi} \sqrt{\frac{k}{m}} = \frac{30}{2\pi} \sqrt{\frac{4800}{50}} \text{ (ft/s)},$$

or about 32 mi/h.

DAMPED FORCED OSCILLATIONS

In real physical systems there is always some damping, from frictional effects if nothing else. The complementary function x_c of the equation

$$mx'' + cx' + kx = F_0 \cos \omega t$$ (17)

CHAPTER 2: Linear Equations of Higher Order

is given by Eq. (18), (19), or (20) of Section 2.4, depending upon whether $c > c_{CR}$, $c = c_{CR}$, or $c < c_{CR}$. The specific form is not important here. What is important is that, in any case, these formulas show that $x_c(t) \to 0$ as $t \to +\infty$. Thus, x_c is a **transient solution** of (17)—one that dies out with the passage of time, leaving only the particular solution x_p.

The method of undetermined coefficients indicates that we should substitute

$$x_p = A \cos \omega t + B \sin \omega t$$

in (17). When we do so, collect terms, and equate coefficients of $\cos \omega t$ and $\sin \omega t$, we obtain the two equations

$$\left.\begin{aligned}
(k - m\omega^2)A + c\omega B &= F_0, \\
-c\omega A + (k - m\omega^2)B &= 0
\end{aligned}\right\} \tag{18}$$

that we solve without difficulty for

$$A = \frac{(k - m\omega^2)F_0}{(k - m\omega^2)^2 + (c\omega)^2}, \qquad B = \frac{(c\omega)F_0}{(k - m\omega^2)^2 + (c\omega)^2}.$$

To simplify the notation, it is convenient to introduce the quantity

$$\rho = \frac{k}{\sqrt{(k - m\omega^2)^2 + (c\omega)^2}} \tag{19}$$

FIGURE 2.24 The angle α.

and the angle α of Fig. 2.24. Then we find that

$$A = \rho \frac{F_0}{k} \cos \alpha, \qquad B = \rho \frac{F_0}{k} \sin \alpha.$$

Hence our particular solution is

$$x_p = A \cos \omega t + B \sin \omega t$$

$$= \rho \frac{F_0}{k} (\cos \omega t \cos \alpha + \sin \omega t \sin \omega \alpha);$$

more concisely,

$$x_p(t) = \rho \frac{F_0}{k} \cos (\omega t - \alpha). \tag{20}$$

Thus we get a **steady periodic solution** that remains after the transient solution has died away. This steady-state solution has amplitude $\rho(F_0/k)$, circular frequency ω, and phase angle α given by

$$\alpha = \tan^{-1} \frac{c\omega}{k - m\omega^2}, \qquad 0 \leq \alpha \leq \pi. \tag{21}$$

Note that α lies in the first or second quadrant, so the formula above does not involve the principal value of the inverse tangent function. If a calculator gives a negative value, we must add π to that value to obtain the actual value of α.

The **amplification factor** ρ, defined in (19), is the amount by which the static displacement F_0/k must be multiplied to get the amplitude of the steady periodic oscillation. Note that when $c > 0$, the amplitude always remains finite (unlike the undamped case). The amplitude may reach a maximum for some value of ω; this is *practical* reasonance. To see when it occurs for various values of the constants m, c, and k, it is useful to express ρ in terms of the dimensionless ratios

$$\tilde{\omega} = \frac{\omega}{\omega_0} = \frac{\omega}{\sqrt{k/m}}, \qquad \tilde{c} = \frac{c}{c_{CR}} = \frac{c}{\sqrt{4km}}.$$

Then (19) is equivalent to

$$\rho = \frac{1}{\sqrt{(1 - \tilde{\omega}^2)^2 + 4\tilde{c}^2\tilde{\omega}^2}}. \tag{19'}$$

Figure 2.25 shows the graph of ρ versus $\tilde{\omega}$ for various values of $\tilde{c}$. It can be shown that if $c \geq c_{CR}/\sqrt{2}$, then ρ steadily decreases as ω increases, but if $c < c_{CR}/\sqrt{2}$, then ρ reaches a maximum value—practical resonance—at some value of ω less than ω_0, and then approaches 0 as $\omega \to +\infty$. It follows that an underdamped system typically will undergo forced oscillations whose amplitude is:

1. Large if ω is close to the critical resonance frequency;
2. Close to F_0/k if ω is very small;
3. Very small if ω is very large.

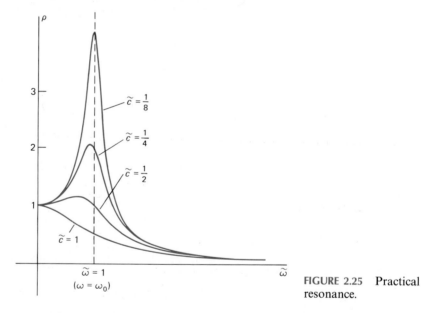

FIGURE 2.25   Practical resonance.

EXAMPLE 5   Find the transient and steady periodic solutions of

$$x'' + 2x' + 2x = 20 \cos 2t, \qquad x(0) = x'(0) = 0.$$

SOLUTION   Instead of applying the general formulas derived above, it is better in a concrete problem to work the problem directly. The roots of the characteristic equation $r^2 + 2r + 2 = 0$ are $-1 \pm i$, so the complementary function is

$$x_c = e^{-t}(c_1 \cos t + c_2 \sin t).$$

CHAPTER 2:   Linear Equations of Higher Order

When we substitute

$$x_p = A \cos 2t + B \sin 2t$$

in the given equation, collect coefficients, and equate coefficients of $\cos 2t$ and $\sin 2t$, we obtain the equations

$$-2A + 4B = 20,$$

$$-4A - 2B = 0,$$

with solution $A = -2$, $B = 4$. Hence the general solution is

$$x(t) = e^{-t}(c_1 \cos t + c_2 \sin t) - 2 \cos 2t + 4 \sin 2t.$$

At this point we impose the initial conditions $x(0) = x'(0) = 0$, and find easily that $c_1 = 2$, $c_2 = -6$. Therefore, the transient solution x_{tr} and the steady periodic solution x_{sp} are given by

$$x_{tr}(t) = e^{-t}(2 \cos t - 6 \sin t)$$

and

$$x_{sp}(t) = -2 \cos 2t + 4 \sin 2t$$

$$= 2\sqrt{5}\left(-\frac{1}{\sqrt{5}} \cos 2t + \frac{2}{\sqrt{5}} \sin 2t\right).$$

The latter can be written in the form

$$x_{sp}(t) = 2\sqrt{5}[\cos(2t - \alpha)]$$

where

$$\alpha = \pi - \tan^{-1}(2) \approx 2.0344.$$

Finally, we indicate how to include damping in the analysis of the vibrations of the car of Example 4 (though we leave the detailed computations to the problems). We use without proof the fact that when $c > 0$, the differential equation is

$$mx'' + cx' + kx = cy' + ky. \tag{22}$$

With $y = a \sin \omega t$ for the road surface, this equation becomes

$$mx'' + cx' + kx = E_0 \cos \omega t + F_0 \sin \omega t, \tag{23}$$

where $E_0 = c\omega a$ and $F_0 = ka$. Substituting the trial solution

$$x_{sp} = A \cos \omega t + B \sin \omega t$$

in (23) to determine the coefficients A and B, we arrive finally at the steady periodic solution

$$x_{sp}(t) = \frac{\sqrt{E_0^2 + F_0^2}}{\sqrt{(k - m\omega^2)^2 + (c\omega)^2}} \cos(\omega t - \alpha - \beta) \tag{24}$$

Velocity v (mi/h)	Amplitude C (in.)
7.5	2.12
15.0	2.54
22.5	3.59
30.0	5.34
37.5	3.62
45.0	2.01
52.5	1.31
60.0	0.95
75.0	0.60
90.0	0.43

FIGURE 2.26 Effect of the washboard road on the car.

where α is defined in (21) and $\beta = \tan^{-1}(F_0/E_0)$. Upon substitution of $E_0 = c\omega a$ and $F_0 = ka$, we see that the amplitude of the steady periodic oscillations of the car is

$$C = \frac{a\sqrt{k^2 + (c\omega)^2}}{\sqrt{(k - m\omega^2)^2 + (c\omega)^2}}. \tag{25}$$

Because $\omega = 2\pi v/L$ when the car is moving with velocity v, this gives C as a function of v. The formula in (25) was used, with the numerical parameters given in Example 4, to calculate the entries in the table shown in Fig. 2.26. As the car accelerates gradually from rest, it initially oscillates with amplitude slightly over 2 in. (the road surface amplitude). Maximum resonance oscillations with amplitude over 5 in. occur around 30 mi/h, but then subside to more tolerable levels at high speeds.

2.8 PROBLEMS

In each of Problems 1–6, express the solution of the given initial value problem as a sum of two oscillations (as in Eq. (8)).

1. $x'' + 9x = 10\cos 2t$; $x(0) = x'(0) = 0$

2. $x'' + 4x = 5\sin 3t$; $x(0) = x'(0) = 0$

3. $x'' + 100x = 15\cos 5t + 20\sin 5t$; $x(0) = 25, x'(0) = 0$

4. $x'' + 25x = 10\cos 4t$; $x(0) = 0, x'(0) = 10$

5. $mx'' + kx = F_0\cos\omega t$ with $\omega \neq \omega_0$; $x(0) = x_0, x'(0) = 0$

6. $mx'' + kx = F_0\sin\omega t$ with $\omega = \omega_0$; $x(0) = 0, x'(0) = v_0$

In each of Problems 7–14, find the steady periodic solution in the form $x_{sp} = C\cos(\omega t - \alpha)$. If initial conditions are given, also find the transient solution.

7. $x'' + 4x' + 4x = 10\cos 3t$

8. $x'' + 3x' + 5x = -4\cos 5t$

9. $2x'' + 2x' + x = 3\sin 10t$

10. $x'' + 3x' + 3x = 8\cos 10t + 6\sin 10t$

11. $x'' + 4x' + 5x = 10\cos 3t$; $x(0) = x'(0) = 0$

12. $x'' + 6x' + 13x = 10\sin 5t$; $x(0) = x'(0) = 0$

13. $x'' + 2x' + 6x = 3\cos 10t$; $x(0) = 10, x'(0) = 0$

14. $x'' + 8x' + 25x = 5\cos t + 13\sin t$; $x(0) = 5, x'(0) = 0$

15. A mass weighing 100 lb is attached to the end of a spring that is stretched 1 in. by a force of 100 lb. A force $F_0\cos\omega t$ acts on the mass. At what frequency (in hertz) will resonance oscillations occur? Neglect damping.

16. A front-loading washing machine is mounted on a thick rubber pad that acts like a spring; the weight of the machine depresses the pad exactly $\frac{1}{4}$ in. When its rotor spins at ω radians per second, the rotor exerts a vertical force $F_0\cos\omega t$ pounds

on the machine. At what speed (in revolutions per minute) will resonance vibrations occur? Neglect damping.

17. See Fig. 2.27 which shows a mass m on the end of a pendulum (of length L) also attached to a horizontal spring (with constant k). Assume small oscillations of m so that the spring remains essentially horizontal, and neglect damping. Find the natural circular frequency ω_0 in terms of L, k, m, and the gravitational constant g.

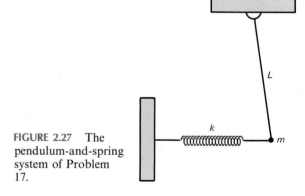

FIGURE 2.27 The pendulum-and-spring system of Problem 17.

18. A mass m hangs on the end of a cord around a pulley of radius a and moment of inertia I, as shown in Fig. 2.28. The rim of the pulley is attached to a spring (with constant k). Assume small oscillations so that the spring remains essentially horizontal, and neglect friction. Find the natural circular frequency of the system in terms of m, a, k, I, and g.

19. A building consists of two floors. The first floor is attached rigidly to the ground, and the second floor is of mass m and

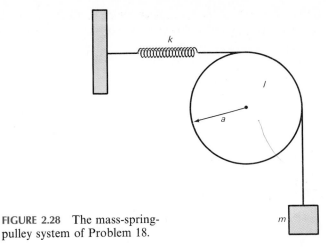

FIGURE 2.28 The mass-spring-pulley system of Problem 18.

weighs 16 tons (32,000 lb). The elastic frame of the building behaves as a spring that resists horizontal displacements of the second floor; it requires a horizontal force of 5 tons to displace the second floor a distance of 1 ft. Assume that in an earthquake the ground oscillates horizontally with amplitude A_0 and circular frequency ω, resulting in an external horizontal force $F(t) = mA_0\omega^2 \sin \omega t$ on the second floor. (a) What is the natural frequency (in hertz) of oscillations of the second floor? (b) If the ground undergoes one oscillation every 2.25 s with an amplitude of 3 in., what is the amplitude of the resulting forced oscillations of the second floor?

20. A mass on a spring without damping is acted upon by the external force $F(t) = F_0 \cos^3 \omega t$. Show that there are *two* values of ω for which resonance occurs, and find both.

21. Derive the steady periodic solution of $mx'' + cx' + kx = F_0 \sin \omega t$. In particular, show that it is what one would expect—the same as the formula in (20) with the same values of ρ and ω, except with $\sin(\omega t - \alpha)$ in place of $\cos(\omega t - \alpha)$.

22. Derive the steady periodic solution in (24) of Eq. (23)—with both sine and cosine forces—by superposition of the steady

periodic solutions separately corresponding to $E_0 \cos \omega t$ and $F_0 \sin \omega t$ (see Problem 21).

23. Recall that the amplification factor ρ is given in terms of the impressed frequency ω by

$$\rho = k[(k - m\omega^2)^2 + (c\omega)^2]^{-1/2}.$$

(a) If $c \geqq c_{CR}/\sqrt{2}$, where $c_{CR} = \sqrt{4km}$, show that ρ steadily decreases as ω increases. (b) If $c < c_{CR}\sqrt{2}$, show that ρ attains a maximum value (practical resonance) when

$$\omega = \omega_m = \sqrt{\frac{k}{m} - \frac{c^2}{2m^2}} < \omega_0 = \sqrt{\frac{k}{m}}.$$

24. Consider the car discussed in this section—with $m = 50$ slugs, $c = 200$ lb-s/ft, $k = 4800$ lb/ft—traveling with velocity v on a washboard road surface described by $y = \frac{1}{6} \sin (2\pi s/30)$. Find the velocity v_m (in miles per hour) at which practical resonance occurs and the amplitude (in inches) of the oscillations of the car at this critical speed (see Problem 23).

25. As indicated by the cart-with-flywheel example discussed in this section, an unbalanced rotating machine part typically results in a force having amplitude proportional to the *square* of the frequency ω. (a) Show that the amplitude of the steady periodic solution of the differential equation

$$mx'' + cx' + kx = mA\omega^2 \cos \omega t$$

(with a forcing term similar to that in Eq. (17)) is $\rho(mA/k)$, where the amplification factor is

$$\rho = k\omega^2[(k - m\omega^2)^2 + (c\omega)^2]^{-1/2}.$$

(b) Suppose that $c^2 < 2mk$. Show that the maximum amplitude occurs at the frequency ω_m given by

$$\omega_m^2 = \frac{k}{m}\left(\frac{2mk}{2mk - c^2}\right).$$

Thus the resonance frequency in this case is *larger* (in contrast to the result of Problem 23) than the natural frequency $\omega_0 = \sqrt{k/m}$. (*Suggestion:* Maximize the *square* of ρ.)

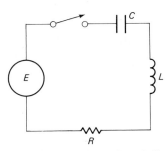

FIGURE 2.29 The series RLC circuit.

*2.9

Electrical Circuits

Here we examine the RLC circuit that is a basic building block in more complicated electrical circuits and networks. As shown in Fig. 2.29 it consists of

A **resistor** with a resistance of R *ohms*,

An **inductor** with an inductance of L *henries*, and

A **capacitor** with a capacitance of C *farads*

in series with a source of electromotive force (such as a battery or a generator) that supplies a voltage of $E(t)$ *volts* at time t. If the switch shown in the circuit of Fig. 2.29 is closed, this results in a current of $I(t)$ *amperes* in the circuit and a charge of $Q(t)$ *coulombs* on the capacitor at time t. The relation between the functions Q and I is

$$\frac{dQ}{dt} = I. \tag{1}$$

We will always use mks electrical units, in which time is measured in seconds.

According to elementary principles of electricity, the **voltage drops** across the three circuit elements are those shown in the table in Fig. 2.30. We can analyze the behavior of the series circuit of Fig. 2.29 with the aid of this table and one of Kirchhoff's laws:

Circuit element	Voltage drop
Inductor	$L \frac{dI}{dt}$
Resistor	RI
Capacitor	$\frac{1}{C} Q$

FIGURE 2.30 Table of voltage drops.

The (algebraic) sum of the voltage drops across the elements in a simple loop of an electrical circuit is equal to the applied voltage.

As a consequence, the current and charge in the simple RLC circuit of Fig. 2.29 satisfy the basic circuit equation

$$L \frac{dI}{dt} + RI + \frac{1}{C} Q = E(t). \tag{2}$$

If we substitute (1) in (2), we get the second order linear differential equation

$$LQ'' + RQ' + \frac{1}{C} Q = E(t) \tag{3}$$

for the charge $Q(t)$, under the assumption that the voltage $E(t)$ is known.

In most practical problems it is the current I rather than the charge Q that is of primary interest, so we differentiate each side of Eq. (3) and substitute $I = Q'$ to obtain

$$LI'' + RI' + \frac{1}{C} I = E'(t). \tag{4}$$

We do *not* assume here a prior familiarity with electrical circuits. It suffices to regard the resistor, inductor, and capacitor in an electrical circuit as "black boxes" that are calibrated by the constants R, C, and L above. A battery or generator is described by the voltage $E(t)$ that it supplies. When the switch is open, no current flows in the circuit; when it is closed, there is a current $I(t)$ in the circuit and a charge $Q(t)$ on the capacitor. All we need to know about these constants and varying quantities is that they satisfy Eqs. (1)–(4)—our mathematical model for the RLC circuit. We can then learn a good deal about electricity by studying the mathematical model.

Note that the equations in (3) and (4) have precisely the same form as the equation

$$mx'' + cx' + kx = F(t) \tag{5}$$

of a mass-spring-dashpot system with external force $F(t)$. The table in Fig. 2.31 details this important **mechanical-electrical analogy**. As a consequence, most of

Mechanical system	Electrical system
Mass m	Inductance L
Damping constant c	Resistance R
Spring constant k	Reciprocal capacitance $1/C$
Position x	Charge Q (using (3)) (or current I using (4))
Force F	Electromotive force E (or its derivative E')

FIGURE 2.31 Mechanical-electrical analogies.

the results derived in Section 2.8 for mechanical systems can be applied at once to electrical circuits. The fact that the same differential equation serves as a mathematical model for such different physical systems is a striking illustration of the unifying role of mathematics in the investigation of natural phenomena. More concretely, the correspondences in Fig. 2.31 can be used to construct an electrical model of a given mechanical system, using inexpensive and readily available circuit elements. The performance of the mechanical system can then be predicted by means of accurate and simple measurements in the electrical circuit. This is especially useful when the actual mechanical system would be expensive to construct and when measurements of displacements and velocities would be inconvenient or inaccurate. This idea is the basis of *analog computers*—electrical models of mechanical systems.

In the typical case of an alternating current voltage $E(t) = E_0 \sin \omega t$, Eq. (4) takes the form

$$LI'' + RI' + \frac{1}{C}I = \omega E_0 \cos \omega t. \tag{6}$$

As in a mass-spring-dashpot system with a simple harmonic external force, the solution of Eq. (6) is the sum of a **transient current** I_{tr} that approaches zero as $t \to +\infty$ (under the assumption that the coefficients in (5) are all positive, so the characteristic roots have negative real parts), and a **steady periodic current** I_{sp}; thus

$$I = I_{tr} + I_{sp}. \tag{7}$$

Recall from Section 2.8 (Eqs. (19)–(21) there) that the steady periodic solution of Eq. (5) with $F(t) = F_0 \cos \omega t$ is

$$x_{sp}(t) = \frac{F_0 \cos (\omega t - \alpha)}{\sqrt{(k - m\omega^2)^2 + (c\omega)^2}}$$

where

$$\alpha = \tan^{-1} \frac{c\omega}{k - m\omega^2}, \qquad 0 \leqq \alpha \leqq \pi.$$

If we make the substitutions L for m, R for c, $1/C$ for k, and ωE_0 for F_0, we get the steady periodic current

$$I_{sp}(t) = \frac{E_0 \cos (\omega t - \alpha)}{\sqrt{R^2 + \left(\omega L - \dfrac{1}{\omega C}\right)^2}} \tag{8}$$

with the phase angle given by

$$\alpha = \tan^{-1} \frac{\omega RC}{1 - LC\omega^2}, \qquad 0 \leqq \alpha \leqq \pi. \tag{9}$$

The quantity

$$Z = \sqrt{R^2 + \left(\omega L - \frac{1}{\omega C}\right)^2} \text{ (ohms)} \tag{10}$$

is called the **impedance** of the circuit. Then the steady periodic current

$$I_{sp}(t) = \frac{E_0}{Z} \cos (\omega t - \alpha) \tag{11}$$

has amplitude

$$I_0 = \frac{E_0}{Z}, \tag{12}$$

reminiscent of Ohm's law $I = E/R$.

Equation (11) gives the steady periodic current as a cosine function, whereas the input voltage $E(t) = E_0 \sin \omega t$ was a sine function. To convert I_{sp} to a sine function, we first introduce the **reactance**

$$S = \omega L - \frac{1}{\omega C}. \tag{13}$$

Then $Z = \sqrt{R^2 + S^2}$, and we see from Eq. (9) that α is as in Fig. 2.32, with delay angle $\delta = \alpha - \pi/2$. Equation (11) now yields

$$I_{sp} = \frac{E_0}{Z} (\cos \alpha \cos \omega t + \sin \alpha \sin \omega t)$$

$$= \frac{E_0}{Z} \left(-\frac{S}{Z} \cos \omega t + \frac{R}{Z} \sin \omega t\right)$$

$$= \frac{E_0}{Z} (\cos \delta \sin \omega t - \sin \delta \cos \omega t).$$

Therefore

$$I_{sp}(t) = \frac{E_0}{Z} \sin (\omega t - \delta), \tag{14}$$

where

$$\delta = \tan^{-1} \frac{S}{R} = \tan^{-1} \frac{LC\omega^2 - 1}{\omega RC}. \tag{15}$$

This finally gives the **time lag** δ/ω (in seconds) of the steady periodic current I_{sp} behind the input voltage (see Fig. 2.33).

When we want to find the transient current, we are usually given the initial values $I(0)$ and $Q(0)$. So we first must find $I'(0)$. To do so, we substitute $t = 0$ in

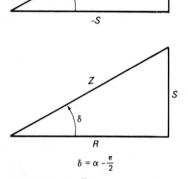

$$\delta = \alpha - \frac{\pi}{2}$$

FIGURE 2.32   Reactance and delay angle.

CHAPTER 2:   Linear Equations of Higher Order

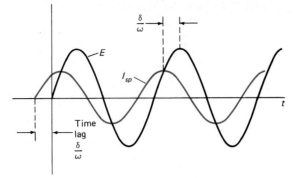

FIGURE 2.33 Time lag of current behind imposed voltage.

Eq. (2) to obtain the equation

$$LI'(0) + RI(0) + \frac{1}{C} Q(0) = E(0) \tag{16}$$

to determine $I'(0)$ in terms of the initial values of current, charge, and voltage.

EXAMPLE 1 Consider an RLC circuit with $R = 50$ ohms, $L = 0.1$ henry (H), and $C = 5 \times 10^{-4}$ farad (F). At time $t = 0$, when both $I(0)$ and $Q(0)$ are zero, the circuit is connected to a 110-V, 60-Hz alternating current generator. Find the current in the circuit and the lag of the steady periodic current behind the voltage.

SOLUTION A frequency of 60 Hz means that $\omega = (2\pi)(60)$ rad/s, approximately 377 rad/s. So we take $E(t) = 110 \sin 377t$, and use equality in place of the symbol for approximate equality in the rest of this discussion. The differential equation in (6) takes the form

$$(0.1)I'' + 50I' + 2000I = (377)(110) \cos 377t.$$

We substitute the given values of R, L, C, and $\omega = 377$ in (10) to find that the impedance is $Z = 59.58$ ohms, so the steady periodic amplitude is

$$I_0 = \frac{110 \text{ (volts)}}{59.58 \text{ (ohms)}} = 1.846 \text{ amperes (A)}.$$

With the same data, Eq. (15) gives the sine phase angle:

$$\delta = \tan^{-1} (0.648) = 0.575.$$

Thus the time lag of current behind voltage is

$$\frac{\delta}{\omega} = \frac{0.575}{377} = 0.0015 \text{ s},$$

and the steady periodic current is $I_{sp} = (1.846) \sin (377t - 0.575)$.
The characteristic equation $(0.1)r^2 + 50r + 2000 = 0$ has roots $r_1 \approx -44$ and $r_2 \approx -456$. With these approximations, the general solution is

$$I(t) = c_1 e^{-44t} + c_2 e^{-456t} + (1.846) \sin (377t - 0.575),$$

with derivative

$$I'(t) = -44c_1e^{-44t} - 456c_2e^{-456t} + 696\cos(377t - 0.575).$$

Because $I(0) = Q(0) = 0$, Eq. (16) gives $I'(0) = 0$ as well. With these initial values substituted, we obtain the equations

$$I(0) = c_1 + c_2 - 1.004 = 0,$$

$$I'(0) = -44c_1 - 456c_2 + 584 = 0;$$

the solution is $c_1 = -0.307$, $c_2 = 1.311$. Thus the transient solution is

$$I_{tr}(t) = (-0.307)e^{-44t} + (1.311)e^{-456t},$$

and it dies out very rapidly, indeed.

EXAMPLE 2 Suppose that the RLC circuit of Example 1, still with $I(0) = Q(0) = 0$, is connected at time $t = 0$ to a battery supplying a constant 110 V. Now find the current in the circuit.

SOLUTION We now have $E(t) = 110$, so Eq. (16) gives

$$I'(0) = \frac{E(0)}{L} = \frac{110}{0.1} = 1100 \text{ A/s},$$

and the differential equation is

$$(0.1)I'' + 50I' + 2000I = E'(t) = 0.$$

Its general solution is the complementary function we found in Example 1: $I(t) = c_1e^{-44t} + c_2e^{-456t}$. We solve the equations

$$I(0) = c_1 + c_2 = 0,$$

$$I'(0) = -44c_1 - 345c_2 = 1100$$

for $c_1 = -c_2 = 2.671$. Therefore $I(t) = (2.671)(e^{-44t} - e^{-456t})$. Note that $I \to 0$ as $t \to +\infty$, even though the voltage is constant.

ELECTRICAL RESONANCE

Consider again the current differential equation in (6) corresponding to a sinusoidal input voltage $E(t) = E_0 \sin \omega t$. We have seen that the amplitude of its steady periodic current is

$$I_0 = \frac{E_0}{Z} = \frac{E_0}{\sqrt{R^2 + \left(\omega L - \dfrac{1}{\omega C}\right)^2}}. \tag{17}$$

For typical values of the constants R, L, C, and E_0, the graph of I_0 as a function of ω resembles the one shown in Fig. 2.34. It reaches a maximum value at $\omega_m = 1/\sqrt{LC}$ and then approaches zero as $\omega \to +\infty$; the critical frequency ω_m is the **resonance frequency** of the circuit.

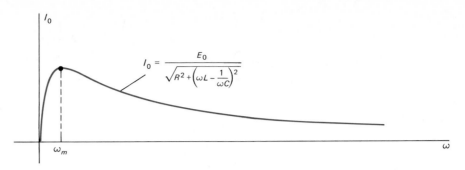

FIGURE 2.34 Effect of frequency on I_0.

In Section 2.8 we emphasized the importance of avoiding resonance in most mechanical systems (the seismograph is an example of a mechanical system in which resonance is *sought*). By contrast, many common electrical devices could not function properly without taking advantage of the phenomenon of resonance. The radio is a familiar example. A highly simplified model of its tuning circuit is the RLC circuit we have discussed. Its inductance L and resistance R are constant, but its capacitance C is varied as one operates the tuning dial (thereby varying the effective area of the plates of its variable capacitor).

Suppose that we want to pick up a particular radio station that is broadcasting at frequency ω, and thereby (in effect) provides an input voltage $E(t) = E_0 \sin \omega t$ to the tuning circuit of the radio. The resulting steady periodic current I_{sp} in the tuning circuit drives its amplifier, and in turn its loudspeaker, with the volume of sound that we hear roughly proportional to the amplitude I_0 of I_{sp}. In order to hear our preferred station (of frequency ω) the loudest (and simultaneously tune out stations broadcasting at other frequencies), we therefore want to choose C to maximize I_0. But examine Eq. (17), thinking of ω as constant and with C the only variable. We see at a glance—no calculus required—that I_0 is maximal when

$$\omega L - \frac{1}{\omega C} = 0;$$

that is, when

$$C = \frac{1}{L\omega^2}. \tag{18}$$

So we merely turn the dial to set the capacitance at this value.

This is the way that the old crystal radios worked, but modern AM radios have a more sophisticated design. A *pair* of variable capacitors are used: The first controls the frequency selected as described above; the second controls the frequency of a signal that the radio itself generates, kept close to 455 kilohertz (kHz) above the desired frequency. The resulting *beat* frequency of 455 kHz, known as the *intermediate frequency*, is then amplified in several stages. This technique has the advantage that the several RLC circuits used in the amplification stages easily can be designed to resonate at 455 kHz and reject other frequencies, resulting in more selectivity of the receiver as well as better amplification of the desired signal.

Problems 1–6 deal with the LR circuit of Fig. 2.35, a series circuit containing an inductor with an inductance of L henries, a resistor with a resistance of R ohms (Ω), and a source of electromotive force (emf), but no capacitor. In this case Eq. (2) reduces to the linear first order equation

$$LI' + RI = E(t).$$

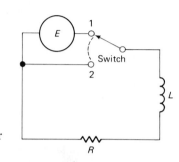

FIGURE 2.35 The circuit for Problems 1–6.

Eq. (3) gives the linear first order differential equation

$$R\frac{dQ}{dt} + \frac{1}{C}Q = E(t)$$

for the charge $Q = Q(t)$ on the capacitor at time t. Note that $I(t) = Q'(t)$.

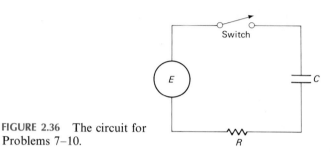

FIGURE 2.36 The circuit for Problems 7–10.

1. In the circuit of Fig. 2.35, suppose that $L = 5$ H, $R = 25\ \Omega$, and that the source E of emf is a battery supplying 100 V to the circuit. Suppose also that the switch has been in position 1 for a long time, so that a steady current of 4 A is flowing in the circuit. At time $t = 0$, the switch is thrown to position 2, so that $I(0) = 4$ and $E = 0$ for $t \geq 0$. Find $I(t)$.

2. Given the same circuit as in Problem 1, suppose that the switch is initially in position 2, but is thrown to position 1 at time $t = 0$, so that $I(0) = 0$ and $E = 100$ for $t \geq 0$. Find $I(t)$, and show that $I(t)$ approaches 4 as $t \to +\infty$.

3. The battery in Problem 2 is replaced with an alternating current generator that supplies a voltage of $E(t) = 100\cos 60t$ volts. With everything else the same, now find $I(t)$.

4. In the circuit of Fig. 2.35, with the switch in position 1, suppose that $L = 2$, $R = 40$, $E = 100e^{-10t}$, and $I(0) = 0$. Find the maximum current in the circuit for $t \geq 0$.

5. In the circuit of Fig. 2.35, with the switch in position 1, suppose that $L = 2$, $R = 20$, $I(0) = 0$, and $E = 100e^{-10t}\cos 60t$. Find $I(t)$.

6. In the circuit of Fig. 2.35, with the switch in position 1, take $L = 1$, $R = 10$, and $E = 30\cos 60t + 40\sin 60t$. (a) Substitute $I_{sp} = A\cos 60t + B\sin 60t$ and then determine A and B to find the steady-state current I_{sp} in the circuit. (b) Write the solution in the form $I_{sp} = C\cos(\omega t - \alpha)$.

Problems 7–10 deal with the RC circuit shown in Fig. 2.36 containing a resistor (R ohms), a capacitor (C farads), a switch, a source E of emf, but no inductor. Substitution of $L = 0$ in

7. (a) Find the charge $Q(t)$ and current $I(t)$ in the RC circuit if $E(t) = E_0$ (a constant voltage supplied by a battery) and the switch is closed at time $t = 0$, so that $Q(0) = 0$. (b) Show that
$$\lim_{t \to +\infty} Q(t) = E_0 C \text{ and that } \lim_{t \to +\infty} I(t) = 0.$$

8. Suppose that in the circuit of Fig. 2.36, we have $R = 10$, $C = 0.02$, $Q(0) = 0$, and $E(t) = 100e^{-5t}$ (volts). (a) Find $Q(t)$ and $I(t)$. (b) What is the maximum charge on the capacitor for $t \geq 0$, and when does it occur?

9. Suppose that in the circuit of Fig. 2.36, $R = 200$, $C = 2.5 \times 10^{-4}$, $Q(0) = 0$, and $E(t) = 100\cos 120t$. (a) Find $Q(t)$ and $I(t)$. (b) What is the amplitude of the steady-state current?

10. An emf of voltage $E(t) = E_0\cos \omega t$ is applied to the RC circuit of Fig. 2.36 at time $t = 0$, and $Q(0) = 0$. Substitute $Q_{sp} = A\cos \omega t + B\sin \omega t$ in the differential equation to show that the steady periodic charge on the capacitor is

$$Q_{sp} = \frac{E_0 C}{(1 + \omega^2 R^2 C^2)^{1/2}}\cos(\omega t - \beta)$$

where $\beta = \tan^{-1}(\omega RC)$.

In each of Problems 11–16, the parameters of an RLC circuit with input voltage $E(t)$ are given. Substitute $I_{sp} = A\cos \omega t + B\sin \omega t$ in Eq. (4), using the appropriate value of ω, to find the steady periodic current in the form $I_{sp} = I_0\sin(\omega t - \delta)$.

11. $R = 30\ \Omega$, $L = 10$ H, $C = 0.02$ F; $E(t) = 50\sin 2t$ volts

12. $R = 200\ \Omega$, $L = 5$ H, $C = 0.001$ F; $E(t) = 100\sin 10t$ volts

13. $R = 20\ \Omega$, $L = 10$ H, $C = 0.01$ F; $E(t) = 200\cos 5t$ volts

14. $R = 50\ \Omega$, $L = 5$ H, $C = 0.005$ F;
$E(t) = 300 \cos 100t + 400 \sin 100t$ volts

15. $R = 100\ \Omega$, $L = 2$ H, $C = 5 \times 10^{-6}$ F;
$E(t) = 110 \sin 60\pi t$ volts

16. $R = 25\ \Omega$, $L = 0.2$ H, $C = 5 \times 10^{-4}$ F;
$E(t) = 120 \cos 377t$ volts

In each of Problems 17–22, an RLC circuit with input voltage $E(t)$ is described. Find the current $I(t)$ given the initial current (in amperes) and charge on the capacitor (in coulombs).

17. $R = 16\ \Omega$, $L = 2$ H, $C = 0.02$ F; $E(t) = 100$ V;
$I(0) = 0$, $Q(0) = 5$

18. $R = 60\ \Omega$, $L = 2$ H, $C = 0.0025$ F; $E(t) = 100e^{-t}$ V;
$I(0) = Q(0) = 0$

19. $R = 60\ \Omega$, $L = 2$ H, $C = 0.0025$ F; $E(t) = 100e^{-10t}$ V;
$I(0) = 0$, $Q(0) = 1$

20. The circuit and input voltage of Problem 11 with $I(0) = 0$ and $Q(0) = 0$.

21. The circuit and input voltage of Problem 13 with $I(0) = 0$ and $Q(0) = 3$.

22. The circuit and input voltage of Problem 15 with $I(0) = 0$ and $Q(0) = 0$.

23. Consider an LC circuit—that is, an RLC circuit with $R = 0$—with input voltage $E(t) = E_0 \sin \omega t$. Show that unbounded oscillations of current occur for a certain resonance frequency; express this frequency in terms of L and C.

24. It was stated in the text that, if R, L, and C are positive, then any solution of $LI'' + RI' + I/C = 0$ is a transient solution—it approaches zero as $t \to +\infty$. Prove this.

25. Prove that the amplitude I_0 of the steady periodic solution of Eq. (6) is maximal at frequency $\omega = 1/\sqrt{LC}$.

*2.10

Endpoint Problems and Eigenvalues

You are now familiar with the fact that a solution of a second order linear differential equation is uniquely determined by two initial conditions. In particular, the only solution of the initial value problem

$$y'' + p(x)y' + q(x)y = 0; \qquad y(a) = 0, \qquad y'(a) = 0 \tag{1}$$

is the trivial solution $y(x) \equiv 0$. Most of Chapter 2 has been based, directly or indirectly, on the uniqueness of solutions of linear initial value problems (as guaranteed by Theorem 2 of Section 2.2).

In this section we will see that the situation is radically different for a problem such as

$$y'' + p(x)y' + q(x)y = 0; \qquad y(a) = 0, \qquad y(b) = 0. \tag{2}$$

The difference between the problems in Eqs. (1) and (2) is that in (2) the two conditions are imposed at two *different* points a and b with (say) $a < b$. In (2) we are to find a solution of the differential equation on the interval (a, b) that satisfies the conditions $y(a) = 0$ and $y(b) = 0$ at the endpoints of the interval. Such a problem is called an **endpoint** or **boundary value** problem. The following two examples illustrate the sorts of complications that can arise in endpoint problems.

EXAMPLE 1 Consider the endpoint problem

$$y'' + 3y = 0; \qquad y(0) = 0, \qquad y(\pi) = 0. \tag{3}$$

The general solution of the differential equation is

$$y(x) = A \cos \sqrt{3}x + B \sin \sqrt{3}x.$$

The endpoint conditions give

$$y(0) = A = 0 \quad \text{and} \quad y(\pi) = B \sin \pi\sqrt{3} \approx (-0.7458)B = 0,$$

so $B = 0$ as well. Thus the only solution of the endpoint problem in (3) is the trivial solution $y(x) \equiv 0$; there is no surprise here.

EXAMPLE 2 Consider the endpoint problem

$$y'' + 4y = 0; \qquad y(0) = 0, \qquad y(\pi) = 0. \qquad (4)$$

The general solution of the differential equation is

$$y(x) = A \cos 2x + B \sin 2x.$$

The endpoint conditions yield $y(0) = A = 0$ and

$$y(\pi) = A \cos 2\pi + B \sin 2\pi = B \cdot 0 = 0.$$

No matter what the value of B, the condition $y(\pi) = 0$ is automatically satisfied because $\sin 2\pi = 0$. Thus the function

$$y(x) = B \sin 2x$$

satisfies the endpoint problem in (4) for *every* value of B. This is an example of an endpoint problem having infinitely many nontrivial solutions!

Rather than being the exceptional cases, these two examples illustrate the typical situation for an endpoint problem as in (2): It may have no nontrivial solution, or it may have infinitely many nontrivial solutions. Note that the problems in (3) and (4) can both be written in the form

$$y'' + p(x)y' + \lambda q(x)y = 0; \qquad y(a) = 0, \qquad y(b) = 0, \qquad (5)$$

with $p(x) \equiv 0$, $q(x) \equiv 1$, $a = 0$, and $b = \pi$. The number λ is a parameter in the problem (nothing to do with the parameters that were varied in Section 2.7). If we take $\lambda = 3$, we get the equations in (3); with $\lambda = 4$, we obtain the equations in (4). Examples 1 and 2 show that the situation in an endpoint problem containing a parameter can (and generally will) depend strongly upon the specific numerical value of the parameter.

An endpoint problem containing a parameter λ, such as the endpoint problem in (5), is called an **eigenvalue problem**. The question we ask in an eigenvalue problem is this: For what (real) values of the parameter λ does there exist a *nontrivial* solution of the endpoint problem? Such a value of λ is called an **eigenvalue** or **characteristic value** of the problem. Thus we saw in Example 2 that $\lambda = 4$ is an eigenvalue of the eigenvalue problem

$$y'' + \lambda y = 0; \qquad y(0) = 0, \qquad y(\pi) = 0. \qquad (6)$$

In Example 1 we saw that $\lambda = 3$ is not an eigenvalue of this problem.

Suppose that λ_* is an eigenvalue of the problem in (5), and that $y_*(x)$ is a nontrivial solution of the problem with this value of λ inserted; that is, that

$$y_*'' + p(x)y_*' + \lambda_* q(x)y_* = 0$$

CHAPTER 2:   Linear Equations of Higher Order

and

$$y_*(a) = 0, \qquad y_*(b) = 0.$$

Then we call y_* an **eigenfunction** associated with the eigenvalue λ_*. Thus we saw in Example 2 that $y_* = \sin 2x$ is an eigenfunction associated with the eigenvalue $\lambda_* = 4$, as is any constant multiple of $\sin 2x$.

More generally, note that the problem in (5) is *homogeneous* in the sense that any constant multiple of an eigenfunction is again an eigenfunction; indeed, one associated with the same eigenvalue. That is, if $y = y_*(x)$ satisfies the problem in (5) with $\lambda = \lambda_*$, then so does any constant multiple $cy_*(x)$. It can be proved under mild restrictions on the coefficient functions p and q that any two eigenfunctions associated with the same eigenvalue must be linearly dependent.

EXAMPLE 3 Determine the eigenvalues and associated eigenfunctions of the problem

$$y'' + \lambda y = 0; \qquad y(0) = 0, \qquad y(L) = 0 \qquad (L > 0). \qquad (7)$$

SOLUTION We must consider all possible (real) values of λ—positive, zero, and negative.

If $\lambda = 0$, then the equation is simply $y'' = 0$, and its general solution is

$$y(x) = Ax + B.$$

Then the endpoint conditions $y(0) = 0 = y(L)$ immediately imply that $A = B = 0$, so the only solution in this case is the trivial function $y(x) \equiv 0$. Therefore, $\lambda = 0$ is *not* an eigenvalue of the problem in (7).

If $\lambda < 0$, let us then write $\lambda = -\alpha^2$ (with $\alpha > 0$) to be specific. Then the differential equation takes the form

$$y'' - \alpha^2 y = 0,$$

and its general solution is

$$y(x) = c_1 e^{\alpha x} + c_2 e^{-\alpha x} = A \cosh \alpha x + B \sinh \alpha x,$$

where $A = c_1 + c_2$ and $B = c_1 - c_2$. (Recall that $\cosh \alpha x = (e^{\alpha x} + e^{-\alpha x})/2$ and $\sinh \alpha x = (e^{\alpha x} - e^{-\alpha x})/2$.) The condition $y(0) = 0$ then gives

$$y(0) = A \cosh 0 + B \sinh 0 = A = 0,$$

so that $y = B \sinh \alpha x$. But now the second endpoint condition, $y(L) = 0$, gives $y(L) = B \sinh \alpha L = 0$. This implies that $B = 0$, because $\alpha \neq 0$, and $\sinh x = 0$ only for $x = 0$ (examine the graphs of $\sinh x$ and $\cosh x$ in Fig. 2.37). Thus the only solution of the problem in (7) in the case $\lambda < 0$ is the trivial solution $y \equiv 0$, and we may therefore conclude that the problem has *no* negative eigenvalues.

The only remaining possibility is that $\lambda = \alpha^2 > 0$ with $\alpha > 0$. In this case the differential equation is

$$y'' + \alpha^2 y = 0,$$

with general solution

$$y(x) = A \cos \alpha x + B \sin \alpha x.$$

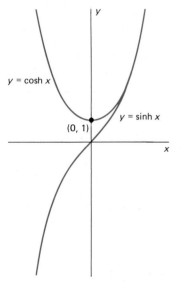

$y = \cosh x$

$y = \sinh x$

$(0, 1)$

FIGURE 2.37 The hyperbolic sine and cosine graphs.

The condition that $y(0) = 0$ implies that $A = 0$, so $y(x) = B \sin \alpha x$. The condition $y(L) = 0$ then gives

$$y(L) = B \sin \alpha L = 0.$$

Can this occur if $B \neq 0$? Yes, but only provided that αL is a (positive) integral multiple of π:

$$\alpha L = \pi, \quad 2\pi, \quad 3\pi, \ldots, \quad n\pi, \ldots;$$

that is, if

$$\lambda = \alpha^2 = \frac{\pi^2}{L^2}, \quad \frac{4\pi^2}{L^2}, \quad \frac{9\pi^2}{L^2}, \ldots, \quad \frac{n^2\pi^2}{L^2}, \ldots.$$

Thus we have discovered that the problem in (7) has an *infinite sequence* of positive eigenvalues,

$$\lambda_n = \frac{n^2\pi^2}{L^2}, \qquad n = 1, 2, 3, \ldots. \tag{8}$$

With $B = 1$, the eigenfunction associated with the eigenvalue λ_n is

$$y_n(x) = \sin \frac{n\pi x}{L}, \qquad n = 1, 2, 3, \ldots. \tag{9}$$

Example 3 illustrates the general situation. According to a theorem whose precise statement we will defer until Section 9.1, under the assumption that $q(x) > 0$ on the interval $[a, b]$, any eigenvalue problem of the form in (5) has a divergent increasing sequence

$$\lambda_1 < \lambda_2 < \lambda_3 < \cdots < \lambda_n < \cdots \to +\infty$$

of eigenvalues, each with an associated eigenfunction. This is also true of the following slightly more general type of eigenvalue problem, in which the endpoint conditions involve values of the derivative y' as well as values of y:

$$y'' + p(x)y' + \lambda q(x)y = 0; \quad a_1 y(a) + a_2 y'(a) = 0, \quad b_1 y(b) + b_2 y'(b) = 0, \quad (10)$$

where a_1, a_2, b_1, and b_2 are given constants. With $a_1 = 1 = b_2$ and $a_2 = 0 = b_1$, we get the problem of the following example (in which $p(x) \equiv 0$ and $q(x) \equiv 1$, as in the previous examples).

EXAMPLE 4 Determine the eigenvalues and eigenfunctions of the problem

$$y'' + \lambda y = 0; \quad y(0) = 0, \quad y'(L) = 0. \tag{11}$$

SOLUTION Virtually the same argument as that used in Example 3 shows that the only possible eigenvalues are positive, so we take $\lambda = \alpha^2 > 0$ to be specific. Then the differential equation is

$$y'' + \alpha^2 y = 0,$$

with general solution

$$y(x) = A \cos \alpha x + B \sin \alpha x.$$

The condition $y(0) = 0$ immediately gives $A = 0$, so

$$y(x) = B \sin \alpha x \quad \text{and} \quad y'(x) = B\alpha \cos \alpha x.$$

The second endpoint condition $y'(L) = 0$ now gives

$$y'(L) = B\alpha \cos \alpha L = 0.$$

This will be so with $B \neq 0$ provided that αL is an odd positive multiple of $\pi/2$:

$$\alpha L = \frac{\pi}{2}, \quad \frac{3\pi}{2}, \ldots, \quad \frac{(2n-1)\pi}{2}, \ldots;$$

that is, if

$$\lambda = \frac{\pi^2}{4L^2}, \quad \frac{9\pi^2}{4L^2}, \ldots, \quad \frac{(2n-1)^2\pi^2}{4L^2}, \ldots.$$

Thus the nth eigenvalue λ_n and associated eigenfunction of the problem in (11) are given by

$$\lambda_n = \frac{(2n-1)^2\pi^2}{4L^2} \quad \text{and} \quad y_n(x) = \sin\frac{(2n-1)\pi x}{2L} \quad \text{for } n = 1, 2, 3, \ldots. \quad (12)$$

A general procedure for determining the eigenvalues of the problem in (10) can be outlined as follows. We first write the general solution of the differential equation in the form

$$y = Ay_1(x, \lambda) + By_2(x, \lambda).$$

We write $y_i(x, \lambda)$ because y_1 and y_2 will depend upon λ, as in Examples 3 and 4, in which

$$y_1 = \cos \alpha x = \cos \sqrt{\lambda} x \quad \text{and} \quad y_2 = \sin \alpha x = \sin \sqrt{\lambda} x.$$

Then we impose the two endpoint conditions, noting that each is linear in y and y', and hence also linear in A and B. When we collect coefficients of A and B in the resulting pair of equations, we therefore get a system of the form

$$\left.\begin{array}{l} \alpha_1(\lambda)A + \beta_1(\lambda)B = 0, \\ \alpha_2(\lambda)A + \beta_2(\lambda)B = 0. \end{array}\right\} \quad (13)$$

Now λ is an eigenvalue if and only if (13) has a nontrivial solution (one with A and B not both zero). But such a homogeneous system of linear equations has a nontrivial solution if and only if the determinant of its coefficients vanishes. We therefore conclude that the eigenvalues of the problem in (10) are the (real) solutions of the equation

$$D(\lambda) = \alpha_1(\lambda)\beta_2(\lambda) - \alpha_2(\lambda)\beta_1(\lambda) = 0. \quad (14)$$

This can be a formidable equation to solve and may require a numerical approximation technique such as Newton's method.

Much of the interest in eigenvalue problems is due to their very diverse physical applications. The remainder of this section is devoted to two such applications. Numerous additional applications are included in Chapters 8 and 9 (on partial differential equations and boundary value problems).

THE WHIRLING STRING

Who of us has not wondered about the shape of a quickly spinning jump rope? Let us consider the shape assumed by a tightly stretched flexible string of length L and constant linear density ρ (mass per unit length) if it is rotated or whirled (like a jump rope) with constant angular speed ω (in radians per second) about its equilibrium position along the x-axis. We assume that the portion of the string to one side of any point exerts a constant tension force T on the portion of the string to the other side of the point, with the direction of T tangential to the string. We further assume that, as the string whirls around the x-axis, each point moves in a circle centered at that point's equilibrium position on the x-axis. Thus the string is elastic, so that as it whirls it also stretches to assume a curved shape. Denote by $y(x)$ the displacement of the string from the axis of rotation. Finally, we assume that the deflection of the string is so slight that $\sin \theta \approx \tan \theta \approx y'(x)$ in Fig. 2.38(c).

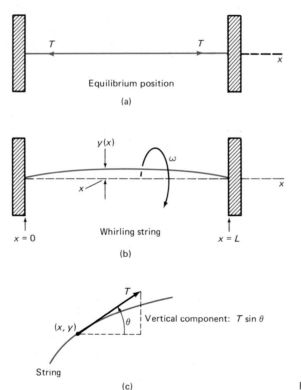

FIGURE 2.38   The whirling string.

We plan to derive a differential equation for $y(x)$ by application of Newton's law $F = ma$ to the piece of string of mass $\rho \, \Delta x$ corresponding to the interval $[x, x + \Delta x]$. The only forces acting on this piece are the tension forces at its two ends. From Fig. 2.39 we see that the net vertical force in the positive y-direction is

$$F = T \sin (\theta + \Delta\theta) - T \sin \theta$$
$$\approx T \tan (\theta + \Delta\theta) - T \tan \theta,$$

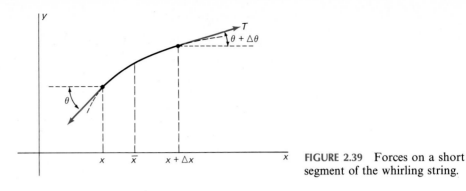

FIGURE 2.39 Forces on a short segment of the whirling string.

so that

$$F \approx Ty'(x + \Delta x) - Ty'(x). \tag{15}$$

Next we recall from elementary calculus or physics the formula $a = r\omega^2$ for the (inward) centripetal acceleration of a body in uniform circular motion (r is the radius of the circle and ω is the angular speed of the body). Here we have $r = y$, so the vertical acceleration of our piece of string is $a = -\omega^2 y$, the minus sign because the inward direction is the negative y-direction. Because $m = \rho \, \Delta x$, substitution of this and (15) in $F = ma$ yields

$$Ty'(x + \Delta x) - Ty'(x) \approx -\rho\omega^2 y \, \Delta x,$$

so that

$$T \cdot \frac{y'(x + \Delta x) - y'(x)}{\Delta x} \approx -\rho\omega^2 y.$$

We now take the limit as $\Delta x \to 0$ to get the differential equation of motion of the string:

$$Ty'' + \rho\omega^2 y = 0. \tag{16}$$

If we write

$$\lambda = \frac{\rho\omega^2}{T} \tag{17}$$

and impose the condition that the ends of the string are fixed, we get finally the eigenvalue problem

$$y'' + \lambda y = 0; \qquad y(0) = 0, \qquad y(L) = 0 \tag{7}$$

that we considered in Example 3. We found there that the eigenvalues of the problem in (7) are

$$\lambda_n = \frac{n^2\pi^2}{L^2}, \qquad n = 1, 2, 3, \ldots, \tag{8}$$

with the eigenfunction $y_n(x) = \sin(n\pi x/L)$ associated with λ_n.

But what does all this mean in terms of our whirling string? It means that unless λ in (17) is one of the eigenvalues in (8), then the only solution of the problem in (7) is the trivial solution $y(x) \equiv 0$. In this case the string remains in its equilibrium position with zero deflection. But if we equate (17) and (8) and solve for the value ω_n corresponding to λ_n,

$$\omega_n = \sqrt{\frac{\lambda_n T}{\rho}} = \frac{n\pi}{L}\sqrt{\frac{T}{\rho}} \tag{18}$$

for $n = 1, 2, 3, \ldots$, we get a sequence of **critical speeds** of angular rotation. Only at these critical angular speeds can the string whirl up out of its equilibrium position. At angular speed ω_n it assumes a shape of the form $y_n = c_n \sin(n\pi x/L)$; our mathematical model is not sufficiently complete to determine the coefficient c_n.

Suppose that we start the string rotating at speed

$$\omega < \omega_1 = \frac{\pi}{L}\sqrt{\frac{T}{\rho}},$$

but gradually increase its speed of rotation. So long as $\omega < \omega_1$, the string remains in its undeflected position $y \equiv 0$. But when $\omega = \omega_1$, the string pops into a whirling position $y = c \sin(\pi x/L)$. And when ω is increased further, the string (theoretically) pops back into its undeflected position along the axis of rotation!

THE BUCKLED ROD

Figure 2.40 shows a uniform rod of length L, hinged at each end, that has been "buckled" by an axial force of compression P applied at each end. We assume this buckling to be so slight that the deflection curve $y = y(x)$ of the rod may be regarded as defined on the interval $0 \leq x \leq L$.

In the theory of elasticity the linear endpoint boundary value problem

$$EIy'' + Py = 0, \qquad y(0) = y(L) = 0 \tag{19}$$

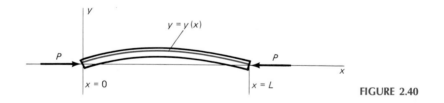

FIGURE 2.40

is used to model the actual (nonlinear) behavior of the rod. As in Section 1.2, E denotes the Young's modulus of the beam material and I denotes the moment of inertia of the beam's cross section about a horizontal line through its centroid.

If we write

$$\lambda = \frac{P}{EI} \tag{20}$$

then the problem in (19) becomes the eigenvalue problem

$$y'' + \lambda y = 0; \qquad y(0) = y(L) = 0 \tag{7}$$

that we considered in Example 3. We found that its eigenvalues $\{\lambda_n\}$ are given by

$$\lambda_n = \frac{n^2\pi^2}{L^2}, \qquad n = 1, 2, 3, \ldots, \tag{8}$$

with the eigenfunction $y_n(x) = \sin(n\pi x/L)$ associated with λ_n. (Thus whirling strings and buckled rods lead to the same eigenvalues and eigenfunctions.)

To interpret this result in terms of our buckled rod, recall from (20) that $P = \lambda EI$. The forces

$$P_n = \lambda_n EI = \frac{n^2\pi^2 EI}{L^2}, \qquad n = 1, 2, 3, \ldots \tag{21}$$

are the *critical buckling forces* of the rod. Only when the compressive force P is one of these critical forces should the rod "buckle" out of its straight (undeflected) shape. The smallest compressive force for which this occurs is

$$P_1 = \frac{\pi^2 EI}{L^2}. \tag{22}$$

This smallest critical force P_1 is called the *Euler buckling force* for the rod; it is the upper bound for those compressive forces to which the rod can safely be subjected without buckling. (In actual practice a rod may fail at a significantly smaller force due to a contribution of factors not taken into account by the mathematical model discussed here.)

For instance, suppose that we want to compute the Euler buckling force for a steel rod 10 ft long having a circular cross section 1 in. in diameter. In cgs units we have

$$E = 2 \times 10^{12} \text{ g/cm-s}^2,$$

$$L = (10 \text{ ft})\left(30.48 \frac{\text{cm}}{\text{ft}}\right) = 304.8 \text{ cm}, \quad \text{and}$$

$$I = \frac{\pi}{4}\left[(0.5 \text{ in.})\left(2.54 \frac{\text{cm}}{\text{in.}}\right)\right]^4 \approx 2.04 \text{ cm}^4.$$

Upon substituting these values in (22) we find that the critical force for this rod is

$$P_1 \approx 4.34 \times 10^8 \text{ dyn} \approx 976 \text{ lb},$$

using the conversion factor 4.448×10^5 dyn/lb.

2.10 PROBLEMS

The eigenvalues in each of Problems 1–5 are all nonnegative. First determine whether $\lambda = 0$ is an eigenvalue; then find the positive eigenvalues and associated eigenfunctions.

1. $y'' + \lambda y = 0$; $y'(0) = 0$, $y(1) = 0$
2. $y'' + \lambda y = 0$; $y'(0) = 0$, $y'(\pi) = 0$
3. $y'' + \lambda y = 0$; $y(-\pi) = 0$, $y(\pi) = 0$
4. $y'' + \lambda y = 0$; $y'(-\pi) = 0$, $y'(\pi) = 0$
5. $y'' + \lambda y = 0$; $y(-2) = 0$, $y'(2) = 0$

6. Consider the eigenvalue problem

$$y'' + \lambda y = 0; \qquad y'(0) = 0, \qquad y(1) + y'(1) = 0.$$

All the eigenvalues are nonnegative, so write $\lambda = \alpha^2$ with $\alpha \geq 0$. (a) Show that $\lambda = 0$ is not an eigenvalue. (b) Show that $y = A\cos\alpha x + B\sin\alpha x$ satisfies the endpoint conditions if and only if $B = 0$ and α is a positive root of the equation $\tan z = 1/z$. These roots $\{\alpha_n\}_1^\infty$ are the abscissas of the points of intersection of the curves $y = \tan z$ and $y = 1/z$, as indicated in Fig. 2.41.

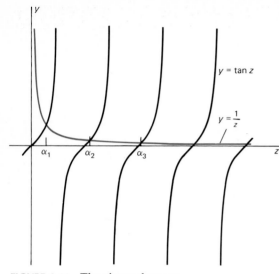

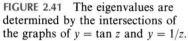

FIGURE 2.41 The eigenvalues are determined by the intersections of the graphs of $y = \tan z$ and $y = 1/z$.

Thus the eigenvalues and eigenfunctions of this problem are the numbers $\{\alpha_n^2\}_1^\infty$ and the functions $\{\cos \alpha_n x\}_1^\infty$, respectively.

7. Consider the eigenvalue problem

$$y'' + \lambda y = 0; \qquad y(0) = 0, \qquad y(1) + y'(1) = 0;$$

all its eigenvalues are nonnegative. (a) Show that $\lambda = 0$ is not an eigenvalue. (b) Show that the eigenfunctions are the functions $\{\sin \alpha_n x\}_1^\infty$, where α_n is the nth positive root of the equation $\tan z = -z$. (c) Draw a sketch indicating the roots $\{\alpha_n\}_1^\infty$ as the points of intersection of the curves $y = z$ and $y = -\tan z$.

8. Consider the eigenvalue problem

$$y'' + \lambda y = 0; \qquad y(0) = 0, \qquad y(1) = y'(1);$$

all its eigenvalues are nonnegative. (a) Show that $\lambda = 0$ is an eigenvalue with associated eigenfunction $y_0(x) = x$. (b) Show that the remaining eigenfunctions are given by $y_n(x) = \sin \alpha_n x$, where α_n is the nth positive root of the equation $\tan z = z$. Draw a sketch showing these roots.

9. Prove that the eigenvalue problem of Example 4 has no negative eigenvalues.

10. Prove that the eigenvalue problem

$$y'' + \lambda y = 0; \qquad y(0) = 0, \qquad y(1) + y'(1) = 0$$

has no negative eigenvalues. (*Suggestion:* Show graphically that the only root of the equation $\tanh z = -z$ is $z = 0$.)

11. Use a method similar to that suggested in Problem 10 to prove that the eigenvalue problem in Problem 6 has no negative eigenvalues.

12. Consider the eigenvalue problem

$$y'' + \lambda y = 0; \qquad y(-\pi) = y(\pi), \qquad y'(-\pi) = y'(\pi),$$

which is not of the type in (10) because the two endpoint conditions are not "separated" between the two endpoints. (a) Show that $\lambda_0 = 0$ is an eigenvalue with eigenfunction $y_0(x) \equiv 1$. (b) Show that there are no negative eigenvalues. (c) Show that the nth positive eigenvalue is n^2, and that it has two linearly independent associated eigenfunctions, $\cos nx$ and $\sin nx$.

13. Consider the eigenvalue problem

$$y'' + 2y' + \lambda y = 0; \qquad y(0) = y(1) = 0.$$

(a) Show that $\lambda = 1$ is not an eigenvalue. (b) Show that there is no eigenvalue λ such that $\lambda < 1$. (c) Show that the nth positive eigenvalue is $\lambda_n = n^2\pi^2 + 1$, with associated eigenfunction $y_n(x) = e^{-x} \sin (n\pi x)$.

14. Consider the eigenvalue problem

$$y'' + 2y' + \lambda y = 0; \qquad y(0) = 0, \qquad y'(1) = 0.$$

Show that the eigenvalues are all positive, and that the nth positive eigenvalue is $\lambda_n = \alpha_n^2 + 1$ with associated eigenfunction $y_n(x) = e^{-x} \sin \alpha_n x$, where α_n is the nth positive root of $\tan z = z$.

15. Consider the eigenvalue problem

$$x^2 y'' + xy' + \lambda y = 0; \qquad y(1) = 0, \qquad y(e) = 0,$$

in which the differential equation is an Euler-Cauchy equation. Show that the eigenvalues are all positive, the nth one being $\lambda_n = n^2\pi^2$, with associated eigenfunction $y_n(x) = \sin (n\pi \ln x)$.

16. Consider the eigenvalue problem

$$x^2 y'' - 3xy' + \lambda y = 0; \qquad y(1) = 0, \qquad y(e) = 0.$$

Show that the eigenvalues all exceed 4, and that the nth one is $\lambda_n = n^2\pi^2 + 4$, with associated eigenfunction $y_n(x) = x^2 \sin (n\pi \ln x)$.

CHAPTER *3*

POWER SERIES SOLUTIONS OF LINEAR EQUATIONS

3.1   INTRODUCTION AND REVIEW OF POWER SERIES

3.2   SERIES SOLUTIONS NEAR ORDINARY POINTS

3.3   REGULAR SINGULAR POINTS

*3.4   METHOD OF FROBENIUS—THE EXCEPTIONAL CASES

3.5   BESSEL'S EQUATION

*3.6   APPLICATIONS OF BESSEL FUNCTIONS

*3.7   APPENDIX ON INFINITE SERIES AND THE ATOM

Introduction and Review of Power Series

In Section 2.3 we saw that solving a homogeneous linear differential equation with constant coefficients can be reduced to the algebraic problem of finding the roots of its characteristic equation. There is no similar procedure for solving linear differential equations with *variable* coefficients, at least not routinely and in finitely many steps. With the exception of special types, such as the Euler-Cauchy equations of Section 2.6, and the occasional equation that can be solved by inspection (perhaps followed by reduction of order), linear equations with variable coefficients generally require the power series techniques of this chapter.

These techniques suffice for many of the nonelementary differential equations that appear most frequently in applications. Perhaps the most important (because of its applications in such areas as acoustics, heat flow, and nuclear reactor design) is **Bessel's equation** of order n:

$$x^2 y'' + xy' + (x^2 - n^2)y = 0.$$

Legendre's equation of order n is important in many applications; it has the form

$$(1 - x^2)y'' - 2xy' + n(n + 1)y = 0.$$

In this section we introduce the **power series method** in its simplest form and, along the way, state (without proof) several theorems that constitute a review of the basic facts about power series. Recall first that a **power series** in (powers of) $x - a$ is an infinite series of the form

$$\sum_{n=0}^{\infty} c_n(x - a)^n = c_0 + c_1(x - a) + c_2(x - a)^2 + \cdots + c_n(x - a)^n + \cdots. \quad (1)$$

If $a = 0$, this is a power series in x:

$$\sum_{n=0}^{\infty} c_n x^n = c_0 + c_1 x + c_2 x^2 + \cdots + c_n x^n + \cdots. \quad (2)$$

We will confine our review mainly to power series in x, but every general property of power series in x can be converted to a general property of power series in $x - a$ by replacement of x by $x - a$.

The power series in (2) **converges** on the interval I provided that the limit

$$\sum_{n=0}^{\infty} c_n x^n = \lim_{N \to \infty} \sum_{n=0}^{N} c_n x^n \quad (3)$$

exists for all x in I. In this case the sum

$$f(x) = \sum_{n=0}^{\infty} c_n x^n \quad (4)$$

is defined on I, and we call the series $\sum c_n x^n$ a **power series representation** of the function f on I. The following power series representations of elementary functions should be familiar to you from introductory calculus.

$$e^x = \sum_{n=0}^{\infty} \frac{x^n}{n!} = 1 + x + \frac{x^2}{2!} + \frac{x^3}{3!} + \cdots; \tag{5}$$

$$\cos x = \sum_{n=0}^{\infty} (-1)^n \frac{x^{2n}}{(2n)!} = 1 - \frac{x^2}{2!} + \frac{x^4}{4!} - \cdots; \tag{6}$$

$$\sin x = \sum_{n=0}^{\infty} (-1)^n \frac{x^{2n+1}}{(2n+1)!} = x - \frac{x^3}{3!} + \frac{x^5}{5!} - \cdots; \tag{7}$$

$$\cosh x = \sum_{n=0}^{\infty} \frac{x^{2n}}{(2n)!} = 1 + \frac{x^2}{2!} + \frac{x^4}{4!} + \cdots; \tag{8}$$

$$\sinh x = \sum_{n=0}^{\infty} \frac{x^{2n+1}}{(2n+1)!} = x + \frac{x^3}{3!} + \frac{x^5}{5!} + \cdots; \tag{9}$$

$$\ln(1+x) = \sum_{n=1}^{\infty} (-1)^{n+1} \frac{x^n}{n} = x - \frac{x^2}{2} + \frac{x^3}{3} - \cdots; \tag{10}$$

$$\frac{1}{1-x} = \sum_{n=0}^{\infty} x^n = 1 + x + x^2 + x^3 + \cdots; \tag{11}$$

and

$$(1+x)^\alpha = 1 + \alpha x + \frac{\alpha(\alpha-1)}{2!} x^2 + \frac{\alpha(\alpha-1)(\alpha-2)}{3!} x^3 + \cdots. \tag{12}$$

In compact summation notation, we observe the usual conventions that $0! = 1$ and that $x^0 = 1$ for all x (including $x = 0$). The series in (5)–(9) converge to the indicated functions for all x, while the series in (10)–(12) converge if $|x| < 1$, but diverge if $|x| > 1$. The series in (12), with α an arbitrary real number, is the **binomial series**; the series in (11) is the **geometric series**.

Power series such as those listed above are often derived as Taylor series. The **Taylor series** with **center** $x = a$ of the function f is the power series

$$\sum_{n=0}^{\infty} \frac{f^{(n)}(a)}{n!} (x-a)^n = f(a) + f'(a)(x-a) + \frac{f''(a)}{2!} (x-a)^2 + \cdots \tag{13}$$

in powers of $x - a$, under the hypothesis that f is infinitely differentiable at a (so that the coefficients in (13) are all defined). If the Taylor series of f converges to $f(x)$ for all x in some open interval containing a, then we say that the function f is **analytic** at $x = a$. For example, every polynomial is analytic everywhere, and every rational function is analytic wherever its denominator is nonzero. More generally, if the two functions f and g are both analytic at $x = a$, then so are their sum $f + g$ and their product $f \cdot g$, as is their quotient f/g wherever g is nonzero.

For instance, the function $\tan x = (\sin x)/(\cos x)$ is analytic at $x = 0$ because $\cos 0 = 1 \neq 0$ and the functions $\sin x$ and $\cos x$ are analytic (by virtue of their convergent power series representations in Eqs. (6) and (7)). It is rather awkward to compute the Taylor series of the function $\tan x$ using (13) because of the way in which its successive derivatives grow in complexity (try it!). Fortunately, power series may be manipulated algebraically in much the same way as polynomials.

For example, if

$$f(x) = \sum_{n=0}^{\infty} a_n x^n \quad \text{and} \quad g(x) = \sum_{n=0}^{\infty} b_n x^n, \tag{14}$$

then

$$f(x) + g(x) = \sum_{n=0}^{\infty} (a_n + b_n) x^n \tag{15}$$

and

$$f(x)g(x) = \sum_{n=0}^{\infty} c_n x^n$$
$$= a_0 b_0 + (a_0 b_1 + a_1 b_0)x + (a_0 b_2 + a_1 b_1 + a_2 b_0)x^2 + \cdots \tag{16}$$

where $c_n = a_0 b_n + a_1 b_{n-1} + \cdots + a_n b_0$. The series in (15) is the result of **termwise addition**, while the series in (16) is the result of **formal multiplication**—multiplying each term of the first series by each term of the second and then collecting coefficients of like powers of x. (Thus the processes strongly resemble addition and multiplication of ordinary polynomials.) The series in (15) and (16) converge to $f(x) + g(x)$ and $f(x)g(x)$, respectively, on any open interval on which both the series in (14) converge. For example,

$$\sin x \cos x = \left(x - \frac{1}{6}x^3 + \frac{1}{120}x^5 - \cdots \right)\left(1 - \frac{1}{2}x^2 + \frac{1}{24}x^4 - \cdots \right)$$

$$= x + \left(-\frac{1}{6} - \frac{1}{2} \right)x^3 + \left(\frac{1}{24} + \frac{1}{12} + \frac{1}{120} \right)x^5 + \cdots$$

$$= x - \frac{4}{6}x^3 + \frac{16}{120}x^5 - \cdots$$

$$= \frac{1}{2}\left[(2x) - \frac{(2x)^3}{3!} + \frac{(2x)^5}{5!} - \cdots \right] = \frac{1}{2}\sin 2x$$

for all x.

Similarly, the quotient of two power series can be computed by long division, as illustrated by the computation shown in Fig. 3.1. This division of the Taylor series for $\cos x$ into that for $\sin x$ yields the first few terms of the series

$$\tan x = x + \frac{1}{3}x^3 + \frac{2}{15}x^5 + \frac{17}{315}x^7 + \cdots. \tag{17}$$

Division of power series is more treacherous than multiplication; the series thus obtained for f/g may fail to converge at some points where the series for f and g both converge. For example, the sine and cosine series converge for all x, but the tangent series in (17) converges only if $|x| < \pi/2$.

The **power series method** for solving a differential equation consists of substituting the power series

$$y = \sum_{n=0}^{\infty} c_n x^n \tag{18}$$

in the differential equation and then attempting to determine what the coefficients $c_0, c_1, c_2, \ldots$ must be in order that the power series will satisfy the differential

$$x + \frac{x^3}{3} + \frac{2x^5}{15} + \frac{17x^7}{315} + \cdots$$

$$1 - \frac{x^2}{2} + \frac{x^4}{24} - \frac{x^6}{720} + \cdots \ \overline{\smash{\big)}\ x - \frac{x^3}{6} + \frac{x^5}{120} - \frac{x^7}{5040} + \cdots}$$

$$x - \frac{x^3}{2} + \frac{x^5}{24} - \frac{x^7}{720} + \cdots$$

$$\frac{x^3}{3} - \frac{x^5}{30} + \frac{x^7}{840} + \cdots$$

$$\frac{x^3}{3} - \frac{x^5}{6} + \frac{x^7}{72} - \cdots$$

$$\frac{2x^5}{15} - \frac{4x^7}{315} + \cdots$$

$$\frac{2x^5}{15} - \frac{x^7}{15} + \cdots$$

$$\frac{17x^7}{315} - \cdots$$

FIGURE 3.1 Obtaining the series for $\tan x$ by division of series.

equation. This is much like the method of undetermined coefficients, but now we have infinitely many coefficients somehow to determine. This method is not always successful, but when it is we obtain an infinite series representation of a solution, in contrast to the "closed form" solutions that our previous methods have yielded.

Before we can substitute the power series in (18) in a differential equation, we must first know what to substitute for the derivatives y', y'', ... The following theorem (stated without proof) tells us that the derivative y' of $y = \sum c_n x^n$ is obtained by the simple procedure of writing the sum of the derivatives of the individual terms of y.

Theorem 1   *Termwise Differentiation of Power Series*

If the power series representation

$$f(x) = \sum_{n=0}^{\infty} c_n x^n = c_0 + c_1 x + c_2 x^2 + c_3 x^3 + \cdots \qquad (19)$$

of the function f converges on the open interval I, then f is differentiable on I, and

$$f'(x) = \sum_{n=1}^{\infty} n c_n x^{n-1} = c_1 + 2c_2 x + 3c_3 x^2 + \cdots \qquad (20)$$

at each point of I.

For example, differentiation of the geometric series

$$\frac{1}{1-x} = \sum_{n=0}^{\infty} x^n = 1 + x + x^2 + x^3 + \cdots \qquad (11)$$

gives the series

$$\frac{1}{(1-x)^2} = \sum_{n=1}^{\infty} nx^{n-1} = 1 + 2x + 3x^2 + 4x^3 + \cdots.$$

The process of determining the coefficients in the series $y = \sum c_n x^n$ so that it will satisfy a given differential equation depends also upon Theorem 2. This theorem—stated without proof—tells us that if two power series represent the same function, then they are the same series. In particular, the Taylor series in (13) is the only power series (in powers of x) that represents the function $f(x)$.

Theorem 2   *Identity Principle*

If

$$\sum_{n=0}^{\infty} a_n x^n = \sum_{n=0}^{\infty} b_n x^n$$

for every point x in some open interval I, then $a_n = b_n$ for all $n \geq 0$.

In particular, if $\sum a_n x^n = 0$ for all x in some open interval, it follows from Theorem 2 that $a_n = 0$ for all n.

EXAMPLE 1   Solve the equation $y' + 2y = 0$.

SOLUTION   We substitute the series

$$y = \sum_{n=0}^{\infty} c_n x^n \quad \text{and} \quad y' = \sum_{n=1}^{\infty} nc_n x^{n-1},$$

and obtain

$$\sum_{n=1}^{\infty} nc_n x^{n-1} + 2 \sum_{n=0}^{\infty} c_n x^n = 0. \tag{21}$$

To compare coefficients here, we need the general term in each to be the term containing x^n. To accomplish this, we shift the index of summation in the first sum. To see how to do this, note that

$$\sum_{n=1}^{\infty} nc_n x^{n-1} = c_1 + 2c_2 x + 3c_3 x^2 + \cdots = \sum_{n=0}^{\infty} (n+1)c_{n+1} x^n.$$

Thus we can replace n by $n+1$ if, at the same time, we start counting one step lower; that is, at $n = 0$ rather than at $n = 1$. This is a shift of $+1$ in the index of summation. The result of making this shift in (21) is the identity

$$\sum_{n=0}^{\infty} (n+1)c_{n+1} x^n + 2 \sum_{n=0}^{\infty} c_n x^n = 0;$$

that is,

$$\sum_{n=0}^{\infty} [(n+1)c_{n+1} + 2c_n] x^n = 0.$$

If this is true on some open interval, then it follows from the identity principle that $(n + 1)c_{n+1} + 2c_n = 0$ for all $n \geq 0$; consequently,

$$c_{n+1} = -\frac{2c_n}{n + 1} \tag{22}$$

for all $n \geq 0$. Equation (22) is a **recurrence relation** from which we can successively compute $c_1, c_2, c_3, \ldots$ in terms of c_0; the latter will turn out to be the arbitrary constant we expect to find in a solution of a first order differential equation.

With $n = 0$, (22) gives

$$c_1 = -\frac{2c_0}{1}.$$

With $n = 1$, (22) gives

$$c_2 = -\frac{2c_1}{2} = +\frac{2^2 c_0}{1 \cdot 2} = +\frac{2^2 c_0}{2!}.$$

With $n = 2$, (22) gives

$$c_3 = -\frac{2c_2}{3} = -\frac{2^3 c_0}{1 \cdot 2 \cdot 3} = -\frac{2^3 c_0}{3!}.$$

By now it should be clear that after n such steps, we will have

$$c_n = (-1)^n \frac{2^n c_0}{n!}, \qquad n \geq 1.$$

(This is easy to prove by induction on n.) Consequently, our solution takes the form

$$y(x) = \sum_{n=0}^{\infty} c_n x^n = \sum_{n=0}^{\infty} (-1)^n \frac{2^n c_0}{n!} x^n$$

$$= c_0 \sum_{n=0}^{\infty} \frac{(-2x)^n}{n!} = c_0 e^{-2x}.$$

In the final step we have used the familiar exponential series in (5) to identify our power series solution as the same solution $y = c_0 e^{-2x}$ we could have obtained immediately by the method of separation of variables.

SHIFT OF INDEX OF SUMMATION

In the solution of Example 1 we wrote

$$\sum_{n=1}^{\infty} nc_n x^{n-1} = \sum_{n=0}^{\infty} (n + 1)c_{n+1} x^n \tag{23}$$

by shifting the index of summation by $+1$ in the series on the left. That is, we simultaneously *increased* the index of summation by 1 (replacing n by $n + 1$, $n \to n + 1$) and *decreased* the starting point by 1, from $n = 1$ to $n = 0$, thereby obtaining the series on the right. This procedure is valid because each infinite series in (23) is simply a compact notation for the single series

$$c_1 + 2c_2 x + 3c_3 x^2 + 4c_4 x^3 + \cdots. \tag{24}$$

More generally, we can shift the index of summation by k in an infinite series by simultaneously *increasing* the summation index by k ($n \rightarrow n + k$) and *decreasing* the starting point by k. For instance, a shift by $+2$ ($n \rightarrow n + 2$) yields

$$\sum_{n=3}^{\infty} a_n x^{n-1} = \sum_{n=1}^{\infty} a_{n+2} x^{n+1}.$$

If k is negative we interpret a "decrease by k" as an increase by $-k = |k|$. Thus a shift by -2 ($n \rightarrow n - 2$) in the index of summation yields

$$\sum_{n=1}^{\infty} n c_n x^{n-1} = \sum_{n=3}^{\infty} (n - 2) c_{n-2} x^{n-3};$$

we have *decreased* the index of summation by 2, but *increased* the starting point by 2, from $n = 1$ to $n = 3$. You should check that the summation on the right is indeed merely another representation of the series in (20).

We know that the power series obtained in Example 1 converges for all x because it is an exponential series. More commonly, a power series solution is not recognizable in terms of the familiar elementary functions. When we get an unfamiliar power series solution, we need a way of ascertaining where it converges. After all, $y = \sum c_n x^n$ is merely an *assumed* form of the solution. The procedure illustrated in Example 1 for determining the coefficients $\{c_n\}$ is only a formal process and may or may not be valid. Its validity—in applying Theorem 1 to compute y' and in applying Theorem 2 to obtain a recurrence relation for the coefficients—depends on the convergence of the initially *unknown* series $y = \sum c_n x^n$. Hence this formal process is justified only if in the end we can show that the power series obtained converges on some open interval. If so, it then represents a solution of the differential equation on that interval. The following theorem (which we state without proof) may be used for this purpose.

Theorem 3 *Radius of Convergence*

Given the power series $\sum c_n x^n$, suppose that the limit

$$\rho = \lim_{n \to +\infty} \left| \frac{c_n}{c_{n+1}} \right| \qquad (25)$$

exists (ρ is finite) or is infinite (in this case, we will write $\rho = \infty$). Then:

(a) If $\rho = 0$, the series diverges for all $x \neq 0$.
(b) If $0 < \rho < \infty$, then $\sum c_n x^n$ converges if $|x| < \rho$ and diverges if $|x| > \rho$.
(c) If $\rho = \infty$, the series converges for all x.

The number ρ in (23) is called the **radius of convergence** of the power series $\sum c_n x^n$. For instance, for the power series obtained in Example 1, we have

$$\rho = \lim_{n \to +\infty} \left| \frac{(-1)^n 2^n c_0 / n!}{(-1)^{n+1} 2^{n+1} c_0 / (n+1)!} \right| = \lim_{n \to +\infty} \frac{n+1}{2} = \infty,$$

and consequently the series we obtained in Example 1 converges for all x. Even if the limit in (25) fails to exist, there always will exist a number ρ such that exactly one of the alternatives (a), (b), or (c) in Theorem 3 holds. This number may be difficult to find, but for the power series we will consider in this chapter, (25) will be quite sufficient for computation of their radii of convergence.

EXAMPLE 2   Solve the equation $(x - 3)y' + 2y = 0$.

SOLUTION   As before, we substitute

$$y = \sum_{n=0}^{\infty} c_n x^n \quad \text{and} \quad y' = \sum_{n=1}^{\infty} n c_n x^{n-1}$$

to obtain

$$(x - 3) \sum_{n=1}^{\infty} n c_n x^{n-1} + 2 \sum_{n=0}^{\infty} c_n x^n = 0,$$

so that

$$\sum_{n=1}^{\infty} n c_n x^n - 3 \sum_{n=1}^{\infty} n c_n x^{n-1} + 2 \sum_{n=0}^{\infty} c_n x^n = 0.$$

In the first sum we can replace $n = 1$ by $n = 0$ with no effect on the sum. In the second sum we shift the index of summation by $+1$. This yields

$$\sum_{n=0}^{\infty} n c_n x^n - 3 \sum_{n=0}^{\infty} (n + 1)c_{n+1} x^n + 2 \sum_{n=0}^{\infty} c_n x^n = 0;$$

that is,

$$\sum_{n=0}^{\infty} [n c_n - 3(n + 1)c_{n+1} + 2c_n]x^n = 0.$$

The identity principle then gives

$$n c_n - 3(n + 1)c_{n+1} + 2c_n = 0,$$

from which we obtain the recurrence relation

$$c_{n+1} = \frac{n + 2}{3(n + 1)} c_n.$$

We apply this formula with $n = 0$, $n = 1$, and $n = 2$ in turn, and find that

$$c_1 = \frac{2}{3} c_0, \qquad c_2 = \frac{3}{3 \cdot 2} c_1 = \frac{3}{3^2} c_0,$$

and

$$c_3 = \frac{4}{3 \cdot 3} c_2 = \frac{4}{3^3} c_0.$$

This is almost enough to make the pattern evident; it is not difficult to show by induction on n that

$$c_n = \frac{n + 1}{3^n} c_0 \qquad \text{if } n \geq 1.$$

Hence our proposed power series solution is

$$y(x) = c_0 \sum_{n=0}^{\infty} \frac{n+1}{3^n} x^n. \tag{26}$$

Its radius of convergence is

$$\rho = \lim_{n \to +\infty} \left| \frac{c_n}{c_{n+1}} \right| = \lim_{n \to +\infty} \frac{3n+3}{n+2} = 3.$$

Thus the series in (26) converges if $-3 < x < 3$, but diverges if $|x| > 3$. In this particular example we can explain why. An elementary solution (obtained by separation of variables) of our differential equation is $y = 1/(3 - x)^2$. If we differentiate termwise the geometric series

$$\frac{1}{3-x} = \frac{\frac{1}{3}}{1-(x/3)} = \frac{1}{3} \sum_{n=0}^{\infty} \frac{x^n}{3^n},$$

we get a constant multiple of the series in (26). Thus this series (with the arbitrary constant c_0 appropriately chosen) represents the solution

$$y(x) = \frac{1}{(3-x)^2}$$

on the interval $-3 < x < 3$, and the singularity at $x = 3$ is the reason why the radius of convergence of our power series solution turned out to be $\rho = 3$.

EXAMPLE 3 Solve the equation $x^2 y' = y - x - 1$.

SOLUTION We make the usual substitutions $y = \sum c_n x^n$ and $y' = \sum n c_n x^{n-1}$, which yield

$$x^2 \sum_{n=1}^{\infty} n c_n x^{n-1} = -1 - x + \sum_{n=0}^{\infty} c_n x^n,$$

so that

$$\sum_{n=1}^{\infty} n c_n x^{n+1} = -1 - x + \sum_{n=0}^{\infty} c_n x^n.$$

Because of the presence of the terms -1 and $-x$ on the right-hand side, we need to split off the first two terms, $c_0 + c_1 x$, of the series on the right for comparison. If we also shift the index of summation in the series on the left by -1 (replace $n = 1$ with $n = 2$ and n with $n - 1$), we get

$$\sum_{n=2}^{\infty} (n-1)c_{n-1} x^n = -1 - x + c_0 + c_1 x + \sum_{n=2}^{\infty} c_n x^n.$$

Because the left-hand side contains neither a constant term nor a term containing x to the first power, the identity principle now yields $c_0 = 1$, $c_1 = 1$, and $c_n = (n-1)c_{n-1}$ for $n \geq 2$. It follows that

$$c_2 = 1 \cdot c_1 = 1!, \qquad c_3 = 2 \cdot c_2 = 2!, \qquad c_4 = 3 \cdot c_3 = 3!,$$

and, in general, that

$$c_n = (n-1)! \quad \text{for } n \geq 2.$$

Thus we obtain the power series

$$y(x) = 1 + x + \sum_{n=2}^{\infty} (n-1)! x^n.$$

But the radius of convergence of this series is

$$\rho = \lim_{n \to +\infty} \frac{(n-1)!}{n!} = \lim_{n \to +\infty} \frac{1}{n} = 0,$$

so it converges only for $x = 0$. What does this mean? Simply that the given differential equation does not have a (convergent) power series solution of the assumed form $y = \sum c_n x^n$. This example serves as a warning that the simple act of writing $y = \sum c_n x^n$ involves an assumption that may be false.

EXAMPLE 4 Solve the equation $y'' + y = 0$.

SOLUTION If we assume a solution of the form

$$y = \sum_{n=0}^{\infty} c_n x^n,$$

we find that

$$y' = \sum_{n=1}^{\infty} n c_n x^{n-1} \quad \text{and} \quad y'' = \sum_{n=2}^{\infty} n(n-1) c_n x^{n-2}.$$

Substitution for y and y'' in the differential equation then yields

$$\sum_{n=2}^{\infty} n(n-1) c_n x^{n-2} + \sum_{n=0}^{\infty} c_n x^n = 0.$$

We shift the index of summation in the left sum by $+2$ (by replacing $n = 2$ with $n = 0$ and n with $n + 2$). This gives

$$\sum_{n=0}^{\infty} (n+2)(n+1) c_{n+2} x^n + \sum_{n=0}^{\infty} c_n x^n = 0.$$

The identity $(n+2)(n+1) c_{n+2} + c_n = 0$ now follows from the identity principle, and thus we obtain the recurrence relation

$$c_{n+2} = -\frac{c_n}{(n+1)(n+2)} \tag{27}$$

for $n \geq 0$. It is evident that this formula will determine the coefficients c_n with even subscripts in terms of c_0, and those of odd subscript in terms of c_1; c_0 and c_1 are not predetermined, and thus will be the two arbitrary constants we expect to find in a general solution of a second order equation.

When we apply the recurrence relation in (27) with $n = 0$, 2, and 4 in turn, we get

$$c_2 = -\frac{c_0}{2!}, \quad c_4 = \frac{c_0}{4!}, \quad \text{and} \quad c_6 = -\frac{c_0}{6!}.$$

Taking $n = 1$, 3, and 5 in turn, we find that

$$c_3 = -\frac{c_1}{3}, \quad c_5 = \frac{c_1}{5!}, \quad \text{and} \quad c_7 = -\frac{c_1}{7!}.$$

Again, the pattern is clear; we leave it for you to show (by induction) that for $k \geq 1$,

$$c_{2k} = \frac{(-1)^k c_0}{(2k)!} \quad \text{and} \quad c_{2k+1} = \frac{(-1)^k c_1}{(2k+1)!}.$$

Thus we get the power series solution

$$y(x) = c_0 \left(1 - \frac{x^2}{2!} + \frac{x^4}{4!} - \frac{x^6}{6!} + \cdots \right)$$

$$+ c_1 \left(x - \frac{x^3}{3!} + \frac{x^5}{5!} - \frac{x^7}{7!} + \cdots \right)$$

—that is, $y = c_0 \cos x + c_1 \sin x$. Note that we have no problem with the radius of convergence here; the Taylor series for the sine and cosine functions converge for all x.

The solution of Example 4 can bear further comment. Suppose that we had never heard of the sine and cosine functions, let alone their Taylor series. We would then have discovered the two power series solutions

$$C(x) = \sum_{n=0}^{\infty} (-1)^n \frac{x^{2n}}{(2n)!} = 1 - \frac{x^2}{2!} + \frac{x^4}{4!} - \cdots \tag{28}$$

and

$$S(x) = \sum_{n=0}^{\infty} (-1)^n \frac{x^{2n+1}}{(2n+1)!} = x - \frac{x^3}{3!} + \frac{x^5}{5!} - \cdots \tag{29}$$

of the differential equation $y'' + y = 0$. It is clear that $C(0) = 1$ and that $S(0) = 0$. After verifying that the two series in (28) and (29) converge for all x, we can differentiate them term by term to find that

$$C'(x) = -S(x) \quad \text{and} \quad S'(x) = C(x). \tag{30}$$

Consequently $C'(0) = 0$ and $S'(0) = 1$. Thus with the aid of the power series method (all the while knowing nothing about the sine and cosine functions), we have discovered that $y = C(x)$ is the unique solution of

$$y'' + y = 0$$

that satisfies the initial conditions $y(0) = 1$ and $y'(0) = 0$, and that $y = S(x)$ is the solution that satisfies the initial conditions $y(0) = 0$ and $y'(0) = 1$. It follows that $C(x)$ and $S(x)$ are linearly independent, and—recognizing the importance of the differential equation $y'' + y = 0$—we can agree to call $C(x)$ the *cosine* function and $S(x)$ the *sine* function. Indeed, all the usual properties of these two functions can be established, using only their initial values (at $x = 0$) and the derivatives in (28); there is no need to refer to triangles or even to angles. (Can you use the series in (26) and (27) to show that $[C(x)]^2 + [S(x)]^2 = 1$ for all x?) This demonstrates that the cosine and sine functions are fully determined by the differential equation $y'' + y = 0$ of which they are the natural linearly independent solutions. Figures 3.2 and 3.3 show how the geometric character of the graphs of $\cos x$ and $\sin x$ is

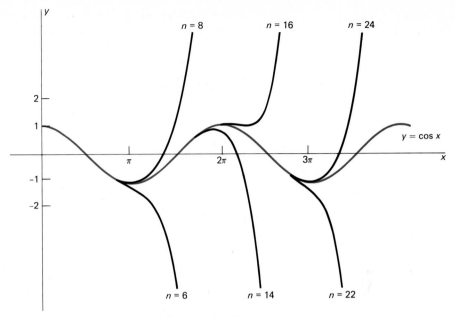

FIGURE 3.2 Taylor polynomial approximations to cos x

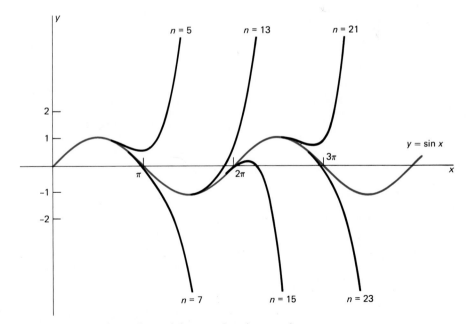

FIGURE 3.3 Taylor polynomial approximations to sin x

revealed by the graphs of the Taylor polynomial approximations that we get by truncating the infinite series in (28) and (29).

This is by no means an uncommon situation. Many important special functions of mathematics occur in the first instance as power series solutions of differential equations, and thus are in practice *defined* by means of these power series. In the remaining sections of this chapter we will see numerous examples of such functions.

3.1 PROBLEMS

In each of Problems 1–10, find a power series solution of the given differential equation. Determine the radius of convergence of the resulting series, and use the series in (5)–(12) to identify the series solution in terms of familiar elementary functions. (Of course, no one can prevent you from checking your work by also solving the equations by the methods of earlier chapters!)

1. $y' = y$ **2.** $y' = 4y$

3. $2y' + 3y = 0$ **4.** $y' + 2xy = 0$

5. $y' = x^2 y$ **6.** $(x - 2)y' + y = 0$

7. $(2x - 1)y' + 2y = 0$ **8.** $2(x + 1)y' = y$

9. $(x - 1)y' + 2y = 0$ **10.** $2(x - 1)y' = 3y$

In each of Problems 11–14, use the method of Example 4 to find two linearly independent power series solutions of the given differential equation. Determine the radius of convergence of each series, and identify the general solution in terms of familiar elementary functions.

11. $y'' = y$ **12.** $y'' = 4y$

13. $y'' + 9y = 0$ **14.** $y'' + y = x$

Show (as in Example 3) that the power series method fails to yield a power series solution of the form $y = \sum c_n x^n$ for the differential equations in Problems 15–18.

15. $xy' + y = 0$ **16.** $2xy' = y$

17. $x^2 y' + y = 0$ **18.** $x^3 y' = 2y$

In each of Problems 19–22, first derive a recurrence relation giving c_n for $n \geq 2$ in terms of c_0 or c_1 (or both). Then apply the given initial conditions to find the values of c_0 and c_1. Next determine c_n (in terms of n, as in the text) and, finally, identify the particular solution in terms of familiar elementary functions.

19. $y'' + 4y = 0$; $y(0) = 0$, $y'(0) = 3$

20. $y'' - 4y = 0$; $y(0) = 2$, $y'(0) = 0$

21. $y'' - 2y' + y = 0$; $y(0) = 0$, $y'(0) = 1$

22. $y'' + y' - 2y = 0$; $y(0) = 1$, $y'(0) = -2$

23. Show that the equation

$$x^2 y'' + x^2 y' + y = 0$$

has no power series solution of the form $y = \sum c_n x^n$.

24. Establish the binomial series in (12) by means of the following steps. (a) Show that $y = (1 + x)^\alpha$ satisfies the initial value problem $(1 + x)y' = \alpha y$, $y(0) = 1$. (b) Show that the power series method gives the binomial series in (12) as the solution of the initial value problem in part (a), and that this series converges if $|x| < 1$. (c) Explain why the validity of the binomial series given in (12) follows from parts (a) and (b).

25. (a) Show that the solution of the initial value problem

$$y' = 1 + y^2, \qquad y(0) = 0$$

is $y = \tan x$. (b) Because $y = \tan x$ is an odd function with $y'(0) = 1$, its Taylor series is of the form

$$y = x + c_3 x^3 + c_5 x^5 + \cdots.$$

Substitute this series in $y' = 1 + y^2$ and equate like powers of x to derive the following relations.

$$3c_3 = 1 \qquad\qquad 5c_5 = 2c_3$$

$$7c_7 = 2c_5 + c_3^2 \qquad 9c_9 = 2c_7 + 2c_3c_5$$

$$11c_{11} = 2c_9 + 2c_3c_7 + c_5^2$$

(c) Conclude that

$$\tan x = x + \frac{1}{3}x^3 + \frac{2}{15}x^5 + \frac{17}{315}x^7$$

$$+ \frac{62}{2835}x^9 + \frac{1382}{155{,}925}x^{11} + \cdots.$$

26. This section introduces the use of infinite series to solve differential equations. Conversely, differential equations can sometimes be used to sum infinite series. For instance, consider the infinite series

$$1 + \frac{1}{1!} - \frac{1}{2!} + \frac{1}{3!} + \frac{1}{4!} - \frac{1}{5!} + \cdots;$$

CHAPTER 3: Power Series Solutions of Linear Equations

note the $++-++-$ pattern of signs superimposed on the terms of the series for the number e. We could evaluate this series if we could obtain a formula for the function

$$f(x) = 1 + x - \frac{1}{2!}x^2 + \frac{1}{3!}x^3 + \frac{1}{4!}x^4 - \frac{1}{5!}x^5 + \cdots,$$

because the sum of the numerical series is simply $f(1)$. (a) It's possible to show that the power series given here converges for all x and is termwise differentiable. Given this fact, show that

$y = f(x)$ satisfies the initial value problem

$$y^{(3)} = y; \qquad y(0) = y'(0) = 1, \qquad y''(0) = -1.$$

(a) Solve this initial value problem to show that

$$f(x) = \frac{1}{3}e^x + \frac{2}{3}e^{-x/2}\left(\cos\frac{\sqrt{3}}{2}x + \sqrt{3}\sin\frac{\sqrt{3}}{2}x\right).$$

For a suggestion, see Problem 42 of Section 2.3. (c) Evaluate $f(1)$ to find the sum of the numerical series given here.

3.2

Series Solutions Near Ordinary Points

The power series method introduced in Section 3.1 can be applied to linear equations of any order, but its most important applications are to homogeneous second order linear differential equations of the form

$$A(x)y'' + B(x)y' + C(x)y = 0, \tag{1}$$

where the coefficients A, B, and C are analytic functions of x. Indeed, in most applications these coefficient functions are simple polynomials.

We saw in Example 3 of Section 3.1 that the series method does not always yield a series solution. To discover when it does succeed, it is best to rewrite Eq. (1) in the form

$$y'' + P(x)y' + Q(x)y = 0 \tag{2}$$

with leading coefficient 1, and with $P = B/A$ and $Q = C/A$. Note that $P(x)$ and $Q(x)$ will generally fail to be analytic at points where $A(x)$ vanishes. For instance, consider the equation

$$xy'' + y' + xy = 0. \tag{3}$$

The coefficient functions in (3) are all analytic everywhere. But in the form of (2) it is the equation

$$y'' + \frac{1}{x}y' + y = 0 \tag{4}$$

with $P(x) = 1/x$ not analytic at $x = 0$.

The point $x = a$ is called an **ordinary point** of Eq. (2)—and of the equivalent Eq. (1)—provided that the functions $P(x)$ and $Q(x)$ are both analytic at $x = a$. Otherwise, $x = a$ is a **singular point**. Thus the only singular point of Eqs. (3) and (4) is $x = 0$. Recall that a quotient of analytic functions is analytic wherever the denominator is nonzero. It follows that, if $A(a) \neq 0$ in Eq. (1) with analytic coefficients, then $x = a$ is an ordinary point. If $A(x)$, $B(x)$, and $C(x)$ are *polynomials* with no common factors, then $x = a$ is an ordinary point if and only if $A(a) \neq 0$.

EXAMPLE 1   The point $x = 0$ is an ordinary point of the equation

$$xy'' + (\sin x)y' + x^2 y = 0,$$

despite the fact that $A(x) = x$ vanishes at $x = 0$. The reason is that

$$P(x) = \frac{\sin x}{x} = \frac{1}{x}\left(x - \frac{x^3}{3!} + \frac{x^5}{5!} - \cdots\right)$$

$$= 1 - \frac{x^2}{3!} + \frac{x^4}{5!} - \cdots$$

is nevertheless analytic at $x = 0$ because the division by x gives a convergent power series.

EXAMPLE 2   The point $x = 0$ is *not* an ordinary point of the equation

$$y'' + x^2 y' + \sqrt{x}\, y = 0.$$

For while $P(x) = x^2$ is analytic at the origin, $Q(x) = \sqrt{x}$ is not. The reason is that $Q(x)$ is not differentiable at $x = 0$ and hence is not analytic there. (Theorem 1 of Section 3.1 implies that an analytic function must be differentiable.)

EXAMPLE 3   The point $x = 0$ is an ordinary point of the equation

$$(1 - x^3)y'' + (7x^2 + 3x^5)y' + (5x - 13x^4)y = 0$$

because the coefficient functions $A(x)$, $B(x)$, and $C(x)$ are polynomials with $A(0) \neq 0$.

Theorem 2 of Section 2.1 implies that Eq. (2) has two linearly independent solutions on any open interval where the coefficient functions $P(x)$ and $Q(x)$ are continuous. The basic fact for our present purpose is that near an *ordinary* point a, these solutions will be power series in powers of $x - a$. A proof of the following theorem can be found in Chapter 3 of Coddington, *An Introduction to Ordinary Differential Equations* (Englewood Cliffs, N.J.: Prentice-Hall, 1961).

> **Theorem**   *Solutions Near an Ordinary Point*
>
> Suppose that a is an ordinary point of the equation
>
> $$A(x)y'' + B(x)y' + C(x)y = 0; \qquad (1)$$
>
> that is, the functions $P = B/A$ and $Q = C/A$ are analytic at $x = a$. Then Eq. (1) has two linearly independent solutions, each of the form
>
> $$y(x) = \sum_{n=0}^{\infty} c_n(x - a)^n. \qquad (5)$$
>
> The radius of convergence of any such series solution is at least as large as the distance from a to the nearest (real or complex) singular point of Eq. (1). The coefficients in the series in (5) can be determined by its substitution in Eq. (1).

CHAPTER 3:   Power Series Solutions of Linear Equations

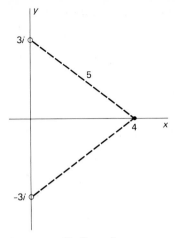

FIGURE 3.4 Radius of convergence as distance to nearest singularity.

EXAMPLE 4 Determine the radius of convergence guaranteed by the above theorem of a series solution of

$$(x^2 + 9)y'' + xy' + x^2y = 0 \tag{6}$$

in powers of x. Repeat for a series in powers of $x - 4$.

SOLUTION This example illustrates the fact that we must take into account complex singular points as well as real ones. Because

$$P(x) = \frac{x}{x^2 + 9} \quad \text{and} \quad Q(x) = \frac{x^2}{x^2 + 9},$$

the only singular points of the equation in (6) are $+3i$ and $-3i$. The distance (in the complex plane) of each from 0 is 3, so a series solution of the form $\sum c_n x^n$ has radius of convergence at least 3. The distance of each singular point from 4 is 5, so a series solution of the form $\sum c_n(x - 4)^n$ has radius of convergence at least 5 (see Fig. 3.4).

EXAMPLE 5 Find the general solution in powers of x of

$$(x^2 - 4)y'' + 3xy' + y = 0. \tag{7}$$

Then find the particular solution with $y(0) = 4$, $y'(0) = 1$.

SOLUTION The only singular points of (7) are $+2$ and -2, so the series we get will have radius of convergence at least 2. Substitution of

$$y = \sum_{n=0}^{\infty} c_n x^n, \qquad y' = \sum_{n=1}^{\infty} nc_n x^{n-1},$$

and

$$y'' = \sum_{n=2}^{\infty} n(n-1)c_n x^{n-2}$$

in Eq. (7) yields

$$\sum_{n=2}^{\infty} n(n-1)c_n x^n - 4\sum_{n=2}^{\infty} n(n-1)c_n x^{n-2} + 3\sum_{n=1}^{\infty} nc_n x^n + \sum_{n=0}^{\infty} c_n x^n = 0.$$

We can begin the first and third summations at $n = 0$ as well, because no nonzero terms are thereby introduced. We shift the index of summation in the second sum by $+2$, replacing n with $n + 2$ and using the initial value $n = 0$. This gives

$$\sum_{n=0}^{\infty} n(n-1)c_n x^n - 4\sum_{n=0}^{\infty} (n+2)(n+1)c_{n+2}x^n + 3\sum_{n=0}^{\infty} nc_n x^n + \sum_{n=0}^{\infty} c_n x^n = 0.$$

After collecting coefficients of c_n and c_{n+2}, we obtain

$$\sum_{n=0}^{\infty} [(n^2 + 2n + 1)c_n - 4(n+2)(n+1)c_{n+2}]x^n = 0.$$

The identity principle yields

$$(n + 1)^2 c_n - 4(n + 2)(n + 1)c_{n+2} = 0,$$

which leads to the recurrence relation

$$c_{n+2} = \frac{(n+1)c_n}{4(n+2)} \tag{8}$$

for $n \geq 0$. With $n = 0$, 2, and 4 in turn, we get

$$c_2 = \frac{c_0}{4 \cdot 2}, \qquad c_4 = \frac{3c_2}{4 \cdot 4} = \frac{3c_0}{4^2 \cdot 2 \cdot 4},$$

and

$$c_6 = \frac{5c_4}{4 \cdot 6} = \frac{3 \cdot 5c_0}{4^3 \cdot 2 \cdot 4 \cdot 6},$$

Continuing in this fashion, we evidently would get

$$c_{2n} = \frac{3 \cdot 5 \cdots (2n-1)}{4^n \cdot 2 \cdot 4 \cdots (2n)} c_0.$$

With the common notation

$$(2n+1)!! = 1 \cdot 3 \cdot 5 \cdots (2n+1) = \frac{(2n+1)!}{2^n \cdot n!}$$

and the observation that $2 \cdot 4 \cdot 6 \cdots (2n) = 2^n \cdot n!$, we finally obtain

$$c_{2n} = \frac{(2n-1)!!}{2^{3n} \cdot n!} c_0. \tag{9}$$

(We also used the fact that $4^n \cdot 2^n = 2^{3n}$.)

With $n = 1$, 3, and 5 in (8), we get

$$c_3 = \frac{2c_1}{4 \cdot 3}, \qquad c_5 = \frac{4c_3}{4 \cdot 5} = \frac{2 \cdot 4c_1}{4^2 \cdot 3 \cdot 5},$$

and

$$c_7 = \frac{6c_5}{4 \cdot 7} = \frac{2 \cdot 4 \cdot 6c_1}{4^3 \cdot 3 \cdot 5 \cdot 7}$$

It is apparent that the pattern is

$$c_{2n+1} = \frac{2 \cdot 4 \cdot 6 \cdots (2n)}{4^n \cdot 3 \cdot 5 \cdots (2n+1)} c_1 = \frac{n!}{2^n \cdot (2n+1)!!} c_1. \tag{10}$$

The formula in (9) gives the coefficients of even subscript in terms of c_0, and the formula in (10) gives the coefficients of odd subscript in terms of c_1. After we separately collect the terms of the series of even and odd degree, we get the general solution

$$y(x) = c_0 \left(1 + \sum_{n=1}^{\infty} \frac{(2n-1)!!}{2^{3n}n!} x^{2n} \right) + c_1 \left(x + \sum_{n=1}^{\infty} \frac{n!}{2^n(2n+1)!!} x^{2n+1} \right). \tag{11'}$$

Alternatively,

$$y(x) = c_0(1 + \tfrac{1}{8}x^2 + \tfrac{3}{128}x^4 + \tfrac{5}{1024}x^6 + \cdots)$$
$$+ c_1(x + \tfrac{1}{6}x^3 + \tfrac{1}{30}x^5 + \tfrac{1}{140}x^7 + \cdots). \tag{11''}$$

Because $y(0) = c_0$ and $y'(0) = c_1$, the given initial conditions imply that $c_0 = 4$ and $c_1 = 1$. Using these values in (11''), the first several terms of the particular solution satisfying $y(0) = 4$ and $y'(0) = 1$ are

$$y(x) = 4 + x + \tfrac{1}{2}x^2 + \tfrac{1}{6}x^3 + \tfrac{3}{32}x^4 + \tfrac{1}{30}x^5 + \cdots. \tag{12}$$

Remark: As in Example 5, substitution of $y = \sum c_n x^n$ in a linear second order equation with $x = 0$ an ordinary point typically leads to a recurrence relation that can be used to express each of the successive coefficients $c_2, c_3, c_4, \ldots$ in terms of the first two, c_0 and c_1. In this event two linearly independent solutions are obtained as follows. Let $y_0(x)$ be the solution obtained with $c_0 = 1$ and $c_1 = 0$, and let $y_1(x)$ be the one obtained with $c_0 = 0$ and $c_1 = 1$. Then

$$y_0(0) = 1, \qquad y_0'(0) = 0$$

while

$$y_1(0) = 0, \qquad y_1'(0) = 1,$$

so it is clear that y_0 and y_1 are linearly independent. In Example 5, $y_0(x)$ and $y_1(x)$ are defined by the two series that appear on the right-hand side in Eq. (11'), which expresses the general solution in the form $y = c_0 y_0 + c_1 y_1$.

   If in Example 5 we had sought a particular solution with given initial values $y(a)$ and $y'(a)$, we would have needed the general solution in the form

$$y(x) = \sum_{n=0}^{\infty} c_n(x - a)^n; \tag{13}$$

that is, in powers of $x - a$ rather than in powers of x. For only with a solution of the form in (13) is it true that

$$y(a) = c_0 \quad \text{and} \quad y'(a) = c_1$$

determine the arbitrary constants c_0 and c_1 in terms of the initial values of y and y'. Consequently, in order to solve an initial value problem, we need a series expansion of the general solution centered at the point where the initial conditions are specified.

EXAMPLE 6   Solve the initial value problem

$$(t^2 - 2t - 3)\frac{d^2 y}{dt^2} + 3(t - 1)\frac{dy}{dt} + y = 0; \qquad y(1) = 4, \qquad y'(1) = 1. \tag{14}$$

SOLUTION   We need a general solution of the form $\sum c_n(t - 1)^n$. But instead of substituting this series in (14) to determine the coefficients, it simplifies the computations if we first make the substitution $x = t - 1$, so that we wind up looking for a series of the form $\sum c_n x^n$ after all. To transform Eq. (14) into one with the new independent variable x, we note that

$$t^2 - 2t - 3 = (x + 1)^2 - 2(x + 1) - 3 = x^2 - 4,$$

$$\frac{dy}{dt} = \frac{dy}{dx}\frac{dx}{dt} = \frac{dy}{dx} = y',$$

and

$$\frac{d^2y}{dt^2} = \left[\frac{d}{dx}\left(\frac{dy}{dt}\right)\right]\frac{dx}{dt} = \frac{d}{dx}(y') = y'',$$

where primes denote differentiation with respect to x. Hence we transform Eq. (14) into

$$(x^2 - 4)y'' + 3xy' + y = 0$$

with initial conditions $y = 4$ and $y' = 1$ at $x = 0$ (corresponding to $t = 1$). This is the initial value problem we solved in Example 5, so the particular solution in (12) is available. We substitute $t - 1$ for x in (12), and thereby obtain the desired particular solution

$$y(t) = 4 + (t - 1) + \tfrac{1}{2}(t - 1)^2 + \tfrac{1}{6}(t - 1)^3$$
$$+ \tfrac{3}{32}(t - 1)^4 + \tfrac{1}{30}(t - 1)^5 + \cdots.$$

This series converges if $-1 < t < 3$. (Why?) A series such as this can be used to estimate numerical values of the solution. For instance,

$$y(0.8) = 4 + (-0.2) + \frac{(-0.2)^2}{2} + \frac{(-0.2)^3}{6} + \frac{3(-0.2)^4}{32} + \frac{(-0.2)^5}{30} + \cdots$$

$$\approx 4 - 0.2 + 0.02 - 0.00133 + 0.00015 - 0.00001 + \cdots,$$

so that $y(0.8) \approx 3.8188$.

The computation above illustrates the fact that series solutions of differential equations are useful not only for establishing general properties of a solution, but also for numerical computations when an expression of the solution in terms of familiar elementary functions is not available.

The formula in Example 5 is an example of a **two-term** recurrence relation; it expresses each coefficient in the series in terms of *one* of the preceding coefficients. A **many-term** recurrence relation expresses each coefficient in the series in terms of two or more preceding coefficients. In the case of a many-term recurrence relation, it is generally inconvenient or even impossible to find a formula that gives the typical c_n in terms of n. The following example shows what we sometimes can do with a three-term recurrence relation.

EXAMPLE 7 Find two linearly independent solutions of

$$y'' - xy' - x^2y = 0. \tag{15}$$

SOLUTION We make the usual substitution of the power series $y = \sum c_n x^n$. This results in the equation

$$\sum_{n=2}^{\infty} n(n-1)c_n x^{n-2} - \sum_{n=1}^{\infty} nc_n x^n - \sum_{n=0}^{\infty} c_n x^{n+2} = 0.$$

We can start the second sum at $n = 0$ without changing anything else. In order to make each sum include x^n in its general term, we shift the index of summation in the first sum by $+2$ (replace n by $n + 2$), and we shift it by -2 in the third sum (replace n by $n - 2$). These shifts yield

$$\sum_{n=0}^{\infty} (n + 2)(n + 1)c_{n+2}x^n - \sum_{n=0}^{\infty} nc_n x^n - \sum_{n=2}^{\infty} c_{n-2}x^n = 0.$$

The common range of these three summations is $n \geq 2$, so we must separate the terms corresponding to $n = 0$ and $n = 1$ in the first two sums before collecting coefficients of x^n. This gives

$$2c_2 + 6c_3 x - c_1 x + \sum_{n=2}^{\infty} [(n + 2)(n + 1)c_{n+2} - nc_n - c_{n-2}]x^n = 0.$$

The identity principle now implies that $2c_2 = 0$, $6c_3 - c_1 = 0$, and

$$(n + 2)(n + 1)c_{n+2} - nc_n - c_{n-2} = 0$$

for $n \geq 2$. Hence we get $c_2 = 0$, $c_3 = \frac{1}{6}c_1$, and the three-term recurrence relation

$$c_{n+2} = \frac{nc_n + c_{n-2}}{(n + 2)(n + 1)} \tag{16}$$

for $n \geq 2$. In particular,

$$c_4 = \frac{2c_2 + c_0}{12}, \qquad c_5 = \frac{3c_3 + c_1}{20}, \qquad c_6 = \frac{4c_4 + c_2}{30},$$

$$c_7 = \frac{5c_5 + c_3}{42}, \qquad c_8 = \frac{6c_6 + c_4}{56}. \tag{17}$$

Thus all values of c_n for $n \geq 4$ are given in terms of the arbitrary constants c_0 and c_1 because $c_2 = 0$ and $c_3 = c_1/6$.

To get our first solution y_1 of Eq. (15), we choose $c_0 = 1$ and $c_1 = 0$, so that $c_2 = c_3 = 0$. Then the formulas in (17) yield

$$c_4 = \tfrac{1}{12}, \qquad c_5 = 0, \qquad c_6 = \tfrac{1}{90}, \qquad c_7 = 0, \qquad c_8 = \tfrac{3}{1120};$$

thus

$$y_1(x) = 1 + \tfrac{1}{12}x^4 + \tfrac{1}{90}x^6 + \tfrac{3}{1120}x^8 + \cdots. \tag{18}$$

Because $c_1 = c_3 = 0$, it is clear from (16) that this series contains only terms of even degree.

To obtain our second linearly independent solution y_2 of (15), we take $c_0 = 0$ and $c_1 = 1$, so that $c_2 = 0$ and $c_3 = \frac{1}{6}$. Then the formulas in (17) yield

$$c_4 = 0, \qquad c_5 = \tfrac{3}{40}, \qquad c_6 = 0, \qquad c_7 = \tfrac{13}{1008},$$

so that

$$y_2(x) = x + \tfrac{1}{6}x^3 + \tfrac{3}{40}x^5 + \tfrac{13}{1008}x^7 + \cdots. \tag{19}$$

Because $c_0 = c_2 = 0$, it is clear from (16) that this series contains only terms of odd degree. The solutions $y_1(x)$ and $y_2(x)$ are linearly independent because $y_1(0) = 1$ and $y_1'(0) = 0$, while $y_2(0) = 0$ and $y_2'(0) = 1$. The general solution of Eq. (15) is a linear combination of the power series in (18) and (19). Equation (15) has no singular points, so the power series representing y_1 and y_2 converge for all x.

*THE LEGENDRE EQUATION

The **Legendre equation** of order α is the second order linear differential equation

$$(1 - x^2)y'' - 2xy' + \alpha(\alpha + 1)y = 0, \tag{20}$$

where the real number α satisfies the inequality $\alpha > -1$. This differential equation has extensive applications, ranging from numerical integration formulas (such as Gaussian quadrature) to the problem of determining the steady state temperature within a solid spherical ball when the temperature at points of its boundary is known. The only singular points of the Legendre equation are at $+1$ and -1, so it has two linearly independent solutions that can be expressed as power series in powers of x with radius of convergence at least 1. The substitution $y = \sum c_m x^m$ in (20) leads (see Problem 31) to the recurrence relation

$$c_{m+2} = -\frac{(\alpha - m)(\alpha + m + 1)}{(m + 1)(m + 2)} c_m \tag{21}$$

for $m \geq 0$. We are using m as the index of summation because we have another role for n to play.

In terms of the arbitrary constants c_0 and c_1, the formula in (21) yields

$$c_2 = -\frac{\alpha(\alpha + 1)}{2!} c_0,$$

$$c_3 = -\frac{(\alpha - 1)(\alpha + 2)}{3!} c_1,$$

$$c_4 = \frac{\alpha(\alpha - 2)(\alpha + 1)(\alpha + 3)}{4!} c_0,$$

$$c_5 = \frac{(\alpha - 1)(\alpha - 3)(\alpha + 2)(\alpha + 4)}{5!} c_1.$$

One can show without much trouble that, for $m > 0$,

$$c_{2m} = (-1)^m \frac{\alpha(\alpha - 2)(\alpha - 4) \cdots (\alpha - 2m + 2)(\alpha + 1)(\alpha + 3) \cdots (\alpha + 2m - 1)}{(2m)!} c_0 \tag{22}$$

and

$$c_{2m+1} = (-1)^m \frac{(\alpha - 1)(\alpha - 3) \cdots (\alpha - 2m + 1)(\alpha + 2)(\alpha + 4) \cdots (\alpha + 2m)}{(2m + 1)!} c_1. \tag{23}$$

Alternatively,

$$c_{2m} = (-1)^m a_{2m} c_0 \quad \text{and} \quad c_{2m+1} = (-1)^m a_{2m+1} c_1,$$

where a_{2m} and a_{2m+1} denote the fractions in (22) and (23), respectively. With this notation, we get two linearly independent power series solutions

$$y_1 = c_0 \sum_{m=0}^{\infty} (-1)^m a_{2m} x^{2m} \quad \text{and} \quad y_2 = c_1 \sum_{m=0}^{\infty} (-1)^m a_{2m+1} x^{2m+1} \tag{24}$$

of Legendre's equation of order α.

Now suppose that $\alpha = n$, a nonnegative *integer*. If $\alpha = n$ is even, we see from (22) that $a_{2m} = 0$ when $2m > n$. In this case y_1 is a *polynomial* of degree n and y_2

is a (nonterminating) infinite series. If $\alpha = n$ is odd, we see from (23) that $a_{2m+1} = 0$ when $2m + 1 > n$. In this case, y_2 is a *polynomial* of degree n and y_1 is a nonterminating infinite series. Thus in either case, one of the two solutions in (24) is a polynomial of degree n and the other is a nonterminating series.

With an appropriate choice (made separately for each n) of the arbitrary constant c_0 (n even) or c_1 (n odd), the nth degree polynomial solution of Legendre's equation of order n,

$$(1 - x^2)y'' - 2xy' + n(n + 1)y = 0, \tag{25}$$

is denoted by $P_n(x)$ and is called the **Legendre polynomial** of degree n. It is customary (for a reason indicated in Problem 32) to choose the arbitrary constant so that the coefficient of x^n in $P_n(x)$ is $(2n)!/[2^n(n!)^2]$. It then turns out that

$$P_n(x) = \sum_{k=0}^{N} \frac{(-1)^k(2n - 2k)!}{2^n k!(n - k)!(n - 2k)!} x^{n-2k} \tag{26}$$

where $N = [\![n/2]\!]$, the integral part of $n/2$. The first six Legendre polynomials are

$$P_0(x) = 1, \qquad\qquad P_1(x) = x,$$
$$P_2(x) = \tfrac{1}{2}(3x^2 - 1), \qquad P_3(x) = \tfrac{1}{2}(5x^3 - 3x),$$
$$P_4(x) = \tfrac{1}{8}(35x^4 - 30x^2 + 3), \qquad P_5(x) = \tfrac{1}{8}(63x^5 - 70x^3 + 15x).$$

3.2 PROBLEMS

Find general solutions in powers of x of the differential equations in Problems 1–15. State the recurrence relation and the guaranteed radius of convergence in each case.

1. $(x^2 - 1)y'' + 4xy' + 2y = 0$

2. $(x^2 + 2)y'' + 4xy' + 2y = 0$

3. $y'' + xy' + y = 0$

4. $(x^2 + 1)y'' + 6xy' + 4y = 0$

5. $(x^2 - 3)y'' + 2xy' = 0$

6. $(x^2 - 1)y'' - 6xy' + 12y = 0$

7. $(x^2 + 3)y'' - 7xy' + 16y = 0$

8. $(2 - x^2)y'' - xy' + 16y = 0$

9. $(x^2 - 1)y'' + 8xy' + 12y = 0$

10. $3y'' + xy' - 4y = 0$

11. $5y'' - 2xy' + 10y = 0$

12. $y'' - x^2y' - 3xy = 0$

13. $y'' + x^2y' + 2xy = 0$

14. $y'' + xy = 0$ (an Airy equation)

15. $y'' + x^2y = 0$

Use power series to solve the initial value problems in Problems 16 and 17.

16. $(1 + x^2)y'' + 2xy' - 2y = 0$; $y(0) = 0$, $y'(0) = 1$

17. $y'' + xy' - 2y = 0$; $y(0) = 1$, $y'(0) = 0$

Solve the initial value problems in Problems 18–22. First make a substitution of the form $t = x - a$, and then find a solution $\sum c_n t^n$ of the transformed differential equation. State the interval of values of x for which the theorem in this section guarantees convergence.

18. $y'' + (x - 1)y' + y = 0$; $y(1) = 2$, $y'(1) = 0$

19. $(2x - x^2)y'' - 6(x - 1)y' - 4y = 0$; $y(1) = 0$, $y'(1) = 1$

20. $(x^2 - 6x + 10)y'' - 4(x - 3)y' + 6y = 0$; $y(3) = 2$, $y'(3) = 0$

21. $(4x^2 + 16x + 17)y'' = 8y$; $y(-2) = 1$, $y'(-2) = 0$

22. $(x^2 + 6x)y'' + (3x + 9)y' - 3y = 0$; $y(-3) = 0$, $y'(-3) = 2$

In each of Problems 23–26, find a three-term recurrence relation for solutions of the form $y = \sum c_n x^n$. Then find the first three nonzero terms in each of two linearly independent solutions.

23. $y'' + (1 + x)y = 0$

24. $(x^2 - 1)y'' + 2xy' + 2xy = 0$

25. $y'' + x^2y' + x^2y = 0$

26. $(1 + x^3)y'' + x^4y = 0$

27. Solve the initial value problem

$$y'' + xy' + (2x^2 + 1)y = 0; \qquad y(0) = 1, \qquad y'(0) = -1.$$

Determine sufficiently many terms to compute $y(\frac{1}{2})$ accurate to four decimal places.

In Problems 28–30, find the first three nonzero terms in each of two linearly independent solutions of the form $y = \sum c_n x^n$. Substitute known Taylor series for the analytic functions and retain enough terms to compute the necessary coefficients.

28. $y'' + e^{-x}y = 0$

29. $(\cos x)y'' + y = 0$

30. $xy'' + (\sin x)y' + xy = 0$

31. Derive the recurrence relation in (21) for the Legendre equation.

32. Follow the steps outlined below to establish **Rodrigues' formula**

$$P_n(x) = \frac{1}{n!2^n}\frac{d^n}{dx^n}(x^2 - 1)^n$$

for the nth degree Legendre polynomial. (a) Show that $v = (x^2 - 1)^n$ satisfies the differential equation

$$(1 - x^2)v' + 2nxv = 0.$$

Differentiate each side of this equation to obtain

$$(1 - x^2)v'' + 2(n - 1)xv' + 2nv = 0.$$

(b) Differentiate each side of the last equation n times in succession to obtain

$$(1 - x^2)v^{(n+2)} - 2xv^{(n+1)} + n(n + 1)v^{(n)} = 0.$$

Thus $u = v^{(n)} = D^n(1 - x^2)^n$ satisfies Legendre's equation of order n.
(c) Show that the coefficient of x^n in u is $(2n)!/n!$; then state why this proves Rodrigues' formula. (Note that the coefficient of x^n in $P_n(x)$ is $(2n)!/[2^n(n!)^2]$.)

33. The **Hermite equation** of order α is

$$y'' - 2xy' + 2\alpha y = 0.$$

(a) Derive the two power series solutions

$$y_1 = 1 + \sum_{n=1}^{\infty}(-1)^m\frac{2^m\alpha(\alpha - 2)\cdots(\alpha - 2m + 2)}{(2m)!}x^{2m}$$

and

$$y_2 = x + \sum_{n=1}^{\infty}(-1)^m\frac{2^m(\alpha - 1)(\alpha - 3)\cdots(\alpha - 2m + 1)}{(2m + 1)!}x^{2m+1}.$$

Show that y_1 is a polynomial if α is an even integer, while y_2 is a polynomial if α is an odd integer. (b) The **Hermite polynomial** of degree n is denoted by $H_n(x)$. It is the nth degree polynomial solution of Hermite's equation, multiplied by a suitable constant so that the coefficient of x^n is 2^n. Show that the first six Hermite polynomials are

$$H_0(x) = 1, \qquad\qquad\qquad H_1(x) = 2x,$$
$$H_2(x) = 4x^2 - 2, \qquad\qquad H_3(x) = 8x^3 - 12x,$$
$$H_4(x) = 16x^4 - 48x^2 + 12, \qquad H_5(x) = 32x^5 - 160x^3 + 120x.$$

(c) The general formula for the Hermite polynomials is

$$H_n(x) = (-1)^n e^{x^2}\frac{d^n}{dx^n}(e^{-x^2}).$$

Verify that this does in fact give an nth degree polynomial.

3.3

Regular Singular Points

We now investigate the solution of the homogeneous second order linear equation

$$A(x)y'' + B(x)y' + C(x)y = 0 \tag{1}$$

near a singular point. Recall that if the functions A, B, and C are polynomials having no common factors, then the singular points of (1) are simply those points where $A(x)$ vanishes. For instance, $x = 0$ is the only singular point of the Bessel equation of order n,

$$x^2y'' + xy' + (x^2 - n^2)y = 0,$$

while the Legendre equation of order n,

$$(1 - x^2)y'' - 2xy' + n(n + 1)y = 0,$$

has the two singular points $x = -1$ and $x = 1$. It turns out that some of the features of the solutions of such equations of the most importance for applications are largely determined by their behavior near their singular points.

We will restrict our attention to the case in which $x = 0$ is a singular point of Eq. (1). A differential equation having $x = a$ as a singular point is easily transformed by the substitution $t = x - a$ into one having a corresponding singular point at 0. For example, let us substitute $t = x - 1$ in the Legendre equation above. Because

$$y' = \frac{dy}{dx} = \frac{dy}{dt}\frac{dt}{dx} = \frac{dy}{dt},$$

$$y'' = \frac{d^2y}{dx^2} = \left[\frac{d}{dt}\left(\frac{dy}{dt}\right)\right]\frac{dt}{dx} = \frac{d^2y}{dt^2},$$

and $1 - x^2 = 1 - (t + 1)^2 = -2t - t^2$, we get the equation

$$-t(t + 2)\frac{d^2y}{dt^2} - 2(t + 1)\frac{dy}{dt} + n(n + 1)y = 0.$$

This new equation has the singular point $t = 0$ corresponding to $x = 1$ in the original equation; it has also the singular point $t = -2$ corresponding to $x = -1$.

A differential equation having a singular point at 0 ordinarily will not have power series solutions of the form $y = \sum c_n x^n$, so the straightforward method of Section 3.2 fails in this case. To investigate the form that a solution of such an equation might take, we assume that Eq. (1) has analytic coefficient functions and rewrite it in the standard form

$$y'' + P(x)y' + Q(x)y = 0 \tag{2}$$

where $P = B/A$ and $Q = C/A$. Recall that $x = 0$ is an ordinary point (rather than a singular point) of the equation in (2) if the functions $P(x)$ and $Q(x)$ are analytic at $x = 0$; that is, if $P(x)$ and $Q(x)$ have convergent power series expansions in powers of x on some open interval containing $x = 0$. Now it can be proved that each of the functions $P(x)$ and $Q(x)$ *either* is analytic at 0 *or* approaches ∞ as $x \to 0$. Consequently $x = 0$ is a singular point of (2) provided that either $P(x)$ or $Q(x)$ (or both) approaches ∞ as $x \to 0$. For instance, if we rewrite the Bessel equation above in the form

$$y'' + \frac{1}{x}y' + \left(1 - \frac{n^2}{x^2}\right)y = 0,$$

we see that $P(x) = 1/x$ and $Q(x) = 1 - (n/x)^2$ each approach infinity as $x \to 0$.

We will see presently that the power series method can be generalized to apply near the singular point $x = 0$ of Eq. (2) provided that $P(x)$ approaches infinity no more rapidly than $1/x$, and $Q(x)$ no more rapidly than $1/x^2$, as $x \to 0$. This is a way of saying that $P(x)$ and $Q(x)$ have only *weak* singularities at $x = 0$. To state it more precisely, we rewrite Eq. (2) in the form

$$y'' + \frac{p(x)}{x}y' + \frac{q(x)}{x^2}y = 0, \tag{3}$$

where

$$p(x) = xP(x) \quad \text{and} \quad q(x) = x^2Q(x). \tag{4}$$

Definition *Regular Singular Point*

The singular point $x = 0$ of Eq. (3) is a **regular singular point** if the functions $p(x)$ and $q(x)$ are both analytic at $x = 0$. Otherwise it is an **irregular singular point**.

In particular, the singular point $x = 0$ is a *regular* singular point if $p(x)$ and $q(x)$ are both polynomials. For instance, we see that $x = 0$ is a regular singular point of Bessel's equation of order n by writing that equation in the form

$$y'' + \frac{1}{x}\, y' + \frac{x^2 - n^2}{x^2}\, y = 0,$$

noting that $p(x) = 1$ and $q(x) = x^2 - n^2$ are both polynomials in x.

By contrast, consider the equation

$$2x^3 y'' + (1 + x)y' + 3xy = 0,$$

which has the singular point $x = 0$. If we write this equation in the form of (3), we get

$$y'' + \frac{(1 + x)/(2x^2)}{x}\, y' + \frac{\frac{3}{2}}{x^2}\, y = 0.$$

Because

$$p(x) = \frac{1 + x}{2x^2} = \frac{1}{2x^2} + \frac{1}{2x} \to \infty$$

as $x \to 0$ (although $q(x) = \frac{3}{2}$ is a polynomial), we see that $x = 0$ is an irregular singular point. We will not discuss the solution of differential equations near irregular singular points; this is a considerably more advanced topic than the solution of differential equations near regular singular points.

EXAMPLE 1 Consider the differential equation

$$x^2(1 + x)y'' + x(4 - x^2)y' + (2 + 3x)y = 0.$$

In the standard form $y'' + Py' + Qy = 0$ it is

$$y'' + \frac{4 - x^2}{x(1 + x)}\, y' + \frac{2 + 3x}{x^2(1 + x)}\, y = 0.$$

Because

$$P(x) = \frac{4 - x^2}{x(1 + x)} \quad \text{and} \quad Q(x) = \frac{2 + 3x}{x^2(1 + x)}$$

both approach ∞ as $x \to 0$, we see that $x = 0$ is a singular point. To determine the nature of this singular point we write the differential equation in the form of Eq. (3):

$$y'' + \frac{(4 - x^2)/(1 + x)}{x} y' + \frac{(2 + 3x)/(1 + x)}{x^2} y = 0.$$

Thus

$$p(x) = \frac{4 - x^2}{1 + x} \quad \text{and} \quad q(x) = \frac{2 + 3x}{1 + x}.$$

Because a quotient of polynomials is analytic wherever the denominator is non-zero, we see that $p(x)$ and $q(x)$ are both analytic at $x = 0$. Hence $x = 0$ is a *regular singular point* of the given differential equation.

It may happen that when we begin with a differential equation in the general form in (1) and rewrite it in the form in (3), the functions $p(x)$ and $q(x)$ as given in (4) are indeterminate forms at $x = 0$. In this case the situation is determined by the limits

$$p_0 = p(0) = \lim_{x \to 0} p(x) = \lim_{x \to 0} xP(x) \tag{5}$$

and

$$q_0 = q(0) = \lim_{x \to 0} q(x) = \lim_{x \to 0} x^2 Q(x). \tag{6}$$

If $p_0 = 0 = q_0$ then $x = 0$ may be an ordinary point. Otherwise, if the limits in both (5) and (6) exist and are *finite*, then $x = 0$ is a regular singular point. If either limit fails to exist or is infinite, then $x = 0$ is an irregular singular point.

Remark: The most common case in applications, for the differential equation written in the form

$$y'' + \frac{p(x)}{x} y' + \frac{q(x)}{x^2} y = 0, \tag{3}$$

is that the functions $p(x)$ and $q(x)$ are *polynomials*. In this case $p_0 = p(0)$ and $q_0 = q(0)$ are simply the constant terms of these polynomials, so the limits in (5) and (6) need not be evaluated.

EXAMPLE 2 To investigate the nature of the point $x = 0$ for the differential equation

$$x^4 y'' + (x^2 \sin x)y' + (1 - \cos x)y = 0,$$

we first rewrite it in the form in (3):

$$y'' + \frac{(\sin x)/x}{x} y' + \frac{(1 - \cos x)/x^2}{x^2} y = 0.$$

Then l'Hôpital's rule gives the values

$$p_0 = \lim_{x \to 0} \frac{\sin x}{x} = \lim_{x \to 0} \frac{\cos x}{1} = 1$$

and

$$q_0 = \lim_{x \to 0} \frac{1 - \cos x}{x^2} = \lim_{x \to 0} \frac{\sin x}{2x} = \frac{1}{2}$$

for the limits in (5) and (6). While not both are zero, we see that $x = 0$ is a regular singular point because each limit is finite. Alternatively, we could write

$$p(x) = \frac{\sin x}{x} = \frac{1}{x}\left(x - \frac{x^3}{3!} + \frac{x^5}{5!} - \cdots\right)$$

$$= 1 - \frac{x^2}{3!} + \frac{x^4}{5!} - \cdots$$

and

$$q(x) = \frac{1 - \cos x}{x^2} = \frac{1}{x^2}\left[1 - \left(1 - \frac{x^2}{2!} + \frac{x^4}{4!} - \frac{x^6}{6!} + \cdots\right)\right]$$

$$= \frac{1}{2!} - \frac{x^2}{4!} + \frac{x^4}{6!} - \cdots.$$

These (convergent) power series show explicitly that $p(x)$ and $q(x)$ are analytic, and moreover that $p_0 = p(0) = 1$ and $q_0 = q(0) = \frac{1}{2}$.

THE METHOD OF FROBENIUS

We now approach the task of actually finding solutions of a second order differential equation near the regular singular point $x = 0$. The simplest such equation is the Euler-Cauchy equation

$$x^2 y'' + p_0 x y' + q_0 y = 0, \tag{7}$$

which we initially rewrite in the form

$$y'' + \frac{p_0}{x} y' + \frac{q_0}{x^2} y = 0 \tag{7'}$$

with $p(x) = p_0$ and $q(x) = q_0$. In Section 2.6 we saw that $y = x^r$ is a solution of Eq. (7) if r is a root of the quadratic equation

$$r(r - 1) + p_0 r + q_0 = 0. \tag{8}$$

In the general case, in which $p(x)$ and $q(x)$ are power series rather than constants, it is a reasonable conjecture that our differential equation might have a solution of the form

$$y = x^r \sum_{n=0}^{\infty} c_n x^n = \sum_{n=0}^{\infty} c_n x^{n+r} \tag{9}$$

$$= c_0 x^r + c_1 x^{r+1} + c_2 x^{r+2} + \cdots$$

—the product of x^r and a power series. This turns out to be a very fruitful conjecture; according to the theorem stated below (following Example 3), every equation of the form in (1) having $x = 0$ as a regular singular point does, indeed, have at least one such solution. This fact is the basis for the *method of Frobenius*, named for the German mathematician Georg Frobenius (1848–1917), who discovered this method in the 1870s.

An infinite series of the form in (9) is called a **Frobenius series**. Note that a Frobenius series is generally *not* a power series. For instance, with $r = -\frac{1}{2}$ the series in (9) takes the form

$$y = c_0 x^{-1/2} + c_1 x^{1/2} + c_2 x^{3/2} + \cdots;$$

it is not a series of *integral* powers of x.

To investigate the possible existence of Frobenius series solutions, we begin with the equation

$$x^2 y'' + x p(x) y' + q(x) y = 0 \tag{10}$$

obtained by multiplying the equation in (3) by x^2. If $x = 0$ is a regular singular point, then $p(x)$ and $q(x)$ are analytic at $x = 0$, so

$$\left. \begin{aligned} p(x) &= p_0 + p_1 x + p_2 x^2 + \cdots, \\ q(x) &= q_0 + q_1 x + q_2 x^2 + \cdots. \end{aligned} \right\} \tag{11}$$

Suppose that Eq. (10) has the Frobenius series solution

$$y = \sum_{n=0}^{\infty} c_n x^{n+r}. \tag{12}$$

We may (and always do) assume that $c_0 \neq 0$ because the series must have a first nonzero term. Termwise differentiation in (12) leads to

$$y' = \sum_{n=0}^{\infty} c_n (n+r) x^{n+r-1} \tag{13}$$

and

$$y'' = \sum_{n=0}^{\infty} c_n (n+r)(n+r-1) x^{n+r-2}. \tag{14}$$

Substitution of the series in (11)–(14) in (10) now yields

$$\begin{aligned} &[r(r-1)c_0 x^r + (r+1)r c_1 x^{r+1} + \cdots] \\ &+ [p_0 x + p_1 x^2 + \cdots][r c_0 x^{r-1} + (r+1)c_1 x^r + \cdots] \\ &+ [q_0 + q_1 x + \cdots][c_0 x^r + c_1 x^{r+1} + \cdots] = 0. \end{aligned} \tag{15}$$

The lowest power of x that appears in (15) is x^r. If (15) is to be satisfied identically, the coefficient $r(r-1)c_0 + p_0 r c_0 + q_0 c_0$ of x^r must vanish. Because $c_0 \neq 0$, it follows that r must satisfy the quadratic equation

$$r(r-1) + p_0 r + q_0 = 0 \tag{16}$$

of precisely the same form as that obtained with the Euler-Cauchy equation. Equation (16) is called the **indicial equation** of the differential equation in (10), and its two roots (possibly equal) are the **exponents** of the differential equation (at the regular singular point $x = 0$).

Our derivation of (16) shows that *if* the Frobenius series $y = x^r \sum c_n x^n$ is to be a solution of the differential equation in (10), *then* the exponent r must be one of the roots r_1 and r_2 of the indicial equation in (16). If $r_1 \neq r_2$ it follows that there are two possible Frobenius series solutions, while if $r_1 = r_2$ there is only one possible Frobenius series solution; the second solution cannot be a Frobenius series. The exponents r_1 and r_2 in the possible Frobenius series solutions are determined (using the indicial equation) by the values $p_0 = p(0)$ and $q_0 = q(0)$ that we have discussed. In practice, particularly when the coefficients in the differential equation in the original form in (1) are polynomials, the simplest way of finding p_0 and q_0 is often to write the equation in the form

$$y'' + \frac{p_0 + p_1 x + \cdots}{x} y' + \frac{q_0 + q_1 x + \cdots}{x^2} y = 0. \tag{17}$$

Then inspection of the series that appear in the two numerators reveals the constants p_0 and q_0.

EXAMPLE 3 Find the exponents in the possible Frobenius series solutions of the equation

$$2x^2(1 + x)y'' + 3x(1 + x)^3 y' - (1 - x^2)y = 0.$$

SOLUTION We divide each term by $2x^2(1 + x)$ to recast the differential equation in the form

$$y'' + \frac{(\frac{3}{2})(1 + 2x + x^2)}{x} y' + \frac{(-\frac{1}{2})(1 - x)}{x^2} y = 0,$$

and thus see that $p_0 = \frac{3}{2}$ and $q_0 = -\frac{1}{2}$. Hence the indicial equation is

$$r(r - 1) + \tfrac{3}{2}r - \tfrac{1}{2} = r^2 + \tfrac{1}{2}r - \tfrac{1}{2} = (r + 1)(r - \tfrac{1}{2}) = 0,$$

with roots $r_1 = \frac{1}{2}$ and $r_2 = -1$. The two possible Frobenius series solutions are then of the forms

$$y_1 = x^{1/2} \sum_{n=0}^{\infty} a_n x^n \quad \text{and} \quad y_2 = x^{-1} \sum_{n=0}^{\infty} b_n x^n.$$

Once the exponents r_1 and r_2 are known, the coefficients in a Frobenius series solution are determined by substitution of the series in (12)–(14) in the differential equation, essentially the same method as was used to determine coefficients in power series solutions in Section 3.2. If the exponents r_1 and r_2 are complex conjugates, then there always exist two independent Frobenius series solutions. We will restrict our attention here to the case in which r_1 and r_2 are both real. We also will seek solutions only for $x > 0$. Once such a solution has been found, we need only replace x^{r_i} by $|x|^{r_i}$ to obtain a solution for $x < 0$. The following theorem is proved in Chapter 4 of Coddington's *An Introduction to Ordinary Differential Equations.*

> **Theorem** *Frobenius Series Solutions*
>
> Suppose that $x = 0$ is a regular singular point of the equation
>
> $$x^2 y'' + xp(x)y' + q(x)y = 0. \qquad (10)$$
>
> Let $\rho > 0$ denote the minimum of the radii of convergence of the power series
>
> $$p(x) = \sum_{n=0}^{\infty} p_n x^n \quad \text{and} \quad q(x) = \sum_{n=0}^{\infty} q_n x^n.$$
>
> Let r_1 and r_2 be the roots, with $r_1 \geqq r_2$, of the indicial equation $r(r-1) + p_0 r + q_0 = 0$. Then:
>
> (a) For $x > 0$, there exists a solution of Eq. (10) of the form
>
> $$y_1 = x^{r_1} \sum_{n=0}^{\infty} a_n x^n \qquad (a_0 \neq 0) \qquad (18)$$
>
>   corresponding to the larger root r_1.
> (b) If $r_1 - r_2$ is neither zero nor a positive integer, then there exists a second linearly independent solution for $x > 0$ of the form
>
> $$y_2 = x^{r_2} \sum_{n=0}^{\infty} b_n x^n \qquad (b_0 \neq 0) \qquad (19)$$
>
>   corresponding to the smaller root r_2.
>
>   The radii of convergence of the power series in (18) and (19) are each at least ρ. The coefficients in these series can be determined by subsituting the series in the differential equation
>
> $$x^2 y'' + xp(x)y' + q(x)y = 0. \qquad (10)$$

We have already seen that if $r_1 = r_2$, then there can exist only one Frobenius series solution. It turns out that, if $r_1 - r_2$ is a positive integer, there may or may not exist a second Frobenius series solution of the form in (19) corresponding to the smaller root r_2. These exceptional cases are discussed in Section 3.4. The examples below illustrate the process of determining the coefficients in those Frobenius series solutions that are guaranteed by the theorem.

EXAMPLE 4 Find the Frobenius series solutions of

$$2x^2 y'' + 3xy' - (x^2 + 1)y = 0. \qquad (20)$$

SOLUTION First we divide each term by $2x^2$ to put the equation in the form in (17):

$$y'' + \frac{\frac{3}{2}}{x} y' + \frac{-(\frac{1}{2}) - (\frac{1}{2})x^2}{x^2} y = 0. \qquad (21)$$

We now see that $x = 0$ is a regular singular point, and that $p_0 = \frac{3}{2}$ and $q_0 = -\frac{1}{2}$. Because $p(x) = \frac{3}{2}$ and $q(x) = -(\frac{1}{2}) - (\frac{1}{2})x^2$ are polynomials, the Frobenius series we obtain will converge for all $x > 0$. The indicial equation is

$$r(r - 1) + \tfrac{3}{2}r - \tfrac{1}{2} = (r - \tfrac{1}{2})(r + 1) = 0,$$

so the exponents are $r_1 = \frac{1}{2}$ and $r_2 = -1$. They do not differ by an integer, so the theorem guarantees the existence of two linearly independent Frobenius series solutions. Rather than separately substituting $y_1 = x^{1/2} \sum a_n x^n$ and $y_2 = x^{-1} \sum b_n x^n$ in the differential equation in (20), it is more efficient to begin by substituting $y = x^r \sum c_n x^n$. We will then get a recurrence relation that depends on r. With the value $r_1 = \frac{1}{2}$ it becomes a recurrence relation for the series for y_1, while with $r_2 = -1$ it becomes a recurrence relation for the series for y_2.

When we substitute

$$y = \sum_{n=0}^{\infty} c_n x^{n+r}, \qquad y' = \sum_{n=0}^{\infty} (n + r)c_n x^{n+r-1},$$

and

$$y'' = \sum_{n=0}^{\infty} (n + r)(n + r - 1)c_n x^{n+r-2}$$

in (20)—the original differential equation, rather than (21)—we get

$$2 \sum_{n=0}^{\infty} (n + r)(n + r - 1)c_n x^{n+r} + 3 \sum_{n=0}^{\infty} (n + r)c_n x^{n+r} - \sum_{n=0}^{\infty} c_n x^{n+r+2} - \sum_{n=0}^{\infty} c_n x^{n+r} = 0. \quad (22)$$

At this stage there are several ways to proceed. A good standard practice is to shift indices so that each exponent will be the same as the smallest one present. In the case above, we shift the index of summation in the third sum by -2 to bring its exponent down from $n + r + 2$ to $n + r$. This gives

$$2 \sum_{n=0}^{\infty} (n + r)(n + r - 1)c_n x^{n+r} + 3 \sum_{n=0}^{\infty} (n + r)c_n x^{n+r} - \sum_{n=2}^{\infty} c_{n-2} x^{n+r} - \sum_{n=0}^{\infty} c_n x^{n+r} = 0. \quad (23)$$

The common range of summation is $n \geq 2$, so we must treat $n = 0$ and $n = 1$ separately. Following our standard practice, the terms corresponding to $n = 0$ will always give the indicial equation

$$[2r(r - 1) + 3r - 1]c_0 = 2(r^2 + \tfrac{1}{2}r - \tfrac{1}{2})c_0 = 0.$$

The terms corresponding to $n = 1$ yield

$$[2(r + 1)r + 3(r + 1) - 1]c_1 = (2r^2 + 5r + 2)c_1 = 0.$$

Because the coefficient $2r^2 + 5r + 2$ of c_1 is nonzero whether $r = \frac{1}{2}$ or $r = -1$, it follows that

$$c_1 = 0 \qquad (24)$$

in either case.

The coefficient of x^{n+r} in (23) is

$$2(n + r)(n + r - 1)c_n + 3(n + r)c_n - c_{n-2} - c_n = 0.$$

We solve for c_n and simplify to obtain the recurrence relation

$$c_n = \frac{c_{n-2}}{2(n+r)^2 + (n+r) - 1} \quad \text{for } n \geqq 2. \tag{25}$$

The case $r_1 = \frac{1}{2}$: We now write a_n in place of c_n and substitute $r = \frac{1}{2}$ in (25). This gives the recurrence relation

$$a_n = \frac{a_{n-2}}{2n^2 + 3n} \quad \text{for } n \geqq 2. \tag{26}$$

With this formula we can determine the coefficients in the first Frobenius solution y_1. In view of (24) we see that $a_n = 0$ whenever n is odd. With $n = 2$, 4, and 6 in (26), we get

$$a_2 = \frac{a_0}{14}, \quad a_4 = \frac{a_2}{44} = \frac{a_0}{616}, \quad \text{and} \quad a_6 = \frac{a_4}{90} = \frac{a_0}{55{,}440}.$$

Hence our first Frobenius series solution is

$$y_1(x) = a_0 x^{1/2}\left(1 + \frac{x^2}{14} + \frac{x^4}{616} + \frac{x^6}{55{,}440} + \cdots\right).$$

The case $r_2 = -1$: We may now write b_n in place of c_n and substitute $r = -1$ in (25). This gives the recurrence relation

$$b_n = \frac{b_{n-2}}{2n^2 - 3n} \quad \text{for } n \geqq 2. \tag{27}$$

Again (24) implies that $b_n = 0$ for n odd. With $n = 2$, 4, and 6 in (27), we get

$$b_2 = \frac{b_0}{2}, \quad b_4 = \frac{b_2}{20} = \frac{b_0}{40}, \quad \text{and} \quad b_6 = \frac{b_4}{54} = \frac{b_0}{2160}.$$

Hence our second Frobenius series solution is

$$y_2(x) = b_0 x^{-1}\left(1 + \frac{1}{2}x^2 + \frac{x^4}{40} + \frac{x^6}{2160} + \cdots\right).$$

EXAMPLE 5 Find a Frobenius series solution of Bessel's equation of order zero:

$$x^2 y'' + xy' + x^2 y = 0. \tag{28}$$

SOLUTION In the form of (17) the equation becomes

$$y'' + \frac{1}{x}y' + \frac{x^2}{x^2}y = 0.$$

Hence $x = 0$ is a regular singular point with $p(x) = 1$ and $q(x) = x^2$, so our series will converge for all $x > 0$. Because $p_0 = 1$ and $q_0 = 0$, the indicial equation is

$$r(r-1) + r = r^2 = 0.$$

Thus we obtain only the single exponent $r = 0$, and so there is only one Frobenius series solution

$$y(x) = x^0 \sum_{n=0}^{\infty} c_n x^n = \sum_{n=0}^{\infty} c_n x^n$$

of (28); it is in fact a power series.

As usual we substitute $y = \sum c_n x^n$ in (28); the result is

$$\sum_{n=0}^{\infty} n(n-1)c_n x^n + \sum_{n=0}^{\infty} nc_n x^n + \sum_{n=0}^{\infty} c_n x^{n+2} = 0.$$

We combine the first two sums and shift the index of summation in the third by -2 to obtain

$$\sum_{n=0}^{\infty} n^2 c_n x^n + \sum_{n=2}^{\infty} c_{n-2} x^n = 0.$$

The term corresponding to x^0 gives $0 = 0$: no information. The term corresponding to x^1 gives $c_1 = 0$, and the term for x^n yields the recurrence relation

$$c_n = -\frac{c_{n-2}}{n^2} \quad \text{for } n \geqq 2. \tag{29}$$

Because $c_1 = 0$, we see that $c_n = 0$ whenever n is odd. Substituting $n = 2$, 4, and 6 in (29), we get

$$c_2 = -\frac{c_0}{2^2}, \quad c_4 = -\frac{c_2}{4^2} = \frac{c_0}{2^2 4^2}, \quad \text{and} \quad c_6 = -\frac{c_4}{6^2} = -\frac{c_0}{2^2 4^2 6^2}.$$

Evidently the pattern is

$$c_{2n} = \frac{(-1)^n c_0}{2^2 \cdot 4^2 \cdots (2n)^2} = \frac{(-1)^n c_0}{2^{2n}(n!)^2}.$$

The choice $c_0 = 1$ gives us one of the most important special functions in mathematics, the **Bessel function of order zero of the first kind**, denoted by $J_0(x)$. Thus

$$J_0(x) = \sum_{n=0}^{\infty} \frac{(-1)^n x^{2n}}{2^{2n}(n!)^2} = 1 - \frac{x^2}{4} + \frac{x^4}{64} - \frac{x^6}{2304} + \cdots. \tag{30}$$

In this example we have not been able to find a second linearly independent solution of Bessel's equation of order zero. We will derive that solution in Section 3.4; it will not be a Frobenius series.

Recall that, if $r_1 - r_2$ is a positive integer, the theorem of this section guarantees only the existence of the Frobenius series solution corresponding to the larger exponent r_1. The following example illustrates the fortunate case in which the series method nevertheless yields a second Frobenius series solution. The case in which the second solution is not a Frobenius series will be discussed in Section 3.4.

EXAMPLE 6   Find the Frobenius series solutions of

$$xy'' + 2y' + xy = 0. \tag{31}$$

SOLUTION In standard form the equation becomes

$$y'' + \frac{2}{x} y' + \frac{x^2}{x^2} y = 0,$$

so we see that $x = 0$ is a regular singular point with $p_0 = 2$ and $q_0 = 0$. The indicial equation

$$r(r - 1) + 2r = r(r + 1) = 0$$

has roots $r_1 = 0$ and $r_2 = -1$, which differ by an integer. In this case when $r_1 - r_2$ is an integer, it is best to depart from the standard procedure of Example 4 and begin our work with the *smaller* exponent. As you will see below, the recurrence relation will then tell us whether or not a second Frobenius series solution exists. If it does exist, our computations will simultaneously yield *both* Frobenius series solutions. If the second solution does not exist, we begin anew with the larger exponent $r = r_1$ to obtain the one Frobenius series solution guaranteed by the theorem of this section.

Hence we begin by substituting

$$y = x^{-1} \sum_{n=0}^{\infty} c_n x^n = \sum_{n=0}^{\infty} c_n x^{n-1}$$

in (31). This gives

$$\sum_{n=0}^{\infty} (n - 1)(n - 2) c_n x^{n-2} + 2 \sum_{n=0}^{\infty} (n - 1) c_n x^{n-2} + \sum_{n=0}^{\infty} c_n x^n = 0.$$

We combine the first two sums and shift the index by -2 in the third to obtain

$$\sum_{n=0}^{\infty} n(n - 1) c_n x^{n-2} + \sum_{n=2}^{\infty} c_{n-2} x^{n-2} = 0. \tag{32}$$

The cases $n = 0$ and $n = 1$ reduce to

$$0 \cdot c_0 = 0 \quad \text{and} \quad 0 \cdot c_1 = 0.$$

Hence we have *two* arbitrary constants c_0 and c_1, and therefore can expect to find a general solution incorporating two linearly independent Frobenius series solutions. If, for $n = 1$, we had obtained an equation such as $0 \cdot c_1 = 3$, which can be satisfied for *no* choice of c_1, this would have told us that no second Frobenius series solution could exist.

Now knowing that all is well, from (32) we read the recurrence relation

$$c_n = -\frac{c_{n-2}}{n(n - 1)} \quad \text{for } n \geq 2. \tag{33}$$

The first few values of n give

$$c_2 = -\frac{1}{2 \cdot 1} c_0, \qquad\qquad c_3 = -\frac{1}{3 \cdot 2} c_1,$$

$$c_4 = -\frac{1}{4 \cdot 3} c_2 = \frac{c_0}{4!}, \qquad c_5 = -\frac{1}{5 \cdot 4} c_3 = \frac{c_1}{5!},$$

$$c_6 = -\frac{1}{6 \cdot 5} c_4 = -\frac{c_0}{6!}, \qquad c_7 = -\frac{1}{7 \cdot 6} c_5 = -\frac{c_1}{7!};$$

evidently the pattern is

$$c_{2n} = \frac{(-1)^n c_0}{(2n)!}, \qquad c_{2n+1} = \frac{(-1)^n c_1}{(2n+1)!}$$

for $n \geq 1$. Therefore our solution is

$$y(x) = x^{-1} \sum_{n=0}^{\infty} c_n x^n$$

$$= \frac{c_0}{x}\left(1 - \frac{x^2}{2!} + \frac{x^4}{4!} - \cdots\right) + \frac{c_1}{x}\left(x - \frac{x^3}{3!} + \frac{x^5}{5!} - \cdots\right)$$

$$= \frac{c_0}{x}\sum_{n=0}^{\infty}\frac{(-1)^n x^{2n}}{(2n)!} + \frac{c_1}{x}\sum_{n=0}^{\infty}\frac{(-1)^n x^{2n+1}}{(2n+1)!};$$

thus

$$y(x) = \frac{1}{x}(c_0 \cos x + c_1 \sin x).$$

We have thus found a general solution expressed as a linear combination of the two Frobenius series solutions

$$y_1(x) = \frac{\cos x}{x} \quad \text{and} \quad y_2(x) = \frac{\sin x}{x}.$$

SUMMARY   When confronted with a linear second order differential equation

$$A(x)y'' + B(x)y' + C(x)y = 0$$

with analytic coefficient functions, in order to investigate the possible existence of series solutions we first write the equation in standard form

$$y'' + P(x)y' + Q(x)y = 0.$$

If $P(x)$ and $Q(x)$ are both analytic at $x = 0$ then $x = 0$ is an ordinary point, and the equation has two linearly independent power series solutions.

Otherwise, $x = 0$ is a singular point, and we next write the differential equation in the form

$$y'' + \frac{p(x)}{x}y' + \frac{q(x)}{x^2}y = 0.$$

If $p(x)$ and $q(x)$ are both analytic at $x = 0$, then $x = 0$ is a regular singular point. In this case we find the two exponents r_1 and r_2 (assumed real, and with $r_1 \geq r_2$) by solving the indicial equation

$$r(r-1) + p_0 r + q_0 = 0,$$

where $p_0 = p(0)$ and $q_0 = q(0)$. There always exists a Frobenius series solution $y_1 = x^{r_1}\sum a_n x^n$ associated with the larger exponent r_1, and if $r_1 - r_2$ is not an integer, the existence of a second Frobenius series solution $y_2 = x^{r_2}\sum b_n x^n$ is also guaranteed.

In each of Problems 1–8, determine whether $x = 0$ is an ordinary point, a regular singular point, or an irregular singular point. If it is a regular singular point, find the exponents of the differential equation at $x = 0$.

1. $xy'' + (x - x^3)y' + (\sin x)y = 0$

2. $xy'' + x^2y' + (e^x - 1)y = 0$

3. $x^2y'' + (\cos x)y' + xy = 0$

4. $3x^3y'' + 2x^2y' + (1 - x^2)y = 0$

5. $x(1 + x)y'' + 2y' + 3xy = 0$

6. $x^2(1 - x^2)y'' + 2xy' - 2y = 0$

7. $x^2y'' + (6 \sin x)y' + 6y = 0$

8. $(6x^2 + 2x^3)y'' + 21xy' + 9(x^2 - 1)y = 0$

If $x = a \neq 0$ is a singular point of a second order linear differential equation, the substitution $t = x - a$ transforms it into a differential equation having $t = 0$ as a singular point. We then attribute to the original equation at $x = a$ the behavior of the new equation at $t = 0$. Classify (as regular or irregular) the singular points of the differential equations in Problems 9–16.

9. $(1 - x)y'' + xy' + x^2y = 0$

10. $(1 - x)^2y'' + (2x - 2)y' + y = 0$

11. $(1 - x^2)y'' - 2xy' + 12y = 0$

12. $(x - 2)^3y'' + 3(x - 2)^2y' + x^3y = 0$

13. $(x^2 - 4)y'' + (x - 2)y' + (x + 2)y = 0$

14. $(x^2 - 9)^2y'' + (x^2 + 9)y' + (x^2 + 4)y = 0$

15. $(x - 2)^2y'' - (x^2 - 4)y' + (x + 2)y = 0$

16. $x^3(1 - x)y'' + (3x + 2)y' + xy = 0$

Find two linearly independent Frobenius series solutions (for $x > 0$) of each of the differential equations in Problems 17–26.

17. $4xy'' + 2y' + y = 0$ **18.** $2xy'' + 3y' - y = 0$

19. $2xy'' - y' - y = 0$ **20.** $3xy'' + 2y' + 2y = 0$

21. $2x^2y'' + xy' - (1 + 2x^2)y = 0$

22. $2x^2y'' + xy' - (3 - 2x^2)y = 0$

23. $6x^2y'' + 7xy' - (x^2 + 2)y = 0$

24. $3x^2y'' + 2xy' + x^2y = 0$

25. $2xy'' + (1 + x)y' + y = 0$

26. $2xy'' + (1 - 2x^2)y' - 4xy = 0$

Use the method of Example 6 to find two linearly independent Frobenius series solutions of each of the differential equations in Problems 27–31.

27. $xy'' + 2y' + 9xy = 0$ **28.** $xy'' + 2y' - 4xy = 0$

29. $4xy'' + 8y' + xy = 0$ **30.** $xy'' - y' + 4x^2y = 0$

31. $4x^2y'' - 4xy' + (3 - 4x^2)y = 0$

In each of Problems 32–34, find the first three nonzero terms of each of two linearly independent Frobenius series solutions.

32. $2x^2y'' + x(x + 1)y' - (2x + 1)y = 0$

33. $(2x^2 + 5x^3)y'' + (3x - x^2)y' - (1 + x)y = 0$

34. $2x^2y'' + (\sin x)y' - (\cos x)y = 0$

35. Note that $x = 0$ is an irregular point of the equation

$$x^2y'' + (3x - 1)y' + y = 0.$$

(a) Show that $y = x^r \sum c_nx^n$ can satisfy this equation only if $r = 0$. (b) Substitute $y = \sum c_nx^n$ to derive the "formal" solution $y = \sum n!x^n$. What is the radius of convergence of this series?

36. (a) Suppose that A and B are nonzero constants. Show that the equation $x^2y'' + Ay' + By = 0$ has at most one solution of the form $y = x^r \sum c_nx^n$. (b) Repeat part (a) with the equation $x^3y'' + Axy' + By = 0$. (c) Show that the equation $x^3y'' + Ax^2y' + By = 0$ has no Frobenius series solution. (*Suggestion:* In each case substitute $y = x^r \sum c_nx^n$ in the given equation to determine the possible values of r.)

37. (a) Use the method of Frobenius to derive the solution $y_1 = x$ of the equation $x^3y'' - xy' + y = 0$. (b) Derive by reduction of order the second solution $y_2 = xe^{-1/x}$. Does y_2 have a Frobenius series representation?

38. Apply the method of Frobenius to Bessel's equation of order $\frac{1}{2}$,

$$x^2y'' + xy' + (x^2 - \tfrac{1}{4})y = 0,$$

to derive its general solution for $x > 0$,

$$y = c_0 \frac{\cos x}{\sqrt{x}} + c_1 \frac{\sin x}{\sqrt{x}}.$$

39. (a) Show that Bessel's equation of order 1,

$$x^2y'' + xy' + (x^2 - 1)y = 0,$$

has exponents $r_1 = 1$ and $r_2 = -1$ at $x = 0$, and that the Frobenius series solution corresponding to $r_1 = 1$ is

$$J_1(x) = \frac{x}{2} \sum_{n=0}^{\infty} \frac{(-1)^nx^{2n}}{n!(n + 1)!2^{2n}}.$$

(b) Show that there is no Frobenius solution corresponding to the smaller exponent $r_2 = -1$; that is, show that it is impossible to determine the coefficients in $y_2 = x^{-1} \sum c_nx^n$.

40. Consider the equation $x^2y'' + xy' + (1 - x)y = 0$. (a) Show that its exponents are $\pm i$, so it has complex-valued Frobenius series solutions $y_+ = x^i \sum p_n x^n$ and $y_- = x^{-i} \sum q_n x^n$ with $p_0 = q_0 = 1$. (b) Show that the recursion formula is

$$c_n = \frac{c_{n-1}}{n^2 + 2rn}.$$

Apply it with $r = i$ to obtain $p_n = c_n$, $r = -i$ to obtain $q_n = c_n$. Conclude that p_n and q_n are complex conjugates: $p_n = a_n + ib_n$ and $q_n = a_n - ib_n$, where the numbers $\{a_n\}$ and $\{b_n\}$ are real. (c) Deduce from part (b) that the differential equation of this problem has real-valued solutions of the form

$$y_1 = A(x)\cos(\ln x) - B(x)\sin(\ln x),$$

$$y_2 = A(x)\sin(\ln x) + B(x)\cos(\ln x)$$

where $A(x) = \sum a_n x^n$ and $B(x) = \sum b_n x^n$.

41. Consider the differential equation

$$x(x-1)(x+1)^2 y'' + 2x(x-3)(x+1)y' - 2(x-1)y = 0$$

that appeared in an advertisement for the MACSYMA© symbolic algebra program (in the March 1984 issue of the *American Mathematical Monthly*). (a) Show that $x = 0$ is a regular singular point with exponents $r_1 = 1$ and $r_2 = 0$. (b) It follows from the theorem on Frobenius series solutions that the differential equation shown here has a power series solution of the form

$$y_1(x) = x + c_2 x^2 + c_3 x^3 + \cdots.$$

Substitute this series (with $c_1 = 1$) in the differential equation to show that $c_2 = -2$, $c_3 = 3$, and

$$c_{n+2} = \frac{n(n-1)c_{n-1} + (n^2 - 5n - 2)c_n - (n^2 + 7n + 4)c_{n+1}}{(n+1)(n+2)}$$

for $n \geq 2$. (c) Use the recurrence relation in part (b) to prove by induction that $c_n = (-1)^{n+1}n$ for $n \geq 1$ (!). Hence deduce (using

the geometric series) that

$$y_1(x) = \frac{x}{(1+x)^2}$$

for $0 < x < 1$. In Problem 53 of Section 2.6 $y_1(x)$ was used to derive by reduction of order a second solution $y_2(x)$.

42. This problem is a brief introduction to Gauss's **hypergeometric equation**

$$x(1-x)y'' + [\gamma - (\alpha + \beta + 1)x]y' - \alpha\beta y = 0, \qquad (34)$$

where α, β, and γ are constants. This famous equation has wide-ranging applications in mathematics and physics. (a) Show that $x = 0$ is a regular singular point of Eq. (34), with exponents 0 and $1 - \gamma$. (b) If γ is not zero or a negative integer, it follows (Why?) that (34) has a power series solution

$$y(x) = x^0 \sum_{n=0}^{\infty} c_n x^n = \sum_{n=0}^{\infty} c_n x^n$$

with $c_0 \neq 0$. Show that the recurrence relation for this series is

$$c_{n+1} = \frac{(\alpha + n)(\beta + n)}{(\gamma + n)(1 + n)} c_n$$

for $n \geq 0$. (c) Conclude that with $c_0 = 1$ the series in part (b) is

$$y(x) = 1 + \sum_{n=0}^{\infty} \frac{\alpha_n \beta_n}{n! \gamma_n} x^n \qquad (35)$$

where $\alpha_n = \alpha(\alpha + 1) \cdots (\alpha + n - 1)$ for $n \geq 1$, and β_n and γ_n are defined similarly. (d) The series in (35) is known as the **hypergeometric series**, and is commonly denoted by $F(\alpha, \beta, \gamma, x)$. Show that

(i) $F(1, 1, 1, x) = \dfrac{1}{1-x}$ (the geometric series)

(ii) $xF(1, 1, 2, -x) = \ln(1 + x)$

(iii) $xF(\frac{1}{2}, 1, \frac{3}{2}, -x^2) = \tan^{-1} x$

(iv) $F(-k, 1, 1, -x) = (1 + x)^k$ (the binomial series)

---

*3.4

Method of Frobenius: The Exceptional Cases

We continue our discussion of the equation

$$y'' + \frac{p(x)}{x}y' + \frac{q(x)}{x^2}y = 0 \qquad (1)$$

where $p(x)$ and $q(x)$ are analytic at $x = 0$, and $x = 0$ is a regular singular point. If the roots r_1 and r_2 of the indicial equation

$$\phi(r) = r(r-1) + p_0 r + q_0 = 0 \qquad (2)$$

do not differ by an integer, then the theorem of Section 3.3 guarantees that Eq. (1) has two linearly independent Frobenius series solutions. We consider now the more complex situation in which $r_1 - r_2$ is an integer. If $r_1 = r_2$, then there is only one exponent available, and thus there can be only one Frobenius series solution. But we saw in Example 6 of Section 3.3 that if $r_1 = r_2 + N$, with N a positive integer, it is possible that a second Frobenius series solution exists. We will also see that it is possible that such a solution does not exist. In fact, the second solution involves $\ln x$ when it is not a Frobenius series. As you will see in Examples 3 and 4, these exceptional cases occur in the solution of Bessel's equation, which for applications is the most important second order linear differential equation with variable coefficients.

THE NONLOGARITHMIC CASE WITH $r_1 = r_2 + N$

In Section 3.3 we derived the indicial equation by substituting the power series $p(x) = \sum p_n x^n$ and $q(x) = \sum q_n x^n$ and the Frobenius series

$$y = x^r \sum_{n=0}^{\infty} c_n x^n = \sum_{n=0}^{\infty} c_n x^{n+r} \qquad (c_0 \neq 0) \tag{3}$$

in the differential equation in the form

$$x^2 y'' + xp(x)y' + q(x)y = 0. \tag{4}$$

The result of this substitution, after collection of the coefficients of like powers of x, is an equation of the form

$$\sum_{n=0}^{\infty} F_n(r)x^{n+r} = 0 \tag{5}$$

with the coefficients depending on r. It turns out that the coefficient of x^r is

$$F_0(r) = [r(r-1) + p_0 r + q_0]c_0 = \phi(r)c_0, \tag{6}$$

which gives the indicial equation because $c_0 \neq 0$ by assumption; also, for $n \geq 1$, the coefficient of x^{n+r} is of the form

$$F_n(r) = \phi(r + n)c_n + L_n(r; c_0, c_1, \ldots, c_{n-1}). \tag{7}$$

Here L_n is a certain linear combination of $c_0, c_1, \ldots, c_{n-1}$. Although the exact formula is not necessary for our purposes, it happens that

$$L_n = \sum_{k=0}^{n-1} [(r + k)p_{n-k} + q_{n-k}]c_k. \tag{8}$$

Because all the coefficients in (5) must vanish for our Frobenius series to be a solution of (4), it follows that the exponent r and the coefficients $c_0, c_1, \ldots, c_n$ must satisfy the equation

$$\phi(r + n)c_n + L_n(r; c_0, c_1, \ldots, c_{n-1}) = 0. \tag{9}$$

This is a *recurrence relation* for c_n in terms of $c_0, c_1, \ldots, c_{n-1}$.

Suppose now that $r_1 = r_2 + N$ with N a *positive* integer. If we use the larger exponent r_1 in (9), then the coefficient $\phi(r_1 + n)$ of c_n will be nonzero for every $n \geq 1$ because $\phi(r) = 0$ only when $r = r_1$ and when $r = r_2 < r_1$. Once $c_0, c_1, \ldots, c_{n-1}$ have been determined, we therefore can solve Eq. (9) for c_n and continue to compute successive coefficients in the Frobenius series solution corresponding to the exponent r_1.

But when we use the smaller exponent r_2, there is a potential difficulty in computing c_N. For in this case $\phi(r_2 + N) = 0$, so Eq. (9) becomes

$$0 \cdot c_N + L_N(r_2; c_0, c_1, \ldots, c_{N-1}) = 0. \tag{10}$$

At this stage $c_0, c_1, \ldots, c_{N-1}$ have already been determined. If it happens that $L_N(r_2; c_0, c_1, \ldots, c_{N-1}) = 0$, then we can choose c_N arbitrarily and continue to determine the remaining coefficients in a second Frobenius series solution. But if it happens that $L_N(r_2; c_0, c_1, \ldots, c_{N-1}) \neq 0$, then Eq. (10) is not satisfied with any choice of c_N; in this case there cannot exist a Frobenius series solution corresponding to the smaller exponent r_2. Examples 1 and 2 illustrate these two possibilities.

EXAMPLE 1 Consider the equation

$$x^2 y'' + (6x + x^2)y' + xy = 0. \tag{11}$$

Here $p_0 = 6$ and $q_0 = 0$, so the indicial equation is

$$\phi(r) = r(r - 1) + 6r = r^2 + 5r = 0 \tag{12}$$

with roots $r_1 = 0$ and $r_2 = -5$; the roots differ by the integer $N = 5$. We substitute the Frobenius series $y = \sum c_n x^{n+r}$ and get

$$\sum_{n=0}^{\infty} (n+r)(n+r-1)c_n x^{n+r} + 6 \sum_{n=0}^{\infty} (n+r)c_n x^{n+r} + \sum_{n=0}^{\infty} (n+r)c_n x^{n+r+1} + \sum_{n=0}^{\infty} c_n x^{n+r+1} = 0.$$

When we combine the first two and also the last two sums, and in the latter shift the index by -1, the result is

$$\sum_{n=0}^{\infty} [(n+r)^2 + 5(n+r)]c_n x^{n+r} + \sum_{n=1}^{\infty} (n+r)c_{n-1}x^{n+r} = 0.$$

The terms corresponding to $n = 0$ give the indicial equation in (12), while for $n \geq 1$ we get the equation

$$[(n+r)^2 + 5(n+r)]c_n + (n+r)c_{n-1} = 0, \tag{13}$$

which in this example corresponds to the general equation in (9). Note that the coefficient of c_n is $\phi(n+r)$.

We now follow the recommendations in Section 3.3 for the case $r_1 = r_2 + N$; we begin with the smaller root $r_2 = -5$. With $r = -5$, Eq. (13) reduces to

$$n(n-5)c_n + (n-5)c_{n-1} = 0. \tag{14}$$

If $n \neq 5$, we can solve this equation for c_n to get the recurrence relation

$$c_n = -\frac{c_{n-1}}{n} \qquad \text{for } n \neq 5. \tag{15}$$

This yields

$$c_1 = -c_0, \qquad c_2 = -\frac{c_1}{2} = \frac{c_0}{2},$$

$$c_3 = -\frac{c_2}{3} = -\frac{c_0}{6}, \quad \text{and} \quad c_4 = -\frac{c_3}{4} = \frac{c_0}{24}. \tag{16}$$

In the case $r_1 = r_2 + N$, it is always the coefficient c_N that requires special consideration. Here $N = 5$, and for $n = 5$ Eq. (14) takes the form $0 \cdot c_5 + 0 = 0$. Hence c_5 is a second arbitrary constant, and we can compute additional coefficients, using the recursion formula in (15):

$$c_6 = -\frac{c_5}{6}, \qquad c_7 = -\frac{c_6}{7} = \frac{c_5}{6 \cdot 7}, \qquad c_8 = -\frac{c_7}{8} = -\frac{c_5}{6 \cdot 7 \cdot 8}, \qquad (17)$$

and so on.

When we combine the results in (16) and (17), we get

$$y = x^{-5} \sum_{n=0}^{\infty} c_n x^n$$

$$= c_0 x^{-5}\left(1 - x + \frac{x^2}{2} - \frac{x^3}{6} + \frac{x^4}{24}\right) + c_5 x^{-5}\left(x^5 - \frac{x^6}{6} + \frac{x^7}{6 \cdot 7} - \frac{x^8}{6 \cdot 7 \cdot 8} + \cdots\right)$$

in terms of the two arbitrary constants c_0 and c_5. Thus we have found the two Frobenius series solutions

$$y_1(x) = x^{-5}\left(1 - x + \frac{x^2}{2} - \frac{x^3}{6} + \frac{x^4}{24}\right)$$

and

$$y_2(x) = 1 + \sum_{n=1}^{\infty} \frac{(-1)^n x^n}{6 \cdot 7 \cdots (n + 5)} = 1 + 120 \sum_{n=1}^{\infty} \frac{(-1)^n x^n}{(n + 5)!}$$

of Eq. (11).

EXAMPLE 2 Determine whether or not the equation

$$x^2 y'' - x y' + (x^2 - 8)y = 0 \qquad (18)$$

has two linearly independent Frobenius series solutions.

SOLUTION Here $p_0 = -1$ and $q_0 = -8$, so the indicial equation is

$$\phi(r) = r(r - 1) - r - 8 = r^2 - 2r - 8 = 0$$

with roots $r_1 = 4$ and $r_2 = -2$ differing by $N = 6$. On substitution of $y = \sum c_n x^{n+r}$ in (18), we get

$$\sum_{n=0}^{\infty} (n + r)(n + r - 1)c_n x^{n+r} - \sum_{n=0}^{\infty} (n + r)c_n x^{n+r} + \sum_{n=0}^{\infty} c_n x^{n+r+2} - 8\sum_{n=0}^{\infty} c_n x^{n+r} = 0.$$

If we shift the index by -2 in the third sum and combine the other three sums, we get

$$\sum_{n=0}^{\infty} [(n + r)^2 - 2(n + r) - 8]c_n x^{n+r} + \sum_{n=2}^{\infty} c_{n-2} x^{n+r} = 0.$$

The coefficient of x^r gives the indicial equation, and the coefficient of x^{r+1} gives

$$[(r + 1)^2 - 2(r + 1) - 8]c_1 = 0.$$

Because the coefficient of c_1 is nonzero both for $r = 4$ and for $r = -2$, it follows that $c_1 = 0$ in each case. For $n \geq 2$ we get the equation

$$[(n + r)^2 - 2(n + r) - 8]c_n + c_{n-2} = 0, \qquad (19)$$

which corresponds in this example to the general equation in (9); note that the coefficient of c_n is $\phi(n + r)$.

We work first with the smaller root $r = r_2 = -2$. Then Eq. (19) becomes

$$n(n - 6)c_n + c_{n-2} = 0 \tag{20}$$

for $n \geq 2$. For $n \neq 6$ we can solve for the recurrence relation

$$c_n = -\frac{c_{n-2}}{n(n - 6)} \qquad (n \geq 2, n \neq 6). \tag{21}$$

Because $c_1 = 0$, this formula gives

$$c_2 = \frac{c_0}{8}, \qquad\qquad c_3 = 0,$$

$$c_4 = \frac{c_2}{8} = \frac{c_0}{64}, \quad \text{and} \quad c_5 = 0.$$

Now Eq. (20) with $n = 6$ reduces to

$$0 \cdot c_6 + \frac{c_0}{64} = 0.$$

But $c_0 \neq 0$ by assumption, and hence there is no way to choose c_6 so that this equation holds. Thus there is *no* Frobenius series solution corresponding to the smaller root $r_2 = -2$.

To find the single Frobenius series solution corresponding to the larger root $r_1 = 4$, we substitute $r = 4$ in (19) to obtain the recurrence relation

$$c_n = -\frac{c_{n-2}}{n(n + 6)} \qquad (n \geq 2). \tag{22}$$

This gives

$$c_2 = -\frac{c_0}{2 \cdot 8}, \qquad c_4 = -\frac{c_2}{4 \cdot 10} = \frac{c_0}{2 \cdot 4 \cdot 8 \cdot 10}.$$

The general pattern is

$$c_{2n} = \frac{(-1)^n c_0}{2 \cdot 4 \cdots (2n) \cdot 8 \cdot 10 \cdots (2n + 6)} = \frac{(-1)^n 48 c_0}{2^{2n+3} n!(n + 3)!}.$$

This yields the Frobenius series solution

$$y_1(x) = x^4 \left(1 + 48 \sum_{n=1}^{\infty} \frac{(-1)^n x^{2n}}{2^{2n+3} n!(n + 3)!} \right)$$

of Eq. (18).

THE LOGARITHMIC CASES

We now investigate the general form of the second solution of the equation

$$y'' + \frac{p(x)}{x} y' + \frac{q(x)}{x^2} y = 0, \tag{1}$$

under the assumption that its exponents r_1 and $r_2 = r_1 - N$ differ by the integer $N \geq 0$. We assume that we have already found the Frobenius series solution

$$y_1(x) = x^{r_1} \sum_{n=0}^{\infty} a_n x^n \qquad (a_0 \neq 0) \qquad (23)$$

for $x > 0$ corresponding to the larger exponent, r_1. Let us write $P(x)$ for $p(x)/x$ and $Q(x)$ for $q(x)/x^2$. Recall from Section 2.6 (reduction of order) that a second solution y_2 is given by

$$y_2(x) = y_1 \int \frac{e^{-\int P(x)\,dx}}{y_1^2}\,dx \qquad (24)$$

on any interval on which y_1 is nonzero.

Because the indicial equation has roots r_1 and $r_2 = r_1 - N$, it can be factored easily:

$$r^2 + (p_0 - 1)r + q_0 = (r - r_1)(r - r_1 + N)$$
$$= r^2 + (N - 2r_1)r + (r_1^2 - r_1 N) = 0,$$

so we see that

$$p_0 - 1 = N - 2r_1;$$

that is,

$$-p_0 - 2r_1 = -1 - N. \qquad (25)$$

In preparation for use of the reduction of order formula in (24), we write

$$P(x) = \frac{p_0 + p_1 x + p_2 x^2 + \cdots}{x} = \frac{p_0}{x} + p_1 + p_2 x + \cdots.$$

Then

$$e^{-\int P(x)\,dx} = \exp\left(-\int \left(\frac{p_0}{x} + p_1 + p_2 x + \cdots\right)dx\right)$$

$$= \exp\left(-p_0 \ln x - p_1 x - \frac{1}{2}p_2 x^2 - \cdots\right)$$

$$= x^{-p_0} \exp\left(-p_1 x - \frac{1}{2}p_1 x^2 - \cdots\right),$$

so that

$$e^{-\int P(x)\,dx} = x^{-p_0}(1 + A_1 x + A_2 x^2 + \cdots). \qquad (26)$$

In the last step we have used the fact that a composition of analytic functions is analytic and therefore has a power series representation; the initial coefficient of that series in (26) is 1 because $e^0 = 1$.

We now substitute (23) and (26) in (24); with the choice $a_0 = 1$ in (23), this yields

$$y_2 = y_1 \int \frac{x^{-p_0}(1 + A_1 x + A_2 x^2 + \cdots)}{x^{2r_1}(1 + a_1 x + a_2 x^2 + \cdots)^2}\,dx.$$

We square the denominator and simplify:

$$y_2 = x_1 \int \frac{x^{-p_0 - 2r_1}(1 + A_1 x + A_2 x^2 + \cdots)}{(1 + B_1 x + B_2 x^2 + \cdots)}\, dx$$

$$= y_1 \int x^{-1-N}(1 + C_1 x + C_2 x^2 + \cdots)\, dx \qquad (27)$$

(we used long division and (25) in the last step). We now consider separately the cases $N = 0$ and $N > 0$. We want to ascertain the general form of y_2 without keeping track of specific coefficients.

Equal Exponents: $r_1 = r_2$
With $N = 0$, (27) gives

$$y_2 = y_1 \int \left(\frac{1}{x} + C_1 + C_2 x + \cdots \right) dx$$

$$= y_1 \ln x + y_1 \left(C_1 x + \frac{1}{2} C_2 x^2 + \cdots \right)$$

$$= y_1 \ln x + x^{r_1}(1 + a_1 x + \cdots)\left(C_1 x + \frac{1}{2} C_2 x^2 + \cdots \right)$$

$$= y_1 \ln x + x^{r_1}(b_0 x + b_1 x^2 + b_2 x^3 + \cdots)$$

Consequently, in the case of equal exponents, the general form of y_2 is

$$y_2 = y_1 \ln x + x^{r_1 + 1} \sum_{n=0}^{\infty} b_n x^n. \qquad (28)$$

Note the logarithmic term; it is always present when $r_1 = r_2$.

Positive Integral Difference: $r_1 = r_2 + N$
With $N > 0$, (27) gives

$$y_2 = y_1 \int x^{-1-N}(1 + C_1 x + C_2 x^2 + \cdots + C_N x^N + \cdots)\, dx$$

$$= y_1 \int \left(\frac{C_N}{x} + \frac{1}{x^{N+1}} + \frac{C_1}{x^N} + \cdots \right) dx$$

$$= C_N y_1 \ln x + y_1 \left(\frac{x^{-N}}{-N} + \frac{C_1 x^{-N+1}}{-N+1} + \cdots \right)$$

$$= C_N y_1 \ln x + x^{r_2 + N}\left(\sum_{n=0}^{\infty} a_n x^n \right) x^{-N}\left(-\frac{1}{N} + \frac{C_1 x}{-N+1} + \cdots \right),$$

so that

$$y_2 = C_N y_1 \ln x + x^{r_2} \sum_{n=0}^{\infty} b_n x^n \qquad (29)$$

where $b_0 = -a_0/N \neq 0$. This gives the general form of y_2 in the case of exponents differing by a positive integer. Note the coefficient C_N that appears in (29) but not

CHAPTER 3:   Power Series Solutions of Linear Equations

in (28). If it happens that $C_N = 0$, then there is no logarithmic term; if so, Eq. (1) has a second Frobenius series solution (as in Example 1).

In our derivation of the formulas in (28) and (29)—which exhibit the general form of the second solution in the cases $r_1 = r_2$ and $r_1 - r_2 = N$, respectively—we have said nothing about the radii of convergence of the various power series that appear. The following theorem is a summation of the discussion above and also tells where the series in (28) and (29) converge. As in the theorem of Section 3.3, we restrict our attention to solutions for $x > 0$. Once a solution for $x > 0$ has been found, we need only replace x^{r_i} by $|x|^{r_i}$ and $\ln x$ by $\ln |x|$ to obtain a solution for $x < 0$.

Theorem *The Exceptional Cases*

Suppose that $x = 0$ is a regular singular point of the equation

$$x^2 y'' + x p(x) y' + q(x) y = 0. \tag{4}$$

Let $\rho > 0$ denote the minimum of the radii of convergence of the power series

$$p(x) = \sum_{n=0}^{\infty} p_n x^n \quad \text{and} \quad q(x) = \sum_{n=0}^{\infty} q_n x^n.$$

Let r_1 and r_2 be the roots, with $r_1 \geq r_2$, of the indicial equation

$$r(r - 1) + p_0 r + q_0 = 0.$$

(a) If $r_1 = r_2$, then (4) has two solutions y_1 and y_2 of the forms

$$y_1 = x^{r_1} \sum_{n=0}^{\infty} a_n x^n \quad (a_0 \neq 0) \tag{30a}$$

and

$$y_2 = y_1 \ln x + x^{r_2 + 1} \sum_{n=0}^{\infty} b_n x^n. \tag{30b}$$

(b) If $r_1 - r_2 = N$, a positive integer, then (4) has two solutions y_1 and y_2 of the forms

$$y_1 = x^{r_1} \sum_{n=0}^{\infty} a_n x^n \quad (a_0 \neq 0) \tag{31a}$$

and

$$y_2 = C y_1 \ln x + x^{r_2} \sum_{n=0}^{\infty} b_n x^n. \tag{31b}$$

In (31b), $b_0 \neq 0$ but C may be either zero or nonzero, so the logarithmic term may or may not actually be present in this case. The radii of convergence of the power series shown above are all at least ρ. The coefficients in these series [and the constant C in (31b)] may be determined by direct substitution of the series in the differential equation in (4).

EXAMPLE 3 We will illustrate the case $r_1 = r_2$ of the theorem by deriving the second solution of Bessel's equation of order zero,

$$x^2 y'' + xy' + x^2 y = 0, \tag{32}$$

for which $r_1 = r_2 = 0$. In Example 5 of Section 3.3 we found the first solution

$$y_1 = J_0(x) = \sum_{n=0}^{\infty} \frac{(-1)^n x^{2n}}{2^{2n}(n!)^2}. \tag{33}$$

According to (30b) the second solution will have the form

$$y_2 = y_1 \ln x + \sum_{n=1}^{\infty} b_n x^n. \tag{34}$$

The first two derivatives of y_2 are

$$y_2' = y_1' \ln x + \frac{y_1}{x} + \sum_{n=1}^{\infty} n b_n x^{n-1}$$

and

$$y_2'' = y_1'' \ln x + \frac{2y_1'}{x} - \frac{y_1}{x^2} + \sum_{n=2}^{\infty} n(n-1) b_n x^{n-2}.$$

We substitute these in Eq. (32) and use the fact that $J_0(x)$ also satisfies this equation to obtain

$$\begin{aligned}
0 &= x^2 y_2'' + xy_2' + x^2 y_2 \\
&= [x^2 y_1'' + xy_1' + x^2 y_1] \ln x + 2xy_1' \\
&\quad + \sum_{n=2}^{\infty} n(n-1) b_n x^n + \sum_{n=1}^{\infty} n b_n x^n + \sum_{n=1}^{\infty} b_n x^{n+2},
\end{aligned}$$

and it follows that

$$0 = 2 \sum_{n=1}^{\infty} \frac{(-1)^n 2n x^{2n}}{2^{2n}(n!)^2} + b_1 x + 2^2 b_2 x^2 + \sum_{n=3}^{\infty} (n^2 b_n + b_{n-2}) x^n. \tag{35}$$

The only term involving x in (35) is $b_1 x$, so $b_1 = 0$. But $n^2 b_n + b_{n-2} = 0$ if n is odd, so it follows that all the coefficients of odd subscript in y_2 vanish.

Now we examine the coefficients with even subscripts in (35). First we see that

$$b_2 = -2 \frac{(-1)(2)}{2^2 2^2 (1!)^2} = \frac{1}{4}. \tag{36}$$

For $n \geq 2$, we read the recurrence relation

$$(2n)^2 b_{2n} + b_{2n-2} = -\frac{(2)(-1)^n (2n)}{2^{2n}(n!)^2} \tag{37}$$

from (35). Nonhomogeneous recurrence relations such as (37) are typical of the exceptional cases of the method of Frobenius, and their solution often requires a bit of ingenuity. The usual strategy depends on detecting the most conspicuous dependence of b_{2n} on n. We note the presence of $2^{2n}(n!)^2$ on the right-hand side in (37); in conjunction with the coefficient $(2n)^2$ on the left-hand side, we are induced to think of b_{2n} as *something* divided by $2^{2n}(n!)^2$. Noting also the alternation of sign, we make the substitution

$$b_{2n} = \frac{(-1)^{n+1}c_{2n}}{2^{2n}(n!)^2},$$ (38)

in the expectation that the recurrence relation for c_{2n} will be simpler than the one for b_{2n}. We chose $(-1)^{n+1}$ rather than $(-1)^n$ because $b_2 = \frac{1}{4} > 0$; with $n = 1$ in (38), we get $c_2 = 1$. Substitution of (38) in (37) gives

$$(2n)^2 \frac{(-1)^{n+1}c_{2n}}{2^{2n}(n!)^2} + \frac{(-1)^n c_{2n-2}}{2^{2n-2}[(n-1)!]^2} = \frac{(-2)(-2)^n(2n)}{2^{2n}(n!)^2},$$

which boils down to the extremely simple recurrence relation

$$c_{2n} = c_{2n-2} + \frac{1}{n}.$$

Thus

$$c_4 = c_2 + \frac{1}{2} = 1 + \frac{1}{2},$$

$$c_6 = c_4 + \frac{1}{3} = 1 + \frac{1}{2} + \frac{1}{3},$$

$$c_8 = c_6 + \frac{1}{4} = 1 + \frac{1}{2} + \frac{1}{3} + \frac{1}{4},$$

and so on. Evidently

$$c_{2n} = 1 + \frac{1}{2} + \frac{1}{3} + \cdots + \frac{1}{n} = H_n,$$ (39)

where by H_n we denote the nth partial sum of the harmonic series $\sum (1/n)$.

Finally, keeping in mind that the coefficients of odd subscript are all zero, we substitute (38) and (39) in (34) to obtain the second solution

$$y_2 = J_0(x) \ln x + \sum_{n=1}^{\infty} \frac{(-1)^{n+1} H_n x^{2n}}{2^{2n}(n!)^2}$$

$$= J_0(x) \ln x + \frac{x^2}{4} - \frac{3x^4}{128} + \frac{11x^6}{13{,}824} - \cdots$$ (40)

of Bessel's equation of order zero. The power series in (40) converges for all x. The most commonly used linearly independent [of $J_0(x)$] second solution is

$$Y_0(x) = \frac{2}{\pi}(\gamma - \ln 2)y_1 + \frac{2}{\pi} y_2;$$

that is,

$$Y_0(x) = \frac{2}{\pi}\left[\left(\gamma + \ln \frac{x}{2}\right)J_0(x) + \sum_{n=1}^{\infty} \frac{(-1)^{n+1} H_n x^{2n}}{2^{2n}(n!)^2}\right],$$ (41)

where γ denotes Euler's constant:

$$\gamma = \lim_{n \to \infty} (H_n - \ln n) \approx 0.57722. \tag{42}$$

This particular combination $Y_0(x)$ is chosen because of its nice behavior as $x \to +\infty$; it is called the **Bessel function of order zero of the second kind**.

EXAMPLE 4 As an alternative to the method of substitution, we illustrate the case $r_1 - r_2 = N$ by employing the technique of reduction of order to derive a second solution of Bessel's equation of order 1,

$$x^2 y'' + x y' + (x^2 - 1)y = 0; \tag{43}$$

the associated indicial equation has roots $r_1 = 1$ and $r_2 = -1$. According to Problem 39 of Section 3.3, one solution of (43) is

$$y_1 = J_1(x) = \frac{x}{2} \sum_{n=0}^{\infty} \frac{(-1)^n x^{2n}}{2^{2n} n! (n + 1)!}$$

$$= \frac{x}{2} - \frac{x^3}{16} + \frac{x^5}{384} - \frac{x^7}{18{,}432} + \cdots. \tag{44}$$

With $P(x) = 1/x$ from (43), the reduction of order formula in (24) yields

$$y_2 = y_1 \int \frac{dx}{x y_1^2}$$

$$= y_1 \int \frac{dx}{x(x/2 - x^3/16 + x^5/384 - x^7/18{,}432 + \cdots)^2}$$

$$= y_1 \int \frac{4\, dx}{x^3(1 - x^2/8 + x^4/192 - x^6/9216 + \cdots)^2}$$

$$= 4y_1 \int \frac{dx}{x^3(1 - x^2/4 + 5x^4/192 - 7x^6/4608 + \cdots)}$$

$$= 4y_1 \int \frac{1}{x^3}\left(1 + \frac{x^2}{4} + \frac{7x^4}{192} + \frac{19x^6}{4608} + \cdots\right) dx$$

$$= 4y_1 \int \left(\frac{1}{4x} + \frac{1}{x^3} + \frac{7x}{192} + \frac{19x^3}{4608} + \cdots\right) dx$$

$$= y_1 \ln x + 4y_1\left(-\frac{1}{2x^2} + \frac{7x^2}{384} + \frac{19x^4}{18{,}432} + \cdots\right).$$

Thus

$$y_2 = y_1 \ln x - \frac{1}{x} + \frac{x}{8} + \frac{x^3}{32} - \frac{11x^5}{4608} + \cdots. \tag{45}$$

Note that the technique of reduction of order readily yields the first several terms of the series, but it does not provide a recurrence relation that can be used to determine the general term of the series.

With a computation similar to that shown in Example 3—but more complicated (see Problem 21)—the method of substitution can be used to derive the solution

$$y_3 = y_1 \ln x - \frac{1}{x} + \sum_{n=1}^{\infty} \frac{(-1)^n (H_n + H_{n-1}) x^{2n-1}}{2^{2n} n! (n-1)!}, \tag{46}$$

where H_n is defined in (39) for $n \geq 1$; $H_0 = 0$. The reader can verify that the terms shown in (45) agree with

$$y_2 = \frac{3}{4} J_1(x) + y_3. \tag{47}$$

The most commonly used linearly independent (of J_1) solution of Bessel's equation of order 1 is the combination

$$Y_1(x) = \frac{2}{\pi} (\gamma - \ln 2) y_1 + \frac{2}{\pi} y_3$$

$$= \frac{2}{\pi} \left[\left(\gamma + \ln \frac{x}{2} \right) J_1(x) - \frac{1}{x} + \sum_{n=1}^{\infty} \frac{(-1)^n (H_n + H_{n-1}) x^{2n-1}}{2^{2n} n! (n-1)!} \right]. \tag{48}$$

Examples 3 and 4 illustrate two methods of finding the solution in the logarithmic cases—direct substitution and reduction of order. A third alternative is outlined in Problem 19.

3.4 PROBLEMS

In each of Problems 1–8, either apply the method of Example 1 to find two linearly independent Frobenius solutions, or find one such series solution and show (as in Example 2) that a second one does not exist.

1. $xy'' + (3 - x)y' - y = 0$

2. $xy'' + (5 - x)y' - y = 0$

3. $xy'' + (5 + 3x)y' + 3y = 0$

4. $5xy'' + (30 + 3x)y' + 3y = 0$

5. $xy'' - (4 + x)y' + 3y = 0$

6. $2xy'' - (6 + 2x)y' + y = 0$

7. $x^2 y'' + (2x + 3x^2)y' - 2y = 0$

8. $x(1 - x)y'' - 3y' + 2y = 0$

In each of Problems 9–14, first find the first four nonzero terms in a Frobenius series solution of the given differential equation. Then use the reduction of order technique (as in Example 4) to find the logarithmic term and the first three nonzero terms in a second linearly independent series solution.

9. $xy'' + y' - xy = 0$

10. $x^2 y'' - xy' + (x^2 + 1)y = 0$

11. $x^2 y'' + (x^2 - 3x)y' + 4y = 0$

12. $x^2 y'' + x^2 y' - 2y = 0$

13. $x^2 y'' + (2x^2 - 3x)y' + 3y = 0$

14. $x^2 y'' + x(1 + x)y' - 4y = 0$

15. Begin with

$$J_0(x) = 1 - \frac{x^2}{4} + \frac{x^4}{64} - \frac{x^6}{2304} + \cdots;$$

by reduction of order, derive the second solution

$$y_2 = J_0(x) \ln x + \frac{x^2}{4} - \frac{3x^4}{128} + \frac{11x^6}{13,824} - \cdots$$

of Bessel's equation of order zero.

16. Find two linearly independent Frobenius series solutions of Bessel's equation of order $\frac{3}{2}$,

$$x^2 y'' + xy' + (x^2 - \tfrac{9}{4})y = 0.$$

17. (a) Verify that $y_1 = xe^x$ is one solution of

$$x^2 y'' - x(1 + x)y' + y = 0.$$

(b) Note that $r_1 = r_2 = 1$. Substitute

$$y_2 = y_1 \ln x + \sum_{n=1}^{\infty} b_n x^{n+1}$$

in the differential equation to deduce that $b_1 = -1$ and that

$$nb_n - b_{n-1} = -\frac{1}{n!} \quad \text{for } n \geq 2.$$

(c) Substitute $b_n = +c_n/n!$ in this recurrence relation, and conclude from the result that $c_n = -H_n$. Thus the second solution is

$$y_2 = xe^x \ln x - \sum_{n=1}^{\infty} \frac{H_n x^{n+1}}{n!}.$$

18. Consider the equation $xy'' - y = 0$, which has exponents $r_1 = 1$ and $r_2 = 0$ at $x = 0$. **(a)** Derive the Frobenius series solution

$$y_1 = \sum_{n=1}^{\infty} \frac{x^n}{n!(n-1)!}.$$

(b) Substitute

$$y_2 = Cy_1 \ln x + \sum_{n=0}^{\infty} b_n x^n$$

in the equation $xy'' - y = 0$ to derive the recurrence relation

$$n(n+1)b_{n+1} - b_n = -\frac{2n+1}{(n+1)!n!} C.$$

Conclude from this result that a second solution is

$$y_2 = y_1 \ln x + 1 - \sum_{n=1}^{\infty} \frac{H_n + H_{n-1}}{n!(n-1)!} x^n.$$

19. Suppose that the differential equation

$$L[y] = x^2 y'' + xp(x)y' + q(x)y = 0 \tag{49}$$

has equal exponents $r_1 = r_2$ at the regular singular point $x = 0$, so that its indicial equation is

$$\phi(r) = (r - r_1)^2 = 0.$$

Let $c_0 = 1$ and define $c_n(r)$ for $n \geq 1$ by using Eq. (9); that is,

$$c_n(r) = -\frac{L_n(r; c_0, c_1, \ldots, c_{n-1})}{\phi(r+n)}. \tag{50}$$

Then define the function $y(x, r)$ of x and r to be

$$y(x, r) = \sum_{n=0}^{\infty} c_n(r) x^{n+r}. \tag{51}$$

(a) Deduce from the discussion preceding Eq. (9) that

$$L[y(x, r)] = x^r (r - r_1)^2. \tag{52}$$

Hence deduce that

$$y_1 = y(x, r_1) = \sum_{n=0}^{\infty} c_n(r_1) x^{n+r_1} \tag{53}$$

is one solution of Eq. (49). **(b)** Differentiate Eq. (52) with respect to r to show that

$$L[y_r(x, r_1)] = \frac{\partial}{\partial r}[x^r(r - r_1)^2]\Big|_{r=r_1} = 0.$$

Deduce that $y_2 = y_r(x, r_1)$ is a second solution of Eq. (49). **(c)** Differentiate Eq. (53) with respect to r to show that

$$y_2 = y_1 \ln x + x^{r_1} \sum_{n=1}^{\infty} c_n'(r_1) x^n. \tag{54}$$

20. Use the method of Problem 19 to derive both the solutions in (33) and (40) of Bessel's equation of order zero. The following steps outline this computation. **(a)** Take $c_0 = 1$; show that Eq. (50) reduces in this case to

$$\left. \begin{array}{l} (r+1)^2 c_1(r) = 0 \\[2mm] \text{and} \\[2mm] c_n(r) = -\dfrac{c_{n-2}(r)}{(n+r)^2} \quad \text{for } n \geq 2. \end{array} \right\} \tag{55}$$

(b) Next show that $c_1(0) = c_1'(0) = 0$, and then deduce from (55) that $c_n(0) = c_n'(0) = 0$ for n odd. Hence you need to compute $c_n(0)$ and $c_n'(0)$ only for n even. **(c)** Deduce from (55) that

$$c_{2n}(r) = \frac{(-1)^n}{(r+2)^2(r+4)^2 \cdots (r+2n)^2}. \tag{56}$$

With $r = r_1 = 0$ in (53), this gives $J_0(x)$. **(d)** Differentiate (56) to show that

$$c_{2n}'(0) = \frac{(-1)^{n+1} H_n}{2^{2n}(n!)^2}.$$

Substitution of this result in (54) gives the second solution in (40).

21. Derive the logarithmic solution in (46) of Bessel's equation of order 1 by the method of substitution. The following steps outline this computation. **(a)** Substitute

$$y_2 = CJ_1(x) \ln x + x^{-1}\left(1 + \sum_{n=1}^{\infty} b_n x^n\right)$$

in Bessel's equation to obtain

$$-b_1 + x + \sum_{n=2}^{\infty} [(n^2 - 1)b_{n+1} + b_{n-1}]x^n$$

$$+ C\left[x + \sum_{n=1}^{\infty} \frac{(-1)^n(2n+1)x^{2n+1}}{2^{2n}(n+1)!n!}\right] = 0. \tag{57}$$

(b) Deduce from Eq. (57) that $C = -1$ and that $b_n = 0$ for n odd. (c) Next deduce the recurrence relation

$$[(2n + 1)^2 - 1]b_{2n+2} + b_{2n} = \frac{(-1)^n(2n + 1)}{2^{2n}(n + 1)!n!} \quad (58)$$

for $n \geq 1$. Note that if b_2 is chosen arbitrarily, then b_{2n} is determined for all $n > 1$. (d) Take $b_2 = \frac{1}{4}$ and substitute

$$b_{2n} = \frac{(-1)^{n+1}c_{2n}}{2^{2n}n!(n - 1)!}$$

in (58) to obtain

$$c_{2n+2} - c_{2n} = \frac{1}{n + 1} + \frac{1}{n}.$$

(e) Note that $c_2 = 1 = H_1 + H_0$ and deduce that $c_{2n} = H_n + H_{n-1}$.

3.5

Bessel's Equation

We have already seen several cases of Bessel's equation of order $p \geq 0$,

$$x^2y'' + xy' + (x^2 - p^2)y = 0. \quad (1)$$

Its solutions are now called Bessel functions of order p. Such functions first appeared in the 1730s in the work of Daniel Bernoulli and Euler on the oscillations of a vertically suspended chain. The equation itself appears in a 1764 article by Euler on the vibrations of a circular drumhead, and Fourier used Bessel functions in his classical treatise on heat (1822). But their general properties were first studied systematically in an 1824 memoir by F. W. Bessel, who was investigating the motion of planets. The standard source of information on Bessel functions is G. N. Watson's *A Treatise on the Theory of Bessel Functions*, 2nd ed. (Cambridge: Cambridge University Press, 1944). Its 36 pages of references, which cover only the period up to 1922, give some idea of the vast literature of this subject.

Bessel's equation in (1) has indicial equation $r^2 - p^2 = 0$, with roots $r = \pm p$. If we substitute $y = \sum c_m x^{m+r}$ in (1), we find in the usual manner that $c_1 = 0$ and that

$$[(m + r)^2 - p^2]c_m + c_{m-2} = 0 \quad (2)$$

for $m \geq 2$. The verification of (2) is left to the reader (see Problem 6).

The Case $r = p > 0$

If we use $r = p$ and write a_m in place of c_m, then (2) yields the recursion formula

$$a_m = \frac{-a_{m-2}}{m(2p + m)}. \quad (3)$$

Because $a_1 = 0$, it follows that $a_m = 0$ for all odd m. The first few even coefficients are

$$a_2 = \frac{-a_0}{2(2p + 2)} = -\frac{a_0}{2^2(p + 1)},$$

$$a_4 = \frac{-a_2}{4(2p + 4)} = \frac{a_0}{2^4 \cdot 2(p + 1)(p + 2)},$$

$$a_6 = \frac{-a_4}{6(2p + 6)} = -\frac{a_0}{2^6 \cdot 2 \cdot 3(p + 1)(p + 2)(p + 3)}.$$

The general pattern is

$$a_{2m} = \frac{(-1)^m a_0}{2^{2m} m!(p+1)(p+2)\cdots(p+m)},$$

so with the larger root $r = p$ we get the solution

$$y_1(x) = a_0 \sum_{m=1}^{\infty} \frac{(-1)^m x^{2m+p}}{2^{2m} m!(p+1)(p+2)\cdots(p+m)}. \qquad (4)$$

If $p = 0$ this is the only Frobenius series solution; with $a_0 = 1$ it is the function $J_0(x)$ we have seen before.

The Case $r = -p < 0$

If we use $r = -p$ and write b_m in place of c_m, Eq. (2) takes the form

$$m(m - 2p)b_m + b_{m-2} = 0 \qquad (5)$$

for $m \geq 2$, while $b_1 = 0$. We see that there is a potential difficulty if it happens that $2p$ is a positive integer—that is, if p is either a positive integer or an odd integral multiple of $\frac{1}{2}$. For then when $m = 2p$, Eq. (5) is simply $0 \cdot b_m + b_{m-2} = 0$. Thus if $b_{m-2} \neq 0$, no value of b_m will satisfy this equation.

But if p is an odd integral multiple of $\frac{1}{2}$, we can circumvent this difficulty. For suppose that $p = k/2$ where k is an odd positive integer. Then we need only choose $b_m = 0$ for all odd values of m. The crucial step is the kth step,

$$k(k - k)b_k + b_{k-2} = 0;$$

and this equation will hold because $b_k = b_{k-2} = 0$.

Hence if p is *not* a positive integer, we take $b_m = 0$ for m odd and define the coefficients of even subscript in terms of b_0 by means of the recursion formula

$$b_m = \frac{-b_{m-2}}{m(m - 2p)}, \qquad m \geq 2. \qquad (6)$$

In comparing (6) with (3), we see that (6) will lead to the same result as that in (4), except with p replaced by $-p$. Thus in this case we obtain the second solution

$$y_2(x) = b_0 \sum_{m=1}^{\infty} \frac{(-1)^m x^{2m-p}}{2^{2m} m!(-p+1)(-p+2)\cdots(-p+m)}. \qquad (7)$$

The series in (4) and (7) converge for all $x > 0$ because $x = 0$ is the only singular point of Bessel's equation. If $p > 0$, the leading term in y_1 is $a_0 x^p$, while the leading term in y_2 is $b_0 x^{-p}$. Hence $y_1(0) = 0$, while $y_2(x) \to \pm\infty$ as $x \to 0$, so it is clear that y_1 and y_2 are linearly independent solutions of Bessel's equation of order $p > 0$.

THE GAMMA FUNCTION

The formulas in (4) and (7) can be simplified by use of the **gamma function** $\Gamma(x)$, which is defined for $x > 0$ by

$$\Gamma(x) = \int_0^{\infty} e^{-t} t^{x-1} \, dt. \qquad (8)$$

It is not difficult to show that this improper integral converges for each $x > 0$. The gamma function is a generalization for $x > 0$ of the factorial function $n!$, which is defined only if n is a nonnegative integer. To see the way in which $\Gamma(x)$ is a generalization of $n!$, we note first that

$$\Gamma(1) = \int_0^\infty e^{-t}\, dt = \lim_{b \to \infty} \left[-e^{-t} \right]_0^b = 1. \tag{9}$$

Then we integrate by parts:

$$\Gamma(x + 1) = \lim_{b \to \infty} \int_0^b e^{-t} t^x\, dt$$

$$= \lim_{b \to \infty} \left(\left[-e^{-t} t^x \right]_0^b + \int_0^b x e^{-t} t^{x-1}\, dt \right)$$

$$= x \lim_{b \to \infty} \int_0^b e^{-t} t^{x-1}\, dt;$$

that is,

$$\Gamma(x + 1) = x\Gamma(x). \tag{10}$$

This is the most important property of the gamma function.

If we combine (9) and (10), we see that

$$\Gamma(2) = 1\Gamma(1) = 1!, \qquad \Gamma(3) = 2\Gamma(2) = 2!, \qquad \Gamma(4) = 3\Gamma(3) = 3!,$$

and in general that

$$\Gamma(n + 1) = n! \quad \text{for } n \geq 0. \tag{11}$$

An important special value of the gamma function is

$$\Gamma(\tfrac{1}{2}) = \int_0^\infty e^{-t} t^{-1/2}\, dt = 2 \int_0^\infty e^{-u^2}\, du = \sqrt{\pi}, \tag{12}$$

where we have substituted u^2 for t in the integral; the fact that

$$\int_0^\infty e^{-u^2}\, du = \frac{\sqrt{\pi}}{2}$$

is known, but is far from obvious.

Although $\Gamma(x)$ is defined in (8) only for $x > 0$, we can use the recursion formula in (10) to define $\Gamma(x)$ whenever x is neither zero nor a negative integer. If $-1 < x < 0$, then

$$\Gamma(x) = \frac{\Gamma(x + 1)}{x};$$

the right-hand term is defined because $0 < x + 1 < 1$. The same formula may then be used to extend the definition of $\Gamma(x)$ to the open interval $(-2, -1)$, then to the open interval $(-3, -2)$, and so on. The graph of the gamma function thus extended is shown in Fig. 3.5. The student who would like to pursue this fascinating topic further should consult Artin's *The Gamma Function* (New York: Holt, Reinhart and Winston, 1964). In only 39 pages, this is one of the finest expositions in the entire literature of mathematics.

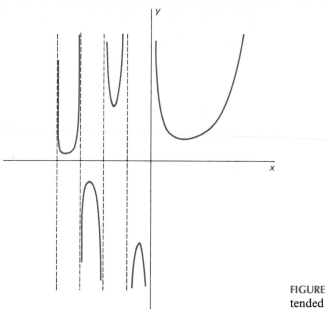

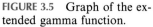

FIGURE 3.5 Graph of the extended gamma function.

BESSEL FUNCTIONS OF THE FIRST KIND

If we choose $a_0 = 1/[2^p\Gamma(p + 1)]$ in (4) and note that

$$\Gamma(p + m + 1) = (p + m)(p + m - 1) \cdots (p + 2)(p + 1)\Gamma(p + 1)$$

by repeated application of (10), we can write the **Bessel function of the first kind of order** p very concisely with the aid of the gamma function:

$$J_p(x) = \sum_{m=0}^{\infty} \frac{(-1)^m}{m!\Gamma(p + m + 1)} \left(\frac{x}{2}\right)^{2m+p}. \tag{13}$$

Similarly, if p is not an integer we choose $b_0 = 1/[2^{-p}\Gamma(-p + 1)]$ in (7) to obtain the linearly independent second solution

$$J_{-p}(x) = \sum_{m=0}^{\infty} \frac{(-1)^m}{m!\Gamma(-p + m + 1)} \left(\frac{x}{2}\right)^{2m-p} \tag{14}$$

of Bessel's equation of order p. If p is not an integer, we have the general solution

$$y(x) = c_1 J_p(x) + c_2 J_{-p}(x) \tag{15}$$

for $x > 0$; x^p must be replaced by $|x|^p$ in Eqs. (13)–(15) to get the correct solutions for $x < 0$.

If $p = n$, a nonnegative integer, then Eq. (13) gives

$$J_n(x) = \sum_{m=0}^{\infty} \frac{(-1)^m}{m!(m + n)!} \left(\frac{x}{2}\right)^{2m+n} \tag{16}$$

for the Bessel functions of the first kind of integral order. Thus

$$J_0(x) = \sum_{m=0}^{\infty} \frac{(-1)^m x^{2m}}{2^{2m}(m!)^2} = 1 - \frac{x^2}{2^2} + \frac{x^4}{2^2 4^2} - \frac{x^6}{2^2 4^2 6^2} + \cdots \tag{17}$$

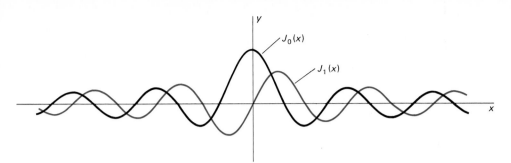

FIGURE 3.6   Graphs of the Bessel functions $J_0(x)$ and $J_1(x)$ (vertical scale exaggerated)

and

$$J_1(x) = \sum_{m=0}^{\infty} \frac{(-1)^m x^{2m+1}}{2^{2m+1} m!(m+1)!} = \frac{x}{2} - \frac{1}{2!}\left(\frac{x}{2}\right)^3 + \frac{1}{2!3!}\left(\frac{x}{2}\right)^5 - \cdots. \qquad (18)$$

The graphs of $J_0(x)$ and $J_1(x)$ are shown in Fig. 3.6. In a general way they resemble damped cosine and sine oscillations, respectively (see Problem 27). Indeed,
if you examine the series in (17), you can see part of the reason why $J_0(x)$ and
$\cos x$ *might* be similar—only minor changes in the denominators in (17) are needed
to produce the Taylor series for $\cos x$. As suggested by Fig. 3.6, the zeros of the
functions $J_0(x)$ and $J_1(x)$ are *interlaced*—between any two consecutive zeros of
$J_0(x)$ there is precisely one zero of $J_1(x)$ (see Problem 26), and vice versa. The
first four zeros of $J_0(x)$ are approximately 2.4048, 5.5201, 8.6537, and 11.7915.
For n large, the nth zero of $J_0(x)$ is approximately $(n - \frac{1}{4})\pi$, while the nth zero
of $J_1(x)$ is approximately $(n + \frac{1}{4})\pi$. Thus the interval between consecutive zeros of
either $J_0(x)$ or of $J_1(x)$ is approximately π—another similarity with $\cos x$ and $\sin x$.
You can see the way the accuracy of these approximations increases with increasing
n by rounding the entries in the table in Fig. 3.7 to two decimal places.

n	nth zero of $J_0(x)$	$(n - \frac{1}{4})\pi$	nth zero of $J_1(x)$	$(n + \frac{1}{4})\pi$
1	2.4048	2.3562	3.8317	3.9270
2	5.5201	5.4978	7.0156	7.0686
3	8.6537	8.6394	10.1735	10.2102
4	11.7915	11.7810	13.3237	13.3518
5	14.9309	14.9226	16.4706	16.4934

FIGURE 3.7   Zeros of $J_0(x)$ and $J_1(x)$

It turns out that $J_p(x)$ is an elementary function if the order p is half an
odd integer. For instance, on substitution of $p = \frac{1}{2}$ and $p = -\frac{1}{2}$ in Eqs. (13) and
(14), respectively, the results can be recognized (Problem 2) as

$$J_{1/2}(x) = \sqrt{\frac{2}{\pi x}} \sin x \quad \text{and} \quad J_{-1/2}(x) = \sqrt{\frac{2}{\pi x}} \cos x. \qquad (19)$$

BESSEL FUNCTIONS OF THE SECOND KIND

The methods of Section 3.4 must be used to find linearly independent second solutions of integral order. A very complicated generalization of Example 3 in that section gives the formula

$$Y_n(x) = \frac{2}{\pi}\left(\gamma + \ln\frac{x}{2}\right)J_n(x) - \frac{1}{\pi}\sum_{m=0}^{n-1}\frac{2^{n-2m}(n-m-1)!}{m!x^{n-2m}}$$

$$-\frac{1}{\pi}\sum_{m=0}^{\infty}\frac{(-1)^m(H_m + H_{m+n})}{m!(m+n)!}\left(\frac{x}{2}\right)^{n+2m}, \tag{20}$$

with the notation used there. If $n = 0$ the first sum in (20) is taken to be zero. Here, $Y_n(x)$ is called the **Bessel function of the second kind of integral order** $n \geq 0$.

The general solution of Bessel's equation of order n is

$$y(x) = c_1 J_n(x) + c_2 Y_n(x). \tag{21}$$

It is important to note that $Y_n(x) \to -\infty$ as $x \to 0$ (see Fig. 3.8). Hence $c_2 = 0$ in (21) if $y(x)$ is continuous at $x = 0$. Thus if $y(x)$ is a *continuous* solution of Bessel's equation of order n, it follows that

$$y(x) = cJ_n(x)$$

for some constant c; because $J_0(0) = 1$ we see in addition that if $n = 0$, then $c = y(0)$. In Section 9.4 we will see that this single fact regarding Bessel functions has numerous physical applications.

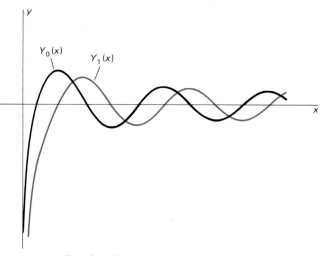

FIGURE 3.8   Graphs of the Bessel functions $Y_0(x)$ and $Y_1(x)$ (vertical scale exaggerated)

BESSEL FUNCTION IDENTITIES

Bessel functions are analogous to trigonometric functions in that they satisfy a large number of standard identities of frequent utility, especially in the evaluation

CHAPTER 3:   Power Series Solutions of Linear Equations

of integrals involving Bessel functions. Differentiation of

$$J_p(x) = \sum_{m=0}^{\infty} \frac{(-1)^m}{m!\,\Gamma(p+m+1)} \left(\frac{x}{2}\right)^{2m+p} \tag{13}$$

in the case that p is a nonnegative integer gives

$$\frac{d}{dx}\left[x^p J_p(x)\right] = \frac{d}{dx} \sum_{m=0}^{\infty} \frac{(-1)^m x^{2m+2p}}{2^{2m+p} m!\,(p+m)!}$$

$$= \sum_{m=0}^{\infty} \frac{(-1)^m x^{2m+2p-1}}{2^{2m+p-1} m!\,(p+m-1)!}$$

$$= x^p \sum_{m=0}^{\infty} \frac{(-1)^m x^{2m+p-1}}{2^{2m+p-1} m!\,(p+m-1)!}$$

and thus we have shown that

$$\frac{d}{dx}\left[x^p J_p(x)\right] = x^p J_{p-1}(x). \tag{22}$$

Similarly,

$$\frac{d}{dx}\left[x^{-p} J_p(x)\right] = -x^{-p} J_{p+1}(x). \tag{23}$$

If we carry out the differentiations in (22) and (23) and then divide the resulting identities by x^p and x^{-p}, respectively, we obtain (see Problem 8) the identities

$$J_p'(x) = J_{p-1}(x) - \frac{p}{x} J_p(x) \tag{24}$$

and

$$J_p'(x) = \frac{p}{x} J_p(x) - J_{p+1}(x). \tag{25}$$

Thus we may express the derivatives of Bessel functions in terms of Bessel functions themselves. Subtraction of (25) from (24) gives the recursion formula

$$J_{p+1}(x) = \frac{2p}{x} J_p(x) - J_{p-1}(x), \tag{26}$$

which can be used to express Bessel functions of high order in terms of Bessel functions of lower orders. In the form

$$J_{p-1}(x) = \frac{2p}{x} J_p(x) - J_{p+1}(x), \tag{27}$$

it can be used to express Bessel functions of large negative order in terms of Bessel functions of numerically smaller negative orders.

The identities in (22)–(27) above hold wherever they are meaningful—that is, whenever no Bessel functions of negative integral order appear. In particular, they hold for all nonintegral values of p.

EXAMPLE 1 With $p = 0$, the identity in (22) gives

$$\int x J_0(x)\, dx = x J_1(x) + C.$$

Similarly, with $p = 0$, the identity in (23) gives

$$\int J_1(x)\, dx = -J_0(x) + C.$$

EXAMPLE 2 Using first $p = 2$ and then $p = 1$ in (26), we get

$$J_3(x) = \frac{4}{x} J_2(x) - J_1(x) = \frac{4}{x}\left[\frac{2}{x} J_1(x) - J_0(x)\right] - J_1(x),$$

so that

$$J_3(x) = -\frac{4}{x} J_0(x) + \left(\frac{8}{x^2} - 1\right) J_1(x).$$

With similar manipulations every Bessel function of positive integral order can be expressed in terms of $J_0(x)$ and $J_1(x)$.

EXAMPLE 3 To evaluate $\int x J_2(x)\, dx$, we first note that $\int x^{-1} J_2(x)\, dx = -x^{-1} J_1(x) + C$ by (23) with $p = 1$. We therefore write

$$\int x J_2(x)\, dx = \int x^2 [x^{-1} J_2(x)]\, dx$$

and integrate by parts with

$$u = x^2, \qquad dv = x^{-1} J_2(x)\, dx,$$

$$du = 2x\, dx, \quad \text{and} \quad v = -x^{-1} J_1(x).$$

This gives

$$\int x J_2(x)\, dx = -x J_1(x) + 2 \int J_1(x)\, dx$$

$$= -x J_1(x) - 2 J_0(x) + C,$$

with the aid of the second result of Example 1.

THE PARAMETRIC BESSEL EQUATION

The **parametric Bessel equation of order** n is

$$x^2 y'' + x y' + (\alpha^2 x^2 - n^2) y = 0, \tag{28}$$

with α a positive parameter. As we will see in Chapter 9, this equation appears in the solution of Laplace's equation in polar coordinates. It is easy to see (Problem 9) that the substitution $t = \alpha x$ transforms (28) into the (standard) Bessel equation

$$t^2 \frac{d^2 y}{dt^2} + t \frac{dy}{dt} + (t^2 - n^2) y = 0 \tag{29}$$

with general solution $y = c_1 J_n(t) + c_2 Y_n(t)$. Hence the general solution of (28) is

$$y(x) = c_1 J_n(\alpha x) + c_2 Y_n(\alpha x). \tag{30}$$

Now consider the eigenvalue problem

$$x^2 y'' + xy' + (\lambda x^2 - n^2)y = 0,$$
$$y(L) = 0$$

(31)

on the interval $[0, L]$. We seek the *positive* values of λ for which there exists a nontrivial solution of (31) that is *continuous* on $[0, L]$. If we write $\lambda = \alpha^2$, then the differential equation in (31) is that in (28), so its general solution is given in (30). Because $Y_n(x) \to -\infty$ as $x \to 0$ while $J_n(0)$ is finite, the continuity of $y(x)$ requires that $c_2 = 0$. Thus $y(x) = c_1 J_n(\alpha x)$. The endpoint condition $y(L) = 0$ now implies that $z = \alpha L$ must be a (positive) root of the equation

$$J_n(z) = 0.$$

(32)

For $n > 1$, $J_n(x)$ oscillates rather like $J_1(x)$ in Fig. 3.6 and hence has an infinite sequence of positive roots $\gamma_{n1}, \gamma_{n2}, \gamma_{n3}, \ldots$. It follows that the kth positive eigenvalue of the problem in (31) is

$$\gamma_k = \alpha_k^2 = \frac{\gamma_{nk}^2}{L^2},$$

(33)

and that its associated eigenfunction is

$$y_k(x) = J_n\left(\frac{\gamma_{nk}x}{L}\right).$$

(34)

The roots γ_{nk} of Eq. (32) for $n \leq 8$ and $k \leq 20$ are tabulated in Table 9.5 of M. Abramowitz and I. A. Stegun, *Handbook of Mathematical Functions* (New York: Dover, 1965).

3.5 PROBLEMS

1. Differentiate termwise the series for $J_0(x)$ to show directly that $J_0'(x) = -J_1(x)$.

2. (a) Deduce from Eqs. (10) and (12) that

$$\Gamma\left(n + \frac{1}{2}\right) = \frac{1 \cdot 3 \cdots (2n-1)}{2^n}\sqrt{\pi}.$$

(b) Use the result of part (a) to verify the formulas in Eq. (19) for $J_{1/2}(x)$ and $J_{-1/2}(x)$.

3. (a) Suppose that m is a positive integer. Show that

$$\Gamma\left(m + \frac{2}{3}\right) = \frac{2 \cdot 5 \cdots (3m-1)}{3^m}\Gamma\left(\frac{2}{3}\right).$$

(b) Conclude from part (a) and the formula in (13) that

$$J_{-1/3}(x) = \frac{(x/2)^{-1/3}}{\Gamma(\frac{2}{3})}\left(1 + \sum_{m=1}^{\infty}\frac{(-1)^m 3^m x^{2m}}{2^{2m}m! \cdot 2 \cdot 5 \cdots (3m-1)}\right).$$

4. Apply the formulas in (19), (26), and (27) to show that

$$J_{3/2}(x) = \sqrt{\frac{2}{\pi x^3}}\,(\sin x - x\cos x)$$

and

$$J_{-3/2}(x) = -\sqrt{\frac{2}{\pi x^3}}\,(\cos x + x\sin x).$$

5. Express $J_4(x)$ in terms of $J_0(x)$ and $J_1(x)$.

6. Derive the recursion formula in Eq. (2) for Bessel's equation.

7. Verify the identity in Eq. (23) by termwise differentiation.

8. Deduce the identities in Eqs. (24) and (25) from those in Eqs. (22) and (23).

9. Verify that the substitution $t = \alpha x$ transforms the parametric Bessel equation in (28) into the equation in (29).

10. Show that

$$4J_p''(x) = J_{p-2}(x) - 2J_p(x) + J_{p+2}(x).$$

11. Use the relation $\Gamma(x + 1) = x\Gamma(x)$ to deduce from Eqs. (13) and (14) that if p is not a negative integer then

$$J_p(x) = \frac{(x/2)^p}{\Gamma(p+1)}\left[1 + \sum_{m=1}^{\infty} \frac{(-1)^m(x/2)^{2m}}{m!(p+1)(p+2)\cdots(p+m)}\right].$$

This form is more convenient for the computation of $J_p(x)$ because only the single value $\Gamma(p + 1)$ of the gamma function is required.

12. Use the series of Problem 11 to find $y(0) = \lim_{x\to 0} y(x)$ if

$$y(x) = x^2\left[\frac{J_{5/2}(x) + J_{-5/2}(x)}{J_{1/2}(x) + J_{-1/2}(x)}\right].$$

Any integral of the form $\int x^m J_n(x)\,dx$ can be evaluated in terms of Bessel functions and the indefinite integral $\int J_0(x)\,dx$. The latter integral cannot be simplified further, but the function $\int_0^x J_0(t)\,dt$ is tabulated in Table 11.1 of Abramowitz and Stegun. Use the identities in Eqs. (22) and (23) to evaluate the integrals in Problems 13–21.

13. $\int x^2 J_0(x)\,dx$ **14.** $\int x^3 J_0(x)\,dx$ **15.** $\int x^4 J_0(x)\,dx$

16. $\int x J_1(x)\,dx$ **17.** $\int x^2 J_1(x)\,dx$ **18.** $\int x^3 J_1(x)\,dx$

19. $\int x^4 J_1(x)\,dx$ **20.** $\int J_2(x)\,dx$ **21.** $\int J_3(x)\,dx$

22. Prove that

$$J_0(x) = \frac{1}{\pi}\int_0^{\pi} \cos(x\sin\theta)\,d\theta$$

by showing that the right-hand side satisfies Bessel's equation of order zero and has the value $J_0(0)$ when $x = 0$. Explain why this constitutes a proof.

23. Prove that

$$J_1(x) = \frac{1}{\pi}\int_0^{\pi} \cos(\theta - x\sin\theta)\,d\theta$$

by showing that the right-hand side satisfies Bessel's equation of order 1 and that its derivative has the value $J_1'(0)$ when $x = 0$. Explain why this constitutes a proof.

24. It can be shown that

$$J_n(x) = \frac{1}{\pi}\int_0^{\pi} \cos(n\theta - x\sin\theta)\,d\theta.$$

With $n \geq 2$, show that the right-hand side satisfies Bessel's equation of order n and also agrees with the values $J_n(0)$ and $J_n'(0)$. Explain why this does *not* suffice to prove the assertion above.

25. Deduce from Problem 22 that

$$J_0(x) = \frac{1}{2\pi}\int_0^{2\pi} e^{ix\sin\theta}\,d\theta.$$

(*Suggestion:* Show first that

$$\int_0^{2\pi} e^{ix\sin\theta}\,d\theta = \int_0^{\pi}(e^{ix\sin\theta} + e^{-ix\sin\theta})\,d\theta;$$

then use Euler's formula.)

26. Use the identities in Eqs. (22) and (23) and Rolle's theorem to prove that between any two consecutive zeros of $J_n(x)$ there is precisely one zero of $J_{n+1}(x)$.

27. (a) Show that the substitution $y = x^{-1/2}z$ in Bessel's equation of order p,

$$x^2 y'' + xy' + (x^2 - p^2)y = 0,$$

yields

$$z'' + \left(1 - \frac{p^2 - \frac{1}{4}}{x^2}\right)z = 0.$$

(b) If x is so large that $(p^2 - \frac{1}{4})/x^2$ is "negligible," then the latter equation reduces to $z'' + z \approx 0$. Explain why this suggests (without proving it) that if $y(x)$ is a solution of Bessel's equation, then

$$y(x) \approx x^{-1/2}(A\cos x + B\sin x) = Cx^{-1/2}\cos(x - \alpha)$$

with C and α constant, and x large.

ASYMPTOTIC APPROXIMATIONS   It is known that the choices $C = \sqrt{2/\pi}$ and $\alpha = (2n + 1)\pi/4$ yield the best approximation to $J_n(x)$ for x large:

$$J_n(x) \approx \sqrt{\frac{2}{\pi x}}\cos\left[x - (2n+1)\frac{\pi}{4}\right]. \tag{35}$$

Similarly,

$$Y_n(x) \approx \sqrt{\frac{2}{\pi x}}\sin\left[x - (2n+1)\frac{\pi}{4}\right]. \tag{36}$$

In particular,

$$J_0(x) \approx \sqrt{\frac{2}{\pi x}}\cos\left(x - \frac{\pi}{4}\right)$$

and

$$Y_0(x) \approx \sqrt{\frac{2}{\pi x}}\sin\left(x - \frac{\pi}{4}\right)$$

if x is large. These are *asymptotic approximations* in that the ratio of the two sides approaches unity as $x \to \infty$.

Applications of Bessel Functions

The importance of Bessel functions stems not only from the frequent appearance of Bessel's equation in applications, but also from the fact that the solutions of many other second order linear differential equations can be expressed in terms of Bessel functions. To see how this comes about, we begin with Bessel's equation of order p in the form

$$z^2 \frac{d^2w}{dz^2} + z \frac{dw}{dz} + (z^2 - p^2)w = 0, \tag{1}$$

and substitute

$$w = x^{-\alpha}y, \qquad z = kx^{\beta}. \tag{2}$$

Then a routine though lengthy transformation of (1) yields the equation

$$x^2y'' + (1 - 2\alpha)xy' + (\alpha^2 - \beta^2p^2 + \beta^2k^2x^{2\beta})y = 0;$$

that is,

$$x^2y'' + Axy' + (B + Cx^q)y = 0, \tag{3}$$

where the constants A, B, C, and q are given by

$$A = 1 - 2\alpha, \qquad B = \alpha^2 - \beta^2p^2, \qquad C = \beta^2k^2, \quad \text{and} \quad q = 2\beta. \tag{4}$$

It is a simple matter to solve the equations in (4) for

$$\alpha = \frac{1 - A}{2} \qquad \beta = \frac{q}{2},$$

$$k = \frac{2\sqrt{C}}{q}, \quad \text{and} \quad p = \frac{\sqrt{(1 - A)^2 - 4B}}{q}. \tag{5}$$

Under the assumption that the square roots in (5) are real, it follows that the general solution of Eq. (3) is

$$y(x) = x^{\alpha}w(z) = x^{\alpha}w(kx^{\beta})$$

where

$$w(z) = c_1 J_p(z) + c_2 J_{-p}(z)$$

(assuming p not an integer) is the general solution of the Bessel equation in (1). This establishes the following result.

Theorem   *Solutions in Bessel Functions*

If $C > 0$, $q \neq 0$, and $(1 - A)^2 \geq 4B$, then the general solution (for $x > 0$) of Eq. (3) is

$$y(x) = x^{\alpha}[c_1 J_p(kx^{\beta}) + c_2 J_{-p}(kx^{\beta})] \tag{6}$$

where α, β, k, and p are given by the equations in (5). If p is an integer, then J_{-p} is to be replaced by Y_p.

EXAMPLE 1 Solve the equation

$$4x^2y'' + 8xy' + (x^4 - 3)y = 0. \tag{7}$$

SOLUTION To compare the equation above with that in (3), we rewrite it as

$$x^2y'' + 2xy' + (-\tfrac{3}{4} + \tfrac{1}{4}x^4)y = 0$$

and see that $A = 2$, $B = -\tfrac{3}{4}$, $C = \tfrac{1}{4}$, and $q = 4$. Then the equations in (5) give $\alpha = -\tfrac{1}{2}$, $\beta = 2$, $k = \tfrac{1}{4}$, and $p = \tfrac{1}{2}$. Thus the general solution in (6) of Eq. (7) is

$$y(x) = x^{-1/2}[c_1 J_{1/2}(\tfrac{1}{4}x^2) + c_2 J_{-1/2}(\tfrac{1}{4}x^2)].$$

If we recall from Eq. (19) of Section 3.5 that

$$J_{1/2}(z) = \sqrt{\frac{2}{\pi z}}\sin z \quad \text{and} \quad J_{-1/2}(z) = \sqrt{\frac{2}{\pi z}}\cos z,$$

we see that a general solution of Eq. (7) can be written in the elementary form

$$y(x) = x^{-3/2}\left[A\cos\frac{x^2}{4} + B\sin\frac{x^2}{4}\right].$$

EXAMPLE 2 Solve the Airy equation

$$y'' + 9xy = 0. \tag{8}$$

SOLUTION First we rewrite the equation in the form

$$x^2y'' + 9x^3y = 0.$$

This is a special case of the equation in (3) with $A = B = 0$, $C = +9$, and $q = 3$. It follows from the equations in (5) that $\alpha = \tfrac{1}{2}$, $\beta = \tfrac{3}{2}$, $k = 2$, and $p = \tfrac{1}{3}$. Thus the general solution of Eq. (8) is

$$y(x) = x^{1/2}[c_1 J_{1/3}(2x^{3/2}) + c_2 J_{-1/3}(2x^{3/2})].$$

BUCKLING OF A VERTICAL COLUMN

For a practical application, we now consider the problem of determining when a uniform vertical column will buckle under its own weight. We take $x = 0$ at the free top end of the column and $x = L$ at its bottom; we assume that the bottom is rigidly imbedded in the ground (or in concrete) (see Fig. 3.9). Denote the angular deflection of the column at the point x by $\theta(x)$. From the theory of elasticity it follows that

$$EI\frac{d^2\theta}{dx^2} + g\rho x\theta = 0, \tag{9}$$

where E is the Young's modulus of the material of the column, I is its cross-sectional moment of inertia, ρ is the linear density of the column, and g is gravitational acceleration. The boundary conditions corresponding to the situation as described are

$$\theta'(0) = 0, \qquad \theta(L) = 0. \tag{10}$$

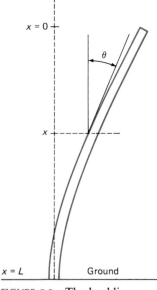

$x = 0$

θ

x

$x = L$ Ground

FIGURE 3.9 The buckling column.

CHAPTER 3: Power Series Solutions of Linear Equations

We will accept (9) and (10) as an appropriate statement of the problem and attempt to solve it in this form. With

$$\lambda = \gamma^2 = \frac{g\rho}{EI},\tag{11}$$

we have the eigenvalue problem

$$\theta'' + \gamma^2 x\theta = 0; \qquad \theta'(0) = 0, \qquad \theta(L) = 0.\tag{12}$$

The column can buckle only if there is a nontrivial solution of (12); otherwise the column will remain in its undeflected vertical position.

The differential equation in (12) is an Airy equation. It has the form of Eq. (3) with $A = 0$, $B = 0$, $C = \gamma^2$, and $q = 3$. The equations in (5) give $\alpha = \frac{1}{2}$, $\beta = \frac{3}{2}$, $k = 2\gamma/3$, and $p = \frac{1}{3}$. So the general solution is

$$\theta(x) = x^{1/2}[c_1 J_{1/3}(\tfrac{2}{3}\gamma x^{3/2}) + c_2 J_{-1/3}(\tfrac{2}{3}\gamma x^{3/2})].\tag{13}$$

In order to apply the initial conditions, we substitute $p = \pm\frac{1}{3}$ in

$$J_p(x) = \sum_{m=0}^{\infty} \frac{(-1)^m}{m!\Gamma(p+m+1)} \left(\frac{x}{2}\right)^{2m+p},$$

and find after some simplification that

$$\theta(x) = \frac{c_1\gamma^{1/3}}{3^{1/3}\Gamma(\tfrac{4}{3})}\left(x - \frac{\gamma^2 x^4}{12} + \frac{\gamma^4 x^7}{504} - \cdots\right)$$
$$+ \frac{c_2 3^{1/3}}{\gamma^{1/3}\Gamma(\tfrac{2}{3})}\left(1 - \frac{\gamma^2 x^3}{6} + \frac{\gamma^4 x^6}{180} - \cdots\right).$$

From this it is clear that the endpoint condition $\theta'(0) = 0$ implies that $c_1 = 0$, so

$$\theta(x) = c_2 x^{1/2} J_{-1/3}(\tfrac{2}{3}\gamma x^{3/2}).\tag{14}$$

The endpoint condition $\theta(L) = 0$ now gives

$$J_{-1/3}(\tfrac{2}{3}\gamma L^{3/2}) = 0.\tag{15}$$

Thus the column will buckle only if $z = \frac{2}{3}\gamma L^{3/2}$ is a root of the equation $J_{-1/3}(z) = 0$. The graph of

$$J_{-1/3}(z) = \frac{(z/2)^{-1/3}}{\Gamma(\tfrac{2}{3})}\left(1 + \sum_{m=1}^{\infty} \frac{(-1)^m 3^m z^{2m}}{2^{2m}(m!2 \cdot 5 \cdots (3m-1)}\right)\tag{16}$$

(see Problem 3 of Section 3.5) is shown in Fig. 3.10. Its zeros $\{z_n\}_1^{\infty}$ are [ignoring

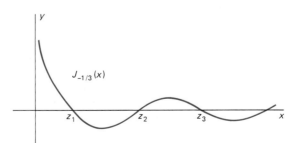

FIGURE 3.10 The graph of $J_{-1/3}(x)$.

the leading factor in (16)] the roots of the equation

$$F(z) = 1 + \sum_{m=1}^{\infty} \frac{(-1)^m 3^m z^{2m}}{2^{2m}(m!)2 \cdot 5 \cdots (3m-1)} = 0. \qquad (17)$$

If $|z|$ is not too large, the series in (17) is an alternating series with terms rapidly approaching zero. Hence the values of $F(z)$ are not difficult to compute. Indeed, if the series is truncated, the error in computing $F(z)$ is numerically less than the first term dropped.

Some typical values of $F(z)$ are given in the table in Fig. 3.11. From these data it is apparent that the least positive root z_1 of $F(z) = 0$ is between $z = 1.5$ and $z = 2.0$. Figure 3.12 shows a BASIC program for an IBM Personal Computer,

z	$F(z)$
0.0	+1.00000
0.5	+0.90799
1.0	+0.65226
1.5	+0.28900
2.0	−0.10257
2.5	−0.43830
3.0	−0.64819

FIGURE 3.11   Values of $F(z)$.

```
100 REM--Program applies Newton's method to find
110 REM--root of Bessel function J sub minus 1/3
120 REM
130       DEFDBL D, F, S, T, X
140       N = 1  :   XNEW = 1.5        'Initial guess
150 REM
160 REM--Newton's method iteration:
170 REM
180       X = XNEW                    'New X becomes old X
190       GOSUB 300                   'To get FCTN  = F(X)
200       GOSUB 400                   'To get DERIV = F'(X)
210       XNEW = X - FCTN/DERIV       'Iterative formula
220       PRINT N, XNEW               'Display new approx
230       IF ABS(X-XNEW) < 1E-08 THEN STOP
240       N = N + 1
250       GOTO 180                    'For another iteration
260 REM
300 REM--Subroutine computes value of function F(X)
310 REM
320 REM--T is typical term, S is partial sum of series
330 REM
340       T = 1  :   S = 1  :   M = 1
350       T = -T*3*X*X/(4*M*(3*M-1))            'New term
360       M = M + 1  :   S = S + T              'New sum
365       IF ABS (T) > 1E-10 THEN GOTO 350
370       FCTN = S    '            F(X) = sum of series
380       RETURN
390 REM
400 REM--Subroutine computes value of derivative F'(X)
410 REM
420 REM--T is typical term, S is partial sum of series
430 REM
440       T = -3*X/4 : S = T : M = 1
450       T = -T*3*X*X/(4*M*(3*M+2))            'New term
460       M = M + 1  :   S = S + T              'New sum
470       IF ABS (T) > 1E-10 THEN GOTO 450
480       DERIV = S   '            F'(X) = sum of series
490       RETURN
500 REM
510       END
```

FIGURE 3.12   Program to compute roots of $J_{-1/3}(z) = 0$.

Cross section of rod	Shortest buckling length L_1
Circular with $r = \frac{1}{2}$ in.	30 ft 6 in.
Circular with $r = 1\frac{1}{2}$ in.	63 ft 5 in.
Annular with $r_{inner} = 1\frac{1}{4}$ in. and $r_{outer} = 1\frac{1}{2}$ in.	75 ft 7 in.

FIGURE 3.13

which we used to compute z_1. This program employs the iteration

$$x_{n+1} = x_n - \frac{F(x_n)}{F'(x_n)} \tag{18}$$

of Newton's method, beginning with the initial guess $x_0 = 1.5$. The subroutine starting at line 300 computes the value of $F(x)$ by summing the series in (17), accurate to nine decimal places. The subroutine starting at line 400 similarly computes the value of the derivative

$$D(x) = F'(x)$$
$$= -\frac{3x}{4} + \sum_{m=1}^{\infty} \frac{(-1)^{m+1} 3^{m+1} x^{2m+1}}{2^{2m+1} \cdot (m!) \cdot 2 \cdot 5 \cdots (3m+2)}. \tag{19}$$

The output of the program was $x_1 = 1.868914$, $x_2 = 1.866350$, and $x_3 = x_4 = 1.866351$. Thus the smallest positive root of $J_{-1/3}(z) = 0$ is $z = 1.86635$, rounded to five decimal places.

The shortest length L_1 for which the column will buckle under its own weight is

$$L_1 = \left(\frac{3z_1}{2\gamma}\right)^{2/3} = \left[\frac{3z_1}{2}\left(\frac{EI}{\rho g}\right)^{1/2}\right]^{2/3}.$$

If we substitute $z_1 \approx 1.86635$ and $\rho = \delta A$, where δ is the volumetric density of the material of the column and A is its cross-sectional area, we finally get

$$L_1 \approx (1.986)\left(\frac{EI}{g\delta A}\right)^{1/3} \tag{20}$$

for the critical buckling length. For example, with a steel column or rod for which $E = 2.8 \times 10^7$ lb/in.^{2} and $g\delta = 0.28$ lb/in.^{3}, the formula in (20) gives the results shown in the table in Fig. 3.13.

We have used the familiar formulas $A = \pi r^2$ and $I = \frac{1}{4}\pi r^4$ for a circular disk. The data in the table show why flagpoles are hollow.

3.6 PROBLEMS

In each of Problems 1–12, express the general solution of the given differential equation in terms of Bessel functions.

1. $x^2 y'' - xy' + (1 + x^2)y = 0$

2. $xy'' + 3y' + xy = 0$

3. $xy'' - y' + 36x^3 y = 0$

4. $x^2 y'' - 5xy' + (8 + x)y = 0$

5. $36x^2y'' + 60xy' + (9x^3 - 5)y = 0$

6. $16x^2y'' + 24xy' + (1 + 144x^3)y = 0$

7. $x^2y'' + 3xy' + (1 + x^2)y = 0$

8. $4x^2y'' - 12xy' + (15 + 16x)y = 0$

9. $16x^2y'' - (5 - 144x^3)y = 0$

10. $2x^2y'' + 3xy' - 2(14 - x^5)y = 0$

11. $y'' + x^4y = 0$

12. $y'' + 4x^3y = 0$

13. Apply the theorem in this section to show that the general solution of

$$xy'' + 2y' + xy = 0$$

is $y = x^{-1}(A \cos x + B \sin x)$.

14. Verify that the substitutions in (2) in Bessel's equation (Eq. (1)) yield Eq. (3).

15. (a) Show that the substitution $y = (-1/u)(du/dx)$ transforms the Ricatti equation $dy/dx = x^2 + y^2$ into $u'' + x^2u = 0$.
(b) Show that the general solution of $y' = x^2 + y^2$ is

$$y(x) = x \frac{J_{3/4}(x^2/2) - cJ_{-3/4}(x^2/2)}{cJ_{1/4}(x^2/2) + J_{-1/4}(x^2/2)}.$$

(*Suggestion:* Apply the identities in Eqs. (22) and (23) of Section 3.5.)

16. (a) Substitute the series of Problem 11 of Section 3.5 in the result of Problem 15 here to show that the solution of the initial value problem

$$\frac{dy}{dx} = x^2 + y^2, \qquad y(0) = 0$$

is

$$y(x) = x \cdot \frac{J_{3/4}(x^2/2)}{J_{-1/4}(x^2/2)}.$$

(b) Deduce similarly that the solution of the initial value problem

$$\frac{dy}{dx} = x^2 + y^2, \qquad y(0) = 1$$

is

$$y(x) = x \cdot \frac{2\Gamma(3/4)J_{3/4}(x^2/2) + \Gamma(1/4)J_{-3/4}(x^2/2)}{2\Gamma(3/4)J_{-1/4}(x^2/2) - \Gamma(1/4)J_{1/4}(x^2/2)}.$$

Some solution curves of the equation $y' = x^2 + y^2$ are shown in Fig. 3.14. The location of the asymptotes where $y(x) \to +\infty$ can be determined by using Newton's method to find the zeros of the denominators in the formula for the solutions as listed above.

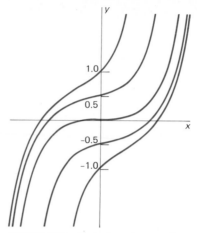

FIGURE 3.14 Solution curves of $y' = x^2 + y^2$ for $y(0) = -1, -0.5, 0.0, 0.5,$ and 1.0.

17. Figure 3.15 shows a linearly tapered rod with circular cross section, subjected to an axial force P of compression. As in Section 2.10, its deflection curve $y = y(x)$ satisfies the endpoint value problem

$$EIy'' + Py = 0, \qquad y(a) = y(b) = 0. \qquad (21)$$

Here, however, the moment of inertia $I = I(x)$ of the cross section at x is given by

$$I(x) = \frac{1}{4}\pi(kx)^4 = I_0\left(\frac{x}{b}\right)^4,$$

where $I_0 = I(b)$, the value of I at $x = b$. Substitution of $I(x)$ in the differential equation in (21) yields the eigenvalue problem

$$x^4y'' + \lambda y = 0, \qquad y(a) = y(b) = 0$$

where

$$\lambda = \mu^2 = \frac{Pb^4}{EI_0}.$$

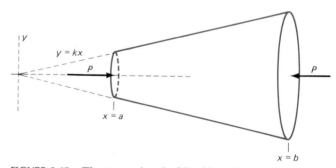

FIGURE 3.15 The tapered rod of Problem 17.

(a) Apply the theorem of this section to show that the general solution of $x^4 y'' + \mu^2 y = 0$ is

$$y(x) = x\left(A \cos \frac{\mu}{x} + B \sin \frac{\mu}{x} \right).$$

(b) Conclude that the *n*th eigenvalue is given by $\mu_n = n\pi ab/L$, where $L = b - a$ is the length of the rod, and hence that the *n*th buckling force is

$$P_n = \frac{n^2 \pi^2}{L^2} \left(\frac{a}{b} \right)^2 EI_0.$$

Note that if $a = b$, this result reduces to Eq. (21) of Section 2.10.

18. Consider a variable length pendulum as indicated in Fig. 3.16. Assume that its length is increasing linearly with time, $L(t) = a + bt$. It can be shown that the oscillations of this pendulum satisfy the differential equation

$$L\theta'' + 2L'\theta' + g\theta = 0$$

under the usual condition that θ is so small that $\sin \theta$ is very well approximated by θ: $\theta \approx \sin \theta$. Substitute $L = a + bt$ to derive the general solution

$$\theta(t) = \frac{1}{\sqrt{L}} \left[A J_1\left(\frac{2}{b}\sqrt{gL} \right) + B Y_1\left(\frac{2}{b}\sqrt{gL} \right) \right].$$

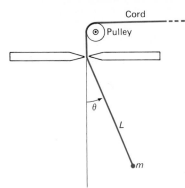

FIGURE 3.16 A variable length pendulum.

For the application of this solution to a discussion of the steadily descending pendulum ("its nether extremity was formed of a crescent of glittering steel, about a foot in length from horn to horn; the horns upward, and the under edge as keen as that of a razor . . . and the whole hissed as it swung through the air . . . down and still down it came") of Edgar Allan Poe's macabre classic "The Pit and the Pendulum," see the article by Borrelli, Coleman, and Hobson in the March 1985 issue of *Mathematics Magazine* (Vol. 58, pp. 78–83).

*3.7

Appendix on Infinite Series and the Atom

Here we illustrate the applications of infinite series methods by briefly outlining one of the great triumphs of modern physics—the interpretation of atomic spectra. During the nineteenth century it was discovered that when a gas is rendered luminous in an electric discharge tube, the light that it emits consists of specific, sharply defined wavelengths that are characteristic of the particular gas used. For example, when the light emitted by hydrogen is analyzed with a spectroscope, we find that its *spectrum* consists of a sequence of sharply defined lines (on the photographic plate), beginning with a red line at wavelength $\lambda = 6563$ angstroms (Å) (1 Å is defined to be 10^{-10} m) and apparently converging toward a limiting line at 3646 Å (see Fig. 3.17). In 1885 J. J. Balmer observed that the wavelengths λ_n of these lines are given by the formula

$$\frac{1}{\lambda_n} = R\left(\frac{1}{2^2} - \frac{1}{n^2} \right) \tag{1}$$

for $n = 3, 4, 5, \ldots$, where $R \approx 109{,}678$ cm^{-1} is the **Rydberg constant**. Thus the formula in (1) fits the **Balmer series** of visible spectral lines of hydrogen. (There are other series of spectral lines outside the visible range, but we shall not discuss them here.)

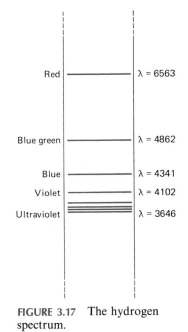

FIGURE 3.17 The hydrogen spectrum.

During the early twentieth century there gradually emerged, from the work of physicists such as Rutherford and Bohr, an understanding that the light emitted by an element originates in the electrons in its individual atoms. In particular, it was found that an electron of an atom can exist in any one of several *states* with well-defined energy levels. When an electron jumps from one state to another with lower energy, it emits a photon of light; the frequency $v = c/\lambda$ of the light (where c denotes the velocity of light) is given by

$$\Delta E = hv, \tag{2}$$

where ΔE is the energy lost by the electron and h is **Planck's constant**. The problem, then, was to construct a mathematical model of the atom consistent with the formulas in (1) and (2).

In the modern quantum theory of atomic structure, the electron in a hydrogen atom does not travel in a well-defined orbit about the nucleus (the proton) of the atom. Instead, it occupies a state called an **orbital**, which may be visualized as a cloud of negative electricity surrounding the nucleus. The density of this "electron cloud" at a particular point is a measure of the probability that, at a given instant, the electron is near that point.

More precisely, an orbital is described by a *probability amplitude function* $\psi(x, y, z)$. This function has the property that the probability $P(x, y, z; \Delta V)$ of finding the electron in a small volume element ΔV surrounding the point (x, y, z) is given approximately by

$$P(x, y, z; \Delta V) \approx [\psi(x, y, z)]^2 \, \Delta V. \tag{3}$$

The error in this approximation is small in comparison with ΔV; that is,

$$\lim_{\Delta V \to 0} \frac{P(x, y, z; \Delta V)}{\Delta V} = [\psi(x, y, z)]^2. \tag{4}$$

If an electron with mass m has constant total energy E—the sum of its kinetic energy and its potential energy $V(x, y, z)$—then its probability amplitude function must satisfy the *time-independent Schrödinger equation*

$$\nabla^2 \psi + \frac{8\pi^2 m}{h^2} (E - V)\psi = 0. \tag{5}$$

Here $\nabla^2 \psi$ denotes the **Laplacian**, defined as follows:

$$\nabla^2 \psi = \frac{\partial^2 \psi}{\partial x^2} + \frac{\partial^2 \psi}{\partial y^2} + \frac{\partial^2 \psi}{\partial z^2}.$$

We want to investigate the possibility of *spherically symmetric* orbitals for which $\psi = \psi(r)$ is a function of the radial coordinate $r = \sqrt{x^2 + y^2 + z^2}$ alone, considering the nucleus of the hydrogen atom (the proton) to be fixed at the origin. In this case

$$\frac{\partial \psi}{\partial x} = \frac{d\psi}{dr} \frac{\partial r}{\partial x} = \frac{x}{r} \frac{d\psi}{dr},$$

so

$$\frac{\partial^2 \psi}{\partial x^2} = \frac{\partial}{\partial x}\left(\frac{x}{r}\frac{d\psi}{dr}\right)$$

$$= \frac{1}{r}\frac{d\psi}{dr} - \frac{x}{r^2}\frac{d\psi}{dr}\frac{\partial r}{\partial x} + \frac{x}{r}\frac{d^2\psi}{dr^2}\frac{\partial r}{\partial x}$$

and thus

$$\frac{\partial^2 \psi}{\partial x^2} = \frac{1}{r}\frac{d\psi}{dr} - \frac{x^2}{r^3}\frac{d\psi}{dr} + \frac{x^2}{r^2}\frac{d^2\psi}{dr^2},$$

with similar formulas for $\partial^2\psi/\partial y^2$ and $\partial^2\psi/\partial z^2$, x being replaced by y and z, respectively. Adding these formulas for the second derivatives of ψ, we get

$$\nabla^2\psi = \frac{3}{r}\frac{d\psi}{dr} - \frac{x^2+y^2+z^2}{r^3}\frac{d\psi}{dr} + \frac{x^2+y^2+z^2}{r^2}\frac{d^2\psi}{dr^2},$$

so

$$\nabla^2\psi = \frac{d^2\psi}{dr^2} + \frac{2}{r}\frac{d\psi}{dr}.$$

Hence the Schrödinger equation in (5) becomes

$$\frac{d^2\psi}{dr^2} + \frac{2}{r}\frac{d\psi}{dr} + \frac{8\pi^2 m}{h^2}(E - V)\psi = 0 \tag{6}$$

in the radially symmetric case.

The proton and the electron have charges $+q$ and $-q$, so in appropriate electrical units the potential energy of the electron orbiting the fixed proton is the *Coulomb potential*

$$V = -\frac{q^2}{r}. \tag{7}$$

Then the total energy of an electron at rest at infinity ($r = \infty$) will be zero, and the total energy of an electron bound to the nucleus will be negative,

$$E = -\alpha^2. \tag{8}$$

On substitution of (7) and (8) in (6) we obtain the ordinary differential equation

$$\frac{d^2\psi}{dr^2} + \frac{2}{r}\frac{d\psi}{dr} + \frac{8\pi^2 m}{h^2}\left(-\alpha^2 - \frac{q^2}{r}\right)\psi = 0. \tag{9}$$

When r is very large, Eq. (9) is approximately the equation

$$\frac{d^2\psi}{dr^2} - \frac{8\pi^2\alpha^2 m}{h^2}\psi = 0,$$

whose solution $u = \psi(r)$ is of the form

$$u = A\exp\left(\sqrt{\frac{8\pi^2\alpha^2 m}{h^2}}\,r\right) + B\exp\left(-\sqrt{\frac{8\pi^2\alpha^2 m}{h^2}}\,r\right);$$

that is,

$$u = Ae^{-x/2} + Be^{+x/2}, \tag{10}$$

where

$$x = \frac{4\pi\alpha r}{h}\sqrt{2m}. \tag{11}$$

Now let us impose the physically motivated condition that we are looking for a solution $\psi(r)$ that is continuous everywhere with $\psi(r) \to 0$ as $r \to \infty$ (so the probability of finding the electron far from the nucleus is negligible). If $\psi = u(x)$ is to be given approximately by (10) when x is large, it is necessary that $B = 0$, so that $\psi \approx Ae^{-x/2}$.

This preliminary analysis suggests that we attempt to simplify Eq. (9) by splitting the factor $e^{-x/2}$ out of its solution. In Problem 1 we ask you to show that the subsitution

$$\psi = e^{-x/2}v(x), \qquad x = \frac{4\pi\alpha r}{h}\sqrt{2m} \tag{12}$$

transforms (9) into the differential equation

$$x\frac{d^2v}{dx^2} + (2 - x)\frac{dv}{dx} + (p - 1)v = 0, \tag{13}$$

where $p = (\pi q^2\sqrt{2m})/\alpha h$. This equation is closely related to **Laguerre's equation** of order p:

$$x\frac{d^2y}{dx^2} + (1 - x)\frac{dy}{dx} + py = 0. \tag{14}$$

Indeed, differentiation of (14) gives

$$xy^{(3)} + (2 - x)y'' + (p - 1)y' = 0,$$

so $v(x) = y'(x)$ is a solution of (13) if $y(x)$ is a solution of Laguerre's equation.

So let us examine Laguerre's equation. First we write it in the form

$$y'' + \frac{1 - x}{x}y' + \frac{p}{x}y = 0.$$

Clearly this equation has $x = 0$ as a regular singular point, and its indicial equation is

$$r(r - 1) + (1)r + 0 = 0;$$

that is, $r^2 = 0$. From the theorem of Section 3.4 we therefore see that Laguerre's equation has two linearly independent solutions of the forms

$$y_1(x) = \sum_{k=1}^{\infty} a_k x^k$$

and

$$y_2(x) = y_1(x)\ln x + \sum_{k=1}^{\infty} b_k x^k.$$

CHAPTER 3:   Power Series Solutions of Linear Equations

But we are searching for a solution that is continuous at $x = 0$, and this condition excludes the singular solution.

In Problem 2 we ask you to verify that substitution of the power series $y = \sum a_k x^k$ in Laguerre's equation yields the recursion formula

$$a_k = -\frac{p - k + 1}{k^2} a_{k-1}. \tag{15}$$

From this it follows by induction that

$$a_k = (-1)^k \frac{p \cdot (p - 1) \cdots (p - k + 1)}{(k!)^2} a_0.$$

Consequently

$$y(x) = a_0 \sum_{k=0}^{\infty} (-1)^k \frac{p \cdot (p - 1) \cdots (p - k + 1)}{(k!)^2} x^k, \tag{16}$$

so the solution of Eq. (13) that we seek is

$$v(x) = y'(x) = \sum_{k=0}^{\infty} b_k x^k \tag{17}$$

where

$$b_k = (-1)^{k+1} a_0 \frac{p \cdot (p - 1) \cdots (p - k)}{(k!)(k + 1)!}. \tag{18}$$

But what is the point to all this? In order to determine what frequency of light photon is emitted when the hydrogen atom's electron jumps from one energy level to a lower one, we need to determine the possible values of the total energy constant $E = -\alpha^2$ that appears in Eq. (9). Equivalently, what are the possible values of $p = (\pi q^2 \sqrt{2m})/\alpha h$?

We will find the answer to this question by imposing the condition that

$$\psi = \frac{v(x)}{e^{x/2}} \to 0 \quad \text{as} \quad x \to \infty. \tag{19}$$

If p is a *nonnegative integer*, then it follows from (18) that $v(x)$ is a polynomial, so the condition in (19) is satisfied. Otherwise, the series in (17) for $v(x)$ is infinite, with

$$\frac{b_{k+1}}{b_k} = \frac{k + 1 - p}{(k + 1)(k + 2)}.$$

Now

$$e^{x/2} = \sum_{n=0}^{\infty} c_k x^k$$

where $c_k = 1/(k! 2^k)$, so

$$\frac{c_{k+1}}{c_k} = \frac{1}{2(k + 1)}.$$

Hence

$$\frac{b_{k+1}/b_k}{c_{k+1}/c_k} = \frac{2(k+1)(k+1-p)}{(k+1)(k+2)} \to 2$$

as $k \to \infty$. It follows that

$$\frac{b_{k+1}}{b_k} > \frac{3}{2}\frac{c_{k+1}}{c_k}$$

for k sufficiently large, say for $k \geq K$. Hence

$$\frac{b_{K+k}}{c_{K+k}} > \left(\frac{3}{2}\right)^k \frac{b_K}{c_K} > 1$$

for k sufficiently large. Thus, if p is *not* a nonnegative integer, then the terms of $v = \sum b_k x^k$ from some point on are positive and larger than those of $e^{x/2} = \sum c_k x^k$, so the condition in (19) cannot be satisfied. We get a nontrivial solution only if p is a *positive integer* because, if $p = 0$, then $y(x) = a_0$, so that $v(x) = y'(x) \equiv 0$.

Thus we finally see that the only values of p for which Eq. (9) has a nontrivial solution satisfying our conditions are the positive integers

$$p = \frac{\pi q^2 \sqrt{2m}}{\alpha h} = n = 1, 2, 3, \ldots.$$

The nth value of α is

$$\alpha_n = \frac{\pi q^2 \sqrt{2m}}{nh},$$

and the energy of the electron in its nth state is

$$E_n = -\alpha_n^2 = -\frac{2\pi^2 m q^4}{n^2 h^2}. \tag{20}$$

If the electron falls from energy level E_n to energy level E_k, where $n > k$, its energy loss is

$$\Delta E = \frac{2\pi^2 m q^4}{h^2}\left(\frac{1}{k^2} - \frac{1}{n^2}\right). \tag{21}$$

Because $\Delta E = h\nu = hc/\lambda$ by Eq. (2), we see that the wavelength λ of the photon emitted is given by

$$\frac{1}{\lambda} = \frac{\Delta E}{hc} = \frac{2\pi^2 m q^4}{ch^3}\left(\frac{1}{k^2} - \frac{1}{n^2}\right). \tag{22}$$

In comparing this result with the formula in (1), we see that the Balmer series of visible spectral lines results from electron transitions from the state n to the state $k = 2$ for $n = 3, 4, 5, \ldots$. We see also that the Rydberg constant is given by

$$R = \frac{2\pi^2 m q^4}{ch^3}. \tag{23}$$

Indeed, when the known values of c, m, q, and h are substituted in (23), the result is the empirically measured value $R \approx 109,678 \text{ cm}^{-1}$. It is the spectacular agreement of theory with experiment that is the great triumph.

CHAPTER 3:   Power Series Solutions of Linear Equations

The polynomial solution of Laguerre's equation,

$$xy'' + (1 - x)y' + ny = 0 \qquad (n \text{ is an integer}),$$

that we found above is called the nth **Laguerre polynomial** $L_n(x)$. In Problem 3 we ask you to deduce from the formula in (16) with $a_0 = 1$ that

$$L_n(x) = \sum_{k=0}^{n} (-1)^k \binom{n}{k} \frac{x^k}{k!}. \tag{24}$$

Thus $L_0(x) = 1$, $L_1(x) = 1 - x$, $L_2(x) = 1 - 2x + \frac{1}{2}x^2$, $L_3(x) = 1 - 3x + \frac{3}{2}x^2 - \frac{1}{6}x^3$, and so on.

3.7 PROBLEMS

1. Verify that the substitution given by the equations in (12) transforms the differential equation in (9) into Eq. (13).

2. Verify the recursion formula in (15) for the Laguerre polynomial of order p.

3. Verify that if $p = n$, a nonnegative integer, then the infinite series in (16) with $a_0 = 1$ reduces to the Laguerre polynomial as given in Eq. (24).

CHAPTER *4*

THE LAPLACE TRANSFORM

4.1 LAPLACE TRANSFORMS AND INVERSE TRANSFORMS

4.2 TRANSFORMATION OF INITIAL VALUE PROBLEMS

4.3 TRANSLATION AND PARTIAL FRACTIONS

4.4 DERIVATIVES, INTEGRALS, AND PRODUCTS OF TRANSFORMS

*4.5 PERIODIC AND PIECEWISE CONTINUOUS FORCING FUNCTIONS

*4.6 IMPULSES AND DELTA FUNCTIONS

TABLE OF LAPLACE TRANSFORMS

Laplace Transforms and Inverse Transforms

In Chapter 2 we saw that linear differential equations with constant coefficients have numerous applications and can be solved systematically. There are common situations, however, in which the alternative method of this chapter is preferable. For example, recall the differential equations

$$mx'' + cx' + kx = F(t)$$

and

$$LI'' + RI' + \frac{1}{C} I = E'(t)$$

corresponding to a mass-spring-dashpot system and a series RLC circuit, respectively. It often happens in practice that the forcing term, $F(t)$ or $E'(t)$, has discontinuities—for instance, when the voltage supplied to an electrical circuit is periodically turned off and on. In this case the methods of Chapter 2 can be quite awkward, and the Laplace transform method is more convenient.

The differentiation operator D can be viewed as a transformation which, when applied to the function $f(t)$, yields the new function $D\{f(t)\} = f'(t)$. The Laplace transformation $\mathscr{L}$ involves the operation of integration, and yields the new function $\mathscr{L}\{f(t)\} = F(s)$ of a new independent variable s. The situation is diagrammed in Fig. 4.1. After learning in this section how to compute the Laplace transform $F(s)$ of a function $f(t)$, we will see in Section 4.2 that the Laplace transformation converts a *differential* equation in the unknown function $f(t)$ into an *algebraic* equation in $F(s)$. Because algebraic equations are generally easier to solve than are differential equations, this is one method that simplifies the problem of finding the solution $f(t)$.

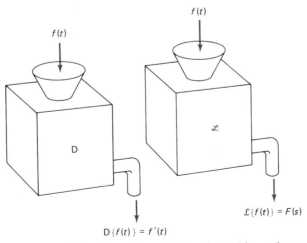

FIGURE 4.1 Transformation of a function: $\mathscr{L}$ in analogy with D.

> **Definition** *The Laplace Transform*
>
> Given a function $f(t)$ defined for all $t \geq 0$, the **Laplace transform** of f is the function F of s defined as follows:
>
> $$\mathcal{L}\{f(t)\} = F(s) = \int_0^\infty e^{-st} f(t)\, dt \qquad (1)$$
>
> for all values of s for which the improper integral converges.

Recall that an **improper integral** over an infinite interval is defined as a limit of integrals over finite intervals; that is,

$$\int_a^\infty g(t)\, dt = \lim_{b \to \infty} \int_a^b g(t)\, dt. \qquad (2)$$

If the limit in (2) exists, then we say that the improper integral **converges**. Otherwise, it **diverges** or fails to exist. Note that the integrand of the improper integral in (1) contains the parameter s in addition to the variable of integration t. Therefore, when the integral in (1) converges, it converges not merely to a number, but to a *function* F of s. As in the following examples, it is typical for the improper integral in the definition of $\mathcal{L}\{f(t)\}$ to converge for some values of s and diverge for others.

EXAMPLE 1 With $f(t) \equiv 1$ for $t \geq 0$, the definition of the Laplace transform in (1) gives

$$\mathcal{L}\{1\} = \int_0^\infty e^{-st}\, dt = \left[-\frac{1}{s} e^{-st} \right]_0^\infty$$

$$= \lim_{b \to \infty} \left[-\frac{1}{s} e^{-bs} + \frac{1}{s} \right],$$

and therefore

$$\mathcal{L}\{1\} = \frac{1}{s} \quad \text{for } s > 0. \qquad (3)$$

As in (3), it's good practice to specify the domain of the Laplace transform—in problems as well as in examples. Also, in this computation we have used the common abbreviation

$$\left[g(t) \right]_a^\infty = \lim_{b \to \infty} \left[g(t) \right]_a^b. \qquad (4)$$

Remark: The limit we computed above would not exist if $s < 0$, for then $(1/s)e^{-bs}$ would become unbounded as $b \to \infty$. Hence $\mathcal{L}\{1\}$ is defined only for $s > 0$. This is typical of Laplace transforms; the domain of a transform is normally of the form $s > a$ for some number a.

EXAMPLE 2 With $f(t) = e^{at}$ for $t \geq 0$, we obtain

$$\mathcal{L}\{e^{at}\} = \int_0^\infty e^{-st}e^{at}\, dt$$

$$= \int_0^\infty e^{-(s-a)t}\, dt = \left[-\frac{e^{-(s-a)t}}{s-a} \right]_0^\infty.$$

If $s - a > 0$, then $e^{-(s-a)t} \to 0$ as $t \to +\infty$, so it follows that

$$\mathcal{L}\{e^{at}\} = \frac{1}{s-a} \quad \text{for } s > a. \tag{5}$$

Note here that the improper integral giving $\mathcal{L}\{e^{at}\}$ diverges if $s \leq a$. It is worth noting also that the formula in (5) holds if a is a complex number. For then, with $a = \alpha + i\beta$,

$$e^{-(s-a)t} = e^{i\beta t}e^{-(s-\alpha)t} \to 0$$

as $t \to +\infty$, provided that $s > \alpha = \mathrm{Re}(a)$; recall that $e^{i\beta t} = \cos \beta t + i \sin \beta t$.

EXAMPLE 3 Suppose that $f(t) = t^a$ for $t \geq 0$, where a is real and $a > -1$. Then

$$\mathcal{L}\{t^a\} = \int_0^\infty e^{-st}t^a\, dt. \tag{6}$$

One could attack this integral using integration by parts (at least if a is an integer), but instead we use the gamma function, which was defined in Section 3.5 for $x > 0$ by the formula

$$\Gamma(x) = \int_0^\infty e^{-t}t^{x-1}\, dt.$$

For an elementary discussion of $\Gamma(x)$, see the subsection on the gamma function in Section 3.5.

If we substitute $u = st$, $t = u/s$, and $dt = du/s$ in (6), we get

$$\mathcal{L}\{t^a\} = \frac{1}{s^{a+1}} \int_0^\infty e^{-u}u^a\, du = \frac{\Gamma(a+1)}{s^{a+1}} \tag{7}$$

for all $s > 0$ (so that $u = st > 0$). Because $\Gamma(n+1) = n!$ if n is a non-negative integer, we see that

$$\mathcal{L}\{t^n\} = \frac{n!}{s^{n+1}} \quad \text{for } s > 0. \tag{8}$$

For instance,

$$\mathcal{L}\{t\} = \frac{1}{s^2}, \quad \mathcal{L}\{t^2\} = \frac{2}{s^3}, \quad \text{and} \quad \mathcal{L}\{t^3\} = \frac{6}{s^4},$$

As in Problems 1 and 2, these formulas can be derived immediately from the definition, without use of the gamma function.

It is not necessary for us to proceed much further in the computation of Laplace transforms directly from the definition. Once we know the Laplace transforms of several functions, we can combine them to obtain transforms of other functions. The reason is that the Laplace transformation is a *linear* operation.

Theorem 1 *Linearity of the Laplace Transformation*

If a and b are constants, then

$$\mathcal{L}\{af(t) + bg(t)\} = a\mathcal{L}\{f(t)\} + b\mathcal{L}\{g(t)\} \tag{9}$$

for all s such that the Laplace transforms of the functions f and g both exist.

The proof of Theorem 1 follows immediately from the linearity of the operations of taking limits and of integration:

$$\mathcal{L}\{af(t) + bg(t)\} = \int_0^\infty e^{-st}[af(t) + bg(t)]\,dt$$

$$= \lim_{c\to\infty} \int_0^c e^{-st}[af(t) + bg(t)]\,dt$$

$$= a\left(\lim_{c\to\infty}\int_0^c e^{-st}f(t)\,dt\right) + b\left(\lim_{c\to\infty}\int_0^c e^{-st}g(t)\,dt\right)$$

$$= a\mathcal{L}\{f(t)\} + b\mathcal{L}\{g(t)\}.$$

EXAMPLE 4 The formulas in (7)–(9) yield

$$\mathcal{L}\{3t^2 + 4t^{3/2}\} = 3\frac{2!}{t^3} + \frac{4\Gamma(\frac{5}{2})}{t^{5/2}}$$

$$= \frac{6}{t^3} + 3\sqrt{\frac{\pi}{t^5}},$$

using the fact that

$$\Gamma(\tfrac{5}{2}) = \tfrac{3}{2}\Gamma(\tfrac{3}{2}) = \tfrac{3}{2}\cdot\tfrac{1}{2}\Gamma(\tfrac{1}{2}) = \tfrac{3}{4}\sqrt{\pi},$$

because

$$\Gamma(x + 1) = x\Gamma(x) \quad\text{and}\quad \Gamma(\tfrac{1}{2}) = \sqrt{\pi}.$$

EXAMPLE 5 Recall that $\cosh kt = (e^{kt} + e^{-kt})/2$. If $k > 0$, then Theorem 1 and Example 2 together give

$$\mathcal{L}\{\cosh kt\} = \frac{1}{2}\mathcal{L}\{e^{kt}\} + \frac{1}{2}\mathcal{L}\{e^{-kt}\}$$

$$= \frac{1}{2}\left(\frac{1}{s-k} + \frac{1}{s+k}\right);$$

CHAPTER 4:   The Laplace Transform

that is,

$$\mathscr{L}\{\cosh kt\} = \frac{s}{s^2 - k^2} \quad \text{for} \quad s > k > 0. \tag{10}$$

Similarly,

$$\mathscr{L}\{\sinh kt\} = \frac{k}{s^2 - k^2} \quad \text{for} \quad s > k > 0. \tag{11}$$

Because $\cos kt = (e^{ikt} + e^{-ikt})/2$, the formula in (5) (with $a = ik$) yields

$$\mathscr{L}\{\cos kt\} = \frac{1}{2}\left(\frac{1}{s - ik} + \frac{1}{s + ik}\right) = \frac{1}{2} \cdot \frac{2s}{s^2 - (ik)^2},$$

and thus

$$\mathscr{L}\{\cos kt\} = \frac{s}{s^2 + k^2} \quad \text{for} \quad s > 0. \tag{12}$$

(The domain follows from $s > \text{Re}(ik) = 0$.) Similarly,

$$\mathscr{L}\{\sin kt\} = \frac{k}{s^2 + k^2} \quad \text{for} \quad s > 0. \tag{13}$$

EXAMPLE 6 Applying linearity, the formula in (12), and a familiar trigonometric identity, we get

$$\mathscr{L}\{3e^{2t} + 2\sin^2 3t\} = \mathscr{L}\{3e^{2t} + 1 - \cos 6t\}$$

$$= \frac{3}{s - 2} + \frac{1}{s} - \frac{s}{s^2 + 36}$$

$$= \frac{3s^3 + 144s - 72}{s(s - 2)(s^2 + 36)} \quad \text{for} \quad s > 0.$$

According to Theorem 3 of this section, no two different functions that are
both continuous for all $t \geq 0$ can have the same Laplace transform. Thus if $F(s)$
is the transform of some continuous function $f(t)$, then $f(t)$ is uniquely determined.
This observation allows us to make the following definition: If $F(s) = \mathscr{L}\{f(t)\}$, then
we call $f(t)$ the **inverse Laplace transform** of $F(s)$ and write

$$f(t) = \mathscr{L}^{-1}\{F(s)\}. \tag{14}$$

Thus

$$\mathscr{L}^{-1}\left\{\frac{1}{s^3}\right\} = \frac{1}{2}t^2, \qquad \mathscr{L}^{-1}\left\{\frac{1}{s + 2}\right\} = e^{-2t},$$

$$\mathscr{L}^{-1}\left\{\frac{2}{s^2 + 9}\right\} = \frac{2}{3}\sin 3t,$$

and so on.

> ***Notation*** *Functions and their Transforms*
>
> Throughout this chapter we denote functions of t by lowercase letters. The transform of a function will always be denoted by the same letter capitalized. Thus $F(s)$ is the Laplace transform of $f(t)$, and $x(t)$ is the inverse Laplace transform of $X(s)$.

A table of Laplace transforms serves a purpose similar to that of a table of known integrals. The table in Fig. 4.2 lists the transforms derived in this section; many additional transforms can be derived from these few, using various general properties of the Laplace transformation (which we will discuss in subsequent sections).

$f(t)$	$F(s)$
1	$\dfrac{1}{s} \quad (s > 0)$
t	$\dfrac{1}{s^2} \quad (s > 0)$
$t^n \ (n \geqslant 0)$	$\dfrac{n!}{s^{n+1}} \quad (s > 0)$
$t^a \ (a > -1)$	$\dfrac{\Gamma(a+1)}{s^{a+1}} \quad (s > 0)$
e^{at}	$\dfrac{1}{s-a} \quad (s > a)$
$\cos kt$	$\dfrac{s}{s^2+k^2} \quad (s > 0)$
$\sin kt$	$\dfrac{k}{s^2+k^2} \quad (s > 0)$
$\cosh kt$	$\dfrac{s}{s^2-k^2} \quad (s > \lvert k \rvert)$
$\sinh kt$	$\dfrac{k}{s^2-k^2} \quad (s > \lvert k \rvert)$
$u(t-a)$	$\dfrac{e^{-as}}{s} \quad (s > 0)$

FIGURE 4.2 A short table of Laplace transforms.

As we remarked at the beginning of this section, we need to be able to handle certain types of discontinuous functions. The function $f(t)$ is said to be **piecewise continuous** on the bounded interval $a \leqq t \leqq b$ provided that $[a, b]$ can be subdivided into finitely many abutting subintervals so that:

1. f is continuous in the interior of each of these subintervals; and
2. $f(t)$ has a finite one-sided limit as t approaches each endpoint of each such subinterval from its interior.

We say that f is piecewise continuous for $t \geq 0$ if it is piecewise continuous on every bounded subinterval of the nonnegative real axis. Thus a piecewise continuous function has only simple discontinuities (if any) at isolated points. At such points the value of the function experiences a finite jump, as indicated in Fig. 4.3. The **jump in $f(t)$ at the point c** is defined to be $f(c+) - f(c-)$, where

$$f(c+) = \lim_{\epsilon \to 0^+} f(c + \epsilon)$$

and

$$f(c-) = \lim_{\epsilon \to 0^+} f(c - \epsilon).$$

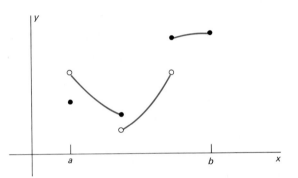

FIGURE 4.3 Graph of a piecewise continuous function; the solid dots indicate values of the function at discontinuities.

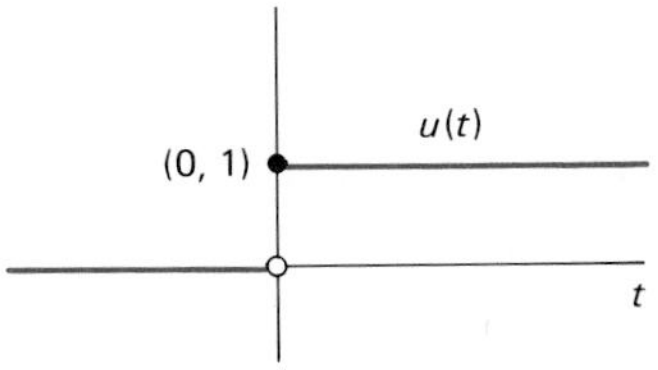

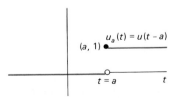

FIGURE 4.4 Graph of the unit step function.

Perhaps the simplest piecewise continuous function is the **unit step function,** shown in Fig. 4.4. It is defined as follows:

$$u(t) = \begin{cases} 0 & \text{for } t < 0, \\ 1 & \text{for } t \geq 0. \end{cases} \tag{15}$$

Because $u(t) = 1$ for $t \geq 0$ and because the Laplace transform involves only the values of a function for $t \geq 0$, we see immediately that

$$\mathcal{L}\{u(t)\} = \frac{1}{s} \quad (s > 0). \tag{16}$$

The unit step function $u_a(t) = u(t - a)$ is shown in Fig. 4.5. Its jump occurs at $t = a$ rather than at $t = 0$; equivalently,

$$u_a(t) = u(t - a) = \begin{cases} 0 & \text{for } t < a, \\ 1 & \text{for } t \geq a. \end{cases} \tag{17}$$

FIGURE 4.5 The unit step function $u_a(t)$—jump at $t = a$.

EXAMPLE 7 Find $\mathcal{L}\{u_a(t)\}$ if $a > 0$.

SOLUTION We begin with the definition of the Laplace transform. We obtain

$$\mathcal{L}\{u_a(t)\} = \int_0^\infty e^{-st} u_a(t)\, dt = \int_a^\infty e^{-st}\, dt = \lim_{b \to \infty} \left[-\frac{e^{-st}}{s} \right]_a^b;$$

consequently,

$$\mathcal{L}\{u_a(t)\} = \frac{e^{-as}}{s} \qquad (s > 0,\ a > 0). \tag{18}$$

It is a familiar fact from calculus that the integral $\int_a^b g(t)\, dt$ exists if g is piecewise continuous on the bounded interval $[a, b]$. Hence if f is piecewise continuous for $t \geq 0$, it follows that the integral $\int_0^b e^{-st}f(t)\, dt$ exists for all $b < \infty$. But in order for $F(s)$—the limit of this last integral as $b \to \infty$—to exist, we need some condition on the rate of growth of $f(t)$ as $t \to \infty$. The function f is said to be of **exponential order** as $t \to \infty$ if there exist nonnegative constants M, c, and T such that

$$|f(t)| \leq M e^{ct} \quad \text{for } t \geq T. \tag{19}$$

Thus a function is of exponential order provided that it grows no more rapidly (as $t \to \infty$) than a constant multiple of some exponential function with a linear exponent. The particular values of M, c, and T are not so important. What is important is that some such values exist so that the condition in (19) is satisfied.

Every polynomial $p(t)$ is of exponential order, with $M = c = 1$ in (19); this follows from the fact that $p(t)/e^t \to 0$ as $t \to \infty$. The function e^{t^2} is an example of one that is *not* of exponential order: Because

$$\lim_{t \to \infty} \frac{e^{t^2}}{e^{ct}} = \lim_{t \to \infty} e^{t^2 - ct} = +\infty,$$

the condition in (19) cannot hold for any (finite) value of M. The condition in (19) merely says that $f(t)/e^{ct}$ is *bounded* for t sufficiently large. In particular, any bounded function—such as $\cos kt$ or $\sin kt$—is of exponential order.

Theorem 2 *Existence of Laplace Transforms*

If the function f is piecewise continuous for $t \geq 0$ and is of exponential order as $t \to \infty$, then its Laplace transform $F(s) = \mathcal{L}\{f(t)\}$ exists. More precisely, if f is piecewise continuous and satisfies the condition in (19), then $F(s)$ exists for all $s > c$.

Proof First we note that we can take $T = 0$ in (19). For by piecewise continuity, $|f(t)|$ is bounded on $[0, T]$. Increasing M in (19) if necessary, we can therefore assume that $|f(t)| \leq M$ if $0 \leq t \leq T$. Because $e^{ct} \geq 1$ for $t \geq 0$, it then follows that $|f(t)| \leq M e^{ct}$ for all $t \geq 0$.

CHAPTER 4: The Laplace Transform

By a standard theorem on convergence of improper integrals—the fact that absolute convergence implies convergence—it suffices for us to prove that the integral $\int_0^\infty |e^{-st}f(t)|\, dt$ exists for $s > c$. To do this, it suffices in turn to show that the value of the integral $\int_0^b |e^{-st}f(t)|\, dt$ remains bounded as $b \to \infty$. But the fact that $|f(t)| \leq Me^{ct}$ for all $t \geq 0$ implies that

$$\int_0^b |e^{-st}f(t)|\, dt \leq \int_0^b |e^{-st}Me^{ct}|\, dt = M \int_0^b e^{-(s-c)t}\, dt$$

$$\leq M \int_0^\infty e^{-(s-c)t}\, dt = \frac{M}{s-c}$$

if $s > c$. This proves Theorem 2. ∎

We have shown, moreover, that

$$|F(s)| \leq \int_0^\infty |e^{-st}f(t)|\, dt \leq \frac{M}{s-c} \tag{20}$$

if $s > c$. When we take limits as $s \to \infty$, we get the following additional result.

Corollary *F(s) for s Large*

If $f(t)$ satisfies the hypotheses of Theorem 2, then

$$\lim_{s \to \infty} F(s) = 0. \tag{21}$$

The condition in (21) severely limits the functions that can be Laplace transforms. For instance, the function $s/(s+1)$ cannot be the Laplace transform of any "reasonable" function because its limit as $s \to \infty$ is 1, not 0. More generally, a rational function—a quotient of two polynomials—can be (and is, as we shall see) a Laplace transform only if the degree of its numerator is less than the degree of its denominator.

On the other hand, the hypotheses of Theorem 2 are sufficient, but not necessary, conditions for existence of the Laplace transform of $f(t)$. For example, the function $f(t) = t^{-1/2} = 1/\sqrt{t}$ fails to be piecewise continuous (at 0), but nevertheless (Example 3 with $a = -\frac{1}{2} > -1$) its Laplace transform

$$\mathscr{L}\{t^{-1/2}\} = \frac{\Gamma(\frac{1}{2})}{s^{1/2}} = \sqrt{\frac{\pi}{s}}$$

both exists and violates the condition in (20), which would imply that $sF(s)$ remains bounded as $s \to \infty$.

The remainder of this chapter will be devoted largely to techniques for solving a differential equation by first finding the Laplace transform of its solution. It is then vital for us to know that this uniquely determines the solution of the differential equation; that is, that the function of s we have found has only one inverse Laplace transform that could be the desired solution. The following theorem is proved in Chapter 6 of Churchill's *Operational Mathematics*, 3rd ed. (New York: McGraw-Hill, 1972).

> **Theorem 3** *Uniqueness of Inverse Laplace Transforms*
>
> Suppose that the functions $f(t)$ and $g(t)$ satisfy the hypotheses of Theorem 2, so that their Laplace transforms $F(s)$ and $G(s)$ exist. If $F(s) = G(s)$ for all $s > c$ (for some c), then $f(t) = g(t)$ wherever f and g are both continuous.

Thus two piecewise continuous functions of exponential order with the same Laplace transform can differ only at their isolated points of discontinuity. This is of no importance in most practical applications, so we may regard inverse Laplace transforms as being essentially unique. In particular, two solutions of a differential equation must both be continuous, and hence must be the same solution if they have the same Laplace transform.

Laplace transforms have an interesting history. The integral in the definition of the Laplace transform probably appeared first in the work of Euler. It is customary in mathematics to name a technique or theorem for the next person after Euler to discover it (or else there would be several hundred different examples of "Euler's theorem"). In this case, the next person was the French mathematician Pierre Simon de Laplace (1749–1827), who employed such integrals in his work on probability theory. The so-called operational techniques for solving differential equations, which are based on Laplace transforms, were not exploited by Laplace. Indeed, they were discovered and popularized by practicing engineers—notably the English electrical engineer Oliver Heaviside (1850–1925). These techniques were successfully and widely applied before they had been rigorously justified, and around the beginning of this century their validity was the subject of considerable controversy.

4.1 PROBLEMS

Apply the definition in (1) to find directly the Laplace transforms of the functions described (by formula or graph) in Problems 1–10.

1. $f(t) = t$ **2.** $f(t) = t^2$

3. $f(t) = e^{3t+1}$ **4.** $f(t) = \cos t$

5. $f(t) = \sinh t$ **6.** $f(t) = \sin^2 t$

7.

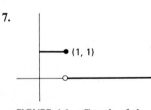

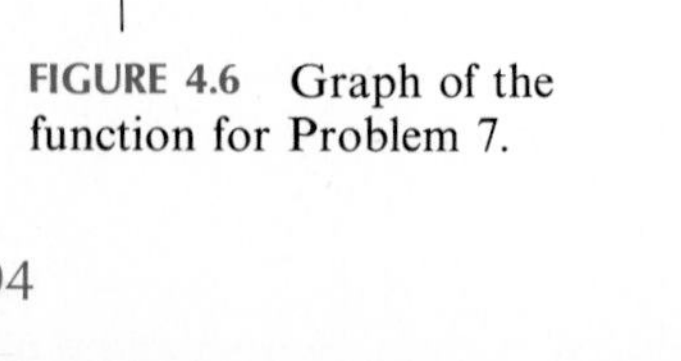

FIGURE 4.6 Graph of the function for Problem 7.

8.

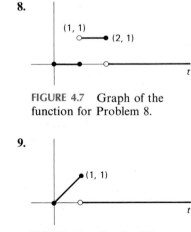

FIGURE 4.7 Graph of the function for Problem 8.

9.

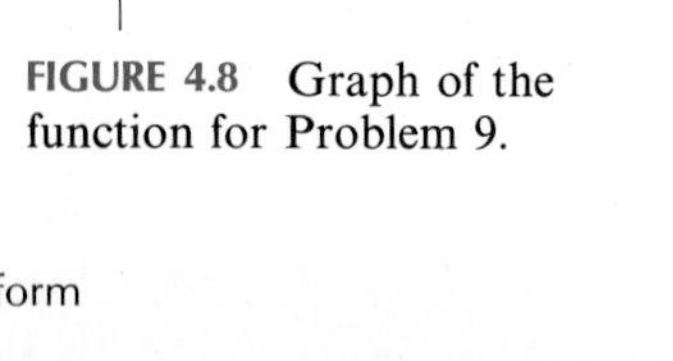

FIGURE 4.8 Graph of the function for Problem 9.

CHAPTER 4:   The Laplace Transform

10.

FIGURE 4.9 Graph of the function for Problem 10.

Use the transforms in Fig. 4.2 to find the Laplace transforms of the functions in Problems 11–22. A preliminary integration by parts may be necessary.

11. $f(t) = \sqrt{t} + 3t$

12. $f(t) = 3t^{5/2} - 4t^3$

13. $f(t) = t - 2e^{3t}$

14. $f(t) = t^{3/2} + e^{-10t}$

15. $f(t) = 1 + \cosh 5t$

16. $f(t) = \sin 2t + \cos 2t$

17. $f(t) = \cos^2 2t$

18. $f(t) = \sin 3t \cos 3t$

19. $f(t) = (1 + t)^3$

20. $f(t) = te^t$

21. $f(t) = t \cos 2t$

22. $f(t) = \sinh^2 3t$

Use the transforms in Fig. 4.2 to find the inverse Laplace transforms of the functions in Problems 23–32.

23. $F(s) = \dfrac{3}{s^4}$

24. $F(s) = s^{-3/2}$

25. $F(s) = \dfrac{1}{s} - \dfrac{2}{s^{5/2}}$

26. $F(s) = \dfrac{1}{s + 5}$

27. $F(s) = \dfrac{3}{s - 4}$

28. $F(s) = \dfrac{3s + 1}{s^2 + 4}$

29. $F(s) = \dfrac{5 - 3s}{s^2 + 9}$

30. $F(s) = \dfrac{9 + s}{4 - s^2}$

31. $F(s) = \dfrac{10s - 3}{25 - s^2}$

32. $F(s) = 2s^{-1}e^{-3s}$

33. Derive the transform of $\sin kt$ by the method used in the text to derive the formula in (12).

34. Derive the transform of $\sinh kt$ by the method used in the text to derive the formula in (10).

35. Use the tabulated integral $\int e^{ax} \cos bx\, dx$ to obtain $\mathscr{L}\{\cos kt\}$ directly from the definition of the Laplace transform.

36. Show that the function $f(t) = \sin(e^{t^2})$ is of exponential order as $t \to \infty$, but that its derivative is not.

37. Let $f(t) = 1$ for $0 \le t \le a$, $f(t) = 0$ for $t > a$ (where $a > 0$). Express f in terms of unit step functions to show that $\mathscr{L}\{f(t)\} = s^{-1}(1 - e^{-as})$.

38. Let $f(t) = 1$ if $a \le t \le b$, $f(t) = 0$ if either $t < a$ or $t > b$ (where $0 < a < b$). Express f in terms of unit step functions to show that $\mathscr{L}\{f(t)\} = s^{-1}(e^{-as} - e^{-bs})$.

39. The unit staircase function is defined as follows:

$$f(t) = n \quad \text{if } n - 1 < t \le n, \, n = 1, 2, 3, \dots.$$

(a) Sketch the graph of f to see why its name is appropriate. (b) Show that

$$f(t) = \sum_{n=0}^{\infty} u(t - n)$$

for all $t > 0$. (c) Assume that the Laplace transform of the infinite series in part (b) can be taken termwise. Apply the geometric series to obtain the result

$$\mathscr{L}\{f(t)\} = \frac{1}{s(1 - e^{-s})}.$$

40. (a) The graph of the function f is shown in Fig. 4.10. Show that f can be written in the form

$$f(t) = \sum_{n=0}^{\infty} (-1)^n u(t - n).$$

(b) Use the method of Problem 39 to show that

$$\mathscr{L}\{f(t)\} = \frac{1}{s(1 + e^{-s})}.$$

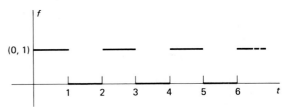

FIGURE 4.10 Graph of the function for Problem 40.

41. The graph of the square wave function $g(t)$ is shown in Fig. 4.11. Express g in terms of the function f of Problem 40 and hence deduce that

$$\mathscr{L}\{g(t)\} = \frac{1 - e^{-s}}{s(1 + e^{-s})} = \frac{1}{s} \tanh \frac{s}{2}.$$

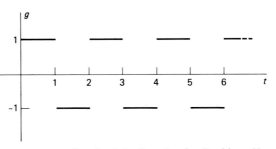

FIGURE 4.11 Graph of the function for Problem 41.

Transformation of Initial Value Problems

We now discuss the application of Laplace transforms to solve a linear differential equation with constant coefficients such as

$$ax''(t) + bx'(t) + cx(t) = f(t), \qquad (1)$$

with given initial conditions $x(0) = x_0$ and $x'(0) = x_0'$. By the linearity of the Laplace transformation, we can transform Eq. (1) by separately taking the Laplace transform of each term in the equation. The transformed equation is

$$a\mathscr{L}\{x''(t)\} + b\mathscr{L}\{x'(t)\} + c\mathscr{L}\{x(t)\} = \mathscr{L}\{f(t)\}; \qquad (2)$$

it involves the transforms of the derivatives x' and x'' of the unknown function $x(t)$. The key to the method is the following theorem, which tells us how to express the transform of the *derivative* of a function in terms of the transform of the function itself.

Theorem 1 *Transforms of Derivatives*

Suppose that the function $f(t)$ is continuous and piecewise smooth for $t \geq 0$ and is of exponential order as $t \to +\infty$, so that there exist nonnegative constants M, c, and T such that

$$|f(t)| \leq Me^{ct} \quad \text{for } t \geq T. \qquad (3)$$

Then $\mathscr{L}\{f'(t)\}$ exists for $s > c$, and

$$\mathscr{L}\{f'(t)\} = s\mathscr{L}\{f(t)\} - f(0) = sF(s) - f(0). \qquad (4)$$

The function f is called **piecewise smooth** on the bounded interval $[a, b]$ if it is piecewise continuous on $[a, b]$ and differentiable except at finitely many points, with $f'(t)$ being piecewise continuous on $[a, b]$. We may assign arbitrary values to $f'(t)$ at the isolated points at which f is not differentiable. We say that f is piecewise smooth for $t \geq 0$ if it is piecewise smooth on every bounded subinterval of the nonnegative real axis. Figure 4.12 indicates how "corners" on the graph of f correspond to discontinuities in its derivative f'.

The main idea of the proof of Theorem 1 is exhibited best by the case when $f'(t)$ is continuous (not merely piecewise continuous) for $t \geq 0$. Then, beginning with the definition of $\mathscr{L}\{f'(t)\}$ and integrating by parts, we get

$$\mathscr{L}\{f'(t)\} = \int_0^\infty e^{-st} f'(t)\, dt$$

$$= \left[e^{-st} f(t)\right]_0^\infty + s \int_0^\infty e^{-st} f(t)\, dt.$$

Because of (3), the integrated term $e^{-st} f(t)$ approaches 0 (when $s > c$) as $t \to +\infty$, and its value at the lower limit $t = 0$ contributes $-f(0)$ to the evaluation of the

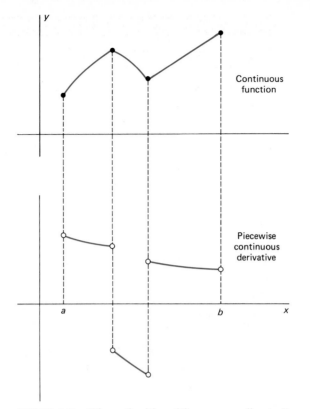

FIGURE 4.12 Discontinuities of f' corresponding to "corners" on the graph of f.

expression above. The integral that remains above is simply $\mathscr{L}\{f(t)\}$; by Theorem 2 of Section 4.1, the integral converges when $s > c$. Thus $\mathscr{L}\{f'(t)\}$ exists when $s > c$, and its value is that given in Eq. (4). We will defer the case in which $f'(t)$ has isolated discontinuities to the end of this section.

In order to transform Eq. (1), we need the transform of the second derivative as well. If we assume that $g(t) = f'(t)$ satisfies the hypotheses of Theorem 1, then that theorem yields

$$\mathscr{L}\{f''(t)\} = \mathscr{L}\{g'(t)\} = s\mathscr{L}\{g(t)\} - g(0)$$
$$= s\mathscr{L}\{f'(t)\} - f'(0) = s[s\mathscr{L}\{f(t)\} - f(0)] - f'(0),$$

and thus

$$\mathscr{L}\{f''(t)\} = s^2 F(s) - sf(0) - f'(0). \tag{5}$$

A repetition of this calculation gives

$$\mathscr{L}\{f'''(t) = s\mathscr{L}\{f''(t)\} - f''(0)$$
$$= s^3 F(s) - s^2 f(0) - sf'(0) - f''(0). \tag{6}$$

After finitely many such steps we obtain the following extension of Theorem 1.

EXAMPLE 1 Solve the initial value problem

$$x'' - x' - 6x = 0; \qquad x(0) = 2, \qquad x'(0) = -1.$$

SOLUTION With the given initial values the formulas in (4) and (5) yield

$$\mathcal{L}\{x'(t)\} = s\mathcal{L}\{x(t)\} - x(0) = sX(s) - 2$$

and

$$\mathcal{L}\{x''(t)\} = s^2 \mathcal{L}\{x(t)\} - sx(0) - x'(0) = s^2 X(s) - 2s + 1,$$

where (according to our convention about notation) $X(s)$ denotes the Laplace transform of the (unknown) function $x(t)$. Hence the transformed equation is

$$[s^2 X(s) - 2s + 1] - [sX(s) - 2] - 6[X(s)] = 0,$$

which we quickly simplify to

$$(s^2 - s - 6)X(s) - 2s + 3 = 0.$$

Thus

$$X(s) = \frac{2s - 3}{s^2 - s - 6} = \frac{2s - 3}{(s - 3)(s + 2)}.$$

By the elementary method of partial fractions, there exist constants A and B such that

$$\frac{2s - 3}{(s - 3)(s + 2)} = \frac{A}{s - 3} + \frac{B}{s + 2},$$

and multiplication of both sides of this equation by $(s - 3)(s + 2)$ yields the identity

$$2s - 3 = A(s + 2) + B(s - 3).$$

If we substitute $s = 3$, we find that $A = \frac{3}{5}$; substitution of $s = -2$ shows that $B = \frac{7}{5}$. Hence

$$\mathcal{L}\{x(t)\} = \frac{\frac{3}{5}}{s - 3} + \frac{\frac{7}{5}}{s + 2}.$$

Because $\mathcal{L}^{-1}\{1/(s - a)\} = e^{at}$, it follows that

$$x(t) = \tfrac{3}{5}e^{3t} + \tfrac{7}{5}e^{-2t}$$

is the solution of the original initial value problem. Note that we did not first find the general solution of the differential equation. The Laplace transform method directly yields the desired particular solution, automatically taking into account— via Theorem 1 and its corollary—the given initial conditions.

EXAMPLE 2 Solve the initial value problem

$$x'' + 4x = \sin 3t; \qquad x(0) = 0 = x'(0).$$

SOLUTION Because both initial values are zero, the formula in (5) yields $\mathscr{L}\{x''(t)\} = s^2 X(s)$. We read the transform of $\sin 3t$ from the table on page 290, and thereby get the transformed equation

$$s^2 X(s) + 4X(s) = \frac{3}{s^2 + 9},$$

and so

$$X(s) = \frac{3}{(s^2 + 4)(s^2 + 9)}.$$

The method of partial fractions calls for

$$\frac{3}{(s^2 + 4)(s^2 + 9)} = \frac{As + B}{s^2 + 4} + \frac{Cs + D}{s^2 + 9}.$$

The fact that there are no terms of odd degree on the left-hand side suggests that we set $A = C = 0$, because nonzero values for A and C would lead to terms of odd degree—but no such terms are present in the numerator on the left. So we replace A and C by zero, then multiply both sides by $(s^2 + 4)(s^2 + 9)$. The result is the identity

$$3 = B(s^2 + 9) + D(s^2 + 4)$$
$$= (B + D)s^2 + (9B + 4D).$$

When we equate coefficients of like powers of s we get the linear equations

$$B + \quad D = 0$$
$$9B + 4D = 3$$

that are readily solved for $B = \frac{3}{5}$ and $D = -\frac{3}{5}$. Hence

$$\mathscr{L}\{x(t)\} = \frac{3}{10} \cdot \frac{2}{s^2 + 4} - \frac{1}{5} \cdot \frac{3}{s^2 + 9}.$$

Because $\mathscr{L}\{\sin 2t\} = 2/(s^2 + 4)$ and $\mathscr{L}\{\sin 3t\} = 3/(s^2 + 9)$, it follows that

$$x(t) = \tfrac{3}{10} \sin 2t - \tfrac{1}{5} \sin 3t.$$

Note that the Laplace transform method again gives the solution directly, without the necessity of first finding the complementary function and a particular solution of the original nonhomogeneous differential equation. Thus nonhomogeneous equations are solved in exactly the same manner as are homogeneous equations.

If we begin with the general linear second order equation

$$ax'' + bx' + cx = f(t) \tag{8}$$

with constant coefficients, the transformed equation is

$$a[s^2 X(s) - sx(0) - x'(0)] + b[sX(s) - x(0)] + cX(s) = F(s). \tag{9}$$

Note that (9) is an *algebraic* equation—indeed, a linear equation—in the "unknown" $X(s)$. This is the source of the power of the Laplace transform method:

> *Linear differential equations are transformed into readily solved algebraic equations.*

If we solve (9) for $X(s)$, we get

$$X(s) = \frac{F(s)}{Z(s)} + \frac{[ax(0)]s + [ax'(0) + bx(0)]}{Z(s)}, \tag{10}$$

where

$$Z(s) = as^2 + bs + c. \tag{11}$$

If Eq. (8) describes the behavior of a physical system under the influence of an external force $f(t)$, then $Z(s)$ depends only on the system itself. For instance, $Z(s) = ms^2 + cs + k$ for the familiar mass-spring-dashpot (mass m, spring constant k, damping constant c) system. Then Eq. (10) presents $\mathcal{L}\{x(t)\}$ as the sum of a term depending only on the external force and one depending only on the initial conditions. In the case of an underdamped system, these two terms are the transforms of the steady periodic solution and the transient solution, respectively. The only potential difficulty in finding these solutions is in finding the inverse Laplace transform of the right-hand side in (10). Much of the remainder of this chapter is devoted to techniques for finding transforms and inverse transforms. In particular, we seek those methods that are sufficiently powerful to enable us to solve problems that—unlike those in Examples 1 and 2—cannot be solved readily by the methods of Chapter 2.

EXAMPLE 3   Show that

$$\mathcal{L}\{te^{at}\} = \frac{1}{(s-a)^2}.$$

SOLUTION   If $f(t) = te^{at}$, then $f(0) = 0$ and $f'(t) = e^{at} + ate^{at}$. Hence Theorem 1 gives

$$\mathcal{L}\{e^{at} + ate^{at}\} = \mathcal{L}\{f'(t)\} = s\mathcal{L}\{f(t)\} = s\mathcal{L}\{te^{at}\}.$$

It follows by linearity of the transform that

$$\mathcal{L}\{e^{at}\} + a\mathcal{L}\{te^{at}\} = s\mathcal{L}\{te^{at}\}.$$

Hence

$$\mathcal{L}\{te^{at}\} = \frac{\mathcal{L}\{e^{at}\}}{s-a} = \frac{1}{(s-a)^2} \tag{12}$$

because $\mathcal{L}\{e^{at}\} = 1/(s-a)$.

EXAMPLE 4 Find $\mathscr{L}\{t \sin kt\}$.

SOLUTION Let $f(t) = t \sin kt$. Then $f(0) = 0$ and

$$f'(t) = \sin kt + kt \cos kt.$$

The derivative involves the new function $t \cos kt$, so we note that $f'(0) = 0$ and differentiate again. The result is

$$f''(t) = 2k \cos kt - k^2 t \sin kt.$$

But $\mathscr{L}\{f''(t)\} = s^2 \mathscr{L}\{f(t)\}$ by the formula in (5) for the transform of the second derivative and $\mathscr{L}\{\cos kt\} = s/(s^2 + k^2)$, so we have

$$\frac{2ks}{s^2 + k^2} - k^2 \mathscr{L}\{t \sin kt\} = s^2 \mathscr{L}\{t \sin kt\}.$$

Finally we solve this equation for

$$\mathscr{L}\{t \sin kt\} = \frac{2ks}{(s^2 + k^2)^2}. \tag{13}$$

This procedure is considerably more pleasant than the alternative of evaluating the integral

$$\mathscr{L}\{t \sin kt\} = \int_0^\infty t e^{-st} \sin kt \, dt.$$

Examples 3 and 4 exploit the fact that, if $f(0) = 0$, then differentiation of f corresponds to multiplication of its transform by s. It is reasonable to expect the inverse operation of integration (antidifferentiation) to correspond to division of the transform by s.

Theorem 2 *Transforms of Integrals*

If $f(t)$ is a piecewise continuous function for $t \geq 0$ and satisfies the condition of exponential order $|f(t)| \leq Me^{ct}$ for $t \geq T$, then

$$\mathscr{L}\left\{\int_0^t f(\tau) \, d\tau\right\} = \frac{1}{s} \mathscr{L}\{f(t)\} = \frac{F(s)}{s} \tag{14}$$

for $s > c$. Equivalently,

$$\mathscr{L}^{-1}\left\{\frac{F(s)}{s}\right\} = \int_0^t f(\tau) \, d\tau. \tag{15}$$

Proof Because $f(t)$ is piecewise continuous, the fundamental theorem of calculus implies that

$$g(t) = \int_0^t f(\tau) \, d\tau$$

is continuous and that $g'(t) = f(t)$ where f is continuous; thus $g(t)$ is continuous and piecewise smooth for $t \geq 0$. Furthermore,

$$|g(t)| \leq \int_0^t |f(\tau)|\, d\tau \leq M \int_0^t e^{c\tau}\, d\tau$$

$$= \frac{M}{c}(e^{ct} - 1) < \frac{M}{c} e^{ct},$$

so $g(t)$ is of exponential order. Hence we can apply Theorem 1 to g; this gives

$$\mathcal{L}\{f(t)\} = \mathcal{L}\{g'(t)\} = s\mathcal{L}\{g(t)\} - g(0).$$

Now $g(0) = 0$, so division by s yields

$$\mathcal{L}\left\{\int_0^t f(\tau)\, d\tau\right\} = \mathcal{L}\{g(t)\} = \frac{\mathcal{L}\{f(t)\}}{s},$$

which completes the proof. ■

EXAMPLE 5 Find the inverse Laplace transform of

$$\frac{1}{s^2(s - a)}.$$

SOLUTION In effect, Eq. (15) means that we can delete a factor of s from the denominator, find the inverse transform of the resulting simpler expression, and finally integrate from 0 to t (to "correct" for the missing factor s). Thus

$$\mathcal{L}^{-1}\left\{\frac{1}{s(s - a)}\right\} = \int_0^t \mathcal{L}^{-1}\left\{\frac{1}{s - a}\right\} d\tau$$

$$= \int_0^t e^{a\tau}\, d\tau = \frac{1}{a}(e^{at} - 1).$$

We now repeat the technique to obtain

$$\mathcal{L}^{-1}\left\{\frac{1}{s^2(s - a)}\right\} = \int_0^t \mathcal{L}^{-1}\left\{\frac{1}{s(s - a)}\right\} d\tau = \int_0^t \frac{1}{a}(e^{a\tau} - 1)\, d\tau$$

$$= \left[\frac{1}{a}\left(\frac{1}{a} e^{a\tau} - \tau\right)\right]_0^t = \frac{1}{a^2}(e^{at} - at - 1).$$

This technique is often a more convenient way than the method of partial fractions for finding an inverse transform of a fraction of the form $p(s)/[s^n q(s)]$.

Proof of Theorem 1 We conclude this section with the proof of Theorem 1 in the general case in which f' is merely piecewise continuous. We need to prove that the limit

$$\lim_{b \to \infty} \int_0^b e^{-st} f'(t)\, dt$$

exists and also need to find its value. With b fixed, let $t_1, t_2, \ldots, t_{k-1}$ be the points interior to the interval $[0, b]$ at which f' is discontinuous. Let $t_0 = 0$ and $t_k = b$.

Then we can integrate by parts on each interval (t_{n-1}, t_n) where f' is continuous. This yields

$$\int_0^b e^{-st}f'(t)\,dt = \sum_{n=1}^k \int_{t_{n-1}}^{t_n} e^{-st}f'(t)\,dt$$

$$= \sum_{n=1}^k \left(\left[e^{-st}f(t) \right]_{t_{n-1}}^{t_n} + s\int_{t_{n-1}}^{t_n} e^{-st}f(t)\,dt \right)$$

$$= -f(0) - \sum_{n=1}^{k-1} e^{(-st_n)}j_f(t_n) + e^{-sb}f(b) + s\int_0^b e^{-st}f(t)\,dt, \quad (16)$$

where

$$j_f(t_n) = f(t_n^+) - f(t_n^-) \tag{17}$$

is the jump in $f(t)$ at $t = t_n$. But because f is continuous, each jump is zero: $j_f(t_n) = 0$ for all n. Moreover, if $b > c$, then $e^{-sb}f(b) \to 0$ as $b \to \infty$. Therefore when we take the limit in (16) as $b \to \infty$, we get the desired result: $\mathcal{L}\{f'(t)\} = s\mathcal{L}\{f(t)\} - f(0)$. ∎

EXTENSION OF THEOREM 1　Suppose that the original function $f(t)$ is itself only piecewise continuous (instead of continuous), with its (finite jump) discontinuities located at the points $t_1, t_2, t_3, \ldots$. Assuming that $\mathcal{L}\{f'(t)\}$ exists, when we take the limit in (16) as $b \to \infty$, we get

$$\mathcal{L}\{f'(t)\} = sF(s) - f(0) - \sum_{n=1}^{\infty} e^{-(st_n)}j_f(t_n). \tag{18}$$

EXAMPLE 6　Let $f(t) = 1 + [\![t]\!]$ be the unit staircase function; its graph is shown in Fig. 4.13. Then $f(0) = 1$, $f'(t) \equiv 0$, and $j_f(n) = 1$ for each $n = 1, 2, 3, \ldots$. Hence

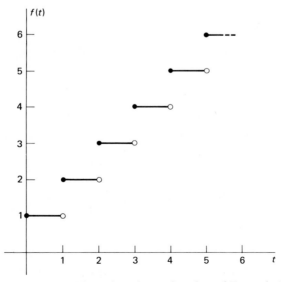

FIGURE 4.13　The unit staircase function of Example 6.

(18) yields

$$0 = sF(s) - 1 - \sum_{n=1}^{\infty} e^{-ns},$$

so the Laplace transform of $f(t)$ is

$$F(s) = \frac{1}{s} \sum_{n=0}^{\infty} e^{-ns} = \frac{1}{s(1 - e^{-s})}.$$

In the last step, we used the formula for the sum of a geometric series,

$$\sum_{n=0}^{\infty} x^n = \frac{1}{1 - x},$$

with $x = e^{-s} < 1$.

4.2 PROBLEMS

Use Laplace transforms to solve the initial value problems in Problems 1–10.

1. $x'' + 4x = 0$; $x(0) = 5$, $x'(0) = 0$
2. $x'' + 9x = 0$; $x(0) = 3$, $x'(0) = 4$
3. $x'' - x' - 2x = 0$; $x(0) = 0$, $x'(0) = 2$
4. $x'' + 8x' + 15x = 0$; $x(0) = 2$, $x'(0) = -3$
5. $x'' + x = \sin 2t$; $x(0) = 0 = x'(0)$
6. $x'' + 4x = \cos t$; $x(0) = 0 = x'(0)$
7. $x'' + x = \cos 3t$; $x(0) = 1$, $x'(0) = 0$
8. $x'' + 9x = 1$; $x(0) = 0 = x'(0)$
9. $x'' + 4x' + 3x = 1$; $x(0) = x'(0) = 0$
10. $x'' + 3x' + 2x = t$; $x(0) = 0$, $x'(0) = 2$

Apply Theorem 2 to find the inverse Laplace transforms of the functions in Problems 11–18.

11. $F(s) = \dfrac{1}{s(s - 3)}$

12. $F(s) = \dfrac{3}{s(s + 5)}$

13. $F(s) = \dfrac{1}{s(s^2 + 4)}$

14. $F(s) = \dfrac{2s + 1}{s(s^2 + 9)}$

15. $F(s) = \dfrac{1}{s^2(s^2 + 1)}$

16. $F(s) = \dfrac{1}{s(s^2 - 9)}$

17. $F(s) = \dfrac{1}{s^2(s^2 - 1)}$

18. $F(s) = \dfrac{1}{s(s + 1)(s + 2)}$

19. Apply Theorem 1 to derive $\mathcal{L}\{\sin kt\}$ from the formula for $\mathcal{L}\{\cos kt\}$.

20. Apply Theorem 1 to derive $\mathcal{L}\{\cosh kt\}$ from the formula for $\mathcal{L}\{\sinh kt\}$.

21. (a) Apply Theorem 1 to show that

$$\mathcal{L}\{t^n e^{at}\} = \frac{n}{s - a} \mathcal{L}\{t^{n-1} e^{at}\}.$$

(b) Deduce that $\mathcal{L}\{t^n e^{at}\} = n!/(s - a)^{n+1}$ for $n = 1, 2, 3, \ldots$.

Apply Theorem 1 as in Example 4 to derive the Laplace transforms in Problems 22–24.

22. $\mathcal{L}\{t \cos kt\} = \dfrac{s^2 - k^2}{(s^2 + k^2)^2}$

23. $\mathcal{L}\{t \sinh kt\} = \dfrac{2ks}{(s^2 - k^2)^2}$

24. $\mathcal{L}\{t \cosh kt\} = \dfrac{s^2 + k^2}{(s^2 - k^2)^2}$

25. Apply the results in Example 4 and Problem 22 to show that

$$\mathcal{L}^{-1}\left\{\frac{1}{(s^2 + k^2)^2}\right\} = \frac{1}{2k^3}(\sin kt - kt \cos kt).$$

Apply the extension of Theorem 1 in (18) to derive the Laplace transforms given in Problems 26–31.

26. $\mathcal{L}\{u(t - a)\} = s^{-1} e^{-as}$ for $a > 0$.

27. If $f(t) = 1$ on the interval $[a, b]$ (where $0 < a < b$) and $f(t) = 0$ otherwise, then $\mathcal{L}\{f(t)\} = s^{-1}(e^{-as} - e^{-bs})$.

28. If $f(t) = (-1)^{[t]}$ is the square wave function shown in Fig. 4.14, then

$$\mathcal{L}\{f(t)\} = \frac{1}{s} \tanh \frac{s}{2}.$$

(*Suggestion:* Use the geometric series.)

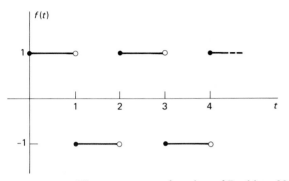

FIGURE 4.14 The square wave function of Problem 28.

29. If $f(t)$ is the unit on-off function shown in Fig. 4.15, then

$$\mathcal{L}\{f(t)\} = \frac{1}{s(1 + e^{-s})}.$$

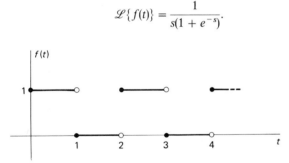

FIGURE 4.15 The off-on function of Problem 29.

30. If $g(t)$ is the triangular wave function shown in Fig. 4.16, then

$$\mathcal{L}\{g(t)\} = \frac{1}{s^2} \tanh \frac{s}{2}.$$

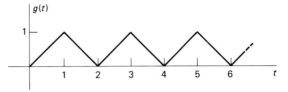

FIGURE 4.16 The triangular wave function of Problem 30.

31. If $f(t)$ is the sawtooth function shown in Fig. 4.17, then

$$\mathcal{L}\{f(t)\} = \frac{1}{s^2} - \frac{e^{-s}}{s(1 - e^{-s})}.$$

(*Suggestion:* Note that $f'(t) \equiv 1$.)

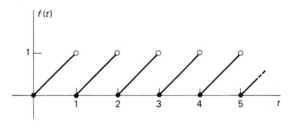

FIGURE 4.17 The sawtooth function of Problem 31.

4.3

Translation and Partial Fractions

As illustrated by Examples 1 and 2 of Section 4.2, the solution of a linear differential equation with constant coefficients can often be reduced to the matter of finding the inverse Laplace transform of a rational function of the form

$$R(s) = \frac{P(s)}{Q(s)} \tag{1}$$

where the degree of $P(s)$ is less than that of $Q(s)$. The technique for finding $\mathscr{L}^{-1}\{R(s)\}$ is based on the same method of partial fractions that we use in elementary calculus to integrate rational functions. The following two rules describe the **partial fraction decomposition** of $R(s)$, in terms of the factorization of the denominator $Q(s)$ into linear factors and irreducible quadratic factors corresponding to the real and complex zeros, respectively, of $Q(s)$.

Rule 1 *Linear Factor Partial Fractions*

The part of the partial fraction decomposition of $R(s)$ corresponding to the linear factor $s - a$ of multiplicity n is a sum of n partial fractions, having the form

$$\frac{A_1}{(s-a)^n} + \frac{A_2}{(s-a)^{n-1}} + \cdots + \frac{A_n}{s-a}, \tag{2}$$

where $A_1, A_2, \ldots,$ and A_n are constants.

Rule 2 *Quadratic Factor Partial Fractions*

The part of the partial fraction decomposition corresponding to the irreducible quadratic factor $(s - a)^2 + b^2$ of multiplicity n is a sum of n partial fractions, having the form

$$\frac{A_1 s + B_1}{[(s-a)^2 + b^2]^n} + \cdots + \frac{A_n s + B_n}{(s-a)^2 + b^2}, \tag{3}$$

where $A_1, A_2, \ldots, A_n, B_1, B_2, \ldots, B_n$ are constants.

Finding $\mathscr{L}^{-1}\{R(s)\}$ involves two steps: First we must find the partial fraction decomposition of $R(s)$, and then we must find the inverse Laplace transform of each of the individual partial fractions of the types that appear in (2) and (3). The latter step is based on the following elementary property of Laplace transforms.

Theorem *Translation on the s-Axis*

If $F(s) = \mathscr{L}\{f(t)\}$ exists for $s > c$, then $\mathscr{L}\{e^{at}f(t)\}$ exists for $s > a + c$, and

$$\mathscr{L}\{e^{at}f(t)\} = F(s - a). \tag{4}$$

Equivalently,

$$\mathscr{L}^{-1}\{F(s - a)\} = e^{at}f(t). \tag{5}$$

Thus the translation $s \to s - a$ in the transform corresponds to multiplication of the original function of t by e^{at}.

CHAPTER 4: The Laplace Transform

Proof   If we simply replace s by $s - a$ in the definition of $F(s) = \mathscr{L}\{f(t)\}$, we obtain

$$F(s - a) = \int_0^\infty e^{-(s-a)t} f(t)\, dt$$

$$= \int_0^\infty e^{-st} [e^{at} f(t)]\, dt = \mathscr{L}\{e^{at} f(t)\}.$$

This is (4), and it is clear that (5) is the same.   ∎

If we apply the translation theorem to the formulas for the Laplace transforms of t^n, $\cos kt$, and $\sin kt$ that we already know—multiplying each of these functions by e^{at} and replacing s by $s - a$ in the transforms—we get the following additions to the table of Fig. 4.2 of Section 4.1.

$f(t)$	$F(s)$		
$e^{at} t^n$	$\dfrac{n!}{(s - a)^{n+1}},$	$s > a$	(6)
$e^{at} \cos kt$	$\dfrac{s - a}{(s - a)^2 + k^2},$	$s > a$	(7)
$e^{at} \sin kt$	$\dfrac{k}{(s - a)^2 + k^2},$	$s > a$	(8)

For ready reference, all the Laplace transforms derived in this chapter are listed in the table of transforms that appears on page 344 at the end of the chapter.

EXAMPLE 1   Consider a mass-spring-dashpot system with $m = \frac{1}{2}$, $k = 17$, and $c = 3$ in mks units. As usual, let $x(t)$ denote the displacement of the mass m from its equilibrium position. If the mass is set in motion with $x(0) = 3$ and $x'(0) = 1$, find $x(t)$ for the resulting damped free oscillations.

SOLUTION   The differential equation of motion is $\frac{1}{2}x'' + 3x' + 17x = 0$, so we need to solve the initial value problem

$$x'' + 6x' + 34x = 0; \qquad x(0) = 3, \qquad x'(0) = 1.$$

We take the Laplace transform of each term of the differential equation; because (obviously) $\mathscr{L}\{0\} = 0$, we get the equation

$$[s^2 X(s) - 3s - 1] + 6[sX(s) - 3] + 34X(s) = 0,$$

which we solve for

$$X(s) = \frac{3s + 19}{s^2 + 6s + 34}$$

$$= 3\,\frac{s + 3}{(s + 3)^2 + 25} + 2\,\frac{5}{(s + 3)^2 + 25}.$$

Applying the formulas in (7) and (8) with $a = -3$ and $k = 5$, we now see that

$$x(t) = e^{-3t}(3\cos 5t + 2\sin 5t).$$

The following example illustrates a useful technique for finding the partial fraction coefficients in the case of nonrepeated linear factors.

EXAMPLE 2   To find the inverse Laplace transform of

$$R(s) = \frac{s^2 + 1}{s^3 - 2s^2 - 8s},$$

we note that the denominator factors as $Q(s) = s(s + 2)(s - 4)$, so,

$$\frac{s^2 + 1}{s^3 - 2s^2 - 8s} = \frac{A}{s} + \frac{B}{s + 2} + \frac{C}{s - 4}.$$

Multiplication of each term of this equation by $Q(s)$ yields

$$s^2 + 1 = A(s + 2)(s - 4) + Bs(s - 4) + Cs(s + 2).$$

When we successively substitute the three zeros $s = 0$, $s = -2$, and $s = 4$ of the denominator $Q(s)$ in this equation, we get the results

$$-8A = 1, \qquad 12B = 5, \quad \text{and} \quad 24C = 17.$$

Thus $A = -\frac{1}{8}$, $B = \frac{5}{12}$, and $C = \frac{17}{24}$, so

$$\frac{s^2 + 1}{s^3 - 2s^2 - 8s} = \frac{-\frac{1}{8}}{s} + \frac{\frac{5}{12}}{s + 2} + \frac{\frac{17}{24}}{s - 4},$$

and hence

$$\mathcal{L}^{-1}\left\{\frac{s^2 + 1}{s^3 - 2s^2 - 8s}\right\} = -\frac{1}{8} + \frac{5}{12} e^{-2t} + \frac{17}{24} e^{4t}.$$

The following example illustrates a differentiation technique for finding the partial fraction coefficients in the case of repeated linear factors.

EXAMPLE 3   Solve the initial value problem

$$y'' + 4y' + 4y = t^2; \qquad y(0) = y'(0) = 0.$$

SOLUTION   The transformed equation is

$$s^2 Y(s) + 4s Y(s) + 4Y(s) = \frac{2}{s^3}.$$

Thus

$$Y(s) = \frac{2}{s^3(s + 2)^2}$$

$$= \frac{A}{s^3} + \frac{B}{s^2} + \frac{C}{s} + \frac{D}{(s + 2)^2} + \frac{E}{s + 2}. \tag{9}$$

To find A, B, and C, we multiply both sides by s^3 to obtain

$$\frac{2}{(s+2)^2} = A + Bs + Cs^2 + s^3 F(s), \tag{10}$$

where $F(s) = D(s+2)^{-2} + E(s+2)^{-1}$ is the sum of the two partial fractions corresponding to $(s+2)^2$. Substitution of $s = 0$ in (10) yields $A = \frac{1}{2}$. To find B and C, we differentiate Eq. (10) twice to obtain

$$\frac{-4}{(s+2)^3} = B + 2Cs + 3s^2 F(s) + s^3 F'(s) \tag{11}$$

and

$$\frac{12}{(s+2)^4} = 2C + 6sF(s) + 6s^2 F'(s) + s^3 F''(s). \tag{12}$$

Now substitution of $s = 0$ in (11) yields $B = -\frac{1}{2}$, and substitution of $s = 0$ in (12) yields $C = \frac{3}{8}$.

To find D and E, we multiply each side in Eq. (9) by $(s+2)^2$ to get

$$\frac{2}{s^3} = D + E(s+2) + (s+2)^2 G(s) \tag{13}$$

where $G(s) = A/s^3 + B/s^2 + C/s$, and then differentiate to obtain

$$\frac{-6}{s^4} = E + 2(s+2)G(s) + (s+2)^2 G'(s). \tag{14}$$

Substitution of $s = -2$ in (13) and (14) now yields $D = -\frac{1}{4}$ and $E = -\frac{3}{8}$. Thus

$$Y(s) = \frac{\frac{1}{2}}{s^3} - \frac{\frac{1}{2}}{s^2} + \frac{\frac{3}{8}}{s} - \frac{\frac{1}{4}}{(s+2)^2} - \frac{\frac{3}{8}}{s+2},$$

so the solution of our initial value problem is

$$y(t) = \tfrac{1}{4}t^2 - \tfrac{1}{2}t + \tfrac{3}{8} - \tfrac{1}{4}te^{-2t} - \tfrac{3}{8}e^{-2t}.$$

The following three examples illustrate techniques for dealing with quadratic factors in partial fraction decompositions.

EXAMPLE 4 Consider the mass-spring-dashpot system as in Example 1, but with initial conditions $x(0) = x'(0) = 0$ and with the imposed external force $F(t) = 15 \sin 2t$. Find the resulting transient motion and steady periodic motion of the mass.

SOLUTION The initial value problem we need to solve is

$$x'' + 6x' + 34x = 30 \sin 2t; \qquad x(0) = x'(0) = 0.$$

The transformed equation is

$$s^2 X(s) + 6sX(s) + 34X(s) = \frac{60}{s^2 + 4}.$$

Hence

$$X(s) = \frac{60}{(s^2 + 4)[(s + 3)^2 + 25]}$$

$$= \frac{As + B}{s^2 + 4} + \frac{Cs + D}{[(s + 3)^2 + 25]}.$$

When we multiply both sides by the common denominator, we get

$$60 = (As + B)[(s + 3)^2 + 25] + (Cs + D)(s^2 + 4). \tag{15}$$

To find A and B, we substitute the zero $s = 2i$ of the quadratic factor $s^2 + 4$ in (15); the result is

$$60 = (2iA + B)[(2i + 3)^2 + 25],$$

which we simplify to

$$60 = (-24A + 30B) + (60A + 12B)i.$$

We now equate real parts and imaginary parts on each side of this equation to obtain the two linear equations

$$-24A + 30B = 60 \quad \text{and} \quad 60A + 12B = 0,$$

which are readily solved for $A = -\frac{10}{29}$ and $B = \frac{50}{29}$.

To find C and D, we substitute the zero $s = -3 + 5i$ of the quadratic factor $[(s + 3)^2 + 25]$ in (15) and get

$$60 = [C(-3 + 5i) + D][(-3 + 5i)^2 + 4],$$

which we simplify to

$$60 = (186C - 12D) + (30C - 30D)i.$$

Again we equate real parts and imaginary parts; this yields the two linear equations

$$186C - 12D = 60 \quad \text{and} \quad 30C - 30D = 0,$$

and we readily find their solution to be $C = D = \frac{10}{29}$.

With these values of the coefficients A, B, C, and D, our partial fraction decomposition of $X(s)$ is

$$X(s) = \frac{1}{29}\left(\frac{-10s + 50}{s^2 + 4} + \frac{10s + 10}{(s + 3)^2 + 25}\right)$$

$$= \frac{1}{29}\left(\frac{-10(s) + (25)(2)}{s^2 + 4} + \frac{10(s + 3) - (4)(5)}{(s + 3)^2 + 25}\right).$$

After we compute the inverse Laplace transforms, we get the position function

$$x(t) = \frac{5}{29}(-2\cos 2t + 5\sin 2t) + \frac{2}{29}e^{-3t}(5\cos 5t - 2\sin 5t).$$

The terms of circular frequency 2 constitute the steady periodic forced oscillations of the mass, while the exponentially damped terms of circular frequency 5 constitute its transient motion. Note that the transient motion is nonzero even though both initial conditions are zero.

CHAPTER 4: The Laplace Transform

The following two inverse Laplace transforms are used in inverting partial fractions that correspond to the case of repeated quadratic factors:

$$\mathcal{L}^{-1}\left\{\frac{s}{(s^2 + k^2)^2}\right\} = \frac{1}{2k}\, t \sin kt \qquad (16)$$

and

$$\mathcal{L}^{-1}\left\{\frac{1}{(s^2 + k^2)^2}\right\} = \frac{1}{2k^3}\, (\sin kt - kt \cos kt). \qquad (17)$$

These follow from Example 4 and Problem 25 of Section 4.2, respectively. Because of the presence in (16) and (17) of the terms $t \sin kt$ and $t \cos kt$, a repeated quadratic factor ordinarily signals the phenomenon of resonance in an undamped mechanical or electrical system.

EXAMPLE 5 Use Laplace transforms to solve the initial value problem

$$x'' + \omega_0^2 x = F_0 \sin \omega t; \qquad x(0) = 0 = x'(0)$$

that determines the undamped forced oscillations of a mass on a spring.

SOLUTION When we transform the differential equation, we get the equation

$$s^2 X(s) + \omega_0^2 X(s) = \frac{F_0 \omega}{s^2 + \omega^2},$$

so

$$X(s) = \frac{F_0 \omega}{(s^2 + \omega^2)(s^2 + \omega_0^2)}$$

If $\omega \neq \omega_0$, we find without difficulty that

$$X(s) = \frac{F_0 \omega}{\omega^2 - \omega_0^2}\left(\frac{1}{s^2 + \omega_0^2} - \frac{1}{s^2 + \omega^2}\right),$$

so it follows that

$$x(t) = \frac{F_0 \omega}{\omega^2 - \omega_0^2}\left(\frac{1}{\omega_0}\sin \omega_0 t - \frac{1}{\omega}\sin \omega t\right).$$

But if $\omega = \omega_0$, we have

$$X(s) = \frac{F_0 \omega_0}{(s^2 + \omega_0^2)^2},$$

so (17) yields the resonance solution

$$x(t) = \frac{F_0}{2\omega_0^2}\, (\sin \omega_0 t - \omega_0 t \cos \omega_0 t).$$

EXAMPLE 6 Solve the initial value problem

$$\begin{cases} y^{(4)} + 2y'' + y = 4te^t; \\ y(0) = y'(0) = y''(0) = y'''(0) = 0. \end{cases}$$

SOLUTION First we observe that $\mathcal{L}\{y''(t)\} = s^2 Y(s)$, $\mathcal{L}\{y^{(4)}(t)\} = s^4 Y(s)$, and $\mathcal{L}\{te^t\} = 1/(s-1)^2$. Hence the transformed equation is

$$(s^4 + 2s^2 + 1)Y(s) = \frac{4}{(s-1)^2}.$$

Thus our problem is to find the inverse transform of

$$Y(s) = \frac{4}{(s-1)^2(s^2+1)^2}$$

$$= \frac{A}{(s-1)^2} + \frac{B}{s-1} + \frac{Cs+D}{(s^2+1)^2} + \frac{Es+F}{s^2+1}. \qquad (18)$$

If we multiply by the common denominator $(s-1)^2(s^2+1)^2$ we get the equation

$$A(s^2+1)^2 + B(s-1)(s^2+1)^2 + Cs(s-1)^2$$
$$+ D(s-1)^2 + Es(s-1)^2(s^2+1) + F(s-1)^2(s^2+1) = 4. \qquad (19)$$

Upon substituting $s = 1$ we find that $A = 1$.

Equation (19) is an identity that holds for all values of s. To find the values of the remaining coefficients, we substitute successively the values $s = 0$, $s = -1$, $s = 2$, $s = -2$, and $s = 3$ in (19). This yields the system

$$\left.\begin{array}{rcl}
-B & + D & + F = 3, \\
-8B - 4C + 4D - 8E + 8F = 0, \\
25B + 2C + D + 10E + 5F = -21, \\
-75B - 18C + 9D - 90E + 45F = -21, \\
200B + 12C + 4D + 120E + 40F = -96
\end{array}\right\} \qquad (20)$$

of five linear equations in B, C, D, E, and F. With the aid of a calculator programmed to solve linear systems, we find that $B = -2$, $C = 2$, $D = 0$, $E = 2$, and $F = 1$.

We now substitute the coefficients we have found in (18), and thus obtain

$$Y(s) = \frac{1}{(s-1)^2} - \frac{2}{s-1} + \frac{2s}{(s^2+1)^2} + \frac{2s+1}{s^2+1}.$$

Recalling (16), the translation property, and the familiar transforms of $\cos t$ and $\sin t$, we see finally that the solution of our initial value problem is

$$y(t) = (t-2)e^t + (t+1)\sin t + 2\cos t.$$

4.3 PROBLEMS

Apply the translation theorem to find the Laplace transforms of the functions in Problems 1–4.

1. $f(t) = t^4 e^{\pi t}$

2. $f(t) = t^{3/2} e^{-4t}$

3. $f(t) = e^{-2t} \sin 3\pi t$

4. $f(t) = e^{-t/2} \cos 2\left(t - \dfrac{\pi}{8}\right)$

Apply the translation theorem to find the inverse Laplace transforms of the functions in Problems 5–10.

5. $F(s) = \dfrac{3}{2s-4}$

6. $F(s) = \dfrac{s-1}{(s+1)^3}$

7. $F(s) = \dfrac{1}{s^2+4s+4}$

8. $F(s) = \dfrac{s+2}{s^2+4s+5}$

9. $F(s) = \dfrac{3s + 5}{s^2 - 6s + 25}$

10. $F(s) = \dfrac{2s - 3}{9s^2 - 12s + 20}$

Use partial fractions to find the inverse Laplace transforms of the functions in Problems 11–22.

11. $F(s) = \dfrac{1}{s^2 - 4}$

12. $F(s) = \dfrac{5s - 6}{s^2 - 3s}$

13. $F(s) = \dfrac{5 - 2s}{s^2 + 7s + 10}$

14. $F(s) = \dfrac{5s - 4}{s^3 - s^2 - 2s}$

15. $F(s) = \dfrac{1}{s^3 - 5s^2}$

16. $F(s) = \dfrac{1}{(s^2 + s - 6)^2}$

17. $F(s) = \dfrac{1}{s^4 - 16}$

18. $F(s) = \dfrac{s^3}{(s - 4)^4}$

19. $F(s) = \dfrac{s^2 - 2s}{s^4 + 5s^2 + 4}$

20. $F(s) = \dfrac{1}{s^4 - 8s^2 + 16}$

21. $F(s) = \dfrac{s^2 + 3}{(s^2 + 2s + 2)^2}$

22. $F(s) = \dfrac{2s^3 - s^2}{(4s^2 - 4s + 5)^2}$

Use the factorization

$$s^4 + 4a^4 = (s^2 - 2as + 2a^2)(s^2 + 2as + 2a^2)$$

to derive the inverse Laplace transforms listed in Problems 23–26.

23. $\mathscr{L}^{-1}\left\{\dfrac{s^3}{s^4 + 4a^4}\right\} = \cosh at \cos at$

24. $\mathscr{L}^{-1}\left\{\dfrac{s}{s^4 + 4a^4}\right\} = \dfrac{1}{2a^2}\sinh at \sin at$

25. $\mathscr{L}^{-1}\left\{\dfrac{s^2}{s^4 + 4a^4}\right\} = \dfrac{1}{2a}(\cosh at \sin at + \sinh at \cos at)$

26. $\mathscr{L}^{-1}\left\{\dfrac{1}{s^4 + 4a^4}\right\} = \dfrac{1}{4a^3}(\cosh at \sin at - \sinh at \cos at)$

Use Laplace transforms to solve the initial value problems in Problems 27–38.

27. $x'' + 6x' + 25x = 0;\quad x(0) = 2,\ x'(0) = 3$

28. $x'' - 6x' + 8x = 2;\quad x(0) = x'(0) = 0$

29. $x'' - 4x = 3t;\quad x(0) = 0 = x'(0)$

30. $x'' + 4x' + 8x = e^{-t};\quad x(0) = x'(0) = 0$

31. $x''' + x'' - 6x' = 0;\quad x(0) = 0,\ x'(0) = x''(0) = 1$

32. $x^{(4)} - x = 0;\quad x(0) = 1,\ x'(0) = x''(0) = x'''(0) = 0$

33. $x^{(4)} + x = 0;\quad x(0) = x'(0) = x''(0) = 0,\ x'''(0) = 1$

34. $x^{(4)} + 13x'' + 36x = 0;\quad x(0) = x''(0) = 0,\ x'(0) = 2,$
$\quad x'''(0) = -13$

35. $x^{(4)} + 8x'' + 16x = 0;\quad x(0) = x'(0) = x''(0) = 0,\ x'''(0) = 1$

36. $x^{(4)} + 2x'' + x = e^{2t};\quad x(0) = x'(0) = x''(0) = x'''(0) = 0$

37. $x'' + 4x' + 13x = te^{-t};\quad x(0) = 0,\ x'(0) = 2$

38. $x'' + 6x' + 18x = \cos 2t;\quad x(0) = 1,\ x'(0) = -1$

Problems 39 and 40 illustrate two types of resonance in a mass-spring-dashpot system with given external force $F(t)$ and with the initial conditions $x(0) = x'(0) = 0$.

39. Suppose that $m = 1$, $k = 9$, $c = 0$, and $F(t) = 6 \cos 3t$. Use the inverse transform given in Eq. (16) to derive the solution $x(t) = t \sin 3t$.

40. Suppose that $m = 1$, $k = 9.04$, $c = 0.4$, and $F(t) = 6e^{-t/5} \cos 3t$. Derive the solution

$$x(t) = te^{-t/5} \sin 3t.$$

Show that the maximum value of the amplitude function $A(t) = te^{-t/5}$ is $A(5) = 5e^{-1}$. Thus (as indicated in Fig. 4.18) the oscillations of the mass increase in amplitude during the first 5 s before being damped out as $t \to \infty$.

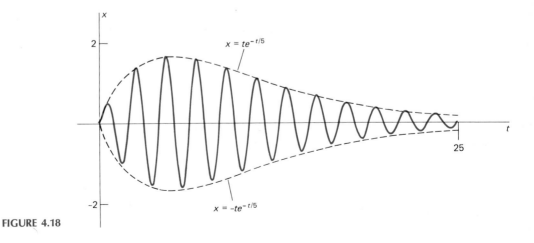

FIGURE 4.18

Derivatives, Integrals, and Products of Transforms

The Laplace transform of the (initially unknown) solution of a differential equation is sometimes recognizable as the product of the transforms of two *known* functions. For example, when we transform the initial value problem

$$x'' + x = \cos t; \qquad x(0) = x'(0) = 0,$$

we get

$$X(s) = \frac{s}{(s^2 + 1)^2} = \frac{s}{s^2 + 1} \cdot \frac{1}{s^2 + 1} = \mathscr{L}\{\cos t\} \cdot \mathscr{L}\{\sin t\}.$$

This strongly suggests that there ought to be a way of combining the two functions $\sin t$ and $\cos t$ to obtain a function $x(t)$ whose transform is the *product* of *their* transforms. But obviously $x(t)$ is *not* simply the product of $\cos t$ and $\sin t$, because

$$\mathscr{L}\{\cos t \sin t\} = \mathscr{L}\{\tfrac{1}{2} \sin 2t\} = \frac{1}{s^2 + 4} \neq \frac{s}{(s^2 + 1)^2}.$$

Thus $\mathscr{L}\{\cos t \sin t\} \neq \mathscr{L}\{\cos t\} \cdot \mathscr{L}\{\sin t\}$.

Theorem 1 tells us that the function

$$h(t) = \int_0^t f(\tau)g(t - \tau)\, d\tau \tag{1}$$

has the desired property that

$$\mathscr{L}\{h(t)\} = H(s) = F(s) \cdot G(s). \tag{2}$$

The new function of t defined as the integral in (1) depends only on f and g and is called the *convolution* of f and g. It is denoted by $f * g$, the idea being that it is a new type of product of f and g, so tailored that its transform is the product of the transforms of f and g.

> **Definition** *The Convolution of Two Functions*
>
> The **convolution** $f * g$ of the piecewise continuous functions f and g is defined for $t \geq 0$ as follows:
>
> $$(f * g)(t) = \int_0^t f(\tau)g(t - \tau)\, d\tau. \tag{3}$$

We also write $f(t) * g(t)$ when convenient. In terms of this convolution product, Theorem 1 of this section says that

$$\mathscr{L}\{f * g\} = \mathscr{L}\{f\} \cdot \mathscr{L}\{g\}.$$

If we make the substitution $u = t - \tau$ in the integral in (3), we see that

$$f(t) * g(t) = \int_0^t f(\tau)g(t - \tau)\,d\tau$$

$$= \int_t^0 f(t - u)g(u)(-du) = \int_0^t g(u)f(t - u)\,du = g(t) * f(t).$$

Thus the convolution is *commutative*: $f * g = g * f$.

EXAMPLE 1   The convolution of $\cos t$ and $\sin t$ is

$$(\cos t) * (\sin t) = \int_0^t \cos \tau \sin (t - \tau)\,d\tau.$$

We apply the trigonometric identity

$$\cos A \sin B = \tfrac{1}{2}[\sin (A + B) - \sin (A - B)]$$

to obtain

$$(\cos t) * (\sin t) = \int_0^t \tfrac{1}{2}[\sin t - \sin (2\tau - t)]\,d\tau$$

$$= \tfrac{1}{2}\Big[\tau \sin t + \tfrac{1}{2} \cos (2\tau - t)\Big]_{\tau = 0}^t;$$

that is,

$$(\cos t) * (\sin t) = \tfrac{1}{2}t \sin t.$$

And we recall from Example 4 of Section 4.2 that the Laplace transform of $\tfrac{1}{2}t \sin t$ is indeed $s/(s^2 + 1)^2$.

The following theorem is proved at the end of this section.

> **Theorem 1**   **The Convolution Property**
>
> Suppose that $f(t)$ and $g(t)$ are piecewise continuous for $t \geq 0$ and that $|f(t)|$ and $|g(t)|$ are bounded by Me^{ct} as $t \to +\infty$. Then the Laplace transform of the convolution $f(t) * g(t)$ exists when $s > c$; moreover,
>
> $$\mathscr{L}\{f(t) * g(t)\} = \mathscr{L}\{f(t)\} \cdot \mathscr{L}\{g(t)\} \tag{4}$$
>
> and
>
> $$\mathscr{L}^{-1}\{F(s) \cdot G(s)\} = f(t) * g(t). \tag{5}$$

Thus we can find the inverse transform of the product $F(s) \cdot G(s)$ provided that we can evaluate the integral

$$\mathscr{L}^{-1}\{F(s) \cdot G(s)\} = \int_0^t f(\tau)g(t - \tau)\,d\tau. \tag{5'}$$

The following example illustrates the fact that convolution often provides a convenient alternative to the use of partial fractions for finding inverse transforms.

EXAMPLE 2 With $f(t) = \sin 2t$ and $g(t) = e^t$, convolution yields

$$\mathscr{L}^{-1}\left\{\frac{2}{(s-1)(s^2+4)}\right\} = (\sin 2t) * e^t$$

$$= \int_0^t e^{t-\tau} \sin 2\tau \, d\tau$$

$$= e^t \int_0^t e^{-\tau} \sin 2\tau \, d\tau$$

$$= e^t\left[\frac{e^{-\tau}}{5}(-\sin 2\tau - 2\cos 2\tau)\right]_0^t,$$

and so

$$\mathscr{L}^{-1}\left\{\frac{2}{(s-1)(s^2+4)}\right\} = \frac{2}{5}e^t - \frac{1}{5}\sin 2t - \frac{2}{5}\cos 2t.$$

DIFFERENTIATION OF TRANSFORMS

According to Theorem 1 of Section 4.2, if $f(0) = 0$ then differentiation of $f(t)$ corresponds to multiplication of its transform $F(s)$ by s. The following theorem (proved at the end of this section) tells us that differentiation of the transform $F(s)$ corresponds to multiplication of the original function $f(t)$ by $-t$.

Theorem 2 *Differentiation of Transforms*

If $f(t)$ is piecewise continuous for $t \geq 0$ and $|f(t)| \leq Me^{ct}$ as $t \to +\infty$, then

$$\mathscr{L}\{-tf(t)\} = F'(s) \tag{6}$$

for $s > c$. Equivalently,

$$f(t) = \mathscr{L}^{-1}\{F(s)\} = -\frac{1}{t}\mathscr{L}^{-1}\{F'(s)\}. \tag{7}$$

Repeated application of (6) yields

$$\mathscr{L}\{t^n f(t)\} = (-1)^n F^{(n)}(s) \tag{8}$$

for $n = 1, 2, 3, \ldots$.

EXAMPLE 3 Find $\mathscr{L}\{t^2 \sin kt\}$.

SOLUTION The formula in (8) gives

$$\mathscr{L}\{t^2 \sin kt\} = (-1)^2 \frac{d^2}{ds^2}\left(\frac{k}{s^2+k^2}\right)$$

$$= \frac{d}{ds}\left[\frac{-2ks}{(s^2+k^2)^2}\right] = \frac{6ks^2 - 2k^3}{(s^2+k^2)^3}. \tag{9}$$

The form of the differentiation property in (7) is often helpful in finding an inverse transform when the *derivative* of the transform is easier to work with than the transform itself.

EXAMPLE 4 Find $\mathscr{L}^{-1}\{\tan^{-1} 1/s\}$.

SOLUTION The derivative of $\tan^{-1}(1/s)$ is a simple rational function, so we apply (7):

$$\mathscr{L}^{-1}\left\{\tan^{-1}\frac{1}{s}\right\} = -\frac{1}{t}\mathscr{L}^{-1}\left\{\frac{d}{ds}\tan^{-1}\frac{1}{s}\right\} = -\frac{1}{t}\mathscr{L}^{-1}\left\{\frac{-1/s^2}{1 + (1/s)^2}\right\}$$

$$= -\frac{1}{t}\mathscr{L}^{-1}\left\{\frac{-1}{s^2 + 1}\right\} = -\frac{1}{t}(-\sin t).$$

Therefore

$$\mathscr{L}^{-1}\left\{\tan^{-1}\frac{1}{s}\right\} = \frac{\sin t}{t}.$$

The formula in (8) can be applied to transform a linear differential equation having polynomial, rather than constant, coefficients. The result will be a differential equation involving the transform; whether this procedure leads to success depends, of course, on whether we can solve the new equation more readily than the old one.

EXAMPLE 5 Let $x(t)$ be the solution of Bessel's equation of order zero,

$$tx'' + x' + tx = 0,$$

such that $x(0) = 1$ and $x'(0) = 0$. Because

$$\mathscr{L}\{x'(t)\} = sX(s) - 1 \quad \text{and} \quad \mathscr{L}\{x''(t)\} = s^2 X(s) - s,$$

and because x and x'' are each multiplied by t, application of (7) yields the transformed equation

$$-\frac{d}{ds}[s^2 X(s) - s] + [sX(s) - 1] - \frac{d}{ds}[X(s)] = 0.$$

The result of differentiation and simplification is the differential equation $(s^2 + 1)X'(s) + sX(s) = 0$. This equation is separable—

$$\frac{X'(s)}{X(s)} = \frac{-s}{s^2 + 1};$$

its general solution is

$$X(s) = \frac{C}{\sqrt{s^2 + 1}}.$$

In Problem 39 we outline the argument that $C = 1$. In Section 3.5 we denoted this solution by $x = J_0(t)$. Because $X(s) = \mathscr{L}\{J_0(t)\}$, it follows that

$$\mathscr{L}\{J_0(t)\} = \frac{1}{\sqrt{s^2 + 1}}. \tag{10}$$

Differentiation of $F(s)$ corresponds to multiplication of $f(t)$ by t (together with a change of sign). It is therefore natural to expect that integration of $F(s)$ will correspond to division of $f(t)$ by t. The following theorem (proved at the end of this section) confirms this, provided that the resulting quotient $f(t)/t$ remains "nice" as $t \to 0$; that is, provided that

$$\lim_{t \to 0^+} \frac{f(t)}{t} \quad \text{exists and is finite.} \tag{11}$$

Theorem 3 *Integration of Transforms*

Suppose that $f(t)$ is piecewise continuous for $t \geq 0$, that $f(t)$ satisfies the condition in (11), and that $|f(t)| \leq Me^{ct}$ as $t \to +\infty$. Then

$$\mathcal{L}\left\{\frac{f(t)}{t}\right\} = \int_s^\infty F(\sigma)\, d\sigma \tag{12}$$

when $s > c$. Equivalently,

$$f(t) = \mathcal{L}^{-1}\{F(s)\} = t\mathcal{L}^{-1}\left\{\int_s^\infty F(\sigma)\, d\sigma\right\}. \tag{13}$$

EXAMPLE 6   Find $\mathcal{L}\{(\sinh t)/t\}$.

SOLUTION   We first verify that

$$\lim_{t=0} \frac{\sinh t}{t} = \lim_{t \to 0} \frac{e^t - e^{-t}}{2t} = \lim_{t \to 0} \frac{e^t + e^{-t}}{2} = 1,$$

with the aid of l'Hôpital's rule. Then the formula in (12), with $f(t) = \sinh t$, yields

$$\mathcal{L}\left\{\frac{\sinh t}{t}\right\} = \int_s^\infty \mathcal{L}\{\sinh t\}\, d\sigma = \int_s^\infty \frac{d\sigma}{\sigma^2 - 1}$$

$$= \frac{1}{2}\int_s^\infty \left(\frac{1}{\sigma - 1} - \frac{1}{\sigma + 1}\right) d\sigma = \frac{1}{2}\left[\ln \frac{\sigma - 1}{\sigma + 1}\right]_s^\infty.$$

Therefore

$$\mathcal{L}\left\{\frac{\sinh t}{t}\right\} = \frac{1}{2}\ln \frac{s + 1}{s - 1},$$

because $\ln 1 = 0$.

The form of the integration property in (13) is often helpful in finding an inverse transform when the indefinite *integral* of the transform is easier to handle than the transform itself.

EXAMPLE 7 Find $\mathscr{L}^{-1}\{2s/(s^2 - 1)^2\}$.

SOLUTION We could use partial fractions, but it is much simpler to apply (13). This gives

$$\mathscr{L}^{-1}\left\{\frac{2s}{(s^2 - 1)^2}\right\} = t\mathscr{L}^{-1}\left\{\int_s^\infty \frac{2\sigma\, d\sigma}{(\sigma^2 - 1)^2}\right\}$$

$$= t\mathscr{L}^{-1}\left\{\left[\frac{-1}{\sigma^2 - 1}\right]_s^\infty\right\} = t\mathscr{L}^{-1}\left\{\frac{1}{s^2 - 1}\right\},$$

and therefore

$$\mathscr{L}^{-1}\left\{\frac{2s}{(s^2 - 1)^2}\right\} = t \sinh t.$$

*PROOFS OF THEOREMS

Proof of Theorem 1 The transforms $F(s)$ and $G(s)$ exist when $s > c$ by Theorem 2 of Section 4.1. The definition of the Laplace transform gives

$$G(s) = \int_0^\infty e^{-su}g(u)\, du$$

$$= \int_{-\tau}^\infty e^{-s(t - \tau)}g(t - \tau)\, dt \qquad (u = t - \tau),$$

and therefore,

$$G(s) = e^{s\tau}\int_0^\infty e^{-st}g(t - \tau)\, dt,$$

because we may *define* $f(t)$ and $g(t)$ to be zero for $t < 0$. Then

$$F(s)G(s) = G(s)\int_0^\infty e^{-s\tau}f(\tau)\, d\tau$$

$$= \int_0^\infty e^{-s\tau}f(\tau)G(s)\, d\tau$$

$$= \int_0^\infty e^{-s\tau}f(\tau)\left(e^{s\tau}\int_0^\infty e^{-st}g(t - \tau)\, dt\right) d\tau$$

$$= \int_0^\infty \left(\int_0^\infty e^{-st}f(\tau)g(t - \tau)\, dt\right) d\tau.$$

Now our hypotheses on f and g imply that the order of integration may be reversed. (The proof of this requires a discussion of uniform convergence of improper integrals, and can be found in Chapter 2 of Churchill's *Operational Mathematics*, 3rd ed. (New York: McGraw-Hill, 1972)). Hence

$$F(s)G(s) = \int_0^\infty \left(\int_0^\infty e^{-st}f(\tau)g(t - \tau)\, d\tau\right) dt$$

$$= \int_0^\infty e^{-st}\left(\int_0^t f(\tau)g(t - \tau)\, d\tau\right) dt$$

$$= \int_0^\infty e^{-st}[f(t) * g(t)]\, dt,$$

and therefore,

$$F(s)G(s) = \mathscr{L}\{f(t) * g(t)\}.$$

We replaced the upper limit of the inner integral by t because $g(t - \tau) = 0$ whenever $\tau > t$. This completes the proof of Theorem 1. ∎

Proof of Theorem 2   Because

$$F(s) = \int_0^\infty e^{-st}f(t)\,dt,$$

differentiation under the integral sign yields

$$F'(s) = \frac{d}{ds}\int_0^\infty e^{-st}f(t)\,dt$$

$$= \int_0^\infty \frac{d}{ds}[e^{-st}f(t)]\,dt = \int_0^\infty e^{-st}[-tf(t)]\,dt;$$

thus

$$F'(s) = \mathscr{L}\{-tf(t)\},$$

which is Eq. (6). We obtain (7) by applying $\mathscr{L}^{-1}$ and then dividing by $-t$. The validity of differentiation under the integral sign depends upon uniform convergence of the resulting integral; this is discussed in Chapter 2 of the book by Churchill just mentioned. ∎

Proof of Theorem 3   By definition,

$$F(\sigma) = \int_0^\infty e^{-\sigma t}f(t)\,dt.$$

So integration of $F(\sigma)$ from s to $+\infty$ gives

$$\int_s^\infty F(\sigma)\,d\sigma = \int_s^\infty \left(\int_0^\infty e^{-\sigma t}f(t)\,dt\right)d\sigma.$$

Under the hypotheses of the theorem, the order of integration may be reversed (cf. Churchill's book again); it follows that

$$\int_s^\infty F(\sigma)\,d\sigma = \int_0^\infty \left(\int_s^\infty e^{-\sigma t}f(t)\,d\sigma\right)dt$$

$$= \int_0^\infty \left[\frac{e^{-\sigma t}}{-t}\right]_{\sigma=s}^\infty f(t)\,dt = \int_0^\infty e^{-st}\frac{f(t)}{t}\,dt$$

$$= \mathscr{L}\left\{\frac{f(t)}{t}\right\}.$$

This verifies (12), and (13) follows upon first applying $\mathscr{L}^{-1}$ and then multiplying by t. ∎

CHAPTER 4:   The Laplace Transform

Find the convolution $f(t) * g(t)$ in each of Problems 1–6.

1. $f(t) = t$, $g(t) = 1$

2. $f(t) = t$, $g(t) = e^{at}$

3. $f(t) = \sin t$, $g(t) = \sin t$

4. $f(t) = t^2$, $g(t) = \cos t$

5. $f(t) = e^{at}$, $g(t) = e^{at}$

6. $f(t) = e^{at}$, $g(t) = e^{bt}$ $(a \neq b)$

Apply the convolution theorem to find the inverse Laplace transforms of the functions in Problems 7–14.

7. $F(s) = \dfrac{1}{s(s-3)}$

8. $F(s) = \dfrac{1}{s(s^2 + 4)}$

9. $F(s) = \dfrac{1}{(s^2 + 9)^2}$

10. $F(s) = \dfrac{1}{s^2(s^2 + k^2)}$

11. $F(s) = \dfrac{s^2}{(s^2 + 4)^2}$

12. $F(s) = \dfrac{1}{s(s^2 + 4s + 5)}$

13. $F(s) = \dfrac{s}{(s-3)(s^2 + 1)}$

14. $F(s) = \dfrac{s}{s^4 + 5s^2 + 4}$

In each of Problems 15–22, apply either Theorem 2 or Theorem 3 to find the Laplace transform of $f(t)$.

15. $f(t) = t \sin 3t$

16. $f(t) = t^2 \cos 2t$

17. $f(t) = te^{2t} \cos 3t$

18. $f(t) = te^{-t} \sin^2 t$

19. $f(t) = \dfrac{\sin t}{t}$

20. $f(t) = \dfrac{1 - \cos 2t}{t}$

21. $f(t) = \dfrac{e^{3t} - 1}{t}$

22. $f(t) = \dfrac{e^t - e^{-t}}{t}$

Find the inverse transforms of the functions in Problems 23–28.

23. $F(s) = \ln \dfrac{s-2}{s+2}$

24. $F(s) = \ln \dfrac{s^2 + 1}{s^2 + 4}$

25. $F(s) = \ln \dfrac{s^2 + 1}{(s+2)(s-3)}$

26. $F(s) = \tan^{-1} \dfrac{3}{s+2}$

27. $F(s) = \ln\left(1 + \dfrac{1}{s^2}\right)$

28. $F(s) = \dfrac{s}{(s^2 + 1)^3}$

In each of Problems 29–34, transform the given differential equation to find a nontrivial solution such that $x(0) = 0$.

29. $tx'' + (t-2)x' + x = 0$

30. $tx'' + (3t - 1)x' + 3x = 0$

31. $tx'' - (4t + 1)x' + 2(2t + 1)x = 0$

32. $tx'' + 2(t - 1)x' - 2x = 0$

33. $tx'' - 2x' + tx = 0$

34. $tx'' + (4t - 2)x' + (13t - 4)x = 0$

35. Apply the convolution theorem to show that

$$\mathcal{L}^{-1}\left\{ \frac{1}{(s-1)\sqrt{s}} \right\} = \frac{2e^t}{\sqrt{\pi}} \int_0^{\sqrt{t}} e^{-u^2} \, du = e^t \operatorname{erf} \sqrt{t}.$$

(*Suggestion:* Substitute $u = \sqrt{\tau}$.)

In each of Problems 36–38, apply the convolution theorem to derive the indicated solution $x(t)$ of the given differential equation with initial conditions $x(0) = x'(0) = 0$.

36. $x'' + 4x = f(t)$; $x(t) = \frac{1}{2} \int_0^t f(t - \tau) \sin 2\tau \, d\tau$

37. $x'' + 2x' + x = f(t)$; $x(t) = \int_0^t \tau e^{-\tau} f(t - \tau) \, d\tau$

38. $x'' + 4x' + 13x = f(t)$; $x(t) = \frac{1}{3} \int_0^t f(t - \tau)e^{-2\tau} \sin 3\tau \, d\tau$

TERMWISE INVERSE TRANSFORMATION OF SERIES   In Chapter 2 of Churchill's *Operational Mathematics*, the following theorem is proved. Suppose that $f(t)$ is continuous for $t \geq 0$, that $f(t)$ is of exponential order as $t \to +\infty$, and that

$$F(s) = \sum_{n=0}^{\infty} \frac{a_n}{s^{n+k+1}}$$

where $0 \leq k < 1$ and the series converges absolutely for $s > c$. Then

$$f(t) = \sum_{n=0}^{\infty} \frac{a_n t^{n+k}}{\Gamma(n + k + 1)}.$$

Apply this result in Problems 39–41.

39. In Example 5 it was shown that

$$\mathcal{L}\{J_0(t)\} = \frac{C}{\sqrt{s^2 + 1}} = \frac{C}{s}\left(1 + \frac{1}{s^2}\right)^{-1/2}.$$

Expand by the binomial series and then compute the inverse transformation term by term to obtain

$$J_0(t) = C \sum_{n=0}^{\infty} \frac{(-1)^n t^{2n}}{2^{2n}(n!)^2}.$$

Finally note that $J_0(0) = 1$ implies that $C = 1$.

40. Expand the function $s^{-1/2}e^{-1/s}$ in powers of s^{-1} to show that

$$\mathcal{L}^{-1}\left\{ \frac{1}{\sqrt{s}} e^{-1/s} \right\} = \frac{1}{\sqrt{\pi t}} \cos 2\sqrt{t}.$$

41. Show that $\mathcal{L}^{-1}\left\{ \dfrac{1}{s} e^{-1/s} \right\} = J_0(2\sqrt{t}).$

Periodic and Piecewise Continuous Forcing Functions

Mathematical models of mechanical or electrical systems often involve functions with discontinuities corresponding to external forces that are turned abruptly on or off. One such simple on-off function is the **unit step function** at $t = a$; its formula is

$$u_a(t) = u(t - a) = \begin{cases} 0 & \text{if } t < a; \\ 1 & \text{if } t \geq a. \end{cases} \tag{1}$$

In Example 7 of Section 4.1 we saw that if $a \geq 0$, then

$$\mathscr{L}\{u(t - a)\} = \frac{e^{-as}}{s}. \tag{2}$$

Because $\mathscr{L}\{u(t)\} = 1/s$, the formula in (2) implies that multiplication of the transform of $u(t)$ by e^{-as} corresponds to the translation $t \to t - a$ in the original independent variable. The following theorem tells us that this fact, when properly interpreted, is a general property of the Laplace transformation.

Theorem 1   *Translation on the t-Axis*

If $\mathscr{L}\{f(t)\} = F(s)$ exists for $s > c$, then

$$\mathscr{L}\{u(t - a)f(t - a)\} = e^{-as}F(s) \tag{3a}$$

and

$$\mathscr{L}^{-1}\{e^{-as}F(s)\} = u(t - a)f(t - a) \tag{3b}$$

when $s > c + a$.

Note that

$$u(t - a)f(t - a) = \begin{cases} 0 & \text{if } t < a, \\ f(t - a) & \text{if } t \geq a. \end{cases} \tag{4}$$

Thus Theorem 1 implies that $\mathscr{L}^{-1}\{e^{-as}F(s)\}$ is the function whose graph for $t \geq a$ is the translation a units to the right of the graph of $f(t)$. Note that the part of the graph of $f(t)$ to the left of $x = 0$ (if any) is cut off and not translated (see Fig. 4.19).

Proof of Theorem 1   From the definition of $\mathscr{L}\{f(t)\}$, we get

$$e^{-as}F(s) = e^{-as} \int_0^\infty e^{-s\tau} f(\tau)\, d\tau = \int_0^\infty e^{-s(\tau + a)} f(\tau)\, d\tau.$$

The substitution $t = \tau + a$ then yields

$$e^{-as}F(s) = \int_a^\infty e^{-st} f(t - a)\, dt.$$

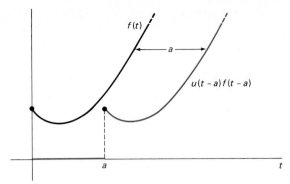

FIGURE 4.19 Translation of $f(t)$ a units to the right.

From (4) we see that this is the same as

$$e^{-as}F(s) = \int_0^\infty e^{-st}u(t-a)f(t-a)\,dt = \mathcal{L}\{u(t-a)f(t-a)\},$$

because $u(t-a)f(t-a) = 0$ for $t < a$. We have therefore completed the proof of Theorem 1. ∎

EXAMPLE 1 With $f(t) = \frac{1}{2}t^2$, Theorem 1 gives

$$\mathcal{L}^{-1}\left\{\frac{e^{-as}}{s^3}\right\} = u(t-a)\frac{1}{2}(t-a)^2 = \begin{cases} 0 & \text{if } t < a, \\ \frac{1}{2}(t-a)^2 & \text{if } t \geq a. \end{cases}$$

EXAMPLE 2 Find $\mathcal{L}\{g(t)\}$ if

$$g(t) = \begin{cases} 0 & \text{if } t < 3, \\ t^2 & \text{if } t \geq 3. \end{cases}$$

SOLUTION Before applying Theorem 1, we must first write $g(t)$ in the form $u(t-3)f(t-3)$. The function $f(t)$ whose translation 3 units to the right agrees (for $t \geq 3$) with $g(t) = t^2$ is $f(t) = (t+3)^2$ because $f(t-3) = t^2$. But then

$$F(s) = \mathcal{L}\{t^2 + 6t + 9\} = \frac{2}{s^3} + \frac{6}{s^2} + \frac{9}{s},$$

so now Theorem 1 yields

$$\mathcal{L}\{g(t)\} = \mathcal{L}\{u(t-3)f(t-3)\}$$

$$= e^{-3s}F(s) = e^{-3s}\left(\frac{2}{s^3} + \frac{6}{s^2} + \frac{9}{s}\right).$$

EXAMPLE 3 Find $\mathcal{L}\{f(t)\}$ if

$$f(t) = \begin{cases} \cos 2t & \text{if } 0 \leq t < 2\pi, \\ 0 & \text{if } t \geq 2\pi. \end{cases}$$

SOLUTION We note first that

$$f(t) = [1 - u(t-2\pi)]\cos 2t$$

$$= \cos 2t - u(t-2\pi)\cos 2(t-2\pi)$$

because of the periodicity of the cosine function. Hence Theorem 1 gives

$$\mathcal{L}\{f(t)\} = \mathcal{L}\{\cos 2t\} - e^{-2\pi s}\mathcal{L}\{\cos 2t\} = \frac{s(1 - e^{-2\pi s})}{s^2 + 4}.$$

EXAMPLE 4 A mass that weighs 32 lb (mass $m = 1$ slug) is attached to the free end of a long light spring that is stretched 1 ft by a force of 4 lb ($k = 4$ lb/ft). The mass is initially at rest in its equilibrium position. Beginning at time $t = 0$ (seconds), an external force $F(t) = \cos 2t$ is applied to the mass, but at time $t = 2\pi$ this force is turned off (abruptly discontinued), and the mass is allowed to continue its motion unimpeded. Find the resulting position function $x(t)$ of the mass.

SOLUTION We need to solve the initial value problem

$$x'' + 4x = f(t); \qquad x(0) = x'(0) = 0$$

where $f(t)$ is the function of Example 3. The transformed equation is

$$(s^2 + 4)X(s) = F(s) = \frac{s(1 - e^{-2\pi s})}{s^2 + 4},$$

so

$$X(s) = \frac{s}{(s^2 + 4)^2} - e^{-2\pi s}\frac{s}{(s^2 + 4)^2}.$$

Because

$$\mathcal{L}^{-1}\left\{\frac{s}{(s^2 + 4)^2}\right\} = \frac{1}{4}t \sin 2t$$

by Eq. (16) of Section 4.3, it follows from Theorem 1 that

$$x(t) = \tfrac{1}{4}t \sin 2t - u(t - 2\pi) \cdot \tfrac{1}{4}(t - 2\pi) \sin 2(t - 2\pi)$$
$$= \tfrac{1}{4}\left[t - u(t - 2\pi) \cdot (t - 2\pi)\right] \sin 2t.$$

If we separate the cases $t < 2\pi$ and $t \geq 2\pi$, we find that the position function may be written in the form

$$x(t) = \begin{cases} \dfrac{1}{4}t \sin 2t & \text{if } t < 2\pi, \\[2mm] \dfrac{\pi}{2} \sin 2t & \text{if } t \geq 2\pi. \end{cases}$$

As indicated by the graph of $x(t)$ shown in Fig. 4.20, the mass oscillates with circular frequency $\omega = 2$ and with linearly increasing amplitude until the force is removed at time $t = 2\pi$. Thereafter, the mass continues to oscillate with the same frequency but with constant amplitude $\pi/2$. The force $F(t) = \cos 2t$ would produce pure resonance if continued indefinitely, but we see that its effect ceases immediately when it is turned off.

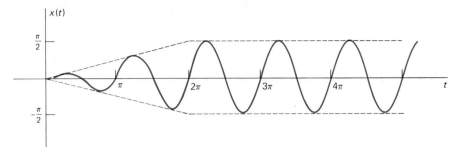

FIGURE 4.20 Graph of the function $x(t)$ of Example 4.

If we were to attack Example 4 with the methods of Chapter 2, we would need to solve one problem for the interval $0 \leq t < 2\pi$, and then a new problem with different initial conditions for the interval $t > 2\pi$. In such a situation the Laplace transform method enjoys the distinct advantage of not requiring the solution of different problems on different intervals.

EXAMPLE 5 Consider the RLC circuit shown in Fig. 4.21, with $R = 110\ \Omega$, $L = 1\ H$, $C = 0.001\ F$, and a battery supplying $E_0 = 90\ V$. Initially, there is no current in the circuit and no charge on the capacitor. At time $t = 0$ the switch is closed and left closed for one second. At time $t = 1$ it is opened and left open. Find the resulting current in the circuit.

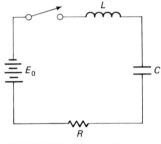

FIGURE 4.21 The series RLC circuit of Example 5.

SOLUTION We recall from Section 2.9 the basic series circuit equation

$$L\frac{di}{dt} + Ri + \frac{1}{C}q = e(t); \tag{5}$$

we use lowercase letters for current and charge and reserve uppercase letters for transforms. With the given circuit elements, Eq. (5) is

$$\frac{di}{dt} + 110i + 1000q = e(t), \tag{6}$$

where $e(t) = 90[1 - u(t - 1)]$, corresponding to the closing and opening of the switch. In Section 2.9 our strategy was to differentiate each side of Eq. (5), then apply the relation

$$i = \frac{dq}{dt} \tag{7}$$

to obtain the second order equation

$$L\frac{d^2i}{dt^2} + R\frac{di}{dt} + \frac{1}{C}i = e'(t).$$

Here we do not use that method, because $e'(t) = 0$ except at $t = 1$, while the jump from $e(t) = 90$ when $t < 1$ to $e(t) = 0$ when $t > 1$ would seem to require that $e'(1) = -\infty$. Thus $e'(t)$ appears to have an "infinite discontinuity" at $t = 1$. This phenomenon will be discussed in Section 4.6. For now, we will simply note that it is an odd situation and circumvent it rather than attempt to deal with it here.

In order to avoid the possible problem at $t = 1$, we observe that the initial value $q(0) = 0$ and the relation in (7) yield, upon integration,

$$q(t) = \int_0^t i(\tau)\, d\tau. \tag{8}$$

We substitute (8) in Eq. (5) to obtain

$$L\frac{di}{dt} + Ri + \frac{1}{C}\int_0^t i(\tau)\, d\tau = e(t). \tag{9}$$

This is the **integrodifferential equation** of a series RLC circuit; it involves both the integral and the derivative of the unknown function $i(t)$. The Laplace transform method works well with such an equation.

In the present example, Eq. (9) is

$$\frac{di}{dt} + 110i + 1000\int_0^t i(\tau)\, d\tau = 90[1 - u(t - 1)]. \tag{10}$$

Because

$$\mathscr{L}\left\{\int_0^t i(\tau)\, d\tau\right\} = \frac{I(s)}{s}$$

by Theorem 2 of Section 4.2 on transforms of integrals, the transformed equation is

$$sI(s) + 110I(s) + 1000\,\frac{I(s)}{s} = \frac{90}{s}(1 - e^{-s}).$$

We solve this equation for $I(s)$ to obtain

$$I(s) = \frac{90(1 - e^{-s})}{s^2 + 110s + 1000}.$$

But

$$\frac{90}{s^2 + 110s + 1000} = \frac{1}{s + 10} - \frac{1}{s + 100},$$

so we have

$$I(s) = \frac{1}{s + 10} - \frac{1}{s + 100} - e^{-s}\left(\frac{1}{s + 10} - \frac{1}{s + 100}\right).$$

We now apply Theorem 1 with $f(t) = e^{-10t} - e^{-100t}$; thus the inverse transform is

$$i(t) = e^{-10t} - e^{-100t} - u(t - 1)[e^{-10(t-1)} - e^{-100(t-1)}].$$

After we separate the cases $t < 1$ and $t \geq 1$, we find that the current in the circuit is given by

$$i(t) = \begin{cases} e^{-10t} - e^{-100t} & \text{if } t < 1, \\ (1 - e^{10})e^{-10t} - (1 - e^{100})e^{-100t} & \text{if } t \geq 1. \end{cases}$$

The portion $e^{-10t} - e^{-100t}$ of the solution would describe the current if the switch were left closed for all t rather than being open for $t \geq 1$.

Periodic forcing functions in practical mechanical or electrical systems often are more complicated than pure sines or cosines. The nonconstant function $f(t)$ defined for $t \geq 0$ is said to be **periodic** if there is a number $p > 0$ such that

$$f(t + p) = f(t) \tag{11}$$

for all $t \geq 0$. The least positive value of p (if any) for which (11) holds is called the **period** of f. Such a function is shown in Fig. 4.22. The following theorem simplifies the computation of the Laplace transform of a periodic function.

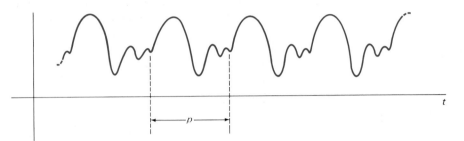

FIGURE 4.22 A function with period p.

Theorem 2   *Transforms of Periodic Functions*

Let $f(t)$ be periodic with period p and piecewise continuous for $t \geq 0$. Then the transform $F(s) = \mathcal{L}\{f(t)\}$ exists for $s > 0$, and is given by

$$F(s) = \frac{1}{1 - e^{-ps}} \int_0^p e^{-st} f(t)\, dt. \tag{12}$$

Proof   The definition of the Laplace transform gives

$$F(s) = \int_0^\infty e^{-st} f(t)\, dt = \sum_{n=0}^\infty \int_{np}^{(n+1)p} e^{-st} f(t)\, dt.$$

The substitution $t = \tau + np$ in the nth integral following the summation sign yields

$$\int_{np}^{(n+1)p} e^{-st} f(t)\, dt = \int_0^p e^{-s(\tau + np)} f(\tau + np)\, d\tau = e^{-nps} \int_0^p e^{-s\tau} f(\tau)\, d\tau$$

because $f(\tau + np) = f(\tau)$ by periodicity. Thus

$$F(s) = \sum_{n=0}^\infty e^{-nps} \int_0^p e^{-s\tau} f(\tau)\, d\tau$$

$$= (1 + e^{-ps} + e^{-2ps} + \cdots) \int_0^p e^{-s\tau} f(\tau)\, d\tau.$$

Consequently,

$$F(s) = \frac{1}{1 - e^{-ps}} \int_0^p e^{-s\tau} f(\tau)\, d\tau.$$

We used the geometric series

$$\frac{1}{1-x} = 1 + x + x^2 + x^3 + \cdots,$$

with $x = e^{-ps} < 1$ (for $s > 0$), to sum the series in the final step. Thus we have derived the formula in (12). ∎

The principal advantage of Theorem 2 is that it enables us to find the Laplace transform of a periodic function without the necessity of an explicit evaluation of an improper integral.

EXAMPLE 6   Figure 4.23 shows the graph of the square wave function $f(t) = (-1)^{[\![at]\!]}$ of period $p = 2a$; $[\![x]\!]$ denotes the greatest integer not exceeding x. By Theorem 2 the Laplace transform of $f(t)$ is

$$F(s) = \frac{1}{1 - e^{-2as}} \int_0^{2a} e^{-st} f(t)\, dt$$

$$= \frac{1}{1 - e^{-2as}} \left(\int_0^a e^{-st}\, dt + \int_a^{2a} (-1)e^{-st}\, dt \right)$$

$$= \frac{1}{1 - e^{-2as}} \left(\left[-\frac{1}{s} e^{-st} \right]_0^a - \left[-\frac{1}{s} e^{-st} \right]_a^{2a} \right)$$

$$= \frac{(1 - e^{-as})^2}{s(1 - e^{-2as})} = \frac{1 - e^{-as}}{s(1 + e^{-as})}.$$

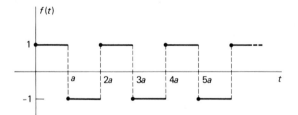

FIGURE 4.23   The square wave function of Example 6.

Therefore,

$$F(s) = \frac{1 - e^{-as}}{s(1 + e^{-as})} \tag{13a}$$

$$= \frac{e^{as/2} - e^{-as/2}}{s(e^{as/2} + e^{-as/2})} = \frac{1}{s} \tanh \frac{as}{2}. \tag{13b}$$

EXAMPLE 7   Figure 4.24 shows the graph of a triangular wave function $g(t)$ of period $p = 2a$. Because the derivative $g'(t)$ is the square wave function of Example 6, it follows from the formula in (13b) and Theorem 2 of Section 4.2 that the transform of this triangular wave function is

$$G(s) = \frac{F(s)}{s} = \frac{1}{s^2} \tanh \frac{as}{2}. \tag{14}$$

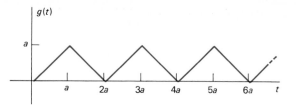

FIGURE 4.24   The triangular wave function of Example 7.

EXAMPLE 8   Consider a mass-spring-dashpot system with $m = 1$, $c = 4$, and $k = 20$ in appropriate units. Suppose that the system is initially at rest at equilibrium $(x(0) = x'(0) = 0)$ and that the mass is acted on by the external force $f(t)$ whose graph is shown in Fig. 4.25: the square wave with amplitude 20 and period 2π. Find the position function $x(t)$.

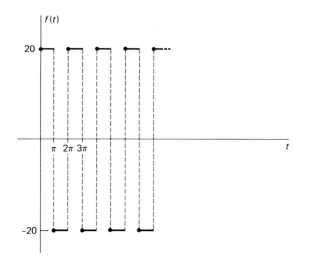

FIGURE 4.25   The external force function of Example 8.

SOLUTION   The initial value problem is

$$x'' + 4x' + 20x = f(t); \qquad x(0) = 0 = x'(0).$$

The transformed equation is

$$s^2 X(s) + 4sX(s) + 20X(s) = F(s). \tag{15}$$

From Example 6 with $a = \pi$ we see that the transform of $f(t)$ is

$$F(s) = \frac{20}{s} \cdot \frac{1 - e^{-\pi s}}{1 + e^{-\pi s}}$$

$$= \frac{20}{s}(1 - e^{-\pi s})(1 - e^{-\pi s} + e^{-2\pi s} - e^{-3\pi s} + \cdots)$$

$$= \frac{20}{s}(1 - 2e^{-\pi s} + 2e^{-2\pi s} - 2e^{-3\pi s} + \cdots),$$

so that

$$F(s) = \frac{20}{s} + \frac{40}{s} \sum_{n=1}^{\infty} (-1)^n e^{-n\pi s}. \tag{16}$$

Substitution of (16) in (15) yields

$$X(s) = \frac{F(s)}{s^2 + 4s + 16}$$

$$= \frac{20}{s[(s + 2)^2 + 16]} + 2 \sum_{n=1}^{\infty} (-1)^n \frac{20e^{-n\pi s}}{s[(s + 2)^2 + 16]}. \tag{17}$$

From the transform in Eq. (8) of Section 4.3, we get

$$\mathcal{L}^{-1} \left\{ \frac{20}{(s + 2)^2 + 16} \right\} = 5e^{-2t} \sin 4t,$$

so by Theorem 2 of Section 4.2 we have

$$g(t) = \mathcal{L}^{-1} \left\{ \frac{20}{s[(s + 2)^2 + 16]} \right\} = \int_0^t 5e^{-2\tau} \sin 4\tau \, d\tau.$$

Using a tabulated formula for $\int e^{at} \sin bt \, dt$, we get

$$g(t) = 1 - e^{-2t}(\cos 4t + \tfrac{1}{2} \sin 4t) = 1 - h(t) \tag{18}$$

where

$$h(t) = e^{-2t}(\cos 4t + \tfrac{1}{2} \sin 4t). \tag{19}$$

Now we apply Theorem 1 to find the inverse transform of the expression in Eq. (17). The result is

$$x(t) = g(t) + 2 \sum_{n=1}^{\infty} (-1)^n u(t - n\pi)g(t - n\pi), \tag{20}$$

and we note that for any fixed value of t the sum above is finite. Moreover,

$$g(t - n\pi) = 1 - e^{-2(t - n\pi)}[\cos 4(t - n\pi) + \tfrac{1}{2} \sin 4(t - n\pi)]$$
$$= 1 - e^{2n\pi}e^{-2t}(\cos 4t + \tfrac{1}{2} \sin 4t).$$

Therefore,

$$g(t - n\pi) = 1 - e^{2n\pi}h(t). \tag{21}$$

Hence if $0 < t < \pi$, then

$$x(t) = 1 - h(t).$$

If $\pi < t < 2\pi$, then

$$x(t) = [1 - h(t)] - 2[1 - e^{2\pi}h(t)]$$
$$= -1 + h(t) - 2h(t)[1 - e^{2\pi}].$$

If $2\pi < t < 3\pi$, then

$$x(t) = [1 - h(t)] - 2[1 - e^{2\pi}h(t)] + 2[1 - e^{4\pi}h(t)]$$
$$= 1 + h(t) - 2h(t)[1 - e^{2\pi} + e^{4\pi}].$$

The general expression for $n\pi < t < (n + 1)\pi$ is

$$x(t) = h(t) + (-1)^n - 2h(t)[1 - e^{2\pi} + \cdots + (-1)^n e^{2n\pi}]$$

$$= h(t) + (-1)^n - 2h(t)\frac{1 + (-1)^n e^{2(n+1)\pi}}{1 + e^{2\pi}}, \tag{22}$$

which we obtained with the aid of the familiar formula for the sum of a finite geometric progression. A rearrangement of (22) finally gives, with the aid of (19),

$$x(t) = \frac{e^{2\pi} - 1}{e^{2\pi} + 1} e^{-2t}\left(\cos 4t + \frac{1}{2}\sin 4t\right) + (-1)^n$$

$$- \frac{(2)(-1)^n e^{2\pi}}{e^{2\pi} + 1} e^{-2(t-n\pi)}\left(\cos 4t + \frac{1}{2}\sin 4t\right) \tag{23}$$

for $n\pi < t < (n + 1)\pi$. The first term in (23) is the transient solution

$$x_{tr}(t) \approx (0.9963)e^{-2t}(\cos 4t + \tfrac{1}{2}\sin 4t) \approx (1.1139)e^{-2t}\cos(4t - 0.4636). \tag{24}$$

The last two terms in (23) give the steady periodic solution x_{sp}. To investigate it, we write $\tau = t - n\pi$ for t in the interval $n\pi < t < (n + 1)\pi$. Then

$$x_{sp}(t) = (-1)^n\left[1 - \frac{2e^{2\pi}}{e^{2\pi} + 1} e^{-2\tau}\left(\cos 4\tau + \frac{1}{2}\sin 4\tau\right)\right]$$

$$\approx (-1)^n[1 - (2.2319)e^{-2\tau}\cos(4\tau - 0.4636)]. \tag{25}$$

Figure 4.26 shows the graph of $x_{sp}(t)$. Its most interesting feature is the appearance of periodically damped oscillations with a frequency *four times* that of the imposed force $f(t)$. In Chapter 8 on Fourier series we will see why a periodic external force sometimes excites oscillations at a higher frequency than the imposed frequency.

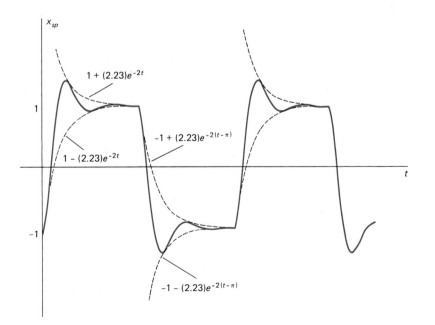

FIGURE 4.26 The steady periodic solution for Example 8; note the "periodically damped" oscillations with frequency four times that of the imposed force.

Find the inverse Laplace transform $f(t)$ of each of the functions given in Problems 1–10, and sketch the graph of $f(t)$.

1. $F(s) = \dfrac{e^{-3s}}{s^2}$

2. $F(s) = \dfrac{e^{-s} - e^{-3s}}{s^2}$

3. $F(s) = \dfrac{e^{-s}}{s+2}$

4. $F(s) = \dfrac{e^{-s} - e^{2-2s}}{s-1}$

5. $F(s) = \dfrac{e^{-\pi s}}{s^2+1}$

6. $F(s) = \dfrac{se^{-s}}{s^2+\pi^2}$

7. $F(s) = \dfrac{1 - e^{-2\pi s}}{s^2+1}$

8. $F(s) = \dfrac{s(1 - e^{-2s})}{s^2+\pi^2}$

9. $F(s) = \dfrac{s(1 + e^{-3s})}{s^2+\pi^2}$

10. $F(s) = \dfrac{2s(e^{-\pi s} - e^{-2\pi s})}{s^2+4}$

Find the Laplace tranform of each of the functions defined in Problems 11–22.

11. $f(t) = 2$ if $0 \le t < 3$; $f(t) = 0$ if $t \ge 3$

12. $f(t) = 3$ if $1 \le t \le 4$; $f(t) = 0$ if $t < 1$ or $t > 4$

13. $f(t) = \sin t$ if $0 \le t \le 2\pi$; $f(t) = 0$ if $t > 2\pi$

14. $f(t) = \cos \pi t$ if $0 \le t \le 2$; $f(t) = 0$ if $t > 2$

15. $f(t) = \sin t$ if $0 \le t \le 3\pi$; $f(t) = 0$ if $t > 3\pi$

16. $f(t) = \sin 2t$ if $\pi \le t \le 2\pi$; $f(t) = 0$ if $t < \pi$ or if $t > 2\pi$

17. $f(t) = \sin \pi t$ if $2 \le t \le 3$; $f(t) = 0$ if $t < 2$ or if $t > 3$

18. $f(t) = \cos \dfrac{\pi t}{2}$ if $3 \le t \le 5$; $f(t) = 0$ if $t < 3$ or $t > 5$

19. $f(t) = 0$ if $t < 1$; $f(t) = t$ if $t \ge 1$

20. $f(t) = t$ if $t \le 1$; $f(t) = 1$ if $t > 1$

21. $f(t) = t$ if $t \le 1$; $f(t) = 2 - t$ if $1 \le t \le 2$; $f(t) = 0$ if $t > 2$

22. $f(t) = t^3$ if $1 \le t \le 2$; $f(t) = 0$ if $t < 1$ or if $t > 2$

23. Apply Theorem 2 with $p = 1$ to verify that $\mathcal{L}\{1\} = 1/s$.

24. Apply Theorem 2 to verify that $\mathcal{L}\{\cos kt\} = s/(s^2 + k^2)$.

25. Apply Theorem 2 to show that the Laplace transform of the square wave function of Fig. 4.27 is

$$\mathcal{L}\{f(t)\} = \dfrac{1}{s(1 + e^{-as})}.$$

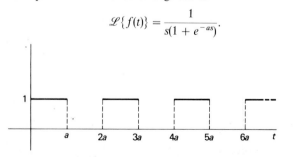

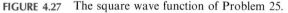

FIGURE 4.27 The square wave function of Problem 25.

26. Apply Theorem 2 to show that the Laplace transform of the sawtooth function $f(t)$ of Fig. 4.28 is

$$F(s) = \dfrac{1}{as^2} - \dfrac{e^{-as}}{s(1 - e^{-as})}.$$

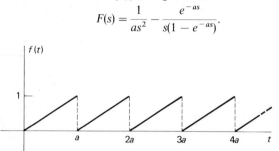

FIGURE 4.28 The sawtooth function of Problem 26.

27. Let $g(t)$ be the staircase function of Fig. 4.29. Show that $g(t) = (t/a) - f(t)$, where $f(t)$ is the sawtooth function of Fig. 4.28, and hence deduce that

$$\mathcal{L}\{g(t)\} = \dfrac{e^{-as}}{s(1 - e^{-as})}.$$

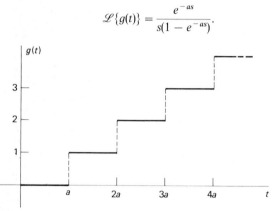

FIGURE 4.29 The staircase function of Problem 27.

28. Suppose that $f(t)$ is a periodic function of period $2a$ with $f(t) = t$ if $0 \le t < a$ and $f(0) = 0$ if $a \le t < 2a$. Find $\mathcal{L}\{f(t)\}$.

29. Suppose that $f(t)$ is the half-wave rectification of $\sin kt$, shown in Fig. 4.30. Show that

$$\mathcal{L}\{f(t)\} = \dfrac{k}{(s^2 + k^2)(1 - e^{-\pi s/k})}.$$

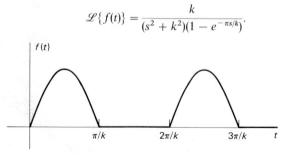

FIGURE 4.30 The half-wave rectification of $\sin kt$.

30. Let $g(t) = u(t - \pi/k)f(t - \pi/k)$, where $f(t)$ is the function of Problem 29. Note that $h(t) = f(t) + g(t)$ is the full-wave rectification of $\sin kt$ shown in Fig. 4.31. Hence deduce from Problem 29 that

$$\mathcal{L}\{h(t)\} = \frac{k}{s^2 + k^2} \coth \frac{\pi s}{2k}.$$

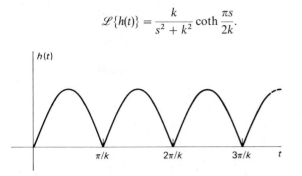

FIGURE 4.31 The full-wave rectification of $\sin kt$.

In each of Problems 31–35, the values of mass m, spring constant k, dashpot resistance c, and force $f(t)$ are given for a mass-spring-dashpot system with external forcing function. Solve the initial value problem

$$\begin{cases} mx'' + cx' + kx = f(t); \\ \qquad x(0) = 0 = x'(0) \end{cases}$$

with the given data.

31. $m = 1$, $k = 4$, $c = 0$; $f(t) = 1$ if $0 \leq t < \pi$, $f(t) = 0$ if $t > \pi$

32. $m = 1$, $k = 4$, $c = 5$; $f(t) = 1$ if $0 \leq t < 2$, $f(t) = 0$ if $t > 2$

33. $m = 1$, $k = 9$, $c = 0$; $f(t) = \sin t$ if $0 \leq t \leq 2\pi$, $f(t) = 0$ if $t > 2\pi$

34. $m = 1$, $k = 1$, $c = 0$; $f(t) = t$ if $0 \leq t < 1$, $f(t) = 0$ if $t \geq 1$

35. $m = 1$, $k = 4$, $c = 4$; $f(t) = t$ if $0 \leq t < 2$, $f(t) = 0$ if $t \geq 2$

In each of Problems 36–40, the values of the elements of an RLC circuit are given. Solve the initial value problem

$$\begin{cases} L\dfrac{di}{dt} + Ri + \dfrac{1}{C}\displaystyle\int_0^t i(\tau)\, d\tau = e(t); \\ \qquad\qquad\qquad i(0) = 0, \end{cases}$$

with the given impressed voltage $e(t)$.

36. $L = 0$, $R = 100$, $C = 10^{-3}$; $e(t) = 100$ if $0 \leq t < 1$, $e(t) = 0$ if $t \geq 1$

37. $L = 1$, $R = 0$, $C = 10^{-4}$; $e(t) = 100$ if $0 \leq t < 2\pi$, $e(t) = 0$ if $t \geq 2\pi$

38. $L = 1$, $R = 0$, $C = 10^{-4}$; $e(t) = 100 \sin 10t$ if $0 \leq t < \pi$, $e(t) = 0$ if $t \geq \pi$

39. $L = 1$, $R = 150$, $C = 2 \times 10^{-4}$; $e(t) = 100t$ if $0 \leq t < 1$, $e(t) = 0$ if $t > 1$

40. $L = 1$, $R = 100$, $C = 4 \times 10^{-4}$; $e(t) = 50t$ if $0 \leq t < 1$, $e(t) = 0$ if $t > 1$

In each of Problems 41 and 42, a mass-spring-dashpot system with external force $f(t)$ is described. Under the assumption that $x(0) = 0 = x'(0)$, use the method of Example 8 to find the transient and steady periodic motions of the mass.

41. $m = 1$, $k = 4$, $c = 0$; $f(t)$ is a square wave function with amplitude 4 and period 2π.

42. $m = 1$, $k = 10$, $c = 2$; $f(t)$ is a square wave function with amplitude 10 and period 2π.

---

*4.6

Impulses and Delta Functions

Consider a force $f(t)$ that acts only during a very short time interval $a \leq t \leq b$, with $f(t) = 0$ outside this interval. A typical example would be the *impulsive force* of a bat striking a ball—the impact is almost instantaneous. A quick surge of voltage (resulting from a lightning bolt, for instance) is an analogous electrical phenomenon. In such a situation it often happens that the principal effect of the force depends only on the value of the integral

$$p = \int_a^b f(t)\, dt \tag{1}$$

and does not depend otherwise on precisely how $f(t)$ varies with t. The number p in (1) is called the **impulse** of the force $f(t)$ over the interval $[a, b]$.

In the case of a force $f(t)$ that acts on a particle of mass m in linear motion, integration of Newton's law

$$f(t) = mv'(t) = \frac{d}{dt}[mv(t)]$$

yields

$$p = \int_a^b \frac{d}{dt}[mv(t)]\, dt = mv(b) - mv(a). \tag{2}$$

Thus the impulse of the force is equal to the change in momentum of the particle. So if change in momentum is the only effect with which we are concerned, we need know only the impulse of the force; we need know neither the precise function $f(t)$ nor even the precise time interval during which it acts. This is fortunate, because in a situation such as that of a batted ball, we are unlikely to have such detailed information about the impulsive force that acts on the ball.

Our strategy for handling such a situation is to set up a reasonable mathematical model in which the unknown force $f(t)$ is replaced with a simple and explicit force that has the same impulse. Suppose for simplicity that $f(t)$ has impulse 1 and acts during some brief time interval beginning at time $t = a \geq 0$. Then we can select a fixed number $\epsilon > 0$ that approximates the length of this time interval and replace $f(t)$ with the specific function

$$d_{a,\epsilon}(t) = \begin{cases} \dfrac{1}{\epsilon} & \text{if } a \leq t < a + \epsilon, \\ 0 & \text{otherwise.} \end{cases} \tag{3}$$

This is a function of t, with a and ϵ being parameters that specify the time interval $[a, a + \epsilon]$. If $b \geq a + \epsilon$, then we see (as in Fig. 4.32) that the impulse of $d_{a,\epsilon}(t)$ over $[a, b]$ is

$$p = \int_a^b d_{a,\epsilon}(t)\, dt = \int_a^{a+\epsilon} \frac{1}{\epsilon}\, dt = 1.$$

Thus $d_{a,\epsilon}(t)$ has a *unit* impulse, whatever the number $\epsilon > 0$ may be. Essentially the same computation gives

$$\int_0^\infty d_{a,\epsilon}(t)\, dt = 1. \tag{4}$$

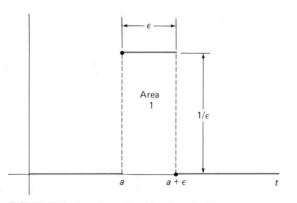

FIGURE 4.32 The impulse function $d_{a,\epsilon}(t)$.

Because the precise time interval during which the force acts seems unimportant, it is tempting to think of an *instantaneous impulse* that occurs precisely at the instant $t = a$. We might try to model such an instantaneous unit impulse by taking the limit as $\epsilon \to 0$, thereby defining

$$\delta_a(t) = \lim_{\epsilon \to 0} d_{a,\epsilon}(t) \tag{5}$$

where $a \geq 0$. If we could also take the limit under the integral sign in (4), then it would follow that

$$\int_0^\infty \delta_a(t) \, dt = 1. \tag{6}$$

But the limit in (5) gives

$$\delta_a(t) = \begin{cases} +\infty & \text{if } t = a, \\ 0 & \text{if } t \neq a. \end{cases} \tag{7}$$

Obviously, no function can satisfy both (6) and (7)—if a function is zero except at a single point, then its integral is not 1 but zero. Nevertheless, the symbol $\delta_a(t)$ is very useful. However it is interpreted, it is called the **Dirac delta function** at a after the British theoretical physicist P. A. M. Dirac (1902–1982), who in the early 1930s introduced a "function" allegedly enjoying the properties in (6) and (7).

The following computation motivates the meaning that we will attach here to the symbol $\delta_a(t)$. If $g(t)$ is a continuous function, then the mean value theorem for integrals implies that

$$\int_a^{a+\epsilon} g(t) \, dt = \epsilon g(\bar{t})$$

for some point $\bar{t}$ in $[a, a + \epsilon]$. It follows that

$$\lim_{\epsilon \to 0} \int_0^\infty g(t) d_{a,\epsilon}(t) \, dt = \lim_{\epsilon \to 0} \int_a^{a+\epsilon} g(t) \cdot \frac{1}{\epsilon} \, dt \tag{8}$$

$$= \lim_{\epsilon \to 0} g(\bar{t}) = g(a)$$

by continuity of g at $t = a$. If $\delta_a(t)$ *were* a function in the strictest sense of the definition, and if we could interchange the limit and the integral in (8), we therefore could conclude that

$$\int_0^\infty g(t) \delta_a(t) \, dt = g(a). \tag{9}$$

We take the equation in (9) as the *definition* (!) of the symbol $\delta_a(t)$. Although we call it the delta function, it is not a genuine function; instead, it specifies the *operation*

$$\int_0^\infty \dots \delta_a(t) \, dt$$

which—when applied to a continuous function $g(t)$—sifts out or selects the value $g(a)$ of this function at the point $a \geq 0$. This idea is shown schematically in Fig. 4.33. Note that we will use the symbol $\delta_a(t)$ only in the context of integrals such as that in (9), or when it will appear subsequently in such an integral.

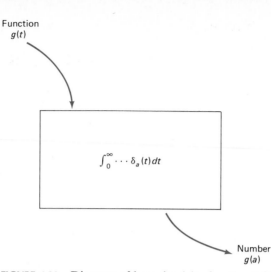

Function
g(t)

$$\int_0^\infty \cdots \delta_a(t)\,dt$$

Number
g(a)

FIGURE 4.33 Diagram of how the delta function "sifts out" the value $g(a)$.

For instance, if we take $g(t) = e^{-st}$ in (9), the result is

$$\int_0^\infty e^{-st}\delta_a(t)\,dt = e^{-as}. \tag{10}$$

We therefore *define* the Laplace transform of the delta function to be

$$\mathscr{L}\{\delta_a(t)\} = e^{-as} \qquad (a \geq 0). \tag{11}$$

If we write

$$\delta(t) = \delta_0(t) \quad \text{and} \quad \delta(t-a) = \delta_a(t), \tag{12}$$

then (11) with $a = 0$ gives

$$\mathscr{L}\{\delta(t)\} = 1. \tag{13}$$

Note that if $\delta(t)$ were an actual function, then (13) would contradict the corollary to Theorem 2 of Section 4.1. There is no problem here; $\delta(t)$ is not a function, and (13) is our *definition* of $\mathscr{L}\{\delta(t)\}$.

Now, finally, suppose that we are given a mechanical system whose response $x(t)$ to the external force $f(t)$ is determined by the differential equation

$$Ax'' + Bx' + Cx = f(t). \tag{14}$$

To investigate the response of this system to a unit impulse at the instant $t - a$, it seems reasonable to replace $f(t)$ with $\delta_a(t)$ and begin with the equation

$$Ax'' + Bx' + Cx = \delta_a(t). \tag{15}$$

But what is meant by a solution of such an equation? We will call $x(t)$ a solution of (15) provided that

$$x(t) = \lim_{\epsilon \to 0} x_\epsilon(t), \tag{16}$$

where $x_\epsilon(t)$ is a solution of

$$Ax'' + Bx' + Cx = d_{a,\epsilon}(t). \tag{17}$$

Because

$$d_{a,\epsilon}(t) = \frac{1}{\epsilon}\left[u_a(t) - u_{a+\epsilon}(t)\right] \tag{18}$$

is an ordinary function, the equation in (17) makes sense. For simplicity suppose the initial conditions to be $x(0) = x'(0) = 0$. When we transform Eq. (17), writing $X_\epsilon = \mathcal{L}\{x_\epsilon\}$, we get the equation

$$(As^2 + Bs + C)X_\epsilon(s) = \frac{1}{\epsilon}\left(\frac{e^{-as}}{s} - \frac{e^{-(a+\epsilon)s}}{s}\right)$$

$$= (e^{-as})\frac{1 - e^{-s\epsilon}}{s\epsilon}.$$

If we take the limit in the last equation as $\epsilon \to 0$, and note that

$$\lim_{\epsilon \to 0} \frac{1 - e^{-s\epsilon}}{s\epsilon} = 1$$

by l'Hôpital's rule, we get the equation

$$(As^2 + Bs + C)X_\epsilon(s) = e^{-as}. \tag{19}$$

Note that this is precisely the same result that we would obtain if we transformed the equation in (15), using the fact that $\mathcal{L}\{\delta_a(t)\} = e^{-as}$.

On this basis it is reasonable to solve a differential equation involving a delta function by employing the Laplace transform method exactly as if $\delta_a(t)$ were an ordinary function. It is important to verify that the solution so obtained agrees with the one defined in (16), but this depends upon a highly technical analysis of the limiting procedures involved; we consider it beyond the scope of the present discussion. The formal method is valid in all the examples of this section and will produce correct results in the subsequent problem set.

EXAMPLE 1 A mass $m = 1$ is attached to a spring with constant $k = 4$; there is no dashpot. The mass is released from rest with $x(0) = 3$. At the instant $t = 2\pi$ the mass is struck with a hammer, providing an impulse $p = 8$. Determine the motion of the mass.

SOLUTION We need to solve the initial value problem

$$\begin{cases} x'' + 4x = 8\delta_{2\pi}(t); \\ x(0) = 3, \quad x'(0) = 0. \end{cases}$$

We apply the Laplace transform to get

$$s^2 X(s) - 3s + 4X(s) = 8e^{-2\pi s},$$

so

$$X(s) = \frac{3s}{s^2 + 4} + \frac{8e^{-2\pi s}}{s^2 + 4}.$$

Recalling the transforms of sine and cosine, as well as the theorem on translations on the t-axis (Theorem 1 of Section 4.5), we see that the inverse transform is

$$x(t) = 3 \cos 2t + 4u(t - 2\pi) \sin 2(t - 2\pi)$$
$$= 3 \cos 2t + 4u_{2\pi}(t) \sin 2t.$$

Because $3 \cos 2t + 4 \sin 2t = 5 \cos(2t - \alpha)$ with $\alpha = \tan^{-1}(4/3) \approx 0.9273$, separation of the cases $t < 2\pi$ and $t > 2\pi$ gives

$$x(t) = \begin{cases} 3 \cos 2t & \text{if } t \leq 2\pi, \\ 5 \cos(2t - 0.9273) & \text{if } t \geq 2\pi. \end{cases}$$

The resulting motion is shown in Fig. 4.34. Note that the impulse at $t = 2\pi$ results in a discontinuity in the velocity at $t = 2\pi$, as it instantaneously increases the amplitude of the oscillations of the mass from 3 to 5.

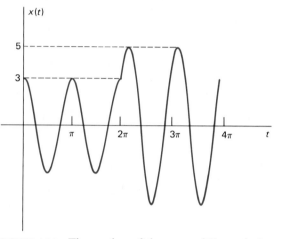

FIGURE 4.34   The motion of the mass of Example 1.

It is useful to regard the delta function $\delta_a(t)$ as the derivative of the unit step function $u_a(t)$. To see why this is reasonable, consider the continuous approximation $u_{a,\epsilon}(t)$ to $u_a(t)$ shown in Fig. 4.35. We readily verify that

$$\frac{d}{dt} u_{a,\epsilon}(t) = d_{a,\epsilon}(t).$$

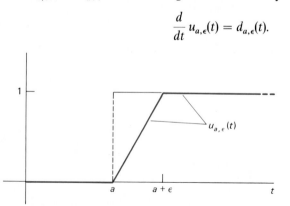

FIGURE 4.35   Approximation of $u_a(t)$ by $u_{a,\epsilon}(t)$.

CHAPTER 4:   The Laplace Transform

Because

$$u_a(t) = \lim_{\epsilon \to 0} u_{a,\epsilon}(t)$$

and

$$\delta_a(t) = \lim_{\epsilon \to 0} d_{a,\epsilon}(t),$$

an interchange of limits and derivatives yields

$$\frac{d}{dt} u_a(t) = \lim_{\epsilon \to 0} \frac{d}{dt} u_{a,\epsilon}(t) = \lim_{\epsilon \to 0} d_{a,\epsilon}(t),$$

and therefore

$$\frac{d}{dt} u_a(t) = \delta_a(t) = \delta(t - a). \tag{20}$$

We may regard this as the *formal definition* of the derivative of the unit step function, although $u_a(t)$ is not differentiable in the ordinary sense at $t = a$.

EXAMPLE 2 We return to the RLC circuit of Example 5 of Section 4.5, with $R = 110\ \Omega$, $L = 1$ H, $C = 0.001$ F, and a battery supplying $E_0 = 90$ V. Suppose that the circuit is initially passive—no current and no charge. At time $t = 0$ the switch is closed, and at time $t = 1$ it is opened and left open. Find the resulting current $i(t)$ in the circuit.

SOLUTION In Section 4.5 we circumvented the discontinuity in the voltage by employing the integrodifferential form of the circuit equation. Now that delta functions are available, we may begin with the ordinary circuit equation

$$Li'' + Ri' + \frac{1}{C} i = e'(t).$$

In this example we have

$$e(t) = 90 - 90u(t - 1) = 90 - 90u_1(t),$$

so $e'(t) = -90\delta(t - 1)$ by Eq. (20). Hence we want to solve the initial value problem

$$i'' + 110i' + 1000i = -90\delta(t - 1); \tag{21}$$

$$i(0) = 0, \qquad i'(0) = 90.$$

The fact that $i'(0) = 90$ comes from substitution of $t = 0$ in the equation

$$Li'(t) + Ri(t) + \frac{1}{C} q(t) = e(t)$$

with the numerical values $i(0) = q(0) = 0$ and $e(0) = 90$.

When we transform the problem in (21), we get the equation

$$s^2 I(s) - 90 + 110sI(s) + 1000I(s) = -90e^{-s}.$$

Hence

$$I(s) = \frac{90(1 - e^{-s})}{s^2 + 110s + 1000}.$$

This is precisely the same transform $I(s)$ we found in Example 5 of Section 4.5, so inversion of $I(s)$ yields the same solution $i(t)$ as recorded there.

EXAMPLE 3 Consider a mass on a spring with $m = k = 1$ and $x(0) = x'(0) = 0$. At each of the instants $t = 0, \pi, 2\pi, \ldots, n\pi, \ldots$, the mass is struck a hammer blow with a unit impulse. Determine the resulting motion.

SOLUTION We need to solve the initial value problem

$$x'' + x = \sum_{n=0}^{\infty} \delta_{n\pi}(t);$$

$$x(0) = 0 = x'(0).$$

Because $\mathcal{L}\{\delta_{n\pi}(t)\} = e^{-n\pi s}$, the transformed equation is

$$s^2 X(s) + X(s) = \sum_{n=0}^{\infty} e^{-n\pi s},$$

so

$$X(s) = \sum_{n=0}^{\infty} \frac{e^{-n\pi s}}{s^2 + 1}.$$

We compute the inverse Laplace transform term by term; the result is

$$x(t) = \sum_{n=0}^{\infty} u(t - n\pi) \sin(t - n\pi).$$

Because $\sin(t - n\pi) = (-1)^n \sin t$ and $u(t - n\pi) = 0$ for $t < n\pi$, we see that if $n\pi < t < (n + 1)\pi$, then

$$x(t) = \sin t - \sin t + \sin t - \cdots + (-1)^n \sin t;$$

that is,

$$x(t) = \begin{cases} \sin t & \text{if } n \text{ is even,} \\ 0 & \text{if } n \text{ is odd.} \end{cases}$$

Hence $x(t)$ is the half-wave rectification of $\sin t$ shown in Fig. 4.36. The physical explanation is that the first hammer blow (at time $t = 0$) starts the mass moving to the right; just as it returns to the origin, the second hammer blow stops it dead; it remains motionless until the third hammer blow starts it moving again, and so on. Of course, if the hammer blows are not perfectly synchronized then the motion of the mass will be quite different.

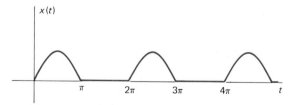

FIGURE 4.36 The half-wave rectification of $\sin t$.

CHAPTER 4: The Laplace Transform

Consider a physical system in which the *output* or *response* $x(t)$ to the *input* function $f(t)$ is described by the differential equation

$$ax'' + bx' + cx = f(t), \tag{22}$$

where the constant coefficients a, b, and c are determined by the physical parameters of the system and are independent of $f(t)$. The mass-spring-dashpot system and the series RLC circuit are familiar examples of this general situation.

For simplicity we assume that the system is initially passive: $x(0) = x'(0) = 0$. Then the transform of Eq. (22) is

$$as^2 X(s) + bsX(s) + cX(s) = F(s),$$

so

$$X(s) = \frac{F(s)}{as^2 + bs + c} = W(s)F(s). \tag{23}$$

The function

$$W(s) = \frac{1}{as^2 + bs + c} \tag{24}$$

is called the **transfer function** of the system. Thus the transform of the response to the input $f(t)$ is the product of $W(s)$ and the transform $F(s)$.

The function

$$w(t) = \mathscr{L}^{-1}\{W(s)\} \tag{25}$$

is called the **weight function** of the system. From (24) we see by convolution that

$$x(t) = \int_0^t w(\tau)f(t - \tau)\, d\tau. \tag{26}$$

This formula is **Duhamel's principle** for the system. What is important is that the weight function $w(t)$ is determined completely by the parameters of the system. Once $w(t)$ has been determined, the integral in (26) gives the response of the system to an arbitrary input function $f(t)$.

EXAMPLE 4 Consider a mass-spring-dashpot system (initially passive) which responds to the external force $f(t)$ in accord with the equation $x'' + 6x' + 10x = f(t)$. Then

$$W(s) = \frac{1}{s^2 + 6s + 10} = \frac{1}{(s + 3)^2 + 1},$$

so the weight function is $w(t) = e^{-3t}\sin t$. Then Duhamel's principle implies that the response $x(t)$ to the force $f(t)$ is

$$x(t) = \int_0^t e^{-3\tau}(\sin \tau)f(t - \tau)\, d\tau.$$

Note that

$$W(s) = \frac{1}{as^2 + bs + c} = \frac{\mathscr{L}\{\delta(t)\}}{as^2 + bs + c}.$$

Consequently, it follows from (23) that the weight function is simply the response of the system to the delta function input $\delta(t)$. For this reason $w(t)$ is sometimes called the **unit impulse response**. A response that is usually easier to measure in practice is the response $h(t)$ to the unit step function $u(t)$; $h(t)$ is the **unit step response**. Because $\mathcal{L}\{u(t)\} = 1/s$, we see from (23) that the transform of $h(t)$ is

$$H(s) = \frac{W(s)}{s}.$$

It follows from the formula for transforms of integrals that

$$h(t) = \int_0^t w(\tau)\, d\tau, \quad \text{so that} \quad w(t) = h'(t). \tag{27}$$

Thus the weight function, or unit impulse response, is the derivative of the unit step response. Substitution of (27) in Duhamel's principle gives

$$x(t) = \int_0^t h'(\tau) f(t - \tau)\, d\tau \tag{28}$$

for the response of the system to the input $f(t)$.

To describe a typical application of (28), suppose that we are given a complex series circuit containing many inductors, resistors, and capacitors. Assume that its circuit equation is a linear equation of the form in (22), though with i in place of x. What if the coefficients a, b, and c are unknown, perhaps only because they are too complicated to compute? We would still want to know the current $i(t)$ corresponding to any input $f(t) = e'(t)$. We connect the circuit to a linearly increasing voltage $e(t) = t$, so that $f(t) = e'(t) = 1 = u(t)$, and measure the response $h(t)$ with an ammeter. We then compute the derivative $h'(t)$, either numerically or graphically. Then according to (28), the output current $i(t)$ corresponding to the input voltage $e(t)$ will be given by

$$i(t) = \int_0^t h'(\tau) e'(t - \tau)\, d\tau$$

(using the fact that $f(t) = e'(t)$).

In conclusion we remark that around 1950, after engineers and physicists had been using delta functions widely and fruitfully for about 20 years without rigorous justification, the French mathematician Laurent Schwartz developed a rigorous mathematical theory of *generalized functions* that supplied the missing logical foundation for delta function techniques. Every piecewise continuous ordinary function is a generalized function, but the delta function is an example of a generalized function that is not an ordinary function.

4.6 PROBLEMS

Solve the initial value problems in Problems 1–8.

1. $x'' + 4x = \delta(t); \quad x(0) = x'(0) = 0$

2. $x'' + 4x = \delta(t) + \delta(t - \pi); \quad x(0) = x'(0) = 0$

3. $x'' + 4x' + 4x = 1 + \delta(t - 2); \quad x(0) = x'(0) = 0$

4. $x'' + 2x' + x = t + \delta(t); \quad x(0) = 0, x'(0) = 1$

5. $x'' + 2x' + 2x = 2\delta(t - \pi); \quad x(0) = 0 = x'(0)$

6. $x'' + 9x = \delta(t - 3\pi) + \cos 3t; \quad x(0) = x'(0) = 0$

7. $x'' + 4x' + 5x = \delta(t - \pi) + \delta(t - 2\pi); \quad x(0) = 0, x'(0) = 2$

8. $x'' + 2x' + x = \delta(t) - \delta(t-2); \quad x(0) = 2 = x'(0)$

Apply Duhamel's principle to write an integral formula for the solution of each of the initial value problems in Problems 9–12.

9. $x'' + 4x = f(t); \quad x(0) = x'(0) = 0$

10. $x'' + 6x' + 9x = f(t); \quad x(0) = x'(0) = 0$

11. $x'' + 6x' + 8x = f(t); \quad x(0) = x'(0) = 0$

12. $x'' + 4x' + 8x = f(t); \quad x(0) = x'(0) = 0$

13. This problem deals with a particle of mass m, initially at rest at the origin, that receives an impulse p at time $t = 0$.
(a) Find the solution $x_\epsilon(t)$ of the problem

$$mx'' = pd_{0,\epsilon}(t); \qquad x(0) = 0 = x'(0).$$

(b) Show that $\lim_{\epsilon \to 0} x_\epsilon(t)$ agrees with the solution of the problem

$$mx'' = p\delta(t); \qquad x(0) = x'(0) = 0.$$

(c) Show that $mv = p$ for $t > 0$ $(v = x')$.

14. Verify that $u'(t-a) = \delta(t-a)$ by solving the problem

$$x' = \delta(t-a); \qquad x(0) = 0$$

to obtain $x(t) = u(t-a)$.

15. This problem deals with a mass m on a spring (with constant k) that receives an impulse $p_0 = mv_0$ at time $t = 0$. Show that the initial value problems

$$mx'' + kx = 0; \qquad x(0) = 0, \qquad x'(0) = v_0$$

and

$$mx'' + kx = p_0\delta(t); \qquad x(0) = 0, \qquad x'(0) = 0$$

have the same solution. Thus the effect of $p_0\delta(t)$ is, indeed, to impart to the particle an initial momentum p_0.

16. This is a generalization of Problem 15. Show that the problems

$$ax'' + bx' + cx = f(t); \qquad x(0) = 0, \qquad x'(0) = v_0$$

and

$$ax'' + bx' + cx = f(t) + av_0\delta(t); \qquad x(0) = x'(0) = 0$$

have the same solution for $t > 0$. Thus the effect of the term $av_0\delta(t)$ is to supply the initial condition $x'(0) = v_0$.

17. Consider an initially passive RC circuit (no inductance) with a battery supplying E_0 volts. (a) If the switch to the battery is closed at time $t = a$ and opened at time $t = b > a$ (and left open thereafter), show that the current $i(t)$ in the circuit satisfies the initial value problem

$$\begin{cases} Ri' + \dfrac{1}{C}i = E_0\delta(t-a) - E_0\delta(t-b); \\[2mm] i(0) = 0. \end{cases}$$

(b) Solve this problem if $R = 100 \, \Omega$, $C = 10^{-4}$ F, $E_0 = 100$ V, $a = 1$ (s), and $b = 2$ (s). Show that $i(t) > 0$ if $1 < t < 2$, and that $i(t) < 0$ if $t > 2$.

18. Consider an initially passive LC circuit (no resistance) with a battery supplying E_0 volts. (a) If the switch is closed at time $t = 0$ and opened at time $t = a$, show that the current $i(t)$ in the circuit satisfies the initial value problem

$$\begin{cases} Li'' + \dfrac{1}{C}i = E_0\delta(t) - E_0\delta(t-a); \\[2mm] i(0) = i'(0) = 0. \end{cases}$$

(b) If $L = 1$ H, $C = 10^{-2}$ F, $E_0 = 10$ V, and $a = \pi$ (s), show that

$$i(t) = \begin{cases} \sin 10t & \text{if } t < \pi, \\ 0 & \text{if } t > \pi. \end{cases}$$

Thus the current oscillates through five cycles and then stops abruptly when the switch is opened.

19. Consider the LC circuit of Problem 18(b), except suppose that the switch is alternately closed and opened at times $t = 0$, $(0.1)\pi$, $(0.2)\pi$, ... (a) Show that $i(t)$ satisfies the initial value problem

$$\begin{cases} i'' + 100i = 10 \displaystyle\sum_{n=0}^{\infty} (-1)^n\delta(t - n\pi/10), \\ i(0) = i'(0) = 0. \end{cases}$$

(b) Solve this initial value problem to show that $i(t) = (n+1)\sin 10t$ if $n\pi/10 < t < (n+1)\pi/10$. Thus a resonance phenomenon occurs.

20. Repeat Problem 19, except suppose that the switch is alternately closed and opened at times $t = 0$, $(0.2)\pi$, $(0.4)\pi$, ..., $n\pi/5$, ... Now show that if $n\pi/5 < t < (n+1)\pi/5$, then

$$i(t) = \begin{cases} \sin 10t & \text{if } n \text{ is even;} \\ 0 & \text{if } n \text{ is odd.} \end{cases}$$

Thus the current in alternate cycles of length $\pi/5$ first executes a sine oscillation during one cycle, then is dormant during the next cycle, and so on.

21. Consider an RLC circuit in series with a battery, with $L = 1$ H, $R = 60 \, \Omega$, $C = 10^{-3}$ F, and $E_0 = 10$ V. (a) Suppose that the switch is alternately closed and opened at times $t = 0$, $(0.1)\pi$, $(0.2)\pi$, ... Show that $i(t)$ satisfies the initial value problem

$$\begin{cases} i'' + 60i' + 1000i = 10 \displaystyle\sum_{n=0}^{\infty} (-1)^n\delta(t - [0.1]n\pi), \\ i(0) = i'(0) = 0. \end{cases}$$

(b) Solve this problem to show that if $(0.1)n\pi < t < (0.1)(n+1)\pi$, then

$$i(t) = \left(\frac{e^{3n\pi + 3\pi} - 1}{e^{3\pi} - 1}\right)e^{-30t}\sin 10t.$$

22. Consider a mass $m = 1$ on a spring with constant $k = 1$, initially at rest, but struck with a hammer at each of the instants $t = 0, 2\pi, 4\pi, \ldots$. Suppose that each hammer blow imparts an impulse of $+1$. Show that the position $x(t)$ of the mass satisfies the initial value problem

$$\begin{cases} x'' + x = \displaystyle\sum_{n=0}^{\infty} \delta(t - 2n\pi), \\ x(0) = x'(0) = 0. \end{cases}$$

Solve this problem to show that if $2n\pi < t < 2(n+1)\pi$, then $x(t) = (n+1)\sin t$. Thus resonance occurs because the mass is struck each time it passes through the origin moving to the right—in contrast to Example 3, in which the mass was struck each time it returned to the origin.

TABLE OF LAPLACE TRANSFORMS

This table summarizes the general properties of Laplace transforms and the Laplace transforms of particular functions derived in this chapter.

Function	Transform	Function	Transform
$f(t)$	$F(s)$	e^{at}	$\dfrac{1}{s-a}$
$af(t) + bg(t)$	$aF(s) + bG(s)$		
$f'(t)$	$sF(s) - f(0)$	$t^n e^{at}$	$\dfrac{n!}{(s-a)^{n+1}}$
$f''(t)$	$s^2 F(s) - sf(0) - f'(0)$		
$f^{(n)}(t)$	$s^n F(s) - s^{n-1}f(0) - \cdots - f^{(n-1)}(0)$	$\cos kt$	$\dfrac{s}{s^2 + k^2}$
$\displaystyle\int_0^t f(\tau)\,d\tau$	$\dfrac{F(s)}{s}$	$\sin kt$	$\dfrac{k}{s^2 + k^2}$
$e^{at}f(t)$	$F(s - a)$		
$u(t-a)f(t-a)$	$e^{-as}F(s)$	$\cosh kt$	$\dfrac{s}{s^2 - k^2}$
$\displaystyle\int_0^t f(\tau)g(t-\tau)\,d\tau$	$F(s)G(s)$		
$tf(t)$	$-F'(s)$	$\sinh kt$	$\dfrac{k}{s^2 - k^2}$
$t^n f(t)$	$(-1)^n F^{(n)}(s)$		
$\dfrac{f(t)}{t}$	$\displaystyle\int_s^\infty F(\sigma)\,d\sigma$	$\dfrac{1}{2k^3}(\sin kt - kt\cos kt)$	$\dfrac{1}{(s^2 + k^2)^2}$
$f(t)$, period p	$\dfrac{1}{1 - e^{-ps}}\displaystyle\int_0^p e^{-st}f(t)\,dt$	$\dfrac{t}{2k}\sin kt$	$\dfrac{s}{(s^2 + k^2)^2}$
1	$\dfrac{1}{s}$	$\dfrac{1}{2k}(\sin kt + kt\cos kt)$	$\dfrac{s^2}{(s^2 + k^2)^2}$
t	$\dfrac{1}{s^2}$	$u(t - a)$	$\dfrac{e^{-as}}{s}$
t^n	$\dfrac{n!}{s^{n+1}}$	$\delta(t - a)$	e^{-as}
$\dfrac{1}{\sqrt{\pi t}}$	$\dfrac{1}{\sqrt{s}}$	$(-1)^{[at]}$ (square wave)	$\dfrac{1}{s}\tanh\dfrac{as}{2}$
t^a	$\dfrac{\Gamma(a + 1)}{s^{a+1}}$	$\left[\dfrac{t}{a}\right]$ (staircase)	$\dfrac{e^{-as}}{s(1 - e^{-as})}$

CHAPTER *5*

LINEAR SYSTEMS OF DIFFERENTIAL EQUATIONS

5.1 INTRODUCTION TO SYSTEMS

5.2 THE METHOD OF ELIMINATION

5.3 LINEAR SYSTEMS AND MATRICES

*__5.4__ MECHANICAL APPLICATIONS OF LINEAR SYSTEMS

5.5 THE EIGENVALUE METHOD FOR HOMOGENEOUS SYSTEMS

5.6 NONHOMOGENEOUS LINEAR SYSTEMS

*__5.7__ MATRIX EXPONENTIALS AND LINEAR SYSTEMS

Introduction to Systems

In the preceding chapters we have discussed methods for solving an ordinary differential equation that involves only one dependent variable. Many applications, however, require the use of two or more dependent variables, each a function of a single independent variable (typically time). Such problems lead naturally to a *system* of simultaneous ordinary differential equations. We will usually denote the independent variable by t and the dependent variables (the unknown functions of t) by $x_1, x_2, x_3, \ldots$, or by $x, y, z, \ldots$. Primes will denote differentiation with respect to t.

We will restrict our attention to systems in which the number of equations is the same as the number of dependent variables (unknown functions). For instance, a system of two first order differential equations in the dependent variables x and y has the general form

$$\left. \begin{aligned} f(t, x, y, x', y') = 0 \\ g(t, x, y, x', y') = 0 \end{aligned} \right\}, \tag{1}$$

where the functions f and g are given. A solution of this system is a *pair* $x(t)$, $y(t)$ of functions of t that satisfy both equations identically over some interval of values of t.

For an example of a second order system, consider a particle of mass m that moves in space under the influence of a force field $\mathbf{F}$ that depends upon time t, the position $(x(t), y(t), z(t))$ of the particle, and its velocity $(x'(t), y'(t), z'(t))$. Applying Newton's law $m\mathbf{a} = \mathbf{F}$ componentwise, we get the system

$$\begin{aligned} mx'' &= F_1(t, x, y, z, x', y', z'), \\ my'' &= F_2(t, x, y, z, x', y', z'), \\ mz'' &= F_3(t, x, y, z, x', y', z'), \end{aligned} \tag{2}$$

where F_1, F_2, and F_3 are the components of the vector function $\mathbf{F}$.

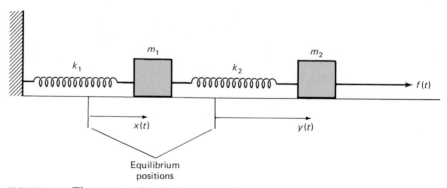

FIGURE 5.1 The mass-and-spring system of Example 1.

CHAPTER 5:   Linear Systems of Differential Equations

The following three examples illustrate further how systems of differential equations arise naturally in scientific problems.

EXAMPLE 1 Consider the system of two masses and two springs shown in Fig. 5.1, with a given external force $f(t)$ acting on the right-hand mass m_2. We denote by $x(t)$ the displacement (to the right) of the mass m_1 from its static position (when the system is motionless and in equilibrium and $f(t) = 0$), and by $y(t)$ the displacement of the mass m_2 from its static position. Thus the two springs are neither stretched nor compressed when x and y are zero.

In the configuration in Fig. 5.1, the first spring is stretched x units and the second by $y - x$ units. We apply Newton's law of motion to the two "free body diagrams" shown in Fig. 5.2; we thereby obtain the system

$$m_1 x'' = -k_1 x + k_2(y - x),$$
$$m_2 y'' = -k_2(y - x) + f(t) \tag{3}$$

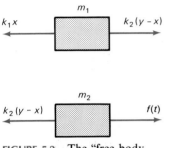

of differential equations that the position functions $x(t)$ and $y(t)$ must satisfy. For instance, if $m_1 = 2$, $m_2 = 1$, $k_1 = 4$, $k_2 = 2$, and $f(t) = 40 \sin 3t$ in appropriate physical units, the system in (3) then reduces to

$$2x'' = -6x + 2y,$$
$$y'' = 2x - 2y + 40 \sin 3t. \tag{4}$$

FIGURE 5.2 The "free body diagrams" for the system of Example 1.

EXAMPLE 2 Consider two brine tanks connected as shown in Fig. 5.3. Tank 1 contains $x(t)$ pounds of salt in 100 gal of brine and Tank 2 contains $y(t)$ pounds of salt in 200 gal of brine. The brine in each tank is kept uniform by stirring, and brine is pumped from each tank to the other at the rates indicated in the figure. In addition, fresh water flows into Tank 1 at 20 gal/min, and the brine in Tank 2 flows out at 20 gal/min (so the total volume of brine in the two tanks remains constant). The salt concentrations in the two tanks are $x/100$ pounds per gallon

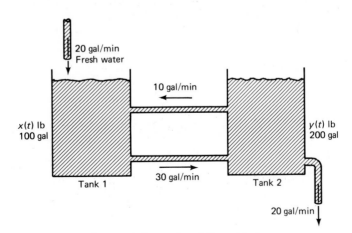

FIGURE 5.3 The two brine tanks of Example 2.

and $y/200$ pounds per gallon, respectively. When we compute the rates of change of the amounts of salt in the two tanks, we therefore get the system of differential equations that $x(t)$ and $y(t)$ must satisfy:

$$x' = -(30)\left(\frac{x}{100}\right) + (10)\left(\frac{y}{200}\right) = -\frac{3}{10}x + \frac{1}{20}y,$$

$$y' = (30)\left(\frac{x}{100}\right) - (10)\left(\frac{y}{200}\right) - (20)\left(\frac{y}{200}\right) = \frac{3}{10}x - \frac{3}{20}y;$$

that is,

$$\begin{aligned} 20x' &= -6x + y, \\ 20y' &= 6x - 3y. \end{aligned} \tag{5}$$

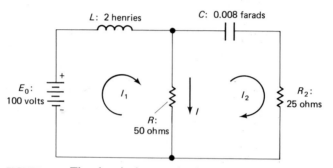

FIGURE 5.4   The electrical network of Example 3.

EXAMPLE 3   Consider the electrical network shown in Fig. 5.4, where $I_1(t)$ denotes the current in the indicated direction through the inductor and $I_2(t)$ denotes the current through the resistor R_2. The current through the resistor R is $I = I_1 - I_2$ in the direction indicated. We recall Kirchhoff's voltage law to the effect that the (algebraic) sum of the voltage drops around any closed loop of such a network is zero. As in Section 2.9, the voltage drops across the three types of circuit elements are those shown in Fig. 5.5. We apply Kirchhoff's law to the left-hand loop of the network to obtain

$$2\frac{dI_1}{dt} + 50(I_1 - I_2) - 100 = 0, \tag{6}$$

because the voltage drop from the negative to the positive pole of the battery is negative. The right-hand loop yields the equation

$$125Q_2 + 25I_2 + 50(I_2 - I_1) = 0 \tag{7}$$

where $Q_2(t)$ is the charge on the capacitor. Because $dQ_2/dt = I_2$, differentiation of each side of Eq. (7) yields

$$-50\frac{dI_1}{dt} + 75\frac{dI_2}{dt} + 125I_2 = 0. \tag{8}$$

Circuit element	Voltage drop
Inductor	$L\frac{dI}{dt}$
Resistor	RI
Capacitor	$\frac{1}{C}Q$

FIGURE 5.5   Voltage drops across common circuit elements.

CHAPTER 5:   Linear Systems of Differential Equations

Dividing Eqs. (6) and (8) by the factors 2 and -25, respectively, we get the system

$$\frac{dI_1}{dt} + 25I_1 - 25I_2 = 50,$$

$$2\frac{dI_1}{dt} - 3\frac{dI_2}{dt} - 5I_2 = 0$$

(9)

of differential equations that the currents $I_1(t)$ and $I_2(t)$ must satisfy.

FIRST ORDER SYSTEMS

Consider a system of differential equations that can be solved for the highest order derivatives of the dependent variables that appear, as explicit functions of t and of lower order derivatives of the dependent variables. For instance, in the case of a system of two second order equations, our assumption is that it can be written in the form

$$x_1'' = f_1(t, x_1, x_2, x_1', x_2'),$$

$$x_2'' = f_2(t, x_1, x_2, x_1', x_2'),$$

(10)

It is of both practical and theoretical importance that any such higher order system can be transformed into an equivalent system of *first order* equations.

To describe how such a tranformation is accomplished, we consider first the "system" consisting of the single nth order equation

$$x^{(n)} = f(t, x, x', \ldots, x^{(n-1)}).$$

(11)

We introduce the dependent variables $x_1, x_2, \ldots, x_n$ defined as follows:

$$x_1 = x, \ x_2 = x', \ x_3 = x'', \ldots, x_n = x^{(n-1)}.$$

(12)

Note that $x_1' = x' = x_2$, $x_2' = x'' = x_3$, and so on. Hence the substitutions of (12) in Eq. (11) yield the system

$$x_1' = x_2,$$

$$x_2' = x_3,$$

$$\vdots$$

$$x_{n-1}' = x_n,$$

$$x_n' = f(t, x_1, x_2, \ldots, x_n)$$

(13)

of n *first order* equations. Evidently this system is equivalent to the original nth order equation in (11), in the sense that $x(t)$ is a solution of (11) if and only if the functions $x_1(t), x_2(t), \ldots, x_n(t)$ defined in (12) satisfy the system of equations in (13).

EXAMPLE 4 The third order equation

$$x''' + 3x'' + 2x' - 5x = \sin 2t$$

is of the form in (11) with

$$f(t, x, x', x'') = 5x - 2x' - 3x'' + \sin 2t.$$

Hence the substitutions

$$x_1 = x, \qquad x_2 = x' = x_1', \qquad x_3 = x'' = x_2'$$

yield the system

$$x_1' = x_2,$$

$$x_2' = x_3,$$

$$x_3' = 5x_1 - 2x_2 - 3x_3 + \sin 2t$$

of three first order equations.

It may appear that the first order system obtained in Example 4 offers little advantage because we could use the methods of Chapter 2 to solve the original (linear) third order equation. But suppose that we were confronted with the nonlinear equation

$$x'' = x^3 + (x')^3,$$

to which none of our earlier methods can be applied. The corresponding first order system is

$$x_1' = x_2,$$

$$x_2' = (x_1)^3 + (x_2)^3, \tag{14}$$

and we will see in Chapter 6 that there exist effective numerical techniques for approximating the solution of essentially any first order system. So in this case the transformation to a first order system *is* advantageous. From a practical viewpoint, large systems of higher order differential equations typically are solved numerically with the aid of a computer, and the first step is to transform such a system into a first order system for which a standard computer program is available.

EXAMPLE 5 The system

$$2x'' = -6x + 2y,$$

$$y'' = \quad 2x - 2y + 40 \sin 3t \tag{4}$$

of second order equations was derived in Example 1. Transform this system into an equivalent first order system.

SOLUTION Motivated by the equations in (12), we define

$$x_1 = x, \qquad x_2 = x' = x_1', \qquad y_1 = y, \qquad y_2 = y' = y_1'.$$

Then the system in (4) yields the system

$$x_1' = x_2,$$

$$2x_2' = -6x_1 + 2y_1,$$

$$y_1' = y_2,$$

$$y_2' = 2x_1 - 2y_1 + 40 \sin 3t \tag{15}$$

of four first order equations in the dependent variables x_1, x_2, y_1, and y_2.

CHAPTER 5: Linear Systems of Differential Equations

In addition to practical advantages for numerical computation, the general theory of systems and systematic solution techniques are more easily and more concisely described for first order systems than for higher order systems. Throughout most of this chapter, we will concentrate our attention on *linear* first order systems of the form

$$x'_1 = p_{11}(t)x_1 + p_{12}(t)x_2 + \cdots + p_{1n}(t)x_n + f_1(t),$$

$$x'_2 = p_{21}(t)x_1 + p_{22}(t)x_2 + \cdots + p_{2n}(t)x_n + f_2(t),$$

$$\vdots$$

$$x'_n = p_{n1}(t)x_1 + p_{n2}(t)x_2 + \cdots + p_{nn}(t)x_n + f_n(t).$$

(16)

We say that this system is **homogeneous** if the functions $f_1, f_2, \ldots, f_n$ are all identically zero; otherwise it is **nonhomogeneous**. Thus the linear system in (5) is homogeneous, while the linear system in (15) is nonhomogeneous. The system in (14) is nonlinear because the right-hand side of the second equation is not a linear function of the dependent variables x_1 and x_2.

A **solution** of the system in (16) is an *n*-tuple of functions $x_1(t), x_2(t), \ldots, x_n(t)$ that (on some interval) identically satisfy each of the equations in (16). We will see that the general theory of a system of *n* linear first order equations has many similarities with the general theory of a single *n*th order linear differential equation. The theorem below (proved in Section 7.6) is analogous to Theorem 2 (existence and uniqueness of the solution of an *n*th order equation) of Section 2.2. It tells us that if the coefficient functions p_{ij} and f_j in (16) are continuous, then the system has a unique solution satisfying given initial conditions.

Theorem *Existence and Uniqueness for Linear Systems*

Suppose that the functions $p_{11}, p_{12}, \ldots, p_{nn}$ and the functions $f_1, f_2, \ldots, f_n$ are continuous on the open interval *I* containing the point *a*. Then, given *n* numbers $b_1, b_2, \ldots, b_n$, the system in (16) has a unique solution on the entire interval *I* that satisfies the *n* initial conditions

$$x_1(a) = b_1, x_2(a) = b_2, \ldots, x_n(a) = b_n.$$

(17)

Thus *n* initial conditions are needed to determine a solution of a system of *n* linear first order equations, and we therefore expect a general solution of such a system to involve *n* arbitrary constants. For instance, we saw in Example 5 that the second order system in (4)—which describes the position functions $x(t)$ and $y(t)$ of the two masses in Example 1—is equivalent to the system of *four* first order linear equations in (15). Hence four initial conditions would be needed to determine the subsequent motions of the two masses in Example 1; typical initial values would be the initial positions $x(0)$ and $y(0)$ and the initial velocities $x'(0)$ and $y'(0)$. On the other hand, we found that the amounts $x(t)$ and $y(t)$ of salt in the two tanks of Example 2 are described by the system in (5) of *two* first order linear equations. Hence the two initial values $x(0)$ and $y(0)$ should suffice to determine the solution. Given a higher order system, we often must transform it into an equivalent first order system to discover how many initial conditions are needed to determine a unique solution; the theorem above tells us that the number of such conditions is precisely the same as the number of equations in the equivalent first order system.

THE LAPLACE TRANSFORM METHOD*

Laplace transforms are frequently used in engineering problems to solve linear systems in which the coefficients are all constants. When initial conditions are specified, the Laplace transform reduces such a linear system of differential equations to a linear system of algebraic equations in which the unknowns are the transforms of the solution functions. As the following example illustrates, the technique for a system is essentially the same as for a single linear differential equation with constant coefficients.

EXAMPLE 6 Solve the system

$$2x'' = -6x + 2y,$$
$$y'' = \quad 2x - 2y + 40 \sin 3t \tag{4}$$

subject to the initial conditions

$$x(0) = x'(0) = y(0) = y'(0) = 0. \tag{18}$$

Thus the force $f(t) = 40 \sin 3t$ is suddenly applied to the second mass of Example 1 at time $t = 0$ when the system is at rest in its equilibrium position.

SOLUTION We write $X(s) = \mathscr{L}\{x(t)\}$ and $Y(s) = \mathscr{L}\{y(t)\}$, exactly as in Chapter 4. Then the initial conditions in (18) imply that

$$\mathscr{L}\{x''(t)\} = s^2 X(s) \quad \text{and} \quad \mathscr{L}\{y''(t)\} = s^2 Y(s).$$

Because $\mathscr{L}\{\sin 3t\} = 3/(s^2 + 9)$, the transforms of the equations in (4) are the equations

$$2s^2 X(s) = -6X(s) + 2Y(s),$$

$$s^2 Y(s) = \quad 2X(s) - 2Y(s) + \frac{120}{s^2 + 9}.$$

Thus the transformed system is

$$(s^2 + 3)X(s) - \quad\quad Y(s) = 0$$

$$-2X(s) + (s^2 + 2)Y(s) = \frac{120}{s^2 + 9}. \tag{19}$$

The determinant of this pair of linear equations in $X(s)$ and $Y(s)$ is

$$(s^2 + 3)(s^2 + 2) - 2 = (s^2 + 1)(s^2 + 4),$$

---

* This subsection can be omitted if Chapter 4 has not been covered.

and we readily solve—using Cramer's rule, for instance—the system in (19) for

$$X(s) = \frac{120}{(s^2 + 1)(s^2 + 4)(s^2 + 9)} = \frac{5}{s^2 + 1} - \frac{8}{s^2 + 4} + \frac{3}{s^2 + 9} \tag{20}$$

and

$$Y(s) = \frac{120(s^2 + 3)}{(s^2 + 1)(s^2 + 4)(s^2 + 9)} = \frac{10}{s^2 + 1} + \frac{8}{s^2 + 4} - \frac{18}{s^2 + 9}. \tag{21}$$

The partial fraction decompositions in (20) and (21) are readily found using the methods of Section 4.4. For instance, if we write

$$\frac{120}{(s^2 + 1)(s^2 + 4)(s^2 + 9)} = \frac{As + B}{s^2 + 1} + H(s),$$

where $H(s)$ denotes the sum of the partial fractions corresponding to the two factors $s^2 + 4$ and $s^2 + 9$, then multiplication of each term by $s^2 + 1$ yields

$$\frac{120}{(s^2 + 4)(s^2 + 9)} = As + B + (s^2 + 1)H(s).$$

Substitution of $s = i$ (one of the zeros of $s^2 + 1$) in this last equation then gives

$$Ai + B = \frac{120}{(i^2 + 4)(i^2 + 9)} = \frac{120}{(3)(8)} = 5,$$

so $A = 0$ and $B = 5$. This gives the first term in (20), and the others can be found similarly.

At any rate, the inverse Laplace transforms of the expressions in (20) and (21) give the solution

$$x(t) = 5 \sin t - 4 \sin 2t + \sin 3t,$$
$$y(t) = 10 \sin t + 4 \sin 2t - 6 \sin 3t. \tag{22}$$

It is of interest to rewrite the solutions in (22) in the form of a single vector equation

$$(x, y) = 5(x_1, y_1) + 4(x_2, y_2) + (x_p, y_p), \tag{23}$$

where

$$x_1(t) = \sin t, \qquad y_1(t) = 2 \sin t; \tag{24a}$$

$$x_2(t) = -\sin 2t, \qquad y_2(t) = \sin 2t; \tag{24b}$$

$$x_p(t) = \sin 3t, \qquad y_p(t) = -6 \sin 3t. \tag{24c}$$

In Problem 25 we ask you to verify that (x_p, y_p) is a particular solution of the nonhomogeneous system in (4), while (x_1, y_1) and (x_2, y_2) are solutions of the associated homogeneous system

$$2x'' = -6x + 2y,$$
$$y'' = 2x - 2y. \tag{25}$$

Thus (x_1, y_1) and (x_2, y_2) represent *free* oscillations of the two masses (no external force), while (x_p, y_p) is a particular forced oscillation in response to the force $f(t) = 40 \sin 3t$ with the imposed frequency $\omega_p = 3$. Note that in the free oscillation

(x_1, y_1) the two masses move in the same direction with the same frequency $\omega_1 = 1$, but with the amplitude of motion of the second mass twice that of the first mass. In the free oscillation (x_2, y_2) the two masses move in opposite directions with equal amplitudes of motion and with frequency $\omega_2 = 2$.

The homogeneous system in (25) has the two additional solutions

$$x_3(t) = \cos t, \qquad y_3(t) = 2 \cos t; \tag{26a}$$

$$x_4(t) = -\cos 2t, \qquad y_4(t) = \cos 2t; \tag{26b}$$

these correspond to (24a) and (24b) with cosines in place of sines. In Problem 25 we ask you to show that if $c_1, c_2, c_3,$ and c_4 are constants, then

$$
\begin{aligned}
x_h &= c_1 x_1 + c_2 x_2 + c_3 x_3 + c_4 x_4, \\
y_h &= c_1 y_1 + c_2 y_2 + c_3 y_3 + c_4 y_4
\end{aligned}
\tag{27}
$$

is a general solution of the homogeneous system in (25), while

$$x = x_h + x_p, \qquad y = y_h + y_p \tag{28}$$

is a general solution of the original nonhomogeneous system in (4). Thus (28) exhibits the general solution of the nonhomogeneous system as the (vector) sum of a particular solution and the general solution of the associated homogeneous system. (This should have a familiar ring!) In Section 5.3 we will see that this situation is typical of linear systems.

EXAMPLE 7 Because the system in (4) involves no first derivatives, we might well have foreseen that it would have a particular solution of the form

$$x_p(t) = A \sin 3t, \qquad y_p(t) = B \sin 3t.$$

Determine A and B by substituting this trial solution in the system.

SOLUTION The substitution in question yields the equations

$$-18A \sin 3t = -6A \sin 3t + 2B \sin 3t,$$

$$-9B \sin 3t = 2A \sin 3t - 2B \sin 3t + 40 \sin 3t.$$

After we remove the common factor $\sin 3t$ and collect coefficients, we see that A and B must satisfy the equations

$$12A + 2B = 0, \qquad -2A - 7B = 40.$$

The solution is $A = 1$ and $B = -6$, and thus $x_p = \sin 3t$ and $y_p = -6 \sin 3t$, exactly as in (24c).

As indicated by Example 6, the Laplace transform method often readily yields the solution of a specific system with given numerical initial conditions. Unfortunately, this method fails to provide an adequate understanding of the general situation for linear systems. The remaining sections of this chapter are devoted to alternative methods for the solution of linear systems that, in addition to yielding solutions, will exhibit more clearly the general nature of the solutions of linear systems.

In each of Problems 1–10, transform the given differential equation or system into an equivalent system of first order differential equations.

1. $x'' + 3x' + 7x = t^2$

2. $x^{(4)} + 6x'' - 3x' + x = \cos 3t$

3. $t^2 x'' + tx' + (t^2 - 1)x = 0$

4. $t^3 x''' - 2t^2 x'' + 3tx' + 5x = \ln t$

5. $x''' = (x')^2 + \cos x$

6. $x'' - 5x + 4y = 0, \ y'' + 4x - 5y = 0$

7. $x'' = \dfrac{-kx}{(x^2 + y^2)^{3/2}}, \ y'' = \dfrac{-ky}{(x^2 + y^2)^{3/2}}$

8. $x'' + 3x' + 4x - 2y = 0, \ y'' + 2y' - 3x + y = \cos t$

9. $x'' = 3x - y + 2z, \ y'' = x + y - 4z, \ z'' = 5x - y - z$

10. $x'' = x(1 - y), \ y'' = y(1 - x)$

In each of Problems 11–13, find a particular solution of the given system, either of the form $x = A \cos \omega t, \ y = B \cos \omega t$ or of the form $x = A \sin \omega t, \ y = B \sin \omega t$.

11. $x'' + 5x - 3y = \sin 2t, \ y'' - 3x + 5y = 0$

12. $x'' + 5x - 4y = 0, \ y'' - 4x + 5y = 2 \cos 4t$

13. $x'' + 3x - y = 2 \cos 3t, \ y'' - 2x + 2y = -\cos 3t$

In each of Problems 14–20, use Laplace transforms to solve the given initial value problem.

14. $x' = x + 2y, \ y' = 2x + y; \quad x(0) = 1, \ y(0) = 0$

15. $x' = 2x + y, \ y' = 6x + 3y; \quad x(0) = 1, \ y(0) = -2$

16. $x' = x + 2y, \ y' = x + e^{-t}; \quad x(0) = y(0) = 0$

17. $x' + 2y' + x = 0, \ x' - y' + y = 0; \quad x(0) = 0, \ y(0) = 1$

18. $x'' + 2x + 4y = 0, \ y'' + x + 2y = 0; \quad x(0) = y(0) = 0, \ x'(0) = y'(0) = -1$

19. $x'' + x' + y' + 2x - y = 0, \ y'' + x' + y' + 4x - 2y = 0; \ x(0) = y(0) = 1, \ x'(0) = y'(0) = 0$

20. $x' = x + z, \ y' = x + y, \ z' = -2x - z; \quad x(0) = 1, \ y(0) = 0, \ z(0) = 0$

21. If each of the two brine tanks in Example 2 initially contains 50 lb of salt, find $x(t)$ and $y(t)$.

22. If no currents are flowing initially in the electrical network of Example 3 and the battery is connected at time $t = 0$, find $I_1(t)$ and $I_2(t)$. Show that $I_1(t)$ approaches 2 (A) as $t \to +\infty$, while $I_2(t)$ approaches zero. You may use the inverse Laplace transform

$$\mathcal{L}^{-1}\left\{ \frac{b(a^2 + b^2)}{s[(s - a)^2 + b^2]} \right\} = b + e^{at}(a \sin bt - b \cos bt).$$

Problems 23 and 24 deal with the linear system

$$\begin{aligned} x' &= ax + by + f(t), \\ y' &= cx + dy + g(t). \end{aligned} \tag{29}$$

23. Suppose that c_1 and c_2 are constants and that x_1, y_1 and x_2, y_2 are two solutions of the homogeneous system associated with (29). Show that $x_h = c_1 x_1 + c_2 x_2, \ y_h = c_1 y_1 + c_2 y_2$ is also a solution of the associated homogeneous system.

24. Suppose that x_p, y_p is a particular solution of (29) and that x_h, y_h is a solution of the associated homogeneous system. Show that $x = x_h + x_p, \ y = y_h + y_p$ is a solution of (29).

25. (a) Verify that the pairs defined in Eqs. (24a), (24b), (26a), and (26b) are all solutions of the associated homogeneous system in (25). (b) Verify that the pair x_h, y_h defined in (27) is a solution of the homogeneous system in (25). (c) Verify that the pair x, y defined in (28) is a solution of the original nonhomogeneous system.

26. Derive the equations

$$\begin{aligned} m_1 x_1'' &= -(k_1 + k_2)x_1 + k_2 x_2, \\ m_2 x_2'' &= k_2 x_1 - (k_2 + k_3)x_2 \end{aligned}$$

for the displacements (from equilibrium) of the two masses shown in Fig. 5.6.

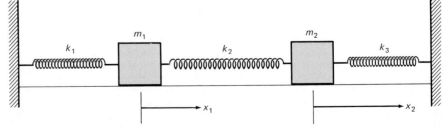

FIGURE 5.6 The system of Problem 26.

27. Two particles each of mass m are attached to a string under (constant) tension T as indicated in Fig. 5.7. Assume that the particles oscillate vertically (that is, parallel to the y-axis) with amplitudes so small that the sines of the angles shown are accurately approximated by their tangents. Show that the displacements y_1 and y_2 of the two masses satisfy the equations

$$ky_1'' = -2y_1 + y_2 \qquad ky_2'' = y_1 - 2y_2$$

where $k = mL/T$.

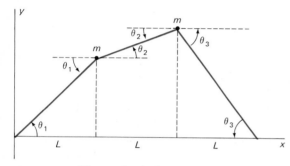

FIGURE 5.7 The mechanical system of Problem 27.

28. Three 100-gal fermentation vats are connected as indicated in Fig 5.8, and the mixtures in each tank are kept uniform by stirring. Denote by $x_i(t)$ the amount (in pounds) of alcohol in tank T_i at time t ($i = 1, 2, 3$). Suppose that the mixture circulates between the tanks at the rate of 10 gal/min. Derive the equations

$$10x_1' = -x_1 \qquad\quad + x_3,$$
$$10x_2' = \quad x_1 - x_2,$$
$$10x_3' = \qquad\quad x_2 - x_3.$$

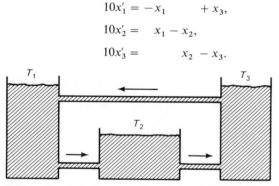

FIGURE 5.8 The fermentation tanks of Problem 28.

29. Set up a system of first order differential equations for the currents in the electrical circuit shown in Fig. 5.9.

30. Repeat Problem 29, except with the generator replaced with a battery supplying an emf of 100 V and with the inductor replaced by a 1-millifarad (mF) capacitor.

31. A particle of mass m moves in the plane with coordinates $(x(t), y(t))$ under the influence of a force that is directed toward

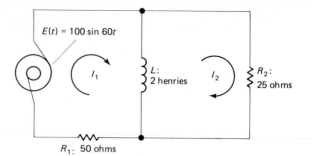

FIGURE 5.9 The electrical circuit of Problem 29.

the origin and has magnitude $k/(x^2 + y^2)$—an inverse-square-law central force field. Show that

$$mx'' = -\frac{kx}{r^3} \quad \text{and} \quad my'' = -\frac{ky}{r^3}$$

where $r = \sqrt{x^2 + y^2}$.

32. Suppose that a projectile of mass m moves in a vertical plane in the atmosphere near the surface of the earth under the influence of two forces: a downward gravitational force of magnitude mg, and a resistive force $\mathbf{F}_R$ that is directed opposite to the velocity vector $\mathbf{v}$ and has magnitude kv^2 (where $v = |\mathbf{v}|$ is the speed of the projectile; see Fig. 5.10). Show that the equations of motion of the projectile are

$$mx'' = -kvx', \qquad my'' = -kvy' - mg$$

where $v = \sqrt{(x')^2 + (y')^2}$.

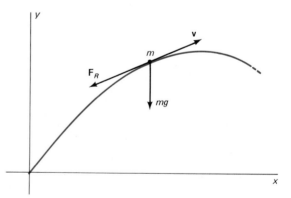

FIGURE 5.10 Trajectory of the projectile of Problem 32.

33. Suppose that a particle with mass m and electrical charge q moves in the xy plane under the influence of the magnetic field $\mathbf{B} = B\mathbf{k}$ (thus a uniform field parallel to the z-axis), so the force on the particle is $\mathbf{F} = q\mathbf{v} \times \mathbf{B}$ if its velocity is $\mathbf{v}$. Show that the equations of motion of the particle are

$$mx'' = +qBy', \qquad my'' = -qBx'.$$

The Method of Elimination

The most elementary approach to linear systems of differential equations with constant coefficients involves the elimination of dependent variables by appropriately combining pairs of equations. The object of this procedure is to eliminate dependent variables in succession until there remains only a single equation containing only one dependent variable. This remaining equation will usually be a linear equation of high order and can frequently be solved by the methods of Chapter 2. After the solution has been found, the other dependent variables can be found in turn, using either the original differential equations or those that have appeared in the elimination process.

This *method of elimination* for linear differential systems is quite similar to the solution of linear algebraic systems by elimination of variables until only one remains. It is most convenient in the case of manageably small systems: those consisting of no more than four equations. For large systems of differential equations, as well as for theoretical discussions, the matrix methods of the subsequent sections are preferable.

EXAMPLE 1   Find the solution of the system

$$x' = 4x - 3y, \qquad y' = 6x - 7y \tag{1}$$

that satisfies the initial conditions $x(0) = 2$ and $y(0) = -1$.

SOLUTION   If we solve the second equation in (1) for x, we get

$$x = \tfrac{1}{6}y' + \tfrac{7}{6}y, \tag{2}$$

so that

$$x' = \tfrac{1}{6}y'' + \tfrac{7}{6}y'. \tag{3}$$

We then substitute these expressions for x and x' in the first equation of the system in (1); this yields

$$\tfrac{1}{6}y'' + \tfrac{7}{6}y' = 4(\tfrac{1}{6}y' + \tfrac{7}{6}y) - 3y,$$

which can be simplified to

$$y'' + 3y' - 10y = 0.$$

This second order equation has characteristic equation

$$r^2 + 3r - 10 = (r - 2)(r + 5) = 0,$$

so its general solution is

$$y(t) = c_1 e^{2t} + c_2 e^{-5t}. \tag{4}$$

Next, substitution of (4) in (2) gives

$$x(t) = \tfrac{1}{6}(2c_1 e^{2t} - 5c_2 e^{-5t}) + \tfrac{7}{6}(c_1 e^{2t} + c_2 e^{-5t});$$

that is,

$$x(t) = \tfrac{3}{2}c_1 e^{2t} + \tfrac{1}{3}c_2 e^{-5t}. \tag{5}$$

Thus (4) and (5) describe the general solution of the system in (1).

The given initial conditions imply that

$$x(0) = \tfrac{3}{2}c_1 + \tfrac{1}{3}c_2 = 2$$

and that

$$y(0) = c_1 + c_2 = -1;$$

these equations are readily solved for $c_1 = 2$ and $c_2 = -3$. Hence the desired solution is

$$x(t) = 3e^{2t} - e^{-5t}, \qquad y(t) = 2e^{2t} - 3e^{-5t}.$$

In Example 1 we used an *ad hoc* procedure to eliminate one of the dependent variables by expressing it in terms of the other. We now describe a systematic elimination procedure. Operator notation is most convenient for these purposes. A **linear differential operator** of order n with *constant* coefficients is one of the form

$$L = a_n D^n + a_{n-1} D^{n-1} + \cdots + a_1 D + a_0, \tag{6}$$

where D denotes differentiation with respect to the independent variable t. If L_1 and L_2 are two such operators, then their product $L_1 L_2$ is defined by means of the equation

$$L_1 L_2 [x] = L_1 [L_2 x]. \tag{7}$$

For instance, if $L_1 = D + a$ and $L_2 = D + b$, then

$$\begin{aligned}
L_1 L_2 [x] &= (D + a)[(D + b)x] \\
&= D[Dx + bx] + a(Dx + bx) = [D^2 + (a + b)D + ab]x.
\end{aligned}$$

This illustrates the fact that two linear operators L_1 and L_2 with constant coefficients can be multiplied as if they were ordinary polynomials in the "variable" D. Because the multiplication of such polynomials is commutative, it follows that

$$L_1 L_2 x = L_2 L_1 x \tag{8}$$

if the necessary derivatives of $x(t)$ exist. By contrast, this property of commutativity generally fails for linear operators with variable coefficients (see Problems 21 and 22).

Any system of two linear differential equations with constant coefficients can be written in the form

$$\begin{aligned}
L_1 x + L_2 y &= f_1(t), \\
L_3 x + L_4 y &= f_2(t),
\end{aligned} \tag{9}$$

where $L_1, L_2, L_3,$ and L_4 are linear differential operators (perhaps of different orders) as in (6), and $f_1(t)$ and $f_2(t)$ are given functions. For instance, the system in (1) of Example 1 can be written in the form

$$(D - 4)x + \qquad 3y = 0,$$
$$-6x + (D + 7)y = 0,$$

(10)

with $L_1 = D - 4$, $L_2 = 3$, $L_3 = -6$, and $L_4 = D + 7$.

To eliminate the dependent variable x from the system in (9), we operate with L_3 on the first equation and with L_1 on the second; thus we obtain the system

$$L_3 L_1 x + L_3 L_2 y = L_3 f_1(t),$$
$$L_1 L_3 x + L_1 L_4 y = L_1 f_2(t).$$

(11)

Subtraction of the first from the second of these equations yields the equation

$$(L_1 L_4 - L_2 L_3)y = L_1 f_2(t) - L_3 f_1(t)$$

(12)

in the single dependent variable y. After solving for y we can substitute the result into either of the original equations in (9) and then solve for x.

Alternatively, we could eliminate similarly the dependent variable y from the original system in (9) and get the equation

$$(L_1 L_4 - L_2 L_3)x = L_4 f_1(t) - L_2 f_2(t),$$

(13)

which can now be solved for x.

Note that the same operator $L_1 L_4 - L_2 L_3$ appears on the left-hand side in both Eq. (12) and Eq. (13). This is the **operational determinant**

$$\begin{vmatrix} L_1 & L_2 \\ L_3 & L_4 \end{vmatrix} = L_1 L_4 - L_2 L_3$$

(14)

of the system in (9). In determinant notation Eqs. (12) and (13) can be rewritten as

$$\begin{vmatrix} L_1 & L_2 \\ L_3 & L_4 \end{vmatrix} x = \begin{vmatrix} f_1(t) & L_2 \\ f_2(t) & L_4 \end{vmatrix},$$

$$\begin{vmatrix} L_1 & L_2 \\ L_3 & L_4 \end{vmatrix} y = \begin{vmatrix} L_1 & f_1(t) \\ L_3 & f_2(t) \end{vmatrix}.$$

(15)

It is important to note that the determinants on the right-hand side in (15) are evaluated by means of the operators operating on the functions. The equations in (15) are strongly reminiscent of Cramer's rule for the solution of two linear equations in two (algebraic) variables and are thereby easily remembered. Indeed, you can solve a system of two linear differential equations either by carrying out the systematic elimination procedure described above or by directly employing the determinant equations in (15). Either process is especially simple if the system is homogeneous ($f_1(t) = f_2(t) = 0$), because in this case the right-hand sides of the equations in (12), (13), and (15) are zero.

EXAMPLE 2 Find the general solution of the system

$$(D - 4)x \qquad + 3y = 0,$$
$$-6x + (D + 7)y = 0.$$

(10)

SOLUTION   The operational determinant of this system is

$$(D - 4)(D + 7) - (3)(-6) = D^2 + 3D - 10. \tag{16}$$

Hence Eqs. (13) and (12) are

$$x'' + 3x' - 10x = 0,$$

$$y'' + 3y' - 10y = 0.$$

The characteristic equation of each is $r^2 + 3r - 10 = (r - 2)(r + 5) = 0$, so their (separate) general solutions are

$$x(t) = a_1 e^{2t} + a_2 e^{-5t},$$

$$y(t) = b_1 e^{2t} + b_2 e^{-5t}. \tag{17}$$

At this point we appear to have *four* arbitrary constants a_1, a_2, b_1, and b_2. But it follows from the theorem of Section 5.1 that the general solution of a system of two first order equations involves only two arbitrary constants. This apparent difficulty demands a resolution.

The explanation is simple: There must be some relation between our four constants. We can discover it by substituting the solutions in (17) into either of the original equations in (10). On substitution in the first equation, we get

$$0 = x' - 4x + 3y$$

$$= (2a_1 e^{2t} - 5a_2 e^{-5t}) - 4(a_1 e^{2t} + a_2 e^{-5t}) + 3(b_1 e^{2t} + b_2 e^{-5t});$$

that is,

$$0 = (-2a_1 + 3b_1)e^{2t} + (-9a_2 + 3b_2)e^{-5t}.$$

But e^{2t} and e^{-5t} are linearly independent functions, and so it follows that $a_1 = \frac{3}{2}b_1$ and $a_2 = \frac{1}{3}b_2$. Therefore the desired general solution is given by

$$x(t) = \tfrac{3}{2}b_1 e^{2t} + \tfrac{1}{3}b_2 e^{-5t}, \qquad y(t) = b_1 e^{2t} + b_2 e^{-5t}.$$

Note that this result is in accord with the general solution (Eqs. (4) and (5)) that we obtained by a different method in Example 1.

As illustrated by Example 2, the elimination procedure ordinarily will introduce a number of interdependent—thus unnecessary and undesirable—arbitrary constants. These must be eliminated by substitution of the proposed general solution into one or more of the original differential equations. If the operational determinant defined in (15) is *not identically zero*, then the number of independent arbitrary constants in a general solution of the system in (9) is equal to the order of the operational determinant—that is, its degree as a polynomial in D. (For a proof of this fact, see pages 144–150 of E. L. Ince's *Ordinary Differential Equations* (New York: Dover, 1956).) Thus the general solution of the system in (10) of Example 2 involves two arbitrary constants because its operational determinant $D^2 + 3D - 10$ is of order 2.

If the operational determinant *is* identically zero, then the system is said to be **degenerate**. A degenerate system may have either no solution or infinitely many independent solutions. For instance, the equations

$$Dx - \quad Dy = 0,$$
$$2Dx - 2Dy = 1$$

with operational determinant zero are obviously inconsistent and thus have no solution. On the other hand, the equations

$$Dx + \quad Dy = t,$$
$$2Dx + 2Dy = 2t$$

with operational determinant zero are obviously redundant; we can substitute *any* (continuously differentiable) function for x and then integrate to obtain y. Roughly speaking, every degenerate system is equivalent to either an inconsistent system or a redundant system.

Although the procedures and results above are described for the case of a system of two equations, they can be generalized readily to systems of three or more equations. For the system

$$L_{11}x + L_{12}y + L_{13}z = f_1(t),$$
$$L_{21}x + L_{22}y + L_{23}z = f_2(t), \tag{18}$$
$$L_{31}x + L_{32}y + L_{33}z = f_3(t)$$

of three linear equations, the dependent variable x satisfies the single linear equation

$$\begin{vmatrix} L_{11} & L_{12} & L_{13} \\ L_{21} & L_{22} & L_{23} \\ L_{31} & L_{32} & L_{33} \end{vmatrix} x = \begin{vmatrix} f_1(t) & L_{12} & L_{13} \\ f_2(t) & L_{22} & L_{23} \\ f_3(t) & L_{32} & L_{33} \end{vmatrix}, \tag{19}$$

with analogous equations for y and z. For most systems of more than three equations, however, the evaluation of operational determinants is too tedious for the method to be practical.

EXAMPLE 3 In Example 1 of Section 5.1 we derived the equations

$$(D^2 + 3)x - \quad\quad y = 0,$$
$$-2x + (D^2 + 2)y = 0 \tag{20}$$

for the displacements of the two masses in Fig. 5.11; here $f(t) = 0$ because we assume no external force. Find the general solution of the system in (20).

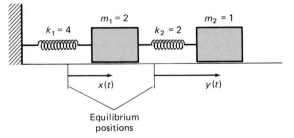

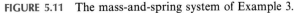

FIGURE 5.11 The mass-and-spring system of Example 3.

SOLUTION The operational determinant of the system in (20) is

$$(D^2 + 3)(D^2 + 2) - (-1)(-2) = D^4 + 5D^2 + 4$$
$$= (D^2 + 1)(D^2 + 4).$$

Hence the equations for x and y are

$$(D^2 + 1)(D^2 + 4)x = 0,$$
$$(D^2 + 1)(D^2 + 4)y = 0. \tag{21}$$

The characteristic equation $(r^2 + 1)(r^2 + 4) = 0$ has roots i, $-i$, $2i$, and $-2i$. So the general solutions of the equations in (21) are

$$x(t) = a_1 \cos t + a_2 \sin t + b_1 \cos 2t + b_2 \sin 2t,$$
$$y(t) = c_1 \cos t + c_2 \sin t + d_1 \cos 2t + d_2 \sin 2t. \tag{22}$$

Because the operational determinant is of order 4, the general solution should involve four (rather than eight) arbitrary constants. When we substitute x and y from (22) in the first equation in (20), we get

$$0 = x'' + 3x - y$$
$$= (-a_1 \cos t - a_2 \sin t - 4b_1 \cos 2t - 4b_2 \sin 2t)$$
$$+ 3(a_1 \cos t + a_2 \sin t + b_1 \cos 2t + b_2 \sin 2t)$$
$$- (c_1 \cos t + c_2 \sin t + d_1 \cos 2t + d_2 \sin 2t);$$

thus

$$0 = (2a_1 - c_1) \cos t + (2a_2 - c_2) \sin t$$
$$+ (-b_1 - d_1) \cos 2t + (-b_2 - d_2) \sin 2t.$$

It follows that

$$c_1 = 2a_1, \qquad c_2 = 2a_2, \qquad d_1 = -b_1, \quad \text{and} \quad d_2 = -b_2.$$

Hence the desired general solution may be written in the form

$$x(t) = a_1 \cos t + a_2 \sin t + b_1 \cos 2t + b_2 \sin 2t,$$
$$y(t) = 2a_1 \cos t + 2a_2 \sin t - b_1 \cos 2t - b_2 \sin 2t. \tag{23}$$

The equations in (23) describe **free oscillations** of the mass-and-spring system of Fig. 5.11—motion subject to *no* external forces. Four initial conditions (typically, initial displacements and velocities) would be needed to determine the values of a_1, a_2, b_1, and b_2. Granted that these computations have been carried out, we can (by the usual trigonometric machinations) write

$$x_1 = a_1 \cos t + a_2 \sin t = A \cos(t - \alpha),$$
$$y_1 = 2a_1 \cos t + 2a_2 \sin t = 2A \cos(t - \alpha) \tag{24}$$

and

$$x_2 = b_1 \cos 2t + b_2 \sin 2t = B \cos(2t - \beta),$$
$$y_2 = -b_1 \cos 2t - b_2 \sin 2t = -B \cos(2t - \beta). \tag{25}$$

The particular solutions (x_1, y_1) and (x_2, y_2) of the system in (20) represent the two **natural modes of oscillation** of the mass-and-spring system, and they exhibit its two (circular) **natural frequencies** $\omega_1 = 1$ and $\omega_2 = 2$. The equations in (23) can be written as a single vector equation

$$(x, y) = (x_1, y_1) + (x_2, y_2). \tag{26}$$

This oscillation represents an arbitrary free oscillation of the mass-and-spring system as a superposition of its two natural modes of oscillation, with the constants A, α, B, and β in Eqs. (24) and (25) determined by the initial conditions. In the natural mode (x_1, y_1), the two masses move in synchrony in the same direction and with the same frequency $\omega_1 = 1$, but with the amplitude of motion of m_2 twice that of m_1. In the natural mode (x_2, y_2), the two masses move in synchrony in opposite directions with the same frequency $\omega_2 = 2$ and with equal amplitudes of oscillation. These remarks generalize the discussion following Example 6 of Section 5.1.

The systematic elimination procedure employed in Examples 2 and 3 is not always optimal. The lesson of the following example is that in solving differential equations, as in doing anything else, it pays to keep your eyes open.

EXAMPLE 4 Find the general solution of the system

$$
\begin{aligned}
x' &= 3x - 2y && + 30t - 1, \\
y' &= -x + 3y - 2z + 2e^{3t}, \\
z' &= \qquad -y + 3z.
\end{aligned} \tag{27}
$$

SOLUTION In operator form, the system in (27) is

$$
\begin{aligned}
(D - 3)x + \quad 2y \qquad\qquad &= 30t - 1, \\
x + (D - 3)y + \quad 2z &= 2e^{3t}, \\
y + (D - 3)z &= 0.
\end{aligned} \tag{28}
$$

The crucial observation is this: If z is found first, then the last equation will give y immediately, and then the second equation will give x. This will involve much less labor than solving separately for x, y, and z and then substituting the results into (two of) the equations in (28) to find six relations between the nine arbitrary constants that would be introduced.

The equation for z analogous to Eq. (19) is

$$
\begin{vmatrix}
D - 3 & 2 & 0 \\
1 & D - 3 & 2 \\
0 & 1 & D - 3
\end{vmatrix} z =
\begin{vmatrix}
D - 3 & 2 & 30t - 1 \\
1 & D - 3 & 2e^{3t} \\
0 & 1 & 0
\end{vmatrix}. \tag{29}
$$

We expand the operational determinant along the first row to obtain

$$
\begin{aligned}
(D - 3)\begin{vmatrix} D - 3 & 2 \\ 1 & D - 3 \end{vmatrix} - 2\begin{vmatrix} 1 & 2 \\ 0 & D - 3 \end{vmatrix} &= (D - 3)[(D - 3)^2 - 2] - 2(D - 3) \\
&= (D - 3)[(D - 3)^2 - 4] \\
&= (D - 3)(D - 1)(D - 5) \tag{30a} \\
&= D^3 - 9D^2 + 23D - 15. \tag{30b}
\end{aligned}
$$

Upon expansion of the right-hand determinant in (29) along the third row, we get

$$(-1)\begin{vmatrix} D-3 & 30t-1 \\ 1 & 2e^{3t} \end{vmatrix} = 30t - 1$$

because $(D-3)e^{3t} = 3e^{3t} - 3e^{3t} = 0$. Thus the equation for z is

$$(D-1)(D-3)(D-5)z = 30t - 1;$$

that is,

$$z''' - 9z'' + 23z' - 15z = 30t - 1. \tag{31}$$

The complementary function of (31) is $z_c = ae^t + be^{3t} + ce^{5t}$ where a, b, and c are arbitrary constants. To find a particular solution, we substitute $z_p = At + B$ in (31) and find routinely that $A = -2$ and $B = -3$. Thus the general solution of (31) is

$$z(t) = ae^t + be^{3t} + ce^{5t} - 2t - 3. \tag{32}$$

Substituting this solution for z in the last equation in (27), we get $y = 3z - z'$; that is,

$$y(t) = 2ae^t - 2ce^{5t} - 6t - 7. \tag{33}$$

We substitute these results in the second equation in (27) to find that

$$x = 3y - 2z - y' + 2e^{3t},$$

and consequently that

$$x(t) = 2ae^t - 2(b-1)e^{3t} + 2ce^{5t} - 14t - 9. \tag{34}$$

Thus we have directly obtained a general solution—Eqs. (32), (33), and (34)—involving the proper number (three) of constants.

5.2   PROBLEMS

Find general solutions of the linear systems in Problems 1–20. If initial conditions are given, find the solution that satisfies them.

1. $x' = -x + 3y,\ y' = 2y$
2. $x' = x - 2y,\ y' = 2x - 3y$
3. $x' = -3x + 2y,\ y' = -3x + 4y;\quad x(0) = 0,\ y(0) = 2$
4. $x' = 3x - y,\ y' = 5x - 3y;\quad x(0) = 1,\ y(0) = -1$
5. $x' = -3x - 4y,\ y' = 2x + y$
6. $x' = x + 9y,\ y' = -2x - 5y;\quad x(0) = 3,\ y(0) = 2$
7. $x' = 4x + y + 2t,\ y' = -2x + y$
8. $x' = 2x + y,\ y' = x + 2y - e^{2t}$
9. $x' = 2x - 3y + 2\sin 2t,\ y' = x - 2y - \cos 2t$
10. $x' + 2y' = 4x + 5y,\ 2x' - y' = 3x;\quad x(0) = 1,\ y(0) = -1$
11. $2y' - x' = x + 3y + e^t,\ 3x' - 4y' = x - 15y + e^{-t}$
12. $x'' = 6x + 2y,\ y'' = 3x + 7y$

13. $x'' = -5x + 2y,\ y'' = 2x - 8y$
14. $x'' = -4x + \sin t,\ y'' = 4x - 8y$
15. $x'' - 3y' - 2x = 0,\ y'' + 3x' - 2y = 0$
16. $x'' + 13y' - 4x = 6\sin t,\ y'' - 2x' - 9y = 0$
17. $x'' + y'' - 3x' - y' - 2x + 2y = 0,$
    $2x'' + 3y'' - 9x' - 2y' - 4x + 6y = 0$
18. $x' = x + 2y + z,\ y' = 6x - y,\ z' = -x - 2y - z$
19. $x' = 4x - 2y,\ y' = -4x + 4y - 2z,\ z' = -4y + 4z$
20. $x' = y + z + e^{-t},\ y' = x + z,\ z' = x + y$
    (*Suggestion:* Solve the characteristic equation by inspection.)
21. Suppose that $L_1 = a_1 D^2 + b_1 D + c_1$ and $L_2 = a_2 D^2 + b_2 D + c_2$, where the coefficients are all constants, and that $x(t)$ is a twice differentiable function. Verify that $L_1 L_2 x = L_2 L_1 x$.
22. Suppose that $L_1 x = t\, Dx + x$ and that $L_2 x = Dx + tx$. Show that $L_1 L_2 \neq L_2 L_1$. Thus linear operators with *variable* coefficients generally do not commute.

Show that each of the systems in Problems 23–25 is degenerate. In each problem determine—by attempting to solve the system—whether it has infinitely many solutions or no solutions.

23. $(D + 2)x + (D + 2)y = e^{-3t}, (D + 3)x + (D + 3)y = e^{-2t}$

24. $(D + 2)x + (D + 2)y = t, (D + 3)x + (D + 3)y = t^2$

25. $(D^2 + 5D + 6)x + D(D + 2)y = 0, (D + 3)x + Dy = 0$

In each of Problems 26–29, first calculate the operational determinant, then attempt to solve the given system in order to determine the number of arbitrary constants that appear in its general solution.

26. $(D^2 + 1)x + D^2y = 2e^{-t}$
$(D^2 - 1)x + D^2y = 0$

27. $(D^2 + 1)x + (D^2 + 2)y = 2e^{-t}$
$(D^2 - 1)x + \quad\quad D^2y = 0$

28. $(D^2 + D)x + \quad\quad D^2y = 2e^{-t}$
$(D^2 - 1)x + (D^2 - D)y = 0$

29. $(D^2 + 1)x - D^2y = 2e^{-t}$
$(D^2 - 1)x + D^2y = 0$

30. Suppose that the salt concentration in each of the two brine tanks of Example 2 of Section 5.1 initially ($t = 0$) is 0.5 lb/gal. Then solve the system in (5) there to find the amounts $x(t)$ and $y(t)$ of salt in the two tanks at time t.

31. Suppose that the electrical network of Example 3 of Section 5.1 is initially open—no currents are flowing. Assume that it is closed at time $t = 0$; solve the system in (9) there to find $I_1(t)$ and $I_2(t)$.

32. Repeat Problem 31, except use the electrical network of Problem 29 of Section 5.1.

33. Repeat Problem 32, except use the electrical network of Problem 30 of Section 5.1. Assume that $I_1(0) = 2$ and $Q(0) = 0$, so that at time $t = 0$ there is no charge on the capacitor.

34. Three 100-gal brine tanks are connected as indicated in Fig. 5.8 of Section 5.1. Assume that the first tank initially contains 100 lb of salt, while the other two are filled with fresh water. Find the amounts of salt in each of the three tanks at time t. (*Suggestion:* Examine the equations to be derived in Problem 28 of Section 5.1.)

35. From Problem 33 of Section 5.1, recall the equations of motion

$$mx'' = qBy', \quad\quad my'' = -qBx'$$

for a particle of mass m and electrical charge q under the influence of the uniform magnetic field $\mathbf{B} = B\mathbf{k}$. Suppose that the initial conditions are $x(0) = r_0$, $y(0) = 0$, $x'(0) = 0$, and $y'(0) = -\omega r_0$ where $\omega = qB/m$. Show that the trajectory of the particle is a circle of radius r_0.

36. If, in addition to the magnetic field $\mathbf{B} = B\mathbf{k}$, the charged particle of Problem 35 moves with velocity $\mathbf{v}$ under the influence of a uniform electric field $\mathbf{E} = E\mathbf{i}$, then the force acting on it is $\mathbf{F} = q(\mathbf{E} + \mathbf{v} \times \mathbf{B})$. Assume that the particle starts from rest at the origin. Show that its trajectory is the cycloid

$$x = a(1 - \cos \omega t), \quad\quad y = -a(\omega t - \sin \omega t)$$

where $a = E/\omega B$ and $\omega = qB/m$. The graph of such a cycloid is shown in Fig. 5.12.

FIGURE 5.12 Cycloidal path of the particle of Problem 36.

37. In the mass-and-spring system of Example 3, suppose instead that $m_1 = 2$, $m_2 = 0.5$, $k_1 = 75$, and $k_2 = 25$. (a) Find the general solution of the equations of motion of the system. In particular, show that its natural frequencies are $\omega_1 = 5$ and $\omega_2 = 5\sqrt{3}$. (b) Describe the natural modes of oscillation of the system.

38. Consider the system of two masses and three springs shown in Fig. 5.13. Derive the equations of motion

$$m_1x'' = -(k_1 + k_2)x + \quad\quad k_2y,$$
$$m_2y'' = \quad\quad k_2x - (k_2 + k_3)y.$$

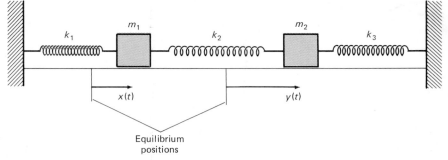

FIGURE 5.13 The mechanical system of Problem 38.

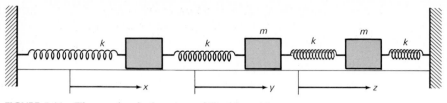

FIGURE 5.14 The mechanical system of Problem 44.

In each of Problems 39–43, find the general solution of the system in Problem 38 with the given masses and spring constants. Find the natural frequencies of the mass-and-spring system, and describe its natural modes of oscillation.

39. $m_1 = 1, m_2 = 1;$ $k_1 = 1, k_2 = 4, k_3 = 1$

40. $m_1 = 1, m_2 = 2;$ $k_1 = 1, k_2 = 2, k_3 = 2$

41. $m_1 = 1, m_2 = 1;$ $k_1 = 1, k_2 = 2, k_3 = 1$

42. $m_1 = 1, m_2 = 1;$ $k_1 = 2, k_2 = 1, k_3 = 2$

43. $m_1 = 1, m_2 = 2;$ $k_1 = 2, k_2 = 4, k_3 = 4$

44. (a) For the system shown in Fig. 5.14, derive the equations of motion

$$mx'' = -2kx + ky,$$
$$my'' = kx - 2ky + kz,$$
$$mz'' = ky - 2kz.$$

(b) Assume that $m = k = 1$. Show that the natural frequencies of oscillation of the system are $\omega_1 = \sqrt{2}, \omega_2 = \sqrt{2 - \sqrt{2}},$ and $\omega_3 = \sqrt{2 + \sqrt{2}}.$

45. Suppose that the trajectory $(x(t), y(t))$ of a particle moving in the plane satisfies the initial value problem

$$x'' - 2y' + 3x = 0,$$
$$y'' + 2x' + 3y = 0;$$
$$x(0) = 4, \qquad y(0) = x'(0) = y'(0) = 0.$$

Solve this problem. You should obtain

$$x(t) = 3 \cos t + \cos 3t,$$
$$y(t) = 3 \sin t - \sin 3t.$$

Verify that these equations describe the *hypocycloid* traced by a point $P(x, y)$ fixed on the circumference of a circle of radius $b = 1$ that rolls around inside a circle of radius $a = 4$. If P begins at $A(a, 0)$ when $t = 0$, then the parameter t represents the angle AOC shown in Fig. 5.15.

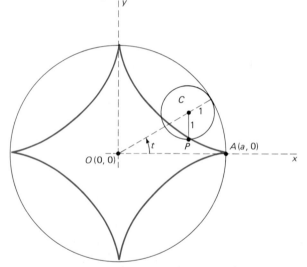

FIGURE 5.15 The hypocycloid of Problem 45.

5.3

Linear Systems and Matrices

Although the simple elimination techniques of Section 5.2 suffice for the solution of small linear systems containing only two or three equations with constant coefficients, the general properties of linear systems—as well as solution methods suitable for larger systems—are most easily and concisely described in matrix notation. It is best that the reader of this section has had some prior exposure to matrices and determinants. If not, we have included—for ready reference—a complete and self-contained account of the notation and terminology that is needed.

REVIEW OF MATRIX NOTATION AND TERMINOLOGY

An $m \times n$ **matrix A** is a rectangular array of mn numbers (or **elements**) arranged in m (horizontal) **rows** and n (vertical) **columns**:

$$\mathbf{A} = \begin{bmatrix} a_{11} & a_{12} & a_{13} & \cdots & a_{1j} & \cdots & a_{1n} \\ a_{21} & a_{22} & a_{23} & \cdots & a_{2j} & \cdots & a_{2n} \\ a_{31} & a_{32} & a_{33} & \cdots & a_{3j} & \cdots & a_{3n} \\ \vdots & \vdots & \vdots & & \vdots & & \vdots \\ a_{i1} & a_{i2} & a_{i3} & \cdots & a_{ij} & \cdots & a_{in} \\ \vdots & \vdots & \vdots & & \vdots & & \vdots \\ a_{m1} & a_{m2} & a_{m3} & \cdots & a_{mj} & \cdots & a_{mn} \end{bmatrix}. \tag{1}$$

We will ordinarily denote matrices by bold-faced capital letters. Sometimes we use the abbreviation $\mathbf{A} = [a_{ij}]$ for the matrix with the element a_{ij} in the ith row and jth column, as above. We denote the **zero matrix**, each entry of which is zero, by

$$\mathbf{0} = \begin{bmatrix} 0 & 0 & \cdots & 0 \\ 0 & 0 & \cdots & 0 \\ \vdots & \vdots & & \vdots \\ 0 & 0 & \cdots & 0 \end{bmatrix}. \tag{2}$$

Actually, for each $m > 0$ and $n > 0$, there is an $m \times n$ zero matrix, but the single symbol $\mathbf{0}$ will suffice for all these zero matrices.

Two $m \times n$ matrices $\mathbf{A} = [a_{ij}]$ and $\mathbf{B} = [b_{ij}]$ are said to be **equal** if corresponding elements are equal; that is, if $a_{ij} = b_{ij}$ for $1 \leq i \leq m$ and $1 \leq j \leq n$. We **add A** and **B** by adding corresponding entries:

$$\mathbf{A} + \mathbf{B} = [a_{ij}] + [b_{ij}] = [a_{ij} + b_{ij}]. \tag{3}$$

Thus the element in row i and column j of $\mathbf{C} = \mathbf{A} + \mathbf{B}$ is $c_{ij} = a_{ij} + b_{ij}$. To multiply the matrix $\mathbf{A} = [a_{ij}]$ by the number c, we simply multiply each of its entries by c:

$$c\mathbf{A} = \mathbf{A}c = [ca_{ij}]. \tag{4}$$

Thus

$$\begin{bmatrix} 2 & -3 \\ 4 & 7 \end{bmatrix} + \begin{bmatrix} -13 & 10 \\ 7 & -5 \end{bmatrix} = \begin{bmatrix} -11 & 7 \\ 11 & 2 \end{bmatrix}$$

and

$$6\begin{bmatrix} 3 & 0 \\ 5 & -7 \end{bmatrix} = \begin{bmatrix} 18 & 0 \\ 30 & -42 \end{bmatrix}.$$

We denote $(-1)\mathbf{A}$ by $-\mathbf{A}$ and define **subtraction** of matrices as follows:

$$\mathbf{A} - \mathbf{B} = \mathbf{A} + (-\mathbf{B}). \qquad (5)$$

The matrix operations just defined have the following properties, each of which is analogous to a familiar algebraic property of the real number system:

$$\mathbf{A} + \mathbf{0} = \mathbf{0} + \mathbf{A} = \mathbf{A}, \qquad \mathbf{A} - \mathbf{A} = \mathbf{0}; \qquad (6)$$

$$\mathbf{A} + \mathbf{B} = \mathbf{B} + \mathbf{A} \qquad \text{(commutativity)}; \qquad (7)$$

$$\mathbf{A} + (\mathbf{B} + \mathbf{C}) = (\mathbf{A} + \mathbf{B}) + \mathbf{C} \qquad \text{(associativity)}; \qquad (8)$$

$$\left.\begin{array}{l} c(\mathbf{A} + \mathbf{B}) = c\mathbf{A} + c\mathbf{B}, \\ (c + d)\mathbf{A} = c\mathbf{A} + d\mathbf{A}. \end{array}\right\} \qquad \text{(distributivity)} \qquad (9)$$

Each of these properties is readily verified by elementwise application of a corresponding property of the real numbers. For example, $a_{ij} + b_{ij} = b_{ij} + a_{ij}$ for all i and j because addition of real numbers is commutative. Consequently

$$\mathbf{A} + \mathbf{B} = \left[a_{ij} + b_{ij}\right] = \left[b_{ij} + a_{ij}\right] = \mathbf{B} + \mathbf{A}.$$

The **transpose** $\mathbf{A}^T$ of the $m \times n$ matrix $\mathbf{A} = \left[a_{ij}\right]$ is the $n \times m$ (note!) matrix whose jth column is the ith row of $\mathbf{A}$ (and, consequently, whose ith row is the jth column of $\mathbf{A}$). Thus $\mathbf{A}^T = \left[a_{ji}\right]$, although this is not notationally perfect; it must be remembered that $\mathbf{A}^T$ will not have the same shape as $\mathbf{A}$ unless $\mathbf{A}$ is a **square** matrix—that is, unless $m = n$.

An $m \times 1$ matrix—one having only a single column—is called a **column vector**, or simply a **vector**. We often denote column vectors by boldface lowercase letters, as in

$$\mathbf{b} = \begin{bmatrix} 3 \\ -7 \\ 0 \end{bmatrix} \quad \text{or} \quad \mathbf{x} = \begin{bmatrix} x_1 \\ x_2 \\ \vdots \\ x_m \end{bmatrix}.$$

Similary, a **row vector** is a $1 \times n$ matrix—one having only a single row, such as $\mathbf{c} = \begin{bmatrix} 5 & 17 & 0 & -3 \end{bmatrix}$. For esthetic and typographical reasons, it is customary to write a column vector as the transpose of a row vector; for example, the two column vectors above may be written in the forms

$$\mathbf{b} = \begin{bmatrix} 3 & -7 & 0 \end{bmatrix}^T \quad \text{and} \quad \mathbf{x} = \begin{bmatrix} x_1 & x_2 & \cdots & x_n \end{bmatrix}^T.$$

Sometimes it is convenient to describe an $m \times n$ matrix in terms of either its m row vectors or its n column vectors. Thus if we write

$$\mathbf{A} = \begin{bmatrix} \mathbf{a}_1 \\ \mathbf{a}_2 \\ \vdots \\ \mathbf{a}_m \end{bmatrix} \quad \text{and} \quad \mathbf{B} = \begin{bmatrix} \mathbf{b}_1 & \mathbf{b}_2 & \cdots & \mathbf{b}_n \end{bmatrix},$$

it is understood that $\mathbf{a}_1, \mathbf{a}_2, \ldots, \mathbf{a}_m$ are the *row* vectors of matrix $\mathbf{A}$, while $\mathbf{b}_1, \mathbf{b}_2, \ldots, \mathbf{b}_n$ are the *column* vectors of the matrix $\mathbf{B}$.

CHAPTER 5:   Linear Systems of Differential Equations

The properties listed in Eqs. (6)–(9) are quite natural and expected. The first surprises in the realm of matrix arithmetic come with multiplication. We define first the **scalar product a · b** of a row vector **a** and a column vector **b**, each having the same number p of entries. If

$$\mathbf{a} = [a_1 \quad a_2 \quad \cdots \quad a_p] \quad \text{and} \quad \mathbf{b} = [b_1 \quad b_2 \quad \cdots \quad b_p]^T,$$

then **a · b** is defined by means of the equation

$$\mathbf{a} \cdot \mathbf{b} = \sum_{k=1}^{p} a_k b_k = a_1 b_1 + a_2 b_2 + \cdots + a_p b_p, \tag{10}$$

exactly as in the scalar or *dot* product of two vectors—a familiar topic from elementary calculus.

The product **AB** of two matrices is defined only if the number of columns of **A** is equal to the number of rows of **B**. If **A** is an $m \times p$ matrix and **B** is a $p \times n$ matrix, then their product **AB** is the $m \times n$ matrix $\mathbf{C} = [c_{ij}]$, where c_{ij} is the scalar product of the ith row vector $\mathbf{a}_i$ of **A** and the jth column vector $\mathbf{b}_j$ of **B**. Thus

$$\mathbf{AB} = [\mathbf{a}_i \cdot \mathbf{b}_j]. \tag{11}$$

In terms of the individual entries of $\mathbf{A} = [a_{ij}]$ and $\mathbf{B} = [b_{ij}]$, Eq. (11) can be recast in the form

$$c_{ij} = \sum_{k=1}^{p} a_{ik} b_{kj}. \tag{12}$$

For purposes of hand computation, the definition in (11) and (12) is easy to remember by visualizing the picture

$$\mathbf{a}_i \rightarrow \begin{bmatrix} a_{11} & a_{12} & \cdots & a_{1p} \\ a_{21} & a_{22} & \cdots & a_{2p} \\ \vdots & \vdots & & \vdots \\ \boxed{a_{i1} \quad a_{i2} \quad \cdots \quad a_{ip}} \\ \vdots & \vdots & & \vdots \\ a_{m1} & a_{m2} & \cdots & a_{mp} \end{bmatrix} \begin{bmatrix} b_{11} & b_{12} & \cdots & \boxed{b_{1j}} & \cdots & b_{1n} \\ b_{21} & b_{22} & \cdots & b_{2j} & \cdots & b_{2n} \\ \vdots & \vdots & & \vdots & & \vdots \\ b_{p1} & b_{p2} & \cdots & \boxed{b_{pj}} & \cdots & b_{pn} \end{bmatrix},$$

in which the row vector $\mathbf{a}_i$ and the column vector $\mathbf{b}_j$ are spotlighted; c_{ij} is simply their scalar product. It may help to think of "pouring the rows of **A** down the columns of **B**." This also reminds us that the number of columns of **A** must be equal to the number of rows of **B**.

You should pause to check your understanding of the definition of matrix multiplication by verifying that

$$\begin{bmatrix} 2 & -3 \\ -1 & 5 \end{bmatrix} \begin{bmatrix} 13 & 9 \\ 4 & 0 \end{bmatrix} = \begin{bmatrix} 14 & 18 \\ 7 & -9 \end{bmatrix}$$

and that

$$\begin{bmatrix} 2 & -3 & 1 \\ 4 & 5 & -2 \\ 6 & -7 & 0 \end{bmatrix} \begin{bmatrix} x \\ y \\ z \end{bmatrix} = \begin{bmatrix} 2x - 3y + z \\ 4x + 5y - 2z \\ 6x - 7y \end{bmatrix}.$$

The BASIC program shown in Fig. 5.16 can be used to compute the product $\mathbf{C} = \mathbf{AB}$ of an $m \times p$ matrix $\mathbf{A}$ and a $p \times n$ matrix $\mathbf{B}$. For instance, if we enter the values $M = 4$, $P = 2$, and $N = 3$ at line 110, insert the data 1, 2, 3, 4, 5, 6, 7, 8 in line 130 and the data 2, 1, 3, -1, 3, -2 in line 140, when we execute this program we find that

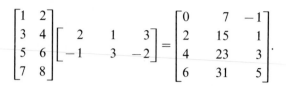

$$
\begin{bmatrix} 1 & 2 \\ 3 & 4 \\ 5 & 6 \\ 7 & 8 \end{bmatrix} \begin{bmatrix} 2 & 1 & 3 \\ -1 & 3 & -2 \end{bmatrix} = \begin{bmatrix} 0 & 7 & -1 \\ 2 & 15 & 1 \\ 4 & 23 & 3 \\ 6 & 31 & 5 \end{bmatrix}.
$$

```
100 REM--Product  C  of matrices  A  and  B
105 REM
110       INPUT "Dimensions m,p,n"; M,P,N
120       DIM A(M,P), B(P,N), C(M,N)
130       DATA [ elements of  A  go here ]
140       DATA [ elements of  B  go here ]
145 REM
150 REM--Load m by p matrix A:
160       FOR I = 1 TO M : FOR J = 1 TO P
170           READ A(I,J)
180       NEXT J : NEXT I
190 REM
200 REM--Load p by n matrix B:
210       FOR I = 1 TO P : FOR J = 1 TO N
220           READ B(I,J)
230       NEXT J : NEXT I
240 REM
250 REM--Calculate matrix C = AB:
260       FOR I = 1 TO M : FOR J = 1 TO N
270           FOR K = 1 TO P
280             C(I,J) = C(I,J) + A(I,K)*B(K,J)
290           NEXT K
300       PRINT C(I,J),
310       NEXT J : PRINT : NEXT I
320 REM
330       END
```

FIGURE 5.16   BASIC program for finding the product of two matrices **A** and **B**.

It can be shown by direct (though lengthy) computation based on its definition that matrix multiplication is associative and is also distributive with respect to matrix addition:

$$\mathbf{A(BC)} = \mathbf{(AB)C} \tag{13}$$

and

$$\mathbf{A(B + C)} = \mathbf{AB} + \mathbf{AC} \tag{14}$$

provided that the matrices are of such sizes that the indicated multiplications and additions are possible.

But matrix multiplication is *not* commutative. That is, if **A** and **B** are both $n \times n$ matrices (so that both the products **AB** and **BA** are defined and have the same dimensions—$n \times n$), then, in general,

$$\mathbf{AB} \neq \mathbf{BA}. \tag{15}$$

Moreover, it can happen that

$$\mathbf{AB} = \mathbf{0} \quad \text{even though} \quad \mathbf{A} \neq \mathbf{0} \quad \text{and} \quad \mathbf{B} \neq \mathbf{0}. \tag{16}$$

Examples illustrating the phenomena in (15) and (16) may be found in the problems, though you can easily construct your own examples using 2×2 matrices with integral entries.

A square $n \times n$ matrix is said to have **order** n. The **identity** matrix of order n is the square matrix

$$\mathbf{I} = \begin{bmatrix} 1 & 0 & 0 & 0 & \cdots & 0 \\ 0 & 1 & 0 & 0 & \cdots & 0 \\ 0 & 0 & 1 & 0 & \cdots & 0 \\ 0 & 0 & 0 & 1 & \cdots & 0 \\ \vdots & \vdots & \vdots & \vdots & & \vdots \\ 0 & 0 & 0 & 0 & \cdots & 1 \end{bmatrix} \tag{17}$$

for which each entry on the **principal diagonal** is 1 and all off-diagonal entries are zero. It is quite easy to verify that

$$\mathbf{AI} = \mathbf{IA} = \mathbf{A} \tag{18}$$

for every square matrix $\mathbf{A}$ of the same order as $\mathbf{I}$.

If $\mathbf{A}$ is a square matrix, then an **inverse** of $\mathbf{A}$ is a square matrix $\mathbf{B}$ of the same order as $\mathbf{A}$ such that *both*

$$\mathbf{AB} = \mathbf{I} \quad \text{and} \quad \mathbf{BA} = \mathbf{I}.$$

It is not difficult to show that if the matrix $\mathbf{A}$ has an inverse, then this inverse is unique. Consequently we may speak of *the* inverse of $\mathbf{A}$ and denote it by $\mathbf{A}^{-1}$. Thus

$$\mathbf{AA}^{-1} = \mathbf{I} = \mathbf{A}^{-1}\mathbf{A}, \tag{19}$$

given the existence of $\mathbf{A}^{-1}$. It is clear that some square matrices do not have inverses; consider any square zero matrix.

In linear algebra it is proved that $\mathbf{A}^{-1}$ exists if and only if the determinant of the square matrix $\mathbf{A}$ (denoted by det $(\mathbf{A})$) is nonzero, in which case the matrix $\mathbf{A}$ is said to be **nonsingular**; if det $(\mathbf{A}) = 0$, then $\mathbf{A}$ is called a **singular** matrix.

We assume that the student has computed 2×2 and 3×3 determinants in earlier courses. If $\mathbf{A} = [a_{ij}]$ is a 2×2 matrix, then its **determinant** det $(\mathbf{A}) = |\mathbf{A}|$ is defined as

$$|\mathbf{A}| = \begin{vmatrix} a_{11} & a_{12} \\ a_{21} & a_{22} \end{vmatrix} = a_{11}a_{22} - a_{12}a_{21}. \tag{20}$$

Determinants of higher order may be defined by induction, as follows. If $\mathbf{A} = [a_{ij}]$ is an $n \times n$ matrix, let $\mathbf{A}_{ij}$ denote the $(n-1) \times (n-1)$ matrix obtained from $\mathbf{A}$ by deleting its ith row and its jth column. The *expansion* of the determinant $|\mathbf{A}|$ along its ith row is given by

$$|\mathbf{A}| = \sum_{j=1}^{n} (-1)^{i+j} a_{ij} |\mathbf{A}_{ij}| \qquad (i \text{ fixed}), \tag{21a}$$

while the expansion along its jth column is given by

$$|\mathbf{A}| = \sum_{i=1}^{n} (-1)^{i+j} a_{ij} |\mathbf{A}_{ij}| \qquad (j \text{ fixed}). \tag{21b}$$

It is shown in linear algebra that whichever row we use in (21a) and whichever column we use in (21b), the results are the same; hence $|\mathbf{A}|$ is well-defined by these formulas. For example, if

$$\mathbf{A} = \begin{bmatrix} 3 & 1 & -2 \\ 4 & 2 & 1 \\ -2 & 3 & 5 \end{bmatrix},$$

then the expansion of $|\mathbf{A}|$ along its second row is

$$|\mathbf{A}| = -(4) \begin{vmatrix} 1 & -2 \\ 3 & 5 \end{vmatrix} + (2) \begin{vmatrix} 3 & -2 \\ -2 & 5 \end{vmatrix} - (1) \begin{vmatrix} 3 & 1 \\ -2 & 3 \end{vmatrix}$$

$$= -(4)(11) + (2)(11) - (1)(11) = -33,$$

while the expansion of $|\mathbf{A}|$ along its third column is

$$|\mathbf{A}| = (-2) \begin{vmatrix} 4 & 2 \\ -2 & 3 \end{vmatrix} - (1) \begin{vmatrix} 3 & 1 \\ -2 & 3 \end{vmatrix} + (5) \begin{vmatrix} 3 & 1 \\ 4 & 2 \end{vmatrix}$$

$$= (-2)(16) - (1)(11) + (5)(2) = -33.$$

MATRIX-VALUED FUNCTIONS

A **matrix-valued function**, or simply matrix function, is a matrix such as

$$\mathbf{x}(t) = \begin{bmatrix} x_1(t) \\ \vdots \\ x_n(t) \end{bmatrix} \tag{22a}$$

or

$$\mathbf{A}(t) = \begin{bmatrix} a_{11}(t) & a_{12}(t) & \cdots & a_{1n}(t) \\ a_{21}(t) & a_{22}(t) & \cdots & a_{2n}(t) \\ \vdots & \vdots & & \vdots \\ a_{m1}(t) & a_{m2}(t) & \cdots & a_{mn}(t) \end{bmatrix} \tag{22b}$$

in which each entry is a function of t. We say that the matrix function $\mathbf{A}(t)$ is continuous (or differentiable) at a point (or on an interval) if each of its entries has the same property. The **derivative** of a differentiable matrix function is defined by elementwise differentiation; that is,

$$\mathbf{A}'(t) = \frac{d\mathbf{A}}{dt} = \left[\frac{da_{ij}}{dt} \right]. \tag{23}$$

Thus if

$$\mathbf{x}(t) = \begin{bmatrix} t \\ t^2 \\ e^{-t} \end{bmatrix} \quad \text{and} \quad \mathbf{A}(t) = \begin{bmatrix} \sin t & 1 \\ t & \cos t \end{bmatrix},$$

then

$$\frac{d\mathbf{x}}{dt} = \begin{bmatrix} 1 \\ 2t \\ -e^{-t} \end{bmatrix} \quad \text{and} \quad \mathbf{A}'(t) = \begin{bmatrix} \cos t & 0 \\ 1 & -\sin t \end{bmatrix}.$$

The differentiation rules

$$\frac{d}{dt}(\mathbf{A} + \mathbf{B}) = \frac{d\mathbf{A}}{dt} + \frac{d\mathbf{B}}{dt} \tag{24}$$

and

$$\frac{d}{dt}(\mathbf{AB}) = \mathbf{A}\frac{d\mathbf{B}}{dt} + \frac{d\mathbf{A}}{dt}\mathbf{B} \tag{25}$$

follow readily by elementwise application of the analogous differentiation rules of elementary calculus for real-valued functions. If c is a (constant) real number and $\mathbf{C}$ is a constant matrix, then

$$\frac{d}{dt}(c\mathbf{A}) = c\frac{d\mathbf{A}}{dt}, \quad \frac{d}{dt}(\mathbf{CA}) = \mathbf{C}\frac{d\mathbf{A}}{dt}, \quad \text{and} \quad \frac{d}{dt}(\mathbf{AC}) = \frac{d\mathbf{A}}{dt}\mathbf{C}. \tag{26}$$

Because of the noncommutativity of matrix multiplication, it is important not to reverse the order of the factors in (25) and (26).

FIRST ORDER LINEAR SYSTEMS

The notation and terminology of matrices and vectors may seem rather elaborate when first encountered, but it is readily assimilated with practice. Our main use for matrix notation will be the simplification of computations with systems of differential equations, especially those computations that would be burdensome in scalar notation.

We discuss here the general system of n first order linear equations

$$\begin{aligned} x_1' &= p_{11}(t)x_1 + p_{12}(t)x_2 + \cdots + p_{1n}(t)x_n + f_1(t), \\ x_2' &= p_{21}(t)x_1 + p_{22}(t)x_2 + \cdots + p_{2n}(t)x_n + f_2(t), \\ &\vdots \\ x_n' &= p_{n1}(t)x_1 + p_{n2}(t)x_2 + \cdots + p_{nn}(t)x_n + f_n(t). \end{aligned} \tag{27}$$

If we introduce the *coefficient matrix*

$$\mathbf{P}(t) = [p_{ij}(t)]$$

and the column vectors

$$\mathbf{x} = [x_i] \quad \text{and} \quad \mathbf{f}(t) = [f_i(t)],$$

then the system in (27) takes the form of a single matrix equation:

$$\mathbf{x}' = \mathbf{P}(t)\mathbf{x} + \mathbf{f}(t). \tag{28}$$

We will see that the general theory of the linear system in (27) closely parallels that of a single nth order linear equation. The matrix notation used in Eq. (28) not only emphasizes this analogy, but also saves a great deal of space.

A **solution** of Eq. (28) on the open interval I is a column vector function $\mathbf{x}(t) = [x_i(t)]$ such that the component functions of $\mathbf{x}$ satisfy the system in (27) identically on I. If the functions $p_{ij}(t)$ and $f_i(t)$ are all continuous on I, then the theorem of Section 5.1 guarantees the existence on I of a unique solution $\mathbf{x}(t)$ satisfying preassigned initial conditions $\mathbf{x}(a) = \mathbf{b}$.

EXAMPLE 1 The first order system

$$x_1' = 4x_1 - 3x_2$$

$$x_2' = 6x_1 - 7x_2$$

can be written as the single matrix equation

$$\mathbf{x}' = \begin{bmatrix} 4 & -3 \\ 6 & -7 \end{bmatrix} \mathbf{x}.$$

To verify that the vector functions

$$\mathbf{x}_1(t) = \begin{bmatrix} 3e^{2t} \\ 2e^{2t} \end{bmatrix} \quad \text{and} \quad \mathbf{x}_2(t) = \begin{bmatrix} e^{-5t} \\ 3e^{-5t} \end{bmatrix}$$

are both solutions of the matrix differential equation with coefficient matrix P, we need only calculate

$$P\mathbf{x}_1 = \begin{bmatrix} 4 & -3 \\ 6 & -7 \end{bmatrix} \begin{bmatrix} 3e^{2t} \\ 2e^{2t} \end{bmatrix} = \begin{bmatrix} 6e^{2t} \\ 4e^{2t} \end{bmatrix} = \mathbf{x}_1'$$

and

$$P\mathbf{x}_2 = \begin{bmatrix} 4 & -3 \\ 6 & -7 \end{bmatrix} \begin{bmatrix} e^{-5t} \\ 3e^{-5t} \end{bmatrix} = \begin{bmatrix} -5e^{-5t} \\ -15e^{-5t} \end{bmatrix} = \mathbf{x}_2'.$$

To investigate the general nature of the solutions of Eq. (28), we consider first the **associated homogeneous equation**

$$\mathbf{x}' = \mathbf{P}(t)\mathbf{x}, \tag{29}$$

which has the form in Eq. (28), though with $\mathbf{f}(t) \equiv \mathbf{0}$. We expect it to have n solutions $\mathbf{x}_1, \mathbf{x}_2, \ldots, \mathbf{x}_n$ that are independent in some appropriate sense, and such that every solution is a linear combination of these n particular solutions. Given n solutions $\mathbf{x}_1, \mathbf{x}_2, \ldots, \mathbf{x}_n$ of (29), let us write

$$\mathbf{x}_j(t) = \begin{bmatrix} x_{1j}(t) \\ \vdots \\ x_{ij}(t) \\ \vdots \\ x_{nj}(t) \end{bmatrix}. \tag{30}$$

Thus $x_{ij}(t)$ denotes the ith component of the vector $\mathbf{x}_j(t)$, so the second subscript refers to the vector function $\mathbf{x}_j(t)$, while the first subscript refers to a component of this function. The following theorem is analogous to Theorem 1 of Section 2.2.

Proof   We know that $\mathbf{x}_i' = \mathbf{P}(t)\mathbf{x}_i$ for each i $(1 \leq i \leq n)$, so it follows immediately that

$$\begin{aligned} \mathbf{x}' &= c_1\mathbf{x}_1' + c_2\mathbf{x}_2' + \cdots + c_n\mathbf{x}_n' \\ &= c_1\mathbf{P}(t)\mathbf{x}_1 + c_2\mathbf{P}(t)\mathbf{x}_2 + \cdots + c_n\mathbf{P}(t)\mathbf{x}_n \\ &= \mathbf{P}(t)(c_1\mathbf{x}_1 + c_2\mathbf{x}_2 + \cdots + c_n\mathbf{x}_n). \end{aligned}$$

That is, $\mathbf{x}' = \mathbf{P}(t)\mathbf{x}$, as desired. The remarkable simplicity of this proof demonstrates clearly the advantages of matrix notation. ∎

For instance, if $\mathbf{x}_1$ and $\mathbf{x}_2$ are the two solutions of

$$\mathbf{x}' = \begin{bmatrix} 4 & -3 \\ 6 & -7 \end{bmatrix}\mathbf{x}$$

discussed in Example 1, then the linear combination

$$\mathbf{x} = c_1\mathbf{x}_1 + c_2\mathbf{x}_2 = c_1\begin{bmatrix} 3e^{2t} \\ 2e^{2t} \end{bmatrix} + c_2\begin{bmatrix} e^{-5t} \\ 3e^{-5t} \end{bmatrix}$$

is also a solution. In scalar form with $\mathbf{x} = (x_1, x_2)$, this gives the solution

$$x_1(t) = 3c_1e^{2t} + c_2e^{-5t}$$

$$x_2(t) = 2c_1e^{2t} + 3c_2e^{-5t}$$

that is equivalent to the general solution we found by the method of elimination in Example 2 of Section 5.2.

Linear independence is defined in the same way for vector-valued functions as for real-valued functions (Section 2.2). The vector-valued functions $\mathbf{x}_1, \mathbf{x}_2, \ldots, \mathbf{x}_n$ are **linearly dependent** on the interval I provided that there exist constants $c_1, c_2, \ldots, c_n$ *not all zero* such that

$$c_1\mathbf{x}_1(t) + c_2\mathbf{x}_2(t) + \cdots + c_n\mathbf{x}_n(t) = \mathbf{0} \qquad (32)$$

for all t in I. Otherwise they are **linearly independent**. Equivalently, they are linearly independent provided that no one of them is a linear combination of the others. For instance, the two solutions $\mathbf{x}_1$ and $\mathbf{x}_2$ of Example 1 are linearly independent, because it is clear that neither is a scalar multiple of the other.

Just as in the case of a single nth order equation, there is a Wronskian determinant that tells us whether or not n given solutions of the homogeneous equation in (29) are linearly dependent. If $\mathbf{x}_1, \mathbf{x}_2, \ldots, \mathbf{x}_n$ are such solutions, then their

Wronskian is the $n \times n$ determinant

$$W = \begin{vmatrix} x_{11}(t) & x_{12}(t) & \cdots & x_{1n}(t) \\ x_{21}(t) & x_{22}(t) & \cdots & x_{2n}(t) \\ \vdots & \vdots & & \vdots \\ x_{n1}(t) & x_{n2}(t) & \cdots & x_{nn}(t) \end{vmatrix}. \tag{33}$$

using the notation in (30) for the components of the solutions. We may write either $W(t)$ or $W(\mathbf{x}_1, \mathbf{x}_2, \ldots, \mathbf{x}_n)$. Note that W is the determinant of the matrix which has as its *column* vectors the solutions $\mathbf{x}_1, \mathbf{x}_2, \ldots, \mathbf{x}_n$. The following theorem is analogous to Theorem 3 of Section 2.2. Moreover, its proof is essentially the same, with the above definition of $W(\mathbf{x}_1, \mathbf{x}_2, \ldots, \mathbf{x}_n)$ substituted for the definition of the Wronskian of n solutions of a single nth order equation (see Problems 30–32).

Theorem 2 *Wronskians of Solutions*

Suppose that $\mathbf{x}_1, \mathbf{x}_2, \ldots, \mathbf{x}_n$ are n solutions of the homogeneous linear equation $\mathbf{x}' = \mathbf{P}(t)\mathbf{x}$ on an open interval I. Suppose also that $\mathbf{P}(t) = [p_{ij}(t)]$ and that each function $p_{ij}(t)$ is continuous on I. Let $W = W(\mathbf{x}_1, \mathbf{x}_2, \ldots, \mathbf{x}_n)$. Then:

(a) If $\mathbf{x}_1, \mathbf{x}_2, \ldots, \mathbf{x}_n$ are linearly dependent on I, then $W \equiv 0$ on I.
(b) If $\mathbf{x}_1, \mathbf{x}_2, \ldots, \mathbf{x}_n$ are linearly independent on I, then $W \neq 0$ at each point of I.

Thus there are only two possibilities for solutions of homogeneous systems: Either $W = 0$ everywhere on I, or $W = 0$ for no point of I.

EXAMPLE 2 It is readily verified directly (as in Example 1) that

$$\mathbf{x}_1 = \begin{bmatrix} 2e^t \\ 2e^t \\ e^t \end{bmatrix}, \qquad \mathbf{x}_2 = \begin{bmatrix} 2e^{3t} \\ 0 \\ -e^{3t} \end{bmatrix}, \quad \text{and} \quad \mathbf{x}_3 = \begin{bmatrix} 2e^{5t} \\ -2e^{5t} \\ e^{5t} \end{bmatrix}$$

are solutions of the equation

$$\mathbf{x}' = \begin{bmatrix} 3 & -2 & 0 \\ -1 & 3 & -2 \\ 0 & -1 & 3 \end{bmatrix} \mathbf{x}. \tag{34}$$

The Wronskian of these solutions is

$$W = \begin{vmatrix} 2e^t & 2e^{3t} & 2e^{5t} \\ 2e^t & 0 & -2e^{5t} \\ e^t & -e^{3t} & e^{5t} \end{vmatrix} = e^{9t} \begin{vmatrix} 2 & 2 & 2 \\ 2 & 0 & -2 \\ 1 & -1 & 1 \end{vmatrix} = -16e^{9t},$$

which is never zero. Hence Theorem 2 implies that the solutions $\mathbf{x}_1$, $\mathbf{x}_2$, and $\mathbf{x}_3$ are linearly independent (on any open interval).

The following theorem is analogous to Theorem 4 of Section 2.2. It says that the **general solution** of the *homogeneous* $n \times n$ system $\mathbf{x}' = \mathbf{P}(t)\mathbf{x}$ is a linear combination

$$\mathbf{x} = c_1\mathbf{x}_1 + c_2\mathbf{x}_2 + \cdots + c_n\mathbf{x}_n \tag{35}$$

of any n given linearly independent solutions $\mathbf{x}_1, \mathbf{x}_2, \ldots, \mathbf{x}_n$.

Theorem 3   *General Solutions of Homogeneous Systems*

Let $\mathbf{x}_1, \mathbf{x}_2, \ldots, \mathbf{x}_n$ be n linearly independent solutions of the homogeneous linear equation $\mathbf{x}' = \mathbf{P}(t)\mathbf{x}$ on an open interval I where the entries $p_{ij}(t)$ of $\mathbf{P}(t)$ are continuous. If $\mathbf{x}(t)$ is any solution whatsoever of the equation $\mathbf{x}' = \mathbf{P}(t)\mathbf{x}$ on I, then there exist numbers $c_1, c_2, \ldots, c_n$ such that

$$\mathbf{x}(t) = c_1\mathbf{x}_1(t) + c_2\mathbf{x}_2(t) + \cdots + c_n\mathbf{x}_n(t) \tag{35}$$

for all t in I.

Proof   Let a be a fixed point of I. We show first that there exist numbers $c_1, c_2, \ldots, c_n$ such that the solution

$$\mathbf{y}(t) = c_1\mathbf{x}_1(t) + c_2\mathbf{x}_2(t) + \cdots + c_n\mathbf{x}_n(t) \tag{36}$$

has the same initial values at $t = a$ as does the given solution $\mathbf{x}(t)$; that is, such that

$$c_1\mathbf{x}_1(a) + c_2\mathbf{x}_2(a) + \cdots + c_n\mathbf{x}_n(a) = \mathbf{x}(a). \tag{37}$$

Let $\mathbf{X}(t)$ be the $n \times n$ matrix with column vectors $\mathbf{x}_1(t), \mathbf{x}_2(t), \ldots, \mathbf{x}_n(t)$, and let $\mathbf{c}$ be the column vector with components $c_1, c_2, \ldots, c_n$. Then Eq. (37) may be written in the form

$$\mathbf{X}(a)\mathbf{c} = \mathbf{x}(a). \tag{38}$$

The Wronskian determinant $W(a) = |\mathbf{X}(a)|$ is nonzero because the solutions $\mathbf{x}_1, \mathbf{x}_2, \ldots, \mathbf{x}_n$ are linearly independent. Hence the matrix $\mathbf{X}(a)$ has an inverse matrix $\mathbf{X}^{-1}(a)$. Therefore, the vector $\mathbf{c} = \mathbf{X}^{-1}(a)\mathbf{x}(a)$ satisfies Eq. (38), as desired.

Finally, note that the given solution $\mathbf{x}(t)$ and the solution $\mathbf{y}(t)$ of Eq. (36)—with the values of c_i determined by the equation $\mathbf{c} = \mathbf{X}^{-1}(a)\mathbf{x}(a)$—have the same initial values (at $t = a$). It follows from the existence-uniqueness theorem of Section 5.1 that $\mathbf{x}(t) = \mathbf{y}(t)$ for all t in I. This establishes Eq. (35).   ∎

EXAMPLE 2 (CONTINUED)   It follows from Theorem 3 that the linear combination

$$\mathbf{x} = c_1\mathbf{x}_1 + c_2\mathbf{x}_2 + c_3\mathbf{x}_3 = c_1\begin{bmatrix} 2e^t \\ 2e^t \\ e^t \end{bmatrix} + c_2\begin{bmatrix} 2e^{3t} \\ 0 \\ -e^{3t} \end{bmatrix} + c_3\begin{bmatrix} 2e^{5t} \\ -2e^{5t} \\ e^{5t} \end{bmatrix}$$

is a general solution of the 3×3 linear system in Eq. (34). In scalar form, this gives the general solution

$$x_1(t) = 2c_1e^t + 2c_2e^{3t} + 2c_3e^{5t}$$
$$x_2(t) = 2c_1e^t \qquad\quad - 2c_3e^{5t}$$
$$x_3(t) = c_1e^t - c_2e^{3t} + c_3e^{5t}.$$

Suppose now that we seek the particular solution satisfying the initial conditions

$$x_1(0) = 0, \qquad x_2(0) = 2, \qquad x_3(0) = 6.$$

When we substitute these values in the three scalar equations just above them, we get the equations

$$2c_1 + 2c_2 + 2c_3 = 0$$
$$2c_1 \qquad\quad - 2c_3 = 2$$
$$c_1 - c_2 + c_3 = 6$$

that are readily solved for $c_1 = 2$, $c_2 = -3$, $c_3 = 1$. Thus the desired particular solution is given by

$$\mathbf{x}(t) = 2\mathbf{x}_1 - 3\mathbf{x}_2 + \mathbf{x}_3 = \begin{bmatrix} 4e^t - 6e^{3t} + 2e^{5t} \\ 4e^t \qquad\quad - 2e^{5t} \\ 2e^t - 3e^{3t} + e^{5t} \end{bmatrix}.$$

Remark: Every $n \times n$ system $\mathbf{x}' = \mathbf{P}(t)\mathbf{x}$ with continuous coefficient matrix does have a set of n linearly independent solutions $\mathbf{x}_1, \mathbf{x}_2, \ldots, \mathbf{x}_n$, as in the hypotheses of Theorem 3. It suffices to choose for $\mathbf{x}_j(t)$ the unique solution such that

$$\mathbf{x}_j(a) = \begin{bmatrix} 0 \\ 0 \\ 0 \\ \vdots \\ 0 \\ 1 \\ 0 \\ \vdots \\ 0 \end{bmatrix} \leftarrow \text{position } j$$

—the column vector with all entries zero except for a 1 in row j. (In other words, $\mathbf{x}_j(a)$ is merely the jth column of the identity matrix.) Then

$$W(\mathbf{x}_1, \mathbf{x}_2, \ldots, \mathbf{x}_n)\big|_{t=a} = |\mathbf{I}| = 1 \neq 0,$$

so the solutions $\mathbf{x}_1, \mathbf{x}_2, \ldots, \mathbf{x}_n$ are linearly independent by Theorem 2. How actually to find these solutions explicitly is another matter—one which we address in Section 5.5 (for the case of constant coefficient matrices).

We now return to the *nonhomogeneous* linear equation

$$\mathbf{x}' = \mathbf{P}(t)\mathbf{x} + \mathbf{f}(t). \tag{28}$$

The following theorem is analogous to Theorem 5 of Section 2.2 and is proved in precisely the same way, substituting the preceding theorems in this section for the analogous theorems of Section 2.2. In brief, Theorem 4 means that the general solution of (28) has the form

$$\mathbf{x} = \mathbf{x}_c + \mathbf{x}_p, \tag{39}$$

where $\mathbf{x}_p(t)$ is a single particular solution of (28) and the **complementary function** $\mathbf{x}_c(t)$ is a general solution of the associated homogeneous equation $\mathbf{x}' = \mathbf{P}(t)\mathbf{x}$.

Theorem 4 *Solutions of Nonhomogeneous Systems*

Let $\mathbf{x}_p$ be a particular solution of the nonhomogeneous linear equation in (28) on an open interval I on which the function entries p_{ij} of $\mathbf{P}(t)$ and f_i of $\mathbf{f}(t)$ are continuous. Let $\mathbf{x}_1, \mathbf{x}_2, \ldots, \mathbf{x}_n$ be linearly independent solutions of the associated homogeneous equation on I. If $\mathbf{x}(t)$ is any solution whatsoever of Eq. (28) on I, then there exist numbers $c_1, c_2, \ldots, c_n$ such that

$$\mathbf{x}(t) = c_1\mathbf{x}_1(t) + c_2\mathbf{x}_2(t) + \cdots + c_n\mathbf{x}_n(t) + \mathbf{x}_p(t) \tag{40}$$

for all t in I.

Thus finding a general solution of a nonhomogeneous linear system involves two separate steps:

1. Finding the general solution $\mathbf{x}_c$ of the associated homogeneous system;
2. Finding a single particular solution $\mathbf{x}_p$ of the nonhomogeneous system.

The sum $\mathbf{x} = \mathbf{x}_c + \mathbf{x}_p$ will then be a general solution of the nonhomogeneous system.

EXAMPLE 3 The nonhomogeneous linear system

$$
\begin{aligned}
x_1' &= 3x_1 - 2x_2   - 9t + 13 \\
x_2' &= -x_1 + 3x_2 - 2x_3 + 7t - 15 \\
x_3' &= - x_2 + 3x_3 - 6t + 7
\end{aligned}
$$

is of the form in (28) with

$$
\mathbf{P}(t) = \begin{bmatrix} 3 & -2 & 0 \\ -1 & 3 & -2 \\ 0 & -1 & 3 \end{bmatrix}, \qquad \mathbf{f}(t) = \begin{bmatrix} -9t + 13 \\ 7t - 15 \\ -6t + 7 \end{bmatrix}.
$$

In Example 2 we saw that a general solution of the associated homogeneous linear system

$$
\mathbf{x}' = \begin{bmatrix} 3 & -2 & 0 \\ -1 & 3 & -2 \\ 0 & -1 & 3 \end{bmatrix}\mathbf{x}
$$

is given by

$$\mathbf{x}_c(t) = \begin{bmatrix} 2c_1e^t + 2c_2e^{3t} + 2c_3e^{5t} \\ 2c_1e^t \qquad\qquad - 2c_3e^{5t} \\ c_1e^t - c_2e^{3t} + c_3e^{5t} \end{bmatrix},$$

and we can verify by substitution that

$$\mathbf{x}_p(t) = \begin{bmatrix} 3t \\ 5 \\ 2t \end{bmatrix}$$

is a particular solution of the original nonhomogeneous system. Consequently Theorem 4 implies that a general solution of the nonhomogeneous system is given by

$$\mathbf{x}(t) = \mathbf{x}_c(t) + \mathbf{x}_p(t);$$

that is, by

$$x_1(t) = 2c_1e^t + 2c_2e^{3t} + 2c_3e^{5t} + 3t$$

$$x_2(t) = 2c_1e^t \qquad\qquad - 2c_3e^{5t} + 5$$

$$x_3(t) = c_1e^t - c_2e^{3t} + c_3e^{5t} + 2t.$$

In Section 5.2 we discussed systematic elimination procedures for solving linear systems of differential equations, and Sections 5.5 and 5.6 are devoted to general matrix methods. But as the following example illustrates, *ad hoc* techniques sometimes are simpler and more efficient.

EXAMPLE 4 Find the displacements $x_1(t)$ and $x_2(t)$ of the two masses in the system shown in Fig. 5.17, given the initial conditions $x_1(0) = x_2(0) = x_1'(0) = x_2'(0) = 0$.

SOLUTION With the numerical values indicated in Fig. 5.17, Example 1 of Section 5.1—in particular, the equations in (3) there—give us the nonhomogeneous system

$$2x_1'' + 100x_1 - 25x_2 = 0,$$

$$\tfrac{1}{2}x_2'' - 25x_1 + 25x_2 = 100 \cos 10t, \qquad (41)$$

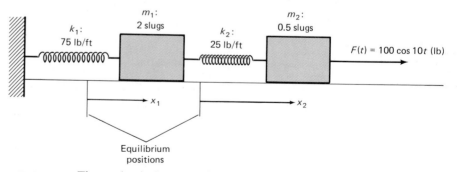

FIGURE 5.17   The mechanical system of Example 4.

CHAPTER 5:   Linear Systems of Differential Equations

The associated homogeneous system is

$$2x_1'' + 100x_1 - 25x_2 = 0,$$
$$\tfrac{1}{2}x_2'' - 25x_1 + 25x_2 = 0.$$

(42)

As in Example 5 of Section 5.1, this system is equivalent to a system of four first order linear equations. Hence its general solution $\mathbf{x}_c$ will be a linear combination of four linearly independent solutions.

To attempt to find these solutions, we note that neither equation in (42) contains a first derivative. Hence it seems likely that there will be solutions of the form

$$x_1 = a_1 \cos \omega t, \qquad x_2 = a_2 \cos \omega t.$$

(43)

For if we substituted the equations in (43) in (42), every term would be a multiple of $\cos \omega t$, so it would merely be a question of whether the constant coefficients sum correctly. When we carry out this substitution, the result is

$$-2a_1\omega^2 \cos \omega t + 100a_1 \cos \omega t - 25a_2 \cos \omega t = 0,$$
$$-\tfrac{1}{2}a_2\omega^2 \cos \omega t - 25a_1 \cos \omega t + 25a_2 \cos \omega t = 0.$$

We cancel $\cos \omega t$ throughout and collect coefficients to obtain

$$(100 - 2\omega^2)a_1 - 25a_2 = 0,$$
$$-25a_1 + (25 - \tfrac{1}{2}\omega^2)a_2 = 0.$$

(44)

But these equations can have a nontrivial solution for a_1 and a_2 only if the determinant of coefficients vanishes:

$$(100 - 2\omega^2)(25 - \tfrac{1}{2}\omega^2) - (25)^2 = 0;$$
$$(50 - \omega^2)^2 = (25)^2;$$
$$\omega^2 = 50 \pm 25.$$

Thus ω must have either the value $\omega_1 = 5$ rad/s or $\omega_2 = 5\sqrt{3}$ rad/s; these are the natural frequencies of the system.

With $\omega_1 = 5$, each equation in (44) reduces to $50a_1 - 25a_2 = 0$, so $a_2 = 2a_1$. With $a_1 = 1$, we get the solution

$$x_1 = \cos 5t, \qquad x_2 = 2 \cos 5t$$

(45)

of the homogeneous system in (42). With $\omega_2 = 5\sqrt{3}$, each equation in (44) reduces to $-50a_1 - 25a_2 = 0$, so that $a_2 = -2a_1$, With $a_1 = 1$ here, we get the second solution

$$x_1 = \cos 5\sqrt{3}t, \qquad x_2 = -2 \cos 5\sqrt{3}t.$$

(46)

To obtain two more solutions, we observe that precisely the same computations would ensue if we had begun with sines rather than with cosines in our trial solution in (43). Hence two additional solutions are

$$x_1 = \sin 5t, \qquad x_2 = 2 \sin 5t$$

(47)

and

$$x_1 = \sin 5\sqrt{3}t, \qquad x_2 = -2 \sin 5\sqrt{3}t.$$

(48)

It can be verified (Problem 33) that these four solutions

$$\mathbf{x}_1(t) = \begin{bmatrix} \cos 5t \\ 2 \cos 5t \end{bmatrix}, \qquad \mathbf{x}_2(t) = \begin{bmatrix} \sin 5t \\ 2 \sin 5t \end{bmatrix},$$

$$\mathbf{x}_3(t) = \begin{bmatrix} \cos 5t\sqrt{3} \\ -2 \cos 5t\sqrt{3} \end{bmatrix}, \qquad \mathbf{x}_4(t) = \begin{bmatrix} \sin 5t\sqrt{3} \\ -2 \sin 5t\sqrt{3} \end{bmatrix}, \qquad (49)$$

are linearly independent. Hence the general solution of the system in (42) is

$$\mathbf{x}_c = c_1\mathbf{x}_1 + c_2\mathbf{x}_2 + c_3\mathbf{x}_3 + c_4\mathbf{x}_4 \qquad (50)$$

where c_1, c_2, c_3, and c_4 are arbitrary constants.

Now we turn our attention to the particular solution of the orginal non-homogeneous system

$$2x_1'' + 100x_1 - 25x_2 = 0,$$

$$\tfrac{1}{2}x_2'' - 25x_1 + 25x_2 = 100 \cos 10t. \qquad (41)$$

Because of the presence of the term $100 \cos 10t$, as well as the absence of any first derivatives, it is reasonable to try

$$x_1 = A_1 \cos 10t, \qquad x_2 = A_2 \cos 10t. \qquad (51)$$

After substitution of (51) in (41) and cancellation of the common term $\cos 10t$, we get the equations

$$-200A_1 + 100A_1 - 25A_2 = 0,$$

$$-50A_2 - 25A_1 + 25A_2 = 100,$$

which are readily solved for $A_1 = \tfrac{4}{3}$ and $A_2 = -\tfrac{16}{3}$. Thus the particular solution is

$$\mathbf{x}_p(t) = \tfrac{4}{3}\begin{bmatrix} \cos 10t \\ -4 \cos 10t \end{bmatrix}. \qquad (52)$$

We now assemble the information in Eqs. (49)–(52): The components of the general solution $\mathbf{x} = \mathbf{x}_c + \mathbf{x}_p$ of the nonhomogeneous system in (41) are

$$x_1 = c_1 \cos 5t + c_2 \sin 5t$$
$$\qquad + c_3 \cos 5\sqrt{3}t + c_4 \sin 5\sqrt{3}t + \tfrac{4}{3} \cos 10t,$$

$$x_2 = 2c_1 \cos 5t + 2c_2 \sin 5t \qquad\qquad (53)$$
$$\qquad - 2c_3 \cos 5\sqrt{3}t - 2c_4 \sin 5\sqrt{3}t - \tfrac{16}{3} \cos 10t.$$

Finally, imposing the initial conditions, we get the equations

$$x_1(0) = c_1 + c_3 + \tfrac{4}{3} = 0,$$

$$x_2(0) = 2c_1 - 2c_3 - \tfrac{16}{3} = 0,$$

$$x_1'(0) = 5c_2 + 5\sqrt{3}c_4 = 0,$$

$$x_2'(0) = 10c_2 - 10\sqrt{3}c_4 = 0.$$

The simultaneous solution of these equations is $c_1 = \tfrac{2}{3}$, $c_2 = 0$, $c_3 = -2$, and $c_4 = 0$. Therefore, the motion of the two masses is described by

$$x_1(t) = \tfrac{2}{3} \cos 5t - 2 \cos 5\sqrt{3}t + \tfrac{4}{3} \cos 10t,$$

$$x_2(t) = \tfrac{4}{3} \cos 5t + 4 \cos 5\sqrt{3}t - \tfrac{16}{3} \cos 10t.$$

CHAPTER 5: Linear Systems of Differential Equations

We have a superposition of two oscillations with the natural frequencies $\omega_1 = 5$ and $\omega_2 = 5\sqrt{3}$ and a forced oscillation with frequency $\omega = 10$. In each of the two natural oscillations the amplitude of motion of m_2 is twice that of m_1, while in the forced oscillation the amplitude of motion of m_2 is four times that of m_1.

5.3 PROBLEMS

1. Let

$$\mathbf{A} = \begin{bmatrix} 2 & -3 \\ 4 & 7 \end{bmatrix} \quad \text{and} \quad \mathbf{B} = \begin{bmatrix} 3 & -4 \\ 5 & 1 \end{bmatrix}.$$

Find (a) $2\mathbf{A} + 3\mathbf{B}$; (b) $3\mathbf{A} - 2\mathbf{B}$; (c) $\mathbf{AB}$; (d) $\mathbf{BA}$.

2. Verify that (a) $\mathbf{A(BC)} = (\mathbf{AB})\mathbf{C}$ and that (b) $\mathbf{A(B + C)} = \mathbf{AB} + \mathbf{AC}$, where $\mathbf{A}$ and $\mathbf{B}$ are the matrices given in Problem 1 and

$$\mathbf{C} = \begin{bmatrix} 0 & 2 \\ 3 & -1 \end{bmatrix}.$$

3. Find $\mathbf{AB}$ and $\mathbf{BA}$ given

$$\mathbf{A} = \begin{bmatrix} 2 & 0 & -1 \\ 3 & -4 & 5 \end{bmatrix} \quad \text{and} \quad \mathbf{B} = \begin{bmatrix} 1 & 3 \\ -7 & 0 \\ 3 & -2 \end{bmatrix}.$$

4. Let $\mathbf{A}$ and $\mathbf{B}$ be the matrices given in Problem 3, and let

$$\mathbf{x} = \begin{bmatrix} 2t \\ e^{-t} \end{bmatrix} \quad \text{and} \quad \mathbf{y} = \begin{bmatrix} t^2 \\ \sin t \\ \cos t \end{bmatrix}.$$

Find $\mathbf{Ay}$ and $\mathbf{Bx}$. Are the products $\mathbf{Ax}$ and $\mathbf{By}$ defined? Explain your answer.

5. Let

$$\mathbf{A} = \begin{bmatrix} 3 & 2 & -1 \\ 0 & 4 & 3 \\ -5 & 2 & 7 \end{bmatrix} \quad \text{and} \quad \mathbf{B} = \begin{bmatrix} 0 & -3 & 2 \\ 1 & 4 & -3 \\ 2 & 5 & -1 \end{bmatrix}.$$

Find (a) $7\mathbf{A} + 4\mathbf{B}$; (b) $3\mathbf{A} - 5\mathbf{B}$; (c) $\mathbf{AB}$; (d) $\mathbf{BA}$; (e) $\mathbf{A} - t\mathbf{I}$.

6. Let

$$\mathbf{A}_1 = \begin{bmatrix} 2 & 1 \\ -3 & 2 \end{bmatrix}, \quad \mathbf{A}_2 = \begin{bmatrix} 1 & 3 \\ -1 & -2 \end{bmatrix}, \quad \mathbf{B} = \begin{bmatrix} 2 & 4 \\ 1 & 2 \end{bmatrix}.$$

(a) Show that $\mathbf{A}_1\mathbf{B} = \mathbf{A}_2\mathbf{B}$ and note that $\mathbf{A}_1 \neq \mathbf{A}_2$. Thus the cancellation law does not hold for matrices; that is, if $\mathbf{A}_1\mathbf{B} = \mathbf{A}_2\mathbf{B}$ and $\mathbf{B} \neq \mathbf{0}$, it does not follow that $\mathbf{A}_1 = \mathbf{A}_2$. (b) Let $\mathbf{A} = \mathbf{A}_1 - \mathbf{A}_2$, and show that $\mathbf{AB} = \mathbf{0}$. Thus the product of two nonzero matrices may be the zero matrix.

7. Compute the determinants of the matrices $\mathbf{A}$ and $\mathbf{B}$ in Problem 6. Are your results consistent with the theorem to the effect that det $(\mathbf{AB}) = [\det (\mathbf{A})][\det (\mathbf{B})]$ for any two square matrices $\mathbf{A}$ and $\mathbf{B}$ of the same order?

8. Suppose that $\mathbf{A}$ and $\mathbf{B}$ are the matrices of Problem 5. Verify that det $(\mathbf{AB}) = \det (\mathbf{BA})$.

In each of Problems 9 and 10, verify the product law for differentiation, $(\mathbf{AB})' = \mathbf{A}'\mathbf{B} + \mathbf{AB}'$.

9. $\mathbf{A}(t) = \begin{bmatrix} t & 2t - 1 \\ t^3 & \dfrac{1}{t} \end{bmatrix}$ and $\mathbf{B}(t) = \begin{bmatrix} 1 - t & 1 + t \\ 3t^2 & 4t^3 \end{bmatrix}$

10. $\mathbf{A}(t) = \begin{bmatrix} e^t & t & t^2 \\ -t & 0 & 2 \\ 8t & -1 & t^3 \end{bmatrix}$ and $\mathbf{B}(t) = \begin{bmatrix} 3 \\ 2e^{-t} \\ 3t \end{bmatrix}$

In each of Problems 11–15, write the given system in the form $\mathbf{x}' = \mathbf{P}(t)\mathbf{x} + \mathbf{f}(t)$; identify $\mathbf{x}$, $\mathbf{P}(t)$, and $\mathbf{f}(t)$.

11. $x' = 3x - 2y$
    $y' = 2x + y$

12. $x' = 2x + 4y + 3e^t$
    $y' = 5x - y - t^2$

13. $x' = \quad tx - e^t y + \cos t$
    $y' = e^{-t}x + t^2 y - \sin t$

14. $x' = 3x - 4y + z + t$
    $y' = x \qquad\quad - 3z + t^2$
    $z' = \qquad\quad 6y - 7z + t^3$

15. $x' = \quad tx - y + e^t z$
    $y' = \quad 2x + t^2 y - z$
    $z' = e^{-t}x + 3ty + t^3 z$

In each of Problems 16–22, first verify that the given vectors are solutions of the given system, and then use the Wronskian to show that they are linearly independent. Finally write the general solution of the system.

16. $\mathbf{x}' = \begin{bmatrix} -3 & 2 \\ -3 & 4 \end{bmatrix} \mathbf{x}; \quad \mathbf{x}_1 = \begin{bmatrix} e^{3t} \\ 3e^{3t} \end{bmatrix}, \mathbf{x}_2 = \begin{bmatrix} 2e^{-2t} \\ e^{-2t} \end{bmatrix}$

17. $\mathbf{x}' = \begin{bmatrix} 3 & -1 \\ 5 & -3 \end{bmatrix} \mathbf{x}; \quad \mathbf{x}_1 = e^{2t}\begin{bmatrix} 1 \\ 1 \end{bmatrix}, \mathbf{x}_2 = e^{-2t}\begin{bmatrix} 1 \\ 5 \end{bmatrix}$

18. $\mathbf{x}' = \begin{bmatrix} 4 & 1 \\ -2 & 1 \end{bmatrix} \mathbf{x}; \quad \mathbf{x}_1 = e^{3t}\begin{bmatrix} 1 \\ -1 \end{bmatrix}, \mathbf{x}_2 = e^{2t}\begin{bmatrix} 2 \\ -1 \end{bmatrix}$

19. $\mathbf{x}' = \begin{bmatrix} 4 & -3 \\ 6 & -7 \end{bmatrix} \mathbf{x}; \quad \mathbf{x}_1 = \begin{bmatrix} 3e^{2t} \\ 2e^{2t} \end{bmatrix}, \mathbf{x}_2 = \begin{bmatrix} e^{-5t} \\ 3e^{-5t} \end{bmatrix}$

20. $\mathbf{x}' = \begin{bmatrix} 3 & -2 & 0 \\ -1 & 3 & -2 \\ 0 & -1 & 3 \end{bmatrix} \mathbf{x}; \quad \mathbf{x}_1 = e^t \begin{bmatrix} 2 \\ 2 \\ 1 \end{bmatrix}, \mathbf{x}_2 = e^{3t} \begin{bmatrix} -2 \\ 0 \\ 1 \end{bmatrix},$

$\mathbf{x}_3 = e^{5t} \begin{bmatrix} 2 \\ -2 \\ 1 \end{bmatrix}$

21. $\mathbf{x}' = \begin{bmatrix} 0 & 1 & 1 \\ 1 & 0 & 1 \\ 1 & 1 & 0 \end{bmatrix} \mathbf{x}; \quad \mathbf{x}_1 = e^{2t} \begin{bmatrix} 1 \\ 1 \\ 1 \end{bmatrix}, \mathbf{x}_2 = e^{-t} \begin{bmatrix} 1 \\ 0 \\ -1 \end{bmatrix},$

$\mathbf{x}_3 = e^{-t} \begin{bmatrix} 0 \\ 1 \\ -1 \end{bmatrix}$

22. $\mathbf{x}' = \begin{bmatrix} 1 & 2 & 1 \\ 6 & -1 & 0 \\ -1 & -2 & -1 \end{bmatrix} \mathbf{x}; \quad \mathbf{x}_1 = \begin{bmatrix} 1 \\ 6 \\ -13 \end{bmatrix},$

$\mathbf{x}_2 = e^{3t} \begin{bmatrix} 2 \\ 3 \\ -2 \end{bmatrix}, \mathbf{x}_3 = e^{-4t} \begin{bmatrix} -1 \\ 2 \\ 1 \end{bmatrix}$

In each of Problems 23–28, find a particular solution of the indicated linear system that satisfies the given initial conditions.

23. The system of Problem 16; $x_1(0) = 0, x_2(0) = 5$

24. The system of Problem 17; $x_1(0) = 5, x_2(0) = -3$

25. The system of Problem 18; $x_1(0) = 11, x_2(0) = -7$

26. The system of Problem 19; $x_1(0) = 8, x_2(0) = 0$

27. The system of Problem 20; $x_1(0) = 0, x_2(0) = 0, x_3(0) = 4$

28. The system of Problem 21; $x_1(0) = 10, x_2(0) = 12,$
$x_3(0) = -1$

29. (a) Show that the vector functions

$$\mathbf{x}_1 = \begin{bmatrix} t \\ t^2 \end{bmatrix} \quad \text{and} \quad \mathbf{x}_2 = \begin{bmatrix} t^2 \\ t^3 \end{bmatrix}$$

are linearly independent on the real line. (b) Why does it follow from Theorem 2 that there is *no* continuous matrix $\mathbf{P}(t)$ such that $\mathbf{x}_1$ and $\mathbf{x}_2$ are both solutions of $\mathbf{x}' = \mathbf{P}(t)\mathbf{x}$?

30. Suppose that one of the vector functions

$$\mathbf{x}_1(t) = \begin{bmatrix} x_{11}(t) \\ x_{21}(t) \end{bmatrix} \quad \text{and} \quad \mathbf{x}_2(t) = \begin{bmatrix} x_{12}(t) \\ x_{22}(t) \end{bmatrix}$$

is a constant multiple of the other on the open interval I. Show that their Wronskian $W(t) = \|[x_{ij}(t)]\|$ must vanish identically on I. This proves part (a) of Theorem 2 in the case $n = 2$.

31. Suppose that the vectors $\mathbf{x}_1(t)$ and $\mathbf{x}_2(t)$ of Problem 30 are solutions of the equation $\mathbf{x}' = \mathbf{P}(t)\mathbf{x}$, where the 2×2 matrix $\mathbf{P}(t)$ is continuous on the open interval I. Show that, if there exists a point a of I at which their Wronskian $W(a)$ is zero, then there exist numbers c_1 and c_2 not both zero such that $c_1\mathbf{x}_1(a) + c_2\mathbf{x}_2(a) = \mathbf{0}$. Then conclude from the uniqueness of solutions of the equation $\mathbf{x}' = \mathbf{P}(t)\mathbf{x}$ that $c_1\mathbf{x}_1(t) + c_2\mathbf{x}_2(t) = \mathbf{0}$ for all t in I; that is, that $\mathbf{x}_1$ and $\mathbf{x}_2$ are linearly dependent. This proves part (b) of Theorem 2 in the case $n = 2$.

32. Generalize Problems 30 and 31 to prove Theorem 2 for n an arbitrary positive integer.

33. Let $\mathbf{x}_1(t), \mathbf{x}_2(t), \ldots, \mathbf{x}_n(t)$ be vector functions whose ith components (for some fixed i) $x_{i1}(t), x_{i2}(t), \ldots, x_{in}(t)$ are linearly independent real-valued functions. Conclude that the vector functions are themselves linearly independent. Does this imply that the solution vectors $\mathbf{x}_1, \mathbf{x}_2, \mathbf{x}_3,$ and $\mathbf{x}_4$ in (49) are linearly independent?

Apply the method of Example 4 to find the general solution of each of the systems in Problems 34–38.

34. $x'' = -5x + 4y, y'' = 4x - 5y$

35. $x_1'' = -3x_1 + 2x_2 + \cos 3t, 2x_2'' = 2x_1 - 4x_2$

36. $x_1'' = -3x_1 + 2x_2 + 2\sin 2t, x_2'' = 2x_1 - 3x_2$

37. $x'' = -3x + y + \cos t, y'' = x - 3y + 2\cos t$

38. $x_1'' = -6x_1 + 4x_2, 2x_2'' = 4x_1 - 8x_2$

---

*5.4

Mechanical Applications of Linear Systems

In this section we apply the theory developed in Section 5.3 to investigate the oscillations of typical mass-and-spring systems having two or more degrees of freedom. Our examples are chosen to illustrate phenomena that are generally characteristic of complex mechanical systems.

Example 1 illustrates the effect of frictional resistance in a mechanical system subjected to a periodic external force that provides a *nonhomogeneous term* in the corresponding system of linear differential equations. On the basis of Theorem 4 of Section 5.3, we know that the solution satisfying given initial conditions will be of the form

$$\mathbf{x}(t) = \mathbf{x}_c(t) + \mathbf{x}_p(t) \tag{1}$$

where $\mathbf{x}_p(t)$ is a particular solution of the nonhomogeneous system and $\mathbf{x}_c(t)$ is a solution of the corresponding homogeneous system. It is typical for the effect of frictional resistance in mechanical systems to damp out the complementary function solution $\mathbf{x}_c(t)$, so that

$$\mathbf{x}_c(t) \to \mathbf{0} \quad \text{as} \quad t \to +\infty. \tag{2}$$

Hence $\mathbf{x}_c(t)$ is a **transient solution** that depends only upon the initial conditions; it dies out with time, leaving the **steady periodic solution** $\mathbf{x}_p(t)$ resulting from the external driving force:

$$\mathbf{x}(t) \to \mathbf{x}_p(t) \quad \text{as} \quad t \to +\infty. \tag{3}$$

As a practical matter, every physical system includes frictional resistance (however small) that damps out transient solutions in this manner.

EXAMPLE 1   We begin with the same system of two masses and two springs that we considered in Example 4 of Section 5.3. But we now include damping or frictional resistance; as indicated in Fig. 5.18, we incorporate dashpots connected to the two masses, each providing resistance proportional to the velocity of the mass to which it is connected. As the figure shows, the constants of proportionality are $c_1 = 4$ lb/ft/s and $c_2 = 1$ lb/ft/s. Also as before, a periodic external force $F(t) = 100 \cos 10t$ (pounds) acts on the second mass. We want to determine the resulting transient and steady periodic motions of the two masses.

SOLUTION   Without the two dashpots—see the equations in (41) of Section 5.3—we derived from $F = ma$ the equations

$$2x_1'' = -100x_1 + 25x_2,$$

$$\tfrac{1}{2}x_2'' = \quad 25x_1 - 25x_2 + 100 \cos 10t$$

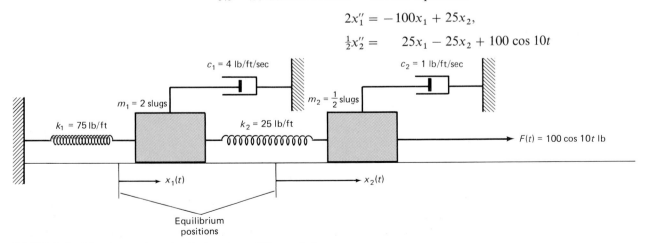

FIGURE 5.18   The damped mechanical system of Example 1.

for the displacements x_1 and x_2 of the two masses from their equilibrium positions at time t. The resistive force of the dashpots is $-c_1 x_1' = -4x_1'$ on the first mass and $-c_2 x_2' = -x_2'$ on the second mass. When we include these additional forces, we obtain the nonhomogeneous system

$$2x_1'' + 4x_1' + 100x_1 - 25x_2 = 0,$$
$$\tfrac{1}{2}x_2'' + x_2' - 25x_1 + 25x_2 = 100 \cos 10t. \tag{4}$$

THE TRANSIENT SOLUTION

First we need to find the general solution of the associated homogeneous system

$$2x_1'' + 4x_1' + 100x_1 - 25x_2 = 0,$$
$$\tfrac{1}{2}x_2'' + x_2' - 25x_1 + 25x_2 = 0; \tag{5}$$

we expect it to give the transient motions of the two masses. Because of the first derivative terms in (5), we can no longer expect the simple sine-cosine solutions we found in Example 4 of Section 5.3. Instead we apply the method of elimination of Section 5.2, in which we first write the system in the operational form

$$(2D^2 + 4D + 100)x_1 - 25x_2 = 0,$$
$$-25x_1 + (\tfrac{1}{2}D^2 + D + 25)x_2 = 0. \tag{6}$$

The operational determinant of this system is

$$(2D^2 + 4D + 100)(\tfrac{1}{2}D^2 + D + 25) - (25)^2 = (D^2 + 2D + 50)^2 - (25)^2.$$

Hence the characteristic equation—for either x_1 or x_2—is

$$(r^2 + 2r + 50)^2 - (25)^2 = 0, \tag{7}$$

which we solve as follows:

$$r^2 + 2r + 50 = \pm 25;$$

$$r^2 + 2r + 50 \pm 25 = 0;$$

$$r = \frac{-2 \pm \sqrt{4 - 4(50 \pm 25)}}{2}.$$

Thus we have two pairs of complex conjugates,

$$r = -1 \pm i\sqrt{24} \quad \text{and} \quad r = -1 \pm i\sqrt{74}. \tag{8}$$

Observe that had the characteristic equation in (7) not assumed so fortuitous a form, we might have needed a computer program for solving a fourth degree equation that has only complex roots.

The characteristic roots in (8) yield the general solutions

$$x_1(t) = e^{-t}(a_1 \cos \sqrt{24}t + a_2 \sin \sqrt{24}t)$$
$$+ e^{-t}(b_1 \cos \sqrt{74}t + b_2 \sin \sqrt{74}t),$$
$$x_2(t) = e^{-t}(c_1 \cos \sqrt{24}t + c_2 \sin \sqrt{24}t)$$
$$+ e^{-t}(d_1 \cos \sqrt{74}t + d_2 \sin \sqrt{74}t). \tag{9}$$

Because we have eight arbitrary constants rather than the four that ought to appear, we substitute x_1 and x_2 in the first equation in (6) to find the relations between these constants. A routine (but long) computation yields

$$(2D^2 + 4D + 100)x_1 = 50e^{-t}(a_1 \cos \sqrt{24}t + a_2 \sin \sqrt{24}t)$$
$$- 50e^{-t}(b_1 \cos \sqrt{74}t + b_2 \sin \sqrt{74}t).$$

In order that this expression be equal to $25x_2$, it is necessary that

$$c_1 = 2a_1, \qquad c_2 = 2a_2, \qquad d_1 = -2b_1, \quad \text{and} \quad d_2 = -2b_2. \qquad (10)$$

Thus the general solution of the homogeneous system in (5) is given by

$$x_1(t) = e^{-t}(a_1 \cos \sqrt{24}t + a_2 \sin \sqrt{24}t)$$
$$+ e^{-t}(b_1 \cos \sqrt{74}t + b_2 \sin \sqrt{74}t),$$

$$x_2(t) = 2e^{-t}(a_1 \cos \sqrt{24}t + a_2 \sin \sqrt{24}t)$$
$$- 2e^{-t}(b_1 \cos \sqrt{74}t + b_2 \sin \sqrt{74}t).$$

$$(11)$$

Because of the presence of the factor e^{-t}, we do indeed have a transient solution. It may be written in the form

$$\mathbf{x}_{tr}(t) = a_1 \begin{bmatrix} 1 \\ 2 \end{bmatrix} e^{-t} \cos t\sqrt{24} + a_2 \begin{bmatrix} 1 \\ 2 \end{bmatrix} e^{-t} \sin t\sqrt{24}$$

$$+ b_1 \begin{bmatrix} 1 \\ -2 \end{bmatrix} e^{-t} \cos t\sqrt{74} + b_2 \begin{bmatrix} 1 \\ -2 \end{bmatrix} e^{-t} \sin t\sqrt{74}. \qquad (12)$$

Note that the two circular frequencies exhibited in (12) are

$$\omega_1 = \sqrt{24} \approx 4.8990 \text{ rad/s} \quad \text{and} \quad \omega_2 = \sqrt{74} \approx 8.6023 \text{ rad/s},$$

as compared with the undamped natural frequencies

$$\omega_1 = 5 \text{ rad/s} \quad \text{and} \quad \omega_2 = 5\sqrt{3} \approx 8.6603 \text{ rad/s}$$

that we found in Example 4 of Section 5.3. Thus the effect of the frictional resistance in this example is both exponential damping of the amplitude of the oscillations and a decrease in their frequencies (just as we observed in the one-dimensional systems of Section 2.6). Equations (11) and (12) describe the "free" damped oscillations that would take place if there were no external force.

THE STEADY PERIODIC SOLUTION

Now we want to find a particular solution of the original nonhomogeneous system

$$2x_1'' + 4x_1' + 100x_1 - 25x_2 = 0,$$
$$\tfrac{1}{2}x_2'' + x_2' - 25x_1 + 25x_2 = 100 \cos 10t. \qquad (4)$$

Again because of the presence of the first derivative terms, we cannot expect a solution of the simple form $x_1 = A \cos 10t$, $x_2 = B \cos 10t$. We will need both sine and cosine terms, so we could substitute

$$x_1 = A_1 \cos 10t + A_2 \sin 10t, \qquad x_2 = B_1 \cos 10t + B_2 \sin 10t$$

in each equation in (4) and attempt to determine the four constants A_1, A_2, B_1, and B_2.

Here it is more efficient to employ an alternative method that involves complex exponentials. We note first that

$$100 \cos 10t = \text{Re}(100e^{10ti}). \tag{13}$$

This observation motivates us to introduce the new nonhomogeneous system

$$2z_1'' + 4z_1' + 100z_1 - 25z_2 = 0,$$
$$\tfrac{1}{2}z_2'' + z_2' - 25z_1 + 25z_2 = 100e^{10ti} \tag{14}$$

in the (new) unknown functions $z_1(t)$ and $z_2(t)$ that we expect to be complex-valued. Because of (13), the *real parts*

$$x_1(t) = \text{Re}(z_1(t)), \qquad x_2(t) = \text{Re}(z_2(t))$$

of any solution $z_1(t)$, $z_2(t)$ of the system in (14) will constitute a solution $x_1(t)$, $x_2(t)$ of the system in (4).

If we substitute in (14) the trial solution

$$z_1(t) = Ae^{10ti}, \qquad z_2(t) = Be^{10ti}, \tag{15}$$

then every term in each equation will be a constant multiple of e^{10ti}. We can therefore hope to determine the (complex) constants A and B, and finally take the real parts of z_1 and z_2 to obtain x_1 and x_2.

When we substitute the expressions in (15) in the equations in (14), we get

$$-200Ae^{10ti} + 40iAe^{10ti} + 100Ae^{10ti} - 25Be^{10ti} = 0,$$

$$-50Be^{10ti} + 10iBe^{10ti} - 25Ae^{10ti} + 25Be^{10ti} = 100e^{10ti}.$$

We cancel the common factor e^{10ti} and collect coefficients to obtain the equations

$$(-100 + 40i)A - 25B = 0,$$

$$-25A + (-25 + 10i)B = 100.$$

The determinant of this system of equations is

$$\Delta = (-100 + 40i)(-25 + 10i) - 625 = 1475 - 2000i.$$

Hence Cramer's rule yields

$$A = \frac{1}{\Delta} \begin{vmatrix} 0 & -25 \\ 100 & -25 + 10i \end{vmatrix} = \frac{2500}{1475 - 2000i}$$

$$\approx \frac{2500}{(2485.08)e^{5.3478i}} \approx (1.0060)e^{-5.3478i}$$

and

$$B = \frac{1}{\Delta} \begin{vmatrix} -100 + 40i & 0 \\ -25 & 100 \end{vmatrix} = \frac{-10,000 + 4000i}{1475 - 2000i}$$

$$\approx \frac{(-10,770.33)e^{-0.3805i}}{(2485.08)e^{5.3478i}} \approx (-4.3340)e^{-5.7283i}.$$

Using the above values for A and B in (15), we get

$$z_1(t) \approx (1.0060)e^{i(10t - 5.3478)}$$

and

$$z_2(t) \approx (-4.3340)e^{i(10t - 5.7283)}.$$

Taking real parts, we finally obtain the components

$$x_1(t) = \text{Re}(z_1(t)) \approx (1.0060) \cos(10t - 5.3478),$$

$$x_2(t) = \text{Re}(z_2(t)) \approx (-4.3340) \cos(10t - 5.7283) \tag{16}$$

of the damped steady periodic solution $\mathbf{x}_p(t)$. It is of interest to compare (16) with the undamped steady periodic solution

$$x_1(t) = \tfrac{4}{3} \cos 10t, \qquad x_2(t) = -\tfrac{16}{3} \cos 10t \tag{17}$$

that we found in Example 4 of Section 5.3. Note that the equations in (16) describe approximately the same situation—with the two masses oscillating approximately $180°$ out of phase and with the amplitude of motion of m_2 about four times that of m_1—but with both amplitudes reduced by the frictional resistance.

Finally, let us combine the particular solution in (16) of the nonhomogeneous system in (4) with the complementary solution in (11) of the associated homogeneous system in (5). We then get the general solution

$$
\begin{aligned}
x_1(t) &= e^{-t}(a_1 \cos\sqrt{24}t + a_2 \sin\sqrt{24}t) \\
&\quad + e^{-t}(b_1 \cos\sqrt{74}t + b_2 \sin\sqrt{74}t) + 1.0060 \cos(10t - 5.3478), \\
x_2(t) &= 2e^{-t}(a_1 \cos\sqrt{24}t + a_2 \sin\sqrt{24}t) \\
&\quad - 2e^{-t}(b_1 \cos\sqrt{74}t + b_2 \sin\sqrt{74}t) - 4.3340 \cos(10t - 5.7283)
\end{aligned}
\tag{18}
$$

of the nonhomogeneous system in (4). If initial values $x_1(0)$, $x_2(0)$, $x_1'(0)$, and $x_2'(0)$ were given, we could then determine the values of the constants a_1, a_2, b_1, and b_2 in (18).

THE TWO-AXLE AUTOMOBILE

In Example 4 of Section 2.8 we investigated the vertical oscillations of a one-axle car—actually a unicycle. Now we can analyze a more realistic model: a car with two axles and with separate front and rear suspension systems. Figure 5.19 represents the suspension system of such a car. We assume that the car body acts as would a solid bar of mass m and length $L = L_1 + L_2$. It has moment of inertia I about its center of mass C, which is at distance L_1 from the front end of the car. The car has front and back suspension springs with Hooke's constants k_1 and k_2, respectively.

When the car is in motion, let x_1 and x_2 denote the elevations of the front and rear ends of the car body, respectively, above their equilibrium positions. Because these are the distances the two suspension springs are stretched or compressed, the two vertical forces on the two ends of the car body are

$$F_1 = -k_1 x_1 \quad \text{(front)} \quad \text{and} \quad F_2 = -k_2 x_2 \quad \text{(rear)}. \tag{19}$$

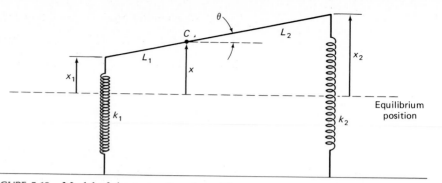

FIGURE 5.19   Model of the two-axle automobile.

If θ denotes the counterclockwise angular displacement (in radians) of the car body from the horizontal, then

$$x_1 = x - L_1\theta, \qquad x_2 = x + L_2\theta \tag{20}$$

where x is the elevation of the center of mass C. We assume that θ is so small that θ itself is an adequate approximation to $\sin\theta$. If we substitute (20) in (19), then Newton's law of motion yields

$$mx'' = F_1 + F_2 = -k_1(x - L_1\theta) - k_2(x + L_2\theta),$$

so that

$$mx'' = -(k_1 + k_2)x + (k_1L_1 - k_2L_2)\theta. \tag{21}$$

The counterclockwise torques of the forces F_1 and F_2 about C are

$$T_1 = -L_1F_1 \quad \text{and} \quad T_2 = +L_2F_2 \tag{22}$$

because their moment arms are L_1 and L_2. Therefore Newton's law of motion of *angular* acceleration yields

$$I\theta'' = T_1 + T_2 = k_1L_1(x - L_1\theta) - k_2L_2(x + L_2\theta);$$

thus

$$I\theta'' = (k_1L_1 - k_2L_2)x - (k_1L_1^2 + k_2L_2^2)\theta. \tag{23}$$

The equations in (21) and (23) constitute a linear system in the car's vertical and angular displacements $x(t)$ and $\theta(t)$, respectively.

EXAMPLE 2   Suppose that $m = 75$ slugs (the car weighs 2400 lb), $L_1 = 7$ ft, $L_2 = 3$ ft (it's a rear-engine car), $k_1 = k_2 = 2000$ lb/ft, and $I = 1000$ ft-lb-s². Then Eqs. (21) and (23) become

$$75x'' + 4000x - \quad 8000\theta = 0,$$
$$1000\theta'' - 8000x + 116{,}000\theta = 0; \tag{24}$$

that is,

$$(3D^2 + 160)x - \quad 320\theta = 0,$$
$$-8x + (D^2 + 116)\theta = 0. \tag{25}$$

The operational determinant of this sytem is

$$(3D^2 + 160)(D^2 + 116) - 2560 = 3D^4 + 508D^2 + 16{,}000. \tag{26}$$

Upon solving the characteristic equation

$$3r^4 + 508r^2 + 16{,}000 = 0 \tag{27}$$

for r^2 and then taking square roots, we get $r \approx \pm(6.4675)i$ and $r \approx \pm(11.2918)i$. Hence the general solution for x is

$$x(t) = a_1 \cos \omega_1 t + b_1 \sin \omega_1 t + a_2 \cos \omega_2 t + b_2 \sin \omega_2 t$$

(and similarly for θ), where the natural frequencies of the system are

$$\begin{aligned} \omega_1 &\approx 6.4675 \text{ rad/s} \approx 1.0293 \text{ Hz}, \\ \omega_2 &\approx 11.2918 \text{ rad/s} \approx 1.7971 \text{ Hz}. \end{aligned} \tag{28}$$

The natural frequencies in (28) are the *two* frequencies at which we would expect the car to be subject to resonance vibrations. For instance, suppose that the car is driven at a speed of v feet per second along a washboard surface shaped like a sine curve with a wavelength of 40 ft. The result is a periodic force on the car with frequency $\omega = 2\pi v/40 = \pi v/20$. Resonance occurs when either $\omega = \omega_1$ or $\omega = \omega_2$; that is, at either of the *two* critical speeds

$$v_1 = \frac{20\omega_1}{\pi} \approx 41 \text{ ft/s} \approx 28 \text{ mi/h}$$

or

$$v_2 = \frac{20\omega_2}{\pi} \approx 72 \text{ ft/s} \approx 49 \text{ mi/h}.$$

At a speed close to either of these two critical speeds, the amplitudes of the car's oscillations (both vertical and angular) would be quite large—see the table in Fig. 5.20—and the ride most uncomfortable for the passengers.

Car speed v (ft/sec)	x–Amplitude (in.)	θ–Amplitude (deg)
10	1.06	0.02
20	1.28	0.10
30	2.01	0.41
40	15.60	7.29
50	−1.56	−1.80
60	−0.33	−1.87
70	2.14	−8.85
80	−0.88	1.71
90	−0.47	0.67
100	−0.32	0.39
110	−0.24	0.26
120	−0.19	0.19

FIGURE 5.20 This table shows the amplitudes of the vertical and angular oscillations that result when the car of Example 2 is tested (at various simulated speeds) on a road test simulator.

Problems 1–5 deal with the mass-and-spring system of Example 1, but with given damping constants c_1 and c_2 and with given external forces $F_1(t)$ and $F_2(t)$ acting on the two masses. The differential equations for $x_1(t)$ and $x_2(t)$ are then

$$2x_1'' + c_1 x_1' + 100x_1 - 25x_2 = F_1(t),$$

$$\tfrac{1}{2}x_2'' + c_2 x_2' - 25x_1 + 25x_2 = F_2(t).$$

Use the complex method to find a steady periodic solution in each problem. For $i = 1$ and 2, first write $F_i(t)$ in the form $F_i(t) = \text{Re}(c_i e^{10ti})$. Then substitute the trial solution $z_1 = Ae^{10ti}$, $z_2 = Be^{10ti}$. Note that $\sin 10t = \text{Re}(-ie^{10ti})$.

1. $c_1 = c_2 = 0;$ $F_1(t) = 50 \cos 10t,\ F_2(t) = 100 \cos 10t$
2. $c_1 = c_2 = 0;$ $F_1(t) = 50 \sin 10t,\ F_2(t) = 100 \cos 10t$
3. $c_1 = c_2 = 0;$ $F_1(t) = 30 \cos 10t + 40 \sin 10t,\ F_2(t) = 0$
4. $c_1 = c_2 = 2;$ $F_1(t) = 100 \sin 10t,\ F_2(t) = 0$
5. $c_1 = 4, c_2 = 1;$ $F_1(t) = 50 \cos 10t,\ F_2(t) = 100 \sin 10t$

Each of Problems 6–10 deals with the undamped system shown in Fig. 5.21, with given fps values for the masses and spring constants. Find the two natural frequencies of the system and describe its two natural modes of free oscillation. Use the equations of motion that are given in Problem 38 of Section 5.2.

6. $m_1 = m_2 = 1;$ $k_1 = 1, k_2 = 4, k_3 = 1$
7. $m_1 = 1, m_2 = 2;$ $k_1 = 1, k_2 = k_3 = 2$
8. $m_1 = m_2 = 1;$ $k_1 = 1, k_2 = 2, k_3 = 1$
9. $m_1 = m_2 = 1;$ $k_1 = 2, k_2 = 1, k_3 = 2$
10. $m_1 = 1, m_2 = 2;$ $k_1 = 2, k_2 = k_3 = 4$

Problems 11 and 12 deal with the mass-and-spring system of Fig. 5.21, except that now the masses m_1 and m_2 are also connected to dashpots with constants c_1 and c_2, respectively. With the given values of the mass-spring-dashpot parameters, find $x_1(t)$ and $x_2(t)$ given that the masses are released from rest with $x_1(0) = x_2(0) = 1$.

11. $m_1 = m_2 = 1;$ $c_1 = c_2 = 2;$ $k_2 = 6, k_1 = k_3 = 5$
12. $m_1 = m_2 = 1;$ $c_1 = c_2 = 2;$ $k_2 = 10, k_1 = k_3 = 17$
13. In the system of Fig. 5.22, assume that $m_1 = 1$ slug, $k_1 = 50$ lb/ft, $k_2 = 10$ lb/ft, $F_0 = 5$ lb, and $\omega = 10$ rad/s. Then find m_2 so that in the resulting steady periodic oscillations, the mass m_1 will remain at rest (!). Thus the effect of the second mass-and-spring pair will be to neutralize the effect of the force on the first mass. This is an example of a *dynamic damper*.
14. In the system of Fig. 5.23, suppose that $m_1 = m_2 = 1$, $c = 1$, $k_2 = 4$, and $k_1 = k_3 = 9$. Set up the system of differential equa-

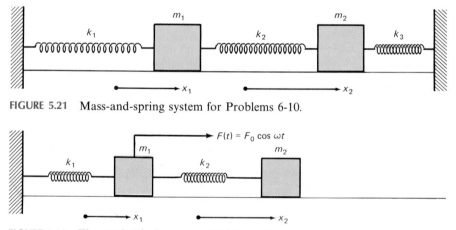

FIGURE 5.21   Mass-and-spring system for Problems 6-10.

$F(t) = F_0 \cos \omega t$

FIGURE 5.22   The mechanical system of Problem 13.

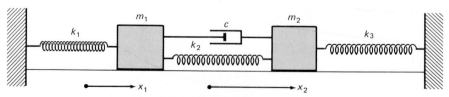

FIGURE 5.23   The mechanical system of Problem 14.

tions for x_1 and x_2 and find its general solution. Describe the two natural modes of oscillation (one is periodic and the other is exponentially damped).

15. Consider a four-story building that is an elastic frame structure with its mass concentrated at the levels of each of its four *floors*. The first floor is anchored to the ground; the second floor weighs 16 tons; the third and fourth floors weigh 8 tons each. The structure, like a spring, resists horizontal shear between any two adjacent floors—the restoring force is 5 tons per foot of (horizontal) relative displacement. Find the natural frequencies and natural modes of horizontal oscillation of the building. Give the ratios of the amplitudes A, B, and C of oscillations of the second, third, and fourth floors in the form $A:B:C$ with $A = 1$.

16. Suppose that the building of Problem 15 is subjected to an earthquake in which the ground undergoes horizontal sinusoidal oscillations with a period of 3 s and an amplitude of 3 in. Find the amplitudes of the resulting steady periodic oscillations of the upper three floors. Assume the fact that a motion $E \sin \omega t$

of the ground, with acceleration $a = -E\omega^2 \sin \omega t$, produces an opposite inertial force $F = -ma = mE\omega^2 \sin \omega t$ on a floor of mass m.

17. Suppose that $k_1 = k_2 = k$ and $L_1 = L_2 = \frac{1}{2}L$ in Fig. 5.19 (the symmetric situation). Then show that every free oscillation is a combination of a vertical oscillation with frequency $\omega_1 = \sqrt{2k/m}$ and an angular oscillation with frequency $\omega_2 = \sqrt{kL^2/2I}$.

In each of Problems 18–20, the system of Fig. 5.19 is taken as a model for an undamped car with the given parameters in fps units. (a) Find the two natural frequencies of oscillation (in hertz). (b) Assume that this car is driven along a sinusoidal washboard surface with a wavelength of 40 ft. Find the two critical speeds.

18. $m = 100$, $I = 800$, $L_1 = L_2 = 5$, $k_1 = k_2 = 2000$

19. $m = 100$, $I = 1000$, $L_1 = 6$, $L_2 = 4$, $k_1 = k_2 = 2000$

20. $m = 100$, $I = 800$, $L_1 = L_2 = 5$, $k_1 = 1000$, $k_2 = 2000$

5.5

The Eigenvalue Method for Homogeneous Systems

We now introduce a powerful alternative to the method of elimination for constructing the general solution of a *homogeneous* first order linear system with *constant* coefficients,

$$
\begin{aligned}
x_1' &= a_{11}x_1 + a_{12}x_2 + \cdots + a_{1n}x_n, \\
x_2' &= a_{21}x_1 + a_{22}x_2 + \cdots + a_{2n}x_n, \\
&\;\;\vdots \\
x_n' &= a_{n1}x_1 + a_{n2}x_2 + \cdots + a_{nn}x_n.
\end{aligned}
\tag{1}
$$

By Theorem 3 of Section 5.3, we know that it suffices to find n linearly independent solution vectors $\mathbf{x}_1, \mathbf{x}_2, \ldots, \mathbf{x}_n$; the linear combination

$$
\mathbf{x} = c_1\mathbf{x}_1 + c_2\mathbf{x}_2 + \cdots + c_n\mathbf{x}_n
\tag{2}
$$

with arbitrary coefficients will then be a general solution of the system in (1).

To search for the n needed linearly independent solution vectors, we proceed by analogy with the characteristic root method for solving a single homogeneous linear equation with constant coefficients (Section 2.3). It is reasonable to anticipate solution vectors of the form

$$
\mathbf{x} = \begin{bmatrix} x_1 \\ x_2 \\ x_3 \\ \vdots \\ x_n \end{bmatrix} = \begin{bmatrix} v_1 e^{\lambda t} \\ v_2 e^{\lambda t} \\ v_3 e^{\lambda t} \\ \vdots \\ v_n e^{\lambda t} \end{bmatrix} = \begin{bmatrix} v_1 \\ v_2 \\ v_3 \\ \vdots \\ v_n \end{bmatrix} e^{\lambda t} = \mathbf{v} e^{\lambda t}
\tag{3}
$$

where $\lambda, v_1, v_2, \ldots, v_n$ are appropriate scalar constants. For if we substitute

$$x_i = v_i e^{\lambda t}, \qquad x_i' = \lambda v_i e^{\lambda t}$$

$(i = 1, 2, \ldots, n)$ in (1), then the factor $e^{\lambda t}$ will cancel throughout. This will leave us with n linear equations which—for appropriate values of λ—we can hope to solve for values of the coefficients $v_1, v_2, \ldots, v_n$ in (3) so that $\mathbf{x}(t) = \mathbf{v} e^{\lambda t}$ is, indeed, a solution of the system in (1).

To investigate this possibility, it is more efficient to write the system in (1) in the matrix form

$$\mathbf{x}' = \mathbf{A}\mathbf{x} \tag{4}$$

where $\mathbf{A} = [a_{ij}]$. When we substitute the trial solution $\mathbf{x} = \mathbf{v} e^{\lambda t}$ with derivative $\mathbf{x}' = \lambda \mathbf{v} e^{\lambda t}$ in Eq. (4), the result is

$$\lambda \mathbf{v} e^{\lambda t} = \mathbf{A}\mathbf{v} e^{\lambda t}.$$

We cancel the nonzero scalar factor $e^{\lambda t}$ to get

$$\mathbf{A}\mathbf{v} = \lambda \mathbf{v}. \tag{5}$$

This means that $\mathbf{x} = \mathbf{v} e^{\lambda t}$ will be a nontrivial solution of (4) provided that $\mathbf{v}$ is a *nonzero* vector and λ is a constant such that (5) holds; that is, *the matrix product $\mathbf{A}\mathbf{v}$ is a scalar multiple of the vector $\mathbf{v}$*. The question now is this: How do we find $\mathbf{v}$ and λ?

To answer this, we rewrite (5) in the form

$$(\mathbf{A} - \lambda \mathbf{I})\mathbf{v} = \mathbf{0}. \tag{6}$$

Given λ, this is a system of n homogeneous linear equations in the unknowns $v_1, v_2, \ldots, v_n$. By a standard theorem of linear algebra, it has a nontrivial solution if and only if the determinant of its coefficient matrix vanishes; that is, if and only if

$$|\mathbf{A} - \lambda \mathbf{I}| = \det(\mathbf{A} - \lambda \mathbf{I}) = 0. \tag{7}$$

In its simplest formulation, the **eigenvalue method** for solving the system $\mathbf{x}' = \mathbf{A}\mathbf{x}$ consists of finding λ so that Eq. (7) holds and next solving (6) with this value of λ to obtain $v_1, v_2, \ldots, v_n$. Then $\mathbf{x} = \mathbf{v} e^{\lambda t}$ will be a solution vector. The name of the method comes from the following definition.

Definition *Eigenvalues and Eigenvectors*

The number λ (either zero or nonzero) is called an **eigenvalue** of the $n \times n$ matrix $\mathbf{A}$ provided that

$$|\mathbf{A} - \lambda \mathbf{I}| = 0. \tag{7}$$

An **eigenvector** associated with the eigenvalue λ is a nonzero vector $\mathbf{v}$ such that $\mathbf{A}\mathbf{v} = \lambda \mathbf{v}$, so that

$$(\mathbf{A} - \lambda \mathbf{I})\mathbf{v} = \mathbf{0}. \tag{6}$$

Note that if **v** is an eigenvector associated with the eigenvalue λ, then so is any nonzero constant scalar multiple $c\mathbf{v}$ of **v**—this follows upon multiplication of each side in Eq. (6) by $c \neq 0$.

The prefix *eigen* is a German word with the approximate translation *characteristic* in this context; the terms *characteristic value* and *characteristic vector* are in common use. For this reason, the equation

$$|\mathbf{A} - \lambda\mathbf{I}| = \begin{vmatrix} a_{11} - \lambda & a_{12} & \cdots & a_{1n} \\ a_{21} & a_{22} - \lambda & \cdots & a_{2n} \\ \vdots & \vdots & & \vdots \\ a_{n1} & a_{n2} & \cdots & a_{nn} - \lambda \end{vmatrix} = 0 \qquad (8)$$

is called the **characteristic equation** of the matrix **A**; its roots are the eigenvalues of **A**. Upon expanding the determinant in (8), we evidently get an nth degree polynomial of the form

$$(-1)^n\lambda^n + b_{n-1}\lambda^{n-1} + \cdots + b_1\lambda + b_0 = 0. \qquad (9)$$

By the fundamental theorem of algebra, this equation has n roots—possibly some are complex, possibly some are repeated—and thus an $n \times n$ matrix has n eigenvalues (counting repetitions, if any). Although we assume that the elements of **A** are real numbers, we allow the possibility of complex eigenvalues and complex-valued eigenvectors.

Our discussion of Eqs. (4)–(7) above provides a proof of the following theorem, which is the basis for the eigenvalue method of solving a first order linear system with constant coefficients.

> **Theorem**   *Eigenvalue Solutions of* $\mathbf{x}' = \mathbf{A}\mathbf{x}$
>
> Let λ be an eigenvalue of the (constant) coefficient matrix **A** of the first order linear system
>
> $$\frac{d\mathbf{x}}{dt} = \mathbf{A}\mathbf{x}.$$
>
> If **v** is an eigenvector associated with λ then
>
> $$\mathbf{x}(t) = \mathbf{v}e^{\lambda t}$$
>
> is a nontrivial solution of the system.

In outline, the eigenvalue method for solving the $n \times n$ system $\mathbf{x}' = \mathbf{A}\mathbf{x}$ proceeds as follows.

1. We first solve the characteristic equation in (8) for the eigenvalues $\lambda_1, \lambda_2, \ldots, \lambda_n$ of the matrix **A**.

2. Next we attempt to find n *linearly independent* eigenvectors $\mathbf{v}_1, \mathbf{v}_2, \ldots, \mathbf{v}_n$ associated with these eigenvalues.

3. Step 2 is not always possible, but when it is, we get n linearly independent solutions

$$\mathbf{x}_1(t) = \mathbf{v}_1 e^{\lambda_1 t}, \quad \mathbf{x}_2(t) = \mathbf{v}_2 e^{\lambda_2 t}, \quad \ldots, \quad \mathbf{x}_n(t) = \mathbf{v}_n e^{\lambda_n t}. \tag{10}$$

In this case the general solution of $\mathbf{x}' = \mathbf{A}\mathbf{x}$ is a linear combination

$$\mathbf{x} = c_1 \mathbf{x}_1 + c_2 \mathbf{x}_2 + \cdots + c_n \mathbf{x}_n$$

of these n solutions.

We will discuss separately the various cases that can occur, depending upon whether the eigenvalues are distinct or repeated, real or complex.

DISTINCT REAL EIGENVALUES

If the eigenvalues $\lambda_1, \lambda_2, \ldots, \lambda_n$ are real and distinct, then we substitute each of them in turn in (6) and solve for the associated eigenvectors $\mathbf{v}_1, \mathbf{v}_2, \ldots, \mathbf{v}_n$. In this case it can be proved that the particular solution vectors given in (10) are always linearly independent. (For instance, see Section 6.2 of Edwards and Penney, *Elementary Linear Algebra* (Englewood Cliffs, N.J.: Prentice-Hall, 1988).) In any particular example such linear independence can always be verified by using the Wronskian determinant of Section 5.3. The following example illustrates the procedure.

EXAMPLE 1   Find a general solution of the system

$$\begin{aligned} x_1' &= 4x_1 + 2x_2, \\ x_2' &= 3x_1 - \quad x_2. \end{aligned} \tag{11}$$

SOLUTION   The matrix form of the system in (11) is

$$\mathbf{x}' = \begin{bmatrix} 4 & 2 \\ 3 & -1 \end{bmatrix} \mathbf{x}. \tag{12}$$

The characteristic equation of the coefficient matrix is

$$\begin{vmatrix} 4 - \lambda & 2 \\ 3 & -1 - \lambda \end{vmatrix} = (4 - \lambda)(-1 - \lambda) - 6$$

$$= \lambda^2 - 3\lambda - 10$$

$$= (\lambda + 2)(\lambda - 5) = 0,$$

so we have the distinct real eigenvalues $\lambda_1 = -2$ and $\lambda_2 = 5$.

For the coefficient matrix $\mathbf{A}$ in (12) the eigenvector equation $(\mathbf{A} - \lambda\mathbf{I})\mathbf{v} = \mathbf{0}$ takes the form

$$\begin{bmatrix} 4 - \lambda & 2 \\ 3 & -1 - \lambda \end{bmatrix} \begin{bmatrix} v_1 \\ v_2 \end{bmatrix} = \begin{bmatrix} 0 \\ 0 \end{bmatrix}. \tag{13}$$

When we substitute the first eigenvalue $\lambda_1 = -2$, we find that the two scalar equations that follow from (13) are

$$6v_1 + 2v_2 = 0 \quad \text{and} \quad 3v_1 + v_2 = 0.$$

These equations are equivalent; we can choose v_1 arbitrary (but nonzero) and solve for v_2. The simplest choice is $v_1 = 1$, which yields $v_2 = -3$, and thus

$$\mathbf{v}_1 = \begin{bmatrix} 1 \\ -3 \end{bmatrix}$$

is an eigenvector associated with $\lambda_1 = -2$ (as is any nonzero constant multiple of $\mathbf{v}_1$).

Remark: If, instead of the "simplest" choice $v_1 = 1$, $v_2 = -3$, we had made another choice, $v_1 = c$, $v_2 = -3c$, we would have obtained the eigenvector

$$\mathbf{v}_1 = \begin{bmatrix} c \\ -3c \end{bmatrix} = c \begin{bmatrix} 1 \\ -3 \end{bmatrix}.$$

Because this is a constant multiple of our previous result, any choice we make leads to the same solution

$$\mathbf{x}_1(t) = \begin{bmatrix} 1 \\ -3 \end{bmatrix} e^{-2t}.$$

Next we substitute in (13) the second eigenvalue $\lambda_2 = 5$ and get the equivalent scalar equations

$$-v_1 + 2v_2 = 0 \quad \text{and} \quad 3v_1 - 6v_2 = 0.$$

With $v_2 = 1$ we obtain $v_1 = 2$, so

$$\mathbf{v}_2 = \begin{bmatrix} 2 \\ 1 \end{bmatrix}$$

is an eigenvector associated with $\lambda_2 = 5$. A different choice $v_1 = 2c$, $v_2 = c$ would merely give a (constant) multiple of $\mathbf{v}_2$.

These two eigenvalues and associated eigenvectors yield the two solutions

$$\mathbf{x}_1 = \begin{bmatrix} 1 \\ -3 \end{bmatrix} e^{-2t} \quad \text{and} \quad \mathbf{x}_2 = \begin{bmatrix} 2 \\ 1 \end{bmatrix} e^{5t}.$$

They are linearly independent because their Wronskian

$$\begin{vmatrix} e^{-2t} & 2e^{5t} \\ -3e^{-2t} & e^{5t} \end{vmatrix} = 7e^{3t}$$

is nonzero. Hence a general solution of the system in (11) is

$$\mathbf{x}(t) = c_1\mathbf{x}_1 + c_2\mathbf{x}_2 = c_1 \begin{bmatrix} 1 \\ -3 \end{bmatrix} e^{-2t} + c_2 \begin{bmatrix} 2 \\ 1 \end{bmatrix} e^{5t};$$

in scalar form,

$$x_1(t) = c_1e^{-2t} + 2c_2e^{5t}, \qquad x_2(t) = -3c_1e^{-2t} + c_2e^{5t}.$$

Remark: As in Example 1, it is convenient when discussing a linear system $\mathbf{x}' = \mathbf{A}\mathbf{x}$ to use *vectors* $\mathbf{x}_1, \mathbf{x}_2, \ldots, \mathbf{x}_n$ to denote different vector-valued solutions of the system, whereas the *scalars* $x_1, x_2, \ldots, x_n$ denote the components of a single vector-valued solution $\mathbf{x}$.

COMPLEX EIGENVALUES

Even if some of the eigenvalues are complex, so long as they are distinct the method described above still yields n linearly independent solutions. The only complication is that the eigenvectors associated with complex eigenvalues are ordinarily complex-valued, so we will have complex-valued solutions.

To obtain real-valued solutions, we note that—because we are assuming that the matrix $\mathbf{A}$ has only real entries—the coefficients in the characteristic equation in (8) all will be real. Consequently any complex eigenvalues must appear in complex conjugate pairs. Suppose then that $\lambda = p + qi$ and $\bar{\lambda} = p - qi$ are such a pair of eigenvalues. If $\mathbf{v}$ is an eigenvector associated with λ, so that

$$(\mathbf{A} - \lambda\mathbf{I})\mathbf{v} = \mathbf{0},$$

then taking complex conjugates in this equation yields

$$(\mathbf{A} - \bar{\lambda}\mathbf{I})\bar{\mathbf{v}} = \mathbf{0}.$$

Thus the conjugate $\bar{\mathbf{v}}$ of $\mathbf{v}$ is an eigenvector associated with $\bar{\lambda}$. Of course the conjugate of a vector is defined componentwise; if

$$\mathbf{v} = \begin{bmatrix} a_1 + b_1 i \\ a_2 + b_2 i \\ \vdots \\ a_n + b_n i \end{bmatrix} = \begin{bmatrix} a_1 \\ a_2 \\ \vdots \\ a_n \end{bmatrix} + \begin{bmatrix} b_1 \\ b_2 \\ \vdots \\ b_n \end{bmatrix} i = \mathbf{a} + \mathbf{b}i \tag{14}$$

then $\bar{\mathbf{v}} = \mathbf{a} - \mathbf{b}i$. The complex-valued solution associated with λ and $\mathbf{v}$ is then

$$\mathbf{x}(t) = \mathbf{v}e^{\lambda t} = \mathbf{v}e^{(p + qi)t}$$
$$= (\mathbf{a} + \mathbf{b}i)e^{pt}(\cos qt + i \sin qt);$$

that is,

$$\mathbf{x}(t) = e^{pt}(\mathbf{a} \cos qt - \mathbf{b} \sin qt) + ie^{pt}(\mathbf{b} \cos qt + \mathbf{a} \sin qt). \tag{15}$$

Because the real and imaginary parts of a complex-valued solution are also solutions, we thus get the two *real-valued* solutions

$$\mathbf{x}_1(t) = \mathrm{Re}(\mathbf{x}(t)) = e^{pt}(\mathbf{a} \cos qt - \mathbf{b} \sin qt),$$
$$\mathbf{x}_2(t) = \mathrm{Im}(\mathbf{x}(t)) = e^{pt}(\mathbf{b} \cos qt + \mathbf{a} \sin qt) \tag{16}$$

associated with the complex conjugate eigenvalues $p \pm qi$; it is easy to check that the same two real-valued solutions result from taking real and imaginary parts of $\bar{\mathbf{v}}e^{\bar{\lambda}t}$. Rather than memorizing the formulas in (16), it is preferable in a specific example to find the complex-valued solution explicitly, and then compute its real and imaginary parts.

EXAMPLE 2 Find a general solution of the system

$$x_1' = 3x_1 + x_3,$$
$$x_2' = 9x_1 - x_2 + 2x_3, \qquad (17)$$
$$x_3' = -9x_1 + 4x_2 - x_3.$$

SOLUTION The characteristic equation of the coefficient matrix in (17) is

$$\begin{bmatrix} 3-\lambda & 0 & 1 \\ 9 & -1-\lambda & 2 \\ -9 & 4 & -1-\lambda \end{bmatrix} = (3-\lambda)[(-1-\lambda)^2 - 8] + (1)[36 - (-9)(-1-\lambda)]$$

$$= (3-\lambda)(\lambda+1)^2 - 24 + 8\lambda + 36 - 9 - 9\lambda$$
$$= (3-\lambda)(\lambda+1)^2 + 3 - \lambda$$
$$= (3-\lambda)[(\lambda+1)^2 + 1] = 0.$$

Thus we have the real eigenvalue $\lambda_1 = 3$ and the conjugate pair $\lambda = -1 + i$, $\bar{\lambda} = -1 - i$ of complex eigenvalues. If we had expanded the determinant in a more slapdash manner, we might have obtained the cubic equation in a form more difficult to solve.

The eigenvectors will be solutions of the eigenvector equation

$$\begin{bmatrix} 3-\lambda & 0 & 1 \\ 9 & -1-\lambda & 2 \\ -9 & 4 & -1-\lambda \end{bmatrix}\begin{bmatrix} v_1 \\ v_2 \\ v_3 \end{bmatrix} = \begin{bmatrix} 0 \\ 0 \\ 0 \end{bmatrix}. \qquad (18)$$

With the real eigenvalue $\lambda_1 = 3$, (18) reduces to the three equations

$$v_3 = 0,$$
$$9v_1 - 4v_2 + 2v_3 = 0,$$
$$-9v_1 + 4v_2 - 4v_3 = 0.$$

Because $v_3 = 0$, we may choose $v_1 = 4$ and $v_2 = 9$ to satisfy the last two equations. This gives the eigenvector and corresponding solution

$$\mathbf{v}_1 = \begin{bmatrix} 4 \\ 9 \\ 0 \end{bmatrix}, \qquad \mathbf{x}_1 = \mathbf{v}_1 e^{3t} = \begin{bmatrix} 4 \\ 9 \\ 0 \end{bmatrix} e^{3t}. \qquad (19)$$

As in Example 1, a different choice $v_1 = 4c$, $v_2 = 9c$ (such as $v_1 = \frac{1}{9}$, $v_2 = \frac{1}{4}$ with $c = \frac{1}{36}$) would merely yield a constant multiple of $\mathbf{v}_1$ and hence would lead to the same solution $\mathbf{x}_1(t)$.

With the complex eigenvalue $\lambda = -1 + i$, Eq. (18) is equivalent to the three simultaneous equations

$$(4 - i)v_1 + v_3 = 0,$$
$$9v_1 - iv_2 + 2v_3 = 0,$$
$$-9v_1 + 4v_2 - iv_3 = 0.$$

The choice $v_1 = 1$ in the first equation gives $v_3 = -4 + i$. Then the third equation gives $v_2 = \frac{1}{4}(9v_1 + iv_3) = 2 - i$. The corresponding complex-valued solution is

$$\mathbf{x}(t) = \begin{bmatrix} 1 \\ 2 - i \\ -4 + i \end{bmatrix} e^{(-1+i)t}$$

$$= \left(\begin{bmatrix} 1 \\ 2 \\ -4 \end{bmatrix} + i \begin{bmatrix} 0 \\ -1 \\ 1 \end{bmatrix} \right) e^{-t}(\cos t + i \sin t),$$

and therefore

$$\mathbf{x}(t) = e^{-t} \left(\begin{bmatrix} 1 \\ 2 \\ -4 \end{bmatrix} \cos t - \begin{bmatrix} 0 \\ -1 \\ 1 \end{bmatrix} \sin t \right) + ie^{-t} \left(\begin{bmatrix} 0 \\ -1 \\ 1 \end{bmatrix} \cos t + \begin{bmatrix} 1 \\ 2 \\ -4 \end{bmatrix} \sin t \right).$$

The real and imaginary parts of $\mathbf{x}$ are the real-valued solutions

$$\mathbf{x}_2 = \mathrm{Re}(\mathbf{x}) = e^{-t} \begin{bmatrix} \cos t \\ 2\cos t + \sin t \\ -4\cos t - \sin t \end{bmatrix},$$

$$\mathbf{x}_3 = \mathrm{Im}(\mathbf{x}) = e^{-t} \begin{bmatrix} \sin t \\ -\cos t + 2\sin t \\ \cos t + 4\sin t \end{bmatrix}.$$

(20)

By computing the Wronskian, it can be verified that the three solution vectors $\mathbf{x}_1$, $\mathbf{x}_2$, and $\mathbf{x}_3$ in (19) and (20) are linearly independent. Hence, putting it all together, we see that the scalar components of the desired general solution

$$\mathbf{x} = c_1 \mathbf{x}_1 + c_2 \mathbf{x}_2 + c_3 \mathbf{x}_3$$

of the system in (17) are given by

$$x_1(t) = 4c_1 e^{3t} + e^{-t}[c_2 \cos t + c_3 \sin t]$$
$$x_2(t) = 9c_1 e^{3t} + e^{-t}[(2c_2 - c_3)\cos t + (c_2 + 2c_3)\sin t]$$
$$x_3(t) = \qquad\quad e^{-t}[(-4c_2 + c_3)\cos t + (-c_2 - 4c_3)\sin t].$$

For instance, with $c_1 = 1$, $c_2 = -4$, and $c_3 = 1$ we obtain the particular solution

$$x_1(t) = 4e^{3t} - e^{-t}(4\cos t - \sin t)$$
$$x_2(t) = 9e^{3t} - e^{-t}(9\cos t + 2\sin t)$$
$$x_3(t) = 17e^{-t}\cos t$$

that satisfies the initial conditions $x_1(0) = 0$, $x_2(0) = 0$, $x_3(0) = 17$.

EXAMPLE 3 Consider the mass-and-spring system of Fig. 5.24. With the values $m_1 = 2$, $m_2 = 1$ and $k_1 = 100$, $k_2 = 50$, the equations of motion of the two masses

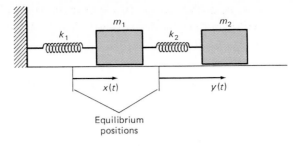

FIGURE 5.24 The mass-and-spring system of Example 3.

simplify to

$$x'' = -75x + 25y$$
$$y'' = 50x - 50y. \tag{21}$$

Apply the eigenvalue method to find the general solution of this system, and thereby determine the natural modes of oscillation of the two masses.

SOLUTION The substitutions

$$x_1 = x, \; x_2 = x', \qquad x_3 = y, \; x_4 = y'$$

transform the second order system in (21) to the first order system

$$
\begin{aligned}
x_1' &= x_2 \\
x_2' &= -75x_1 + 25x_3 \\
x_3' &= x_4 \\
x_4' &= 50x_1 - 50x_3.
\end{aligned} \tag{22}
$$

When we calculate the determinant of the matrix

$$
\mathbf{A} - \lambda\mathbf{I} =
\begin{bmatrix}
-\lambda & 1 & 0 & 0 \\
-75 & -\lambda & 25 & 0 \\
0 & 0 & -\lambda & 1 \\
50 & 0 & -50 & -\lambda
\end{bmatrix}, \tag{23}
$$

we find that the characteristic equation of the coefficient matrix $\mathbf{A}$ of the system in (22) is

$$\lambda^4 + 125\lambda^2 + 2500 = (\lambda^2 + 25)(\lambda^2 + 100) = 0.$$

Thus the eigenvalues of $\mathbf{A}$ are $\lambda = \pm 5i$ and $\lambda = \pm 10i$.

With $\lambda = -5i$ in the matrix in Eq. (23), the eigenvector equation $(\mathbf{A} - \lambda\mathbf{I})\mathbf{v} = \mathbf{0}$ takes the form

$$
\begin{bmatrix}
5i & 1 & 0 & 0 \\
-75 & 5i & 25 & 0 \\
0 & 0 & 5i & 1 \\
50 & 0 & -50 & 5i
\end{bmatrix}
\begin{bmatrix}
v_1 \\
v_2 \\
v_3 \\
v_4
\end{bmatrix}
=
\begin{bmatrix}
0 \\
0 \\
0 \\
0
\end{bmatrix}. \tag{24}
$$

The first and third scalar equations in (24) give $v_2 = -5iv_1$ and $v_4 = -5iv_3$, respectively, and then the fourth equation yields $v_3 = 2v_1$. Consequently, the choice $v_1 = 1$

leads to the eigenvector $\mathbf{v} = \begin{bmatrix} 1 & -5i & 2 & -10i \end{bmatrix}^T$ associated with $\lambda = -5i$. The corresponding solution $\mathbf{x} = \mathbf{v}e^{\lambda t}$ of the system in (22) is then given by

$$\mathbf{x}(t) = \begin{bmatrix} 1 \\ -5i \\ 2 \\ -10i \end{bmatrix} (\cos 5i - i \sin 5t) = \mathbf{x}_1(t) - i\mathbf{x}_2(t),$$

where

$$\mathbf{x}_1(t) = \begin{bmatrix} \cos 5t \\ -5 \sin 5t \\ 2 \cos 5t \\ -10 \sin 5t \end{bmatrix} \quad \text{and} \quad \mathbf{x}_2(t) = \begin{bmatrix} \sin 5t \\ 5 \cos 5t \\ 2 \sin 5t \\ 10 \cos 5t \end{bmatrix} \tag{25}$$

are real-valued solutions of the system associated with the conjugate eigenvalues $\lambda = \pm 5i$.

Beginning instead with $\lambda = -10i$, a very similar computation leads to the real-valued solutions

$$\mathbf{x}_3(t) = \begin{bmatrix} \cos 10t \\ -10 \sin 10t \\ -\cos 10t \\ 10 \sin 10t \end{bmatrix} \quad \text{and} \quad \mathbf{x}_4(t) = \begin{bmatrix} \sin 10t \\ 10 \cos 10t \\ -\sin 10t \\ -10 \cos 5t \end{bmatrix} \tag{26}$$

of the first-order system in (22) associated with the conjugate eigenvalues $\lambda = \pm 10i$.

Because $x = x_1$ and $y = x_3$, we get a general solution of the original second order system in (21) by extracting the first and third components of the general solution $\mathbf{x} = c_1\mathbf{x}_1 + c_2\mathbf{x}_2 + c_3\mathbf{x}_3 + c_4\mathbf{x}_4$. Therefore, a general solution of (21) is given by

$$\begin{aligned} x(t) &= c_1 \cos 5t + c_2 \sin 5t + c_3 \cos 10t + c_4 \sin 10t, \\ y(t) &= 2c_1 \cos 5t + 2c_2 \sin 5t - c_3 \cos 10t - c_4 \sin 10t. \end{aligned} \tag{27}$$

Thus the natural frequencies of oscillation of the system are $\omega_1 = 5$ and $\omega_2 = 10$. In the natural mode with frequency $\omega_1 = 5$ the two masses move synchronously in the same direction, with the amplitude of oscillation of m_2 twice that of m_1. In the natural mode with frequency $\omega_2 = 10$ the two masses move in opposite directions with equal amplitudes of oscillation.

REPEATED EIGENVALUES

Suppose that the $n \times n$ matrix $\mathbf{A}$ has $m < n$ distinct eigenvalues $\lambda_1, \lambda_2, \ldots, \lambda_n$. Then at least one of these eigenvalues is a repeated root of the characteristic equation

$$|\mathbf{A} - \lambda\mathbf{I}| = 0. \tag{7}$$

An eigenvalue is of **multiplicity** k if it is a k-fold root of Eq. (7). For each eigenvalue λ_i, the eigenvector equation

$$(\mathbf{A} - \lambda_i\mathbf{I})\mathbf{v} = \mathbf{0} \tag{28}$$

CHAPTER 5: Linear Systems of Differential Equations

has at least one nontrivial solution, so there is at least one eigenvector associated with λ_i. But if some eigenvalue λ_i is of multiplicity $k > 1$, and Eq. (28) has *fewer* than k linearly independent solutions—so there are fewer than k linearly independent eigenvectors associated with λ_i—then the eigenvalue method as described thusfar will produce *fewer* than the needed n linearly independent solutions of the system $\mathbf{x}' = \mathbf{Ax}$.

We may call an eigenvalue of multiplicity k **complete** if it has k linearly independent associated eigenvectors and **incomplete** if there are fewer than k. There are then two cases that we must consider:

1. Every eigenvalue of $\mathbf{A}$ is complete;
2. Some eigenvalue of $\mathbf{A}$ is incomplete.

In case 1, the matrix $\mathbf{A}$ has n linearly independent eigenvectors $\mathbf{v}_1, \mathbf{v}_2, \ldots, \mathbf{v}_n$ associated with the eigenvalues $\lambda_1, \lambda_2, \ldots, \lambda_n$ (each repeated with its multiplicity), and the general solution of $\mathbf{x}' = \mathbf{Ax}$ is

$$\mathbf{x}(t) = c_1\mathbf{v}_1 e^{\lambda_1 t} + c_2\mathbf{v}_2 e^{\lambda_2 t} + \cdots + c_n\mathbf{v}_n e^{\lambda_n t}. \tag{29}$$

This case always occurs when the matrix $\mathbf{A}$ is **symmetric**; that is, when $\mathbf{A} = \mathbf{A}^T$. Indeed, it is proved in most linear algebra texts that each eigenvalue of a symmetric (real) matrix is both real and complete. (For instance, see Section 6.4 of Edwards and Penney, *Elementary Linear Algebra* (Englewood Cliffs, N.J.: Prentice-Hall, 1988).) The following example illustrates the approach in case 1.

EXAMPLE 4   Find a general solution of the system

$$
\begin{aligned}
x_1' &= 2x_2 + 2x_3, \\
x_2' &= 2x_1 + 2x_3, \\
x_3' &= 2x_1 + 2x_2.
\end{aligned} \tag{30}
$$

SOLUTION   The characteristic equation is

$$
0 = \begin{vmatrix} -\lambda & 2 & 2 \\ 2 & -\lambda & 2 \\ 2 & 2 & -\lambda \end{vmatrix}
$$

$$
\begin{aligned}
&= -\lambda(\lambda^2 - 4) - 2(-2\lambda - 4) + 2(4 + 2\lambda) \\
&= -\lambda(\lambda + 2)(\lambda - 2) + 8\lambda + 16 \\
&= (\lambda + 2)(-\lambda^2 + 2\lambda + 8);
\end{aligned}
$$

that is,

$$-(\lambda + 2)(\lambda + 2)(\lambda - 4) = 0.$$

Thus we have the unrepeated eigenvalue $\lambda_1 = 4$ and the eigenvalue $\lambda_2 = -2$ of multiplicity 2.

When we substitute $\lambda_1 = 4$ in the eigenvector equation

$$(\mathbf{A} - \lambda\mathbf{I})\mathbf{v} = \begin{bmatrix} -\lambda & 2 & 2 \\ 2 & -\lambda & 2 \\ 2 & 2 & -\lambda \end{bmatrix}\begin{bmatrix} v_1 \\ v_2 \\ v_3 \end{bmatrix} = \begin{bmatrix} 0 \\ 0 \\ 0 \end{bmatrix}, \tag{31}$$

we obtain three scalar equations that are readily solved (to within a constant multiple) for the eigenvector

$$\mathbf{v}_1 = \begin{bmatrix} 1 \\ 1 \\ 1 \end{bmatrix}$$

associated with $\lambda_1 = 4$.

When we substitute the repeated eigenvalue $\lambda_2 = -2$ in (31), the three scalar equations reduce to the single equation

$$2v_1 + 2v_2 + 2v_3 = 0.$$

We can choose v_1 and v_2 arbitrarily and solve for v_3. With $v_1 = 1$ and $v_2 = 0$, we get $v_3 = -1$, which we also get with $v_1 = 0$ and $v_2 = 1$. Thus we find the *two* linearly independent eigenvectors

$$\mathbf{v}_2 = \begin{bmatrix} 1 \\ 0 \\ -1 \end{bmatrix} \quad \text{and} \quad \mathbf{v}_3 = \begin{bmatrix} 0 \\ 1 \\ -1 \end{bmatrix}$$

associated with $\lambda_2 = -2$.

The three eigenvectors we have found yield the general solution

$$\mathbf{x}(t) = c_1 \begin{bmatrix} 1 \\ 1 \\ 1 \end{bmatrix} e^{4t} + c_2 \begin{bmatrix} 1 \\ 0 \\ -1 \end{bmatrix} e^{-2t} + c_3 \begin{bmatrix} 0 \\ 1 \\ -1 \end{bmatrix} e^{-2t}, \tag{32}$$

because the Wronskian can be used to verify that the three solutions of (30) displayed in (32) are indeed linearly independent.

Remark: The "arbitrariness" in Example 4 is a bit more interesting than that in Examples 1 and 2. Suppose that, after finding the first two eigenvectors $\mathbf{v}_1$ and $\mathbf{v}_2$ as above, we had chosen $v_1 = a$ and $v_2 = b$ (not both zero) in the equation $2v_1 + 2v_2 + 2v_3 = 0$, so that $v_3 = -a - b$. Then our third eigenvector would have been

$$\mathbf{v} = \begin{bmatrix} a \\ b \\ -a - b \end{bmatrix} = a \begin{bmatrix} 1 \\ 0 \\ -1 \end{bmatrix} + b \begin{bmatrix} 0 \\ 1 \\ -1 \end{bmatrix}.$$

It would follow that $\mathbf{v} = a\mathbf{v}_2 + b\mathbf{v}_3$. Because $\mathbf{v}$ is a linear combination of $\mathbf{v}_2$ and $\mathbf{v}_3$, our new general solution

$$\mathbf{x}(t) = c_1 \mathbf{v}_1 e^{4t} + c_2 \mathbf{v}_2 e^{-2t} + c\mathbf{v} e^{-2t}$$

would be equivalent to the one in (32). Thus we need not worry about making the "right" choice; any choice will do.

In case 2, we need to tell how to find the missing solutions corresponding to an incomplete eigenvalue λ of multiplicity $k > 1$. We will discuss in detail only the case $k = 2$. Suppose we have found that there is only one eigenvector $\mathbf{v}$ associated with λ, so we have the single solution

$$\mathbf{x}_1(t) = \mathbf{v}e^{\lambda t}.$$

By analogy with the case of a repeated characteristic root for a single differential equation (Section 2.3), we might expect to find a second solution of the form

$$\mathbf{x} = \mathbf{w}te^{\lambda t}. \tag{33}$$

But when we substitute (33) in $\mathbf{x}' = \mathbf{A}\mathbf{x}$, we get the equation

$$\mathbf{w}e^{\lambda t} + \mathbf{w}te^{\lambda t} = \mathbf{A}\mathbf{w}te^{\lambda t}. \tag{34}$$

Because the coefficients of both $e^{\lambda t}$ and $te^{\lambda t}$ must balance, it follows that $\mathbf{w} = \mathbf{0}$. Thus there is *no* nontrivial solution of the form in (33).

Still trying to make this idea work, we turn to the next most plausible possibility: a second solution of the form

$$\mathbf{x}_2(t) = \mathbf{v}te^{\lambda t} + \mathbf{w}e^{\lambda t}. \tag{35}$$

When we substitute (35) in $\mathbf{x}' = \mathbf{A}\mathbf{x}$, we get the equation

$$\mathbf{v}e^{\lambda t} + \lambda\mathbf{v}te^{\lambda t} + \lambda\mathbf{w}e^{\lambda t} = \mathbf{A}\mathbf{v}te^{\lambda t} + \mathbf{A}\mathbf{w}e^{\lambda t}. \tag{36}$$

We equate coefficients of $te^{\lambda t}$ and of $e^{\lambda t}$ in (36); thus we obtain the two equations

$$(\mathbf{A} - \lambda\mathbf{I})\mathbf{v} = \mathbf{0} \tag{37}$$

and

$$(\mathbf{A} - \lambda\mathbf{I})\mathbf{w} = \mathbf{v} \tag{38}$$

that the vectors $\mathbf{v}$ and $\mathbf{w}$ must satisfy in order for (35) to be a solution. (Note that Eq. (37) merely means that $\mathbf{v}$ is an eigenvector associated with λ.) Having found $\mathbf{v}$, we must then attempt to solve (38) for $\mathbf{w}$. Despite the initially discouraging fact that $|\mathbf{A} - \lambda\mathbf{I}| = 0$, it turns out that a nontrivial solution $\mathbf{w}$ of Eq. (38) can always be found, as in the following example.

EXAMPLE 5 Find a general solution of the system

$$\begin{align} x_1' &= x_1 - 3x_2, \\ x_2' &= 3x_1 + 7x_2. \end{align} \tag{39}$$

SOLUTION The characteristic equation is

$$0 = \begin{vmatrix} 1 - \lambda & -3 \\ 3 & 7 - \lambda \end{vmatrix} = (1 - \lambda)(7 - \lambda) + 9,$$

so that $0 = \lambda^2 - 8\lambda + 16 = (\lambda - 4)^2$. So we have the single eigenvalue $\lambda_1 = 4$ of multiplicity 2. When we substitute $\lambda_1 = 4$ in the eigenvector equation

$$(\mathbf{A} - \lambda\mathbf{I})\mathbf{v} = \begin{bmatrix} 1 - \lambda & -3 \\ 3 & 7 - \lambda \end{bmatrix}\begin{bmatrix} v_1 \\ v_2 \end{bmatrix} = \begin{bmatrix} 0 \\ 0 \end{bmatrix},$$

we obtain the equivalent scalar equations

$$-3v_1 - 3v_2 = 0 = 3v_1 + 3v_2.$$

We choose $v_1 = 1$ and $v_2 = -1$; this yields the eigenvector and solution

$$\mathbf{v} = \begin{bmatrix} 1 \\ -1 \end{bmatrix} \quad \text{and} \quad \mathbf{x}_1(t) = \begin{bmatrix} 1 \\ -1 \end{bmatrix} e^{4t} \qquad (40)$$

With this eigenvector $\mathbf{v}$ and eigenvalue $\lambda = 4$, Eq. (38) takes the form

$$\begin{bmatrix} -3 & -3 \\ 3 & 3 \end{bmatrix}\begin{bmatrix} w_1 \\ w_2 \end{bmatrix} = \begin{bmatrix} 1 \\ -1 \end{bmatrix},$$

so that $3w_1 + 3w_2 = -1$. A nontrivial solution is $w_1 = 0$, $w_2 = -\frac{1}{3}$. Then the second solution given by (35) is

$$\mathbf{x}_2(t) = \mathbf{v}te^{4t} + \mathbf{w}e^{4t}$$

$$= \begin{bmatrix} 1 \\ -1 \end{bmatrix} te^{4t} + \begin{bmatrix} 0 \\ -\frac{1}{3} \end{bmatrix} e^{4t} = \begin{bmatrix} t \\ -t - \frac{1}{3} \end{bmatrix} e^{4t}. \qquad (41)$$

The Wronskian of the two solutions $\mathbf{x}_1$ and $\mathbf{x}_2$ is

$$W = \begin{vmatrix} e^{4t} & te^{4t} \\ -e^{4t} & -(t + \frac{1}{3})e^{4t} \end{vmatrix} = -\frac{1}{3}e^{8t} \neq 0,$$

so $\mathbf{x}_1$ and $\mathbf{x}_2$ are linearly independent. Therefore a general solution of (39) is

$$\mathbf{x} = c_1\mathbf{x}_1 + c_2\mathbf{x}_2 = c_1\begin{bmatrix} 1 \\ -1 \end{bmatrix} e^{4t} + c_2\begin{bmatrix} t \\ -t - \frac{1}{3} \end{bmatrix} e^{4t}$$

$$= \begin{bmatrix} c_1 + c_2 t \\ -c_1 - c_2(t + \frac{1}{3}) \end{bmatrix} e^{4t}. \qquad (42)$$

The situation is more complicated in the case of an incomplete eigenvalue λ of multiplicity $k > 2$. If $k = 3$, for instance, there may be either one or two linearly independent eigenvectors associated with λ, and thus either two or one missing solutions to be found. A general discussion of systems having incomplete eigenvalues of higher multiplicity is beyond the scope of our discussion, but some of the possibilities are illustrated in the problems that follow. For instance, Problem 34 illustrates the fact that if λ is an eigenvalue of $\mathbf{A}$ of multiplicity $k = 3$ having only one associated eigenvector $\mathbf{u}$, then three independent solutions of $\mathbf{x}' = \mathbf{A}\mathbf{x}$ are given by

$$\mathbf{x}_1(t) = \mathbf{u}e^{\lambda t}$$

$$\mathbf{x}_2(t) = \mathbf{u}te^{\lambda t} + \mathbf{v}e^{\lambda t} \qquad (43)$$

$$\mathbf{x}_3(t) = \tfrac{1}{2}\mathbf{u}t^2 e^{\lambda t} + \mathbf{v}te^{\lambda t} + \mathbf{w}e^{\lambda t},$$

where the nonzero vectors $\mathbf{u}$, $\mathbf{v}$, and $\mathbf{w}$ are obtained by solving in turn the systems

$$(\mathbf{A} - \lambda\mathbf{I})\mathbf{u} = \mathbf{0}$$

$$(\mathbf{A} - \lambda\mathbf{I})\mathbf{v} = \mathbf{u} \qquad (44)$$

$$(\mathbf{A} - \lambda\mathbf{I})\mathbf{w} = \mathbf{v}.$$

Problem 35 illustrates the case of a triple eigenvalue that has only two linearly independent associated eigenvectors.

CHAPTER 5: Linear Systems of Differential Equations

In each of Problems 1–20, apply the eigenvalue method of this section to find a general solution of the given system. If initial values are given, find also the corresponding particular solution.

1. $x_1' = x_1 + 2x_2$
$x_2' = 2x_1 + x_2$

2. $x_1' = 2x_1 + 3x_2$
$x_2' = 2x_1 + x_2$

3. $x_1' = 3x_1 + 4x_2$
$x_2' = 3x_1 + 2x_2$
$x_1(0) = x_2(0) = 1$

4. $x_1' = 4x_1 + x_2$
$x_2' = 6x_1 - x_2$

5. $x_1' = 6x_1 - 7x_2$
$x_2' = x_1 - 2x_2$

6. $x_1' = 9x_1 + 5x_2$
$x_2' = -6x_1 - 2x_2$
$x_1(0) = 1, x_2(0) = 0$

7. $x_1' = -3x_1 + 4x_2$
$x_2' = 6x_1 - 5x_2$

8. $x_1' = x_1 - 5x_2$
$x_2' = x_1 - x_2$

9. $x_1' = 2x_1 - 5x_2$
$x_2' = 4x_1 - 2x_2$
$x_1(0) = 2, x_2(0) = 3$

10. $x_1' = -3x_1 - 2x_2$
$x_2' = 9x_1 + 3x_2$

11. $x_1' = x_1 - 2x_2$
$x_2' = 2x_1 + x_2$
$x_1(0) = 0, x_2(0) = 4$

12. $x_1' = x_1 - 5x_2$
$x_2' = x_1 + 3x_2$

13. $x_1' = 5x_1 - 9x_2$
$x_2' = 2x_1 - x_2$

14. $x_1' = 3x_1 - 4x_2$
$x_2' = 4x_1 + 3x_2$

15. $x_1' = 7x_1 - 5x_2$
$x_2' = 4x_1 + 3x_2$

16. $x_1' = 3x_1 - x_2$
$x_2' = x_1 + x_2$

17. $x_1' = x_1 - 2x_2$
$x_2' = 2x_1 + 5x_2$

18. $x_1' = 3x_1 - x_2$
$x_2' = x_1 + 5x_2$

19. $x_1' = 7x_1 + x_2$
$x_2' = -4x_1 + 3x_2$

20. $x_1' = x_1 - 4x_2$
$x_2' = 4x_1 + 9x_2$

In each of Problems 21–29, the eigenvalues of the coefficient matrix are complete and can be found by inspection and factoring. Apply the eigenvalue method to find a general solution of each system.

21. $x_1' = 4x_1 + x_2 + 4x_3$
$x_2' = x_1 + 7x_2 + x_3$
$x_3' = 4x_1 + x_2 + 4x_3$

22. $x_1' = x_1 + 2x_2 + 2x_3$
$x_2' = 2x_1 + 7x_2 + x_3$
$x_3' = 2x_1 + x_2 + 7x_3$

23. $x_1' = 4x_1 + x_2 + x_3$
$x_2' = x_1 + 4x_2 + x_3$
$x_3' = x_1 + x_2 + 4x_3$

24. $x_1' = 5x_1 + x_2 + 3x_3$
$x_2' = x_1 + 7x_2 + x_3$
$x_3' = 3x_1 + x_2 + 5x_3$

25. $x_1' = 5x_1 - 6x_3$
$x_2' = 2x_1 - x_2 - 2x_3$
$x_3' = 4x_1 - 2x_2 - 4x_3$

26. $x_1' = 3x_1 + 2x_2 + 2x_3$
$x_2' = -5x_1 - 4x_2 - 2x_3$
$x_3' = 5x_1 + 5x_2 + 3x_3$

27. $x_1' = 3x_1 + x_2 + x_3$
$x_2' = -5x_1 - 3x_2 - x_3$
$x_3' = 5x_1 + 5x_2 + 3x_3$

28. $x_1' = 2x_1 + x_2 - x_3$
$x_2' = -4x_1 - 3x_2 - x_3$
$x_3' = 4x_1 + 4x_2 + 2x_3$

29. $x_1' = 5x_1 + 5x_2 + 2x_3$
$x_2' = -6x_1 - 6x_2 - 5x_3$
$x_3' = 6x_1 + 6x_2 + 5x_3$

30. Solve the system described in Example 3 of this section with the values $m_1 = 2$, $m_2 = 1$ and $k_1 = 4$, $k_2 = 2$.

31. Solve the system described in Example 3 of this section with the values $m_1 = 2$, $m_2 = 0.5$ and $k_1 = 75$, $k_2 = 25$. Compare your results with those of Example 4 of Section 5.3.

32. Show that the coefficient matrix of the system

$$x_1' = 2x_1 + 3x_2 + 3x_3,$$
$$x_2' = -x_2 - 3x_3,$$
$$x_3' = 2x_3$$

has the eigenvalue $\lambda = 2$ of multiplicity 2. Find two linearly independent eigenvectors associated with $\lambda = 2$, and then find a general solution of the system.

33. Show that the coefficient matrix of the system

$$x_1' = 3x_1 + 2x_2 + x_3,$$
$$x_2' = -x_1 - x_3,$$
$$x_3' = x_1 + x_2 + 2x_3$$

has the eigenvalue $\lambda = 2$ of multiplicity 2, but with only one linearly independent eigenvector associated with it. Use the method of Example 5 to find a general solution of this system.

34. (a) Show that the coefficient matrix **A** of the system

$$x_1' = 3x_1 + x_2,$$
$$x_2' = -x_1 - x_3,$$
$$x_3' = x_1 + 2x_2 + 3x_3$$

has the single eigenvalue $\lambda = 2$ of multiplicity 3, but with only one linearly independent eigenvector **u** associated with it. Then one solution is $\mathbf{x}_1(t) = \mathbf{u}e^{2t}$. (b) Find a second solution of the form $\mathbf{x}_2(t) = \mathbf{u}te^{2t} + \mathbf{v}e^{2t}$ where **v** is a nontrivial solution of $(\mathbf{A} - 2\mathbf{I})\mathbf{v} = \mathbf{u}$. (c) Find the third solution of the form

$$\mathbf{x}_3(t) = \tfrac{1}{2}\mathbf{u}t^2 e^{2t} + \mathbf{v}te^{2t} + \mathbf{w}e^{2t}$$

where **w** is a nontrivial solution of $(\mathbf{A} - 2\mathbf{I})\mathbf{w} = \mathbf{v}$.

35. (a) Show that $\lambda = 2$ is a triple eigenvalue of the coefficient matrix **A** of the system

$$x_1' = 3x_1 + x_2,$$
$$x_2' = -x_1 + x_2,$$
$$x_3' = x_1 + x_2 + 2x_3,$$

and that there are only two linearly independent eigenvectors $\mathbf{v}_1 = (1 \;\; -1 \;\; 0)^T$ and $\mathbf{v}_2 = (1 \;\; -1 \;\; 1)^T$ associated with this eigenvalue. Then $\mathbf{x}_1(t) = \mathbf{v}_1 e^{2t}$ and $\mathbf{x}_2(t) = \mathbf{v}_2 e^{2t}$ are two linearly independent solutions. (b) Show that if $\mathbf{x}_3(t) = \mathbf{v}te^{2t} + \mathbf{w}e^{2t}$ is a third solution, then $\mathbf{v}$ and $\mathbf{w}$ must satisfy the equations

$$(\mathbf{A} - 2\mathbf{I})\mathbf{v} = \mathbf{0}, \tag{45a}$$

$$(\mathbf{A} - 2\mathbf{I})\mathbf{w} = \mathbf{v}. \tag{45b}$$

(c) Every solution of Eq. (45a) is of the form $\mathbf{v} = c_1\mathbf{v}_1 + c_2\mathbf{v}_2$, where c_1 and c_2 are constants. According to a fundamental theorem of linear algebra, Eq. (45b) has a nontrivial solution $\mathbf{w}$ if and only if the vector $\mathbf{v}$ is orthogonal to every solution of the system $(\mathbf{A}^T - 2\mathbf{I})\mathbf{y} = \mathbf{0}$. Show that the general solution of this system is

$$\mathbf{y} = \alpha \begin{bmatrix} 1 \\ 0 \\ -1 \end{bmatrix} + \beta \begin{bmatrix} 0 \\ 1 \\ 1 \end{bmatrix}.$$

(d) Show that $\mathbf{v} \cdot \mathbf{y} = c_1\alpha - c_1\beta$, so $\mathbf{v}$ and $\mathbf{y}$ are orthogonal provided that $c_1 = 0$. Thus, taking $c_2 = 1$, we can solve Eq. (45b) with $\mathbf{v} = \mathbf{v}_2$. (e) Solve the equation $(\mathbf{A} - 2\mathbf{I})\mathbf{w} = \mathbf{v}_2$ for $\mathbf{w} = (1 \;\; 0 \;\; 0)^T$. Conclude that

$$\mathbf{x}_3(t) = \begin{bmatrix} t + 1 \\ -t \\ t \end{bmatrix} e^{2t}$$

is a third independent solution of the given system.

36. Show that the coefficient matrix of the system

$$x_1' = \qquad\quad x_2 - x_3 - x_4,$$

$$x_2' = \qquad\quad -x_2,$$

$$x_3' = -2x_1 + 2x_2 + x_3 - 2x_4,$$

$$x_4' = -x_1 - x_2 + x_3$$

has the repeated eigenvalue $\lambda = -1$ with two linearly independent eigenvectors and the repeated eigenvalue $\lambda = +1$ with only one linearly independent eigenvector. Use the method of Example 5 to find a general solution.

37. The coefficient matrix $\mathbf{A}$ of the 4×4 system

$$x_1' = 4x_1 + \quad x_2 + \quad x_3 + 7x_4$$

$$x_2' = \quad x_1 + 4x_2 + 10x_3 + \quad x_4$$

$$x_3' = \quad x_1 + 10x_2 + 4x_3 + \quad x_4$$

$$x_4' = 7x_1 + \quad x_2 + \quad x_3 + 4x_4$$

has eigenvalues $\lambda_1 = -3$, $\lambda_2 = -6$, $\lambda_3 = 10$, and $\lambda_4 = 15$. Find the particular solution of this system that satisfies the initial conditions

$$x_1(0) = 3, \qquad x_2(0) = x_3(0) = 1, \qquad x_4(0) = 3.$$

5.6

Nonhomogeneous Linear Systems

In Sections 2.5 and 2.7 we discussed two methods of finding a particular solution of a single nonhomogeneous nth order linear differential equation—undetermined coefficients and variation of parameters. Each of these methods can be generalized to nonhomogeneous linear systems.

Given the nonhomogeneous first order linear system

$$\mathbf{x}' = \mathbf{P}(t)\mathbf{x} + \mathbf{f}(t) \tag{1}$$

with $\mathbf{P}(t)$ and $\mathbf{f}(t)$ continuous, we know from Theorem 4 of Section 5.3 that a general solution of (1) has the form

$$\mathbf{x}(t) = \mathbf{x}_c(t) + \mathbf{x}_p(t), \tag{2}$$

where $\mathbf{x}_c$ denotes a general solution of the associated homogeneous system $\mathbf{x}' = \mathbf{P}(t)\mathbf{x}$, and $\mathbf{x}_p$ is a single particular solution of the equation in (1). Section 5.5 was devoted to $\mathbf{x}_c$; our task now is to find $\mathbf{x}_p$.

UNDETERMINED COEFFICIENTS

Just as in the case of a single differential equation, the method of undetermined coefficients must be restricted to linear systems of equations with *constant* coefficients. Let us then consider the system

$$\mathbf{x}' = \mathbf{A}\mathbf{x} + \mathbf{f}(t) \tag{3}$$

where $\mathbf{A}$ is a matrix of constants, and $\mathbf{f}(t)$ is a linear combination (with constant vector coefficients) of products of polynomials, exponential functions, and sines and cosines. The method is essentially the same for systems as for single differential equations—we make an intelligent guess as to the general form of a particular solution $\mathbf{x}_p$ and then attempt to determine the coefficients in $\mathbf{x}_p$ by substitution in (3). Moreover, the choice of the general form of $\mathbf{x}_p$ is essentially the same as given in the table in Fig. 2.14 in Section 2.5; we modify it only by using undetermined *vector* coefficients rather than undetermined scalars. We will therefore confine the present discussion mainly to illustrative examples.

EXAMPLE 1 Find a general solution of the system

$$\begin{aligned} x' &= 4x + 2y - 8t, \\ y' &= 3x - y + 2t + 3. \end{aligned} \tag{4}$$

SOLUTION In Example 1 of Section 5.5, we found the general solution

$$\mathbf{x}_c(t) = \begin{bmatrix} x_c \\ y_c \end{bmatrix} = c_1 \begin{bmatrix} 1 \\ -3 \end{bmatrix} e^{-2t} + c_2 \begin{bmatrix} 2 \\ 1 \end{bmatrix} e^{5t} \tag{5}$$

of the associated homogeneous system. Because there is no duplication between the terms of $\mathbf{x}_c$ and the nonhomogeneous terms in (4), we assume a trial solution of the form

$$\mathbf{x}_p = \begin{bmatrix} x_p \\ y_p \end{bmatrix} = \mathbf{a}t + \mathbf{b} = \begin{bmatrix} a_1 \\ a_2 \end{bmatrix} t + \begin{bmatrix} b_1 \\ b_2 \end{bmatrix}. \tag{6}$$

Upon substitution of (6) in (4) we get

$$\begin{aligned} \begin{bmatrix} a_1 \\ a_2 \end{bmatrix} &= \begin{bmatrix} 4 & 2 \\ 3 & -1 \end{bmatrix} \begin{bmatrix} a_1 t + b_1 \\ a_2 t + b_2 \end{bmatrix} + \begin{bmatrix} -8t \\ 2t + 3 \end{bmatrix} \\ &= \begin{bmatrix} 4a_1 + 2a_2 - 8 \\ 3a_1 - a_2 + 2 \end{bmatrix} t + \begin{bmatrix} 4b_1 + 2b_2 \\ 3b_1 - b_2 + 3 \end{bmatrix}. \end{aligned} \tag{7}$$

When we equate the coefficients of t and the constant terms (in both x- and y-components) in (7), we get the equations

$$\begin{aligned} 4a_1 + 2a_2 &= 8, \\ 3a_1 - a_2 &= -2, \\ 4b_1 + 2b_2 &= a_1, \\ 3b_1 - b_2 &= a_2 - 3. \end{aligned} \tag{8}$$

We solve the first two equations in (8) for $a_1 = \frac{2}{5}$ and $a_2 = \frac{16}{5}$ and then solve the last two equations for $b_1 = \frac{2}{25}$ and $b_2 = \frac{1}{25}$. Thus our particular solution is

$$\mathbf{x}_p(t) = \begin{bmatrix} \dfrac{2}{5}t + \dfrac{2}{25} \\[2mm] \dfrac{16}{5}t + \dfrac{1}{25} \end{bmatrix},$$

and the general solution of the system in (4) is

$$\begin{aligned} x(t) &= c_1 e^{-2t} + 2c_2 e^{5t} + \tfrac{2}{5}t + \tfrac{2}{25}, \\ y(t) &= -3c_1 e^{-2t} + c_2 e^{5t} + \tfrac{16}{5}t + \tfrac{1}{25}. \end{aligned} \tag{9}$$

In the case of duplicate expressions in the complementary function and the nonhomogeneous terms, there is one difference between the method of undetermined coefficients for systems and for single equations (Rule 2 of Section 2.5). For a system, the usual choice for a trial solution must be multiplied not only by the smallest integral power of t that will eliminate the duplication, but also by all lower (nonnegative integral) powers as well; all the resulting terms must be included in the actual trial solution. For instance, if $\lambda = 0$ had been an (unrepeated) eigenvalue of the associated homogeneous system in Example 1, our trial solution would have been not $\mathbf{a}t^2 + \mathbf{b}t$ (the "usual" choice), but instead $\mathbf{a}t^2 + \mathbf{b}t + \mathbf{c}$.

EXAMPLE 2   Find a general solution of the sytem

$$\begin{aligned} x' &= 4x + 2y, \\ y' &= 3x - y + e^{-2t}. \end{aligned} \tag{10}$$

SOLUTION   Because of duplication of the term e^{-2t} in the nonhomogeneous system above and in the complementary function

$$\mathbf{x}_c(t) = \begin{bmatrix} x_c \\ y_c \end{bmatrix} = c_1 \begin{bmatrix} 1 \\ -3 \end{bmatrix} e^{-2t} + c_2 \begin{bmatrix} 2 \\ 1 \end{bmatrix} e^{5t} \tag{5}$$

previously found, we assume a trial solution of the form

$$\mathbf{x}_p = \mathbf{a}t e^{-2t} + \mathbf{b}e^{-2t} = \begin{bmatrix} a_1 t + b_1 \\ a_2 t + b_2 \end{bmatrix} e^{-2t} \tag{11}$$

(rather than $\mathbf{a}t e^{-2t}$ alone). Then

$$\mathbf{x}_p' = (\mathbf{a} - 2\mathbf{b})e^{-2t} - 2\mathbf{a}t e^{-2t}. \tag{12}$$

Substitution of (11) and (12) in (10), and cancellation of e^{-2t} throughout, yields

$$\begin{bmatrix} a_1 - 2b_1 - 2a_1 t \\ a_2 - 2b_2 - 2a_2 t \end{bmatrix} = \begin{bmatrix} 4 & 2 \\ 3 & -1 \end{bmatrix} \begin{bmatrix} a_1 t + b_1 \\ a_2 t + b_2 \end{bmatrix} + \begin{bmatrix} 0 \\ 1 \end{bmatrix}$$

$$= \begin{bmatrix} (4a_1 + 2a_2)t + 4b_1 + 2b_2 \\ (3a_1 - a_2)t + 3b_1 - b_2 + 1 \end{bmatrix}.$$

We equate the coefficients of t and the constant terms to get the equations

$$6a_1 + 2a_2 = 0,$$
$$3a_1 + a_2 = 0,$$
$$6b_1 + 2b_2 = a_1,$$
$$3b_1 + b_2 = a_2 - 1.$$

(13)

The first two equations in (13) imply that $a_2 = -3a_1$. In order for the last two equations to be consistent, we must have

$$a_1 = 2(a_2 - 1) = 2(-3a_1 - 1) = -6a_1 - 2.$$

It follows that $a_1 = -\frac{2}{7}$, and so $a_2 = \frac{6}{7}$. Each of the last two equations in (13) now reduces to $6b_1 + 2b_2 = -\frac{2}{7}$, so $b_2 = -3b_1 - \frac{1}{7}$. Thus our particular solution is

$$\mathbf{x}_p(t) = \begin{bmatrix} -\frac{2}{7} \\ \frac{6}{7} \end{bmatrix} te^{-2t} + b_1 \begin{bmatrix} 1 \\ -3 \end{bmatrix} e^{-2t} + \begin{bmatrix} 0 \\ -\frac{1}{7} \end{bmatrix} e^{-2t}.$$

(14)

The middle term on the right-hand side is a solution of the associated homogeneous system and can therefore be absorbed into the complementary function in (5). Hence the general solution of the system in (10) is given by

$$x(t) = c_1 e^{-2t} + 2c_2 e^{5t} - \tfrac{2}{7} te^{-2t},$$
$$y(t) = -3c_1 e^{-2t} + c_2 e^{5t} + \tfrac{6}{7} te^{-2t} - \tfrac{1}{7} e^{-2t}.$$

(15)

FUNDAMENTAL MATRICES

In order to generalize the method of variation of parameters so it applies to the problem of finding a particular solution of a nonhomogeneous linear system, we need the concept of a fundamental matrix of a homogeneous system. Suppose that $\mathbf{x}_1(t), \mathbf{x}_2(t), \ldots, \mathbf{x}_n(t)$ are n linearly independent solutions on some open interval of the homogeneous system

$$\mathbf{x}' = \mathbf{P}(t)\mathbf{x}$$

(16)

of n linear equations. Then the $n \times n$ matrix

$$\mathbf{\Phi}(t) = \begin{bmatrix} x_{11}(t) & x_{12}(t) & \cdots & x_{1n}(t) \\ x_{21}(t) & x_{22}(t) & \cdots & x_{2n}(t) \\ \vdots & \vdots & & \vdots \\ x_{n1}(t) & x_{n2}(t) & \cdots & x_{nn}(t) \end{bmatrix},$$

(17)

having as *column* vectors the solution vectors $\mathbf{x}_1, \mathbf{x}_2, \ldots, \mathbf{x}_n$, is called a **fundamental matrix** for the system. Because its column vectors are linearly independent, it follows that the matrix $\mathbf{\Phi}(t)$ is nonsingular and therefore has an inverse matrix $\mathbf{\Phi}^{-1}(t)$.

In terms of the fundamental matrix $\mathbf{\Phi}(t)$, the general solution

$$\mathbf{x}(t) = c_1 \mathbf{x}_1(t) + c_2 \mathbf{x}_2(t) + \cdots + c_n \mathbf{x}_n(t)$$

(18)

of the system $\mathbf{x}' = \mathbf{P}(t)\mathbf{x}$ can be written in the form

$$\mathbf{x}(t) = \boldsymbol{\Phi}(t)\mathbf{c} \tag{19}$$

where

$$\mathbf{c} = \begin{bmatrix} c_1 \\ c_2 \\ c_3 \\ \vdots \\ c_n \end{bmatrix}.$$

In order that the solution $\mathbf{x}(t)$ satisfy the initial condition

$$\mathbf{x}(a) = \mathbf{b} \tag{20}$$

where $b_1, b_2, \ldots, b_n$ are given, it therefore will suffice for the coefficient vector $\mathbf{c}$ in (19) to satisfy $\boldsymbol{\Phi}(a)\mathbf{c} = \mathbf{b}$; that is,

$$\mathbf{c} = \boldsymbol{\Phi}^{-1}(a)\mathbf{b}. \tag{21}$$

When we substitute (21) in (19) we get the conclusion of the following theorem.

Theorem 1 *Fundamental Matrix Solutions*

Let $\boldsymbol{\Phi}(t)$ be a fundamental matrix for the homogeneous linear system $\mathbf{x}' = \mathbf{P}(t)\mathbf{x}$ on an open interval containing the point $t = a$. Then the (unique) solution of the initial value problem

$$\mathbf{x}' = \mathbf{P}(t)\mathbf{x}, \qquad \mathbf{x}(a) = \mathbf{b} \tag{22}$$

is given by

$$\mathbf{x}(t) = \boldsymbol{\Phi}(t)\boldsymbol{\Phi}^{-1}(a)\mathbf{b} = \boldsymbol{\Phi}(t)\boldsymbol{\Phi}^{-1}(a)\mathbf{x}(a). \tag{23}$$

Section 5.5 tells us how to find a fundamental matrix for the system

$$\mathbf{x}' = \mathbf{A}\mathbf{x} \tag{24}$$

with constant $n \times n$ coefficient matrix $\mathbf{A}$, at least in the case where $\mathbf{A}$ has a complete set of n linearly independent eigenvectors $\mathbf{v}_1, \mathbf{v}_2, \ldots, \mathbf{v}_n$ associated with the (not necessarily distinct) eigenvalues $\lambda_1, \lambda_2, \ldots, \lambda_n$, respectively. In this event the corresponding solution vectors of (24) are given by

$$\mathbf{x}_i(t) = \mathbf{v}_i e^{\lambda_i t}$$

for $i = 1, 2, \ldots, n$. Therefore, the $n \times n$ matrix

$$\boldsymbol{\Phi}(t) = \begin{bmatrix} \Big| & \Big| & & \Big| \\ \mathbf{v}_1 e^{\lambda_1 t} & \mathbf{v}_2 e^{\lambda_2 t} & \cdots & \mathbf{v}_n e^{\lambda_n t} \\ \Big| & \Big| & & \Big| \end{bmatrix} \tag{25}$$

CHAPTER 5: Linear Systems of Differential Equations

having the solutions $\mathbf{x}_1, \mathbf{x}_2, \ldots, \mathbf{x}_n$ as column vectors is a fundamental matrix for the system $\mathbf{x}' = \mathbf{Ax}$ (on the entire real line).

In order to apply the formula in (23), we must be able to compute the inverse matrix $\mathbf{\Phi}^{-1}(a)$. The inverse of the nonsingular 2×2 matrix

$$\mathbf{A} = \begin{bmatrix} a & b \\ c & d \end{bmatrix}$$

is

$$\mathbf{A}^{-1} = \frac{1}{\Delta} \begin{bmatrix} d & -b \\ -c & a \end{bmatrix} \tag{26}$$

where $\Delta = \det(\mathbf{A}) = ad - bc \neq 0$. The inverse of the nonsingular 3×3 matrix $\mathbf{A} = (a_{ij})$ is given by

$$\mathbf{A}^{-1} = \frac{1}{\Delta} \begin{bmatrix} +A_{11} & -A_{12} & +A_{13} \\ -A_{21} & +A_{22} & -A_{23} \\ +A_{31} & -A_{32} & +A_{33} \end{bmatrix}^T \tag{27}$$

where $\Delta = \det(\mathbf{A}) \neq 0$, and A_{ij} denotes the determinant of the 2×2 submatrix of $\mathbf{A}$ obtained by deleting the ith row and jth column of A. (Do not overlook the symbol T for *transpose* in Eq. (27).) The formula in (27) is also valid upon generalization to $n \times n$ matrices, but in practice inverses of larger matrices are usually computed instead by row reduction methods (see any linear algebra text).

EXAMPLE 3 Find a fundamental matrix for the system

$$\begin{aligned} x' &= 4x + 2y, \\ y' &= 3x - y \end{aligned} \tag{28}$$

and use it to find the solution of (28) that satisfies the initial conditions $x(0) = 1$, $y(0) = -1$.

SOLUTION The linearly independent solutions

$$\mathbf{x}_1(t) = \begin{bmatrix} e^{-2t} \\ -3e^{-2t} \end{bmatrix} \quad \text{and} \quad \mathbf{x}_2(t) = \begin{bmatrix} 2e^{5t} \\ e^{5t} \end{bmatrix}$$

displayed in (5) yield the fundamental matrix

$$\mathbf{\Phi}(t) = \begin{bmatrix} e^{-2t} & 2e^{5t} \\ -3e^{-2t} & e^{5t} \end{bmatrix}. \tag{29}$$

Then

$$\mathbf{\Phi}(0) = \begin{bmatrix} 1 & 2 \\ -3 & 1 \end{bmatrix},$$

and the formula in (26) gives the inverse matrix

$$\mathbf{\Phi}^{-1}(0) = \tfrac{1}{7} \begin{bmatrix} 1 & -2 \\ 3 & 1 \end{bmatrix}. \tag{30}$$

Hence the formula in (23) gives the solution

$$\mathbf{x}(t) = \begin{bmatrix} e^{-2t} & 2e^{5t} \\ -3e^{-2t} & e^{5t} \end{bmatrix} (\tfrac{1}{7}) \begin{bmatrix} 1 & -2 \\ 3 & 1 \end{bmatrix} \begin{bmatrix} 1 \\ -1 \end{bmatrix}$$

$$= \tfrac{1}{7} \begin{bmatrix} e^{-2t} & 2e^{5t} \\ -3e^{-2t} & e^{5t} \end{bmatrix} \begin{bmatrix} 3 \\ 2 \end{bmatrix},$$

and so

$$\mathbf{x}(t) = \tfrac{1}{7} \begin{bmatrix} 3e^{-2t} + 4e^{5t} \\ -9e^{-2t} + 2e^{5t} \end{bmatrix}.$$

Thus the solution of the original initial value problem is given by

$$x(t) = \tfrac{3}{7}e^{-2t} + \tfrac{4}{7}e^{5t},$$

$$y(t) = -\tfrac{9}{7}e^{-2t} + \tfrac{2}{7}e^{5t}.$$

Remark: An advantage of the fundamental matrix approach is this: Once we know the fundamental matrix $\mathbf{\Phi}(t)$ and the inverse matrix $\mathbf{\Phi}^{-1}(a)$, we can calculate rapidly by matrix multiplication the solutions corresponding to different initial conditions. For example, suppose we seek the solution of the system in (28) satisfying the new initial conditions $x(0) = 77$, $y(0) = 49$. Then substitution of (29) and (30) in (23) gives the new solution

$$\mathbf{x}(t) = \tfrac{1}{7} \begin{bmatrix} e^{-2t} & 2e^{5t} \\ -3e^{-2t} & e^{5t} \end{bmatrix} \begin{bmatrix} 1 & -2 \\ 3 & 1 \end{bmatrix} \begin{bmatrix} 77 \\ 49 \end{bmatrix}$$

$$= \tfrac{1}{7} \begin{bmatrix} e^{-2t} & 2e^{5t} \\ -3e^{-2t} & e^{5t} \end{bmatrix} \begin{bmatrix} -21 \\ 280 \end{bmatrix} = \begin{bmatrix} -3e^{-2t} + 80e^{5t} \\ 9e^{-2t} + 40e^{5t} \end{bmatrix}.$$

VARIATION OF PARAMETERS

Recall from Section 2.7 that the method of variation of parameters may be applied to a linear differential equation with variable coefficients, and is not restricted to nonhomogeneous terms involving only polynomials, exponentials, and sinusoidal functions. The method of variation of parameters for systems enjoys the same flexibility, and has a concise matrix formulation that is convenient for both practical and theoretical purposes.

We want to find a particular solution $\mathbf{x}_p$ of the nonhomogeneous linear system

$$\mathbf{x}' = \mathbf{P}(t)\mathbf{x} + \mathbf{f}(t), \tag{31}$$

given that we have found already a general solution

$$\mathbf{x}_c = c_1\mathbf{x}_1 + c_2\mathbf{x}_2 + \cdots + c_n\mathbf{x}_n \tag{32}$$

of the associated homogeneous system

$$\mathbf{x}' = \mathbf{P}(t)\mathbf{x}. \tag{33}$$

We first use the fundamental matrix $\mathbf{\Phi}(t)$ with column vectors $\mathbf{x}_1, \mathbf{x}_2, \ldots, \mathbf{x}_n$ to rewrite the complementary function in (32) as

$$\mathbf{x}_c(t) = \mathbf{\Phi}(t)\mathbf{c}. \tag{34}$$

Our idea is to replace the vector "parameter" $\mathbf{c}$ with a variable vector $\mathbf{u}(t)$, and thus we seek a particular solution of the form

$$\mathbf{x}_p(t) = \mathbf{\Phi}(t)\mathbf{u}(t). \tag{35}$$

We must determine $\mathbf{u}(t)$ so that $\mathbf{x}_p$ does, indeed, satisfy Eq. (31).

The derivative of $\mathbf{x}_p(t)$ is (by the product rule)

$$\mathbf{x}_p'(t) = \mathbf{\Phi}'(t)\mathbf{u}(t) + \mathbf{\Phi}(t)\mathbf{u}'(t). \tag{36}$$

Hence substitution of (35) and (36) in (31) yields

$$\mathbf{\Phi}'(t)\mathbf{u}(t) + \mathbf{\Phi}(t)\mathbf{u}'(t) = \mathbf{P}(t)\mathbf{\Phi}(t)\mathbf{u}(t) + \mathbf{f}(t). \tag{37}$$

But

$$\mathbf{\Phi}'(t) = \mathbf{P}(t)\mathbf{\Phi}(t) \tag{38}$$

because each column vector of $\mathbf{\Phi}(t)$ satisfies Eq. (33). Therefore, (37) reduces to

$$\mathbf{\Phi}(t)\mathbf{u}'(t) = \mathbf{f}(t). \tag{39}$$

Thus it suffices to choose $\mathbf{u}(t)$ so that

$$\mathbf{u}'(t) = \mathbf{\Phi}^{-1}(t)\mathbf{f}(t); \tag{40}$$

that is, so that

$$\mathbf{u}(t) = \int \mathbf{\Phi}^{-1}(t)\mathbf{f}(t)\,dt. \tag{41}$$

Upon substitution of (41) in (35), we finally obtain the desired particular solution, as stated in the following theorem.

Theorem: *Variation of Parameters*

If $\mathbf{\Phi}(t)$ is a fundamental matrix for the homogeneous system $\mathbf{x}' = \mathbf{P}(t)\mathbf{x}$ on some interval where $\mathbf{P}(t)$ and $\mathbf{f}(t)$ are continuous, then a particular solution of the nonhomogeneous system

$$\mathbf{x}' = \mathbf{P}(t)\mathbf{x} + \mathbf{f}(t)$$

is given by

$$\mathbf{x}_p(t) = \mathbf{\Phi}(t)\int \mathbf{\Phi}^{-1}(t)\mathbf{f}(t)\,dt. \tag{42}$$

This is the **variation of parameters formula** for first order linear systems. If we add this particular solution and the complementary function in (34), we get the general solution

$$\mathbf{x}(t) = \mathbf{\Phi}(t)\mathbf{c} + \mathbf{\Phi}(t)\int \mathbf{\Phi}^{-1}(t)\mathbf{f}(t)\,dt \tag{43}$$

of the nonhomogeneous system in (31).

The choice of the constant of integration in (42) is immaterial, for we need only a single particular solution. In solving initial value problems it often is convenient to choose the constant of integration so that $\mathbf{x}_p(a) = \mathbf{0}$, and thus integrate from a to t:

$$\mathbf{x}_p(t) = \mathbf{\Phi}(t) \int_a^t \mathbf{\Phi}^{-1}(s)\mathbf{f}(s) \, ds. \tag{42'}$$

If we add this particular solution and the homogeneous solution in (23), we get the solution

$$\mathbf{x}(t) = \mathbf{\Phi}(t)\mathbf{\Phi}^{-1}(a)\mathbf{b} + \mathbf{\Phi}(t) \int_a^t \mathbf{\Phi}^{-1}(s)\mathbf{f}(s) \, ds \tag{43'}$$

of the initial value problem $\mathbf{x}' = \mathbf{P}(t)\mathbf{x} + \mathbf{f}(t)$, $\mathbf{x}(a) = \mathbf{b}$.

EXAMPLE 4 Solve the initial value problem

$$\mathbf{x}' = \begin{bmatrix} 4 & 2 \\ 3 & -1 \end{bmatrix}\mathbf{x} - \begin{bmatrix} 15 \\ 4 \end{bmatrix} te^{-2t}, \qquad \mathbf{x}(0) = \begin{bmatrix} 1 \\ -1 \end{bmatrix}. \tag{44}$$

SOLUTION If we were to use the method of undetermined coefficients, our trial solution would be of the form

$$\mathbf{x}_p(t) = \mathbf{a}t^2 e^{-2t} + \mathbf{b}te^{-2t} + \mathbf{c}e^{-2t},$$

and we would have six scalar coefficients to determine. Here it will be simpler to use the method of variation of parameters, for we already know from Example 3 the fundamental matrix

$$\mathbf{\Phi}(t) = \begin{bmatrix} e^{-2t} & 2e^{5t} \\ -3e^{-2t} & e^{5t} \end{bmatrix} \tag{45}$$

of the associated homogeneous system.

The determinant of $\mathbf{\Phi}(t)$ is $\Delta = 7e^{3t}$, so the formula in (26) gives the inverse matrix

$$\mathbf{\Phi}^{-1}(t) = \tfrac{1}{7}e^{-3t} \begin{bmatrix} e^{5t} & -2e^{5t} \\ 3e^{-2t} & e^{-2t} \end{bmatrix}. \tag{46}$$

Then the formula in (42') gives the particular solution

$$\begin{aligned}
\mathbf{x}_p(t) &= \mathbf{\Phi}(t) \int_0^t \tfrac{1}{7}e^{-3s} \begin{bmatrix} e^{5s} & -2e^{5s} \\ 3e^{-2s} & e^{-2s} \end{bmatrix}\begin{bmatrix} -15se^{-2s} \\ -4se^{-2s} \end{bmatrix} ds \\
&= \mathbf{\Phi}(t) \int_0^t \tfrac{1}{7}e^{-3s} \begin{bmatrix} -7se^{3s} \\ -49se^{-4s} \end{bmatrix} ds \\
&= \mathbf{\Phi}(t) \int_0^t \begin{bmatrix} -s \\ -7se^{-7s} \end{bmatrix} ds \\
&= \mathbf{\Phi}(t) \left[\begin{bmatrix} -\tfrac{1}{2}s^2 \\ se^{-7s} + \tfrac{1}{7}e^{-7s} \end{bmatrix} \right]_{s=0}^{s=t} \\
&= \begin{bmatrix} e^{-2t} & 2e^{5t} \\ -3e^{-2t} & e^{5t} \end{bmatrix}\begin{bmatrix} -\tfrac{1}{2}t^2 \\ te^{-7t} + \tfrac{1}{7}e^{-7t} - \tfrac{1}{7} \end{bmatrix}.
\end{aligned}$$

Therefore

$$\mathbf{x}_p(t) = \tfrac{1}{14}e^{-2t}\begin{bmatrix} 4 + 28t - 7t^2 \\ 2 + 14t + 21t^2 \end{bmatrix} - \tfrac{1}{7}e^{5t}\begin{bmatrix} 2 \\ 1 \end{bmatrix}. \tag{47}$$

This is a particular solution such that $\mathbf{x}_p(0) = \mathbf{0}$. In Example 3 we found the solution

$$\mathbf{x}_c(t) = \tfrac{1}{7}\begin{bmatrix} 3e^{-2t} + 4e^{5t} \\ -9e^{-2t} + 2e^{5t} \end{bmatrix} \tag{48}$$

of the associated homogeneous system such that

$$\mathbf{x}_c(0) = \begin{bmatrix} 1 \\ -1 \end{bmatrix}.$$

Upon adding the solutions in (47) and (48), we find that the solution of the initial value problem in (44) is given by

$$x(t) = \tfrac{1}{14}(10 + 28t - 7t^2)e^{-2t} + \tfrac{2}{7}e^{5t},$$
$$y(t) = \tfrac{1}{14}(-16 + 14t + 21t^2)e^{-2t} + \tfrac{1}{7}e^{5t}.$$

5.6 PROBLEMS

Apply the method of undetermined coefficients to find a particular solution of each of the systems in Problems 1–14. If initial conditions are given, find the particular solution that satisfies these conditions.

1. $x' = x + 2y + 3$, $y' = 2x + y - 2$
2. $x' = 2x + 3y + 5$, $y' = 2x + y - 2t$
3. $x' = 3x + 4y$, $y' = 3x + 2y + t^2$; $x(0) = y(0) = 0$
4. $x' = 4x + y + e^t$, $y' = 6x - y - e^t$; $x(0) = y(0) = 1$
5. $x' = 6x - 7y + 10$, $y' = x - 2y - 2e^{-t}$
6. $x' = 9x + y + 2e^t$, $y' = -8x - 2y + te^t$
7. $x' = -3x + 4y + \sin t$, $y' = 6x - 5y$; $x(0) = 1$, $y(0) = 0$
8. $x' = x - 5y + 2\sin t$, $y' = x - y - 3\cos t$
9. $x' = x - 5y + \cos 2t$, $y' = x - y$
10. $x' = x - 2y$, $y' = 2x - y + e^t \sin t$
11. $x' = 2x + 4y + 2$, $y' = x + 2y + 3$; $x(0) = 1$, $y(0) = -1$
12. $x' = x + y + 2t$, $y' = x + y - 2t$
13. $x' = 2x + y + 2e^t$, $y' = x + 2y - 3e^t$
14. $x' = 2x + y + 1$, $y' = 4x + 2y + e^{4t}$

Find a fundamental matrix of each of the systems in Problems 15–22 and then apply the formula in Eq. (23) to find a solution satisfying the given initial conditions.

15. $\mathbf{x}' = \begin{bmatrix} 2 & 1 \\ 1 & 2 \end{bmatrix}\mathbf{x}$, $\mathbf{x}(0) = \begin{bmatrix} 3 \\ -2 \end{bmatrix}$

16. $\mathbf{x}' = \begin{bmatrix} 2 & -1 \\ -4 & 2 \end{bmatrix}\mathbf{x}$, $\mathbf{x}(0) = \begin{bmatrix} 2 \\ -1 \end{bmatrix}$

17. $\mathbf{x}' = \begin{bmatrix} 2 & -5 \\ 4 & -2 \end{bmatrix}\mathbf{x}$, $\mathbf{x}(0) = \begin{bmatrix} 0 \\ 1 \end{bmatrix}$

18. $\mathbf{x}' = \begin{bmatrix} 3 & -1 \\ 1 & 1 \end{bmatrix}\mathbf{x}$, $\mathbf{x}(0) = \begin{bmatrix} 1 \\ 0 \end{bmatrix}$

19. $\mathbf{x}' = \begin{bmatrix} -3 & -2 \\ 9 & 3 \end{bmatrix}\mathbf{x}$, $\mathbf{x}(0) = \begin{bmatrix} 1 \\ -1 \end{bmatrix}$

20. $\mathbf{x}' = \begin{bmatrix} 7 & -5 \\ 4 & 3 \end{bmatrix}\mathbf{x}$, $\mathbf{x}(0) = \begin{bmatrix} 2 \\ 0 \end{bmatrix}$

21. $\mathbf{x}' = \begin{bmatrix} 5 & 0 & -6 \\ 2 & -1 & -2 \\ 4 & -2 & -4 \end{bmatrix}\mathbf{x}$, $\mathbf{x}(0) = \begin{bmatrix} 2 \\ 1 \\ 0 \end{bmatrix}$

22. $\mathbf{x}' = \begin{bmatrix} 3 & 2 & 2 \\ -5 & -4 & -2 \\ -5 & 5 & 3 \end{bmatrix}\mathbf{x}$, $\mathbf{x}(0) = \begin{bmatrix} 1 \\ 0 \\ -1 \end{bmatrix}$

In each of Problems 23–32, apply the method of variation of parameters to find a particular solution of the given system.

23. $\mathbf{x}' = \begin{bmatrix} 6 & -7 \\ 1 & -2 \end{bmatrix}\mathbf{x} + \begin{bmatrix} 2t \\ 3 \end{bmatrix}$

24. $\mathbf{x}' = \begin{bmatrix} 1 & 2 \\ 2 & -2 \end{bmatrix} \mathbf{x} + \begin{bmatrix} 3e^{2t} \\ 5t \end{bmatrix}$

25. $\mathbf{x}' = \begin{bmatrix} 4 & -1 \\ 5 & -2 \end{bmatrix} \mathbf{x} + \begin{bmatrix} e^{-t} \\ 2e^{-t} \end{bmatrix}$

26. $\mathbf{x}' = \begin{bmatrix} 3 & -1 \\ 9 & -3 \end{bmatrix} \mathbf{x} + \begin{bmatrix} \dfrac{1}{t^2} \\ \dfrac{1}{t^3} \end{bmatrix}$

27. $\mathbf{x}' = \begin{bmatrix} 2 & -4 \\ 5 & -2 \end{bmatrix} \mathbf{x} + \begin{bmatrix} \cos 3t \\ \sin 3t \end{bmatrix}$

28. $\mathbf{x}' = \begin{bmatrix} 3 & -5 \\ 5 & -3 \end{bmatrix} \mathbf{x} + \begin{bmatrix} \cos 4t \\ \sin 4t \end{bmatrix}$

29. $\mathbf{x}' = \begin{bmatrix} 2 & -1 \\ 5 & -2 \end{bmatrix} \mathbf{x} + \begin{bmatrix} \sec t \\ \tan t \end{bmatrix}$

30. $\mathbf{x}' = \begin{bmatrix} 2 & -1 \\ 1 & 2 \end{bmatrix} \mathbf{x} + \begin{bmatrix} e^{2t} \tan t \\ 0 \end{bmatrix}$

31. $\mathbf{x}' = \begin{bmatrix} 3 & -2 \\ 2 & 3 \end{bmatrix} \mathbf{x} + \begin{bmatrix} 0 \\ e^{3t} \cos 2t \end{bmatrix}$

32. $\mathbf{x}' = \begin{bmatrix} 2 & -4 \\ 1 & -2 \end{bmatrix} \mathbf{x} + \begin{bmatrix} \ln t \\ t \end{bmatrix}$

Find the particular solution such that $x_1(1) = x_2(1) = 0$.

*5.7
Matrix Exponentials and Linear Systems

We have seen that exponential functions play a central role in the solution of linear differential equations and systems, ranging from the scalar equation $x' = kx$ with solution $x(t) = x_0 e^{kt}$ to the vector solution $\mathbf{x}(t) = \mathbf{v}e^{kt}$ of the linear system $\mathbf{x}' = \mathbf{A}\mathbf{x}$ whose coefficient matrix $\mathbf{A}$ has eigenvalue λ with associated eigenvector $\mathbf{v}$. In this section we define exponentials of matrices in such a way that

$$\mathbf{X}(t) = e^{\mathbf{A}t}$$

is a matrix solution of the matrix differential equation

$$\mathbf{X}' = \mathbf{A}\mathbf{X}$$

with $n \times n$ coefficient matrix $\mathbf{A}$.

The exponential e^z of a complex number z may be defined (as in Section 2.3) by means of the exponential series

$$e^z = 1 + z + \frac{z^2}{2!} + \cdots + \frac{z^n}{n!} + \cdots \tag{1}$$

Similarly, if $\mathbf{A}$ is a $n \times n$ matrix, then the **exponential matrix** $e^{\mathbf{A}}$ is the $n \times n$ matrix defined by the series

$$e^{\mathbf{A}} = \mathbf{I} + \mathbf{A} + \frac{\mathbf{A}^2}{2!} + \cdots + \frac{\mathbf{A}^n}{n!} + \cdots \tag{2}$$

where $\mathbf{I}$ is the identity matrix. The meaning of the infinite series on the right in (2) is given by

$$\sum_{n=0}^{\infty} \frac{\mathbf{A}^n}{n!} = \lim_{k \to \infty} \left(\sum_{n=0}^{k} \frac{\mathbf{A}^n}{n!} \right) \tag{3}$$

where $\mathbf{A}^0 = \mathbf{I}$, $\mathbf{A}^2 = \mathbf{AA}$, $\mathbf{A}^3 = \mathbf{AA}^2$, and so on; inductively, $\mathbf{A}^{n+1} = \mathbf{AA}^n$ if $n \geq 0$. It can be shown that the limit in (3) exists for every $n \times n$ matrix $\mathbf{A}$. That is, the exponential matrix $e^{\mathbf{A}}$ is defined (by (2)) for every square matrix $\mathbf{A}$.

EXAMPLE 1 Consider the 2×2 diagonal matrix

$$\mathbf{A} = \begin{bmatrix} a & 0 \\ 0 & b \end{bmatrix}.$$

Then it is apparent that

$$\mathbf{A}^n = \begin{bmatrix} a^n & 0 \\ 0 & b^n \end{bmatrix}$$

for each integer $n \geq 1$. It therefore follows that

$$e^{\mathbf{A}} = \mathbf{I} + \mathbf{A} + \frac{\mathbf{A}^2}{2!} + \cdots$$

$$= \begin{bmatrix} 1 & 0 \\ 0 & 1 \end{bmatrix} + \begin{bmatrix} a & 0 \\ 0 & b \end{bmatrix} + \begin{bmatrix} a^2/2! & 0 \\ 0 & b^2/2! \end{bmatrix} + \cdots$$

$$= \begin{bmatrix} 1 + a + a^2/2! + \cdots & 0 \\ 0 & 1 + b + b^2/2! + \cdots \end{bmatrix}.$$

Thus

$$e^{\mathbf{A}} = \begin{bmatrix} e^a & 0 \\ 0 & e^b \end{bmatrix},$$

so the exponential of the *diagonal* 2×2 matrix $\mathbf{A}$ is obtained simply by exponentiating each diagonal element of $\mathbf{A}$.

The $n \times n$ analogue of the 2×2 result in Example 1 is established in the same way. The exponential of the $n \times n$ diagonal matrix

$$\mathbf{D} = \begin{bmatrix} a_1 & 0 & \cdots & 0 \\ 0 & a_2 & \cdots & 0 \\ \vdots & \vdots & & \vdots \\ 0 & 0 & \cdots & a_n \end{bmatrix} \tag{4}$$

is the $n \times n$ diagonal matrix

$$e^{\mathbf{D}} = \begin{bmatrix} e^{a_1} & 0 & \cdots & 0 \\ 0 & e^{a_2} & \cdots & 0 \\ \vdots & \vdots & & \vdots \\ 0 & 0 & \cdots & e^{a_n} \end{bmatrix} \tag{5}$$

obtained by exponentiating each diagonal element of D.

The matrix exponential $e^{\mathbf{A}}$ satisfies most of the exponential relations that are familiar in the case of scalar exponents. For instance, if $\mathbf{O}$ is the $n \times n$ zero matrix, then (2) yields

$$e^{\mathbf{O}} = \mathbf{I}. \tag{6}$$

In Problem 21 we ask you to show that a useful law of exponents holds for $n \times n$ matrices that commute:

$$\text{If } \mathbf{AB} = \mathbf{BA}, \quad \text{then} \quad e^{\mathbf{A}+\mathbf{B}} = e^{\mathbf{A}}e^{\mathbf{B}}. \tag{7}$$

In Problem 22 we ask you to conclude that

$$(e^{\mathbf{A}})^{-1} = e^{-\mathbf{A}}. \tag{8}$$

In particular, the matrix $e^{\mathbf{A}}$ is nonsingular for every $n \times n$ matrix $\mathbf{A}$ (reminiscent of the fact that $e^t \neq 0$ for all t). It follows by elementary linear algebra that the column vectors of $e^{\mathbf{A}}$ are always linearly independent.

If t is a scalar variable, then substitution of $\mathbf{A}t$ for $\mathbf{A}$ in (2) gives

$$e^{\mathbf{A}t} = \mathbf{I} + \mathbf{A}t + \mathbf{A}^2 \frac{t^2}{2!} + \cdots + \mathbf{A}^n \frac{t^n}{n!} + \cdots. \tag{9}$$

(Of course, $\mathbf{A}t$ is obtained simply by multiplying each element of $\mathbf{A}$ by t.) It happens that term-by-term differentiation of the series in (9) is valid, with the result

$$\frac{d}{dt}(e^{\mathbf{A}t}) = \mathbf{A} + \mathbf{A}^2 t + \mathbf{A}^3 \frac{t^2}{2!} + \cdots$$

$$= \mathbf{A}\left(\mathbf{I} + \mathbf{A}t + \mathbf{A}^2 \frac{t^2}{2!} + \cdots\right);$$

that is,

$$\frac{d}{dt}(e^{\mathbf{A}t}) = \mathbf{A}e^{\mathbf{A}t}, \tag{10}$$

in analogy to the formula $D_t(e^{kt}) = ke^{kt}$ from elementary calculus. Thus the matrix-valued function

$$\mathbf{X}(t) = e^{\mathbf{A}t} \tag{11}$$

satisfies the matrix differential equation

$$\mathbf{X}' = \mathbf{AX}. \tag{12}$$

It follows from the results in (11) and (12) that each column vector of $e^{\mathbf{A}t}$ is a solution vector for the linear system

$$\mathbf{x}' = \mathbf{Ax}. \tag{13}$$

Because the n column vectors of $e^{\mathbf{A}t}$ are automatically linearly independent, it also follows that the $n \times n$ matrix

$$\mathbf{\Phi}(t) = e^{\mathbf{A}t} \tag{14}$$

is a *fundamental matrix* for the linear system in (13). Therefore, Theorem 1 of Section 5.6 implies the following result.

CHAPTER 5: Linear Systems of Differential Equations

> **Theorem 1 Exponential Matrix Solutions**
>
> If $\mathbf{A}$ is an $n \times n$ matrix, then the (unique) solution of the initial value problem
>
> $$\mathbf{x}' = \mathbf{A}\mathbf{x}, \qquad \mathbf{x}(0) = \mathbf{x}_0 \tag{15}$$
>
> is given by
>
> $$\mathbf{x}(t) = e^{\mathbf{A}t}\mathbf{x}_0. \tag{16}$$

Thus the solution of homogeneous linear systems of differential equations reduces to the task of computing exponential matrices. This task is easiest in the case of an $n \times n$ matrix $\mathbf{A}$ having a complete set $\mathbf{v}_1, \mathbf{v}_2, \ldots, \mathbf{v}_n$ of n linearly independent eigenvectors corresponding (in order) to the n eigenvalues $\lambda_1, \lambda_2, \ldots, \lambda_n$ (not necessarily distinct). In this event we define the **eigenvector matrix P** and the diagonal **eigenvalue matrix D** of $\mathbf{A}$ by

$$\mathbf{P} = \begin{bmatrix} | & | & & | \\ \mathbf{v}_1 & \mathbf{v}_2 & \cdots & \mathbf{v}_n \\ | & | & & | \end{bmatrix} \quad \text{and} \quad \mathbf{D} = \begin{bmatrix} \lambda_1 & 0 & \cdots & 0 \\ 0 & \lambda_2 & \cdots & 0 \\ \vdots & \vdots & & \vdots \\ 0 & 0 & \cdots & \lambda_n \end{bmatrix}. \tag{17}$$

Then it is not difficult to show by direct computation (for instance, see Section 6.2 of Edwards and Penney, *Elementary Linear Algebra* (Englewood Cliffs, N.J.: Prentice-Hall, 1988)) that

$$\mathbf{A} = \mathbf{P}\mathbf{D}\mathbf{P}^{-1}. \tag{18}$$

The matrix factorization in (18) facilitates the calculation of $e^{\mathbf{A}t}$. If we substitute $\mathbf{A} = \mathbf{P}\mathbf{D}\mathbf{P}^{-1}$ in the series in (9) then a formal computation gives

$$e^{\mathbf{A}t} = e^{(\mathbf{P}\mathbf{D}\mathbf{P}^{-1})t} = e^{\mathbf{P}(\mathbf{D}t)\mathbf{P}^{-1}}$$

$$= \mathbf{I} + \mathbf{P}(\mathbf{D}t)\mathbf{P}^{-1} + \frac{1}{2!}\mathbf{P}(\mathbf{D}t)\mathbf{P}^{-1} \cdot \mathbf{P}(\mathbf{D}t)\mathbf{P}^{-1} + \cdots$$

$$= \mathbf{I} + \mathbf{P}(\mathbf{D}t)\mathbf{P}^{-1} + \frac{1}{2!}\mathbf{P}(\mathbf{D}t)^2\mathbf{P}^{-1} + \cdots$$

$$= \mathbf{P}\left[\mathbf{I} + (\mathbf{D}t) + \frac{1}{2!}(\mathbf{D}t)^2 + \cdots\right]\mathbf{P}^{-1}.$$

Thus

$$e^{\mathbf{A}t} = \mathbf{P}e^{\mathbf{D}t}\mathbf{P}^{-1}. \tag{19}$$

If we define

$$\boldsymbol{\Lambda}_t = e^{\mathbf{D}t} = \begin{bmatrix} e^{\lambda_1 t} & 0 & \cdots & 0 \\ 0 & e^{\lambda_2 t} & \cdots & 0 \\ \vdots & \vdots & & \vdots \\ 0 & 0 & \cdots & e^{\lambda_n t} \end{bmatrix} \tag{20}$$

using (5), then we have the following result.

EXAMPLE 2 Find the fundamental matrix $\mathbf{\Phi}(t) = e^{\mathbf{A}t}$ and the general solution $\mathbf{x}(t) = \mathbf{\Phi}(t)\mathbf{c}$ for the linear system

$$\begin{aligned} x'_1 &= 4x_1 + 2x_2 \\ x'_2 &= 3x_1 - x_2. \end{aligned} \qquad (22)$$

SOLUTION In Example 1 of Section 5.5 we saw that the coefficient matrix

$$\mathbf{A} = \begin{bmatrix} 4 & 2 \\ 3 & -1 \end{bmatrix}$$

has eigenvalues $\lambda_1 = -2$ and $\lambda_2 = 5$ with associated eigenvectors

$$\mathbf{v}_1 = \begin{bmatrix} 1 \\ -3 \end{bmatrix} \quad \text{and} \quad \mathbf{v}_2 = \begin{bmatrix} 2 \\ 1 \end{bmatrix},$$

so

$$\mathbf{P} = \begin{bmatrix} 1 & 2 \\ -3 & 1 \end{bmatrix} \quad \text{and} \quad \mathbf{\Lambda}_t = \begin{bmatrix} e^{-2t} & 0 \\ 0 & e^{5t} \end{bmatrix}.$$

Because

$$\mathbf{P}^{-1} = \tfrac{1}{7}\begin{bmatrix} 1 & -2 \\ 3 & 1 \end{bmatrix},$$

it follows from Eq. (21) that

$$\begin{aligned} \mathbf{\Phi}(t) = e^{\mathbf{A}t} &= \tfrac{1}{7}\begin{bmatrix} 1 & 2 \\ -3 & 1 \end{bmatrix}\begin{bmatrix} e^{-2t} & 0 \\ 0 & e^{5t} \end{bmatrix}\begin{bmatrix} 1 & -2 \\ 3 & 1 \end{bmatrix} \\ &= \tfrac{1}{7}\begin{bmatrix} 1 & 2 \\ -3 & 1 \end{bmatrix}\begin{bmatrix} e^{-2t} & -2e^{-2t} \\ 3e^{5t} & e^{5t} \end{bmatrix} \\ &= \tfrac{1}{7}\begin{bmatrix} e^{-2t} + 6e^{5t} & -2e^{-2t} + 2e^{5t} \\ -3e^{-2t} + 3e^{5t} & 6e^{-2t} + e^{5t} \end{bmatrix}. \end{aligned}$$

Note that the column vectors $\mathbf{x}_1(t)$ and $\mathbf{x}_2(t)$ of the fundamental matrix $\mathbf{\Phi}(t)$ satisfy the initial conditions

$$\mathbf{x}_1(0) = \begin{bmatrix} 1 \\ 0 \end{bmatrix} \quad \text{and} \quad \mathbf{x}_2(0) = \begin{bmatrix} 0 \\ 1 \end{bmatrix}.$$

The general solution $\mathbf{x}(t) = \boldsymbol{\Phi}(t)\mathbf{c} = c_1\mathbf{x}_1 + c_2\mathbf{x}_2$ is

$$\mathbf{x}(t) = \frac{1}{7}\begin{bmatrix} (c_1 - 2c_2)e^{-2t} + (6c_1 + 2c_2)e^{5t} \\ (-3c_1 + 6c_2)e^{-2t} + (3c_1 + c_2)e^{5t} \end{bmatrix}.$$

EXAMPLE 3 Use the exponential matrix to find a general solution of

$$\mathbf{x}' = \begin{bmatrix} 2 & 1 \\ 0 & 2 \end{bmatrix}\mathbf{x}. \tag{23}$$

SOLUTION The 2×2 coefficient matrix $\mathbf{A}$ in (23) has only the single repeated eigenvalue $\lambda = 2$ and the single associated eigenvector

$$\mathbf{v} = \begin{bmatrix} 1 \\ 0 \end{bmatrix},$$

so Theorem 2 does not apply for the computation of $e^{\mathbf{A}t}$. We instead note that $\mathbf{A} = 2\mathbf{I} + \mathbf{B}$ where

$$\mathbf{B} = \begin{bmatrix} 0 & 1 \\ 0 & 0 \end{bmatrix},$$

so $e^{\mathbf{A}t} = e^{2\mathbf{I}t}e^{\mathbf{B}t}$ by (7). Now

$$e^{2\mathbf{I}t} = \begin{bmatrix} e^{2t} & 0 \\ 0 & e^{2t} \end{bmatrix}$$

by (5), and $\mathbf{B}^2 = \mathbf{0}$ by direct multiplication, so it follows from the series in (9) that

$$e^{\mathbf{B}t} = \mathbf{I} + \mathbf{B}t = \begin{bmatrix} 1 & t \\ 0 & 1 \end{bmatrix}.$$

Hence the fundamental matrix $\boldsymbol{\Phi}(t) = e^{\mathbf{A}t}$ of the system in (23) is given by

$$\boldsymbol{\Phi}(t) = \begin{bmatrix} e^{2t} & 0 \\ 0 & e^{2t} \end{bmatrix}\begin{bmatrix} 1 & t \\ 0 & 1 \end{bmatrix} = \begin{bmatrix} e^{2t} & te^{2t} \\ 0 & e^{2t} \end{bmatrix},$$

and so a general solution is

$$\mathbf{x}(t) = \boldsymbol{\Phi}(t)\mathbf{c} = c_1\begin{bmatrix} e^{2t} \\ 0 \end{bmatrix} + c_2\begin{bmatrix} te^{2t} \\ e^{2t} \end{bmatrix}.$$

Remark: The simplicity of the computations in Example 3 may be a bit misleading. Only because of the special (so-called *upper triangular*) form of the matrix $\mathbf{A}$ was the computation of $e^{\mathbf{A}t}$ relatively easy. For a general $n \times n$ matrix $\mathbf{A}$ not having a complete set of n linearly independent eigenvalues, the computation of the exponential matrix $e^{\mathbf{A}t}$ can be a formidable task, one that involves the *canonical forms* of matrices that are studied in advanced linear algebra.

Finally, let us consider the nonhomogeneous linear system

$$\mathbf{x}' = \mathbf{A}\mathbf{x} + \mathbf{f}(t) \tag{24}$$

with constant $n \times n$ coefficient matrix $\mathbf{A}$. We know that a fundamental matrix of the associated homogeneous system $\mathbf{x}' = \mathbf{A}\mathbf{x}$ is $\mathbf{\Phi}(t) = e^{\mathbf{A}t}$, so $\mathbf{\Phi}^{-1}(t) = e^{-\mathbf{A}t}$ by (8). Consequently, the variation of parameters formula (Eq. (43) of Section 5.6) yields the formula

$$\mathbf{x}(t) = e^{\mathbf{A}t}\mathbf{c} + e^{\mathbf{A}t} \int e^{-\mathbf{A}t}\mathbf{f}(t)\, dt \tag{25}$$

for a general solution of the nonhomogeneous system in (24).

EXAMPLE 4 Apply the formula in (25) with $\mathbf{c} = \mathbf{0}$ to find a particular solution of the system

$$
\begin{aligned}
x_1' &= 2x_1 + x_2 + 3e^{2t} \\
x_2' &= 2x_2 + 4e^{2t}.
\end{aligned}
\tag{26}
$$

SOLUTION With $\mathbf{A}$ denoting the 2×2 coefficient matrix in (26), we found in Example 3 that

$$e^{\mathbf{A}t} = \begin{bmatrix} e^{2t} & te^{2t} \\ 0 & e^{2t} \end{bmatrix},$$

so

$$e^{-\mathbf{A}t} = \begin{bmatrix} e^{-2t} & -te^{-2t} \\ 0 & e^{-2t} \end{bmatrix}.$$

Hence with $\mathbf{c} = \mathbf{0}$ the formula in (25) gives the particular solution

$$
\begin{aligned}
\mathbf{x}(t) &= \begin{bmatrix} e^{2t} & te^{2t} \\ 0 & e^{2t} \end{bmatrix} \int \begin{bmatrix} e^{-2t} & -te^{-2t} \\ 0 & e^{-2t} \end{bmatrix} \begin{bmatrix} 3e^{2t} \\ 4e^{2t} \end{bmatrix} dt \\
&= \begin{bmatrix} e^{2t} & te^{2t} \\ 0 & e^{2t} \end{bmatrix} \int \begin{bmatrix} 3 - 4t \\ 4 \end{bmatrix} dt \\
&= \begin{bmatrix} e^{2t} & te^{2t} \\ 0 & e^{2t} \end{bmatrix} \begin{bmatrix} 3t - 2t^2 \\ 4t \end{bmatrix}.
\end{aligned}
$$

Therefore

$$\mathbf{x}(t) = \begin{bmatrix} (3t + 2t^2)e^{2t} \\ 4te^{2t} \end{bmatrix}.$$

Thus a particular solution of Eq. (26) in scalar form is

$$x_1(t) = (3t + 2t^2)e^{2t}, \qquad x_2(t) = 4te^{2t}.$$

5.7 PROBLEMS

Compute the fundamental matrix $\mathbf{\Phi}(t) = e^{\mathbf{A}t}$ for each system $\mathbf{x}' = \mathbf{A}\mathbf{x}$ given in Problems 1–12.

1. $\begin{aligned} x_1' &= 5x_1 - 4x_2 \\ x_2' &= 2x_1 - x_2 \end{aligned}$

2. $\begin{aligned} x_1' &= 6x_1 - 6x_2 \\ x_2' &= 4x_1 - x_2 \end{aligned}$

3. $\begin{aligned} x_1' &= 5x_1 - 3x_2 \\ x_2' &= 2x_1 \end{aligned}$

4. $\begin{aligned} x_1' &= 5x_1 - 4x_2 \\ x_2' &= 3x_1 - 2x_2 \end{aligned}$

5. $\begin{aligned} x_1' &= 9x_1 - 8x_2 \\ x_2' &= 6x_1 - 5x_2 \end{aligned}$

6. $\begin{aligned} x_1' &= 10x_1 - 6x_2 \\ x_2' &= 12x_1 - 7x_2 \end{aligned}$

7. $x_1' = 6x_1 - 10x_2$
$\quad\ x_2' = 2x_1 - 3x_2$

8. $x_1' = 11x_1 - 15x_2$
$\quad\ x_2' = 6x_1 - 8x_2$

9. $x_1' = 3x_1 + x_2$
$\quad\ x_2' = x_1 + 3x_2$

10. $x_1' = 4x_1 + 2x_2$
$\quad\ x_2' = 2x_1 + 4x_2$

11. $x_1' = 9x_1 + 2x_2$
$\quad\ x_2' = 2x_1 + 6x_2$

12. $x_1' = 13x_1 + 4x_2$
$\quad\ x_2' = 4x_1 + 7x_2$

Apply the method of Example 3 to find the fundamental matrix $\Phi(t) = e^{At}$ for each system $\mathbf{x}' = \mathbf{Ax}$ given in Problems 13–16.

13. $\mathbf{x}' = \begin{bmatrix} 1 & 2 \\ 0 & 1 \end{bmatrix} \mathbf{x}$

14. $\mathbf{x}' = \begin{bmatrix} 3 & 5 \\ 0 & 3 \end{bmatrix} \mathbf{x}$

15. $\mathbf{x}' = \begin{bmatrix} 1 & 1 & 0 \\ 0 & 1 & 1 \\ 0 & 0 & 1 \end{bmatrix} \mathbf{x}$

16. $\mathbf{x}' = \begin{bmatrix} 2 & 3 & 4 \\ 0 & 2 & 3 \\ 0 & 0 & 2 \end{bmatrix} \mathbf{x}$

Apply the formula in (25) with $\mathbf{c} = \mathbf{0}$ to find a particular solution of each of the nonhomogeneous systems in Problems 17–20.

17. $\mathbf{x}' = \begin{bmatrix} 1 & 2 \\ 0 & 1 \end{bmatrix} \mathbf{x} + \begin{bmatrix} 12e^t \\ 6e^t \end{bmatrix}$

18. $\mathbf{x}' = \begin{bmatrix} 1 & 2 \\ 0 & 1 \end{bmatrix} \mathbf{x} + \begin{bmatrix} 12t^2 \\ 6t \end{bmatrix}$

19. $\mathbf{x}' = \begin{bmatrix} 1 & 1 & 0 \\ 0 & 1 & 1 \\ 0 & 0 & 1 \end{bmatrix} \mathbf{x} + \begin{bmatrix} 6 \\ 12 \\ 24 \end{bmatrix} te^t$

20. $\mathbf{x}' = \begin{bmatrix} 2 & 3 & 4 \\ 0 & 2 & 3 \\ 0 & 0 & 2 \end{bmatrix} \mathbf{x} + \begin{bmatrix} 8t^3 \\ 6t^2 \\ 4t \end{bmatrix} e^{2t}$

21. Suppose that the matrices $\mathbf{A}$ and $\mathbf{B}$ commute; that is, that $\mathbf{AB} = \mathbf{BA}$. Prove that $e^{\mathbf{A}+\mathbf{B}} = e^{\mathbf{A}}e^{\mathbf{B}}$. (*Suggestion:* Group the terms in the product of the two series on the right-hand side to obtain the series on the left.)

22. Deduce from the result of Problem 21 that, for every square matrix $\mathbf{A}$, the matrix $e^{\mathbf{A}}$ is nonsingular with $(e^{\mathbf{A}})^{-1} = e^{-\mathbf{A}}$.

23. Suppose that

$$\mathbf{A} = \begin{bmatrix} 0 & 1 \\ 1 & 0 \end{bmatrix}.$$

Show that $\mathbf{A}^{2n} = \mathbf{I}$ while $\mathbf{A}^{2n+1} = \mathbf{A}$. Conclude that $e^{\mathbf{A}t} = \mathbf{I}\cosh t + \mathbf{A}\sinh t$, and apply this fact to find a general solution of $\mathbf{x}' = \mathbf{Ax}$. Verify that it is equivalent to the general solution found by the eigenvalue method.

24. Suppose that

$$\mathbf{A} = \begin{bmatrix} 0 & 2 \\ -2 & 0 \end{bmatrix}.$$

Show that $e^{\mathbf{A}t} = \mathbf{I}\cos 2t + \frac{1}{2}\mathbf{A}\sin 2t$. Apply this fact to find a general solution of $\mathbf{x}' = \mathbf{Ax}$, and verify that it is equivalent to the one found by the eigenvalue method.

25. Apply the technique of Example 3 to find a general solution of

$$\mathbf{x}' = \begin{bmatrix} 1 & 2 & 0 \\ 0 & 1 & 2 \\ 0 & 0 & 1 \end{bmatrix} \mathbf{x}.$$

CHAPTER *6*

NUMERICAL METHODS

6.1 INTRODUCTION: EULER'S METHOD

6.2 A CLOSER LOOK AT THE EULER METHOD, AND IMPROVEMENTS

6.3 THE RUNGE-KUTTA METHOD

6.4 SYSTEMS OF DIFFERENTIAL EQUATIONS

Introduction: Euler's Method

It is the exception rather than the rule when a differential equation of the general form

$$\frac{dy}{dx} = f(x, y)$$

can be solved exactly and explicitly by elementary methods like those discussed in Chapter 1. For example, consider the simple equation

$$\frac{dy}{dx} = e^{-x^2}. \tag{1}$$

A solution of (1) is simply an antiderivative of e^{-x^2}. But it is known that every antiderivative of $f(x) = e^{-x^2}$ is a **nonelementary** function—one that cannot be expressed as a finite combination of the familiar functions of introductory calculus. Hence no particular solution of Eq. (1) is finitely expressible in terms of elementary functions.

Suppose that we need the solution of (1) satisfying an initial condition $y(0) = y_0$, in a form that will enable us to compute numerical values of the solution. One approach would be to apply the methods of Chapter 3 to derive an infinite series solution. This might well suffice in the case of Eq. (1), but recall that in Chapter 3 we discussed only *linear* equations (such as Bessel's equation). Infinite series methods are poorly adapted to nonlinear equations. For example, you can easily solve the differential equation $dy/dx = 4x^3y^3$ by separating the variables. But try the method of infinite series, and note the difficulties that result when you must cube an infinite series.

Finding a particular solution of a differential equation is, in most applications, merely a means to an end. The goal is to be able to predict—with a reasonable degree of accuracy—the shape of a column, the trajectory of a particle, or the modes of vibration of a mechanical system. Suppose that the situation under investigation is modeled by the initial value problem

$$\frac{dy}{dx} = f(x, y), \qquad y(a) = y_0. \tag{2}$$

For purposes such as those mentioned above, it might be quite adequate to have a table of values of the unknown solution $y(x)$ at selected points of some interval $[a, b]$. In this section we discuss **Euler's method** for computing a table of numerical approximations to the solution of the initial value problem in (2).

To describe Euler's method, we first choose a fixed **step size** $h > 0$ and consider the points

$$x_0 = a, \quad x_1, \quad x_2, \quad \ldots, \quad x_n, \quad \ldots$$

where $x_n = a + nh$, so that

$$x_{n+1} = x_n + h$$

for $n = 0, 1, 2, \ldots$. Our goal is to find suitable *approximations*

$$y_1, \quad y_2, \quad y_3, \quad \ldots, \quad y_n, \quad \ldots$$

to the *true values*

$$y(x_1), \quad y(x_2), \quad y(x_3), \quad \ldots, \quad y(x_n), \quad \ldots$$

of the solution $y(x)$ of (2) at the points $x_1, x_2, x_3, \ldots$. Thus we seek reasonably accurate approximations

$$y_n \approx y(x_n) \tag{3}$$

for $n = 1, 2, 3, \ldots$. The question is this: How do we "step" from the approximate value y_n at x_n to the approximate value y_{n+1} at x_{n+1}?

When $x = x_0$, the rate of change of y with respect to x is $y' = f(x_0, y_0)$. If y continued to change at this same rate from $x = x_0$ to $x = x_1 = x_0 + h$, the change in y would be exactly $hf(x_0, y_0)$. We therefore take

$$y_1 = y_0 + hf(x_0, y_0) \tag{4}$$

as our approximation to the true value $y(x_1)$ of the solution at $x = x_1$. Similarly, we take

$$y_2 = y_1 + hf(x_1, y_1) \tag{5}$$

as our approximation to $y(x_2)$. In general, having reached the nth approximate value $y_n \approx y(x_n)$, we take

$$y_{n+1} = y_n + hf(x_n, y_n) \tag{6}$$

as our approximation to the true value $y(x_{n+1})$.

The formula in (6) tells how to make the typical step from y_n to y_{n+1}, and is the heart of Euler's method. Note that the formulas in (4) and (5) are the first two cases, $n = 0$ and $n = 1$, of the general iterative formula in (6).

Algorithm *The Euler Method*

Given the initial value problem

$$y' = f(x, y), \qquad y(a) = y_0, \tag{2}$$

Euler's method with step size h consists in applying the iterative formula

$$y_{n+1} = y_n + hf(x_n, y_n) \qquad (n \geqq 0) \tag{6}$$

to compute successively approximations $y_1, y_2, y_3, \ldots$ to the (true) values $y(x_1), y(x_2), y(x_3), \ldots$ of the (exact) solution $y = y(x)$ at the points $x_1, x_2, x_3, \ldots$, respectively.

Although the most important practical applications of Euler's method are to nonlinear differential equations, we will first illustrate the method with the linear initial value problem

$$\frac{dy}{dx} = x + y, \qquad y(0) = 1, \tag{7}$$

in order that we may compare our approximate solution with the exact solution

$$y(x) = 2e^x - x - 1. \tag{8}$$

(This solution is readily found using the method of Section 1.5.)

EXAMPLE 1 Apply Euler's method with step size $h = 0.1$ to approximate the solution of the initial value problem in (7) on the interval $0 \leq x \leq 1$.

SOLUTION Here $f(x, y) = x + y$, so the iterative formula in (6) is

$$y_{n+1} = y_n + h(x_n + y_n). \tag{9}$$

Beginning with $x_0 = 0$, $y_0 = 1$ and rounding values to four decimal places, the first three approximate values given by (9) are

$$y_1 = 1.0000 + (0.1)(0.0 + 1.0000) = 1.1000,$$

$$y_2 = 1.1000 + (0.1)(0.1 + 1.1000) = 1.2200,$$

$$y_3 = 1.2200 + (0.1)(0.2 + 1.2200) = 1.3620.$$

The table in Fig. 6.1 shows the approximate values obtained in all ten steps, together with the actual value $y(x_n)$ given by (8) and the error

$$y_{\text{actual}} - y_{\text{approx}} = y(x_n) - y_n$$

n	x_n	Approximate y_n	Actual $y(x_n)$	Error $y(x_n) - y_n$	Percent error
0	0.0	1.0000	1.0000	0.0000	0.00%
1	0.1	1.1000	1.1103	0.0103	0.93%
2	0.2	1.2200	1.2428	0.0228	1.84%
3	0.3	1.3620	1.3997	0.0377	2.69%
4	0.4	1.5282	1.5836	0.0554	3.50%
5	0.5	1.7210	1.7974	0.0764	4.25%
6	0.6	1.9431	2.0442	0.1011	4.95%
7	0.7	2.1974	2.3275	0.1301	5.59%
8	0.8	2.4872	2.6511	0.1639	6.18%
9	0.9	2.8159	3.0192	0.2033	6.73%
10	1.0	3.1875	3.4366	0.2491	7.25%

FIGURE 6.1 Using Euler's method to approximate the solution of Eq. (7).

at each step. Observe that the error in y_n increases as n increases—that is, as x_n gets further and further from the starting point x_0. The final column of the table shows the percentage error $100(y_{\text{actual}} - y_{\text{approx}})/y_{\text{actual}}$ in each approximate value; these percentage errors increase from 0.93% at $x_1 = 0.1$ to 7.25% at $x_{10} = 1.0$.

There are several sources of error in Euler's method that may make the approximation y_n to $y(x_n)$ unreliable for large values of n, those for which x_n is not sufficiently close to x_0. The error in the linear approximation formula

$$y(x_{n+1}) \approx y_n + hf(x_n, y_n) = y_{n+1} \tag{10}$$

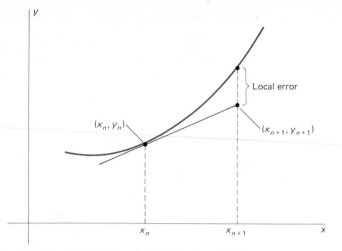

FIGURE 6.2 The local error in Euler's method.

is the amount by which the tangent line at (x_n, y_n) departs from the solution curve through (x_n, y_n), as illustrated in Fig. 6.2. This error, introduced at each step in the process, is called the **local error** in Euler's method.

The local error indicated in Fig. 6.2 *would be* the total error in y_{n+1} *if* the starting point y_n in (10) were an exact value, rather than merely an approximation to the actual value $y(x_n)$. But y_n itself suffers from the accumulated effects of all the local errors introduced at the previous steps. Thus the tangent line in Fig. 6.2 is tangent to the "wrong" solution curve—the one through (x_n, y_n) rather than the actual solution curve through the initial point (x_0, y_0). Figure 6.3 illustrates the **cumulative error** in Euler's method; it is the amount by which the polygonal step-

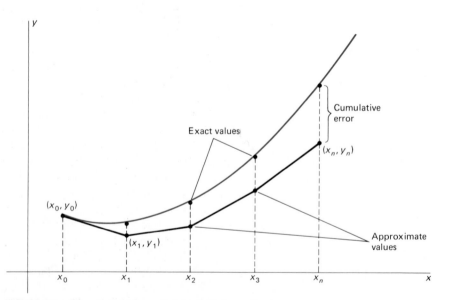

FIGURE 6.3 The cumulative error in Euler's method.

CHAPTER 6: Numerical Methods

wise path from (x_0, y_0) to (x_n, y_n) departs from the actual solution curve $y = y(x)$ through (x_0, y_0).

The usual way of attempting to reduce the cumulative error in Euler's method is to decrease the step size h. The table in Fig. 6.4 shows the results obtained in approximating the solution of the initial value problem

$$y' = x + y, \qquad y(0) = 1$$

of Example 1, using the successively smaller step sizes $h = 0.1$, $h = 0.02$, $h = 0.005$, and $h = 0.001$. We show computed values only at intervals of $\Delta x = 0.1$. For instance, with $h = 0.001$ the computation required 1000 Euler steps, but the value y_n is shown only when n is a multiple of 100, so that x_n is an integral multiple of 0.1.

x	y with $h = 0.1$	y with $h = 0.02$	y with $h = 0.005$	y with $h = 0.001$	Actual y
0.1	1.1000	1.1082	1.1098	1.1102	1.1103
0.2	1.2200	1.2380	1.2416	1.2426	1.2428
0.3	1.3620	1.3917	1.3977	1.3993	1.3997
0.4	1.5282	1.5719	1.5807	1.5831	1.5836
0.5	1.7210	1.7812	1.7933	1.7966	1.7974
0.6	1.9461	2.0227	2.0388	2.0431	2.0442
0.7	2.1974	2.2998	2.3205	2.3261	2.3275
0.8	2.4872	2.6161	2.6422	2.6493	2.6511
0.9	2.8159	2.9757	3.0082	3.0170	3.0192
1.0	3.1875	3.3832	3.4230	3.4338	3.4266

FIGURE 6.4   Approximating the solution of $y' = x + y$, $y(0) = 1$ with successively smaller step sizes.

By scanning the columns in Fig. 6.4 we observe that, for each fixed step size h, the error $y_{\text{actual}} - y_{\text{approx}}$ increases as x gets further from the starting point $x_0 = 0$. But by scanning the rows of the table we see that, for each fixed x, the error decreases as the step size h is reduced. The percentage errors at the final point $x = 1$ range from 7.25% with $h = 0.1$ down to only 0.08% with $h = 0.001$. Thus the smaller is the step size h, the more slowly does the error grow with increasing distance from the starting point.

The column of data for $h = 0.1$ in Fig. 6.4 requires only 10 steps, so Euler's method can be carried out (as in Example 1) with a hand-held calculator. But 50 steps are required to reach $x = 1$ with $h = 0.02$, 200 steps with $h = 0.005$, and 1000 steps with $h = 0.001$. A computer is almost always used to implement Euler's method when more than 10 or 20 steps are required. Once an appropriate computer program has been written, one step size is—in principle—just as convenient as another; after all, the computer hardly cares how many steps it is asked to carry out.

Why, then, do we not simply choose an exceedingly small step size (such as $h = 10^{-9}$), with the expectation that very great accuracy will result? There are two reasons for not doing so. The first is obvious: the time required for the computation. For example, the data in Fig. 6.4 were obtained using a·Tandy PC-6 pocket computer that carried out about nine Euler steps per second. Thus it required slightly over a second to approximate $y(1)$ with $h = 0.1$, about 1 min 50 s with $h = 0.001$. But with $h = 10^{-9}$ it would require over 3 years!

The second reason is more subtle. In addition to the local and cumulative errors discussed previously, the computer itself will contribute **roundoff error** at each stage because only finitely many significant digits can be used in each calculation. An Euler's method computation with $h = 0.0001$ will introduce roundoff errors 1000 times as often as one with $h = 0.1$. Hence with certain differential equations, $h = 0.1$ could actually produce more accurate results than those obtained with $h = 0.0001$, because the cumulative effect of roundoff error in the latter case might exceed combined cumulative and roundoff error in the case $h = 0.1$.

The "best" choice of h is difficult to determine in practice as well as in theory. It depends on the nature of the function $f(x, y)$ in the initial value problem in (2), on the exact code in which the program is written, and on the specific computer used. With a step size that is too large, the approximations inherent in Euler's method may not be sufficiently accurate, while if h is too small, then roundoff errors may accumulate to an unacceptable degree or the program may require too much time to be practical. The subject of *error propagation* in numerical algorithms is treated in numerical analysis courses and textbooks.

PROGRAMMING

One's understanding of a numerical algorithm is enhanced by considering how it can be implemented by means of a computer program. The programs displayed in this chapter (except where specifically noted) are written in a portable version of BASIC that should run (perhaps with trivial modifications, such as replacement of commas by semicolons) on any computer (from pocket to mainframe) that accepts BASIC. These programs are structured to make them intelligible to the reader who has little or no programming experience.

Each program in this chapter was executed on either a pocket computer (a Casio FX-700P or a Tandy PC-6) or a microcomputer (an IBM PC or a TRS-80 Model II), or both. When essentially equivalent programs are executed on two different computers, it is normal for the numerical results to differ slightly—perhaps in the last digit or two—because of differing internal methods of performing simple arithmetic.

Figure 6.5 lists the program used to compute the data shown in Fig. 6.4. Line 140 defines the function $f(x, y) = x + y$ corresponding to the equation

```
100 REM--Program EULER
110 REM
120 REM--Initialization:
130 REM
140      DEF FNF(X,Y) = X + Y
150      INPUT "Initial x,y"; X,Y
160      INPUT "Step size h"; H
170      INPUT "Number of steps"; K
180      INPUT "Print step p"; P
190 REM
200 REM--Euler iteration:
210 REM
220      FOR N = 1 TO K
230          Y = Y + H*FNF(X,Y)
240          X = X + H
250          IF INT(N/P) = N/P THEN PRINT X,Y
260      NEXT N
270 REM
280      END
```

FIGURE 6.5 Listing of Program EULER.

$y' = x + y$, and this is the only line of the program that must be changed in order to approximate a solution of a different differential equation. Lines 150–170 call for the user to enter the initial values of x and y, the step size h, and the total number of steps to be computed. The FOR–NEXT loop in lines 220–260 carries out the Euler iteration. Observe that line 230 is the equation $y_{n+1} = y_n + hf(x_n, y_n)$ written in BASIC notation. Line 250 directs that the data be printed only after every pth step. For instance, to produce the data in the $h = 0.005$ column in Fig. 6.4, we entered the values $x = 0$, $y = 1$, $h = 0.005$, $k = 200$, and $p = 20$.

Figure 6.6 lists a pocket computer version of the Euler program with the remark lines deleted. Pocket calculator BASIC generally does not support explicit function definition (as in line 140 of Fig. 6.5), so the value of the function $f(x, y) = x + y$ is computed in line 160 as the value of the variable F; the Euler iterative formula follows immediately in line 170.

```
100 INPUT "X", X
110 INPUT "Y", Y
120 INPUT "H", H
130 INPUT "K", K
140 INPUT "P", P
150 FOR N = 1 TO K
160    F = X + Y
170    Y = Y + H*F
180    X = X + H
190    IF INT(N/P) = N/P THEN PRINT X,Y
200 NEXT N
210 END
```

FIGURE 6.6   Pocket version of Program EULER.

The data shown in Fig. 6.4 indicate that Euler's method works well in approximating the solution of $y' = x + y$, $y(0) = 1$ on the interval $[0, 1]$. That is, for each fixed x it appears that the approximate values approach the actual value of $y(x)$ as the step size h is decreased. For instance, the approximate values in the rows corresponding to $x = 0.3$ and $x = 0.5$ suggest that $y(0.3) \approx 1.40$ and $y(0.5) \approx 1.80$, in agreement with the actual values shown in the final column of the table.

The following example, in contrast, shows that some initial value problems are not so well-behaved.

EXAMPLE 2   Use Euler's method to approximate the solution of the initial value problem

$$\frac{dy}{dx} = x^2 + y^2, \qquad y(0) = 1. \tag{11}$$

SOLUTION   Here $f(x, y) = x^2 + y^2$, so the iterative formula of Euler's method is

$$y_{n+1} = h(x_n^2 + y_n^2). \tag{12}$$

With step size $h = 0.1$ we obtain

$$y_1 = 1 + (0.1)[(0)^2 + (1)^2] = 1.1,$$

$$y_2 = 1.1 + (0.1)[(0.1)^2 + (1.1)^2] = 1.222,$$

$$y_3 = 1.222 + (0.1)[(0.2)^2 + (1.222)^2] \approx 1.3753,$$

x	y with h = 0.1	y with h = 0.02	y with h = 0.005
0.1	1.1000	1.1088	1.1108
0.2	1.2220	1.2458	1.2512
0.3	1.3753	1.4243	1.4357
0.4	1.5735	1.6658	1.6882
0.5	1.8371	2.0074	2.0512
0.6	2.1995	2.5201	2.6104
0.7	2.7193	3.3612	3.5706
0.8	3.5078	4.9601	5.5763
0.9	4.8023	9.0000	12.2061
1.0	7.1895	30.9167	1502.2090

FIGURE 6.7 Attempting to approximate the solution of $y' = x^2 + y^2$, $y(0) = 1$.

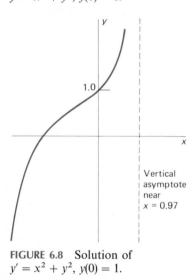

FIGURE 6.8 Solution of $y' = x^2 + y^2$, $y(0) = 1$.

and so forth. Rounded to four decimal places, the first ten values obtained in this manner are

$$y_1 = 1.1000 \qquad y_6 = 2.1995$$

$$y_2 = 1.2220 \qquad y_7 = 2.7193$$

$$y_3 = 1.3753 \qquad y_8 = 3.5078$$

$$y_4 = 1.5735 \qquad y_9 = 4.8023$$

$$y_5 = 1.8371 \qquad y_{10} = 7.1895$$

Instead of unwarily accepting these results as accurate approximations, let us repeat the computation with smaller values of h. In order to use program EULER (listed in Fig. 6.5), we need only replace the function definition in line 140 with DEF FNF(X, Y) = X*X + Y*Y. The table in Fig. 6.7 shows the results obtained with step sizes $h = 0.1$, $h = 0.02$, and $h = 0.005$. Observe that now the "stability" of our previous tables is missing. Indeed, it seems obvious that something is going wrong near $x = 1$.

Here is the explanation. The exact solution of the initial value problem in (11) was found in Problem 16 of Section 3.6. Using this solution (involving Bessel functions), it can be shown that $y(x) \to +\infty$ as x approaches 0.969811 (approximately), so the graph of the solution looks as indicated in Fig. 6.8. Although Euler's method gives values (albeit spurious ones) at $x = 1$, the actual solution does not exist on the entire interval $[0, 1]$. Moreover, Euler's method is unable to "keep up" with the rapid changes in $y(x)$ that occur as x approaches the infinite discontinuity near 0.969811.

The moral of Example 2 is that there are pitfalls in the numerical solution of certain types of initial value problems. Certainly it's pointless to attempt to approximate a solution on an interval where it doesn't even exist (or where it's not unique, in which case there's no generally applicable way to predict which way the numerical approximation will branch at a point of nonuniqueness). One should never accept as accurate the results of applying Euler's method with a single fixed step size h. A second "run" with smaller step size ($h/2$, say, or $h/5$, or $h/10$) may give seemingly consistent results, thereby suggesting their accuracy, or it may—as in Example 2—reveal the presence of some hidden difficulty in the problem. Many problems simply require the more accurate and powerful methods that are discussed in subsequent sections of this chapter.

6.1 PROBLEMS

A hand-held calculator will suffice for Problems 1–10. In each problem, find the exact solution of the given initial value problem. Then apply Euler's method twice to approximate (to four decimal places) this solution on the interval $[0, 0.5]$, first with step size $h = 0.1$, then with $h = 0.05$. Make a table showing the approximate values with $h = 0.1$, the approximate values with $h = 0.05$, and the actual values of the solution at the points $x = 0.1, 0.2, 0.3, 0.4$, and 0.5.

1. $y' = -y$, $y(0) = 2$
2. $y' = 2y$, $y(0) = \frac{1}{2}$
3. $y' = y + 1$, $y(0) = 1$
4. $y' = x - y$, $y(0) = 1$
5. $y' = y - x - 1$, $y(0) = 1$
6. $y' = -2xy$, $y(0) = 2$

7. $y' = -3x^2y$, $y(0) = 3$ **8.** $y' = e^{-y}$, $y(0) = 0$

9. $y' = \frac{1}{4}(1 + y^2)$, $y(0) = 1$ **10.** $y' = 2xy^2$, $y(0) = 1$

A programmable calculator or a computer will be useful for Problems 11–16. In each problem find the exact solution of the given initial value problem. Then apply Euler's method twice to approximate (to four decimal places) this solution on the given interval, first with step size $h = 0.01$, then with step size $h = 0.005$. Make a table showing the approximate values and the actual value, together with the percentage error in the more accurate approximation, for x an integral multiple of 0.2.

11. $y' = y - 2$, $y(0) = 1$; $0 \le x \le 1$

12. $y' = \frac{1}{2}(y - 1)^2$, $y(0) = 2$; $0 \le x \le 1$

13. $yy' = 2x^3$, $y(1) = 3$; $1 \le x \le 2$

14. $xy' = y^2$, $y(1) = 1$; $1 \le x \le 2$

15. $xy' = 3x - 2y$, $y(2) = 3$; $2 \le x \le 3$

16. $y^2y' = 2x^5$, $y(2) = 3$; $2 \le x \le 3$

A computer with a printer is required for Problems 17–24. In each of these initial value problems, use Euler's method with step sizes $h = 0.1$, 0.02, 0.004, and 0.0008 to approximate to four decimal places the values of the solution at ten equally spaced points of the given interval. Print the results in tabular form with appropriate headings to make it easy to gauge the effect of varying the step size h.

17. $y' = x^2 + y^2$, $y(0) = 0$; $0 \le x \le 1$

18. $y' = x^2 - y^2$, $y(0) = 1$; $0 \le x \le 2$

19. $y' = x + \sqrt{y}$, $y(0) = 1$; $0 \le x \le 2$

20. $y' = x + y^{1/3}$, $y(0) = -1$; $0 \le x \le 2$

21. $y' = \ln y$, $y(1) = 2$; $1 \le x \le 2$

22. $y' = x^{2/3} + y^{2/3}$, $y(0) = 1$; $0 \le x \le 2$

23. $y' = \sin x + \cos y$, $y(0) = 0$; $0 \le x \le 1$

24. $y' = \frac{x}{1 + y^2}$, $y(-1) = 1$; $-1 \le x \le 1$

25. Consider the initial value problem

$$7xy' + y = 0, \qquad y(-1) = 1.$$

(a) Solve this problem for the exact solution

$$y(x) = -\frac{1}{x^{1/7}},$$

which has an infinite discountinuity at $x = 0$. (b) Apply Euler's method with step size $h = 0.15$ to approximate this solution on the interval $-1 \le x \le 0.5$. Note that, from these data alone, you might not suspect any difficulty near $x = 0$. The reason is that the numerical approximation "jumps across the discontinuity" to another solution of $7xy' + y = 0$ for $x > 0$. (c) Finally, apply Euler's method again with step sizes $h = 0.03$ and $h = 0.006$, but still printing results only at the original points $x = -1.00, -0.85, -0.70, \ldots, 1.20, 1.35, 1.50$. Would you now suspect a discontinuity in the exact solution?

26. Apply Euler's method with successively smaller step sizes on the interval $0 \le x \le 2$ to verify empirically that the solution of the initial value problem

$$y' = x^2 + y^2, \qquad y(0) = 0$$

has a vertical asymptote near $x = 2.003147$. (Contrast this with Example 2, in which $y(0) = 1$.)

27. The general solution of the equation

$$y' = (1 + y^2) \cos x$$

is $y(x) = \tan (C + \sin x)$. With the initial condition $y(0) = 0$ the solution $y(x) = \tan (\sin x)$ is well-behaved. But with $y(0) = 1$ the solution $y(x) = \tan (\frac{1}{4}\pi + \sin x)$ has a vertical asymptote at $x = \sin^{-1} (\pi/4) \approx 0.90334$. Use Euler's method to verify this fact empirically.

6.2

A Closer Look at the Euler Method, and Improvements

The Euler method as presented in Section 6.1 is not often used in practice, mainly because more accurate methods are available. But Euler's method has the advantage of simplicity, and a careful study of this method yields insights into the workings of the more accurate methods, because many of the latter are extensions or refinements of the Euler method. In order to compare two different numerical methods of approximation, we need some way to measure the accuracy of each. The following theorem tells what degree of accuracy we can expect when we use Euler's method.

> **Theorem** *The Error in the Euler Method*
>
> Suppose that the initial value problem
>
> $$y' = f(x, y), \qquad y(a) = y_0 \tag{1}$$
>
> has a unique solution $y(x)$ on the closed interval $[a, b]$, and assume that $y(x)$ has a continuous second derivative on $[a, b]$. (This would follow from the assumption that f, f_x, and f_y are all continuous for $a \leq x \leq b$ and $c \leq y \leq d$, where $c \leq y(x) \leq d$ for all x in $[a, b]$.) Then there exists a constant C such that the following is true: If the approximations $y_1, y_2, y_3, \ldots, y_k$ to the actual values $y(x_1), y(x_2), y(x_3), \ldots, y(x_k)$ at points of $[a, b]$ are computed using Euler's method with step size $h > 0$, then
>
> $$\left| y_n - y(x_n) \right| \leq Ch \tag{2}$$
>
> for each $n = 1, 2, 3, \ldots, k$.

Remark: The **error**

$$y_{\text{actual}} - y_{\text{approx}} = y(x_n) - y_n$$

in (2) denotes the (cumulative) error in Euler's method after n steps in the approximation, *exclusive* of roundoff error (as though we were using a perfect machine that made no roundoff errors). The theorem can be summarized by saying that *the error in Euler's method is of order h*; that is, the error is bounded by a (predetermined) constant C multiplied by the step size h. It follows, for instance, that (on a given closed interval) halving the step size cuts the maximum error in half; similarly, with step size $h/10$ we get 10 times the accuracy (that is, $1/10$ the maximum error) as with step size h. Consequently, we can (in principle) get any degree of accuracy we want by choosing h sufficiently small.

We will omit the proof of this theorem, but one can be found in Chapter 7 of G. Birkhoff and G.-C. Rota, *Ordinary Differential Equations*, 2nd ed. (New York: John Wiley, 1969). The constant C deserves some comment. Because C tends to increase as the maximum value of $|y''(x)|$ on $[a, b]$ increases, it follows that C must depend in a fairly complicated way upon y, and actual computation of a value of C such that (2) holds is usually impractical. In practice, the following type of procedure is commonly employed.

1. Apply Euler's method to the initial value problem in (1) with a reasonable value of h.

2. Repeat with $h/2$, $h/4$, and so forth, at each stage halving the step size for the next application of Euler's method.

3. Continue until the results obtained at one stage agree—to an appropriate number of significant digits—with those obtained at the previous stage. Then the approximate values obtained at the final stage are considered likely to be accurate to the indicated number of significant digits.

h	Approximate $y(1)$	Actual $y(1)$	$\frac{\lvert \text{Error} \rvert}{h}$
0.04	0.50451	0.50000	0.11
0.02	0.50220	0.50000	0.11
0.01	0.50109	0.50000	0.11
0.005	0.50054	0.50000	0.11
0.0025	0.50027	0.50000	0.11
0.00125	0.50013	0.50000	0.10
0.000625	0.50007	0.50000	0.11
0.0003125	0.50003	0.50000	0.10

FIGURE 6.9 Table of values in Example 1.

EXAMPLE 1 Carry out this procedure with the initial value problem

$$\frac{dy}{dx} = -\frac{2xy}{1 + x^2}, \qquad y(0) = 1 \tag{3}$$

to approximate accurately the value $y(1)$ of the solution at $x = 1$.

SOLUTION Using Program EULER listed in Section 6.1, we began with a step size $h = 0.04$ requiring $k = 25$ steps to reach $x = 1$. The table in Fig. 6.9 shows the approximate values of $y(1)$ obtained with successively smaller values of h. The data suggest that the true value of $y(1)$ is exactly 0.5. Indeed, the exact solution of the initial value problem in (3) is $y(1) = 1/(1 + x^2)$, so $y(1) = 1/2$.

The final column of the table in Fig. 6.9 displays the ratio of the magnitude of the error to h; that is, $\lvert y_{\text{actual}} - y_{\text{approx}} \rvert / h$. Observe how the data in this column substantiate the theorem above—in this computation, the error bound in (2) appears to hold with a value of C slightly larger than $1/10$.

AN IMPROVEMENT IN EULER'S METHOD

As Fig. 6.10 shows, Euler's method is rather unsymmetrical. It uses the predicted slope of the graph of the solution at the left-hand end point of the interval $[x, x + h]$ as if it were the actual slope of the solution over that entire interval. We now turn our attention to a way in which increased accuracy can easily be obtained; it is known as the *improved Euler method*.

Given the initial value problem

$$y' = f(x, y), \qquad y(x_0) = y_0, \tag{4}$$

suppose that after carrying out n steps with step size h we have computed the approximation y_n to the actual value $y(x_n)$ of the solution at $x_n = x_0 + nh$. We can use the Euler method to obtain a first estimate—which we now call u_n rather than y_n—of the value of the solution at $x_{n+1} = x_n + h$. Thus

$$u_{n+1} = y_n + hf(x_n, y_n).$$

Now that $u_{n+1} \approx y(x_{n+1})$ has been computed, we can take

$$m_{n+1} = f(x_{n+1}, u_{n+1})$$

as a prediction of the slope of the solution curve $y = y(x)$ at $x = x_{n+1}$.

(x + h, y(x + h))

Solution
$y = y(x)$

Error

Predicted
y-value

Slope $y'(x)$

x $x + h$

FIGURE 6.10 True and predicted values in Euler's method.

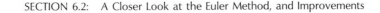

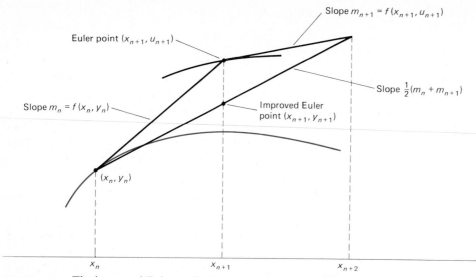

FIGURE 6.11 The improved Euler method: Average the slopes of the tangent lines at (x_n, y_n) and (x_{n+1}, u_{n+1}).

Of course, the approximate slope $m_n = f(x_n, y_n)$ at $x = x_n$ has already been calculated. Why not *average* these two slopes to obtain a more accurate estimate of the average slope of the solution curve over the entire subinterval $[x_n, x_{n+1}]$? This idea is the essence of the *improved* Euler method. Figure 6.11 shows the geometry behind this method; here it is in algorithmic terms.

Algorithm *The Improved Euler Method*

Given the initial value problem

$$y' = f(x, y), \qquad y(x_0) = y_0$$

in (4), suppose that x_n and y_n have already been computed. Then compute $x_{n+1} = x_n + h$ and y_n as follows:

$$\text{Let } m_n \quad = f(x_n, y_n).$$
$$\text{Let } u_{n+1} = y_n + hm_n.$$
$$\text{Let } m_{n+1} = f(x_{n+1}, u_{n+1}). \tag{5}$$
$$\text{Let } \bar{m}_{n+1} = \tfrac{1}{2}(m_n + m_{n+1}).$$
$$\text{Let } y_{n+1} = y_n + h\bar{m}_{n+1}.$$

The improved Euler method is one of a class of numerical techniques known as predictor-corrector methods. First a **predictor** u_{n+1} of the next y-value is computed; then it is used to **correct** itself. Thus the **improved Euler method** with step size h consists in using the predictor

$$u_{n+1} = y_n + hf(x_n, y_n) \tag{6}$$

and the corrector

$$y_{n+1} = y_n + \frac{h}{2}\left[f(x_n, y_n) + f(x_{n+1}, u_{n+1})\right] \qquad (7)$$

iteratively to compute successive approximations $y_1, y_2, y_3, \ldots$ to the values $y(x_1)$, $y(x_2)$, $y(x_3)$, $\ldots$ of the actual solution of the initial value problem in (4).

Under the assumption that the exact solution $y = y(x)$ of the initial value problem in (4) has a continuous third derivative, it can be proved—see Chapter 7 of Birkhoff and Rota—that the error in the improved Euler method is of order h^2. This means that on a given bounded interval $[a, b]$, each approximate value y_n satisfies the inequality

$$\left|y(x_n) - y_n\right| \leqq Ch^2 \qquad (8)$$

where the constant C does not depend on h. Because h^2 is much smaller than h if h is small, this means that the improved Euler method is more accurate than Euler's method itself. This advantage is offset by the fact that about twice as many computations are required. But the factor h^2 in (8) means that halving the step size results in 1/4 the maximum error, while with step size $h/10$ we get 100 times the accuracy (that is, 1/100 the maximum error) as with step size h.

```
100 REM--Program IMPEULER
110 REM
120 REM--Initialization:
130 REM
140       DEF FNF(X,Y) = X + Y
150       INPUT "Initial x,y"; X,Y
160       INPUT "Step size h"; H
170       INPUT "Number of steps"; K
180       INPUT "Print step p"; P
190 REM
200 REM--Improved Euler iteration:
210 REM
220       FOR N = 1 TO K
230            F0 = FNF(X,Y)
240            U = Y + H*F0
250            X = X + H
260            F1 = FNF(X,U)
270            Y = Y + (H/2)*(F0 + F1)
280            IF INT(N/P) = N/P THEN PRINT X,Y
290       NEXT N
300 REM
310       END
```

FIGURE 6.12 Listing of Program IMPEULER.

Figures 6.12 and 6.13 show two versions of an improved Euler program. Note its overall similarity to the Euler program in Section 6.1. The variables F0 and F1 denote the *two* values of the function $f(x, y)$ that must now be calculated at each step. Note the appearance of the improved Euler predictor-corrector formulas in lines 240 and 270 of this program.

```
100 INPUT "X"; X
110 INPUT "Y"; Y
120 INPUT "H"; H
130 INPUT "K"; K
140 INPUT "P"; P
150 FOR N = 1 TO K
160    L = X + Y
170    U = Y + H*L
180    X = X + H
190    M = X + U
200    Y = Y + (H/2)*(L + M)
210    IF INT(N/P) = N/P THEN PRINT X,Y
220 NEXT N
230 END
```

FIGURE 6.13 Pocket version of Program IMPEULER.

EXAMPLE 2 In Example 1 of Section 6.1 we applied Euler's method to the initial value problem

$$y' = x + y, \qquad y(0) = 1 \qquad (9)$$

with exact solution $y(x) = 2e^x - x - 1$. With $f(x, y) = x + y$ in (6) and (7), the predictor-corrector formulas for the improved Euler method are

$$u_{n+1} = y_n + h(x_n + y_n),$$

$$y_{n+1} = y_n + \frac{h}{2}\left[(x_n + y_n) + (x_{n+1} + u_{n+1})\right].$$

With step size $h = 0.1$ we calculate

$$u_1 = 1 + (0.1)(0 + 1) = 1.1,$$

$$y_1 = 1 + (0.05)[(0 + 1) + (0.1 + 1.1)] = 1.11,$$

$$u_2 = 1.11 + (0.1)(0.1 + 1.11) = 1.231,$$

$$y_2 = 1.11 + (0.05)[(0.1 + 1.11) + (0.2 + 1.231)] = 1.24205,$$

and so forth. The table in Fig. 6.14 compares the results obtained using the improved Euler method with those obtained previously using the "unimproved" Euler method. When the same step size $h = 0.1$ is used, the error in the Euler approximation to $y(1)$ is 7.25%, whereas the error in the improved Euler approximation is only 0.24%.

x	Euler method, $h = 0.1$, values of y	Euler method, $h = 0.005$, values of y	Improved Euler, $h = 0.1$, values of y	Actual y
0.1	1.1000	1.1098	1.1100	1.1103
0.2	1.2200	1.2416	1.2421	1.2428
0.3	1.3620	1.3977	1.3985	1.3997
0.4	1.5282	1.5807	1.5818	1.5836
0.5	1.7210	1.7933	1.7949	1.7974
0.6	1.9431	2.0388	2.0409	2.0442
0.7	2.1974	2.3205	2.3231	2.3275
0.8	2.4872	2.6422	2.6456	2.6511
0.9	2.8159	3.0082	3.0124	3.0192
1.0	3.1875	3.4230	3.4282	3.4366

FIGURE 6.14   Euler and improved Euler approximations to the solution of $y' = x + y$, $y(0) = 1$.

Indeed, the improved Euler method with $h = 0.1$ is more accurate (in this example) than the original Euler method with $h = 0.005$. The latter requires 200 evaluations of the function $f(x, y)$, while the former requires only 20 such evaluations, so in this case the improved Euler method yields greater accuracy with only about one-tenth the work!

Figure 6.15 shows the results obtained when the improved Euler method is applied to the initial value problem in (9) using step size $h = 0.005$. Accuracy of five significant figures is apparent in the table. This suggests that, in contrast with the original Euler method, the improved Euler method is sufficiently accurate for certain types of practical applications.

x	Improved Euler, approximate y	Actual y
0.0	1.00000	1.00000
0.1	1.11034	1.11034
0.2	1.24280	1.24281
0.3	1.39971	1.39972
0.4	1.58364	1.58365
0.5	1.79744	1.79744
0.6	2.04423	2.04424
0.7	2.32749	2.32751
0.8	2.65107	2.65108
0.9	3.01919	3.01921
1.0	3.43654	3.43656

FIGURE 6.15   Improved Euler approximations to the solution of Eq. (9) with step size $h = 0.005$.

A double-precision version of the improved Euler program listed in Fig. 6.12 was used to compute approximations to the exact value $y(1) = 0.5$ of the solution

$y(x) = 1/(1 + x^2)$ of the initial value problem

$$y' = -\frac{2xy}{1 + x^2}, \qquad y(0) = 1 \tag{3}$$

of Example 1. The results obtained by successively halving the step size appear in the table in Fig. 6.16. Note that the final column of this table impressively corroborates the form of the error bound in (8), and that each halving of the step size reduces the error by a factor of almost exactly 4, as should happen if the error is proportional to h^2.

h	Improved Euler approximation to $y(1)$	Error	$\frac{|\text{Error}|}{h^2}$
0.04	0.500195903	-0.000195903	0.12
0.02	0.500049494	-0.000049494	0.12
0.01	0.500012437	-0.000012437	0.12
0.005	0.500003117	-0.000003117	0.12
0.0025	0.500000780	-0.000000780	0.12
0.00125	0.500000195	-0.000000195	0.12
0.000625	0.500000049	-0.000000049	0.12
0.0003125	0.500000012	-0.000000012	0.12

FIGURE 6.16   Improved Euler approximation to $y(1)$ for $y' = -2xy/(1 + x^2)$, $y(0) = 1$.

TAYLOR SERIES METHODS

Another way to improve Euler's method of approximating the solution of $y' = f(x, y)$, $y(x_0) = y_0$ is to use the Taylor series

$$y(a + h) = y(a) + hy'(a) + \frac{h^2}{2!} y''(a) + \frac{h^3}{3!} y'''(a) + \cdots \tag{10}$$

at each step. If we substitute $a = x_n$, $a + h = x_{n+1}$ and retain only the two terms on the right-hand side, then (10) yields the linear approximation

$$y(x_{n+1}) \approx y(x_n) + hy'(x_n).$$

With $y(x_n) \approx y_n$ and $y'(x_n) \approx f(x_n, y_n)$ this is simply the iterative formula

$$y_{n+1} = y_n + hf(x_n, y_n)$$

of the original Euler method.

The idea now is to improve the accuracy at each step by retaining more terms of the Taylor series in (10). Retention of *three* terms gives

$$y(x_{n+1}) \approx y(x_n) + hy'(x_n) + \tfrac{1}{2}h^2 y''(x_n). \tag{11}$$

Because $y'(x) = f(x, y)$ is our differential equation, the second derivative on the right in (11) is given by the chain rule:

$$y''(x) = \frac{d}{dx} [f(x, y)] = f_x(x, y) + f_y(x, y)f(x, y). \tag{12}$$

When the indicated derivatives are evaluated at the point (x_n, y_n), Eq. (11) therefore yields the approximation

$$y_{n+1} = y_n + hf(x_n, y_n) + \tfrac{1}{2}h^2[f_x(x_n, y_n) + f_y(x_n, y_n)f(x_n, y_n)] \tag{13}$$

to the actual value $y(x_{n+1})$ of the solution.

The use of the iterative formula in (13), to compute successive approximations $y_1, y_2, y_3, \ldots$ to the actual values $y(x_1), y(x_2), y(x_3), \ldots$ of the solution of the initial value problem in (4), constitutes the **three-term Taylor series method** with step size h. Like the improved Euler method on a given bounded interval $[a, b]$, this method has a (cumulative) error of order h^2. For this reason, and also because it results from retaining terms through h^2 in the Taylor series, it is sometimes called the **Taylor series method of order 2**. The Taylor series method of order 1 is simply Euler's method.

Figure 6.17 shows a three-term Taylor series method program. The function $f(x, y)$ and its partial derivatives f_x and f_y must be edited into lines 140–160 before the program is executed. The iterative formula in (13) appears in line 270. Figure 6.18 shows a pocket computer version of this program.

```
100 INPUT "X"; X
110 INPUT "Y"; Y
120 INPUT "H"; H
130 INPUT "K"; K
140 INPUT "P"; P
150 FOR N = 1 TO K
160    F = 2*X*Y
170    R = 2*Y
180    S = 2*X
190    Y = Y + H*F + (H*H/2)*(R + S*F)
200    X = X + H
210    IF INT(N/P) = N/P THEN PRINT X,Y
220 NEXT N
230 END
```

FIGURE 6.18 Pocket version of Program TAYLOR3T.

```
100 REM--Program TAYLOR3T
110 REM
120 REM--Initialization:
130 REM
140        DEF FNF(X,Y) = 2*X*Y
150        DEF FNFX(X,Y) = 2*Y
160        DEF FNFY(X,Y) = 2*X
170        INPUT "Initial x,y"; X,Y
180        INPUT "Step size h"; H
190        INPUT "Number of steps"; K
200        INPUT "Print step p"; P
210 REM
220 REM--Three term Taylor series iteration:
230 REM
240        FOR N = 1 TO K
250            F = FNF(X,Y)
260            R = FNFX(X,Y) : S = FNFY(X,Y)
270            Y = Y + H*F + (H*H/2)*(R + S*F)
280            X = X + H
290            IF INT(N/P) = N/P THEN PRINT X,Y
300        NEXT N
310 REM
320        END
```

FIGURE 6.17 Listing of Program TAYLOR3T.

EXAMPLE 3 Compare the improved Euler method with the three-term Taylor series method when each is applied to the initial value problem

$$y' = 2xy, \qquad y(0) = 1 \tag{14}$$

on the interval $[0, 1]$.

SOLUTION The table in Fig. 6.19 shows the results obtained using these two methods, each with step size $h = 0.005$. The final column shows values of the exact solution $y = e^{-x^2}$. In this particular problem the improved Euler method is slightly more accurate than the three-term Taylor series method, though in another problem the reverse might be true.

x	Improved Euler y	Three-term Taylor y	Actual y
0.1	1.01005	1.01005	1.01005
0.2	1.04081	1.04081	1.04081
0.3	1.09417	1.09417	1.09417
0.4	1.17351	1.17351	1.17351
0.5	1.28402	1.28402	1.28403
0.6	1.43333	1.43332	1.43333
0.7	1.63231	1.63229	1.63232
0.8	1.89647	1.89644	1.89648
0.9	2.24790	2.24785	2.24791
1.0	2.71826	2.71819	2.71828

FIGURE 6.19 Approximating the solution of $y' = 2xy$, $y(0) = 1$ with $h = 0.005$.

If the function $f(x, y)$ is sufficiently simple that the calculation of higher order partial derivatives is feasible, then more than three terms of the Taylor series

$$y(x_{n+1}) = y(x_n) + hy'(x_n) + \frac{h^2}{2!} y''(x_n) + \frac{h^3}{3!} y'''(x_n) + \cdots \tag{15}$$

can be retained at each step. The **Taylor series method of order** k consists of using at each step all terms in (15) through the one involving h^k, and it is known that (on a fixed bounded interval) the (cumulative) error in the Taylor series method of order k is of order h^k; that is, $|y(x_n) - y_n| < C|h|^k$ for some constant C. The disadvantage of these methods is that the higher order derivatives in (15) may be very tedious to compute analytically. The following example illustrates the great accuracy that higher order Taylor series methods offer when they are practical, and also the use of a *negative* step size to step backward rather than forward.

EXAMPLE 4 Consider the initial value problem

$$y' = y^2, \qquad y(1) = 1, \tag{16}$$

whose exact solution for $x < 2$ is $y(x) = 1/(2 - x)$. To obtain the Taylor polynomial of degree 4, for instance, we begin with the differential equation in (16) and compute the following derivatives:

$$\begin{aligned}
y' &= y^2, \\
y'' &= 2yy' = 2y^3, \\
y^{(3)} &= 6y^2y' = 6y^4, \\
y^{(4)} &= 24y^3y' = 24y^5.
\end{aligned} \tag{17}$$

Thus

$$y(x + h) \approx y + hy^2 + h^2y^3 + h^3y^4 + h^4y^5,$$

with y in place of $y(x)$ on the right-hand side. Hence the iterative formula for the Taylor series method of order 4 is

$$y_{n+1} = y_n + hy_n^2 + h^2y_n^3 + h^3y_n^4 + h^4y_n^5;$$

that is,

$$y_{n+1} = y_n(1 + hy_n\{1 + hy_n[1 + hy_n(1 + hy_n)]\}) \tag{18}$$

in the "stacked" form that is more convenient for automated computation.

It is clear from the pattern in (17) that $y^{(m)} = m!y^{m+1}$ for all $m > 0$. It follows that the iterative formula for the Taylor series method of order m for the problem in (16) is the mth degree analogue of the 4th degree (in h) stacked polynomial in (18). Figure 6.20 shows an *ad hoc* program written to carry out Taylor series approximations to the solution of (16) when the desired order m is entered at line 145. The loop in lines 170–190 then calculates the stacked polynomial like the one in Eq. (18). Figure 6.21 shows the results of applying the Taylor series method with orders $m = 4$, $m = 6$, $m = 8$, and step size $h = -0.05$, so that only 20 steps are required to reach $x = 0$, where $y(1) = 1/2$. With $m = 8$ the results are correct to 10 decimal places on the entire interval.

```
100 INPUT "Initial x"; X
110 INPUT "Initial y"; Y
120 INPUT "Step size"; H
130 INPUT "No of steps"; K
140 INPUT "Print steps"; P
145 INPUT "Order"; M
150 FOR N = 1 TO K
160     U = H*Y  :  V = 0
170     FOR I = 1 TO M
180         V = U*(1 + V)
190     NEXT I
200     Y = Y*(1 + V)
210     X = X + H
220     IF INT(N/P) = N/P THEN PRINT X,Y
230 NEXT N
240 END
```

FIGURE 6.20 Program for Example 4.

x	Order $m = 4$ approximate y	Order $m = 6$ approximate y	Order $m = 8$ approximate y	Actual y
1.0	1.0000000000	1.0000000000	1.0000000000	1.0000000000
0.8	0.8333340327	0.8333333349	0.8333333333	0.8333333333
0.6	0.7142864839	0.7142857158	0.7142857143	0.7142857143
0.4	0.6250006974	0.6250000013	0.6250000000	0.6250000000
0.2	0.5555561577	0.5555555566	0.5555555556	0.5555555556
0.0	0.5000005139	0.5000000009	0.5000000000	0.5000000000

FIGURE 6.21 Using the Taylor series methods with step size $h = -0.05$ to approximate the solution of the initial value problem in Eq. (16).

6.2 PROBLEMS

A hand-held calculator will suffice for Problems 1–10. In each problem, find the exact solution of the given initial value problem. Then apply both (a) the improved Euler method, and (b) the three-term Taylor series method to approximate (to four decimal places) this solution on the interval $[0, 0.5]$ with step size $h = 0.1$. Make a table showing the approximate values and the actual values of the solution at the points $x = 0.1, 0.2, 0.3, 0.4, 0.5$.

1. $y' = -y$, $y(0) = 2$
2. $y' = 2y$, $y(0) = \frac{1}{2}$
3. $y' = y + 1$, $y(0) = 1$
4. $y' = x - y$, $y(0) = 1$
5. $y' = y - x - 1$, $y(0) = 1$
6. $y' = -2xy$, $y(0) = 2$
7. $y' = -3x^2y$, $y(0) = 3$
8. $y' = e^{-y}$, $y(0) = 0$
9. $y' = \frac{1}{4}(1 + y^2)$, $y(0) = 1$
10. $y' = 2xy^2$, $y(0) = 1$

A programmable calculator or a computer will be useful for Problems 11–16. In each problem, find the exact solution of the given initial value problem. Then apply the improved Euler method twice to approximate (to five decimal places) this solution on the given interval, first with step size $h = 0.01$, then with step size $h = 0.005$. Make a table showing the approximate values and the actual value, together with the percentage error in the more accurate approximation, when x is an integral multiple of 0.2.

11. $y' = y - 2$, $y(0) = 1$; $0 \leq x \leq 1$
12. $y' = \frac{1}{2}(y - 1)^2$, $y(0) = 2$; $0 \leq x \leq 1$
13. $yy' = 2x^3$, $y(1) = 3$; $1 \leq x \leq 2$
14. $xy' = y^2$, $y(1) = 1$; $1 \leq x \leq 2$
15. $xy' = 3x - 2y$, $y(2) = 3$; $2 \leq x \leq 3$
16. $y^2y' = 2x^5$, $y(2) = 3$; $2 \leq x \leq 3$

A computer with a printer is required for Problems 17–24. In each of these initial value problems, use the improved Euler

method with step sizes $h = 0.1, 0.02, 0.004$, and 0.0008 to approximate (to five decimal places) the values of the solution at ten equally spaced points of the given interval. Print the results in tabular form to make it easy to gauge the effect of varying the step size.

17. $y' = x^2 + y^2$, $y(0) = 0$; $0 \leq x \leq 1$
18. $y' = x^2 - y^2$, $y(0) = 1$; $0 \leq x \leq 2$
19. $y' = x + \sqrt{y}$, $y(0) = 1$; $0 \leq x \leq 2$
20. $y' = x + y^{1/3}$, $y(0) = -1$; $0 \leq x \leq 2$
21. $y' = \ln y$, $y(1) = 2$; $1 \leq x \leq 2$
22. $y' = x^{2/3} + y^{2/3}$, $y(0) = 1$; $0 \leq x \leq 2$
23. $y' = \sin x + \cos y$, $y(0) = 0$; $0 \leq x \leq 1$
24. $y' = \dfrac{x}{1 + y^2}$, $y(-1) = 1$; $-1 \leq x \leq 1$

For each of the differential equations in Problems 25–27, write the iterative formula (for y_{n+1} in terms of y_n) for the Taylor series method of order 3.

25. $y' = 1 + y^2$
26. $y' = y + y^2$
27. $y' = (x + y)^2$

28. Show that for the initial value problem

$$y' = y, \qquad y(0) = 1$$

the iterative formula for the Taylor series method of order m is

$$y_{n+1} = \left(1 + h + \frac{h^2}{2!} + \cdots + \frac{h^m}{m!}\right)y_n.$$

Use this formula with various combinations of $h = 0.1, 0.2$ and $m = 4, 6, 8$ to approximate (to nine decimal places) the number $y(1) = e \approx 2.718281828$.

29. For the initial value problem

$$y' = xy, \qquad y(0) = 1$$

show that the successive derivatives of the solution are given by

$$y^{(k+1)} = xy^{(k)} + ky^{(k-1)}.$$

Then write a program to apply the Taylor series method of order m (with the value of m to be entered when the program is executed) to approximate the actual solution $y = \exp\left(\frac{1}{2}x^2\right)$, $0 \le x \le 1$.

30. For the initial value problem

$$y' = e^{-y}, \qquad y(0) = 0$$

with exact solution $y(x) = \ln(1+x)$ $(x > -1)$, show that successive derivatives are given by

$$y^{(k)} = (-1)^{k+1}(k-1)!e^{-ky}.$$

Hence conclude that the iterative formula for the Taylor series method of order m is

$$y_{n+1} = y_n + he^{-y_n} - \frac{1}{2}h^2e^{-2y_n} + \cdots + (-1)^{m-1}\frac{h^m}{m}e^{-my_n}.$$

31. Here is one idea that might yield an improvement in both the improved Euler method and the Taylor series method, and it depends upon combining the two. Suppose that x_n and y_n have both been computed for some value of n (and, of course, all previous values) in the initial value problem

$$y' = f(x, y), \qquad y(a) = y_0. \tag{1}$$

Use the differential equation in (1) to compute estimates of $y'(x_n)$ and $y''(x_n)$, then use the Taylor series method to obtain a first estimate u_{n+1} of the solution at x_{n+1}. Next see the differential equation in (1) to compute estimates of $y'(x_{n+1})$ and $y''(x_{n+1})$. Then average the approximate first derivatives at the points x_n and x_{n+1}, and also average the approximate second derivatives there, to obtain (possibly better) estimates of y' and y'' over the interval $[x_n, x_{n+1}]$. Write a flowchart for this new algorithm, write a program to implement it, and apply it to the initial value problem

$$y' = 2xy, \qquad y(0) = 1 \tag{14}$$

on $[0, 1]$ discussed in Example 3 of this section. Comment on the results. (*Note:* Data for the improved Euler method and the three-term Taylor series method are shown in Fig. 6.19.)

6.3

The Runge-Kutta Method

We now discuss a method for approximating the solution $y = y(x)$ of the initial value problem

$$y' = f(x, y), \qquad y(x_0) = y_0 \tag{1}$$

that is considerably more accurate than the improved Euler method, and is more widely used in practice than any of the numerical methods discussed in Sections 6.1 and 6.2. It is called the *Runge-Kutta method* and is named for the German mathematicians Carl Runge (1856–1927) and Wilhelm Kutta (1867–1944).

With the usual notation, suppose that we have computed the approximations $y_1, y_2, \ldots, y_n$ to the actual values $y(x_1), y(x_2), \ldots, y(x_n)$ and now want to compute $y_{n+1} \approx y(x_{n+1})$. Then

$$y(x_{n+1}) - y(x_n) = \int_{x_n}^{x_{n+1}} y'(x)\, dx = \int_{x_n}^{x_n+h} y'(x)\, dx \tag{2}$$

by the fundamental theorem of calculus, and Simpson's rule for numerical integration therefore yields

$$y(x_{n+1}) - y(x_n) \approx \frac{h}{6}\left[y'(x_n) + 4y'\left(x_n + \frac{1}{2}h\right) + y'(x_{n+1})\right]. \tag{3}$$

Hence we want to define y_{n+1} so that

$$y_{n+1} \approx y_n + \frac{h}{6}\left[y'(x_n) + 2y'\left(x_n + \frac{1}{2}h\right) + 2y'\left(x_n + \frac{1}{2}h\right) + y'(x_{n+1})\right]; \tag{4}$$

we have split $4y'(x_n + \frac{1}{2}h)$ into a sum of two terms because we intend to approximate the slope $y'(x + \frac{1}{2}h)$ at the midpoint $x_n + \frac{1}{2}h$ of the interval $[x_n, x_{n+1}]$ in two different ways.

On the right-hand side in (4), we replace the (true) slopes $y'(x_n)$, $y'(x_n + \frac{1}{2}h)$, $y'(x_n + \frac{1}{2}h)$, and $y'(x_{n+1})$ (respectively) by the following estimates.

$$k_{n1} = f(x_n, y_n) \tag{5a}$$

—This is the Euler method slope at x_n.

$$k_{n2} = f(x_n + \tfrac{1}{2}h, y_n + \tfrac{1}{2}hk_{n1}) \tag{5b}$$

—This is an estimate of the slope at the midpoint of the interval $[x_n, x_{n+1}]$, using the Euler method to predict the ordinate there.

$$k_{n3} = f(x_n + \tfrac{1}{2}h, y_n + \tfrac{1}{2}hk_{n2}) \tag{5c}$$

—This is an improved Euler value for the slope at the midpoint.

$$k_{n4} = f(x_{n+1}, y_n + hk_{n3}) \tag{5d}$$

—This is the Euler method slope at x_{n+1}, using the improved slope k_{n3} at the midpoint to step to x_{n+1}.

When these substitutions are made in (4), the result is the iterative formula

$$y_{n+1} = y_n + \frac{h}{6}(k_{n1} + 2k_{n2} + 2k_{n3} + k_{n4}). \tag{6}$$

The use of this formula to compute the approximations $y_1, y_2, y_3, \ldots$ successively constitutes the **Runge-Kutta method**.

It happens that this method is equivalent in accuracy to the five-term (fourth degree) Taylor formula that uses in place of (6) the iterative formula

$$y_{n+1} = y_n + y_n'h + \frac{y_n''}{2!}h^2 + \frac{y_n^{(3)}}{3!}h^3 + \frac{y_n^{(4)}}{4!}h^4. \tag{7}$$

The Runge-Kutta method is, however, simpler to apply in practice because the computation of the numbers k_{n1}, k_{n2}, k_{n3}, and k_{n4} in (5a)–(5d) involves only evaluation of the original function $f(x, y)$ in (1), whereas use of the Taylor formula in (7) requires high-order partial derivatives of f.

The Runge-Kutta method is a *fourth order* method—it can be proved that cumulative error on a bounded interval $[a, b]$ with $a = x_0$ is of order h^4. (Thus the iteration in (6) is sometimes called the *fourth order* Runge-Kutta method because it is possible to develop Runge-Kutta methods of other orders.) That is,

$$|y(x_n) - y_n| \leq Ch^4, \tag{8}$$

where the constant C depends on the function $f(x, y)$ and the interval $[a, b]$, but does not depend on the step size h. The following example illustrates this high accuracy in comparison with the lower-order accuracy of our previous numerical methods.

EXAMPLE 1 We first apply the Runge-Kutta method to the illustrative initial value problem

$$y' = x + y, \qquad y(0) = 1 \tag{9}$$

that we considered in Example 1 of Section 6.1 and in Example 2 of Section 6.2, with exact solution $y(x) = 2e^x - x - 1$. To make a point we use $h = 0.5$, a larger step size than in any previous example, so only two steps are required to go from $x = 0$ to $x = 1$.

In the first step we use the formulas in (5) and (6) to calculate

$$k_1 = (0) + (1) = 1,$$

$$k_2 = (0 + 0.25) + (1 + (0.25)(1)) = 1.5,$$

$$k_3 = (0 + 0.25) + (1 + (0.25)(1.5)) = 1.625,$$

$$k_4 = (0.5) + (1 + (0.5)(1.625)) = 2.3125,$$

and thence

$$y_1 = 1 + \frac{0.5}{6} \left[1 + (2)(1.5) + (2)(1.625) + 2.3125 \right] \approx 1.7969.$$

Similarly, the second step yields $y_2 \approx 3.4347$.

Figure 6.22 presents these results together with the results (from Fig. 6.14) of applying the improved Euler method with step size $h = 0.1$. We see that even with the larger step size, the Runge-Kutta method gives (for this problem) four to five times the accuracy (in terms of relative percentage errors) of the improved Euler method.

	Improved Euler		Runge-Kutta		
x	y with h = 0.1	Percent error	y with h = 0.5	Percent error	Actual y
0.0	1.0000	0.00%	1.0000	0.00%	1.0000
0.5	1.7949	0.14%	1.7969	0.03%	1.7974
1.0	3.4282	0.24%	3.4347	0.05%	3.4366

FIGURE 6.22 Runge-Kutta and improved Euler results for the initial value problem $y' = x + y$, $y(0) = 1$.

It is customary to measure the computational labor involved in solving $y' = f(x, y)$ numerically by counting the number of evaluations of the function $f(x, y)$ that are required. In Example 1, the Runge-Kutta method required eight evaluations of $f(x, y) = x + y$ (four at each step), whereas the improved Euler method required 20 such evaluations (two for each of 10 steps). Thus the Runge-Kutta method gave over four times the accuracy with only 40% of the labor.

Figure 6.23 is a listing of a program that implements the Runge-Kutta method. The four evaluations of $f(x, y)$ in each step occur at lines 240–270, and line 280 is the Simpson's rule formula in (6). The version shown in Fig. 6.24 stores the four slopes k_{n1}, k_{n2}, k_{n3}, k_{n4} in an array K(1), K(2), K(3), K(4), and will run on a pocket computer that accepts only single-character variable names; a subroutine is used for the repeated evaluation of $f(x, y)$.

Figure 6.25 shows the results obtained by applying the improved Euler and Runge-Kutta methods to the problem $y' = x + y$, $y(0) = 1$ with the same step size $h = 0.1$. The relative error in the improved Euler value at $x = 1$ is about 0.24%,

```
100 REM--Program RNGEKTTA
110 REM
120 REM--Initialization:
130 REM
140     DEFDBL F,H,K,X,Y  :  DEFINT M,N,P
150     DEF FNF(X,Y) = X + Y
160     INPUT "Initial x,y"; X,Y
170     INPUT "Step size h"; H
180     INPUT "Number of steps"; M
190     INPUT "Print step p"; P
200 REM
210 REM--Runge-Kutta iteration:
220 REM
230     FOR N = 1 TO M
240         K1 = FNF(X,Y)
250         K2 = FNF(X + H/2,Y + H*K1/2)
260         K3 = FNF(X + H/2,Y + H*K2/2)
270         K4 = FNF(X + H,   Y + H*K3)
280         Y = Y + (H/6)*(K1 + 2*K2 + 2*K3 + K4)
290         X = X + H
300         IF INT(N/P) = N/P THEN PRINT X,Y
310     NEXT N
320 REM
330     END
```

FIGURE 6.23 Listing of Program RNGEKTTA.

```
100 INPUT "X"; X
110 INPUT "Y"; Y
120 INPUT "H"; H
130 INPUT "M"; M
140 INPUT "P"; P
150 DIM K(4)
160 FOR N = 1 TO M
170   Z = Y
180   GOSUB 330
190   K(1) = F
200   X = X + H/2  :  Y = Z + (H/2)*K(1)
210   GOSUB 330
220   K(2) = F
230   Y = Z + (H/2)*K(2)
240   GOSUB 330
250   K(3) = F
260   X = X + H/2  :  Y = Z + H*K(3)
270   GOSUB 330
280   K(4) = F
290   Y = Z + (H/6)*(K(1)+2*K(2)+2*K(3)+K(4))
300   IF INT(N/P) = N/P THEN PRINT X,Y
310 NEXT N
320 STOP
330 F = X + Y
340 RETURN
350 END
```

FIGURE 6.24 Pocket version of Program RNGEKTTA.

x	Improved Euler y	Runge-Kutta y	Actual y
0.1	1.1100	1.110342	1.110342
0.2	1.2421	1.242805	1.242806
0.3	1.3985	1.399717	1.399718
0.4	1.5818	1.583648	1.583649
0.5	1.7949	1.797441	1.797443
0.6	2.0409	2.044236	2.044238
0.7	2.3231	2.327503	2.327505
0.8	2.6456	2.651079	2.651082
0.9	3.0124	3.019203	3.019206
1.0	3.4282	3.436559	3.436564

FIGURE 6.25 Runge-Kutta and improved Euler results for the initial value problem $y' = x + y$, $y(0) = 1$ with the same step size $h = 0.1$.

whereas for the Runge-Kutta value it is 0.00012%. In this comparison the Runge-Kutta method is about 2000 times as accurate, but requires only twice as many function evaluations as does the improved Euler method.

The error bound

$$|y(x_n) - y_n| \leq Ch^4 \tag{8}$$

for the Runge-Kutta method results in a rapid decrease in errors when the step size h is reduced (except for the possibility that very small step sizes may result in unacceptable roundoff errors). It follows from Eq. (8) that (on a fixed bounded interval) halving the step size decreases the absolute error by a factor of $(1/2)^4 = 1/16$. Consequently, the common practice of successively halving the step size until the computed results "stabilize" is particularly effective with the Runge-Kutta method.

EXAMPLE 2 In Example 2 of Section 6.1 we saw that Euler's method is not adequate to approximate the solution $y(x)$ of the initial value problem

$$y' = x^2 + y^2, \qquad y(0) = 1 \tag{10}$$

as it approaches the infinite discontinuity at $x \approx 0.969811$ (see Fig. 6.8). Now we apply the Runge-Kutta method to this problem.

Figure 6.26 shows Runge-Kutta results on the interval $[0, 0.9]$, computed with step sizes $h = 0.1$, $h = 0.05$, and $h = 0.025$. There is still some difficulty near $x = 0.9$, but it seems safe to conclude from these data that $y(0.5) \approx 2.0670$.

We therefore start afresh and apply the Runge-Kutta method to the initial value problem

$$y' = x^2 + y^2, \qquad y(0.5) = 2.0670. \tag{11}$$

Figure 6.27 shows results on the interval $[0.5, 0.9]$, obtained with steps sizes $h = 0.01$, $h = 0.005$, and $h = 0.0025$. We now conclude that $y(0.9) \approx 14.3049$.

x	y with $h = 0.1$	y with $h = 0.05$	y with $h = 0.025$
0.1	1.1115	1.1115	1.1115
0.3	1.4397	1.4397	1.4397
0.5	2.0670	2.0670	2.0670
0.7	3.6522	3.6529	3.6529
0.9	14.0218	14.2712	14.3021

FIGURE 6.26 Approximating the solution of the problem in Eq. (10).

x	y with $h = 0.01$	y with $h = 0.005$	y with $h = 0.0025$
0.5	2.0670	2.0670	2.0670
0.6	2.6440	2.6440	2.6440
0.7	3.6529	3.6529	3.6529
0.8	5.8486	5.8486	5.8486
0.9	14.3048	14.3049	14.3049

FIGURE 6.27 Approximating the solution of the problem in Eq. (11).

x	y with $h = 0.002$	y with $h = 0.001$	y with $h = 0.0005$
0.90	14.3049	14.3049	14.3049
0.91	16.7024	16.7024	16.7024
0.92	20.0617	20.0617	20.0617
0.93	25.1073	25.1073	25.1073
0.94	33.5363	33.5363	33.5363
0.95	50.4722	50.4723	50.4723

FIGURE 6.28 Approximating the solution of the problem in Eq. (12).

Finally, Fig. 6.28 shows results on the interval $[0.90, 0.95]$ for the initial value problem

$$y' = x^2 + y^2, \qquad y(0.9) = 14.3049 \tag{12}$$

obtained using step sizes $h = 0.002$, $h = 0.001$, and $h = 0.0005$. Our final approximate result is $y(0.95) \approx 50.4723$. The actual value of the solution at $x = 0.95$ is $y(0.95) \approx 50.471867$. Our slight overestimate results mainly from the fact that the four-place initial value in (12) is (in effect) the result of rounding *up* the actual value $y(0.9) \approx 14.304864$; such errors are magnified considerably as we approach the singularity in the solution.

EXAMPLE 3 A skydiver jumps from an aircraft at an initial altitude of 10,000 ft. Assume that she falls vertically with initial velocity zero, weighs 128 lb, and experiences an upward force F_R of air resistance given in terms of her velocity v (in feet per second) by

$$F_R = (0.01)v + (0.001)v^2 + (0.0001)v^3$$

(in pounds, and with the coordinate axis directed downward so that $v > 0$). If she does not open her parachute, what will be her terminal velocity? How fast will she be falling after 5 s have elapsed? After 10 s? After 20 s?

SOLUTION With $g = 32$ ft/s^2, the skydiver's mass is $m = 4$ slugs, so Newton's law $F = ma$ yields

$$m \frac{dv}{dt} = mg - F_R;$$

that is,

$$4 \frac{dv}{dt} = 128 - (0.01)v - (0.001)v^2 - (0.0001)v^3. \tag{13}$$

Thus the velocity $v(t)$ satisfies the initial value problem

$$\frac{dv}{dt} = f(v), \qquad v(0) = 0 \tag{14}$$

where

$$f(v) = 32 - (0.000025)(100v + 10v^2 + v^3). \tag{15}$$

The skydiver reaches her terminal velocity when the forces of gravity and air resistance balance, so $f(v) = 0$. We therefore can calculate her terminal velocity immediately by the simple expedient of solving the equation

$$32 - (0.000025)(100v + 10v^2 + v^3) = 0. \tag{16}$$

Using Newton's method we find that her terminal velocity is approximately 105.046 ft/s, approximately 71.6 mi/h.

Figure 6.29 shows the results of Runge-Kutta approximations to the solution of the initial value problem in (14); the step sizes $h = 0.2$ and $h = 0.1$ yield the same results (to three decimal places). Observe that the terminal velocity is attained in 16 s. But the skydiver's velocity is 95.35% of her terminal velocity after only 5 s, and 99.94% after 10 s.

t (s)	v (ft/s)
1	31.675
2	60.264
3	81.423
4	93.933
5	100.163
6	102.972
7	104.178
8	104.685
9	104.897
10	104.984
11	105.020
12	105.035
13	105.042
14	105.044
15	105.045
16	105.046
17	105.046
18	105.046
19	105.046
20	105.046

FIGURE 6.29 Data in Example 3.

x	Runge-Kutta y with $h = 0.2$	Runge-Kutta y with $h = 0.1$	Runge-Kutta y with $h = 0.05$	Actual y
0.4	0.66880	0.67020	0.67031	0.67032
0.8	0.43713	0.44833	0.44926	0.44933
1.2	0.21099	0.29376	0.30067	0.30199
1.6	-0.46019	0.14697	0.19802	0.20190
2.0	-4.72142	-0.27026	0.10668	0.13534
2.4	-35.53415	-2.90419	-0.12102	0.09072
2.8	-261.25023	-22.05352	-1.50367	0.06081
3.2	-1,916.69395	-163.25077	-11.51868	0.04076
3.6	-14,059.35494	-1205.71249	-85.38156	0.02732
4.0	-103,126.5270	-8903.12866	-631.03934	0.01832

FIGURE 6.30 Runge-Kutta attempts to solve numerically the initial value problem in Eq. (17).

CHAPTER 6: Numerical Methods

The final example in this section constitutes a *warning*: For certain types of initial value problems, the numerical methods we have discussed are not nearly so successful as in the previous examples.

EXAMPLE 4 Consider the seemingly innocuous initial value problem

$$\frac{dy}{dx} = 5y - 6e^{-x}, \qquad y(0) = 1 \tag{17}$$

whose exact solution is $y(x) = e^{-x}$. The table in Fig. 6.30 shows the results obtained by applying the Runge-Kutta method on the interval $0 \leq x \leq 4$ with step sizes $h = 0.2$, $h = 0.1$, and $h = 0.05$. Obviously, these attempts are spectacularly unsuccessful. Although $y(x) = e^{-x} \to 0$ as $x \to +\infty$, it appears that our numerical approximations are headed toward $-\infty$ rather than zero.

The explanation lies in the fact that the general solution of the equation $y' = 5y - 6e^{-x}$ is

$$y(x) = e^{-x} + Ce^{5x}. \tag{18}$$

The particular solution of (17) satisfying the initial condition $y(0) = 1$ is obtained with $C = 0$. But any departure from the exact solution $y(x) = e^{-x}$, however small— even if due only to roundoff error—introduces (in effect) a nonzero value of C in (18). And as indicated in Fig. 6.31, all solution curves of the form in (18) with

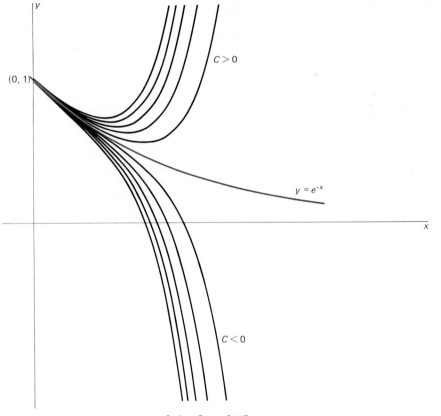

FIGURE 6.31 Solution curves of $y' = 5y - 6e^{-x}$.

$C \neq 0$ diverge strongly away from the one with $C = 0$, even if their initial values are close to 1.

Difficulties of the sort illustrated by Example 4 sometimes are unavoidable, but one can at least hope to recognize such a problem when it appears. Approximate values whose order of magnitude varies with changing step size are a common indicator of such instability. These difficulties are discussed in numerical analysis textbooks and are the object of current research in the field.

6.3 PROBLEMS

A hand-held calculator will suffice for Problems 1–10. In each problem, find the exact solution of the given initial value problem. Then apply the Runge-Kutta method with step size $h = 0.25$ to approximate (to five decimal places) this solution on the interval $[0, 0.5]$. Make a table showing the approximate and actual values of the solution at the points $x = 0.25$ and $x = 0.50$.

1. $y' = -y$, $y(0) = 2$
2. $y' = 2y$, $y(0) = \frac{1}{2}$
3. $y' = y + 1$, $y(0) = 1$
4. $y' = x - y$, $y(0) = 1$
5. $y' = y - x - 1$, $y(0) = 1$
6. $y' = -2xy$, $y(0) = 2$
7. $y' = -3x^2 y$, $y(0) = 3$
8. $y' = e^{-y}$, $y(0) = 0$
9. $y' = \frac{1}{4}(1 + y^2)$, $y(0) = 1$
10. $y' = 2xy^2$, $y(0) = 1$

A programmable calculator or a computer will be useful for Problems 11–16. In each problem find the exact solution of the given initial value problem. Then apply the Runge-Kutta method twice to approximate (to five decimal places) this solution on the given interval, first with step size $h = 0.2$, then with step size $h = 0.1$. Make a table showing the approximate values and the actual value, together with the percentage error in the more accurate approximation, for x an integral multiple of 0.2.

11. $y' = y - 2$, $y(0) = 1$; $0 \leq x \leq 1$
12. $y' = \frac{1}{2}(y - 1)^2$, $y(0) = 2$; $0 \leq x \leq 1$
13. $yy' = 2x^3$, $y(1) = 3$; $1 \leq x \leq 2$
14. $xy' = y^2$, $y(1) = 1$; $1 \leq x \leq 2$
15. $xy' = 3x - 2y$, $y(2) = 3$; $2 \leq x \leq 3$
16. $y^2 y' = 2x^5$, $y(2) = 3$; $2 \leq x \leq 3$

A computer with a printer is required for Problems 17–24. In each of these initial value problems, use the Runge-Kutta method with step sizes $h = 0.2$, 0.1, 0.05, and 0.025 to approximate (to six decimal places) the values of the solution at ten equally spaced points of the given interval. Print the results in tabular form with appropriate headings to make it easy to gauge the effect of varying the step size h.

17. $y' = x^2 + y^2$, $y(0) = 0$; $0 \leq x \leq 1$

18. $y' = x^2 - y^2$, $y(0) = 1$; $0 \leq x \leq 2$
19. $y' = x + \sqrt{y}$, $y(0) = 1$; $0 \leq x \leq 2$
20. $y' = x + y^{1/3}$, $y(0) = -1$; $0 \leq x \leq 2$
21. $y' = \ln y$, $y(1) = 2$; $1 \leq x \leq 2$
22. $y' = x^{2/3} + y^{2/3}$, $y(0) = 1$; $0 \leq x \leq 2$
23. $y' = \sin x + \cos y$, $y(0) = 0$; $0 \leq x \leq 1$
24. $y' = \dfrac{x}{1 + y^2}$, $y(-1) = 1$; $-1 \leq x \leq 1$

VELOCITY-ACCELERATION PROBLEMS

In each of Problems 25–30, the linear acceleration $a = dv/dt$ of a moving particle is given by a formula

$$\frac{dv}{dt} = f(t, v), \tag{19}$$

where the velocity $v = ds/dt$ is the time derivative of the distance function s. The velocity $v(t)$ is approximated by using the Runge-Kutta method to solve numerically the initial value problem

$$v' = f(t, v), \qquad v(0) = v_0. \tag{20}$$

Suppose that we also want to compute the distance $s(t)$ traveled by the particle, beginning with $s(0) = s_0$. We can do this by calculating

$$s_{n+1} = s_n + v_n h + \tfrac{1}{2} a_n h^2 \tag{21}$$

at each step, where $a_n = f(t_n, v_n)$. (The formula in (21) would give the correct increment in the distance s if the acceleration a_n remained *constant* during the time interval $[t_n, t_{n+1}]$.) Thus at each step we compute

$$a_n = f(t_n, v_n),$$

$$s_{n+1} = s_n + v_n h + \tfrac{1}{2} a_n h^2,$$

$$v_{n+1} = v_n + \frac{h}{6}(k_1 + 2k_2 + 2k_3 + k_4), \tag{22}$$

$$t_{n+1} = t_n + h.$$

Here, k_1, k_2, k_3, and k_4 are given by Eqs. (5a)–(5d) with t in place of x and v in place of y.

25. If the skydiver of Example 3 jumped from an altitude of 10,000 ft and there were *no* air resistance, then (by the formula $s = \frac{1}{2}gt^2$) she would fall 1600 ft in the first 10 s, 6400 ft in the first 20 s, and (if the parachute remains unopened) would hit the ground in a total of 25 s. Use the equations in (22) with $h = 0.2$ to find the distance fallen after 10 and 20 s and the time elapsed at impact with the ground, under the assumption that her acceleration due to gravity *and* air resistance is that given by the formula in Eq. (15).

26. Suppose that a skydiver steps out of an airplane at an altitude of 10,000 ft and that his downward acceleration is given by

$$\frac{dv}{dt} = 32 - (0.005)v^2.$$

(a) Show that his velocity and distance fallen are given (exactly) by

$$v(t) = 80 \tanh{(0.4t)}, \qquad s(t) = 200 \ln{(\cosh{(0.4t)})}.$$

(b) Use the method outlined in (22) to approximate (with $h = 0.1$) the distance fallen after 5, 10, and 15 s. Compare with the actual values. (c) What terminal velocity will the skydiver attain and when will he hit the ground (with no parachute)?

27. Suppose that a car starts from rest and its engine provides an acceleration of 10 ft/s². Assume also that it experiences a deceleration of $(0.1)v$ ft/s² ($v = v(t)$ denotes its velocity at time t seconds) due to friction and air resistance, so that its net acceleration is given by

$$\frac{dv}{dt} = 10 - (0.1)v.$$

Find Runge-Kutta approximations (with $h = 0.2$) to $v(t)$ and $s(t)$ in order to answer the following questions. (a) What is the limiting velocity of the car? (b) How far will it travel in the first 10 s, and what will its velocity be then? (c) How long will it take to reach a velocity of 60 mi/h (88 ft/s) and how far will it travel during that time?

28. Repeat Problem 27, except use acceleration

$$\frac{dv}{dt} = 10 - (0.001)v^2.$$

29. Suppose that a ball is thrown straight upward from the ground with initial velocity $v_0 = 160$ ft/s. If there were no air resistance it would reach a maximum height of 400 ft in 5 s, then fall back to the ground in another 5 s. But assume, more realistically, that the effect of air resistance is to give the ball acceleration

$$\frac{dv}{dt} = -32 - (0.1)v.$$

Find Runge-Kutta approximations (with $h = 0.1$) to the velocity $v(t)$ and height $s(t)$ of the ball in order to answer the following questions. (a) What is the maximum height that the ball reaches, and at what time t is that height attained? (b) How long will it take the ball to fall from its maximum height back to the ground, and with what velocity will it strike the ground?

30. Repeat Problem 29, but take

$$\frac{dv}{dt} = -32 - (0.001)v|v|$$

as the acceleration of the ball. Note that $v' = -32 - (0.001)v^2$ during the ascent of the ball whereas $v' = -32 + (0.001)v^2$ during its descent. The answers should be comparable to those found in Problem 29.

6.4

Systems of Differential Equations

We now discuss the numerical approximation of solutions of systems of differential equations. Our goal is to apply the methods of earlier sections to the initial value problem

$$\mathbf{x}' = f(t, \mathbf{x}), \qquad \mathbf{x}(t_0) = \mathbf{x}_0 \tag{1}$$

for a system of m first order differential equations. In (1) the independent variable is the scalar t, whereas $\mathbf{x} = (x_1, x_2, \ldots, x_m)$ and $\mathbf{f} = (f_1, f_2, \ldots, f_m)$ are vector-valued functions. If the component functions of $\mathbf{f}$ and their first order partial derivatives are all continuous in a neighborhood of the point $(t_0, \mathbf{x}_0)$, then Theorems 3 and 4 of Section 7.6 guarantee the existence and uniqueness of a solution $\mathbf{x} = \mathbf{x}(t)$

of (1) on some subinterval (of the t-axis) containing t_0. With this assurance we can proceed to discuss the numerical approximation of this solution.

Beginning with step size h, we want to approximate the values of $\mathbf{x}(t)$ at the points $t_1, t_2, t_3, \ldots$, where $t_{n+1} = t_n + h$ for $n \geq 0$. Suppose that we have already computed the *approximations*

$$\mathbf{x}_1, \quad \mathbf{x}_2, \quad \mathbf{x}_3, \quad \ldots$$

to the *actual values*

$$\mathbf{x}(t_1), \quad \mathbf{x}(t_2), \quad \mathbf{x}(t_3), \quad \ldots$$

of the exact solution of the system in (1). We can then make the step from $\mathbf{x}_n$ to the next approximation $\mathbf{x}_{n+1} \approx \mathbf{x}(t_{n+1})$ by any one of the methods of Sections 6.1 through 6.3. Essentially all that is required is to write the iterative formula of the selected method in the vector notation of the present discussion.

For example, the iterative formula of Euler's method for systems is

$$\mathbf{x}_{n+1} = \mathbf{x}_n + h\mathbf{f}(t, \mathbf{x}_n). \tag{2}$$

To examine the case $m = 2$ of a pair of first order differential equations, let us write

$$\mathbf{x} = \begin{bmatrix} x \\ y \end{bmatrix} \quad \text{and} \quad \mathbf{f} = \begin{bmatrix} f \\ g \end{bmatrix}.$$

Then the initial value problem in (1) is

$$\begin{aligned} x' &= f(t, x, y), & x(t_0) &= x_0, \\ y' &= g(t, x, y), & y(t_0) &= y_0 \end{aligned} \tag{3}$$

and the scalar components of the vector formula in (2) are

$$\begin{aligned} x_{n+1} &= x_n + hf(t_n, x_n, y_n), \\ y_{n+1} &= y_n + hg(t_n, x_n, y_n). \end{aligned} \tag{4}$$

Note that each iterative formula in (4) has the form of a single Euler iteration, but with y_n inserted like a parameter in the first formula (for x_{n+1}) and with x_n inserted like a parameter in the second formula (for y_{n+1}). The generalization to the system in (3) of each of the other methods in Sections 6.1 through 6.3 follows a similar pattern.

The improved Euler method for systems consists at each step of calculating first the predictor

$$\mathbf{u}_{n+1} = \mathbf{x}_n + h\mathbf{f}(t_n, \mathbf{x}_n) \tag{5}$$

and then the corrector

$$\mathbf{x}_{n+1} = \mathbf{x}_n + \frac{h}{2}\left[\mathbf{f}(t_n, \mathbf{x}_n) + \mathbf{f}(t_{n+1}, \mathbf{u}_{n+1})\right]. \tag{6}$$

For the case of the two-dimensional initial value problem in (3), the scalar components of the formulas in (5) and (6) are

$$\begin{aligned} u_{n+1} &= x_n + hf(t_n, x_n, y_n), \\ v_{n+1} &= y_n + hg(t_n, x_n, y_n) \end{aligned} \tag{7}$$

and

$$x_{n+1} = x_n + \frac{h}{2} \left[f(t_n, x_n, y_n) + f(t_{n+1}, u_{n+1}, v_{n+1}) \right],$$

$$y_{n+1} = y_n + \frac{h}{2} \left[g(t_n, x_n, y_n) + g(t_{n+1}, u_{n+1}, v_{n+1}) \right].$$

(8)

EXAMPLE 1 Consider the initial value problem

$$\begin{aligned} x' &= 3x - 2y, & x(0) &= 3 \\ y' &= 5x - 4y, & y(0) &= 6 \end{aligned}$$

(9)

whose exact solution is given by

$$x(t) = 2e^{-2t} + e^t, \qquad y(t) = 5e^{-2t} + e^t.$$

(10)

Here we have $f(x, y) = 3x - 2y$ and $g(x, y) = 5x - 4y$ in (3), so the Euler iterative formulas in (4) are

$$x_{n+1} = x_n + h(3x_n - 2y_n), \qquad y_{n+1} = y_n + h(5x_n - 4y_n).$$

With step size $h = 0.1$ we calculate

$$x_1 = 3 + (0.1)[(3)(3) - (2)(6)] = 2.7,$$
$$y_1 = 6 + (0.1)[(5)(3) - (4)(6)] = 5.1$$

and

$$x_2 = 2.7 + (0.1)[(3)(2.7) - (2)(5.1)] = 2.49,$$
$$y_2 = 5.1 + (0.1)[(5)(2.7) - (4)(5.1)] = 4.41.$$

The actual values at $t_2 = 0.2$ given by (10) are $x(0.2) \approx 2.562$ and $y(0.2) \approx 4.573$.

To compute the improved Euler approximations to $x(0.2)$ and $y(0.2)$ with a single step of size $h = 0.2$, we first calculate the predictors

$$u_1 = 3 + (0.2)[(3)(3) - (2)(6)] = 2.4,$$
$$v_1 = 6 + (0.2)[(5)(3) - (4)(6)] = 4.2.$$

Then the corrector formulas in (8) yield

$$x_1 = 3 + (0.1)\{[(3)(3) - (2)(6)] + [(3)(2.4) - (2)(4.2)]\} = 2.58,$$
$$y_1 = 6 + (0.1)\{[(5)(3) - (4)(6)] + [(5)(2.4) - (4)(4.2)]\} = 4.62.$$

As we would expect, a single improved Euler step gives better accuracy than two ordinary Euler steps.

THE RUNGE-KUTTA METHOD AND SECOND ORDER EQUATIONS

The vector version of the iterative formula for the Runge-Kutta method is

$$\mathbf{x}_{n+1} = \mathbf{x}_n + \frac{h}{6} (\mathbf{k}_1 + 2\mathbf{k}_2 + 2\mathbf{k}_3 + \mathbf{k}_4)$$

(11)

where the vectors $\mathbf{k}_1$, $\mathbf{k}_2$, $\mathbf{k}_3$, $\mathbf{k}_4$ are defined (in analogy with Eqs. (5a)–(5d) of Section 6.3) as follows:

$$\mathbf{k}_1 = \mathbf{f}(t_n, \mathbf{x}_n)$$
$$\mathbf{k}_2 = \mathbf{f}(t_n + \tfrac{1}{2}h, \mathbf{x}_n + \tfrac{1}{2}h\mathbf{k}_1)$$
$$\mathbf{k}_3 = \mathbf{f}(t_n + \tfrac{1}{2}h, \mathbf{x}_n + \tfrac{1}{2}h\mathbf{k}_2) \tag{12}$$
$$\mathbf{k}_4 = \mathbf{f}(t_n + h, \mathbf{x}_n + h\mathbf{k}_3).$$

To describe in scalar notation the Runge-Kutta method for the two-dimensional initial value problem

$$x' = f(t, x, y), \qquad x(t_0) = x_0$$
$$y' = g(t, x, y), \qquad y(t_0) = y_0, \tag{3}$$

let us write

$$\mathbf{x} = \begin{bmatrix} x \\ y \end{bmatrix}, \qquad \mathbf{f} = \begin{bmatrix} f \\ g \end{bmatrix}, \quad \text{and} \quad \mathbf{k}_i = \begin{bmatrix} F_i \\ G_i \end{bmatrix}.$$

Then the Runge-Kutta iterative formulas for the step from (x_n, y_n) to the next approximation $(x_{n+1}, y_{n+1}) \approx (x(t_{n+1}), y(t_{n+1}))$ are

$$x_{n+1} = x_n + \frac{h}{6}(F_1 + 2F_2 + 2F_3 + F_4),$$

$$y_{n+1} = y_n + \frac{h}{6}(G_1 + 2G_2 + 2G_3 + G_4) \tag{13}$$

where the values F_1, F_2, F_3, F_4 of the function f are

$$F_1 = f(t_n, x_n, y_n)$$
$$F_2 = f(t_n + \tfrac{1}{2}h, x_n + \tfrac{1}{2}hF_1, y_n + \tfrac{1}{2}hG_1)$$
$$F_3 = f(t_n + \tfrac{1}{2}h, x_n + \tfrac{1}{2}hF_2, y_n + \tfrac{1}{2}hG_2) \tag{14}$$
$$F_4 = f(t_n + h, x_n + hF_3, y_n + hG_3);$$

G_1, G_2, G_3, and G_4 are the similarly defined values of the function g.

The program listed in Fig. 6.32 is a straightforward implementation of the two-dimensional Runge-Kutta method. Note that lines 260–330 carry out the computations in (14), whereas lines 340–350 carry out those in (13). The functions $f(t, x, y)$ and $g(t, x, y)$ that specify the system to be solved numerically must be edited into lines 150 and 160 of the program before its execution.

Perhaps the most common appplication of the two-dimensional Runge-Kutta method is to the numerical solution of second order initial value problems of the form

$$x'' = g(t, x, x'),$$

$$x(t_0) = x_0, \qquad x'(t_0) = y_0. \tag{15}$$

If we introduce the auxiliary variable $y = x'$, then the problem in (15) translates

```
100 REM--Program RK2DIM
110 REM
120 REM--Initialization:
130 REM
140       DEFDBL F-H,T-Y : DEFINT M-P
150       DEF FNF(T,X,Y) = Y
160       DEF FNG(T,X,Y) = - X
170       INPUT "Initial t"; T
180       INPUT "Initial x,y"; X,Y
190       INPUT "Step size h"; H
200       INPUT "Number of steps"; M
210       INPUT "Print step p"; P
220 REM
230 REM--Runge-Kutta iteration:
240 REM
250       FOR N = 1 TO M
260           F1 = FNF(T,X,Y)
270           G1 = FNG(T,X,Y)
280           F2 = FNF(T + H/2,X + H*F1/2,Y + H*G1/2)
290           G2 = FNG(T + H/2,X + H*F1/2,Y + H*G1/2)
300           F3 = FNF(T + H/2,X + H*F2/2,Y + H*G2/2)
310           G3 = FNG(T + H/2,X + H*F2/2,Y + H*G2/2)
320           F4 = FNF(T + H,   X + H*F3,   Y + H*G3)
330           G4 = FNG(T + H,   X + H*F3,   Y + H*G3)
340           X   = X + (H/6)*(F1 + 2*F2 + 2*F3 + F4)
350           Y   = Y + (H/6)*(G1 + 2*G2 + 2*G3 + G4)
360           T = T + H
370           IF INT(N/P) = N/P THEN PRINT T,X,Y
380       NEXT N
390 REM
400       END
```

FIGURE 6.32 Listing of Program RK2DIM.

into the two-dimensional first order problem

$$x' = y, \qquad x(t_0) = x_0$$
$$y' = g(t, x, y), \qquad y(t_0) = y_0. \tag{16}$$

This is a problem of the form in (3) with $f(t, x, y) = y$.

EXAMPLE 2 The exact solution of the initial value problem

$$x'' = -x; \qquad x(0) = 0, \qquad x'(0) = 1 \tag{17}$$

is $x(t) = \sin t$. The substitution $x' = y$ translates (17) into the two-dimensional problem

$$x' = y, \qquad x(0) = 0$$
$$y' = -x, \qquad y(0) = 1 \tag{18}$$

which has the form in (3) with $f(t, x, y) = y$ and $g(t, x, y) = -x$. The table in Fig. 6.33 shows the results produced for $0 \leq t \leq 5$ (radians) using Program RK2DIM with step size $h = 0.05$. The values shown for $x = \sin t$ and $y = \cos t$ are all accurate to five decimal places.

EXAMPLE 3 In Example 3 of Section 1.9 we considered a lunar lander that initially is falling freely toward the surface of the moon at a speed of 1000 mi/h. Its retrorockets, when fired in free space, provide a deceleration of 33,000 mi/h²; in addition, the lander is subject to the gravitational attraction of the moon. We found that a soft touchdown ($v = 0$ at impact) is achieved by firing the rockets

t	$x = \sin t$	$y = \cos t$
0.5	+0.47943	+0.87758
1.0	+0.84147	+0.54030
1.5	+0.99749	+0.07074
2.0	+0.90930	-0.41615
2.5	+0.59847	-0.80114
3.0	+0.14112	-0.98999
3.5	-0.35078	-0.93646
4.0	-0.75680	-0.65364
4.5	-0.97753	-0.21080
5.0	-0.95892	+0.28366

FIGURE 6.33 Runge-Kutta values (with $h = 0.05$) for the problem in Eq. (18).

t	x	v
0.004	+1.10080	-0.91788
0.008	+1.09730	-0.83610
0.012	+1.09412	-0.75462
0.016	+1.09126	-0.67342
0.020	+1.08873	-0.59248
0.024	+1.08652	-0.51175
0.028	+1.08463	-0.43122
0.032	+1.08307	-0.35086
0.036	+1.08183	-0.27062
0.040	+1.08090	-0.19049
0.044	+1.08030	-0.11044
0.048	+1.08002	-0.03043
0.052	+1.08006	+0.04958

FIGURE 6.34 The lander's descent to the lunar surface.

t	x	v
0.0482	+1.08001	-0.02643
0.0484	+1.08001	-0.02243
0.0486	+1.08001	-0.01843
0.0488	+1.08000	-0.01443
0.0490	+1.08000	-0.01043
0.0492	+1.08000	-0.00643
0.0494	+1.08000	-0.00243
0.0496	+1.08000	+0.00157
0.0498	+1.08000	+0.00557
0.0500	+1.08000	+0.00957

FIGURE 6.35 Focusing on the lunar lander's soft touchdown.

beginning at time $t = 0$ at a height of approximately 25 miles above the lunar surface.

Here we want to compute the *descent time* of the lunar lander. Let the distance $x(t)$ of the lander from the center of the moon be measured in kilomiles and measure time t in hours. According to the analysis in Section 1.9, $x(t)$ satisfies the initial value problem

$$\frac{d^2x}{dt^2} = 33 - \frac{15.1632}{x^2};$$

$$x(0) = 1.10464, \qquad x'(0) = -1. \tag{19}$$

(Actually, in Section 1.9 we used y where here we use x; we also use more precise values of the constants to take advantage of the accuracy of the Runge-Kutta method.) We seek the value of t when $x = 1.080$ (the radius of the moon is 1080 mi).

The problem in (19) is equivalent to the first order system

$$x' = y, \qquad\qquad x(0) = 1.10464;$$

$$y' = 33 - \frac{15.1632}{x^2}, \qquad y(0) = -1. \tag{20}$$

The table in Fig. 6.34 shows the results of a run of Program RK2DIM with step size $h = 0.002$. Evidently, touchdown ($x = 1.080$) occurs sometime between $t = 0.048$ and $t = 0.052$. The table in Fig. 6.35 shows the result of a second run with $t_0 = 0.048$, $x_0 = 1.08002$, $y_0 = v_0 = -0.03043$, and $h = 0.0002$. Now it is apparent that the lander's time of descent to the lunar surface is about $t = 0.0495$ h—approximately 2 min 58 s.

HIGHER ORDER SYSTEMS

As we saw in Section 5.1, any system of higher order differential equations can be replaced with an equivalent system of first order differential equations. For example, consider the system

$$x'' = F(t, x, y, x', y'),$$
$$y'' = G(t, x, y, x', y') \tag{21}$$

of two second order equations. If we define the two auxiliary functions $p(t) = x'(t)$ and $q(t) = y'(t)$, we get the equivalent system

$$x' = p$$
$$y' = q$$
$$p' = F(t, x, y, p, q)$$
$$q' = G(t, x, y, p, q) \tag{22}$$

of four first order equations in the unknown functions $x(t)$, $y(t)$, $p(t)$, and $q(t)$. It would be a routine matter to write a four-dimensional version of Program RK2DIM for the purpose of solving numerically such a system. Indeed, there are widely available software packages containing sophisticated Runge-Kutta programs that will accommodate an arbitrary number of input differential equations.

Such a general purpose program might be used, for example, to model numerically the major components of the solar system: the sun and the nine (known) major planets. If m_i denotes the mass and $\mathbf{r}_i = (x_i, y_i, z_i)$ denotes the position vector of the ith one of these 10 bodies, then by Newton's laws the equation of motion of m_i is

$$m_i \mathbf{r}_i'' = \sum_{j \neq i} \frac{Gm_i m_j}{r_{ij}^3} (\mathbf{r}_j - \mathbf{r}_i), \tag{23}$$

where $r_{ij} = |\mathbf{r}_j - \mathbf{r}_i|$ denotes the distance between m_i and m_j. For each $i = 1, 2, \ldots, 10$, the summation in (23) is over all values of $j \neq i$ from 1 to 10. The 10 vector equations in (23) constitute a system of 30 second order scalar equations, and the equivalent first order system consists of 60 differential equations in the coordinates and velocity components of the 10 major (known) bodies in the solar system. Mathematical models that involve this many (or more) differential equations—and require sophisticated software and hardware for their numerical analysis—are quite common in science, engineering, and applied technology.

Of course, many interesting applications—especially those in a personal computing environment—are best treated in an *ad hoc* manner utilizing the numerical methods introduced in this chapter. The following example illustrates the possibility of combining two different techniques in a single problem.

EXAMPLE 4   Suppose that a batted baseball starts at $x_0 = 0$, $y_0 = 0$ with initial velocity $v_0 = 160$ ft/s and with initial angle of inclination $\theta = 30°$. If air resistance is ignored, we find by the elementary methods of Section 1.2 that the baseball travels a (horizontal) distance of $400\sqrt{3}$ ft (approximately 693 ft) in 5 s before striking the ground. Now suppose that in additional to a downward gravitational acceleration ($g = 32$ ft/s^2), the baseball experiences an acceleration due to air resistance of $(0.0025)v^2$ feet per second per second, directed opposite to its instantaneous direction of motion. Determine how far the baseball will travel horizontally under these conditions.

SOLUTION   According to Problem 32 of Section 5.1, the equations of motion of the baseball are

$$\begin{aligned} x'' &= -cvx', \\ y'' &= -cvy' - g \end{aligned} \tag{24}$$

where $v = \sqrt{(x')^2 + (y')^2}$ is the speed of the ball, and where $c = 0.0025$ and $g = 32$ in fps units. We convert to a first order system as in (22) and thereby obtain

$$\begin{aligned} x' &= p \\ y' &= q \\ p' &= -cp\sqrt{p^2 + q^2} \\ q' &= -cq\sqrt{p^2 + q^2} - g, \end{aligned} \tag{25}$$

four first order equations with initial conditions

$$x_0 = y_0 = 0,$$

$$p_0 = 80\sqrt{3}, \qquad q_0 = 80.$$

Note that $p(t)$ and $q(t)$ are simply the x- and y-components of the velocity vector of the baseball, so that $v = \sqrt{p^2 + q^2}$.

To approximate the solution of the system in (25), we use a three-term Taylor series technique to update the values of x and y at each step, combined with a Runge-Kutta technique to update the values of p and q. Having reached the approximations x_n, y_n, p_n, q_n, we increment t_n by the step size h to move from these values to the approximations $x_{n+1}, y_{n+1}, p_{n+1}, q_{n+1}$. In the notation of Eqs. (13) and (14), we first compute the values

$$p'_n = \tfrac{1}{6}(F_1 + 2F_2 + 2F_3 + F_4),$$

$$q'_n = \tfrac{1}{6}(G_1 + 2G_2 + 2G_3 + G_4)$$

(26)

```
100 REM--Program BASEBALL
110 REM
120 REM--Initialization:
130 REM
140     DEFINT K-N : DEFDBL A-H,P-Z
150     G = 32  :  PI = 3.141593  :  C = .0025
160     DEF FNF(P,Q) = - C*P*SQR(P*P + Q*Q)
170     DEF FNG(P,Q) = - C*Q*SQR(P*P + Q*Q)  - G
180     INPUT "Initial t"; T
190     INPUT "Initial x,y"; X,Y
200     INPUT "Initial velocity"; V
210     INPUT "Initial angle (deg)"; A
220     INPUT "Step size h"; H
230     INPUT "Number of steps"; M
240     INPUT "Print step k"; K
250     PRINT T,X,Y,V,A
260     A = PI*A/180
270     P = V*COS(A)  :  Q = V*SIN(A)
280 REM
290 REM--Runge-Kutta iteration:
300 REM
310     FOR N = 1 TO M
320         F1 = FNF(P,Q)
330         G1 = FNG(P,Q)
340         F2 = FNF(P + H*F1/2,Q + H*G1/2)
350         G2 = FNG(P + H*F1/2,Q + H*G1/2)
360         F3 = FNF(P + H*F2/2,Q + H*G2/2)
370         G3 = FNG(P + H*F2/2,Q + H*G2/2)
380         F4 = FNF(P + H*F3,  Q + H*G3)
390         G4 = FNG(P + H*F3,  Q + H*G3)
400         DP = (F1 + 2*F2 + 2*F3 + F4)/6
410         DQ = (G1 + 2*G2 + 2*G3 + G4)/6
420         X   = X + P*H + .5*DP*H*H
430         Y   = Y + Q*H + .5*DQ*H*H
440         P   = P + DP*H
450         Q   = Q + DQ*H
460         T = T + H
470         IF INT(N/K) = N/K THEN PRINT
                T,X,Y,SQR(P*P + Q*Q),180*ATN(Q/P)/PI
480     NEXT N
490 REM
500     END
```

FIGURE 6.36 Listing of Program BASEBALL.

CHAPTER 6:   Numerical Methods

using the functions

$$f(p, q) = -cp\sqrt{p^2 + q^2},$$
$$g(p, q) = -cq\sqrt{p^2 + q^2} - q \tag{27}$$

that appear in (25). With these values of p_n' and q_n', the Runge-Kutta iteration is given by

$$p_{n+1} = p_n + hp_n',$$
$$q_{n+1} = q_n + hq_n'. \tag{28}$$

The three-term Taylor series formulas for x_{n+1} and y_{n+1} are then

$$x_{n+1} = x_n + hx_n' + \tfrac{1}{2}h^2 x_n'' = x_n + hp_n + \tfrac{1}{2}h^2 p_n',$$
$$y_{n+1} = y_n + hy_n' + \tfrac{1}{2}h^2 y_n'' = y_n + hq_n + \tfrac{1}{2}h^2 q_n'. \tag{29}$$

Figure 6.36 lists a program implementing the iterative computation just described. Observe that lines 400–410 correspond to the equations in (26), lines 420–430 correspond to the equations in (29), and lines 440–450 correspond to the equations in (28). For convenience in interpreting the results, the printed output at each selected step consists of the horizontal and vertical coordinates x and y of the baseball, its velocity v, and the angle of inclination α of its velocity vector (in degrees measured from the horizontal).

To test the validity of the program, it was first executed with step size $h = 0.1$ and with $c = 0$ (no air resistance) in line 150. The results, shown in Fig. 6.37, agree with the exact solution when $c = 0$. The ball travels a horizontal distance of $400\sqrt{3} \approx 692.82$ ft in exactly 5 s, having reached a maximum height of 100 ft after 2.5 s. Note also that it strikes the ground at the same angle and with the same speed as its initial angle and initial speed.

t	x	y	v	α
0.0	0.00	0.00	160.00	+30
0.5	69.28	36.00	152.63	+25
1.0	138.56	64.00	146.64	+19
1.5	207.85	84.00	142.21	+13
2.0	277.13	96.00	139.48	+7
2.5	346.41	100.00	138.56	+0
3.0	415.69	96.00	139.48	-7
3.5	484.97	84.00	142.21	-13
4.0	554.26	64.00	146.64	-19
4.5	623.54	36.00	152.63	-25
5.0	692.82	0.00	160.00	-30

FIGURE 6.37 The batted baseball with no air resistance ($c = 0$).

t	x	y	v	α
0.0	0.00	0.00	160.00	+30
0.5	63.25	32.74	127.18	+24
1.0	117.12	53.20	104.86	+17
1.5	164.32	63.60	89.72	+8
2.0	206.48	65.30	80.17	-3
2.5	244.61	59.23	75.22	-15
3.0	279.29	46.06	73.99	-27
3.5	310.91	26.41	75.47	-37
4.0	339.67	0.91	78.66	-46

FIGURE 6.38 The batted baseball with air resistance ($c = 0.0025$).

Figure 6.38 shows the results obtained with the fairly realistic value of $c = 0.0025$ for the air resistance for a batted baseball. To within a hundredth of a foot in either direction, the same results are obtained with step sizes $h = 0.05$ and $h = 0.025$. We now see that with air resistance the ball travels a distance well under

t	x	y	v	α
1.5	164.32	63.60	89.72	+8
1.6	173.12	64.60	87.40	+5
1.7	181.72	65.27	85.29	+3
1.8	190.15	65.60	83.39	+1
1.9	198.40	65.61	81.68	−1
2.0	206.48	65.30	80.17	−3
.	.	.	.	.
.	.	.	.	.
.	.	.	.	.
3.8	328.50	11.77	77.24	−42
3.9	334.14	6.45	77.93	−44
4.0	339.67	0.91	78.66	−46
4.1	345.10	−4.84	79.43	−47
4.2	350.42	−10.79	80.22	−49

—Apex

—Impact

FIGURE 6.39 The batted ball's apex and its impact with the ground.

400 ft in just over 4 s. The more refined data in Fig. 6.39 show that the ball now travels horizontally only about 340 ft and that its maximum height is only about 66 ft. As illustrated in Fig. 6.40, air resistance has converted a massive home run into a routine fly ball (if hit straightaway to center field). Note also that when the ball strikes the ground, it has slightly under *half* its initial speed (only about 79 ft/s) and is falling at a steeper angle (about 46°). Every baseball fan has observed empirically these aspects of the trajectory of a fly ball.

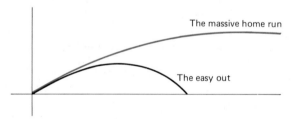

The massive home run

The easy out

FIGURE 6.40 An "easy out" or a home run?

6.4 PROBLEMS

A hand-held calculator will suffice for Problems 1–8. In each problem an initial value problem and its exact solution are given. Approximate the values of $x(0.2)$ and $y(0.2)$ in three ways: (a) by the Euler method with two steps of size $h = 0.1$, (b) by the improved Euler method with a single step of size $h = 0.2$, and (c) by the Runge-Kutta method with a single step of size $h = 0.2$. Compare the approximate values with the actual values $x(0.2)$ and $y(0.2)$.

1. $x' = x + 2y, \; x(0) = 0$
$y' = 2x + y, \; y(0) = 2$
$x(t) = e^{3t} - e^{-t}, \; y(t) = e^{3t} + e^{-t}$

2. $x' = 2x + 3y, \; x(0) = 1$
$y' = 2x + y, \; y(0) = -1$
$x(t) = e^{-t}, \; y(t) = -e^{-t}$

3. $x' = 3x + 4y, \; x(0) = 1$
$y' = 3x + 2y, \; y(0) = 1$
$x(t) = \frac{1}{7}(8e^{6t} - e^{-t}), \; y(t) = \frac{1}{7}(6e^{6t} + e^{-t})$

4. $x' = 9x + 5y, \; x(0) = 1$
$y' = -6x - 2y, \; y(0) = 0$
$x(t) = -5e^{3t} + 6e^{4t}, \; y(t) = 6e^{3t} - 6e^{4t}$

5. $x' = 2x - 5y$, $x(0) = 2$
$y' = 4x - 2y$, $y(0) = 3$
$x(t) = 2 \cos 4t - \frac{11}{4} \sin 4t$, $y(t) = 3 \cos 4t + \frac{1}{2} \sin 4t$

6. $x' = x - 2y$, $x(0) = 0$
$y' = 2x + y$, $y(0) = 4$
$x(t) = -4e^t \sin 2t$, $y(t) = 4e^t \cos 2t$

7. $x' = 3x - y$, $x(0) = 2$
$y' = x + y$, $y(0) = 1$
$x(t) = (t + 2)e^{2t}$, $y(t) = (t + 1)e^{2t}$

8. $x' = 5x - 9y$, $x(0) = 0$
$y' = 2x - y$, $y(0) = -1$
$x(t) = 3e^{2t} \sin 3t$, $y(t) = e^{2t}(\sin 3t - \cos 3t)$

A computer will be required for the remaining problems in this section. In Problems 9–12, an initial value problem and its exact solution are given. In each of these four problems, use the Runge-Kutta method with step sizes $h = 0.1$ and $h = 0.05$ to approximate (to five decimal places) the values $x(1)$ and $y(1)$. Compare the approximate values with the actual values.

9. $x' = 2x - y$, $x(0) = 1$
$y' = x + 2y$, $y(0) = 0$
$x(t) = e^{2t} \cos t$, $y(t) = e^{2t} \sin t$

10. $x' = x + 2y$, $x(0) = 0$
$y' = x + e^{-t}$, $y(0) = 0$
$x(t) = \frac{1}{9}(2e^{2t} - 2e^{-t} - 6te^{-t})$, $y(t) = \frac{1}{9}(e^{2t} - e^{-t} + 6te^{-t})$

11. $x' = -x + y - (1 + t^3)e^{-t}$, $x(0) = 0$
$y' = -x - y - (t - 3t^2)e^{-t}$, $y(0) = 1$
$x(t) = e^{-t}(\sin t - t)$, $y(t) = e^{-t}(\cos t + t^3)$

12. $x'' + x = \sin t$, $x(0) = x'(0) = 0$
$x(t) = \frac{1}{2}(\sin t - t \cos t)$

13. Suppose that a crossbow bolt is shot straight upward with initial velocity 288 ft/s. If its deceleration due to air resistance is $(0.04)v$, then its height $x(t)$ satisfies the initial value problem

$$x'' = -32 - (0.04)x'; \qquad x(0) = 0, \qquad x'(0) = 288.$$

Find the maximum height that the bolt attains and the time required for it to reach this height.

14. Repeat Problem 13, but assume that the deceleration of the bolt due to air resistance is $(0.0002)v^2$.

15. Suppose that a projectile is fired straight upward with initial velocity v_0 from the surface of the earth. Then its height $x(t)$ at time t satisfies the initial value problem

$$\frac{dx^2}{dt^2} = -\frac{gR^2}{(x + R)^2}; \qquad x(0) = 0, \qquad x'(0) = v_0.$$

Use the values $g = 32.15$ ft/s$^2 \approx 0.006089$ mi/s^2 for the gravitational acceleration of the earth at its surface and $R = 3960$ mi as the radius of the earth. If $v_0 = 1$ mi/s, find the maximum

height attained by the projectile and its time of ascent to this height.

Problems 16–18 deal with the batted baseball of Example 4, having initial velocity 160 ft/s and air resistance coefficient $c = 0.0025$.

16. Find the *range*—the horizontal distance the ball travels before it hits the ground—and its total time of flight with initial inclination angles 40°, 45°, and 50°.

17. Find (to the nearest degree) the initial inclination that maximizes the range. If there were no air resistance it would be exactly 45°, but your answer should be less than 45°.

18. Find (to the nearest half degree) the initial inclination angle greater than 45° for which the range is 300 ft.

19. Find the initial velocity of a baseball hit by Babe Ruth (with $c = 0.0025$ and initial inclination 40°) if it hit the bleachers at a point 50 ft high and 500 horizontal feet from home plate.

20. Consider the crossbow bolt of Problem 14, fired with the same initial velocity of 288 ft/s and with the air resistance deceleration $(0.0002)v^2$ directed opposite its direction of motion. Suppose that this bolt is fired from ground level at an initial angle of 45°. Find how high vertically and how far horizontally it goes, and how long it remains in the air.

21. Suppose that an artillery projectile is fired from ground level with initial velocity 3000 ft/s and initial inclination angle 40°. Assume that its air resistance deceleration is $(0.0001)v^2$. (a) What is the range of the projectile and what is its total time of flight? What is its speed at impact with the ground? (b) What is the maximum altitude of the projectile, and when is that altitude attained? (c) You will find that the projectile is still losing speed at the apex of its trajectory. What is the *minimum* speed it attains during its descent?

22. This problem is a project to investigate numerically Kepler's laws of planetary (or satellite) motion. Consider a satellite in elliptical orbit about a planet of mass M, and suppose that physical units are chosen so that $GM = 1$ (where G is the gravitational constant). If the planet is located at the origin in the xy-plane, then the equations of motion of the satellite are

$$\frac{d^2x}{dt^2} = -\frac{x}{(x^2 + y^2)^{3/2}}, \qquad \frac{d^2y}{dt^2} = -\frac{y}{(x^2 + y^2)^{3/2}}. \qquad (30)$$

Let T denote the period of revolution of the satellite. Kepler's third law says that the *square* of T is proportional to the *cube* of the major semiaxis a of its elliptical orbit. In particular, if $GM = 1$, then

$$T^2 = 4\pi^2 a^3. \qquad (31)$$

(For details, see Section 13-7 of Edwards and Penney, *Calculus and Analytic Geometry*, 2nd ed. (Englewood Cliffs, N.J.:

Prentice-Hall, 1986).) (a) If the satellite's x- and y-components of velocity, $p = x'$ and $q = y'$, are introduced, then the system in (30) translates into a system of four first order differential equations of the form of those in Eq. (22) of this section. Alter Program BASEBALL to obtain a program—call it KEPLER— to solve this 4×4 system when the user enters the initial conditions

$$x(0) = x_0, \qquad y(0) = y_0,$$

$$p(0) = p_0, \qquad q(0) = q_0,$$

and the step size h. (b) Test Program KEPLER by running it with the initial conditions

$$x_0 = 1, \qquad y_0 = 0, \qquad p_0 = 0, \qquad q_0 = 1$$

that correspond to a circular orbit of radius $a = 1$, so Eq. (30) gives $T = 2\pi$. With step size $h = \pi/1000$ your results should show reasonable agreement with the following values.

t	x	y
$\pi/4$	$\frac{1}{2}\sqrt{2}$	$\frac{1}{2}\sqrt{2}$
$\pi/2$	0	1
$3\pi/4$	$-\frac{1}{2}\sqrt{2}$	$\frac{1}{2}\sqrt{2}$
π	-1	0

(c) Test Program KEPLER by running it with the initial conditions

$$x_0 = 1, \qquad y_0 = 0,$$

$$p_0 = 0, \qquad q_0 = \frac{1}{2}\sqrt{6}$$

that correspond to an elliptical orbit with major semiaxis $a = 2$, so (30) gives $T = 4\pi\sqrt{2}$. After 2000 steps of size $h = (\pi\sqrt{2})/1000$, the satellite should be near the point $(-3, 0)$.

CHAPTER *7*

QUALITATIVE PROPERTIES AND EXISTENCE OF SOLUTIONS

7.1 INTRODUCTION TO STABILITY

7.2 STABILITY AND THE PHASE PLANE

7.3 LINEAR AND ALMOST LINEAR SYSTEMS

7.4 NONLINEAR MECHANICAL SYSTEMS

7.5 ECOLOGICAL APPLICATIONS—PREDATORS AND COMPETITORS

7.6 EXISTENCE AND UNIQUENESS OF SOLUTIONS

Introduction to Stability

It is often difficult, if not impossible, to solve a given differential equation explicitly, especially one that is nonlinear. Therefore, it is important to determine whether qualitative information about the solutions of a differential equation can be obtained without the necessity of obtaining an explicit solution. For example, we may be able to establish that every solution $x(t)$ grows without bound as $t \to +\infty$, or approaches a finite limit, or is a periodic function of t. In this section we introduce—by consideration of simple differential equations that *can* be solved explicitly—some of the more important qualitative questions that can sometimes be answered for less tractable equations.

The question of whether a population $x(t)$ is bounded or unbounded as $t \to +\infty$ is of evident interest. In Section 1.8 we introduced the general population equation

$$\frac{dx}{dt} = (\beta - \delta)x \tag{1}$$

where β and δ are the birth and death rates, respectively, in births or deaths per individual per unit of time. If β and δ are constants, we have the case of natural population growth, and the solution of Eq. (1) is

$$x(t) = x_0 e^{kt}$$

where $k = \beta - \delta$ and $x_0 = x(0)$.

If $k > 0$, then it is clear that $x(t) \to +\infty$ as $t \to +\infty$ if $x_0 > 0$, while $x(t) \to -\infty$ if $x_0 < 0$. (Although an actual population would not be negative, we can discuss the initial value problem $x' = kx$, $x(0) = x_0 < 0$.) On the other hand, if $k < 0$, then $x(t) \to 0$ as $t \to +\infty$ for every value of x_0. In either case we see that, given k, the qualitative behavior as $t \to +\infty$ of a solution $x(t)$ of $x' = kx$ is determined by the initial condition $x(0) = x_0$.

In more interesting situations β and δ are (known) functions of x. Then Eq. (1) takes the form

$$\frac{dx}{dt} = f(x). \tag{2}$$

This is an **autonomous** first order differential equation—one in which the independent variable t does not appear explicitly. A **critical point** of Eq. (2) is a root of $f(x) = 0$. If $x = c$ is a critical point of (2), then the differential equation has the constant solution $x(t) \equiv c$. A constant solution of a differential equation is sometimes called an **equilibrium solution** (one may think of a population that remains constant because it is in "equilibrium" with its environment). Thus the critical point $x = c$, a number, corresponds to the equilibrium solution $x(t) \equiv c$, a constant-valued function.

The following example illustrates the fact that the qualitative behavior (as t increases) of the solutions of an autonomous first order equation can be described in terms of its critical points.

EXAMPLE 1 Suppose that the death rate $\delta = \delta_0$ in Eq. (1) is constant but that, because of cultural sophistication or a limited food supply or for some other reason, the birth rate is a linearly decreasing function $\beta = \beta_0 - \beta_1 x$ of the population x. Then Eq. (1) becomes the autonomous equation

$$\frac{dx}{dt} = ax - bx^2 \tag{3}$$

with $a = \beta_0 - \delta_0$ and $b = \beta_1$; we assume that a and b are each positive. When we rewrite (3) in the form

$$\frac{dx}{dt} = bx(M - x) \qquad \left(M = \frac{a}{b} > 0 \right), \tag{4}$$

we recognize it as the **logistic equation** that we introduced in Section 1.8 in connection with limited population growth.

For a specific numerical example let us take $a = 4$ and $b = 1$ in (3), so our differential equation is

$$\frac{dx}{dt} = 4x - x^2. \tag{5}$$

It has two critical points—the roots $x = 0$ and $x = 4$ of the equation

$$f(x) = 4x - x^2 = x(4 - x) = 0.$$

By separating the variables in (5) we readily find the solution

$$x(t) = \frac{4x_0}{x_0 + (4 - x_0)e^{-4t}} \tag{6}$$

that satisfies the initial condition $x(0) = x_0$. Note that (6) includes the equilibrium solutions $x(t) \equiv 0$ (if $x_0 = 0$) and $x(t) \equiv 4$ (if $x_0 = 4$).

If $x_0 > 0$, then the denominator on the right-hand side in (6) is positive for all $t > 0$ and approaches 4 as $t \to +\infty$, so

$$\lim_{t \to \infty} x(t) = 4 \qquad \text{if } x_0 > 0. \tag{7a}$$

That is, if $x_0 > 0$, then $x(t)$ approaches the critical point $x = 4$ as $t \to +\infty$. On the other hand, if $x_0 < 0$, then the denominator in (6) is initially positive but approaches zero as

$$t \to t_0 = \frac{1}{4} \ln \frac{x_0 - 4}{x_0} > 0,$$

so

$$\lim_{t \to t_0^-} x(t) = -\infty \qquad \text{if } x_0 < 0. \tag{7b}$$

From (7a) and (7b) we see that the solution curves of Eq. (5) appear as shown in Fig. 7.1.

This figure illustrates the concept of stability. A critical point $x = c$ of an autonomous first order equation is called *stable* provided that, if the initial value

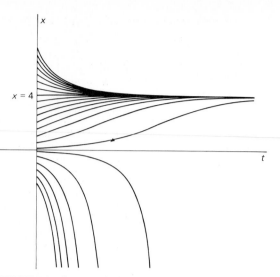

FIGURE 7.1 Some solution curves of $x' = 4x - x^2$.

x_0 is sufficiently close to c, then $x(t)$ is close to c for all $t > 0$. More precisely, the critical point c is **stable** if, given $\epsilon > 0$, there exists $\delta > 0$ such that

$$|x_0 - c| < \delta \quad \text{implies} \quad |x(t) - c| < \epsilon \tag{8}$$

for all $t > 0$. The critical point $x = c$ is **unstable** if it is not stable.

In Fig. 7.1 we see that solution curves of $x' = 4x - x^2$ that start near $x = 4$ remain near $x = 4$, so the critical point $x = 4$ is stable; indeed, the condition in (8) is satisfied with $\delta = \epsilon$. In contrast, solution curves that start near $x = 0$ do not remain near $x = 0$, so the critical point $x = 0$ is unstable.

Remark: The critical point $x = 4$ in Example 1 not only has the property that $x(t)$ remains close to 4 if x_0 is sufficiently close to 4, it has the stronger property that $x(t)$ *approaches* 4 (as $t \to +\infty$) if x_0 is sufficiently close to 4. The stable critical point $x = c$ is said to be **asymptotically stable** if there exists $\delta > 0$ such that

$$|x_0 - c| < \delta \quad \text{implies} \quad \lim_{t \to \infty} x(t) = c. \tag{8'}$$

In Section 7.2 we will see examples of critical points that are stable but not asymptotically stable.

In Problem 14 we ask you to carry out the analysis of Example 1 for the original logistic population equation

$$\frac{dx}{dt} = ax - bx^2 \tag{3}$$

with $a, b > 0$. The critical point $x = M = a/b$ is stable (M is the *limiting population*), whereas the critical point $x = 0$ is unstable. The solution curves of (3) look just like those shown in Fig. 7.1 (with the equilibrium solution $x(t) \equiv 4$ replaced with $x(t) \equiv M = a/b$). We can summarize the behavior of solutions of (3)—in terms of

0            $M = \dfrac{a}{b}$ x

Unstable   Stable

FIGURE 7.2   The phase portrait shows a stable critical point and an unstable critical point.

their initial values—by means of the *phase portrait* shown in Fig. 7.2. It indicates that $x(t) \to M$ as $t \to +\infty$ if either $x_0 > M$ or $0 < x_0 < M$, whereas $x(t) \to -\infty$ as t increases if $x_0 < 0$. The fact that M is a stable critical point would be important, for instance, if we wished to conduct an experiment with a population of M bacteria. It is impossible to count precisely M bacteria for M large, but any initially positive population will approach M with increasing t.

Here is another aspect of stability. The coefficients a and b in Eq. (3) are unlikely to be known precisely for an actual population. But if they are replaced with close approximations a^* and b^*, then the approximate limiting population $M^* = a^*/b^*$ will be close to the actual limiting population $M = a/b$. Thus the limiting population that Eq. (3) predicts is stable with respect to small perturbations of its constant coefficients.

EXAMPLE 2   The autonomous differential equation

$$\frac{dx}{dt} = ax - bx^2 - k \tag{9}$$

(with a, b, and k all positive) may be considered to describe a logistic population *with harvesting*. For instance, we might think of the population of fish in a lake from which k fish per year are removed by fishing. If $a = 4$ and $b = 1$, as in Example 1, and $k = 3$, then Eq. (9) is

$$\frac{dx}{dt} = 4x - x^2 - 3 \tag{10}$$

with

$$f(x) = 4x - x^2 - 3 = -(x - 1)(x - 3),$$

so it has the two critical points $x = 1$ and $x = 3$.

Separation of variables in (10) yields the solution

$$x(t) = \frac{3(x_0 - 1) - (x_0 - 3)e^{-2t}}{(x_0 - 1) - (x_0 - 3)e^{-2t}} \tag{11}$$

in terms of the initial value $x(0) = x_0$. In order to determine the stability of the two critical points $x = 1$ and $x = 3$, we analyze the behavior of $x(t)$ for different initial values—in the open intervals $x_0 > 3$, $1 < x_0 < 3$, and $x_0 < 1$ into which the critical points separate the x-axis.

If $x_0 > 3$, then each of the quantities within parentheses in (11) is positive. Hence it follows (because of the negative exponents) that

$$\lim_{t \to \infty} x(t) = 3 \qquad \text{if } x_0 > 3. \tag{12}$$

If $1 < x_0 < 3$, then we write (11) in the form

$$x(t) = \frac{(3x_0 - 3) + (3 - x_0)e^{-2t}}{(x_0 - 1) + (3 - x_0)e^{-2t}} \tag{11'}$$

with each of the quantities within parentheses positive. Now it also follows that

$$\lim_{t \to \infty} x(t) = 3 \qquad \text{if } 1 < x_0 < 3. \qquad (12')$$

If $x_0 < 1$, then we write (11) in the form

$$x(t) = \frac{(3 - x_0)e^{-2t} - (3 - 3x_0)}{(3 - x_0)e^{-2t} - (1 - x_0)} \qquad (11'')$$

with each quantity within parentheses positive. The numerator and denominator in (11'') both are positive initially, but the denominator approaches zero as

$$t \to t_0 = \frac{1}{2}\ln\frac{3 - x_0}{1 - x_0},$$

and the numerator approaches zero as

$$t \to t_1 = \frac{1}{2}\ln\frac{3 - x_0}{3(1 - x_0)} < t_0.$$

Hence it follows that

$$\lim_{t \to t_0} x(t) = -\infty \qquad \text{if } x_0 < 1. \qquad (12'')$$

From (12), (12'), and (12'') we see that the solution curves of Eq. (10) appear as shown in Fig. 7.3. Hence we conclude that the critical point $x = 3$ is stable, whereas the critical point $x = 1$ is unstable.

For a concrete application of our stability conclusions in Example 2, suppose that $x(t)$ in Eq. (10) denotes the fish population in hundreds after t years

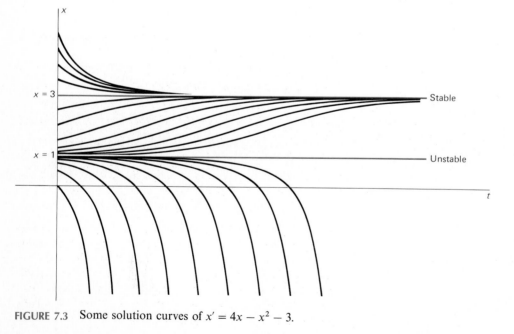

FIGURE 7.3 Some solution curves of $x' = 4x - x^2 - 3$.

CHAPTER 7: Qualitative Properties and Existence of Solutions

(so 300 fish per year are removed by fishing). If the lake is stocked initially with more than 100 fish, then as t increases the fish will approach a stable population of 300 fish. But if the lake is stocked initially with fewer than 100 fish, the lake will be "fished out" and the fish will disappear in a finite period of time.

7.1 PROBLEMS

In each of Problems 1–12, first solve explicitly for $x(t)$ in terms of t and the initial value $x_0 = x(0)$. Then construct a sketch like the ones in Figs. 7.1 and 7.3, showing the nature of the trajectories for a wide variety of possible values of x_0. Finally, determine the stability or instability of each critical point.

1. $\dfrac{dx}{dt} = -x$

2. $\dfrac{dx}{dt} = x - 2$

3. $\dfrac{dx}{dt} = x - x^2$

4. $\dfrac{dx}{dt} = x^2 - 2x$

5. $\dfrac{dx}{dt} = x + x^2$

6. $\dfrac{dx}{dt} = (x - 1)^2$

7. $\dfrac{dx}{dt} = 1 - x^2$

8. $\dfrac{dx}{dt} = x^2 - 4$

9. $\dfrac{dx}{dt} = 2 - x - x^2$

10. $\dfrac{dx}{dt} = x - x^3$

11. $\dfrac{dx}{dt} = x^2 - 5x + 6$

12. $\dfrac{dx}{dt} = (2 - x)^3$

13. Consider the equation

$$\frac{dx}{dt} = kx - x^3.$$

(a) Show that the single critical point $x = 0$ is stable if $k \leq 0$.
(b) Suppose that $k > 0$. Show that the critical point $x = 0$ is now unstable, but that the critical points $x = \pm\sqrt{k}$ are stable. The value $k = 0$ of the parameter, for which the qualitative nature of the solutions of $x' = kx - x^3$ changes as k increases, is called a **bifurcation point** for the differential equation.

14. Solve the logistic equation

$$\frac{dx}{dt} = ax - bx^2, \qquad x(0) = x_0$$

explicitly (with $a, b > 0$) to show that the critical point $x = 0$ is unstable whereas the critical point $x = a/b$ is stable.

15. Consider the initial value problem

$$\frac{dx}{dt} = bx^2 - ax, \qquad x(0) = x_0$$

(with $a, b > 0$) for a population of unsophisticated animals in which deaths occur at a natural linear rate, but births occur as a result of chance encounters (as in the "doomsday versus extinction" example of Section 1.8). Show that the critical point $x = 0$ is stable while the critical point $x = a/b$ is unstable (the reverse of the situation for the logistic equation of Problem 14).

16. Suppose that the equation $x' = ax - bx^2$ of Example 1 describes the fish population $x(t)$ in a lake after t months in which no fishing occurs. Now suppose that, because of fishing, fish are removed from the lake at the rate of kx fish per month.
(a) What is the new limiting population in the case $0 < k < a$?
(b) Show that if $k \geq a$, then $x(t) \to 0$ as $t \to +\infty$.

Problems 17–19 deal with the phenomenon of harvesting a logistic population (as illustrated in Example 2 of this section). If k is the (constant) rate at which individuals are removed from the population by harvesting, then the differential equation is

$$\frac{dx}{dt} = ax - bx^2 - k. \tag{13}$$

The long-range situation depends on the values of the positive parameters a, b, and k in (13).

17. Suppose that $4bk = a^2$. Show that $x(t) \to a/2b$ (half the limiting population in the logistic case $k = 0$) as $t \to +\infty$.

18. Suppose that $4bk > a^2$. Show that $x(t) = 0$ after a finite period of time. In this case the harvesting rate k is too large, and extinction of the population results.

19. Suppose that $4bk < a^2$. Show that $x(t)$ approaches the limiting population

$$M_k = \frac{a + (a^2 - 4bk)^{1/2}}{2b}$$

as $t \to +\infty$. Note that $M_k \to M = a/b$ as $k \to 0$. (*Suggestion:* Complete the square on the right-hand side in (13). The results of this and the previous two problems show that when the harvesting rate k is gradually increased, the limiting population M_k gradually decreases to half its original ($k = 0$) value $M = a/b$, until k reaches the critical value $k = a^2/4b$, above which extinction of the population occurs.)

20. This problem deals with the initial value problem

$$\frac{dx}{dt} = bx^2 - ax - k, \qquad x(0) = x_0 \qquad (14)$$

(with $a, b, k > 0$) that describes the harvesting of an unsophisticated population (such as alligators) of the type mentioned in Problem 15. The two critical points of the equation in (14) are

$$\alpha = \frac{a + \sqrt{a^2 + 4bk}}{2b} > 0$$

and

$$\beta = \frac{a - \sqrt{a^2 + 4bk}}{2b} < 0.$$

Show that the critical point $x = \alpha$ is unstable, whereas $x = \beta$ is stable. In particular, show that if $x_0 > \alpha$, then $x(t) \to +\infty$ in a finite period of time, whereas if $x_0 < \alpha$, then $x(t) \to 0$ in a finite period of time. Thus $x_0 = \alpha$ is a threshold value that increases when the harvesting rate k is increased—naturally, the more alligators are killed annually, the greater must be their initial number to prevent extinction from resulting.

7.2

Stability and the Phase Plane

An autonomous second order differential equation is one of the form

$$x'' = G(x, x') \qquad (1)$$

in which the independent variable t does not appear explicitly. If we introduce the new dependent variable $y = x'$, we obtain the equivalent system

$$\left.\begin{aligned} \frac{dx}{dt} &= y, \\[2mm] \frac{dy}{dt} &= G(x, y) \end{aligned}\right\} \qquad (2)$$

—that is, two first order equations neither of which involves explicitly the independent variable t. This system is a special case of the general **autonomous system** of two first order equations:

$$\left.\begin{aligned} \frac{dx}{dt} &= F(x, y), \\[2mm] \frac{dy}{dt} &= G(x, y); \end{aligned}\right\} \qquad (3)$$

again, the independent variable t does not appear explicitly. We assume that the functions F and G are continuously differentiable in some region R in the xy-plane, which is called the **phase plane** for the system in (3). Then, according to the existence and uniqueness theorems of Section 7.6, given t_0 and any point (x_0, y_0) of R, there is a *unique* solution $x = x(t)$, $y = y(t)$ of (3) that is defined on some open interval $a < t_0 < b$ and satisfies the initial conditions

$$x(t_0) = x_0, \qquad y(t_0) = y_0. \qquad (4)$$

The equations $x = x(t)$, $y = y(t)$ then describe a parametrized solution curve in the phase plane. Any such solution curve is called a **trajectory** of the system in (3), and precisely one trajectory passes through each point of the region R (see

Problem 29). A **critical point** of the system in (4) is a point (x_*, y_*) such that

$$F(x_*, y_*) = G(x_*, y_*) = 0. \tag{5}$$

If (x_*, y_*) is a critical point of the system, then the constant-valued functions

$$x(t) \equiv x_*, \qquad y(t) \equiv y_* \tag{6}$$

satisfy the equations in (3). Such a constant-valued solution is called an **equilibrium solution** of the system. Note that the trajectory of the equilibrium solution in (5) consists of the single point (x_*, y_*).

In some practical situations these very simple solutions and trajectories are the ones of most interest. For example, suppose that the system $x' = F(x, y)$, $y' = G(x, y)$ models two populations $x(t)$ and $y(t)$ of animals that cohabit the same environment, and perhaps compete for the same food or prey on one another; $x(t)$ might denote the number of foxes and $y(t)$ the number of rabbits present at time t. Then a critical point (x_*, y_*) of the system specifies a *constant* population x_* of foxes and a *constant* population y_* of rabbits that can coexist with one another in the environment. If (x_0, y_0) is *not* a critical point of the system, it is *not* possible for constant populations of x_0 foxes and y_0 rabbits to coexist; one or both must change with time.

EXAMPLE 1   Find the critical points of the system

$$\left.\begin{array}{l} x' = 60x - 3x^2 - 4xy, \\ y' = 42y - 3y^2 - 2xy. \end{array}\right\} \tag{7}$$

SOLUTION   When we look at the equations

$$60x - 3x^2 - 4xy = x(60 - 3x - 4y) = 0,$$

$$42y - 3y^2 - 2xy = y(42 - 3y - 2x) = 0$$

that a critical point (x, y) must satisfy, we see that either

$$x = 0 \quad \text{or} \quad 60 - 3x - 4y = 0 \tag{8a}$$

and either

$$y = 0 \quad \text{or} \quad 42 - 3y - 2x = 0. \tag{8b}$$

If $x = 0$ and $y \neq 0$, then the second equation in (8b) gives $y = 14$. If $y = 0$ and $x \neq 0$, the second equation in (8a) gives $x = 20$. If x and y are each nonzero, we solve the simultaneous equations

$$3x + 4y = 60, \qquad 2x + 3y = 42$$

for $x = 12$, $y = 6$. Thus the system in (7) has the four critical points $(0, 0)$, $(0, 14)$, $(20, 0)$, and $(12, 6)$. If $x(t)$ and $y(t)$ denote the number of foxes and the number of rabbits, respectively, and if both populations are *constant*, it follows that the equations in (7) allow only three nontrivial possibilities: either 0 foxes and 14 rabbits, or 20 foxes and 0 rabbits, or 12 foxes and 6 rabbits. In particular, the critical point $(12, 6)$ describes the *only* possibility for the coexistence of constant nonzero populations of both species.

If the initial point (x_0, y_0) is not a critical point, the corresponding trajectory is a curve in the xy-plane along which the point $(x(t), y(t))$ moves as t increases. It turns out that any trajectory not consisting of a single point is a nondegenerate curve with no self-intersections (see Problem 30). We can demonstrate qualitatively the behavior of solutions of the system in (3) by constructing its **phase portrait**—a phase plane picture of its critical points and typical nondegenerate trajectories. The behavior of the trajectories near an isolated critical point is of particular interest. In the remainder of this section we illustrate with simple examples some of the most common possibilities.

EXAMPLE 2   Consider the autonomous linear system

$$\left. \begin{array}{l} x' = -x, \\[2mm] y' = -ky \quad (k \text{ constant}) \end{array} \right\} \tag{9}$$

that has the origin $(0, 0)$ as its only critical point. The solution with initial point (x_0, y_0) is

$$x(t) = x_0 e^{-t}, \qquad y(t) = y_0 e^{-kt}. \tag{10}$$

If $x_0 \neq 0$, we can write

$$y = y_0 e^{-kt} = \frac{y_0}{x_0^k} (x_0 e^{-t})^k = bx^k \tag{11}$$

where $b = y_0 / x_0^k$.

If $k = 1$, then each curve in (11) is a straight line through the origin. Each trajectory is an open ray along which the point $(x(t), y(t)) = (x_0 e^{-t}, y_0 e^{-t})$ approaches the origin as $t \to +\infty$. This type of critical point, shown in Fig. 7.4, is

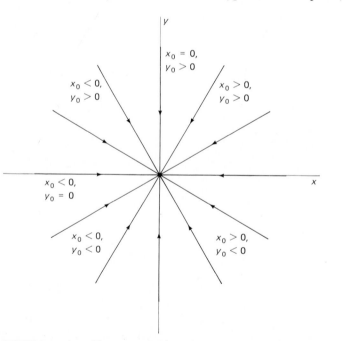

FIGURE 7.4   A stable proper node.

CHAPTER 7:   Qualitative Properties and Existence of Solutions

called a **proper node**. Note the arrows that indicate the orientations of the trajectories; they point in the direction of increasing t.

In general, the critical point (x_0, y_0) of the autonomous system in (3) is called a **node** provided that *either* every trajectory approaches (x_0, y_0) as $t \to +\infty$ *or* every trajectory recedes from (x_0, y_0) as $t \to +\infty$, *and* every trajectory is tangent at (x_0, y_0) to some straight line through the critical point. Note that the critical point $(0, 0)$ in Fig. 7.4—where the trajectories *are* straight lines, not merely tangents to straight lines—would still be a node if all the arrows were reversed to make the trajectories recede from the critical point rather than approach it. This node is said to be *proper* because no two different pairs of "opposite" trajectories are tangent to the same straight line. In the following paragraph we consider a case in which the critical point is a node that is not proper.

If $k = 2$ and neither x_0 nor y_0 is zero in (11), then each curve is a parabola $y = bx^2$ tangent to the x-axis at the origin. The solution curve in (10) is half of the x-axis if $y_0 = 0$, and half of the y-axis if $x_0 = 0$. The trajectories are the semiaxes and the right and left halves of the parabolas shown in Fig. 7.5. Along each trajectory the point $(x(t), y(t))$ approaches the origin as $t \to +\infty$. Thus all trajectories except for a single pair approach the origin tangent to the same line—the x-axis. This sort of critical point is called an **improper node**.

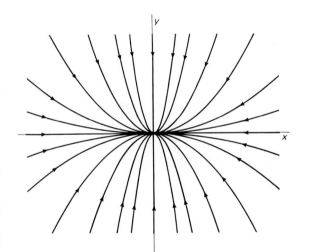

FIGURE 7.5   A stable improper node.

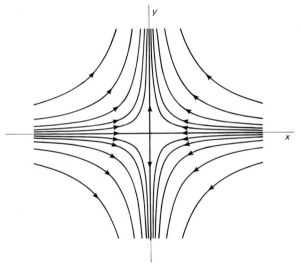

FIGURE 7.6   A saddle point.

If $k = -1$, then $x(t) = x_0 e^{-t}$ and $y(t) = y_0 e^{t}$, so $xy = x_0 y_0 = b$. If neither x_0 nor y_0 is zero, the trajectory is one branch of the rectangular hyperbola $xy = b$, and $y \to \pm\infty$ as $t \to +\infty$. If $x_0 = 0$ or $y_0 = 0$, the trajectory is a semiaxis of the hyperbola. The point $(x(t), y(t))$ approaches the origin along the x-axis, but recedes from it along the y-axis as $t \to +\infty$. Thus there are two trajectories that approach the critical point $(0, 0)$, and all others are unbounded as $t \to +\infty$. This type of critical point, shown in Fig. 7.6, is called a **saddle point**.

A critical point (x_*, y_*) of the autonomous system in (3) is said to be *stable* provided that if the initial point (x_0, y_0) is sufficiently close to (x_*, y_*), then $(x(t), y(t))$ remains close to (x_*, y_*) for all $t > 0$. In vector notation, with $\mathbf{x}(t) = (x(t), y(t))$, the distance between the initial point $\mathbf{x}_0 = (x_0, y_0)$ and the critical point $\mathbf{x}_* = (x_*, y_*)$ is

$$|\mathbf{x}_0 - \mathbf{x}_*| = [(x_0 - x_*)^2 + (y_0 - y_*)^2]^{1/2}.$$

Then the critical point $\mathbf{x}_*$ is **stable** provided that, for each $\epsilon > 0$, there exists $\delta > 0$ such that

$$|\mathbf{x}_0 - \mathbf{x}_*| < \delta \quad \text{implies} \quad |\mathbf{x}(t) - \mathbf{x}_*| < \epsilon \tag{12}$$

for all $t > 0$. The critical point (x_*, y_*) is called **unstable** if it is not stable. In each of Figs. 7.4 and 7.5, the origin $(0, 0)$ is a stable critical point. The saddle point at $(0, 0)$ shown in Fig. 7.6 is an unstable critical point.

If the signs on the right-hand sides in (9) are changed to obtain the system

$$\left. \begin{array}{l} x' = x, \\ y' = ky, \end{array} \right\} \tag{13}$$

the solution is $x(t) = x_0 e^t$, $y(t) = y_0 e^{kt}$. Then with $k = 1$ and $k = 2$, the trajectories are the same as those shown in Figs. 7.4 and 7.5, respectively, but with the arrows reversed. In Fig. 7.4 with the arrows reversed, the origin is an unstable proper node; in Fig. 7.5 with the arrows reversed, the origin is an unstable improper node.

If (x_*, y_*) is a critical point, the equilibrium solution $x(t) \equiv x_*$, $y(t) \equiv y_*$ is called stable or unstable depending on the nature of the critical point. In applications the stability of an equilibrium solution often is a crucial matter. For instance, suppose in Example 1 that $x(t)$ and $y(t)$ denote the fox and rabbit populations, respectively, *in hundreds*. We will see in Section 7.5 that the critical point $(12, 6)$ is stable. It follows that if we begin with *close to* 1200 foxes and 600 rabbits—rather than exactly these equilibrium values—then for all future time there will remain close to 1200 foxes and close to 600 rabbits. Thus the practical consequence of stability is that slight changes (perhaps due to random births and deaths) in the equilibrium populations will not so upset the equilibrium as to result in large deviations from the equilibrium solution.

It is possible for trajectories to remain near a stable critical point without approaching it, as the following example shows.

EXAMPLE 3 Consider a mass m that oscillates without damping on a spring with Hooke's constant k, so that its position function $x(t)$ satisfies the differential equation $x'' + \omega^2 x = 0$ (where $\omega^2 = k/m$). If we introduce its velocity $y = x'$, we get the system

$$\left. \begin{array}{l} x' = y, \\ y' = -\omega^2 x \end{array} \right\} \tag{14}$$

with general solution

$$x(t) = \quad A \cos \omega t + \quad B \sin \omega t, \tag{15a}$$

$$y(t) = -A\omega \sin \omega t + B\omega \cos \omega t. \tag{15b}$$

With $C = (A^2 + B^2)^{1/2}$ and $\alpha = \tan^{-1}(B/A)$, we can rewrite (15) in the form

$$x(t) = \quad C \cos(\omega t - \alpha), \tag{16a}$$

$$y(t) = -\omega C \sin(\omega t - \alpha), \tag{16b}$$

so it becomes clear that each trajectory other than the critical point $(0, 0)$ is an ellipse with equation of the form

$$\frac{x^2}{C^2} + \frac{y^2}{\omega^2 C^2} = 1. \tag{17}$$

Each point $(x_0, y_0) \neq (0, 0)$ in the xy-plane lies on exactly one of these ellipses, and a solution $(x(t), y(t))$ with initial point (x_0, y_0) traverses the ellipse containing (x_0, y_0) in the clockwise direction with *period* $T = 2\pi/\omega$. (It is clear from (15) that $x(t + T) = x(t)$ and $y(t + T) = y(t)$ for all t.) Thus each nontrivial solution of the system in (14) is periodic and its trajectory is a simple closed curve.

It is clear in Fig. 7.7, in which $\omega < 1$, that if the distance from (x_0, y_0) to $(0, 0)$ is less than $\delta = \omega\epsilon$, then for all t the distance from $(x(t), y(t))$ is less than ϵ. Hence $(0, 0)$ is a stable critical point of the system $x' = y$, $y' = -\omega^2 x$. Unlike the situation shown in Figs. 7.4 and 7.5, though, no single trajectory approaches $(0, 0)$. A critical point surrounded by simple closed trajectories representing periodic solutions is called a **(stable) center**.

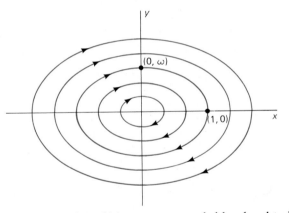

FIGURE 7.7 A (stable) center surrounded by closed trajectories.

The critical point (x_*, y_*) is called **asymptotically stable** if it is stable and, moreover, every trajectory that begins sufficiently close to (x_*, y_*) also approaches (x_*, y_*) as $t \to +\infty$. That is, there exists $\delta > 0$ such that

$$|\mathbf{x}_0 - \mathbf{x}_*| < \delta \quad \text{implies} \quad \lim_{t \to \infty} \mathbf{x}(t) = \mathbf{x}_*, \tag{18}$$

where $\mathbf{x}_0 = (x_0, y_0)$, $\mathbf{x}_* = (x_*, y_*)$, and $\mathbf{x}(t) = (x(t), y(t))$ is a solution with $\mathbf{x}(0) = \mathbf{x}_0$.

The stable nodes shown in Figs. 7.4 and 7.5 are asymptotically stable because every trajectory approaches the critical point $(0, 0)$ as $t \to +\infty$. The center $(0, 0)$ in Fig. 7.7 is stable but not asymptotically stable, because however small an elliptical trajectory we consider, a point moving around this ellipse does not approach the origin. Thus asymptotic stability is a stronger condition than mere stability.

Now suppose that $x(t)$ and $y(t)$ denote coexisting populations for which (x_*, y_*) is an asymptotically stable critical point. Then if the initial populations x_0 and y_0 are sufficiently close to x_* and y_*, respectively, it follows that both

$$\lim_{t \to \infty} x(t) = x_* \quad \text{and} \quad \lim_{t \to \infty} y(t) = y_*. \tag{19}$$

That is, $x(t)$ and $y(t)$ actually approach the equilibrium populations x_* and y_* as $t \to +\infty$, rather than merely remaining close to those values.

For a mechanical system as in Example 3, a critical point represents an *equilibrium state* of the system—if the velocity $y = x'$ and the acceleration $y' = x''$ vanish simultaneously, the mass remains at rest with no force acting on it. Stability of a critical point concerns the question whether, when the mass is displaced slightly from its equilibrium, it (1) moves back toward the equilibrium point as $t \to +\infty$, (2) merely remains near the equilibrium point without approaching it, or (3) moves farther away from equilibrium. In Case 1 the critical (equilibrium) point is asymptotically stable; in Case 2 it is stable but not asymptotically so; in Case 3 it is an unstable critical point. A marble balanced on the top of a basketball is an example of an unstable equilibrium state. A mass on a spring with damping illustrates the case of asymptotic stability of a mechanical system. The mass-and-spring without damping in Example 3 is an example of a system that is stable but not asymptotically stable.

EXAMPLE 4   Suppose that $m = 1$ and $k = 2$ for the mass and spring of Example 3 and that the mass is attached also to a dashpot with damping constant $c = 2$. Then its displacement function $x(t)$ satisfies the second order equation

$$x''(t) + 2x'(t) + 2x(t) = 0. \tag{20}$$

With $y = x'$ we obtain the equivalent first order system

$$\left. \begin{array}{l} \dfrac{dx}{dt} = y, \\[2mm] \dfrac{dy}{dt} = -2x - 2y \end{array} \right\} \tag{21}$$

with critical point $(0, 0)$. The characteristic equation $r^2 + 2r + 2 = 0$ of (20) has roots $-1 + i$ and $-1 - i$, so the general solution of the system in (21) is given by

$$\begin{aligned} x(t) &= e^{-t}(A \cos t + B \sin t) \\ &= Ce^{-t} \cos (t - \alpha), \end{aligned} \tag{22a}$$

$$\begin{aligned} y(t) &= e^{-t}[(B - A) \cos t - (A + B) \sin t] \\ &= -C\sqrt{2}e^{-t} \sin\left(t - \alpha + \frac{\pi}{4}\right), \end{aligned} \tag{22b}$$

where $C = \sqrt{A^2 + B^2}$ and $\alpha = \tan^{-1}(B/A)$. We see that $x(t)$ and $y(t)$ oscillate between positive and negative values and that each approaches zero as $t \to +\infty$. Consequently, a typical trajectory spirals toward the origin, as indicated in Fig. 7.8.

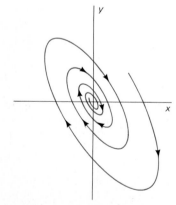

FIGURE 7.8   A stable spiral point and one nearby trajectory.

478

It is clear that the critical point (0, 0) is asymptotically stable; this type of critical point is called a (stable) **spiral point**. In the case of the mass-spring-dashpot system, a spiral point is the manifestation in the phase plane of the damped oscillations that occur because of resistance.

The trajectory illustrated in Fig. 7.8 spirals into a critical point as $t \to +\infty$. The next example shows that it is also possible for a trajectory to spiral into a closed trajectory. A **closed** trajectory is a simple closed solution curve representing a periodic solution (like the ellipses in Fig. 7.7).

EXAMPLE 5   Consider the standard system

$$\left. \begin{aligned} x' &= -y + x(1 - x^2 - y^2), \\ y' &= x + y(1 - x^2 - y^2). \end{aligned} \right\} \tag{23}$$

In Problem 21 we ask you to show that (0, 0) is its only critical point. This system can be solved explicitly by introducing polar coordinates $x = r \cos \theta$, $y = r \sin \theta$, as follows. First note that

$$\frac{d\theta}{dt} = \frac{d}{dt} \left(\arctan \frac{y}{x} \right) = \frac{xy' - yx'}{x^2 + y^2}.$$

Then substitute the expressions given in (23) for x' and y' to obtain

$$\frac{d\theta}{dt} = \frac{x^2 + y^2}{x^2 + y^2} = 1,$$

so it follows that

$$\theta(t) = t + \theta_0, \quad \text{where} \quad \theta_0 = \theta(0). \tag{24}$$

Then differentiation of $r^2 = x^2 + y^2$ yields

$$2r \frac{dr}{dt} = 2x \frac{dx}{dt} + 2y \frac{dy}{dt}$$

$$= 2(x^2 + y^2)(1 - x^2 - y^2) = 2r^2(1 - r^2),$$

so $r = r(t)$ satisfies the differential equation

$$\frac{dr}{dt} = r(1 - r^2). \tag{25}$$

In Problem 22 we ask you to derive the solution

$$r(t) = \frac{r_0}{[r_0^2 + (1 - r_0^2)e^{-2t}]^{1/2}} \tag{26}$$

where $r_0 = r(0)$. Thus the typical solution of Eq. (23) is described by

$$x(t) = r(t) \cos (t + \theta_0), \qquad y(t) = r(t) \sin (t + \theta_0). \tag{27}$$

If $r_0 = 1$, then (26) gives $r(t) \equiv 1$, so (27) yields the closed trajectory $r = 1$ (the unit circle). Otherwise, if $r_0 > 0$, then (26) implies that $r(t) \to 1$ as $t \to +\infty$. Hence the trajectory defined by (27) spirals in toward the unit circle if $r_0 > 1$ and spirals out toward this closed trajectory if $0 < r_0 < 1$ (see Fig. 7.9).

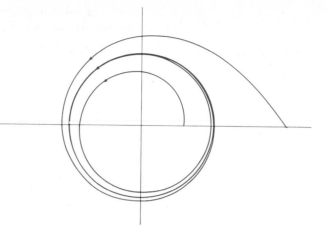

FIGURE 7.9 A closed trajectory and two nonclosed trajectories spiraling toward it.

Under rather general hypotheses it can be shown that there are four possibilities for a nondegenerate trajectory of the autonomous system

$$x' = F(x, y), \qquad y' = G(x, y).$$

The four possibilities are these:

1. $(x(t), y(t))$ approaches a critical point as $t \to +\infty$.
2. $(x(t), y(t))$ is unbounded with increasing t.
3. $(x(t), y(t))$ is a periodic solution with a closed trajectory.
4. $(x(t), y(t))$ spirals toward a closed trajectory as $t \to +\infty$.

As a consequence, the qualitative nature of the phase plane picture of the trajectories of an autonomous system is determined largely by the locations of its critical points and by the behavior of its trajectories near its critical points. We will see in Section 7.3 that, subject to mild restrictions on the functions F and G, each isolated critical point of the system $x' = F(x, y)$, $y' = G(x, y)$ resembles qualitatively one of the examples of this section—it is either a node (proper or improper), a saddle point, a center, or a spiral point.

7.2   PROBLEMS

Find the critical points of the two-dimensional systems in Problems 1–8.

1. $\dfrac{dx}{dt} = 3x - y, \quad \dfrac{dy}{dt} = x + 3y$

2. $\dfrac{dx}{dt} = x - 2y, \quad \dfrac{dy}{dt} = 2x - 4y$

3. $\dfrac{dx}{dt} = 3x - 2y, \quad \dfrac{dy}{dt} = 4x - 3y + 1$

4. $\dfrac{dx}{dt} = xy - x^2, \quad \dfrac{dy}{dt} = xy + y^2$

5. $\dfrac{dx}{dt} = 2x - xy, \quad \dfrac{dy}{dt} = xy - 3y$

6. $\dfrac{dx}{dt} = x - 3x^2 + xy, \quad \dfrac{dy}{dt} = 4y - y^2 - 2xy$

7. $\dfrac{dx}{dt} = y, \quad \dfrac{dy}{dt} = -\sin x$

CHAPTER 7:   Qualitative Properties and Existence of Solutions

8. $\dfrac{dx}{dt} = x + y - x(x^2 + y^2), \quad \dfrac{dy}{dt} = -x + y - y(x^2 + y^2)$

In each of Problems 9–12, find all equilibrium solutions (of the form $x(t) \equiv x_0$, a constant) of the given second order differential equation.

9. $x'' + 4x - x^3 = 0$

10. $x'' + 2x' + x + 4x^3 = 0$

11. $x'' + 3x' + 4 \sin x = 0$

12. $x'' + (x^2 - 1)x' + x = 0$

Solve each of the linear systems in Problems 13–20 to determine whether the critical point $(0, 0)$ is stable, asymptotically stable, or unstable. Sketch typical trajectories, and indicate the direction of motion with increasing t. Identify the critical point as a node, a saddle point, a center, or a spiral point.

13. $x' = -2x, \; y' = -2y$

14. $x' = 2x, \; y' = -2y$

15. $x' = -2x, \; y' = -y$

16. $x' = x, \; y' = 3y$

17. $x' = y, \; y' = -x$

18. $x' = -y, \; y' = 4x$

19. $x' = 2y, \; y' = -2x$

20. $x' = y, \; y' = -5x - 4y$

21. Verify that $(0, 0)$ is the only critical point of the system in Example 5.

22. Separate variables in (25) to derive the solution given in (26).

In each of Problems 23–26, a system $x' = F(x, y), \; y' = G(x, y)$ is given. Solve the equation

$$\frac{dy}{dx} = \frac{G(x, y)}{F(x, y)}$$

to find the trajectories of the given system.

23. $x' = y, \; y' = -x$

24. $x' = y(1 + x^2 + y^2), \; y' = x(1 + x^2 + y^2)$

25. $x' = 4y(1 + x^2 + y^2), \; y' = -x(1 + x^2 + y^2)$

26. $x' = y^3 e^{x+y}, \; y' = -x^3 e^{x+y}$

27. Let $(x(t), y(t))$ be a nontrivial solution of the non-autonomous system

$$x' = y, \qquad y' = tx.$$

Suppose that $\phi(t) = x(t + \gamma)$ and $\psi(t) = y(t + \gamma)$ where $\gamma \neq 0$. Show that $(\phi(t), \psi(t))$ is *not* a solution of the system.

Problems 28–30 deal with the system

$$\frac{dx}{dt} = F(x, y), \qquad \frac{dy}{dt} = G(x, y)$$

in a region where the functions F and G are continuously differentiable, so for each number a and point (x_0, y_0), there is a unique solution with $x(a) = x_0$ and $y(a) = y_0$.

28. Suppose that $(x(t), y(t))$ is a solution of the autonomous system and that $\gamma \neq 0$. Define $\phi(t) = x(t + \gamma)$ and $\psi(t) = y(t + \gamma)$. Then show (in contrast with the situation in Problem 27) that $(\phi(t), \psi(t))$ is also a solution of the system. Thus autonomous systems have the simple but important property that a "t-translate" of a solution is again a solution.

29. Let $(x_1(t), y_1(t))$ and $(x_2(t), y_2(t))$ be two solutions having trajectories that meet at the point (x_0, y_0); thus $x_1(a) = x_2(b) = x_0$ and $y_1(a) = y_2(b) = y_0$ for some values a and b of t. Define

$$x_3(t) = x_2(t + \gamma) \quad \text{and} \quad y_3(t) = y_2(t + \gamma)$$

where $\gamma = b - a$, so $(x_2(t), y_2(t))$ and $(x_3(t), y_3(t))$ have the same trajectory. Apply the uniqueness theorem to show that $(x_1(t), y_1(t))$ and $(x_3(t), y_3(t))$ are identical solutions. Hence the original two trajectories are identical. Thus no two different trajectories of an autonomous system can intersect.

30. Suppose that the solution $(x_1(t), y_1(t))$ is defined for all t and that its trajectory has an apparent self-intersection:

$$x_1(a) = x_1(a + T) = x_0, \qquad y_1(a) = y_1(a + T) = y_0$$

for some $T > 0$. Introduce the solution

$$x_2(t) = x_1(t + T), \qquad y_2(t) = y_1(t + T)$$

and then apply the uniqueness theorem to show that

$$x_1(t) = x_1(t + T) \quad \text{and} \quad y_1(t) = y_1(t + T)$$

for *all* t. Thus the solution $(x_1(t), y_1(t))$ is periodic with period T and has a closed trajectory. Consequently a solution of an autonomous system either is periodic with a closed trajectory, or else its trajectory never passes through the same point twice.

7.3

Linear and Almost Linear Systems

We now discuss the behavior of solutions of the autonomous system

$$\frac{dx}{dt} = F(x, y), \qquad \frac{dy}{dt} = G(x, y) \tag{1}$$

near an isolated critical point (x_0, y_0) at which $F(x_0, y_0) = G(x_0, y_0) = 0$. A critical point is called **isolated** if some neighborhood of it contains no other critical point. We assume throughout that the functions F and G are continuously differentiable in a neighborhood of (x_0, y_0).

We can assume without loss of generality that $x_0 = y_0 = 0$. Otherwise, we make the substitutions $u = x - x_0$, $v = y - y_0$. Then $dx/dt = du/dt$ and $dy/dt = dv/dt$, so (1) is equivalent to the system

$$\left. \begin{aligned} \frac{du}{dt} &= F(u + x_0, v + y_0) = F_1(u, v), \\[2mm] \frac{dv}{dt} &= G(u + x_0, v + y_0) = G_1(u, v) \end{aligned} \right\} \tag{2}$$

that has $(0, 0)$ as an isolated critical point.

EXAMPLE 1 The system

$$\left. \begin{aligned} \frac{dx}{dt} &= 3x - x^2 - xy = x(3 - x - y), \\[2mm] \frac{dy}{dt} &= y + y^2 - 3xy = y(1 - 3x + y) \end{aligned} \right\} \tag{3}$$

has $(1, 2)$ as one of its critical points. We substitute $u = x - 1$, $v = y - 2$; that is, $x = u + 1$, $y = v + 2$. Then

$$3 - x - y = 3 - (u + 1) - (v + 2) = -u - v$$

and

$$1 - 3x + y = 1 - 3(u + 1) + (v + 2) = -3u + v,$$

so the system in (3) takes the form

$$\left. \begin{aligned} \frac{du}{dt} &= (u + 1)(-u - v) = -u - v - u^2 - uv, \\[2mm] \frac{dv}{dt} &= (v + 2)(-3u + v) = -6u + 2v + v^2 - 3uv \end{aligned} \right\} \tag{4}$$

that has $(0, 0)$ as a critical point. If we can determine the trajectories of the system in (4) near $(0, 0)$, then their translates under the rigid motion that carries $(0, 0)$ to $(1, 2)$ will be the trajectories near $(1, 2)$ of the original system in (3).

We therefore assume hereafter that $(0, 0)$ is an isolated critical point of the autonomous system in (1). It then follows from Taylor's formula for functions of two variables that (1) can be written in the form

$$\left. \begin{aligned} \frac{dx}{dt} &= ax + by + f(x, y), \\[2mm] \frac{dy}{dt} &= cx + dy + g(x, y), \end{aligned} \right\} \tag{5}$$

where $a = F_x(0, 0)$, $b = F_y(0, 0)$, $c = G_x(0, 0)$, and $d = G_y(0, 0)$, and the functions $f(x, y)$ and $g(x, y)$ have the property that

$$\lim_{(x, y) \to (0, 0)} \frac{f(x, y)}{\sqrt{x^2 + y^2}} = \lim_{(x, y) \to (0, 0)} \frac{g(x, y)}{\sqrt{x^2 + y^2}} = 0. \tag{6}$$

That is, when (x, y) is near $(0, 0)$, the quantities $f(x, y)$ and $g(x, y)$ are small in comparison with $r = \sqrt{x^2 + y^2}$, which itself is small. Thus, when (x, y) is near $(0, 0)$, the nonlinear system in (5) is in some sense "near" the **linearized** system

$$\left.\begin{aligned} \frac{dx}{dt} &= ax + by, \\[2mm] \frac{dy}{dt} &= cx + dy. \end{aligned}\right\} \tag{7}$$

Under the assumption that $(0, 0)$ is also an isolated critical point of this linear system, the autonomous system in (5) is therefore called **almost linear** provided that f and g satisfy the condition in (6). It turns out that in most (but not all) cases, the trajectories near $(0, 0)$ of the almost linear system in (5) strongly resemble—qualitatively—those of its "linearization" in (7). Consequently, the first step toward understanding general autonomous systems is to characterize the critical points of linear systems.

LINEAR SYSTEMS

The linear system in (7) is readily solved by the elimination method of Section 5.2 or by the eigenvalue method of Section 5.5. Using the latter method, we substitute $x = Ae^{\lambda t}$ and $y = Be^{\lambda t}$ in (7). Upon dividing the resulting equations by $e^{\lambda t}$, we get the two homogeneous linear equations

$$\begin{aligned} (a - \lambda)A + \quad bB &= 0, \\ cA + (d - \lambda)B &= 0 \end{aligned} \tag{8}$$

that the coefficients A and B must satisfy. In order for these equations to have a nontrivial solution, the determinant of coefficients must vanish:

$$\begin{aligned} \Delta &= (a - \lambda)(d - \lambda) - bc \\ &= \lambda^2 - (a + d)\lambda + (ad - bc) = 0. \end{aligned} \tag{9}$$

The nature of the solution of the system in (7) is determined by the nature of the roots of the quadratic **characteristic equation** in (9). Note that this equation is simply the characteristic equation of the coefficient matrix

$$\begin{bmatrix} a & b \\ c & d \end{bmatrix}$$

of the linear system in (7).

We assume that $(0, 0)$ is an isolated critical point of the system in (7). It follows that $ad - bc \neq 0$, so $\lambda = 0$ cannot be a root of (9). The reason is that the constant term $ad - bc$ in (9) is the determinant of the system of linear equations

$$\left.\begin{array}{l} ax + by = 0, \\ cx + dy = 0. \end{array}\right\} \tag{10}$$

If $ad - bc = 0$, then the equations in (10) are the equations of the same line through $(0, 0)$, so the system has a whole line of critical points rather than an isolated critical point at $(0, 0)$.

Consequently the characteristic equation in (9) has either:

1. real unequal roots of the same sign;
2. real unequal roots of opposite sign;
3. real nonzero equal roots;
4. complex conjugate roots; or
5. pure imaginary roots.

These five cases are discussed separately below. In each case the critical point $(0, 0)$ resembles one of those we saw in the examples of Section 7.2—a node (proper or improper), a saddle point, a center, or a spiral point.

REAL UNEQUAL ROOTS OF THE SAME SIGN   In this case the general solution of (7) is of the form

$$\begin{aligned} x(t) &= A_1 e^{\lambda_1 t} + A_2 e^{\lambda_2 t}, \\ y(t) &= B_1 e^{\lambda_1 t} + B_2 e^{\lambda_2 t} \end{aligned} \tag{11}$$

where only two of the four coefficients are arbitrary. If both roots are negative $(\lambda_2 < \lambda_1 < 0)$, it is clear from (11) that x and y approach 0 as $t \to +\infty$, so the critical point $(0, 0)$ is asymptotically stable. But if both roots are positive $(0 < \lambda_2 < \lambda_1)$, then x and y are unbounded as $t \to +\infty$, so the critical point $(0, 0)$ is unstable.

The system $x' = -x$, $y' = -2y$ considered in Example 2 of Section 7.2, and the improper node shown in Fig. 7.5 there, are typical of this case. In general, there is a linear change of coordinates—$u = a_1 x + a_2 y$, $v = b_1 x + b_2 y$—that transforms (11) into

$$u = A e^{\lambda_1 t}, \qquad v = B e^{\lambda_2 t}. \tag{12}$$

In the (oblique) uv-coordinate system, the trajectories of the system in (12) are the straight lines $u = 0$ and $v = 0$ and curves of the form $v = Cu^k$, where $k = \lambda_2/\lambda_1 > 0$. The trajectories in the xy-coordinate system are the images of these curves under the inverse transformation from uv- to xy-coordinates. Thus $(0, 0)$ is an **improper node** like the one shown in Fig. 7.10, where $k = 2$, so the trajectories are parabolas in the oblique uv-system. The distinguishing feature of an improper node is that one pair of trajectories lies on the line $u = 0$, while all other trajectories are tangent to the line $v = 0$ at $(0, 0)$. If λ_1 and λ_2 are negative, then all trajectories approach $(0, 0)$ as $t \to 0$, while if λ_1 and λ_2 are positive, then the arrows in Fig. 7.10 would be reversed.

REAL UNEQUAL ROOTS OF OPPOSITE SIGN   In this case the situation is the same as in the previous case, except that $\lambda_2 < 0 < \lambda_1$ in (12). In the oblique uv-coordinate system, the trajectories determined by (12) are the straight lines $u = 0$ and $v = 0$

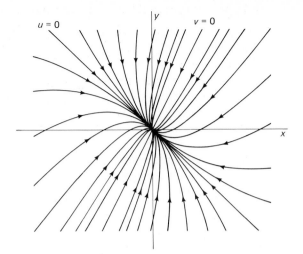

FIGURE 7.10 An improper node; both roots of the characteristic equation are negative.

and curves of the form $uv^k = C$, where $k = -\lambda_1/\lambda_2 > 0$. The general situation resembles that shown in Fig. 7.11, where $k = 1$, so the nonlinear trajectories are hyperbolas in the uv-system. As $t \to +\infty$, the trajectories $u = Ae^{\lambda_1 t}$, $v = 0$ go to infinity because $\lambda_1 > 0$, while the trajectories $u = 0$, $v = Be^{\lambda_2 t}$ approach the origin because $\lambda_2 < 0$. The critical point $(0, 0)$ is therefore an unstable **saddle point** in this case.

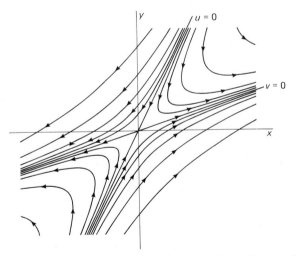

FIGURE 7.11 A saddle point; the roots of the characteristic equation are real and of opposite sign.

REAL EQUAL ROOTS   In this case, with $\lambda_1 = \lambda_2 = \lambda$, the general solution of the system in (7) is of the form

$$x(t) = (A_1 + A_2 t)e^{\lambda t}, \qquad y(t) = (B_1 + B_2 t)e^{\lambda t} \tag{13}$$

where only two of the four coefficients are arbitrary. If $\lambda < 0$, then it is clear from (13) that $x \to 0$ and $y \to 0$ as $t \to +\infty$, so the critical point $(0, 0)$ is asymptotically stable. But if $\lambda > 0$, then x and y increase without bound as $t \to +\infty$, so $(0, 0)$ is an unstable critical point.

The nature of the trajectories depends on whether the term $te^{\lambda t}$ is actually present in (13). The system $x' = -x$, $y' = -y$ considered in Example 2 of Section 7.2 illustrates the case in which $A_2 = B_2 = 0$, so (13) reduces to

$$x = Ae^{\lambda t}, \qquad y = Be^{\lambda t}. \tag{14}$$

Then $y = (B/A)x$ if $A \neq 0$, so all trajectories lie on straight lines through the critical point $(0, 0)$, which is therefore a **proper node** (see Fig. 7.12).

The situation when the term $te^{\lambda t}$ is present is illustrated by the system $x' = -x$, $y' = x - y$, which has the general solution

$$x = Ae^{-t}, \qquad y = Be^{-t} + Ate^{-t}. \tag{15}$$

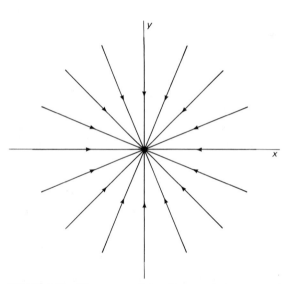

FIGURE 7.12 The roots of the characteristic equation are real and equal, resulting in a proper node.

After we solve the first equation for $t = -(1/A)\ln x$ and substitute this in the second equation, we find that

$$y = \frac{B}{A}x - x \ln \frac{x}{A} \tag{16}$$

and

$$\frac{dy}{dx} = \frac{y'}{x'} = \frac{-Be^{-t} + Ae^{-t} - Ate^{-t}}{-Ae^{-t}}$$

$$= \frac{1}{A}(B - A + At) \tag{17}$$

if $A \neq 0$. It is clear from (17) that $dy/dx \to +\infty$ as $t \to +\infty$, so all the trajectories are tangent to the y-axis at $(0, 0)$. In this case the critical point $(0, 0)$ is an **improper**

CHAPTER 7: Qualitative Properties and Existence of Solutions

node as in Fig. 7.13. For a more general improper node, all trajectories are tangent to an arbitrary fixed line through the critical point, as in Fig. 7.14.

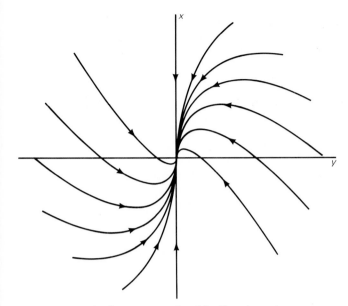

FIGURE 7.13   An improper node with all trajectories tangent to the y-axis.

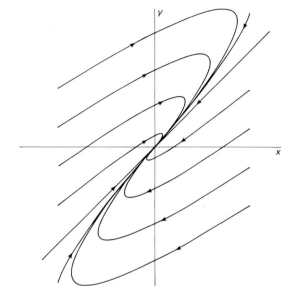

FIGURE 7.14   An improper node with all trajectories tangent to the graph of $y = x$.

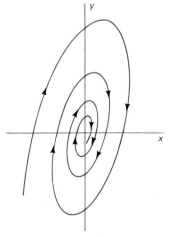

FIGURE 7.15   A spiral point; the roots of the characteristic equation are complex conjugates.

COMPLEX CONJUGATE ROOTS   In this case, with $\lambda_1 = p + qi$ and $\lambda_2 = p - qi$ with p and q nonzero, the general solution has the form

$$
x(t) = e^{pt}(A_1 \cos qt + A_2 \sin qt),
$$
$$
y(t) = e^{pt}(B_1 \cos qt + B_2 \sin qt).
$$
(18)

Thus $x(t)$ and $y(t)$ oscillate between positive and negative values as t increases, and the critical point $(0, 0)$ is a **spiral point** as in Example 4 of Section 7.2. If the real part p of λ_1 and λ_2 is negative, then it is clear from (18) that (x, y) approaches $(0, 0)$ as $t \to +\infty$, so $(0, 0)$ is an asymptotically stable critical point, as shown in Fig. 7.15. But if $p > 0$, then the critical point is unstable.

PURE IMAGINARY ROOTS   If $\lambda_1 = qi$ and $\lambda_2 = -qi$ with $q \neq 0$, then the general solution is of the form

$$
x(t) = A_1 \cos qt + A_2 \sin qt, \qquad y(t) = B_1 \cos qt + B_2 \sin qt. \quad (19)
$$

As in Example 3 of Section 7.2, the trajectories are (rotated) ellipses (see Fig. 7.16), so the critical point $(0, 0)$ is a **center**: stable but not asymptotically stable.

For the linear system in (7) with $ad - bc \neq 0$, the table in Fig. 7.17 lists the type of critical point at $(0, 0)$ found in the five cases discussed above, according to the nature of the roots λ_1 and λ_2 of the characteristic equation in (9). Our discussion of these cases shows also that the stability of the critical point $(0, 0)$ is

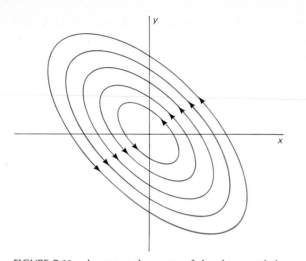

FIGURE 7.16 A center; the roots of the characteristic equation are pure imaginary.

Roots of characteristic equation	Type of critical point
Real, unequal, same sign	Improper node
Real, unequal, opposite sign	Saddle point
Real and equal	Proper or improper node
Complex conjugate	Spiral point
Pure imaginary	Center

FIGURE 7.17 Type of critical point $(0, 0)$ of the linear system in Eq. (7).

determined by the real parts of the characteristic roots λ_1 and λ_2, as summarized in Theorem 1. If λ_1 and λ_2 are real, then they are themselves their real parts.

Theorem 1 *Stability of Linear Systems*

Let λ_1 and λ_2 be the roots of the characteristic equation

$$\lambda^2 - (a + d)\lambda + (ad - bc) = 0 \tag{9}$$

of the linear system

$$\left.\begin{aligned} \frac{dx}{dt} &= ax + by, \\[2mm] \frac{dy}{dt} &= cx + dy \end{aligned}\right\} \tag{7}$$

with $ad - bc \neq 0$. Then the critical point $(0, 0)$ is:

(a) asymptotically stable if the real parts of λ_1 and λ_2 are both negative;
(b) stable but not asymptotically stable if the real parts of λ_1 and λ_2 are both zero (so that $\lambda_1, \lambda_2 = \pm qi$);
(c) unstable if either λ_1 or λ_2 has a positive real part.

It is worthwhile to consider the effect of small perturbations in the coefficients a, b, c, and d of the linear system in (7), which result in small perturbations of the characteristic roots λ_1 and λ_2. If these perturbations are sufficiently small, then positive real parts (of λ_1 and λ_2) remain positive, and negative real parts remain negative. Hence an asymptotically stable critical point remains asympto-

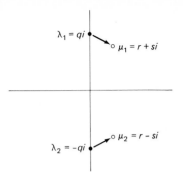

FIGURE 7.18 Effect of perturbation of pure imaginary roots.

tically stable, and an unstable critical point remains unstable. Part (b) of Theorem 1 is therefore the only one in which arbitrarily small perturbations can affect the stability of the critical point $(0, 0)$. In this case pure imaginary roots $\lambda_1, \lambda_2 = \pm qi$ can be changed to nearby complex roots $\mu_1, \mu_2 = r \pm si$, with r either positive or negative (see Fig. 7.18). Consequently, a small perturbation of the coefficients of the linear system in (7) can change a stable center to a spiral point that is either unstable or asymptotically stable.

There is one other exceptional case in which the type, though not the stability, of the critical point $(0, 0)$ of a linear system can be altered by a small perturbation of its coefficients. This is the case with $\lambda_1 = \lambda_2$, equal real roots that (under a small perturbation of the coefficients) can split into two roots μ_1 and μ_2, which are either complex conjugates or unequal real roots (see Fig. 7.19). In either case, the sign of the real parts of the roots is preserved, so the stability of the critical point is unaltered. Its nature may change, however; the table in Fig. 7.17 shows that a node with $\lambda_1 = \lambda_2$ can either remain a node (if μ_1 and μ_2 are real) or change to a spiral point (if μ_1 and μ_2 are complex conjugates).

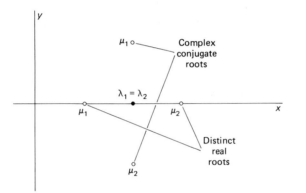

FIGURE 7.19 Effects of perturbation of real equal roots.

Suppose that the linear system in (7) is used to model a physical situation. It is unlikely that the coefficients in (7) can be measured with total accuracy, so let the unknown precise linear model be

$$\left. \begin{aligned} \frac{dx}{dt} &= a^*x + b^*y, \\[2mm] \frac{dy}{dt} &= c^*x + d^*y. \end{aligned} \right\} \tag{7*}$$

If the coefficients in (7) are sufficiently close to those in (7*), it then follows from the discussion in the preceding paragraph that the origin $(0, 0)$ is an asymptotically stable critical point for (7) if it is an asymptotically stable critical point for (7*) and is an unstable critical point for (7) if it is an unstable critical point for (7*). Thus in this case the approximate model in (7) and the precise model in (7*) predict the same qualitative behavior (with regard to asymptotic stability versus instability.)

ALMOST LINEAR SYSTEMS

We now return to the almost linear system

$$\left.\begin{array}{l}\dfrac{dx}{dt} = ax + by + f(x, y), \\[3mm] \dfrac{dy}{dt} = cx + dy + g(x, y)\end{array}\right\} \qquad (5)$$

having $(0, 0)$ as an isolated critical point with $ad - bc \neq 0$. Theorem 2, which we state without proof, essentially implies that—with regard to the type and stability of the critical point $(0, 0)$—the effect of the small nonlinear terms $f(x, y)$ and $g(x, y)$ is equivalent to the effect of a small perturbation in the coefficients of the associated *linear* system in (7).

Theorem 2 *Stability of Almost Linear Systems*

Let λ_1 and λ_2 be the characteristic roots of the linear system in (7) associated with the almost linear system in (5). Then:

(a) If $\lambda_1 = \lambda_2$ are real equal roots, then the critical point $(0, 0)$ of (5) is either a node or a spiral point, and is asymptotically stable if $\lambda_1 = \lambda_2 < 0$, unstable if $\lambda_1 = \lambda_2 > 0$.
(b) If λ_1 and λ_2 are pure imaginary, then $(0, 0)$ is either a center or a spiral point, which may be either asymptotically stable, stable, or unstable.
(c) Otherwise—that is, unless λ_1 and λ_2 are either real equal or pure imaginary—the critical point $(0, 0)$ of the almost linear system in (5) is of the same type and stability as the critical point $(0, 0)$ of the associated linear system in (7).

Thus, if $\lambda_1 \neq \lambda_2$ and $\mathrm{Re}(\lambda_1) \neq 0$, then the type and stability of the critical point of the almost linear system in (5) can be determined by analysis of its associated linear system in (7), and only in the case of pure imaginary characteristic roots is the stability of $(0, 0)$ not determined by the linear system. Except in the sensitive cases $\lambda_1 = \lambda_2$ and $\mathrm{Re}(\lambda_i) = 0$, the trajectories near $(0, 0)$ will resemble qualitatively those of the associated linear system—they enter or leave the critical point in the same way, but may be "deformed" in a nonlinear manner.

An important consequence of the classification of cases in Theorem 2 is that *a critical point of an almost linear system is asymptotically stable if it is an asymptotically stable critical point of the linearization of the system.* Moreover, a critical point of the almost linear system is unstable if it is an unstable critical point of the linearized system. If an almost linear system is used to model a physical situation, then—apart from the sensitive cases mentioned in the preceding paragraph—it follows that the qualitative behavior of the system near a critical point can be determined by examining its linearization.

EXAMPLE 2 Determine the type and stability of the critical point $(0, 0)$ of the almost linear system

$$\left.\begin{aligned} \frac{dx}{dt} &= 4x + 2y + 2x^2 - 3y^2, \\ \frac{dy}{dt} &= 4x - 3y + 7xy. \end{aligned}\right\} \tag{20}$$

SOLUTION The characteristic equation of the associated linear system (obtained simply by deleting the quadratic terms in (20)) is

$$(4 - \lambda)(-3 - \lambda) - 8 = (\lambda - 5)(\lambda + 4) = 0,$$

so the roots $\lambda_1 = 5$ and $\lambda_2 = -4$ are real, unequal, and have opposite sign. By our discussion of this case we know that $(0, 0)$ is an unstable saddle point of the linear system, and hence by Part (c) of Theorem 2, it is also an unstable saddle point of the almost linear system in (20). The trajectories of the linear system near $(0, 0)$ are shown in Fig. 7.20, while those of the nonlinear system in (20) are shown in Fig. 7.21.

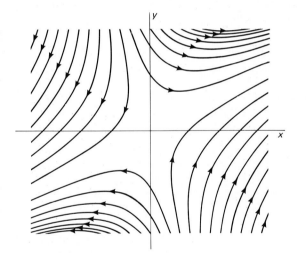

FIGURE 7.20 Trajectories of the linearized system of Example 2; here we have $|x| \leq 0.5$, $|y| \leq 0.4$.

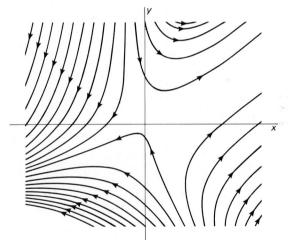

FIGURE 7.21 Trajectories of the nonlinear system of Example 2; $|x| \leq 0.5$, $|y| \leq 0.4$.

EXAMPLE 3 Determine the type and stability of the critical point $(4, 3)$ of the almost linear system

$$\left.\begin{aligned} x' &= 33 - 10x - 3y + x^2, \\ y' &= -18 + 6x + 2y - xy. \end{aligned}\right\} \tag{21}$$

SOLUTION The substitution $u = x - 4$, $v = y - 3$ (that is, $x = u + 4$, $y = v + 3$) in (21) yields the almost linear system

$$\left.\begin{aligned} u' &= -2u - 3v + u^2, \\ v' &= 3u - 2v - uv \end{aligned}\right\} \tag{22}$$

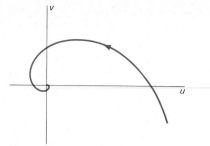

FIGURE 7.22 A spiral trajectory of the linear system in Eq. (23).

having (0, 0) as the corresponding critical point. The associated linear system

$$\left.\begin{array}{rl} u' &= -2u - 3v, \\ v' &= 3u - 2v \end{array}\right\} \tag{23}$$

has characteristic equation $(\lambda + 2)^2 + 9 = 0$, with complex conjugate roots $\lambda = -2 \pm 3i$. Hence (0, 0) is an asymptotically stable spiral point of the linear system in (23), so Theorem 2 implies that (4, 3) is an asymptotically stable spiral point of the original almost linear system in (21). Figure 7.22 shows a typical trajectory of the linear system in (23), and Fig. 7.23 shows a corresponding trajectory of the almost linear system.

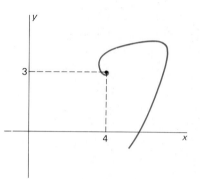

FIGURE 7.23 A spiral trajectory of the almost linear system in Eq. (21).

7.3 PROBLEMS

In each of Problems 1–10, determine the type of the critical point (0, 0) and whether it is asymptotically stable, stable, or unstable.

1. $x' = -2x + y, \ y' = x - 2y$
2. $x' = 4x - y, \ y' = 2x + y$
3. $x' = x + 2y, \ y' = 2x + y$
4. $x' = 3x + y, \ y' = 5x - y$
5. $x' = x - 2y, \ y' = 2x - 3y$
6. $x' = 5x - 3y, \ y' = 3x - y$
7. $x' = 3x - 2y, \ y' = 4x - y$
8. $x' = x - 3y, \ y' = 6x - 5y$
9. $x' = 2x - 2y, \ y' = 4x - 2y$
10. $x' = x - 2y, \ y' = 5x - y$

Each of the systems in Problems 11–18 has a single critical point (x_0, y_0). Classify it as to type and stability. Begin by making the substitutions $u = x - x_0$, $v = y - y_0$, as in Example 1.

11. $x' = x - 2y, \ y' = 3x - 4y - 2$
12. $x' = x - 2y - 8, \ y' = x + 4y + 10$
13. $x' = 2x - y - 2, \ y' = 3x - 2y - 2$
14. $x' = x + y - 7, \ y' = 3x - y - 5$
15. $x' = x - y, \ y' = 5x - 3y - 2$
16. $x' = x - 2y + 1, \ y' = x + 3y - 9$
17. $x' = x - 5y - 5, \ y' = x - y - 3$
18. $x' = 4x - 5y + 3, \ y' = 5x - 4y + 6$

In each of Problems 19–28, investigate the type and stability of the critical point (0, 0) of the given almost linear system.

19. $x' = x - 3y + 2xy, \ y' = 4x - 6y - xy$
20. $x' = 6x - 5y + x^2, \ y' = 2x - y + y^2$
21. $x' = x + 2y + x^2 + y^2, \ y' = 2x - 2y - 3xy$
22. $x' = x + 4y - xy^2, \ y' = 2x - y + x^2y$
23. $x' = 2x - 5y + x^3, \ y' = 4x - 6y + y^4$
24. $x' = 5x - 5y + x(x^2 + y^2), \ y' = 5x - 3y + y(x^2 + y^2)$

25. $x' = x - 2y + 3xy$, $y' = 2x - 3y - x^2 - y^2$

26. $x' = 3x - 2y - x^2 - y^2$, $y' = 2x - y + 3xy$

27. $x' = x - y + x^4 - y^2$, $y' = 2x - y + y^4 - x^2$

28. $x' = 3x - y + x^3 + y^3$, $y' = 13x - 3y + 3xy$

In each of Problems 29–32, find all critical points of the given system and investigate the type and stability of each.

29. $x' = x - y$, $y' = x^2 - y$

30. $x' = y - 1$, $y' = x^2 - y$

31. $x' = y^2 - 1$, $y' = x^3 - y$

32. $x' = xy - 2$, $y' = x - 2y$

Problems 33 and 34 illustrate the sensitive cases in which a small perturbation in the coefficients of a linear system can change the type or stability (or both) of the critical point $(0, 0)$.

33. Consider the linear system

$$\frac{dx}{dt} = hx - 4y, \qquad \frac{dy}{dt} = x + hy.$$

(a) Show that $(0, 0)$ is a center if $h = 0$. (b) Show that $(0, 0)$ is an unstable spiral point if $h > 0$. (c) Show that $(0, 0)$ is an asymptotically stable spiral point if $h < 0$. Thus small perturbations of the system $x' = -4y$, $y' = x$ can change both the type and the stability of the critical point $(0, 0)$.

34. Consider the linear system

$$\frac{dx}{dt} = -x + hy, \qquad \frac{dy}{dt} = x - y.$$

(a) Show that $(0, 0)$ is an asymptotically stable node if $h = 0$. (b) Show that $(0, 0)$ is an asymptotically stable spiral point if $h < 0$. (c) Show that $(0, 0)$ is an asymptotically stable node if $0 < h < 1$. Thus small perturbations of the system $x' = -x$, $y' = x - y$ can change the type of the critical point $(0, 0)$ without affecting its stability.

35. This problem deals with the almost linear system

$$\frac{dx}{dy} = y + hx(x^2 + y^2), \qquad \frac{dy}{dt} = -x + hy(x^2 + y^2),$$

in illustration of the sensitive case of Theorem 2, in which the theorem provides no information about the stability of the critical point $(0, 0)$. (a) Show that $(0, 0)$ is a center of the linear system obtained by setting $h = 0$. (b) Suppose that $h \neq 0$. Let $r^2 = x^2 + y^2$, and then apply the fact that $x(dx/dt) + y(dy/dt) = r(dr/dt)$ to show that $dr/dt = hr^3$. (c) Suppose that $h = -1$. Integrate the differential equation in (b); then show that $r \to 0$ as $t \to +\infty$. Thus $(0, 0)$ is an asymptotically stable critical point

of the almost linear system in this case. (d) Suppose that $h = +1$. Show that $r \to +\infty$ as t increases, so $(0, 0)$ is an unstable critical point in this case.

36. In the case of a two-dimensional system that is *not* almost linear, the trajectories near an isolated critical point can exhibit a considerably more complicated structure than those near the nodes, centers, saddle points, and spiral points discussed in this section. For example, consider the system

$$\frac{dx}{dt} = x(x^3 - 2y^3),$$

$$\frac{dy}{dt} = y(2x^3 - y^3) \tag{24}$$

having $(0, 0)$ as an isolated critical point. This system is not almost linear because $(0, 0)$ is not an *isolated* critical point of the trivial associated linear system $x' = 0$, $y' = 0$. Solve the homogeneous first order equation

$$\frac{dy}{dx} = \frac{y(2x^3 - y^3)}{x(x^3 - 2y^3)}$$

to show that the trajectories of the system in (24) are folia of Descartes of the form

$$x^3 + y^3 = 3cxy$$

where c is an arbitrary constant (see Fig. 7.24).

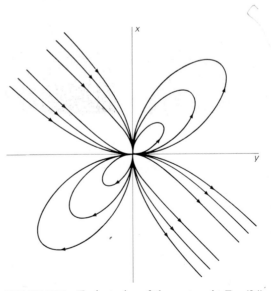

FIGURE 7.24 Trajectories of the system in Eq. (24).

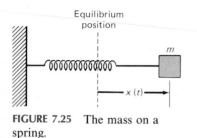

Equilibrium
position

m

$x(t)$

FIGURE 7.25 The mass on a spring.

Nonlinear Mechanical Systems

Now we apply the qualitative methods of Sections 7.2 and 7.3 to the analysis of simple mechanical systems like the mass-on-a-spring system shown in Fig. 7.25. Let m denote the mass in a suitable system of units and let $x(t)$ denote the displacement of the mass at time t from its equilibrium position (where the spring is unstretched). Previously, we always have assumed that the force $F(x)$ exerted by the spring on the mass is a *linear* function of x: $F(x) = -kx$ (Hooke's law). In reality, however, every spring in nature actually is nonlinear (even if only slightly so). Here we are interested specifically in the effects of nonlinearity.

So now we allow the force function $F(x)$ to be nonlinear. Because $F(0) = 0$ at the equilibrium position $x = 0$, we may assume that F has a power series expansion of the form

$$F(x) = -kx + \alpha x^2 + \beta x^3 + \cdots. \tag{1}$$

We take $k > 0$ so that the reaction of the spring is directed opposite to the displacement when x is sufficiently small. If we assume also that the reaction of the spring is symmetric with respect to positive and negative displacements by the same distance, then $F(-x) = -F(x)$, so F is an *odd* function. In this case it follows that the coefficient of x^n in (1) is zero if n is even, so the first nonlinear term is the one involving x^3.

For a simple mathematical model of a nonlinear spring we therefore take

$$F(x) = -kx + \beta x^3, \tag{2}$$

ignoring all terms in (1) of degree greater than 3. The equation of motion of the mass m is then

$$mx'' = -kx + \beta x^3. \tag{3}$$

Introducing the velocity

$$y = \frac{dx}{dt}, \tag{4}$$

we get the equivalent first order system

$$\left. \begin{array}{l} \dfrac{dx}{dt} = y, \\[2mm] m\dfrac{dy}{dt} = -kx + \beta x^3. \end{array} \right\} \tag{5}$$

We can solve explicitly for the phase plane trajectories of this system by writing

$$\frac{dy}{dx} = \frac{dy/dt}{dx/dt} = \frac{-kx + \beta x^3}{my},$$

whence

$$my \, dy + (kx - \beta x^3) \, dx = 0.$$

Integration then yields

$$\tfrac{1}{2}my^2 + \tfrac{1}{2}kx^2 - \tfrac{1}{4}\beta x^4 = E \tag{6}$$

for the equation of a typical trajectory. We write E for the arbitrary constant because $\tfrac{1}{2}my^2$ is the kinetic energy of the mass, and it is natural to define

$$V(x) = \tfrac{1}{2}kx^2 - \tfrac{1}{4}\beta x^4 \tag{7}$$

as the potential energy of the spring. Then Eq. (6) expresses conservation of energy for the (undamped) motion of a mass on a spring.

There are two cases to consider. The spring is called *hard* if $\beta < 0$, *soft* if $\beta > 0$. If $\beta < 0$, then the only critical point of the system in (5) is the origin $(0, 0)$. and each trajectory

$$\tfrac{1}{2}my^2 + \tfrac{1}{2}kx^2 + \tfrac{1}{4}|\beta|x^4 = E > 0 \tag{8}$$

is an oval like those shown in Fig. 7.26. Each of these oval trajectories represents a periodic oscillation of the mass back and forth about its equilibrium position, and the critical point $(0, 0)$ is a stable center. Thus the behavior of a mass on a hard spring resembles qualitatively that of a mass on a linear spring with $\beta = 0$ (as in Example 3 of Section 7.2); the phase plane trajectories in Fig. 7.26 resemble the elliptical trajectories in Fig. 7.7. But one difference between the linear and nonlinear situations is that, whereas the period $T = 2\pi\sqrt{m/k}$ of oscillation of a mass on a linear spring is independent of the initial conditions, the period of a mass on a nonlinear spring depends on its initial position $x(0)$ and initial velocity $y(0)$ (see Problem 15).

As the following example illustrates, there is a greater range of possible behavior for a mass on a soft spring with $\beta > 0$.

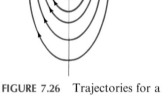

FIGURE 7.26 Trajectories for a mass on a hard spring with $\beta < 0$.

EXAMPLE 1 If $m = 1$, $k = 4$, and $\beta = 1$, then the equation of motion of the mass is

$$x'' + 4x - x^3 = 0 \tag{9}$$

and Eq. (6) gives the trajectories in the form

$$\tfrac{1}{2}y^2 + 2x^2 - \tfrac{1}{4}x^4 = E. \tag{10}$$

After solving for

$$y = \pm\sqrt{2E - 4x^2 + \tfrac{1}{2}x^4}$$

we can plot for various values of the (constant) energy E the trajectories shown in Fig. 7.27.

The emphasized curves obtained with $E = 4$ are called *separatrices*—they separate regions of different behavior—and intersect the x-axis at the critical points $(-2, 0)$ and $(2, 0)$. These, together with the other critical point $(0, 0)$, are the only points at which the mass can remain at rest.

The nature of the motion of the mass is determined by which type of trajectory its initial conditions place it on. The simple closed trajectories encircling $(0, 0)$ in the region bounded by the separatrices correspond to energies in the range $0 < E < 4$. These closed trajectories represent *periodic* oscillations of the mass back and forth around the equilibrium point $x = 0$.

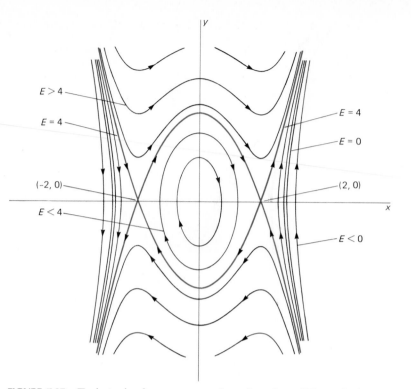

FIGURE 7.27 Trajectories for a mass on the soft spring of Example 1.

The unbounded trajectories lying in the regions above and below the separatrices correspond to values of E greater than 4. These represent motions in which the mass approaches $x = 0$ with sufficient energy that it continues on through the equilibrium point, never to return again.

The unbounded trajectories opening to the right and left correspond to values of E smaller than 4, including negative values. These represent motions in which the mass initially is headed toward the equilibrium point $x = 0$, but with insufficient energy to reach it. At some point the mass reverses direction and heads back whence it came.

Although in this example Eq. (10) makes it possible to construct the phase portrait in Fig. 7.27, it nevertheless is instructive to examine the critical points $(0, 0)$, $(-2, 0)$, and $(2, 0)$ of the equivalent first order system

$$\left.\begin{array}{l} \dfrac{dx}{dt} = y, \\[2ex] \dfrac{dy}{dt} = -4x + x^3. \end{array}\right\} \tag{11}$$

At $(0, 0)$ we find that the linearized system

$$\left.\begin{array}{l} x' = y, \\ y' = -4x \end{array}\right\}$$

CHAPTER 7: Qualitative Properties and Existence of Solutions

has a stable center and elliptical trajectories given by

$$x(t) = A \cos 2t + B \sin 2t, \\ y(t) = 2B \cos 2t - 2A \sin 2t. \tag{12}$$

Note that the direction of motion along these ellipses is consistent with the clockwise orientation of the closed trajectories in Fig. 7.27.

At $(2, 0)$ the substitution $u = x - 2$, $v = y$ in (11) yields the system

$$u' = v, \\ v' = 8u + 6u^2 + u^3 \tag{13}$$

having $(0, 0)$ as the corresponding critical point. We find that the linearized system

$$u' = v, \\ v' = 8u \tag{ }$$

has a saddle point at $(0, 0)$, with hyperbolic trajectories given by

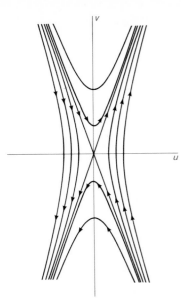

$$u(t) = A \cosh t\sqrt{8} + B \sinh t\sqrt{8}, \\ v(t) = B\sqrt{8} \cosh t\sqrt{8} + A\sqrt{8} \sinh t\sqrt{8}. \tag{14}$$

FIGURE 7.28 Linearized trajectories of the system in Eq. (13).

These equations yield the trajectories shown in Fig. 7.28. Note that in appearance and direction they are consistent with the picture in Fig. 7.27 near $(2, 0)$. The analysis of the critical point $(-2, 0)$ is similar.

DAMPED NONLINEAR VIBRATIONS

Suppose now that the mass on a spring is connected also to a dashpot that provides a force of resistance proportional to the velocity $y = x'$ of the mass. If the spring is still assumed nonlinear as in (2), then the equation of motion of the mass is

$$mx'' = -cx' - kx + \beta x^3 \tag{15}$$

where $c > 0$ is the dashpot constant. If $\beta > 0$, then the equivalent first order system

$$x' = y, \\ my' = -kx - cy + \beta x^3 \tag{16}$$

has critical points $(0, 0)$ and $(\pm\sqrt{k/\beta}, 0)$. If we take $m = 1$ for convenience, then at $(0, 0)$ the characteristic equation of the corresponding linear system is

$$\lambda(\lambda + c) + k = \lambda^2 + c\lambda + k = 0. \tag{17}$$

It follows that $(0, 0)$ is a stable node if $c^2 > 4k$, a stable spiral point if $c^2 < 4k$. The next example illustrates the latter case.

EXAMPLE 2 Suppose that $m = 1$, $k = 5$, $\beta = \frac{5}{4}$, and $c = 2$. Then the nonlinear system in (16) is

$$x' = y, \\ y' = -5x - 2y + \frac{5}{4}x^3. \tag{18}$$

Now no explicit solution for the phase plane trajectories is available, so we proceed to investigate the critical points $(0, 0)$, $(2, 0)$, and $(-2, 0)$.

At $(0, 0)$ the linearized system

$$\left. \begin{array}{l} x' = y, \\ y' = -5x - 2y \end{array} \right\}$$

has characteristic equation $\lambda^2 + 2\lambda + 5 = 0$ with roots $\lambda = -1 \pm 2i$. Hence $(0, 0)$ is a stable spiral point of (18), and the linearized position function of the mass is of the form

$$x(t) = e^{-t}(A \cos 2t + B \sin 2t),$$

an exponentially damped oscillation about $x = 0$.

At $(2, 0)$ the substitution $u = x - 2$, $v = y$ in (18) yields the system

$$\left. \begin{array}{l} u' = v, \\ v' = 10u - 2v + \dfrac{15}{2} u^2 + \dfrac{5}{4} u^3 \end{array} \right\} \tag{19}$$

with corresponding critical point $(0, 0)$. The linearized system

$$\left. \begin{array}{l} u' = v, \\ v' = 10u - 2v \end{array} \right\}$$

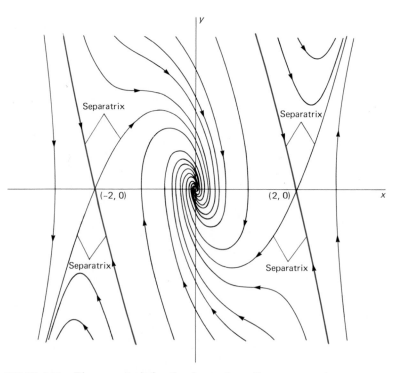

FIGURE 7.29 Phase portrait for the damped nonlinear system in Example 2.

has characteristic equation $\lambda^2 + 2\lambda - 10 = 0$ with roots $\lambda_1 = -1 - \sqrt{11} < 0$ and $\lambda_2 = -1 + \sqrt{11} > 0$. It follows that $(2, 0)$ is an unstable saddle point of the original system in (18). A similar analysis shows that $(-2, 0)$ is also an unstable saddle point.

The phase portrait in Fig. 7.29 shows trajectories of (18) plotted using the numerical methods of Section 6.4, and exhibits a stable spiral point at $(0, 0)$ and unstable saddle points at $(-2, 0)$ and $(2, 0)$. The periodic oscillations of the undamped case (Fig. 7.27) are now replaced by damped oscillations about the equilibrium point $x = 0$. The emphasized branches of the separatrices divide the phase plane into regions of stability and instability. Every trajectory whose initial point lies in the region between these emphasized branches must spiral into the origin as $t \to +\infty$, and is therefore a "stable trajectory" representing a damped oscillation. The trajectories that start outside this region are unstable ones along which $x(t) \to \pm\infty$ as $t \to +\infty$.

THE NONLINEAR PENDULUM

In Section 2.4 we derived the equation

$$\frac{d^2\theta}{dt^2} + \frac{g}{L} \sin \theta = 0 \tag{20}$$

for the undamped oscillations of the simple pendulum shown in Fig. 7.30. There we used the approximation $\sin \theta \approx \theta$ for θ small to replace (20) with the linear model

$$\theta'' + \omega^2\theta = 0 \tag{21}$$

where $\omega^2 = g/L$. The general solution

$$\theta(t) = A \cos \omega t + B \sin \omega t \tag{22}$$

of (21) describes oscillations about the equilibrium position $\theta = 0$ with circular frequency ω and amplitude $C = \sqrt{A^2 + B^2}$.

The linear model does not adequately describe the possible motions of the pendulum for large values of θ. For instance, the equilibrium solution $\theta(t) \equiv \pi$ of (20), with the pendulum standing straight up, does not satisfy the linear equation in (21). Nor does (22) include the situation in which the pendulum "goes over the top" repeatedly, so that $\theta(t)$ is a steadily increasing rather than an oscillatory function of t. To investigate these phenomena we must analyze the nonlinear equation $\theta'' + \omega^2 \sin \theta = 0$ rather than merely its linearization. We also want to include the possibility of resistance proportional to velocity, so we consider the general nonlinear pendulum equation

$$\theta'' + c\theta' + \omega^2 \sin \theta = 0. \tag{23}$$

We examine first the undamped case, in which $c = 0$. With $x = \theta(t)$ and $y = \theta'(t)$ the equivalent first order system is

$$\left.\begin{aligned} x' &= y, \\ y' &= -\omega^2 \sin x. \end{aligned}\right\} \tag{24}$$

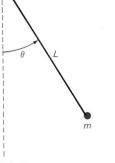

FIGURE 7.30 The simple pendulum.

Upon substituting the Taylor series for $\sin x$ we get

$$\left.\begin{array}{l} x' = y, \\[2mm] y' = -\omega^2 x + \dfrac{\omega^2 x^3}{3!} - \dfrac{\omega^2 x^5}{5!} + \cdots \end{array}\right\} \qquad (25)$$

Thus (24) is an almost linear system of the form

$$x' = ax + by + f(x, y), \qquad y' = cx + dy + g(x, y)$$

with $f(x, y) \equiv 0$ and $g(x, y) = \omega^2(x^3/3! - x^5/5! + \cdots)$.

The critical points of the system in (24) are the points $(n\pi, 0)$ with n an integer. The nature of the critical point $(n\pi, 0)$ depends on whether n is even or odd.

If $n = 2m$ is even then, because $\sin(u + 2m\pi) = \sin u$, the substitution $u = x - 2m\pi$, $v = y$ in (24) yields the system

$$\left.\begin{array}{l} u' = v, \\[2mm] v' = -\omega^2 \sin u \end{array}\right\} \qquad (26)$$

having $(0, 0)$ as the corresponding critical point. Just as in (25), the linearization of (26) is the system

$$\left.\begin{array}{l} u' = v, \\[2mm] v' = -\omega^2 u \end{array}\right\} \qquad (27)$$

for which $(0, 0)$ is the familiar stable center with elliptical trajectories shown in Fig. 7.31 (see Example 3 of Section 7.2). Although this is the delicate case in which Theorem 2 of Section 7.3 does not settle the matter, we will soon see that $(2m\pi, 0)$ is also a stable center for the original system in (24).

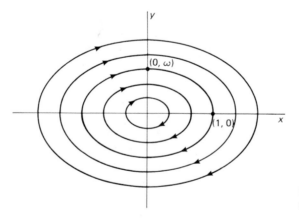

FIGURE 7.31   Trajectories of the system in (27).

If $n = 2m + 1$ is odd, then, because $\sin(u + (2m + 1)\pi) = -\sin u$, the substitution $u = x - (2m + 1)\pi$, $v = y$ in (24) yields the system

$$\left.\begin{array}{l} u' = v, \\[2mm] v' = +\omega^2 \sin u \end{array}\right\} \qquad (28)$$

having $(0, 0)$ as the corresponding critical point. The linearization

$$\left. \begin{array}{l} u' = v, \\ v' = +\omega^2 u \end{array} \right\} \tag{29}$$

of (28) has general solution

$$u(t) = \quad A \cosh \omega t + \quad B \sinh \omega t,$$

$$v(t) = B\omega \cosh \omega t + A\omega \sinh \omega t,$$

so we see that $(0, 0)$ is the unstable saddle point shown in Fig. 7.32. It follows from Theorem 2 of Section 7.3 that $((2m + 1)\pi, 0)$ is a similar unstable saddle point for the original almost linear system in (24).

In the undamped case we can see how these centers and saddle points fit together by solving explicitly for the trajectories of (24). If we write

$$\frac{dy}{dx} = \frac{y'}{x'} = -\frac{\omega^2 \sin x}{y}$$

and separate the variables,

$$y\,dy + \omega^2 \sin x\,dx = 0,$$

then integration from $x = 0$ to $x = x$ yields

$$\tfrac{1}{2}y^2 + \omega^2(1 - \cos x) = E. \tag{30}$$

We write E for the arbitrary constant because if physical units are chosen so that $m = L = 1$, then the first term on the left is the kinetic energy and the second term is the potential energy of the mass on the end of the pendulum. Then E is the total energy; Eq. (30) thus expresses conservation of energy for the undamped pendulum. If we solve (30) for y and use a half-angle identity, we get the equation

$$y = \pm\sqrt{2E - 4\omega^2 \sin^2 \tfrac{1}{2}x} \tag{31}$$

for the trajectories. Figure 7.33 shows the result of plotting these trajectories for

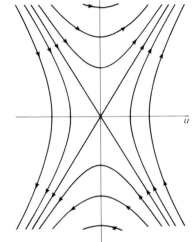

FIGURE 7.32   Trajectories of the system in Eq. (29).

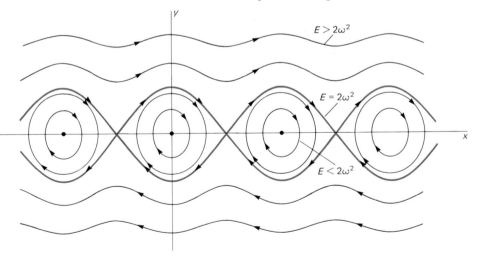

FIGURE 7.33   Phase plane trajectories for the undamped nonlinear pendulum. There are unstable saddles at the odd multiples of π and stable centers at the even multiples of π.

various values of E. Arrows have been inserted consistent with the critical point analysis given above.

The separatrices, which are emphasized in Fig. 7.33, correspond to the critical value $E = 2\omega^2$ of the energy; they enter and leave the unstable critical points $(n\pi, 0)$ with n an odd integer. Following the arrow along a separatrix, the pendulum theoretically approaches a balanced vertical position $\theta = x = (2m + 1)\pi$ with just enough energy to reach it but not enough to "go over the top." The instability of this equilibrium position indicates that this behavior will never be observed in practice!

The simple closed trajectories encircling the stable critical points—all of which correspond to the downward position $\theta = 2m\pi$ of the pendulum—represent periodic oscillations of the pendulum back and forth about the stable equilibrium position $\theta = 0$. These correspond to energies $E < 2\omega^2$ that are insufficient for the pendulum to ascend to the vertical upward position. The unbounded trajectories with $E > 2\omega^2$ represent whirling motions of the pendulum in which it goes over the top repeatedly.

If the pendulum is released from rest with initial conditions

$$x(0) = \theta(0) = \alpha, \qquad y(0) = \theta'(0) = 0, \tag{32}$$

then Eq. (30) with $t = 0$ reduces to

$$\omega^2(1 - \cos \alpha) = E. \tag{33}$$

Hence $E < 2\omega^2$ if $0 < \alpha < \pi$, so a periodic oscillation of the pendulum ensues. To determine the *period* of this oscillation, we subtract Eq. (33) from Eq. (30) and write the result (with $x = \theta$ and $y = d\theta/dt$) in the form

$$\frac{1}{2}\left(\frac{d\theta}{dt}\right)^2 = \omega^2(\cos \theta - \cos \alpha). \tag{34}$$

The period T of time required for one complete oscillation is four times the amount of time required for θ to decrease from $\theta = \alpha$ to $\theta = 0$, one-fourth of an oscillation. Hence we solve (34) for $dt/d\theta$ and integrate to get

$$T = \frac{4}{\omega\sqrt{2}} \int_0^\alpha \frac{d\theta}{\sqrt{\cos \theta - \cos \alpha}}. \tag{35}$$

To attempt to evaluate this integral we first use the identity $\cos \theta = 1 - 2 \sin^2 (\theta/2)$ and get

$$T = \frac{2}{\omega} \int_0^\alpha \frac{d\theta}{\sqrt{k^2 - \sin^2 (\theta/2)}} \tag{36}$$

where

$$k = \sin \frac{\alpha}{2}. \tag{37}$$

Next, the substitution $u = (1/k) \sin (\theta/2)$ yields

$$T = \frac{4}{\omega} \int_0^1 \frac{du}{\sqrt{(1 - u^2)(1 - k^2 u^2)}}. \tag{38}$$

CHAPTER 7: Qualitative Properties and Existence of Solutions

Finally, the substitution $u = \sin \phi$ gives

$$T = \frac{4}{\omega} \int_0^{\pi/2} \frac{d\phi}{\sqrt{1 - k^2 \sin^2 \phi}}. \tag{39}$$

The integral in (39) is the *elliptic integral of the first kind* that often is denoted by $F(k, \pi/2)$. It can be evaluated numerically as follows. First we use the binomial series

$$\frac{1}{\sqrt{1 - x}} = 1 + \sum_{n=1}^{\infty} \frac{1 \cdot 3 \cdots (2n - 1)}{2 \cdot 4 \cdots (2n)} x^n \tag{40}$$

with $x = k^2 \sin^2 \phi < 1$ to expand the integrand in (39). Then we integrate termwise using the tabulated integral formula

$$\int_0^{\pi/2} \sin^{2n} \phi \, d\phi = \frac{\pi}{2} \cdot \frac{1 \cdot 3 \cdots (2n - 1)}{2 \cdot 4 \cdots (2n)}. \tag{41}$$

The final result is the formula

$$T = \frac{2\pi}{\omega} \left[1 + \sum_{n=1}^{\infty} \left(\frac{1 \cdot 3 \cdots (2n - 1)}{2 \cdot 4 \cdots (2n)} \right)^2 k^{2n} \right]$$

$$= T_0 \left[1 + \left(\frac{1}{2} \right)^2 k^2 + \left(\frac{1 \cdot 3}{2 \cdot 4} \right)^2 k^4 + \left(\frac{1 \cdot 3 \cdot 5}{2 \cdot 4 \cdot 6} \right)^2 k^6 + \cdots \right] \tag{42}$$

for the period T of the nonlinear pendulum released from rest with initial angle $\theta(0) = \alpha$, in terms of the linearized period $T_0 = 2\pi/\omega$ and $k = \sin(\alpha/2)$.

The infinite series within the brackets in (42) gives the factor T/T_0 by which the nonlinear period T is longer than the linearized period. The table in Fig. 7.34, obtained by summing this series numerically, shows how T/T_0 increases as α is increased. Thus T is 0.19% greater than T_0 if $\alpha = 10°$, whereas T is 18.03% greater if $\alpha = 90°$. But even a 0.19% discrepancy is significant—the calculation

$$(0.0019) \times 3600 \, \frac{\text{seconds}}{\text{hour}} \times 24 \, \frac{\text{hours}}{\text{day}} \times 7 \, \frac{\text{days}}{\text{week}}$$

$$\approx 1149 \text{ (seconds)}$$

shows that the linearized model is quite inadequate for a pendulum clock; a discrepancy of 19 min 9 s after only one week is unacceptable.

Finally, we discuss briefly the *damped* nonlinear pendulum. The almost linear first order system equivalent to Eq. (23) is

$$\left. \begin{aligned} x' &= y, \\ y' &= -\omega^2 \sin x - cy, \end{aligned} \right\} \tag{43}$$

and again the critical points are of the form $(n\pi, 0)$ with n an integer. In Problem 9 we ask you to verify that:

- If n is odd, then $(n\pi, 0)$ is an unstable saddle point of (43), just as in the undamped case; but
- If n is even and $c^2 > 4\omega^2$, then $(n\pi, 0)$ is a stable node; while
- If n is even and $c^2 < 4\omega^2$, then $(n\pi, 0)$ is a stable spiral point.

α	T/T_0
10°	1.0019
20°	1.0077
30°	1.0174
40°	1.0313
50°	1.0498
60°	1.0732
70°	1.1021
80°	1.1375
90°	1.1803

FIGURE 7.34 Dependence of the period T of a nonlinear pendulum on its initial angle α.

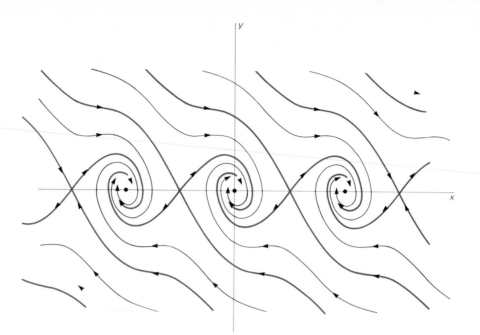

FIGURE 7.35 Phase plane trajectories for the underdamped nonlinear pendulum. There are unstable saddles at the odd multiples of π and spiral points at the even multiples of π.

Figure 7.35 shows the phase plane trajectories for the more interesting under-damped case, $c^2 < 4\omega^2$. Other than the physically unattainable separatrix trajectories that enter unstable saddle points, every trajectory eventually is "trapped" by one of the stable spiral points $(n\pi, 0)$ with n an even integer. What this means is that even if the pendulum starts with enough energy to go over the top, after a certain (finite) number of revolutions it has lost enough energy that thereafter it undergoes damped oscillations about its stable (lower) equilibrium position.

7.4 PROBLEMS

In each of Problems 1–4, show that the given system is almost linear with (0, 0) as a critical point, and classify this critical point as to type and stability.

1. $x' = 1 - e^x + 2y$
$\quad y' = -x - 4 \sin y$

2. $x' = 2 \sin x + \sin y$
$\quad y' = \sin x + 2 \sin y$

3. $x' = e^x + 2y - 1$
$\quad y' = 8x + e^y - 1$

4. $x' = \sin x \cos y - 2y$
$\quad y' = 4x - 3 \cos x \sin y$

Find and classify each of the critical points of the almost linear systems in Problems 5–8.

5. $x' = -x + \sin y$
$\quad y' = 2x$

6. $x' = y$
$\quad y' = \sin \pi x - y$

7. $x' = 1 - e^{x-y}$
$\quad y' = 2 \sin x$

8. $x' = 3 \sin x + y$
$\quad y' = \sin x + 2y$

9. For the damped pendulum system in (43), verify the classification of critical points of the form $(n\pi, 0)$ that is stated in the text.

In each of Problems 10–14, a second order equation of the form

$$\frac{d^2x}{dt^2} + f(x, x') = 0$$

corresponding to a certain mass-and-spring system is given. Find and classify the critical points of the equivalent first order system.

10. $x'' + 20x - 5x^3 = 0$: Verify that the critical points look like the ones shown in Fig. 7.27.

11. $x'' + 2x' + 20x - 5x^3 = 0$: Verify that the critical points look like the ones shown in Fig. 7.29.

12. $x'' - 8x + 2x^3 = 0$: Here the linear part of the force is repulsive rather than attractive (as for an ordinary spring). Verify that the critical points look like the ones shown in Fig. 7.36. Thus there are two stable equilibrium points and three types of periodic oscillations.

13. $x'' + 4x - x^2 = 0$: Here the force function is nonsymmetric. Verify that the critical points look like those shown in Fig. 7.37.

14. $x'' + 4x - 5x^3 + x^5 = 0$: The idea here is that terms through the fifth degree in an odd force function have been retained. Verify that the critical points look like those shown in Fig. 7.38.

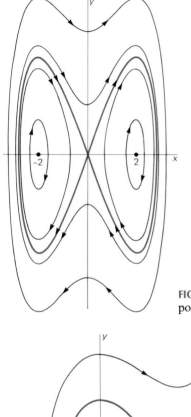

FIGURE 7.36   The phase portrait for Problem 12.

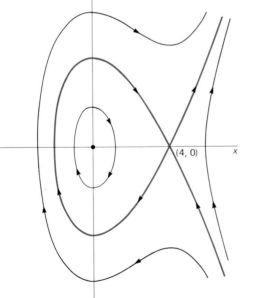

FIGURE 7.37   The phase portrait for Problem 13.

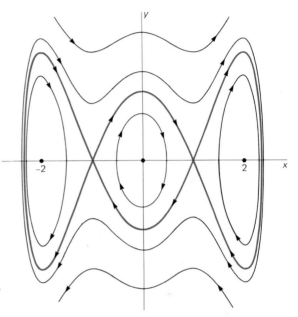

FIGURE 7.38   The phase portrait for Problem 14.

15. This problem outlines an investigation of the period T of oscillation of a mass on a nonlinear spring with equation of motion

$$x'' + \phi(x) = 0. \tag{44}$$

If $\phi(x) = kx$ with $k > 0$, then the spring actually is linear with period $T_0 = 2\pi/\sqrt{k}$. (a) Integrate once (as in Eq. (6)) to derive the energy equation

$$\tfrac{1}{2}y^2 + V(x) = E \tag{45}$$

where $y = dx/dt$ and

$$V(x) = \int_0^x \phi(u)\, du. \tag{46}$$

(b) If the mass is released from rest with initial conditions $x(0) = x_0$, $y(0) = 0$ and periodic oscillations ensue, conclude from Eq. (45) that $E = V(x_0)$ and that the time T required for one complete oscillation is

$$T = \frac{4}{\sqrt{2}} \int_0^{x_0} \frac{du}{\sqrt{V(x_0) - V(u)}}. \tag{47}$$

(c) If $\phi(x) = kx - \beta x^3$ as in the text, deduce from Eqs. (46) and (47) that

$$T = 4\sqrt{2} \int_0^{x_0} \frac{dx}{\sqrt{(x_0^2 - u^2)(2k - \beta x_0^2 - \beta u^2)}}. \tag{48}$$

(d) Substitute $u = x_0 \cos \phi$ in (48) to show that

$$T = \frac{2T_0}{\pi \sqrt{1 - \epsilon}} \int_0^{\pi/2} \frac{d\phi}{\sqrt{1 - \mu \sin^2 \phi}} \tag{49}$$

where $T_0 = 2\pi/\sqrt{k}$ is the linear period and

$$\epsilon = \frac{\beta}{k} x_0^2, \qquad \mu = -\frac{1}{2} \cdot \frac{\epsilon}{1 - \epsilon}. \tag{50}$$

(e) Finally, use the binomial series in (40) and the integral formula in (41) to evaluate the elliptic integral in (49) and thereby show that the period T of oscillation is given by

$$T = \frac{T_0}{\sqrt{1 - \epsilon}} \left(1 + \frac{1}{4} \mu + \frac{9}{64} \mu^2 + \frac{25}{256} \mu^3 + \cdots \right). \tag{51}$$

(f) If $\epsilon = \beta x_0^2/k$ is sufficiently small that ϵ^2 is negligible, deduce from Eqs. (50) and (51) that

$$T \approx T_0 \left(1 + \frac{3}{8} \epsilon \right) = T_0 \left(1 + \frac{3\beta}{8k} x_0^2 \right). \tag{52}$$

It follows that:

- If $\beta > 0$, so the spring is *soft*, then $T > T_0$, and increasing x_0 increases T, so the larger ovals in Fig. 7.27 correspond to smaller frequencies.

- If $\beta < 0$, so the spring is *hard*, then $T < T_0$, and increasing x_0 decreases T, so the larger ovals in Fig. 7.26 correspond to larger frequencies.

7.5

Ecological Applications: Predators and Competitors

Some of the most interesting and important applications of stability theory involve the interactions between two or more biological populations occupying the same environment. We consider first a **predator-prey** situation involving two species. One species—the **predators**—feeds on the other species—the **prey**—which in turn feeds on some third food item readily available in the environment. A standard example is a population of foxes and rabbits in a woodland; the foxes (predators) eat rabbits (the prey), while the rabbits eat certain vegetation in the woodland. Other examples are sharks (predators) and food fish (prey), bass (predators) and sunfish (prey), ladybugs (predators) and aphids (prey), and beetles (predators) and scale insects (prey).

The classical mathematical model of a predator-prey situation was developed in the 1920s by the Italian mathematician Vito Volterra (1860–1940) in order to analyze the cyclic variations observed in the shark and food fish populations in the Adriatic Sea. To construct such a model, we denote the number of prey by $x(t)$ and the number of predators at time t by $y(t)$, and make the following simplifying assumptions:

1. In the absence of predators, the prey population would grow at a natural rate, with $dx/dt = ax$, $a > 0$.

2. In the absence of prey, the predator population would decline at a natural rate, with $dy/dt = -cy$, $c > 0$.

3. When both predator and prey are present, there occurs, in combination with these natural rates of growth and decline, a decline in the prey population and a growth in the predator population, each at a rate proportional to the frequency of encounters between individuals of the two species. We assume further that the frequency of such encounters is proportional to the product xy, reasoning that doubling either population alone should double the fre-

quency of encounters, while doubling both populations ought to quadruple the frequency of encounters. Consequently, the effect of predators eating prey is an *interaction rate* of decline $-bxy$ in the prey population x, and an interaction rate of growth dxy of the predator population y, with b and d positive constants.

When we add the natural and interaction rates described above, we obtain the **predator-prey equations**

$$\left.\begin{aligned}
\frac{dx}{dt} &= ax - bxy = x(a - by), \\[2mm]
\frac{dy}{dt} &= -cy + dxy = y(-c + dx),
\end{aligned}\right\} \tag{1}$$

with the constants a, b, c and d all positive. This is an almost linear system with two critical points, $(0, 0)$ and $(c/d, a/b)$. The point $(0, 0)$ is a saddle point, but the corresponding equilibrium solution $x(t) \equiv 0$, $y(t) \equiv 0$ merely describes simultaneous extinction of both species.

The critical point $(c/d, a/b)$ is of greater interest; $x(t) \equiv c/d$ and $y(t) \equiv a/b$ are the nonzero constant prey and predator populations, respectively, that can coexist in equilibrium. We would like to know, if the initial populations x_0 and y_0 are near these critical populations, whether $(x(t), y(t))$ remains near $(c/d, a/b)$ for all $t > 0$. That is, is the critical point $(c/d, a/b)$ stable? To attempt to answer this question, we substitute $u = x - c/d$, $v = y - a/b$ in (1). We thereby obtain the almost linear system

$$\left.\begin{aligned}
\frac{du}{dt} &= -\frac{bc}{d} v - buv, \\[2mm]
\frac{dv}{dt} &= \frac{ad}{b} u - duv,
\end{aligned}\right\} \tag{2}$$

which has $(0, 0)$ as its critical point corresponding to the critical point $(c/d, a/b)$ of the system in (1). The corresponding linear system

$$\left.\begin{aligned}
\frac{du}{dt} &= -\frac{bc}{d} v, \\[2mm]
\frac{dv}{dt} &= \frac{ad}{b} u
\end{aligned}\right\} \tag{3}$$

has characteristic equation $\lambda^2 + ac = 0$ with pure imaginary roots $\lambda_1, \lambda_2 = \pm i\sqrt{ac}$. Hence $(0, 0)$ is a stable center of (3), and the trajectories are ellipses. In fact, if we divide the second equation in (3) by the first, we obtain

$$\frac{dv}{du} = -\frac{ad/b}{bc/d}\frac{u}{v} = -\frac{ad^2 u}{cb^2 v},$$

$$ad^2 u\,du + cb^2 v\,dv = 0,$$

$$ad^2 u^2 + cb^2 v^2 = C, \tag{4}$$

$$\frac{u^2}{A^2} + \frac{v^2}{B^2} = 1,$$

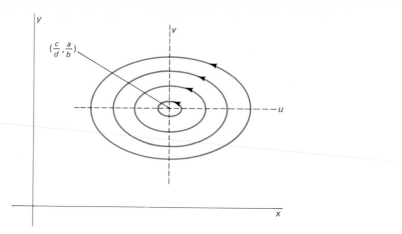

FIGURE 7.39 Linearized trajectories of the system in (2); the directions indicated by the arrows are determined by the signs in Eq. (3).

where C is a constant of integration, $A^2 = C/ad^2$, and $B^2 = C/cb^2$. In terms of x and y, the trajectories of the linearized system are, therefore, ellipses of the form

$$\frac{1}{A^2}\left(x - \frac{c}{d}\right)^2 + \frac{1}{B^2}\left(y - \frac{a}{b}\right)^2 = 1 \tag{5}$$

centered at the critical point $(c/d, a/b)$. Some of these ellipses are shown in Fig. 7.39.

Unfortunately, this analysis does not settle the question of the stability of the critical point $(c/d, a/b)$ of the original nonlinear system in (1), because a stable center represents the indeterminate case of Theorem 2 of Section 7.3, in which the critical point can (aside from a center) be either an unstable or an asymptotically stable spiral point. We can in this case, however, find the trajectories explicitly by dividing the second equation in (1) by the first to obtain

$$\frac{dy}{dx} = \frac{y(-c + dx)}{x(a - by)}.$$

We separate the variables to get

$$\frac{c - dx}{x}\, dx + \frac{a - by}{y}\, dy = 0,$$

and thus

$$c \ln x - dx + a \ln y - by = C, \tag{6}$$

where C is a constant of integration (and dx in Eq. (6) is not a differential, but the product of the positive constant d with x). In any case, the trajectories of the system in (1) near the critical point $(c/d, a/b)$ are the level curves of the function $f(x, y)$ that appears on the left-hand side in Eq. (6). It can be shown that these trajectories are simple closed curves enclosing $(c/d, a/b)$, which is therefore a stable center. Figure 7.40 shows a numerical plot of these trajectories for the case $a = b = c = d = 1$. It follows from Problem 30 of Section 7.2 that $x(t)$ and $y(t)$ are both periodic functions of t; this explains the fluctuations that are experimentally observed in predator-prey populations. If we follow a single trajectory in Fig. 7.40,

CHAPTER 7: Qualitative Properties and Existence of Solutions

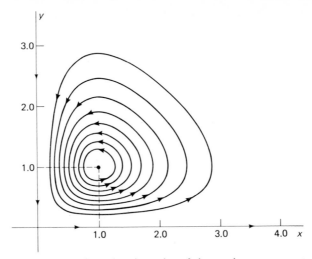

FIGURE 7.40   Actual trajectories of the predator-prey system $x' = x - xy$, $y' = -y + xy$.

beginning at a point where the prey population x is maximal and $y = a/b$ (Why is $y = a/b$ when x is maximal?), we see that x decreases and y increases until $x = c/d$ and the predator population y is maximal. Then both decrease until x is minimal and $y = a/b$ again. And so on around the trajectory back to the initial point. In particular, we see that if both $x_0 > 0$ and $y_0 > 0$, then both $x(t) > 0$ and $y(t) > 0$ for all t, so both populations survive in coexistence with each other. Exception: If the fluctuations are so wide that $x(t)$ is nearly zero, there is a possibility that the last few prey will be devoured, resulting in their immediate extinction and the consequent eventual extinction of the predators. This will certainly take place if ever $x(t) < 2$ in a population of mammals!

COMPETING SPECIES

Now we consider two species (of animals, plants, or bacteria, for instance) with populations $x(t)$ and $y(t)$ and which compete with each other for the food available in their common environment. This is in marked contrast to the case in which one species preys upon the other. To construct a mathematical model that is as realistic as possible, let us assume that in the absence of either species, the other would have a bounded (logistic) population like that considered in Example 1 of Section 7.1. In the absence of any interaction or competition between the two species, their populations would then satisfy the differential equations

$$\left.\begin{aligned} \frac{dx}{dt} &= a_1 x - b_1 x^2, \\[2mm] \frac{dy}{dt} &= a_2 y - b_2 y^2, \end{aligned}\right\} \tag{7}$$

each of the form of Eq. (3) of Section 7.1. But in addition, we assume that competition has the effect of a rate of decline in each population that is proportional to their product xy. We insert such terms with *negative* proportionality constants in the equations in (7) to obtain the **competition equations**

$$\frac{dx}{dt} = a_1 x - b_1 x^2 - c_1 xy = x(a_1 - b_1 x - c_1 y),$$

$$\frac{dy}{dt} = a_2 y - b_2 y^2 - c_2 xy = y(a_2 - b_2 y - c_2 x),$$

(8)

where the coefficients a_1, a_2, b_1, b_2, c_1, and c_2 are all positive.

The almost linear system in (8) has four critical points. Upon setting the right-hand sides of the two equations equal to zero, we see that if $x = 0$, then either $y = 0$ or $y = a_2/b_2$, while if $y = 0$, then either $x = 0$ or $x = a_1/b_1$. This gives the three critical points $(0, 0)$, $(0, a_2/b_2)$, and $(a_1/b_1, 0)$. The fourth is at the intersection of the two lines

$$b_1 x + c_1 y = a_1,$$

$$c_2 x + b_2 y = a_2.$$

(9)

We assume that these two lines are not parallel and that they intersect at a point in the first quadrant. (The other cases will be explored in the exercises.) Then this point (x_E, y_E) is the fourth critical point, and it represents the possibility of peaceful coexistence of the two species, with stable populations $x(t) \equiv x_E$ and $y(t) \equiv y_E$.

We are interested in the stability of the critical point (x_E, y_E). This turns out to depend upon the relative orientation of the two lines in (9). The two possibilities are shown in Fig. 7.41, with the first line in (9) solid and the second dashed. Comparing the slopes of the two lines, we see that Fig. 7.41(a) corresponds to the condition that

$$\frac{a_2/b_2}{a_2/c_2} < \frac{a_1/c_1}{a_1/b_1}; \quad \text{that is,} \quad c_1 c_2 < b_1 b_2.$$

(10)

Similarly, Fig. 7.41(b) corresponds to the condition that

$$\frac{a_2/b_2}{a_2/c_2} > \frac{a_1/c_1}{a_1/b_1}; \quad \text{that is,} \quad c_1 c_2 > b_1 b_2.$$

(11)

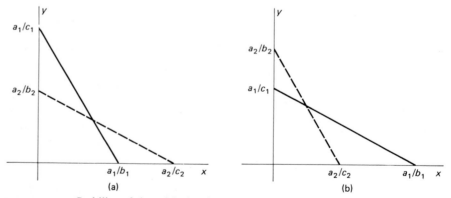

FIGURE 7.41   Stability of the critical point (x_E, y_E).

The conditions in (10) and (11) have a natural interpretation. In the equations in (7), we see that b_1 and b_2 represent the inhibiting effect of each population on its own growth (possibly due to limitations of food or space). On the other hand, c_1 and c_2 represent the effect of competition between the two populations. Thus $b_1 b_2$ is a measure of *inhibition*, while $c_1 c_2$ is a measure of *competition*. A general analysis of the system in (8) shows the following:

1. If $c_1 c_2 < b_1 b_2$, so that competition is small in comparison with inhibition, then (x_E, y_E) is an asymptotically stable critical point that is approached by each solution as $t \to +\infty$. Thus the two species can and do coexist in this case.

2. If $c_1 c_2 > b_1 b_2$, so that competition is large in comparison with inhibition, then (x_E, y_E) is an unstable critical point, and either $x(t)$ or $y(t)$ approaches zero as $t \to +\infty$. Thus the two species cannot coexist in this case; one survives and the other becomes locally extinct.

Rather than carrying out the general analysis that leads to the conclusions stated above, we present two examples that illustrate these two possibilities.

EXAMPLE 1 **(Survival of a single species):** Suppose that the populations x and y satisfy the equations

$$\left.\begin{aligned} \frac{dx}{dt} &= 60x - 4x^2 - 3xy = x(60 - 4x - 3y), \\[2mm] \frac{dy}{dt} &= 42y - 2y^2 - 3xy = y(42 - 3x - 2y). \end{aligned}\right\} \tag{12}$$

These are the competition equations with $a_1 = 60$, $a_2 = 42$, $b_1 = 4$, $b_2 = 2$, and $c_1 = c_3 = 3$. Note that $c_1 c_2 = 9 > 8 = b_1 b_2$, so we should expect the results in Case 2. The four critical points are $(0, 0)$, $(0, 21)$, $(15, 0)$, and $(6, 12)$. We will analyze them individually.

THE CRITICAL POINT $(0, 0)$ We linearize the system in (12) by dropping the quadratic terms; the result is the linear system

$$x' = 60x, \qquad y' = 42y, \tag{13}$$

in which primes denote, as usual, derivatives with respect to time t. The general solution of this system is

$$x = Ae^{60t}, \qquad y = Be^{42t}, \tag{14}$$

so $(0, 0)$ is an unstable node for the linearized system in (13). Because the equations in (14) yield $y = Cx^{7/10}$, all the trajectories other than the x-axis are tangent to the y-axis at the origin. Figure 7.42 shows some of these trajectories near the critical point $(0, 0)$.

THE CRITICAL POINT $(0, 21)$ To linearize the equations in (12) near the critical point $(0, 21)$, we substitute $u = x$ and $v = y - 21$. The result is the almost linear

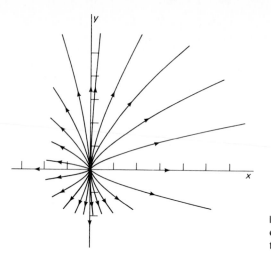

FIGURE 7.42 Linearized trajectories of (12) near (0, 0); the common tangent is the y-axis.

system

$$\left.\begin{array}{l} \dfrac{du}{dt} = -3u - 4u^2 - 3uv, \\[2mm] \dfrac{dv}{dt} = -63u - 42v - 3uv - 2v^2 \end{array}\right\} \tag{15}$$

with critical point (0, 0). The corresponding linear system

$$u' = -3u, \qquad v' = -63u - 42v \tag{16}$$

has characteristic roots $\lambda_1 = -3$ and $\lambda_2 = -42$. So (0, 21) is an asymptotically stable node. The general solution of (16) has the form

$$\left.\begin{array}{l} u = Ae^{-3t}, \\[2mm] v = -\dfrac{21}{13} Ae^{-3t} + Be^{-42t}, \end{array}\right\} \tag{17}$$

where A and B are arbitrary constants. As $t \to +\infty$, (u, v) approaches (0, 0), so (x, y) approaches (0, 21). With $A = 0$, (17) gives the vertical line through the critical point. The slope of any trajectory with $A \neq 0$ is

$$\frac{dy}{dx} = \frac{dy/dt}{dx/dt} = \frac{dv/dt}{du/dt} = \frac{\frac{63}{13}Ae^{-3t} - 42Be^{-42t}}{-3Ae^{-3t}} = -\frac{21}{13} + \frac{14B}{A}e^{-39t}.$$

Hence $dy/dx \to -\frac{21}{13}$ as $t \to +\infty$ and (x, y) approaches (0, 21). Consequently, the trajectories near the critical point (0, 21) resemble those shown in Fig. 7.43.

THE CRITICAL POINT (15, 0) The substitution $u = x - 15$, $v = y$ in the equations in (12) yields

$$\left.\begin{array}{l} \dfrac{du}{dt} = -60u - 45v - 4u^2 - 3uv, \\[2mm] \dfrac{dv}{dt} = -3v - 3uv - 2v^2. \end{array}\right\} \tag{18}$$

CHAPTER 7: Qualitative Properties and Existence of Solutions

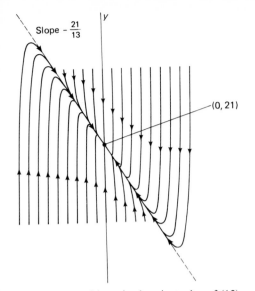

FIGURE 7.43 Linearized trajectories of (12) near the improper node (0, 21); the trajectories are drawn accurately by computer. Through they appear to "center" the line with slope $-\frac{21}{13}$, they actually are all tangent to it at the single point (0, 21).

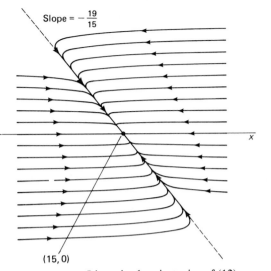

FIGURE 7.44 Linearized trajectories of (12) near the improper node (15, 0); the trajectories are drawn accurately by computer. Though they appear to "enter" the line with slope $-\frac{19}{15}$, they actually are all tangent to it at the single point (15, 0).

The corresponding linear system

$$u' = -60u - 45v, \qquad v' = -3v \tag{19}$$

has characteristic roots $\lambda_1 = -3$ and $\lambda_2 = -60$, so this critical point is also an asymptotically stable note. The general solution is of the form

$$\left. \begin{aligned} u &= -\frac{15}{19} Ae^{-3t} + Be^{-60t}, \\ v &= Ae^{-3t}. \end{aligned} \right\} \tag{20}$$

As $t \to +\infty$, $(u, v) \to (0, 0)$ and so $(x, y) \to (15, 0)$. With $A = 0$ we get the horizontal line through the critical point. The slope of any trajectory with $A \neq 0$ is

$$\frac{dy}{dx} = \frac{dy/dt}{dx/dt} = \frac{dv/dt}{du/dt} = \frac{-3Ae^{-3t}}{\frac{45}{19}Ae^{-3t} - 60Be^{-60t}} = \frac{-A}{\frac{15}{19}A - 20Be^{-57t}}.$$

Hence $dy/dx \to -\frac{19}{15}$ as $t \to +\infty$ and (x, y) approaches (15, 0). The trajectories near (15, 0) therefore resemble those shown in Fig. 7.44.

THE CRITICAL POINT (6, 12)   To linearize the equations in (12) near the critical point (6, 12) (the one representing the possibility of coexistence of the two species), we substitute $u = x - 6$ and $v = y - 12$. The result is the almost linear system

$$\left.\begin{aligned} \frac{du}{dt} &= -24u - 18v - 4u^2 - 3uv, \\ \frac{dv}{dt} &= -36u - 24v - 3uv - 2v^2. \end{aligned}\right\} \quad (21)$$

The associated linear system is

$$u' = -24u - 18v, \qquad v' = -36u - 24v. \quad (22)$$

The characteristic equation associated with Eq. (22) is

$$(-24 - \lambda)^2 - (-36)(-18) = (\lambda + 24)^2 - 2(18)^2 = 0$$

with roots $\lambda_1 = -24 + 18\sqrt{2} > 0$ and $\lambda_2 = -24 - 18\sqrt{2} < 0$. Thus this critical point is an unstable saddle point. We find that the general solution of (22) is

$$\left.\begin{aligned} u &= Ae^{\lambda_1 t} + Be^{\lambda_2 t}, \\ v &= -\sqrt{2}Ae^{\lambda_1 t} + \sqrt{2}Be^{\lambda_2 t}, \end{aligned}\right\} \quad (23)$$

where A and B are arbitrary constants. The slope of a trajectory in the linearized system is given by

$$\frac{dy}{dx} = \frac{dv/dt}{du/dt} = \sqrt{2}\left(\frac{-\lambda_1 Ae^{\lambda_1 t} + \lambda_2 Be^{\lambda_2 t}}{\lambda_1 Ae^{\lambda_1 t} + \lambda_2 Be^{\lambda_2 t}}\right). \quad (24)$$

If $A = 0$ then $(u, v) \to (0, 0)$ and so $(x, y) \to (6, 12)$ as $t \to +\infty$ because $\lambda_2 < 0$. From (23) and (24) we see that (x, y) then approaches the critical point $(6, 12)$ along the straight line of slope $\sqrt{2}$. If $A \neq 0$ then from (24) we find that

$$\frac{dy}{dx} = \sqrt{2}\left(\frac{-\lambda_1 A + \lambda_2 Be^{-36\sqrt{2}t}}{\lambda_1 A + \lambda_2 Bc^{-36\sqrt{2}t}}\right). \quad (25)$$

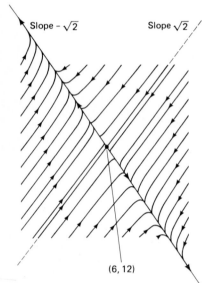

FIGURE 7.45 Linearized trajectories of (12) near the saddle point (6, 12); the trajectories are drawn accurately by computer. Though they appear to "enter" the line with slope $-\sqrt{2}$, they actually have the two lines with slope $\pm\sqrt{2}$ as asymptotes.

Slope $-\sqrt{2}$ *Slope* $\sqrt{2}$

(6, 12)

With $B = 0$ we have the straight line of slope $-\sqrt{2}$ along which (x, y) leaves the saddle point as t increases, and from (25) we see that $dy/dx \to -\sqrt{2}$ along any other trajectory as $t \to +\infty$. Hence the trajectories near the critical point $(6, 12)$ resemble those shown in Fig. 7.45.

Now that we have completed our local analysis of the four critical points of the almost linear system in (12), we want to assemble this information into a coherent whole—to construct a phase plane picture (phase portrait) of the global behavior of the trajectories in the first quadrant (where both populations are nonnegative). If we accept the facts that (1) near each critical point the trajectories look qualitatively like the linearized trajectories shown in Figs. 7.42 through 7.45, and (2) as $t \to +\infty$, each trajectory either approaches a critical point or diverges toward infinity somewhere in the first quadrant, then it follows that the phase plane picture should look qualitatively similar to the linearized trajectories shown in the schematic Fig. 7.46.

The two trajectories that approach the saddle point $(6, 12)$ together with that saddle point, form a separatrix. It plays a crucial role in determining the long-term behavior of the two populations. If the initial point (x_0, y_0) lies precisely on the separatrix, then $(x(t), y(t))$ approaches $(6, 12)$ as $t \to +\infty$. Of course, random events

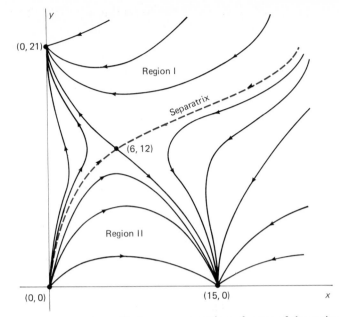

FIGURE 7.46 A qualitative representation of some of the trajectories of the system in Eq. (12).

make it extremely unlikely that $(x(t), y(t))$ will remain on the separatrix. If not, peaceful coexistence of the two species is impossible. If (x_0, y_0) lies in Region I above the separatrix, then $(x(t), y(t))$ approaches $(0, 21)$ as $t \to +\infty$, so the population $x(t)$ decreases to zero. Alternatively, if (x_0, y_0) lies in Region II below the separatrix, then $(x(t), y(t))$ approaches $(15, 0)$ as $t \to +\infty$, so the population $y(t)$ dies out. In short, whichever population has the initial competitive advantage survives, while the other faces extinction.

EXAMPLE 2 **(Peaceful coexistence):** Now suppose that the two populations satisfy the equations

$$\left.\begin{aligned} \frac{dx}{dt} &= 60x - 3x^2 - 4xy = x(60 - 3x - 4y), \\[2mm] \frac{dy}{dt} &= 42y - 3y^2 - 2xy = y(42 - 2x - 3y). \end{aligned}\right\} \tag{26}$$

Here $a_1 = 60$, $a_2 = 42$, $b_1 = b_2 = 3$, $c_1 = 4$, and $c_2 = 2$. So $c_1 c_2 = 8 < b_1 b_2 = 9$: Competition is smaller than inhibition. The analysis of the system in (26) follows step by step that of the system in (12) of Example 1. We will present only the results of this analysis, leaving the details to Problems 2–6. There are four critical points—$(0, 0)$, $(0, 14)$, $(20, 0)$, and $(12, 6)$.

The critical point $(0, 0)$ is an unstable node of the linearized system, just as in Example 1.

The critical point $(0, 14)$ is an unstable saddle point of the linearized system. The entering trajectories lie along the y-axis, while the departing trajectories lie along the line through $(0, 14)$ with slope $-\frac{14}{23}$.

The critical point (20, 0) is also an unstable saddle point of the linearized system. The entering trajectories lie along the x-axis, while the departing trajectories lie along the straight line through (20, 0) with slope $-\frac{31}{40}$.

The critical point (12, 6) represents the possibility of peaceful coexistence because it is an asymptotically stable node. One pair of trajectories lie along the line through (12, 6) with slope $(-3 + \sqrt{73})/16 \approx 0.35$. All other trajectories of the linearized system enter the node and are tangent to the line with slope $(-3 - \sqrt{73})/16 \approx -0.72$.

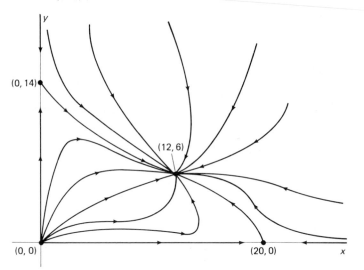

FIGURE 7.47 A qualitative representation of some of the trajectories of the system in Eq. (26).

Figure 7.47 shows a reasonable phase portrait that is qualitatively consistent with the results of our critical point analysis. One especially notable feature of this system is that for *any* positive initial values x_0 and y_0, $(x(t), y(t))$ approaches (12, 6) as $t \to +\infty$, so the two species both survive in stable (peaceful) coexistence.

Examples 1 and 2 illustrate the power of elementary critical point analysis. But we must conclude with a word of caution. Ecological systems in nature are rarely so simple as in these examples; they normally involve far more than two species, and the rates of growth of these populations and their interactions are almost always more complex than those discussed in this section.

7.5 PROBLEMS

1. Let $x(t)$ be a harmful insect population (aphids?) that under natural conditions is held somewhat in check by a benign predator insect population $y(t)$ (ladybugs?). Assume that $x(t)$ and $y(t)$ satisfy the predator-prey equations in (1), so that the stable equilibrium populations are $x_E = c/d$ and $y_E = a/b$. Now suppose that an insecticide is employed that kills (per unit time) the same fraction $f < a$ of each species of insect. Show that the harmful population x_E is increased, while the benign population y_E is decreased, so the use of the insecticide is counterproductive. This is an instance in which mathematical analysis reveals undesirable consequences of a well-intentioned interference with nature.

CHAPTER 7: Qualitative Properties and Existence of Solutions

Problems 2–6 provide the details cited in Example 2 for the two competing populations $x(t)$ and $y(t)$ that satisfy the equations

$$\left.\begin{aligned}\frac{dx}{dt} &= 60x - 3x^2 - 4xy = x(60 - 3x - 4y),\\[2mm]\frac{dy}{dt} &= 42y - 3y^2 - 2xy = y(42 - 2x - 3y).\end{aligned}\right\} \quad (26)$$

2. Show that the critical points of the system in (26) are $(0, 0)$, $(0, 14)$, $(20, 0)$, and $(12, 6)$.

3. Show that the critical point $(0, 0)$ is an unstable node at which all but two trajectories are tangent to the y-axis.

4. (a) To investigate the critical point $(0, 14)$, substitute $u = x$, $v = y - 14$; then show that the corresponding linear system is $u' = 4u$, $v' = -28u - 42v$. (b) Show that the general solution of this linear system is $u = Ae^{4t}$, $v = -\frac{14}{23}Ae^{4t} + Be^{-42t}$. (c) Then conclude that $(0, 14)$ is an unstable saddle point at which the entering and departing trajectories lie on the y-axis and on the line through $(0, 14)$ with slope $-\frac{14}{23}$.

5. (a) To investigate the critical point $(20, 0)$, substitute $u = x - 20$, $v = y$; then show that the corresponding linear system is $u' = -60u - 80v$, $v' = 2v$. (b) Show that the general solution of this linear system is $u = Ae^{-60t} - \frac{40}{31}Be^{2t}$, $v = Be^{2t}$. (c) Then conclude that $(20, 0)$ is an unstable saddle point at which entering and departing trajectories lie on the x-axis and on the line through $(20, 0)$ with slope $-\frac{31}{40}$.

6. (a) To investigate the critical point $(12, 6)$, substitute $u = x - 12$, $v = y - 6$; then show that the corresponding linear system is $u' = -36u - 48v$, $v' = -12u - 18v$. (b) Show that the characteristic roots of this linear system are $\lambda_1 = -27 + 3\sqrt{73} < 0$ and $\lambda_2 = -27 - 3\sqrt{73} < 0$, and that its general solution is

$$u = Ae^{\lambda_1 t} + Be^{\lambda_2 t}, \qquad v = Ce^{\lambda_1 t} + De^{\lambda_2 t}$$

where $C = (-3 - \sqrt{73})A/16$ and $D = (-3 + \sqrt{73})B/16$. (c) Hence conclude that $(12, 6)$ is an asymptotically stable node, where two trajectories lie on the line through $(12, 6)$ with slope $(-3 + \sqrt{73})/16$ and the others are tangent to the line through $(12, 6)$ with slope $(-3 - \sqrt{73})/16$.

Problems 7–11 deal with the predator-prey system that is modeled by the equations

$$\left.\begin{aligned}\frac{dx}{dt} &= 5x - x^2 - xy = x(5 - x - y),\\[2mm]\frac{dy}{dt} &= xy - 2y = y(x - 2).\end{aligned}\right\} \quad (27)$$

In contrast with the system in (1) discussed in the text, the prey population $x(t)$ would—in the absence of any predators—be a bounded population described by the logistic equation $dx/dt = 5x - x^2$.

7. Show that the critical points of the system in (27) are $(0, 0)$, $(5, 0)$, and $(2, 3)$.

8. Show that the critical point $(0, 0)$ is an unstable saddle point of the linearized system, with trajectories entering along the y-axis and departing along the x-axis.

9. Show that the critical point $(5, 0)$ is an unstable saddle point of the linearized system, with trajectories entering along the x-axis and departing along the line through $(5, 0)$ with slope $-\frac{8}{5}$.

10. Show that the critical point $(2, 3)$ is an asymptotically stable spiral point of the linearized system.

11. Show that the results of Problems 7–10 are consistent with the global phase portrait for the original nonlinear system in (27) shown in Fig. 7.48. Conclude that the prey and predators coexist with stable equilibrium populations $x_E = 2$ and $y_E = 3$.

Problems 12–17 deal with the predator-prey system modeled by the equations

$$\left.\begin{aligned}\frac{dx}{dt} &= x^2 - 2x - xy = x(x - y - 2),\\[2mm]\frac{dy}{dt} &= y^2 - 4y + xy = y(x + y - 4).\end{aligned}\right\} \quad (28)$$

In this system, each population—the prey population $x(t)$ and the predator population $y(t)$—is an unsophisticated population

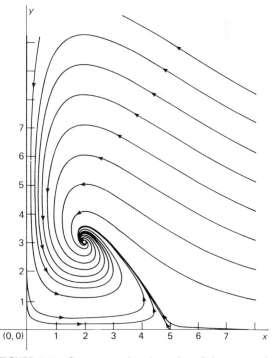

FIGURE 7.48 Some actual trajectories of the system in (27).

(similar to the one considered in Problem 15 of Section 7.1) for each of which the only alternatives (in the absence of the other) are doomsday and extinction.

12. Show that the critical points of the system in (28) are (0, 0), (0, 4), (2, 0), and (3, 1).

13. Show that the critical point (0, 0) is an asymptotically stable node of the linearized system, at which one pair of trajectories lie along the y-axis and the others are tangent to the x-axis.

14. Show that the critical point (0, 4) is an unstable saddle point of the linearized system, with trajectories departing along the y-axis and entering along the line through (0, 4) with slope $-\frac{2}{5}$.

15. Show that the critical point (2, 0) is an unstable saddle point of the linearized system, with trajectories departing along the x-axis and entering along the line through (2, 0) with slope 2.

16. Show that the critical point (3, 1) is an unstable spiral point of the linearized system.

17. Show that the results of Problems 12–16 are consistent with the global phase portrait for the original nonlinear system in (28) shown in Fig. 7.49. This is a two-dimensional version of doomsday versus extinction. If the initial point (x_0, y_0) lies

in Region I, then both populations increase without bound (until doomsday), while if it lies in Region II, then both populations will decrease to zero.

The next two problems deal with the competition equations in (8) and the associated linearized system. The discussion in the text covered only the case in which the critical point of interest (x_E, y_E) lay in the first quadrant. You may now explore the other possibilities.

18. Determine the behavior of the linearized system associated with the system in (8) in the case that (x_E, y_E) exists but does not lie in the first quadrant. Because the populations are never negative, there are only three critical points to examine. There are two cases, depending upon whether (x_E, y_E) lies in the second quadrant or the fourth quadrant.

19. Determine the behavior of the linearized system associated with the system in (8) in the case that (x_E, y_E) does not exist because the two lines having the equations in (9) are parallel. There will be two cases, depending upon which line is above the other. (You may take for granted that the improbable—coincidence of the two lines—does not occur.)

20. A lake is initially stocked with 100 bass and 600 redear. There is ample food for the redear. Because bass prey on redear, the population of bass will increase at a rate proportional to the number of encounters between the species; bass will also die at a rate proportional to the bass population. The redear multiply at a rate proportional to their population and die off at a rate proportional to the number of encounters between the two species. This implies that the populations $B(t)$ of bass and $R(t)$ of redear satisfy a system of differential equations of the following form:

$$B'(t) = pBR - qB,$$

$$R'(t) = uR - vBR.$$

Suppose it is known that $p = 0.00004$, $q = 0.02$, $u = 0.05$, and $v = 0.0004$, with t measured in days. Then we have a predator-prey system as in Eq. (1) of this section, and the populations of the two species oscillate periodically with the same period. Approximate this period by solving the system numerically with the given initial conditions, using the Runge-Kutta method of Section 6.4.

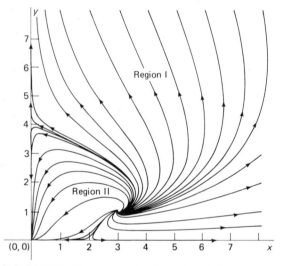

FIGURE 7.49 Some actual trajectories of the predator-prey system in (28).

Existence and Uniqueness of Solutions

In Chapter 1 we saw that an initial value problem of the form

$$\frac{dy}{dx} = f(x, y), \qquad y(a) = b \tag{1}$$

can fail (on a given interval containing the point $x = a$) to have a unique solution. For instance, in Example 4 of Section 1.3, we saw that the initial value problem

$$x^2 \frac{dy}{dx} + y^2 = 0, \qquad y(0) = b \tag{2}$$

has no solution at all unless $b = 0$, in which case there are infinitely many different solutions. According to Problem 37 of Section 1.3, the initial value problem

$$\frac{dy}{dx} = -\sqrt{1 - y^2}, \qquad y(0) = 1 \tag{3}$$

has the two distinct solutions $y_1(x) \equiv 1$ and $y_2(x) = \cos x$ on the interval $0 \leq x \leq \pi$. In this final section of Chapter 7 we investigate conditions on the function $f(x, y)$ that suffice to guarantee that the initial value problem in (1) has one and only one solution, and then proceed to establish appropriate versions of the existence-uniqueness theorems that were stated without proof in Sections 1.3, 2.1, 2.2, and 5.1.

EXISTENCE OF SOLUTIONS

The approach we employ is the **method of successive approximations**, which was developed by the French mathematician Emile Picard (1856–1941). This method is based on the fact that the function $f(x)$ satisfies the initial value problem in (1) on the open interval I containing $x = a$ if and only if it satisfies the integral equation

$$y(x) = b + \int_a^x f(t, y(t)) \, dt \tag{4}$$

for all x in I. In particular, if $y(x)$ satisfies (4), then clearly $y(a) = b$, and differentiation of both sides in (4)—using the fundamental theorem of calculus—yields the differential equation $y'(x) = f(x, y(x))$.

To attempt to solve Eq. (4), we begin with the initial function

$$y_0(x) \equiv b, \tag{5}$$

and then define iteratively a sequence $y_1, y_2, y_3, \ldots$, of functions that we hope will converge to the solution. Specifically, we let

$$y_1(x) = b + \int_a^x f(t, y_0(t)) \, dt$$

and $\tag{6}$

$$y_2(x) = b + \int_a^x f(t, y_1(t)) \, dt.$$

In general, y_{n+1} is obtained by substitution of y_n for y in the right-hand side in (4):

$$y_{n+1}(x) = b + \int_a^x f(t, y_n(t)) \, dt. \tag{7}$$

Suppose we know that each of these functions $\{y_n(x)\}_0^\infty$ is defined on some open interval (the same for each n) containing $x = a$, and that the limit

$$y(x) = \lim_{n \to \infty} y_n(x) \tag{8}$$

exists at each point of this interval. Then it will follow that

$$y(x) = \lim_{n \to \infty} y_{n+1}(x)$$

$$= \lim_{n \to \infty} \left[b + \int_a^x f(t, y_n(t)) \, dt \right]$$

$$= b + \lim_{n \to \infty} \int_a^x f(t, y_n(t)) \, dt \tag{9}$$

$$= b + \int_a^x f(t, \lim_{n \to \infty} y_n(t)) \, dt \tag{10}$$

and hence that

$$y(x) = b + \int_a^x f(t, y(t)) \, dt,$$

provided that we can validate the interchange of limit operations involved in passing from (9) to (10). It is therefore reasonable to expect that, under favorable conditions, the sequence $\{y_n(x)\}$ defined iteratively in (5) and (7) will converge to a solution $y(x)$ of the integral equation in (4), and hence to a solution of the original initial value problem in (1).

EXAMPLE 1 To apply the method of successive approximations to the initial value problem

$$\frac{dy}{dx} = y, \qquad y(0) = 1, \tag{11}$$

we write Eqs. (5) and (7), thereby obtaining

$$y_0(x) \equiv 1, \qquad y_{n+1}(x) = 1 + \int_0^x y_n(t) \, dt. \tag{12}$$

The iteration formula in (12) yields

$$y_1(x) = 1 + \int_0^x 1 \, dt = 1 + x,$$

$$y_2(x) = 1 + \int_0^x (1 + t) \, dt$$

$$= 1 + x + \tfrac{1}{2}x^2,$$

$$y_3(x) = 1 + \int_0^x (1 + t + \tfrac{1}{2}t^2) \, dt$$

$$= 1 + x + \tfrac{1}{2}x^2 + \tfrac{1}{6}x^3,$$

and

$$y_4(x) = 1 + \int_0^x (1 + t + \tfrac{1}{2}t^2 + \tfrac{1}{6}t^3) \, dt$$

$$= 1 + x + \tfrac{1}{2}x^2 + \tfrac{1}{6}x^3 + \tfrac{1}{24}x^4.$$

It is clear that we are generating the sequence of partial sums of a power series solution; indeed, we immediately recognize the series as that of $y(x) = e^x$. There is no difficulty in demonstrating that the exponential function is indeed the solution

CHAPTER 7:   Qualitative Properties and Existence of Solutions

of the initial value problem in (11); moreover, a diligent student can verify (using a proof by induction on n) that $y_n(x)$, obtained in the manner above, is indeed the nth partial sum for the Taylor series with center zero for $y(x) = e^x$.

EXAMPLE 2 To apply the method of successive approximations to the initial value problem

$$\frac{dy}{dx} = 4xy, \qquad y(0) = 3, \tag{13}$$

we write Eqs. (5) and (7) as in Example 1. Now we obtain

$$y_0(x) \equiv 3, \qquad y_{n+1}(x) = 3 + \int_0^x 4t y_n(t)\, dt \tag{14}$$

The iteration formula in (14) yields

$$y_1(x) = 3 + \int_0^x (4t)(3)\, dt = 3 + 6x^2,$$

$$y_2(x) = 3 + \int_0^x 4t(3 + 6t^2)\, dt$$

$$= 3 + 6x^2 + 6x^4,$$

$$y_3(x) = 3 + \int_0^x 4t(3 + 6t^2 + 6t^4)\, dt$$

$$= 3 + 6x^2 + 6x^4 + 4x^6,$$

and

$$y_4(x) = 3 + \int_0^x 4t(3 + 6t^2 + 6t^4 + 4t^6)\, dt$$

$$= 3 + 6x^2 + 6x^4 + 4x^6 + 2x^8.$$

It is again clear that we are generating partial sums of a power series solution. It is not quite so obvious what function has such a power series representation, but the initial value problem in (13) is readily solved by separation of variables:

$$y(x) = 3e^{2x^2} = 3 \sum_{n=0}^{\infty} \frac{(2x^2)^n}{n!}$$

$$= 3 + 6x^2 + 6x^4 + 4x^6 + 2x^8 + \frac{4}{5}x^{10} + \cdots.$$

In some cases it may be necessary to compute a much larger number of terms, either in order to identify the solution or to use a partial sum of its series with large subscript to approximate the solution accurately for x near its initial value. Fortunately, there are computer programs available (even for microcomputers) that will perform the symbolic integrations (as opposed to numerical integrations) of the sort in the examples above. If necessary, you could generate the first hundred terms in Example 2 in a matter of minutes.

In general, of course, we apply Picard's method because we cannot find a solution by elementary methods. Suppose that we have produced a large number of terms of what we believe to be the correct power series expansion of the solution. We *must* have conditions under which the sequence $\{y_n(x)\}$ provided by the method of successive approximation is guaranteed in advance to converge to a solution. It is just as convenient to discuss the intial value problem

$$\frac{d\mathbf{x}}{dt} = \mathbf{f}(\mathbf{x}, t), \qquad \mathbf{x}(a) = \mathbf{b} \tag{15}$$

for a system of m first order equations, where

$$\mathbf{x} = \begin{bmatrix} x_1 \\ x_2 \\ x_3 \\ \vdots \\ x_m \end{bmatrix}, \qquad \mathbf{f} = \begin{bmatrix} f_1 \\ f_2 \\ f_3 \\ \vdots \\ f_m \end{bmatrix}, \qquad \text{and} \qquad \mathbf{b} = \begin{bmatrix} b_1 \\ b_2 \\ b_3 \\ \vdots \\ b_m \end{bmatrix}.$$

It turns out that with the aid of this vector notation (which we introduced in Section 5.3), most results concerning a single (scalar) equation $x' = f(x, t)$ can be generalized readily to analogous results for a system of m first order equations, as abbreviated in (15). Consequently, the effort of using vector notation is amply justified by the generality it affords.

The method of successive approximations for the system in (15) calls for us to compute the sequence $\{\mathbf{x}_n(t)\}_0^\infty$ of vector-valued functions of t,

$$\mathbf{x}_n(t) = \begin{bmatrix} x_{1n}(t) \\ x_{2n}(t) \\ x_{3n}(t) \\ \vdots \\ x_{mn}(t) \end{bmatrix},$$

defined iteratively by

$$\mathbf{x}_0(t) \equiv \mathbf{b}, \qquad \mathbf{x}_{n+1}(t) = \mathbf{b} + \int_a^t \mathbf{f}(\mathbf{x}_n(s), s)\, ds. \tag{16}$$

Recall that vector-valued functions are integrated componentwise.

EXAMPLE 3 Consider the m-dimensional initial value problem

$$\frac{d\mathbf{x}}{dt} = \mathbf{A}\mathbf{x}, \qquad \mathbf{x}(0) = \mathbf{b} \tag{17}$$

for a homogeneous linear system with $m \times m$ constant coefficient matrix $\mathbf{A}$. The equations in (16) take the form

$$\mathbf{x}_0(t) = \mathbf{b}, \qquad \mathbf{x}_{n+1}(t) = \mathbf{b} + \int_0^t \mathbf{A}\mathbf{x}_n(s)\, ds. \tag{18}$$

Thus

$$\mathbf{x}_1(t) = \mathbf{b} + \int_0^t \mathbf{A}\mathbf{b}\, ds$$

$$= \mathbf{b} + \mathbf{A}\mathbf{b}t = (\mathbf{I} + \mathbf{A}t)\mathbf{b};$$

$$\mathbf{x}_2(t) = \mathbf{b} + \int_0^t \mathbf{A}(\mathbf{b} + \mathbf{A}\mathbf{b}s)\, ds$$

$$= \mathbf{b} + \mathbf{A}\mathbf{b}t + \tfrac{1}{2}\mathbf{A}^2\mathbf{b}t^2$$

$$= (\mathbf{I} + \mathbf{A}t + \tfrac{1}{2}\mathbf{A}^2t^2)\mathbf{b},$$

and

$$\mathbf{x}_3(t) = \mathbf{b} + \int_0^t \mathbf{A}(\mathbf{b} + \mathbf{A}\mathbf{b}s + \tfrac{1}{2}\mathbf{A}^2\mathbf{b}s^2)\, ds$$

$$= (\mathbf{I} + \mathbf{A}t + \tfrac{1}{2}\mathbf{A}^2 t^2 + \tfrac{1}{6}\mathbf{A}^3 t^3)\mathbf{b}.$$

We have therefore obtained the first several partial sums of the exponential series solution

$$\mathbf{x}(t) = e^{\mathbf{A}t}\mathbf{b} = \left(\sum_{n=0}^{\infty} \frac{(\mathbf{A}t)^n}{n!} \right) \mathbf{b} \tag{19}$$

of (17), which was derived earlier in Section 5.7.

The key to establishing convergence in the method of successive approximations is an appropriate condition on the rate at which $\mathbf{f}(\mathbf{x}, t)$ changes when $\mathbf{x}$ varies but t is held fixed. If R is a region in $(m + 1)$-dimensional $(\mathbf{x}, t)$-space, then the function $\mathbf{f}(\mathbf{x}, t)$ is said to be **Lipschitz continuous** in R if there exists a constant $k > 0$ such that

$$|\mathbf{f}(\mathbf{x}_1, t) - \mathbf{f}(\mathbf{x}_2, t)| \leq k|\mathbf{x}_1 - \mathbf{x}_2| \tag{20}$$

if $(\mathbf{x}_1, t)$ and $(\mathbf{x}_2, t)$ are points of R. Recall that the norm of an m-dimensional point or vector $\mathbf{x}$ is defined to be

$$|\mathbf{x}| = (x_1^2 + x_2^2 + \cdots + x_m^2)^{1/2}. \tag{21}$$

Then $|\mathbf{x}_1 - \mathbf{x}_2|$ is simply the Euclidean distance between the points $\mathbf{x}_1$ and $\mathbf{x}_2$.

EXAMPLE 4 Let $f(x, t) = x^2 e^{-t^2} \sin t$ and let R be the strip $0 \leq x \leq 2$ in the xt-plane. If (x_1, t) and (x_2, t) are both points of R, then

$$|f(x_1, t) - f(x_2, t)| = |e^{-t^2} \sin t| |x_1 + x_2| |x_1 - x_2|$$

$$\leq 4|x_1 - x_2|,$$

because $|e^{-t^2} \sin t| \leq 1$ for all t and $|x_1 + x_2| \leq 4$ if x_1 and x_2 are both in the interval $[0, 2]$. Thus f satisfies the Lipschitz condition in (20) with $k = 4$ and is therefore Lipschitz continuous in the strip R.

EXAMPLE 5 Let $f(x, t) = t\sqrt{x}$ on the rectangle R consisting of the points (x, t) in the plane for which $0 \leq x \leq 1$ and $0 \leq t \leq 1$. Then, taking $x_1 = x$, $x_2 = 0$, and $t = 1$, we find that

$$|f(x, 1) - f(0, 1)| = \sqrt{x} = \frac{1}{\sqrt{x}} |x - 0|.$$

Because $x^{-1/2} \to +\infty$ as $x \to 0^+$, we see that the Lipschitz condition in (20) cannot be satisfied by any (finite) constant $k > 0$. Thus the function f, though obviously continuous on R, is *not* Lipschitz continuous on R.

Suppose, however, that the function $f(x, t)$ has a continuous partial derivative $f_x(x, t)$ on the closed rectangle R in the xt-plane, and denote by k the maximum value of $\left|f_x(x, t)\right|$ on R. Then the mean value theorem of differential calculus yields

$$\left|f(x_1, t) - f(x_2, t)\right| = \left|f_x(\bar{x}, t)(x_1 - x_2)\right|$$

for some $\bar{x}$ in (x_1, x_2), so it follows that

$$\left|f(x_1, t) - f(x_2, t)\right| \leq k\left|x_1 - x_2\right|$$

because $\left|f_x(\bar{x}, t)\right| \leq k$. Thus a continuously differentiable function $f(x, t)$ defined on a closed rectangle *is* Lipschitz continuous there. More generally, the multivariable mean value theorem of advanced calculus can be used similarly to prove that *a vector-valued function* $\mathbf{f}(\mathbf{x}, t)$ *with continuously differentiable component functions on a closed rectangular region R in $(\mathbf{x}, t)$-space is Lipschitz continuous in R.*

EXAMPLE 6 The function $f(x, t) = x^2$ is Lipschitz continuous on any closed (bounded) rectangle in the xt-plane. But consider this function on the infinite strip R consisting of the points (x, t) for which $0 \leq t \leq 1$ and x is arbitrary. Then

$$\begin{aligned}\left|f(x_1, t) - f(x_2, t)\right| &= \left|x_1^2 - x_2^2\right| \\ &= \left|x_1 + x_2\right|\left|x_1 - x_2\right|.\end{aligned}$$

Because $\left|x_1 + x_2\right|$ can be made arbitrarily large, it follows that f is *not* Lipschitz continuous on the infinite strip R.

If I is an interval on the t-axis, then the set of all points $(\mathbf{x}, t)$ with t in I is an infinite strip or slab in $(m + 1)$-space (as indicated in Fig. 7.50). Example 6 shows that Lipschitz continuity of $\mathbf{f}(\mathbf{x}, t)$ on such an infinite slab is a very strong condition. Nevertheless, the existence of a solution of the initial value problem

$$\frac{d\mathbf{x}}{dt} = \mathbf{f}(\mathbf{x}, t), \qquad \mathbf{x}(a) = \mathbf{b} \tag{15}$$

under the hypothesis of Lipschitz continuity of $\mathbf{f}$ in such a slab is of considerable importance.

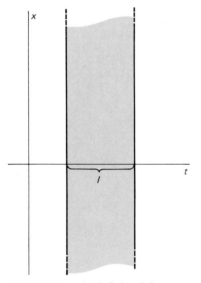

FIGURE 7.50 An infinite slab in $(m + 1)$-space.

> **Theorem 1** *Global Existence of Solutions*
>
> Let $\mathbf{f}$ be a vector-valued function (with m components) of $m + 1$ real variables, and let I be a (bounded or unbounded) open interval containing $t = a$. If $\mathbf{f}(\mathbf{x}, t)$ is continuous and satisfies the Lipschitz condition in (20) for all t in I and for all $\mathbf{x}_1$ and $\mathbf{x}_2$, then the initial value problem in (15) has a solution on the (entire) interval I.

Proof We want to show that the sequence $\{\mathbf{x}_n(t)\}_0^\infty$ of successive approximations determined iteratively by

$$\mathbf{x}_0(t) \equiv \mathbf{b}, \qquad \mathbf{x}_{n+1}(t) = \mathbf{b} + \int_a^t \mathbf{f}(\mathbf{x}_n(s), s)\, ds \tag{16}$$

converges to a solution $\mathbf{x}(t)$ of (15). We see that each of these functions in turn is continuous on I, being an (indefinite) integral of a continuous function.

We can assume that $a = 0$, because the transformation $t \to t + a$ converts (15) into an equivalent problem with initial point $t = 0$. Also, we will consider only the portion $t \geq 0$ of the interval I; the details for the case $t \leq 0$ are very similar.

The main part of the proof consists in showing that if $[0, T]$ is a closed (and bounded) interval contained in I, then the sequence $\{\mathbf{x}_n(t)\}$ converges *uniformly* on I to a limit function $\mathbf{x}(t)$. This means that, given $\epsilon > 0$, there exists an integer N such that

$$|\mathbf{x}_n(t) - \mathbf{x}(t)| < \epsilon \tag{22}$$

for all $n \geq N$ and all t in $[0, T]$. For ordinary (perhaps nonuniform) convergence the integer N, for which (22) holds for all $n \geq N$, may depend upon t, with no single value of N working for all t in I. Once this uniform convergence of the sequence $\{\mathbf{x}_n(t)\}$ has been established, the following conclusions will follow from standard theorems of advanced calculus (see pages 620–622 of A. E. Taylor and W. R. Mann, *Advanced Calculus*, 3rd ed., (New York: John Wiley, 1983)):

1. The limit function $\mathbf{x}(t)$ is continuous on $[0, T]$.

2. If N is chosen so that the inequality in (22) holds for $n \geq N$, then the Lipschitz continuity of $\mathbf{f}$ implies that

$$|\mathbf{f}(\mathbf{x}_n(t), t) - \mathbf{f}(\mathbf{x}(t), t)| \leq k|\mathbf{x}_n(t) - \mathbf{x}(t)| < k\epsilon$$

for all t in $[0, T]$ and $n \geq N$, so it follows that the sequence $\{\mathbf{f}(\mathbf{x}_n(t), t)\}_0^\infty$ converges uniformly to $\mathbf{f}(\mathbf{x}(t), t)$ on $[0, T]$.

3. But a uniformly convergent sequence or series can be integrated termwise, so it follows that, upon taking limits in the iterative formula in (16),

$$\mathbf{x}(t) = \lim_{n \to \infty} \mathbf{x}_{n+1}(t)$$

$$= \mathbf{b} + \lim_{n \to \infty} \int_0^t \mathbf{f}(\mathbf{x}_n(s), s)\, ds$$

$$= \mathbf{b} + \int_0^t \lim_{n \to \infty} \mathbf{f}(\mathbf{x}_n(s), s)\, ds;$$

thus

$$\mathbf{x}(t) = \mathbf{b} + \int_0^t \mathbf{f}(\mathbf{x}(s), s)\, ds. \tag{23}$$

4. Because the function $\mathbf{x}(t)$ is continuous on $[0, T]$, the integral equation in (23) (analogous to the one-dimensional case in (4)) implies that $\mathbf{x}'(t) = \mathbf{f}(\mathbf{x}(t), t)$ on $[0, T]$. But if this is true on every closed subinterval of the open interval I, then it is true on the entire interval I.

It therefore remains only to prove that the sequence $\{\mathbf{x}_n(t)\}_0^\infty$ converges uniformly on the closed interval $[0, T]$. Let M be the maximum value of $|\mathbf{f}(\mathbf{b}, t)|$ for t in $[0, T]$. Then

$$|\mathbf{x}_1(t) - \mathbf{x}_0(t)| = \left| \int_0^t \mathbf{f}(\mathbf{x}_0(s), s)\, ds \right| \leq \int_0^t |\mathbf{f}(\mathbf{b}, s)|\, ds = Mt. \tag{24}$$

Next,

$$\left| \mathbf{x}_2(t) - \mathbf{x}_1(t) \right| = \left| \int_0^t \left[\mathbf{f}(\mathbf{x}_1(s), s) - \mathbf{f}(\mathbf{x}_0(s), s) \right] ds \right|$$

$$\leq k \int_0^t \left| \mathbf{x}_1(s) - \mathbf{x}_0(s) \right| ds,$$

and hence

$$\left| \mathbf{x}_2(t) - \mathbf{x}_1(t) \right| \leq k \int_0^t M s\, ds = \tfrac{1}{2} k M t^2. \tag{25}$$

We now proceed by induction: Assume that

$$\left| \mathbf{x}_n(t) - \mathbf{x}_{n-1}(t) \right| \leq \frac{M}{k} \cdot \frac{(kt)^n}{n!}. \tag{26}$$

It then follows that

$$\left| \mathbf{x}_{n+1}(t) - \mathbf{x}_n(t) \right| = \left| \int_0^t \left[\mathbf{f}(\mathbf{x}_n(s), s) - \mathbf{f}(\mathbf{x}_{n-1}(s), s) \right] ds \right|$$

$$\leq k \int_0^t \left| \mathbf{x}_n(s) - \mathbf{x}_{n-1}(s) \right| ds;$$

consequently,

$$\left| \mathbf{x}_{n+1}(t) - \mathbf{x}_n(t) \right| \leq k \int_0^t \frac{M}{k} \cdot \frac{(ks)^n}{n!} ds.$$

It follows on evaluating this integral that

$$\left| \mathbf{x}_{n+1}(t) - \mathbf{x}_n(t) \right| \leq \frac{M}{k} \cdot \frac{(kt)^{n+1}}{(n+1)!}.$$

Thus (26) holds on $[0, T]$ for all $n \geq 1$.

Hence the terms of the infinite series

$$\mathbf{x}_0(t) + \sum_{n=1}^{\infty} \left[\mathbf{x}_n(t) - \mathbf{x}_{n-1}(t) \right] \tag{27}$$

are dominated (in magnitude on the interval $[0, T]$) by the terms of the convergent series

$$\sum_{n=1}^{\infty} \frac{M}{k} \cdot \frac{(kT)^{n+1}}{(n+1)!} = \frac{M}{k} (e^{kT} - 1), \tag{28}$$

which is a series of positive constants. It therefore follows (from the Weierstrass M-test on pages 618–619 of Taylor and Mann) that the series in (27) converges uniformly on $[0, T]$. But the sequence of partial sums of this series is simply our original sequence $\{\mathbf{x}_n(t)\}_0^{\infty}$ of successive approximations, so the proof of Theorem 1 is finally complete. ∎

LINEAR SYSTEMS   An important application of the global existence theorem just given is to the initial value problem

$$\frac{d\mathbf{x}}{dt} = \mathbf{A}(t)\mathbf{x} + \mathbf{g}(t), \qquad \mathbf{x}(a) = \mathbf{b} \tag{29}$$

for a linear system, where the $m \times m$ matrix-valued function $\mathbf{A}(t)$ and the vector-valued function $\mathbf{g}(t)$ are continuous on a (bounded or unbounded) open interval I containing the point $t = a$. In order to apply Theorem 1 to the linear system in (29), we note first that the proof of Theorem 1 requires only that, for each closed and bounded subinterval J of I, there exists a Lipschitz constant k such that

$$\left| \mathbf{f}(\mathbf{x}_1, t) - \mathbf{f}(\mathbf{x}_2, t) \right| \leq k \left| \mathbf{x}_1 - \mathbf{x}_2 \right| \tag{20}$$

for all t in J (and all $\mathbf{x}_1$ and $\mathbf{x}_2$). Thus we do not need a single Lipschitz constant for the entire open interval I.

In (29) we have $\mathbf{f}(\mathbf{x}, t) = \mathbf{Ax} + \mathbf{g}$, so

$$\mathbf{f}(\mathbf{x}_1, t) - \mathbf{f}(\mathbf{x}_2, t) = \mathbf{A}(t)(\mathbf{x}_1 - \mathbf{x}_2). \tag{30}$$

It therefore suffices to show that, if $\mathbf{A}(t)$ is continuous on the closed and bounded interval J, then there is a constant k such that

$$\left| \mathbf{A}(t)\mathbf{x} \right| \leq k \left| \mathbf{x} \right| \tag{31}$$

for all t in J. But this follows from the fact (see Problem 17) that

$$\left| \mathbf{Ax} \right| \leq \left\| \mathbf{A} \right\| \cdot \left| \mathbf{x} \right|, \tag{32}$$

where the **norm** $\left\| \mathbf{A} \right\|$ of the matrix $\mathbf{A}$ is defined to be

$$\left\| \mathbf{A} \right\| = \left(\sum_{i,j=1}^{m} a_{ij}^2 \right)^{1/2}. \tag{33}$$

Because $\mathbf{A}(t)$ is continuous on the closed and bounded interval J, the norm $\left\| \mathbf{A}(t) \right\|$ is bounded on J, so Eq. (31) follows, as desired. Thus we have the following global existence theorem for the linear initial value problem in (29).

Theorem 2 *Existence for Linear Systems*

Let the $m \times m$ matrix-valued function $\mathbf{A}(t)$ and the vector-valued function $\mathbf{g}(t)$ be continuous on the (bounded or unbounded) open interval I containing the point $t = a$. Then the initial value problem

$$\frac{d\mathbf{x}}{dt} = \mathbf{A}(t)\mathbf{x} + \mathbf{g}(t), \qquad \mathbf{x}(a) = \mathbf{b} \tag{29}$$

has a solution on the (entire) interval I.

As we saw in Section 5.1, the mth order initial value problem

$$\left. \begin{array}{l} x^{(m)} + a_1(t)x^{(m-1)} + \cdots + a_{m-1}(t)x' + a_m(t)x = p(t), \\ x(a) = b_0, \qquad x'(a) = b_1, \ldots, x^{(m-1)}(a) = b_{m-1} \end{array} \right\} \tag{34}$$

is easily transformed into an equivalent $m \times m$ system of the form in (29). It therefore follows from Theorem 2 that if the functions $a_1(t), a_2(t), \ldots, a_m(t)$, and $p(t)$ in (34) are all continuous on the (bounded or unbounded) open interval I containing $t = a$, then the initial value problem in (34) has a solution on the (entire) interval I.

LOCAL EXISTENCE   In the case of a *nonlinear* initial value problem

$$\frac{d\mathbf{x}}{dt} = \mathbf{f}(\mathbf{x}, t), \qquad \mathbf{x}(a) = \mathbf{b}, \tag{35}$$

the hypothesis in Theorem 1 that $\mathbf{f}$ satisfies a Lipschitz condition on a slab $(\mathbf{x}, t)$ (t in I, all $\mathbf{x}$) is unrealistic and rarely satisfied. This is illustrated by the following simple example.

EXAMPLE 7   Consider the initial value problem

$$\frac{dx}{dt} = x^2, \qquad x(0) = b > 0. \tag{36}$$

As we saw in Example 6, the equation $x' = x^2$ does not satisfy a "strip Lipschitz condition." When we solve (36) by separation of variables, we get

$$x(t) = \frac{b}{1 - bt}. \tag{37}$$

Because the denominator vanishes for $t = 1/b$, (37) provides a solution of the initial value problem in (36) only for $t < 1/b$, despite the fact that the differential equation $x' = x^2$ "looks nice" on the whole real line—certainly the function appearing on the right-hand side of the equation is continuous everywhere. In particular, if b is large, then we have a solution only on a very small interval to the right of $t = 0$.

Although Theorem 2 assures us that *linear* equations have global solutions, Example 7 shows that, in general, even a "nice" nonlinear differential equation can be expected to have a solution only on a small interval about the initial point $t = a$, and that the length of this interval of existence can depend on the initial value $\mathbf{x}(a) = \mathbf{b}$, as well as on the differential equation itself. The reason is this: If $\mathbf{f}(\mathbf{x}, t)$ is continuously differentiable in a neighborhood of the point $(\mathbf{b}, a)$ in $(m + 1)$-space, then—as indicated in the discussion preceding Example 6—we can conclude that $\mathbf{f}(\mathbf{x}, t)$ satisfies a Lipschitz condition on some rectangular region R centered at $(\mathbf{b}, a)$, of the form

$$|t - a| < A, \qquad |x_i - b_i| < B_i \tag{38}$$

($i = 1, 2, \ldots, m$). In the proof of Theorem 1, we need to apply the Lipschitz condition on the function $\mathbf{f}$ in analyzing the iterative formula

$$\mathbf{x}_{n+1}(t) = \mathbf{b} + \int_a^t \mathbf{f}(\mathbf{x}_n(s), s) \, ds. \tag{39}$$

The potential difficulty is that unless the values of t are suitably restricted, the points $(\mathbf{x}_n(t), t)$ appearing in the integrand in (39) may not lie in the region R where $\mathbf{f}$ is known to satisfy a Lipschitz condition. On the other hand, it can be shown that—on a sufficiently small open interval J containing the point $t = a$—the graphs of the functions $\{\mathbf{x}_n(t)\}$ given iteratively by the formula in (39) remain within the region R, so the proof of convergence can then be carried out as in the proof of Theorem 1. A proof of the following *local* existence theorem can be found in Chapter 6 of G. Birkhoff and G.-C. Rota, *Ordinary Differential Equations*, 2nd. ed. (New York: John Wiley, 1969).

> **Theorem 3   Local Existence of Solutions**
>
> Let $\mathbf{f}$ be a vector-valued function (with m components) of the $m+1$ real values $x_1, x_2, \ldots, x_m$, and t. If the first order partial derivatives of $\mathbf{f}$ all exist and are continuous in some neighborhood of the point $\mathbf{x} = \mathbf{b}$, $t = a$, then the initial value problem
>
> $$\frac{d\mathbf{x}}{dt} = \mathbf{f}(\mathbf{x}, t), \qquad \mathbf{x}(a) = \mathbf{b}, \tag{35}$$
>
> has a solution on some open interval containing the point $t = a$.

UNIQUENESS OF SOLUTIONS

It is possible to establish the existence of solutions of the initial value problem in (35) under the much weaker hypothesis that $\mathbf{f}(\mathbf{x}, t)$ is merely continuous; techniques other than those used in this section are required. By contrast, the Lipschitz condition that we used in proving Theorem 1 is the key to *uniqueness* of solutions. In particular, the solution provided by Theorem 3 is unique near the point $t = a$.

> **Theorem 4   Uniqueness of Solutions**
>
> Suppose that on some region R in $(m+1)$-space, the function $\mathbf{f}$ in (35) is continuous and satisfies the Lipschitz condition
>
> $$\left|\mathbf{f}(\mathbf{x}_1, t) - \mathbf{f}(\mathbf{x}_2, t)\right| \leq k\left|\mathbf{x}_1 - \mathbf{x}_2\right|. \tag{20}$$
>
> If $\mathbf{x}_1(t)$ and $\mathbf{x}_2(t)$ are two solutions of the initial value problem in (35) on some open interval I containing $t = a$, such that the solution curves $(\mathbf{x}_1(t), t)$, and $(\mathbf{x}_2(t), t)$ both lie in R for all t in I, then $\mathbf{x}_1(t) = \mathbf{x}_2(t)$ for all t in I.

We will outline the proof of Theorem 4 for the one-dimensional case in which x is a real variable. A generalization of this proof to the multivariable case can be found in Chapter 6 of Birkhoff and Rota.

Let us consider the function

$$\phi(t) = [x_1(t) - x_2(t)]^2 \tag{40}$$

for which $\phi(a) = 0$, because $x_1(a) = x_2(a) = b$. We want to show that $\phi(t) \equiv 0$, so that $x_1(t) \equiv x_2(t)$. We will consider only the case $t \geqq a$; the details are similar for the case $t \leqq a$.

If we differentiate each side in (40), we find that

$$\begin{aligned}
|\phi'(t)| &= |2[x_1(t) - x_2(t)][x_1'(t) - x_2'(t)]| \\
&= |2[x_1(t) - x_2(t)][f(x_1(t), t) - f(x_2(t), t)]| \\
&\leq 2k|x_1(t) - x_2(t)|^2 = 2k\phi(t),
\end{aligned}$$

using the Lipschitz condition on f. Hence

$$\phi'(t) \leq 2k\phi(t). \tag{41}$$

Now let us temporarily ignore the fact that $\phi(a) = 0$ and compare $\phi(t)$ with the solution of the differential equation

$$\Phi'(t) = 2k\Phi(t) \tag{42}$$

such that $\Phi(a) = \phi(a)$; clearly

$$\Phi(t) = \phi(a)e^{2k(t-a)}. \tag{43}$$

In comparing (41) with (42), it seems inevitable that

$$\phi(t) \leq \Phi(t) \quad \text{for } t \geq a, \tag{44}$$

and this is easily proved (see Problem 18). Hence

$$0 \leq [x_1(t) - x_2(t)]^2 \leq [x_1(a) - x_2(a)]^2 e^{2k(t-a)}.$$

On taking square roots, we get

$$0 \leq |x_1(t) - x_2(t)| \leq |x_1(a) - x_2(a)|e^{k(t-a)}. \tag{45}$$

But $x_1(a) - x_2(a) = 0$, so (45) implies that $x_1(t) \equiv x_2(t)$.

EXAMPLE 8   The initial value problem

$$\frac{dx}{dt} = 3x^{2/3}, \qquad x(0) = 0 \tag{46}$$

has both the obvious solution $x_1(t) \equiv 0$ and the solution $x_2(t) = t^3$ that is easily found by separation of variables. Hence the function $f(x, t) = 3x^{2/3}$ must *fail* to satisfy a Lipschitz condition near $(0, 0)$. Indeed, the mean value theorem yields

$$|f(x, 0) - f(0, 0)| = |f_x(\bar{x}, 0)| |x - 0|$$

for some $\bar{x}$ between 0 and x. But $f_x(x, 0) = 2x^{-1/3}$ is unbounded as $x \to 0$, so no Lipschitz condition can be satisfied.

WELL POSED PROBLEMS AND MODELS

In addition to uniqueness, another consequence of the inequality in (45) is the fact that solutions of the differential equation

$$\frac{dx}{dt} = f(x, t) \tag{47}$$

depend *continuously* on the initial value $x(a)$; that is, if $x_1(t)$ and $x_2(t)$ are two solutions of (47) on the interval $a \leq t \leq T$ such that the initial values $x_1(a)$ and $x_2(a)$ are sufficiently close to one another, then the values $x_1(t)$ and $x_2(t)$ remain close to one another. In particular, if $|x_1(a) - x_2(a)| \leq \delta$, then (45) implies that

$$|x_1(t) - x_2(t)| \leq \delta e^{k(T-a)} = \epsilon \tag{48}$$

for all t with $a \leq t \leq T$. Obviously, we can make ϵ as small as we wish by choosing δ sufficiently close to zero.

This continuity of solutions of (47) with respect to initial values is important in practical applications where we are unlikely to know the initial value $x_0 = x(a)$ with absolute precision. For example, suppose that the initial value problem

$$\frac{dx}{dt} = f(x, t), \qquad x(a) = x_0 \tag{49}$$

models a population for which we know only that the initial population is within $\delta > 0$ of the assumed value x_0. Then even if the function $f(x, t)$ is accurate, the solution $x(t)$ of (49) will be only an approximation to the actual population. But (45) implies that the actual population at time t will be within $\delta e^{k(t-a)}$ of the approximate population $x(t)$. Thus, on a given closed interval $a \leq t \leq T$, $x(t)$ will be a close approximation to the actual population provided that $\delta > 0$ is sufficiently small.

An initial value problem is usually considered *well posed* as a mathematical model for a real-world situation only if the differential equation has unique solutions that are continuous with respect to initial values. Otherwise it is unlikely that the initial value problem adequately mirrors the real-world situation.

An even stronger "continuous dependence" of solutions is often desirable. In addition to possible inaccuracy in the initial value, the function $f(x, t)$ may not model precisely the physical situation. For instance, it may involve physical parameters (such as resistance coefficients) whose values cannot be measured with absolute precision. Birkhoff and Rota generalize the proof of Theorem 4 to establish the following result.

Theorem 5 *Continuous Dependence of Solutions*

Let $\mathbf{x}(t)$ and $\mathbf{y}(t)$ be solutions of the equations

$$\frac{d\mathbf{x}}{dt} = \mathbf{f}(\mathbf{x}, t) \quad \text{and} \quad \frac{d\mathbf{y}}{dt} = \mathbf{g}(\mathbf{y}, t) \tag{50}$$

on the closed interval $a \leq t \leq T$. Let $\mathbf{f}$ and $\mathbf{g}$ be continuous for $a \leq t \leq T$ and for $\mathbf{x}$ and $\mathbf{y}$ in a common region D of n-space and assume that $\mathbf{f}$ satisfies the Lipschitz condition in (20) in the region D. If

$$\left| \mathbf{f}(\mathbf{z}, t) - \mathbf{g}(\mathbf{z}, t) \right| \leq \mu \tag{51}$$

for all t in the interval $[a, T]$ and all $\mathbf{z}$ in D, it then follows that

$$\left| \mathbf{x}(t) - \mathbf{y}(t) \right| \leq \left| \mathbf{x}(a) - \mathbf{y}(a) \right| e^{k(t-a)} + \frac{\mu}{k} \left[e^{k(t-a)} - 1 \right] \tag{52}$$

on the interval $a \leq t \leq T$.

If $\mu > 0$ is small, then Eq. (51) implies that the functions $\mathbf{f}$ and $\mathbf{g}$ appearing in the two differential equations, though different, are "close" to each other. If

$\epsilon > 0$ is given, then it is apparent from (52) that

$$|\mathbf{x}(t) - \mathbf{y}(t)| \leq \epsilon \qquad (53)$$

for all t in $[a, T]$ if both $|\mathbf{x}(a) - \mathbf{y}(a)|$ and μ are sufficiently small. Thus Theorem 5 says (roughly) that if both the two initial values and the two differential equations in (50) are close to each other, then the two solutions remain close to each other for $a \leq t \leq T$.

For example, suppose that a falling body is subject both to constant gravitational acceleration g and to resistance proportional to some power of its velocity, so (with the positive axis directed downward) its velocity v satisfies the differential equation

$$\frac{dv}{dt} = g - cv^\rho. \qquad (54)$$

Assume, however, that only an approximation $\bar{c}$ to the actual resistance coefficient c and an approximation $\bar{\rho}$ to the actual exponent ρ are known. Then our mathematical model is based on the differential equation

$$\frac{du}{dt} = g - \bar{c}u^{\bar{\rho}} \qquad (55)$$

instead of the actual equation in (54). Thus if we solve Eq. (55), we obtain only an approximation $u(t)$ to the actual velocity $v(t)$. But if the parameters $\bar{c}$ and $\bar{\rho}$ are sufficiently close to the actual values c and ρ, then the right-hand sides in (54) and (55) will be close to each other; if so, then Theorem 5 implies that the actual and approximate velocity functions $v(t)$ and $u(t)$ are close to each other. In this case the approximation in (55) will be a good model of the actual physical situation.

7.6 PROBLEMS

In each of Problems 1–8, apply the successive approximations formula to compute $y_n(x)$ for $n \leq 4$. Then write the exponential series for which these approximations are partial sums (perhaps minus the first term or two; for example,

$$e^x - 1 = x + \tfrac{1}{2}x^2 + \cdots).$$

1. $y' = y$, $y(0) = 3$

2. $y' = -2y$, $y(0) = 4$

3. $y' = -2xy$, $y(0) = 1$

4. $y' = 3x^2y$, $y(0) = 2$

5. $y' = 2y + 2$, $y(0) = 0$

6. $y' = x + y$, $y(0) = 0$

7. $y' = 2x(1 + y)$, $y(0) = 0$

8. $y' = 4x(y + 2x^2)$, $y(0) = 0$

In each of Problems 9–12, compute the successive approximations $y_n(x)$ for $n \leq 3$; then compare them with the appropriate partial sums of the Taylor series of the exact solution.

9. $y' = x + y$, $y(0) = 1$

10. $y' = y + e^x$, $y(0) = 0$

11. $y' = y^2$, $y(0) = 1$

12. $y' = \tfrac{1}{2}y^3$, $y(0) = 1$

13. Apply the iterative formula in (16) to compute the first three successive approximations to the solution of the initial value problem

$$x' = 2x - y, \qquad x(0) = 1;$$
$$y' = 3x - 2y, \qquad y(0) = -1.$$

14. Apply the matrix exponential series in (19) to solve (in closed form) the initial value problem

$$\mathbf{x}'(t) = \begin{bmatrix} 1 & 1 \\ 0 & 1 \end{bmatrix}\mathbf{x}, \qquad \mathbf{x}(0) = \begin{bmatrix} 1 \\ 1 \end{bmatrix}.$$

(*Suggestion:* Show first that

$$\begin{bmatrix} 1 & 1 \\ 0 & 1 \end{bmatrix}^n = \begin{bmatrix} 1 & n \\ 0 & 1 \end{bmatrix}$$

for each positive integer n.)

15. For the initial value problem $dy/dx = 1 + y^3$, $y(1) = 1$, show that the second Picard approximation is

$$y_2(x) = 1 + 2(x - 1) + 3(x - 1)^2 + 4(x - 1)^3 + 2(x - 1)^4.$$

Then compute $y_2(1.1)$ and $y_2(1.2)$. The fourth-order Runge-Kutta method with step size $h = 0.005$ yields $y(1.1) \approx 1.2391$ and $y(1.2) \approx 1.6269$.

16. For the initial value problem $dy/dx = x^2 + y^2$, $y(0) = 0$, show that the third Picard approximation is

$$y_3(x) = \frac{1}{3} x^3 + \frac{1}{63} x^7 + \frac{2}{2079} x^{11} + \frac{1}{59,535} x^{15}.$$

Compute $y_3(1)$. The fourth-order Runge-Kutta method yields $y(1) \approx 0.350232$, both with step size $h = 0.05$ and with step size $h = 0.025$.

17. Prove as follows the inequality $|\mathbf{Ax}| \leq \|\mathbf{A}\| \cdot |\mathbf{x}|$, where $\mathbf{A}$ is an $m \times m$ matrix with row vectors $\mathbf{a}_1, \mathbf{a}_2, \ldots, \mathbf{a}_m$, and $\mathbf{x}$ is an m-dimensional vector. First note that the components of the vector $\mathbf{Ax}$ are $\mathbf{a}_1 \cdot \mathbf{x}, \mathbf{a}_2 \cdot \mathbf{x}, \ldots, \mathbf{a}_m \cdot \mathbf{x}$, so

$$|\mathbf{Ax}| = \left[\sum_{i=1}^{m} (\mathbf{a}_i \cdot \mathbf{x})^2 \right]^{1/2}.$$

Then use the Cauchy-Schwarz inequality $(\mathbf{a} \cdot \mathbf{x})^2 \leq |\mathbf{a}|^2 |\mathbf{x}|^2$ for the dot product.

18. Suppose that $\phi(t)$ is a differentiable function with

$$\phi'(t) \leq k\phi(t) \qquad (k > 0)$$

for $t \geq a$. Multiply both sides by e^{-kt}; then transpose to show that

$$\frac{d}{dt} \left[\phi(t) e^{-kt} \right] \leq 0$$

for $t \geq a$. Then apply the mean value theorem to conclude that

$$\phi(t) \leq \phi(a) e^{k(t-a)}$$

for $t \geq a$.

REFERENCES
FOR FURTHER STUDY

The literature of the theory and applications of differential equations is vast. The following list includes a selection of books that might be useful to readers who wish to pursue further the topics introduced in this book.

1. M. ABRAMOWITZ and I. A. STEGUN, *Handbook of Mathematical Functions*. New York: Dover, 1965. The comprehensive collection of tables to which frequent reference is made in the text.
2. G. BIRKHOFF and G.-C. ROTA, *Ordinary Differential Equations* (2nd ed.). New York: John Wiley, 1969. An intermediate-level text that includes more complete treatment of existence and uniqueness theorems, Sturm-Liouville problems, and eigenfunction expansions.
3. F. BRAUER and J. NOHEL, *Qualitative Theory of Ordinary Differential Equations*. New York: W. A. Benjamin, 1969. A more complete treatment of linear systems and of qualitative properties of solutions.
4. M. BRAUN, *Differential Equations and Their Applications* (3rd ed.). New York: Springer-Verlag, 1983. An introductory text at a slightly higher level than this one; it includes several interesting "case study" applications.
5. R. V. CHURCHILL, *Operational Mathematics* (3rd ed.). New York: McGraw-Hill, 1972. Standard reference for theory and applications of Laplace transforms, starting at about the same level as Chapter 4 of this book.
6. R. V. CHURCHILL and J. W. BROWN, *Fourier Series and Boundary Value Problems* (3rd ed.). New York: McGraw-Hill, 1978. At about the same level as Chapters 8 and 9 of this book.
7. E. A. CODDINGTON, *An Introduction to Ordinary Differential Equations*. Englewood Cliffs, N.J.: Prentice-Hall, 1961. An intermediate-level introduction; Chapters 3 and 4 include proofs of the theorems on power series and Frobenius series solutions stated in Chapter 3 of this book.
8. E. A. CODDINGTON and N. LEVINSON, *Theory of Ordinary Differential Equations*. New York: McGraw-Hill, 1955. An advanced theoretical text; Chapter 5 discusses solution near an irregular singular point.
9. S. D. CONTE and C. DeBOOR, *Elementary Numerical Analysis* (3rd ed.). New York: McGraw-Hill, 1980. An introductory text; Chapter 8 treats numerical solution of differential equations, and includes FORTRAN programs.

10. P. HENRICI, *Discrete Variable Methods in Ordinary Differential Equations.* New York: John Wiley, 1962. A more complete and somewhat advanced treatment of numerical methods of solution.

11. M. W. HIRSCH and S. SMALE, *Differential Equations, Dynamical Systems, and Linear Algebra.* New York: Academic Press, 1974. Intermediate-level treatment of linear systems and of qualitative properties of solutions.

12. E. L. INCE, *Ordinary Differential Equations.* New York: Dover, 1956. First published in 1926, this is the classic older reference work on the subject.

13. N. N. LEBEDEV, *Special Functions and Their Applications.* New York: Dover, 1972. A comprehensive account of Bessel functions and the other special functions of mathematical physics.

14. N. N. LEBEDEV, I. P. SKALSKAYA, and Y. S. UFLYAND. *Worked Problems in Applied Mathematics.* New York: Dover, 1979. A large collection of applied examples and problems similar to those discussed in Chapter 9 of this book.

15. N. W. MCLACHLAN, *Bessel Functions for Engineers* (2nd ed.). London: Oxford University Press, 1955. Includes numerous physical applications of Bessel functions.

16. N. W. MCLACHLAN, *Ordinary Non-Linear Differential Equations in Engineering and Physical Sciences.* London: Oxford University Press, 1956. A concrete introduction to the effects of nonlinearity in physical systems.

17. B. NOBLE, *Applied Linear Algebra* (2nd ed.). Englewood Cliffs, N.J.: Prentice-Hall, 1977. An introduction to linear algebra and its applications.

18. E. RAINVILLE, *Intermediate Differential Equations* (2nd ed.). New York: Macmillan, 1964. Chapters 3 and 4 include proofs of the theorems on power series and Frobenius series solutions stated in Chapter 3 of this book.

19. J. R. RICE, *Numerical Methods, Software, and Analysis: IMSL Reference Edition.* New York: McGraw-Hill, 1983. Chapter 9 contains detailed discussion of sophisticated numerical algorithms and their mainframe implementation.

20. H. SAGAN, *Boundary and Eigenvalue Problems in Mathematical Physics.* New York: John Wiley, 1961. Discusses the classical boundary value problems and the variational approach to Sturm-Liouville problems, eigenvalues, and eigenfunctions.

21. G. F. SIMMONS, *Differential Equations.* New York: McGraw-Hill, 1972. An introductory text with interesting historical notes and fascinating applications and with the most eloquent preface in any mathematics book currently in print.

22. W. G. STRANG, *Linear Algebra and Its Applications* (2nd ed.). New York: Academic Press, 1980. An introductory treatment of linear algebra with motivating applications.

23. G. P. TOLSTOV, *Fourier Series.* New York: Dover, 1976. An introductory text including detailed discussion of both convergence and applications of Fourier series.

24. H. F. WEINBERGER, *A First Course in Partial Differential Equations.* New York: Blaisdell, 1965. Includes separation of variables, Sturm-Liouville methods, and applications of Laplace transform methods to partial differential equations.

25. R. WEINSTOCK, *Calculus of Variations.* New York: Dover, 1974. Includes variational derivations of the partial differential equations of vibrating strings and membranes, rods, and bars.

ANSWERS
TO SELECTED
PROBLEMS

SECTION 1.1

11. If $y = y_1 = x^{-2}$, then $y' = -2x^{-3}$ and $y'' = 6x^{-4}$, so $x^2 y'' + 5xy' + 4y = x^2(6x^{-4}) + 5x(-2x^{-3}) + 4(x^{-2}) = 6x^{-2} - 10x^{-2} + 4x^{-2} = 0$. If $y = y_2 = x^{-2} \ln x$, then $y' = x^{-3} - 2x^{-3} \ln x$ and $y'' = -5x^{-4} + 6x^{-4} \ln x$, so $x^2 y'' + 5xy' + 4y = x^2(-5x^{-4} + 6x^{-4} \ln x) + 5x(x^{-3} - 2x^{-3} \ln x) + 4(x^{-2} \ln x) = (-5x^{-2} + 5x^{-2}) + (6x^{-2} - 10x^{-2} + 4x^{-2}) \ln x = 0$.

13. $r = \frac{2}{3}$ **14.** $r = \pm\frac{1}{2}$ **15.** $r = -2, 1$ **17.** $C = 2$ **18.** $C = 3$ **19.** $C = 6$ **20.** $C = 11$ **21.** $C = 7$ **22.** $C = 1$
23. $C = -56$ **24.** $C = 17$ **25.** $C = \pi/4$ **26.** $C = -\pi$ **27.** $y' = x + y$ **28.** $y' = 2y/x$ **29.** $y' = x/(1 - y)$
31. $y' = (y - x)/(x + y)$ **32.** $dP/dt = k\sqrt{P}$ **33.** $dv/dt = kv^2$ **35.** $dN/dt = k(P - N)$ **37.** $y = 1$ or $y = x$ **39.** $y = x^2$
41. $y = \frac{1}{2}e^x$ **42.** $y = \cos x$ or $y = \sin x$
43. (a) $C = 10.1$; (b) No such C, but the constant function $y(x) = 0$ satisfies the conditions $y' = y^2$ and $y(0) = 0$.

SECTION 1.2

1. $y(x) = x^2 + x + 3$ **3.** $y = (2x^{3/2} - 16)/3$ **5.** $y = 2\sqrt{x + 2} - 5$ **7.** $y = 10 \tan^{-1} x$ **9.** $y = \sin^{-1} x$
11. $x(t) = 25t^2 + 10t + 20$ **13.** $x = \frac{1}{2}t^3 + 5t$ **15.** $x = \frac{1}{3}(t + 3)^4 - 37t - 26$ **17.** $x(t) = \frac{1}{2}\{(t + 1)^{-1} + t - 1\}$
19. The car stops when $t \approx 2.78$ (s), so the distance traveled before stopping is $x(2.78) \approx 38.59$ (m).
21. $y_0 \approx 178.59$ (m) **23.** $v_0 = 544/3$ (ft/s) **25.** $v_0 = 150$ (ft/s) ≈ 102 (mi/h) **27.** $v = -20\sqrt{10} \approx 63.25$ (ft/s) **31.** 25 (mi)
32. 1:10 P.M.

SECTION 1.3

1.

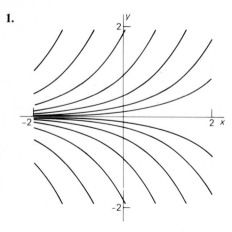

2.

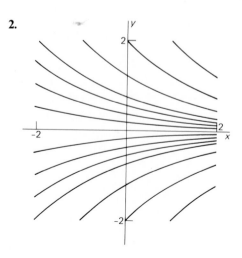

3.

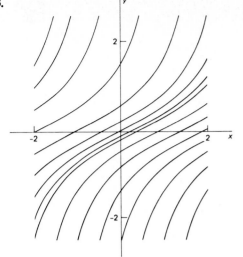

4.

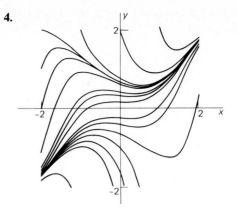

5.

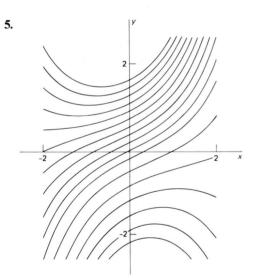

6.

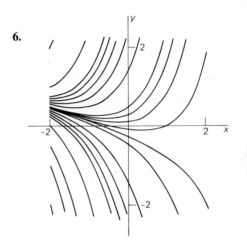

7.

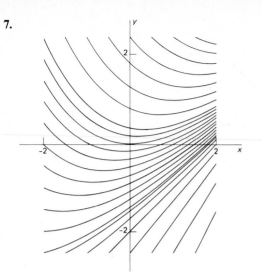

8.

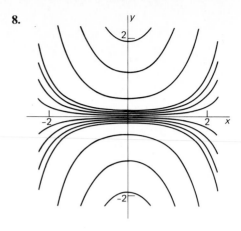

9.

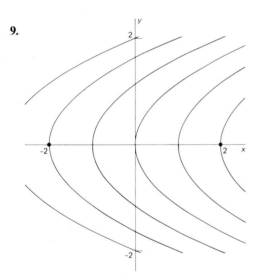

10.

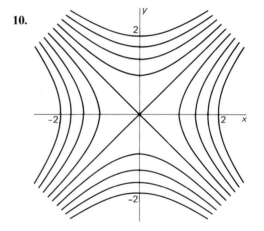

Answers to Selected Problems

11.

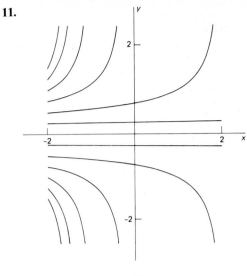

12.

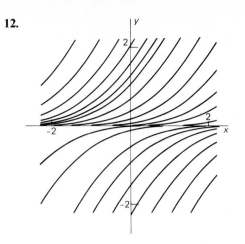

13.

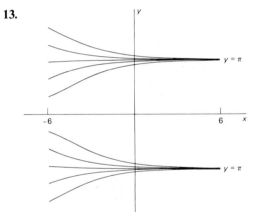

14. Of course, no solutions of this differential equation branch, and the solution curves fill the entire xy-plane. Nevertheless, it is interesting how certain solutions appear to be "stable" solutions that "nearby" solutions approach for x near zero. (All these solution curves were plotted using the improved Euler method of Chapter 6.)

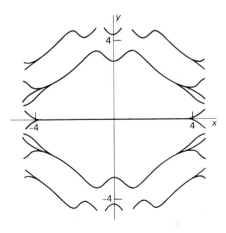

15. **16.**

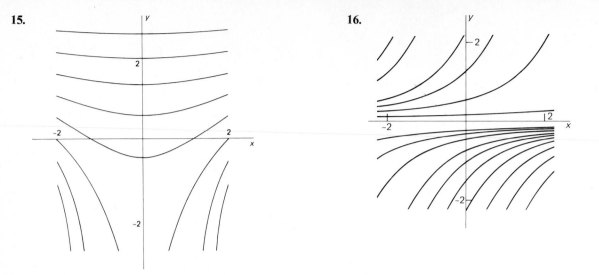

17. Each isocline is a vertical straight line.   **19.** Each isocline is a horizontal straight line.
21. Each isocline is a straight line through the origin.
23. Each isocline $xy = C$ is a rectangular hyperbola that opens along the line $y = x$ if $C > 0$, along the line $y = -x$ if $C < 0$.
25. Each isocline is a parabola symmetric about the y-axis.   **27.** A unique solution exists in some neighborhood of $x = 1$.
29. A unique solution exists in some neighborhood of $x = 0$.
31. Existence, but not uniqueness, is guaranteed in some neighborhood of $x = 2$.
33. A unique solution exists in some neighborhood of $x = 0$.   **35.** A unique solution exists in some neighborhood of $x = 0$.
37. $\partial f/\partial y$ is not continuous when $y = 1$, so the theorem does not guarantee uniqueness.

SECTION 1.4

1. $y = C \exp(-x^2)$ **3.** $y = C \exp(-\cos x)$ **5.** $y = \sin(C + \sqrt{x})$ **7.** $y = (2x^{4/3} + C)^{3/2}$ **9.** $y = C(1 + x)/(1 - x)$
11. $y = (C - x^2)^{-1/2}$ **13.** $\ln(y^4 + 1) = C + 4\sin x$ **15.** $1/(3y^3) - 2/y = 1/x + \ln|x| + C$ **17.** $\ln|1 + y| = x + \frac{1}{2}x^2 + C$
19. $y = 2\exp(e^x)$ **21.** $y^2 = 1 + (x^2 - 16)^{1/2}$ **23.** $\ln(2y - 1) = 2(x - 1)$ **25.** $\ln y = x^2 - 1 + \ln x$ **27.** About 51,840 persons
29. About 14,735 years   **31.** \$21,103.48   **33.** 2585 mg   **35.** About 4.86 billions years ago
37. After a total of about 63 min have elapsed   **39.** (a) 0.495 m; (b) $(8.32 \times 10^{-7})I_0$; (c) 3.29 m
41. (a) $dA/dt = rA + Q$; (b) $Q = 0.70482$, so you make deposits at the rate of \$704.82 per year   **43.** After about 66 min 40 s
45. After about 46 days   **47.** 972 s   **49.** Approximately 869 s   **52.** At time $t = 2048/1562 \approx 1.31$ (in hours)
53. At approximately 10:29 A.M.

SECTION 1.5

1. $y = 2(1 - e^{-x})$ **2.** $\rho = e^{-2x}$ **3.** $y = e^{-3x}(x^2 + C)$ **4.** $\rho = \exp(-x^2)$ **5.** $y = x + 4/x^2$ **6.** $\rho = x^5$ **7.** $y = 5x^{1/2} + Cx^{-1/2}$
8. $\rho = x^{1/3}$ **9.** $y = x(7 + \ln x)$ **10.** $\rho = x^{-3/2}$ **11.** $y(x) = 0$ **12.** $\rho = x^3$ **13.** $y = (e^x + e^{-x})/2$ **14.** $\rho = x^{-3}$
15. $y = \{1 - 5\exp(-x^2)\}/2$ **16.** $\rho = \exp(\sin x)$ **17.** $y = (1 + \sin x)/(1 + x)$ **18.** $\rho = x^{-2}$ **19.** $y = \frac{1}{2}\sin x + C\csc x$
20. $\rho = \exp(-x - \frac{1}{2}x^2)$ **21.** $y = x^3\sin x$ **22.** $\rho = \exp(-x^2)$ **23.** $y = x^3(2 + Ce^{-2x})$ **24.** $\rho = (x^2 + 4)^{3/2}$
25. $y = \{\exp(-3x^2/2)\}\{3(x^2 + 2)^{3/2} - 2\}$ **26.** $\rho(y) = y^4$ **27.** $x = e^y(C + \frac{1}{2}y^2)$ **28.** $\rho(y) = (1 + y^2)^{-1}$
29. $y = \{\exp(x^2)\}\{C + \frac{1}{2}\sqrt{\pi}\,\text{erf}(x)\}$ **33.** After about 7 min 41 s   **34.** About 22.2 days   **35.** After about 10.85 years
37. 393.75 lb   **39.** (b) $y_{\max} = 100e^{-1} \approx 36.79$ (gal)   **41.** (b) Approximately \$1,308,283

SECTION 1.6

1. $x^2 - 2xy - y^2 = C$ **3.** $y = x(C + \ln|x|)^2$ **5.** $\ln|xy| = C + x/y$ **7.** $y^3 = 3x^3(C + \ln|x|)$ **9.** $y = x/(C - \ln|x|)$
11. $y = C(x^2 + y^2)$ **13.** $y + (x^2 + y^2)^{1/2} = Cx^2$ **15.** $x^2(2xy + y^2) = C$ **17.** $y = -4x + 2\tan(2x + C)$ **19.** $y^2 = x/(2 + Cx^5)$
20. Substitute $v = y^3$ **21.** $y^2 = 1/(Ce^{-2x} - 1)$ **22.** Substitute $v = y^{-3}$ **23.** $y = (x + Cx^2)^{-3}$ **24.** Substitute $v = y^{-2}$

25. $2x^3y^3 = 3(1 + x^4)^{1/2} + C$ **26.** Substitute $v = y^3$ **27.** $y = (x^4 + Cx)^{1/3}$ **28.** Substitute $v = e^y$ **29.** $\sin^2 y = Cx^3 - 4x^2$
34. Substitute $v = \ln y$ **35.** $x^2 - 2xy - y^2 - 2x - 6y = C$ **37.** $x = \tan(x - y) + \sec(x - y) + C$ **41.** $y = x + (C - x)^{-1}$
45. Approximately 3.68 mi

SECTION 1.7

1. $x^2 + 3xy + y^2 = C$ **3.** $x^3 + 2xy^2 + 2y^3 = C$ **5.** $3x^4 + 4y^3 + 12y \ln x = C$ **7.** $\sin x + x \ln y + e^y = C$
9. $5x^3y^3 + 5xy^4 + y^5 = C$ **11.** $x^2y^{-1} + y^2x^{-3} + 2y^{1/2} = C$ **13.** $2x = y^3 + Cy$ **15.** $xy - x - 2y = C$ **17.** $x^3y + e^y = C$
19. $x + \arctan(y/x) = C$ **21.** $\rho(x) = x^3;\ x^4y = C$ **23.** $\rho(y) = y^{-2};\ x^2 + y^2 = Cy$ **25.** $\rho(y) = y^{-1};\ e^x + x \ln y + \sin y = C$
27. $\rho(y) = \sin y;\ x^2 \sin y = C$ **29.** $\rho(x, y) = x^2y;\ x^7y^2 - x^3y^9 = C$ **31.** $\arctan(x/y) + (x^2 + y^2)^{1/2} = C$

SECTION 1.8

1. 20 weeks   **2.** (b) 256   **3.** (b) At 6 months   **4.** It approaches zero.   **6.** Approximately 4.87 million
7. (a) Approximately 271.15 million; (b) Approximately 282.22 million   **9.** (a) 200 g; (b) (1.25) $\ln 3$ (s)   **10.** (20 $\ln 4$)/3 days
11. (a) $M = 100$ and $k = 0.0002$; (b) In the year 2035   **12.** 50 $\ln(9/8)$ months   **13.** (a) 100 $\ln(9/5)$ months; (b) 100 $\ln 2$ months
16. $\alpha \approx 0.3915;\ 2.15 \times 10^6$ cells

SECTION 1.9

1. Approximately 31.5 s   **3.** 400/($\ln 2$) ft   **5.** 400 $\ln 7$ ft   **8.** (b) About 71 mi/h; about 102 mi/h   **9.** 50 ft/s   **10.** 5 min 47 s
11. Time of fall: about 12.5 s   **12.** Approximately 649 ft   **15.** Approximately 30.46 ft/s; exactly 40 ft/s
18. Approximately 1044.14 ft in approximately 7.8003 s   **19.** Approximately 8.362 s; about 233.72 ft/s
20. Approximately 20.66 ft/s; about 484.57 s   **22.** (a) About 0.88 cm; (b) About 2.91 k
24. (b) About 11 min 33 s; about 11,230 mi/h   **26.** Height: about 56,711 ft; velocity: about 7858 ft/s
27. If its exhaust velocity is greater than 800 ft/s, it lifts off immediately. If $160 < c \leq 800$, then the rocket lifts off sometime during the 20 s its fuel is burning. If $c \leq 160$, then the rocket never leaves the ground.
32. Velocity: about 5337 ft/s; height: about 136,925 ft

CHAPTER 1 REVIEW PROBLEMS

1. Linear: $y = x^3(C + \ln x)$ **2.** Separable: $y = x/(3 - Cx - x \ln x)$ **3.** Homogeneous: $y = x/(C - \ln x)$
4. Exact: $x^2y^3 + e^x - \cos y = C$ **5.** Separable: $y = C \exp(x^{-3} - x^{-2})$ **6.** Separable: $y = x/(1 + Cx + 2x \ln x)$
7. Linear: $y = x^{-2}(C + \ln x)$ **8.** Homogeneous: $y = 3Cx/(C - x^3) = 3x/(1 + Kx^3)$ **9.** Bernoulli: $y = (x^2 + Cx^{-1})^2$
10. Separable: $y = \tan(C + x + \frac{1}{3}x^3)$ **11.** Homogeneous: $y = x/(C - 3 \ln x)$ **12.** Exact: $3x^2y^3 + 2xy^4 = C$
13. Separable: $y = 1/(C + 2x^2 - x^5)$ **14.** Homogeneous: $y^2 = x^2/(C + 2 \ln x)$ **15.** Linear: $y = (x^3 + C)e^{-3x}$
16. Substitution: $v = y - x$; solution: $y(1 + Ce^{2x}) = 1 + x + C(1 - x)e^{2x}$ **17.** Exact: $e^x + e^y + e^{xy} = C$
18. Homogeneous: $y^2 = Cx^2(x^2 - y^2)$ **19.** Separable: $y = x^2/(x^5 + Cx^2 + 1)$ **20.** Linear: $y = 2x^{-3/2} + Cx^{-3}$
21. Linear: $y = \{C + \ln(x - 1)\}/(x + 1)$ **22.** Bernoulli: $y = (2x^4 + Cx^2)^3$ **23.** Exact: $xe^y + y \sin x = C$
24. Separable: $y = x^{1/2}/(6x^2 + Cx^{1/2} + 2)$ **25.** Linear: $y = (x + 1)^{-2}(x^3 + 3x^2 + 3x + C) = x + 1 + K/(x + 1)^2$
26. Exact: $3x^{3/2}y^{4/3} - 5x^{6/5}y^{3/2} = C$ **27.** Bernoulli: $y = x^{-1}(C + \ln x)^{-1/3}$ **28.** Linear: $y = x^{-1}(C + e^{2x})$
29. Linear: $y = (x^2 + x + C)(2x + 1)^{-1/2}$ **30.** Substitution: $v = x + y$; solution: $x = 2(x + y)^{1/2} - 2 \ln\{1 + (x + y)^{1/2}\} + C$

SECTION 2.1

1. $y = \frac{5}{2}e^x - \frac{5}{2}e^{-x}$ **3.** $y = 3 \cos 2x + 4 \sin 2x$ **5.** $y = 2e^x - e^{2x}$ **7.** $y = 6 - 8e^{-x}$ **9.** $y = 2e^{-x} + xe^{-x}$ **11.** $y = 5e^x \sin x$
13. $y = 5x - 2x^2$ **15.** $y = 7x - 5x \ln x$ **21.** Linearly independent   **23.** Linearly independent   **25.** Linearly independent
29. There is no contradiction because if the given differential equation is divided by x^2 to get the form in Eq. (8), then the resulting functions $p(x) = -4/x$ and $q(x) = 6/x^2$ are not continuous at $x = 0$.

SECTION 2.2

1. $\frac{5}{2}(2x) - \frac{8}{3}(3x^2) - (5x - 8x^2) = 0$ **3.** $1(0) + 0(\sin x) + 0(e^x) = 0$ **5.** $17 - 34(\cos^2 x) + 17(\cos 2x) = 0$ **13.** $y = \frac{4}{3}e^x - \frac{1}{3}e^{-2x}$
15. $y = (2 - 2x + x^2)e^x$ **17.** $y = \frac{1}{9}(29 - 2 \cos 3x - 3 \sin 3x)$ **19.** $y = x + 2x^2 + 3x^3$ **21.** $y = 2 \cos x - 5 \sin x + 3x$
23. $y = e^{-x} + 4e^{3x} - 2$

SECTION 2.3

1. $y = c_1 e^{2x} + c_2 e^{-2x}$ **3.** $y = c_1 e^{2x} + c_2 e^{-5x}$ **5.** $y = c_1 e^{-3x} + c_2 x e^{-3x}$ **7.** $y = c_1 e^{(3x/2)} + c_2 x e^{(3x/2)}$
9. $y = e^{-4x}(c_1 \cos 3x + c_2 \sin 3x)$ **11.** $y = c_1 + c_2 x + c_3 e^{4x} + c_4 x e^{4x}$ **13.** $y = c_1 + c_2 e^{-(2x/3)} + c_3 x e^{-(2x/3)}$
15. $y = c_1 e^{2x} + c_2 x e^{2x} + c_3 e^{-2x} + c_4 x e^{-2x}$ **17.** $y = c_1 \cos (x/\sqrt{2}) + c_2 \sin (x/\sqrt{2}) + c_3 \cos (2x/\sqrt{3}) + c_4 \sin (2x/\sqrt{3})$
19. $y = c_1 e^x + c_2 e^{-x} + c_3 x e^{-x}$ **21.** $y = 5e^x + 2e^{3x}$ **23.** $y = e^{3x}(3 \cos 4x - 2 \sin 4x)$ **25.** $y = \frac{1}{4}(-13 + 6x + 9e^{-2x/3})$
27. $y = c_1 e^x + c_2 e^{-2x} + c_3 x e^{-2x}$ **29.** $y = c_1 e^{-3x} + e^{3x/2}(c_1 \cos \frac{3}{2}\sqrt{3}x + c_2 \sin \frac{3}{2}\sqrt{3}x)$
31. $y = c_1 e^{0.6527x} + c_2 e^{2.8974x} + c_3 e^{-0.5321x}$ **33.** $y = c_1 e^{1.2115x} + c_2 e^{-1.9631x} + e^{0.3758x}(c_1 \cos 1.6739x + c_2 \sin 1.6739x)$
35. $y = c_1 e^{-x/2} + c_2 e^{-x/3} + c_3 \cos 2x + c_4 \sin 2x$ **39.** $y = c_1 e^{-ix} + c_2 e^{3ix}$ **41.** $y = c_1 \exp ([1 + i\sqrt{3}]x) + c_2 \exp (-[1 + i\sqrt{3}]x)$
43. $y = 2e^{2x} - 5e^{-x} + 3 \cos x - 9 \sin x$

SECTION 2.4

1. Frequency: 2 rad/s $= 1/\pi$ Hz; period: π s **3.** Amplitude: 2 m; frequency: 5 rad/s; period: $2\pi/5$ s **7.** About 10,450 ft
11. About 3.8 in. **13.** $x = 4e^{-2t} - 2e^{-4t}$; overdamped **15.** $x = 5e^{-4t} + 10te^{-4t}$; critically damped
17. $x = \frac{1}{3}\sqrt{313}e^{-5t/2} \cos (6t - 0.8254)$; underdamped
19. $x = e^{-5t}(6 \cos 10t + 8 \sin 10t) \approx 10e^{-5t} \cos (10t - 0.9273)$; underdamped
20. (b) The time-varying amplitude is $2/\sqrt{3}$ ft, the frequency is $4\sqrt{3}$ rad/s, and the phase angle is $\pi/6$.
21. (a) $k \approx 7018$ lb/ft; (b) After about 2.47 s **32.** Damping constant: $c \approx 11.5$ lb/ft; spring constant: $k \approx 189.68$ lb/ft

SECTION 2.5

1. $y_p = \frac{1}{25}e^x$ **3.** $y_p = \frac{1}{39}(\cos 3x - 5 \sin 3x)$ **5.** $y_p = \frac{1}{26}(13 + 3 \cos 2x - 2 \sin 2x)$ **7.** $y_p = -\frac{1}{6}(e^x - e^{-x}) = -\frac{1}{3}\sinh x$
9. $y_p = -\frac{1}{3} + \frac{1}{16}(2x^2 - x)e^x$ **11.** $y_p = \frac{1}{8}(3x^2 - 2x)$ **13.** $y_p = \frac{1}{65}e^x(7 \sin x - 4 \cos x)$ **15.** $y_p = -17$
17. $y_p = \frac{1}{10}(x^2 \sin x - 4x \cos x)$ **19.** $y_p = \frac{1}{8}(10x^2 - 4x^3 + x^4)$ **21.** $y_p = xe^x(A \cos x + B \sin x)$
23. $y_p = Ax \cos 2x + Bx \sin 2x + Cx^2 \cos 2x + Dx^2 \sin 2x$ **25.** $y_p = Axe^{-x} + Bx^2 e^{-x} + Cxe^{-2x} + Dx^2 e^{-2x}$
27. $y_p = Ax \cos x + Bx \sin x + Cx \cos 2x + Dx \sin 2x$ **29.** $y_p = Ax^3 e^x + Bx^4 e^x + Cxe^{2x} + Dxe^{-2x}$
31. $y_p = \cos 2x + \frac{3}{4} \sin 2x + \frac{1}{2}x$ **33.** $y = \cos 3x - \frac{2}{15} \sin 3x + \frac{1}{5} \sin 2x$ **35.** $y = e^x(2 \cos x - \frac{5}{2} \sin x) + \frac{1}{2}x + 1$
37. $y = 2 - 2e^x + 2xe^x - \frac{1}{2}x^2 e^x + \frac{1}{6}x^3 e^x$ **39.** $y = -3 + 3x - \frac{1}{2}x^2 + \frac{1}{6}x^3 + 4e^{-x} + xe^{-x}$
41. $y_p = 255 - 450x + 30x^2 + 20x^3 + 10x^4 - 4x^5$ **43.** (b) $y = c_1 \cos 2x + c_2 \sin 2x + \frac{1}{4} \cos x - \frac{1}{20} \cos 3x$
45. $y = c_1 \cos 3x + c_2 \sin 3x + \frac{1}{24} - \frac{1}{10} \cos 2x - \frac{1}{56} \cos 4x$ **47.** $y_p = Ax + Bx^2 + Cx^3$ **49.** $y_p = Axe^{5x} + Bx^2 e^{5x}$
51. $y_p = Ax^2 \cos 2x + Bx^2 \sin 2x$

SECTION 2.6

1. $y_2 = xe^{2x}$ **3.** $y_2 = x^{1/2} \ln x$ **5.** $y_2 = 2x^2 + 2x + 1$ **7.** $y_2 = x \sin x$ **9.** $y_2 = 2x + 1$ **11.** $y_2 = xe^{x/2}$ **13.** $y_2 = 1/x^3$
15. $y_2 = xe^x$ **17.** $y_2 = x^2 \cos x$ **21.** $y = c_1 + c_2 \ln x$ **23.** $y = c_1 x^3 + c_2 x^{-4}$ **25.** $y = c_1 x^{1/2} + c_2 x^{-3/2}$
27. $y = x[c_1 \cos (\ln x) + c_2 \sin (\ln x)]$ **29.** $y = c_1 \cos ([\frac{3}{2}]^{1/2} \ln x) + c_2 \sin ([\frac{3}{2}]^{1/2} \ln x)$ **31.** $y = c_1 + c_2 x^2 + c_3 x^2 \ln x$
33. $y = c_1 + c_2 \ln x + c_3 (\ln x)^2$ **35.** $y = c_1 x^{-1} + x^{1/2}\{c_2 \cos (\frac{1}{2}\sqrt{3} \ln x) + c_3 \sin (\frac{1}{2}\sqrt{3} \ln x)\}$ **37.** $y = Ax^2 + B$
39. $y = A \cos 2x + B \sin 2x$ **41.** $y = A - \ln |x + B|$ **43.** $y = (A + Be^x)^{1/2}$ **45.** $y^3 + 3x + Ay + B = 0$

47. $y = A \tan (Ax + B)$ **63.** $y(x) = \frac{1}{2}\left(\frac{1}{2c}x^2 - c \ln x\right) + A$

SECTION 2.7

1. $y_p = \frac{2}{3}e^x$ **3.** $y_p = x^2 e^{2x}$ **5.** $y_p = -\frac{1}{4}(\cos 2x \cos x - \sin 2x \sin x) + \frac{1}{20}(\cos 5x \cos 2x + \sin 5x \sin 2x) = -\frac{1}{5} \cos 3x$ (!)
7. $y_p = \frac{2}{3}x \sin 3x + \frac{2}{9}(\cos 3x) \ln |\cos 3x|$ **9.** $y_p = \frac{1}{8}(1 - x \sin 2x)$ **11.** $y_p = -e^x(1 + \ln x)$ **13.** $y_p = \frac{1}{4}x^4$ **15.** $y_p = \ln x$
17. $y_p = \frac{1}{6}x^3 e^{-x}$ **19.** $y_p = \frac{1}{2}e^x(1 - x^{-1}) - \frac{1}{2}e^{-x}\int_1^x t^{-2}e^{2t} dt$ **21.** $y_p = \frac{1}{2}\int_0^x \exp(-t^2) \sin 2(x - t) dt$
23. $y_p = \frac{3}{4}x^2 + (\frac{1}{2}x^2 - 2x) \ln x + e^x \int_1^x e^{-t} \ln t \, dt$ **25.** $y_p = \frac{1}{90}x^4$ **27.** $y_p = -x^2 + x \ln |(x + 1)/(x - 1)| + \frac{1}{2}(1 + x^2) \ln |x^2 - 1|$
29. $y_p = e^x\left(e^{-1} + \frac{1}{2}\int_1^x t^{-1}e^{-t} dt - \frac{1}{2}e^{-2x}\int_1^x t^{-1}e^t dt\right)$ **31.** This is a special case of Problem 37.

SECTION 2.8

1. $x(t) = 2 \cos 2t - 2 \cos 3t$

3. $x(t) = \frac{1}{15}\sqrt{138,388} \cos(10t + \alpha) + \frac{1}{3}\cos(5t - \beta)$ where $\alpha = \tan^{-1}(\frac{1}{186}) \approx 0.0054$ and $\beta = \tan^{-1}(\frac{4}{3}) \approx 0.9273$

5. $x(t) = (x_0 - C) \cos \omega_0 t + C \cos \omega t$ where $C = F_0/(k - m\omega^2)$ 7. $x_{sp} = -\frac{10}{13}\cos(3t + \alpha)$ where $\alpha = \tan^{-1}(\frac{12}{5}) \approx 1.1760$

9. $x_{sp} = -(3/\sqrt{40,001}) \cos(10t - \alpha)$ where $\alpha = \tan^{-1}(199/20) \approx 1.4706$

11. $x_{sp} = \frac{1}{4}\sqrt{10}\cos(3t - \alpha)$ where $\tan \alpha = -3$, $-\pi/2 < \alpha < \pi$ (so $\alpha \approx 1.8925$); $x_{tr} = \frac{5}{4}\sqrt{2}e^{-2t}\cos(t - \beta)$ where $\tan \beta = -7$, $3\pi/2 < \beta < 2\pi$ (so $\beta \approx 4.8543$)

13. $x_{sp} = -(3/\sqrt{9236}) \cos(10t + \alpha)$ where $\alpha = \tan^{-1}(\frac{10}{47}) \approx 0.2096$; $x_{tr} \approx (10.9761)e^{-t}\cos(t\sqrt{5} - \beta)$ where $\beta \approx 0.4181$

15. $\omega = \sqrt{384}$ rad/s (approximately 3.12 Hz)   17. $\omega_0 = \sqrt{g/L + k/m}$

19. (a) Natural frequency: $\sqrt{10}$ rad/s ≈ 0.50 Hz; (b) amplitude: approximately $10\frac{5}{8}$ in.

SECTION 2.9

1. $I(t) = 4e^{-5t}$ 3. $I(t) = \frac{4}{145}(\cos 60t + 12 \sin 60t - e^{-5t})$ 5. $I(t) = \frac{5}{6}e^{-10t}\sin 60t$

7. (a) $Q(t) = E_0 C(1 - e^{-t/RC})$; $I(t) = (E_0/R)e^{-t/RC}$

9. (a) $Q(t) = \frac{1}{1480}(\cos 120t + 6 \sin 120t - e^{-20t})$; $I(t) = \frac{1}{74}(36 \cos 120t - 6 \sin 120t + e^{-20t})$; (b) Steady-state amplitude: $3/\sqrt{37}$

11. $I_{sp} = (10/\sqrt{37}) \sin(2t - \delta)$ where $\delta = 2\pi - \tan^{-1}(\frac{1}{6}) \approx 6.1180$

13. $I_{sp} = (20/\sqrt{13}) \sin(5t - \delta)$ where $\delta = 2\pi - \tan^{-1}(\frac{2}{3}) \approx 5.6952$

15. $I_0 = 33\pi[(1000 - 36\pi^2)^2 + (30\pi)^2]^{-1/2} \approx 0.1591$; $\delta = 2\pi - \tan^{-1}[(1000 - 36\pi^2)/(30\pi)] \approx 4.8576$ 17. $I(t) = -25e^{-4t}\sin 3t$

19. $I(t) = 10e^{-20t} - 10e^{-10t} - 50te^{-10t}$ 21. $I(t) = \frac{20}{13}(2 \cos 5t + 3 \sin 5t) - \frac{10}{39}e^{-t}(12 \cos 3t + 47 \sin 3t)$

23. Critical frequency: $\omega_0 = 1/\sqrt{LC}$

SECTION 2.10

1. Eigenvalues: $\{(2n + 1)^2\pi^2/4\}$; associated eigenfunctions: $\{\cos(2n + 1)\pi x/2\}$ ($n \geq 0$)

3. Eigenvalues: $\{n^2/4\}$ for $n > 0$; associated eigenfunctions: $y_n(x) = \cos(nx/2)$ for n odd and $y_n(x) = \sin(nx/2)$ for n even

5. Eigenvalues $\{(2n + 1)^2\pi^2/64\}$ for $n \geq 0$; associated eigenfunctions given by $y_n(x) = \cos([2n + 1]\pi x/8) + (-1)^n \sin([2n + 1]\pi x/8)$

SECTION 3.1

1. $y = c_0(1 + x + x^2/2! + x^3/3! + \cdots) = c_0 e^x$; $\rho = \infty$ 2. $c_{n+1} = 4c_n/(n + 1)$

3. $y = c_0(1 - 3x/2 + 3^2 x^2/(2!2^2) - 3^3 x^3/(3!2^3) + 3^4 x^4/(4!2^4) - \cdots)$; $\rho = \infty$ 4. $c_1 = 0$ and $c_{n+2} = -2c_n/(n + 2)$

5. $y = c_0(1 + x^3/3 + x^6/(2!3^2) + x^9/(3!3^3) + \cdots) = c_0 \exp(x^3/3)$; $\rho = \infty$ 6. $c_{n+1} = c_n/2$

7. $y = c_0(1 + 2x + 4x^2 + 8x^3 + \cdots) = c_0/(1 - 2x)$; $\rho = \frac{1}{2}$ 8. $y = c_0(1 + x)^{1/2}$

9. $y = c_0(1 + 2x + 3x^2 + 4x^3 + \cdots) = c_0/(1 - x)^2$; $\rho = 1$ 10. $y = c_0(1 - x)^{3/2}$

11. $y = c_0(1 + x^2/2! + x^4/4! + x^6/6! + \cdots) + c_1(x + x^3/3! + x^5/5! + x^7/7! + \cdots) = c_0 \cosh x + c_1 \sinh x$; $\rho = \infty$

12. $c_{n+2} = 4c_n/((n + 1)(n + 2))$

13. $y = c_0(1 - 3^2 x^2/2! + 3^4 x^4/4! - 3^6 x^6/6! + \cdots) + (c_1/3)(3x - 3^3 x^3/3! + 3^5 x^5/5! - 3^7 x^7/7! + \cdots) = c_0 \cos 3x + (c_1/3) \sin 3x$; $\rho = \infty$

14. $y = x + c_0 \cos x + (c_1 - 1) \sin x$ 15. $(n + 1)c_n = 0$ for all $n \geq 0$, so $c_n = 0$ for all $n \geq 0$

16. $2nc_n = c_n$ for all $n \geq 0$, so $c_n = 0$ for all $n \geq 0$ 17. $c_0 = c_1 = 0$ and $c_{n+1} = -nc_n$ for $n \geq 1$, thus $c_n = 0$ for all $n \geq 0$

18. $c_n = 0$ for all $n \geq 0$ 19. $c_n = -4c_{n-2}/(n(n - 1))$ for $n \geq 2$; $c_0 = 0$ and $c_1 = 3$; $y = (3/2) \sin 2x$ 20. $y = 2 \cosh 2x$

21. $c_{n+1} = (2nc_n - c_{n-1})/(n(n + 1))$ for $n \geq 1$; $c_0 = 0$, $c_1 = 1$, and $y = xe^x$ 22. $y = e^{-2x}$

23. Because $c_0 = c_1 = 0$ and $(n^2 - n + 1)c_n + (n - 1)c_{n-1} = 0$ for $n \geq 2$, $c_n = 0$ for all $n \geq 0$

SECTION 3.2

1. $c_{n+2} = c_n$; $y = c_0 \sum_{n=0}^{\infty} x^{2n} + c_1 \sum_{n=0}^{\infty} x^{2n+1} = (c_0 + c_1 x)/(1 - x^2)$

2. $c_{n+2} = -\frac{1}{2}c_n$; $y = c_0 \sum_{n=0}^{\infty} (-1)^n x^{2n}/2^n + c_1 \sum_{n=0}^{\infty} (-1)^n x^{2n+1}/2^n$

3. $c_{n+2} = -c_n/(n+2)$; $y = c_0 \sum_{n=0}^{\infty} (-1)^n x^{2n}/(n!2^n) + c_1 \sum_{n=0}^{\infty} (-1)^n x^{2n+1}/(2n+1)!!$

4. $c_{n+2} = -[(n+4)/(n+2)]c_n$; $y = c_0 \sum_{n=0}^{\infty} (-1)^n (n+1)x^{2n} + \frac{1}{3}c_1 \sum_{n=0}^{\infty} (-1)^n (2n+3)x^{2n+1}$

5. $c_{n+2} = nc_n/[3(n+2)]$; $y = c_0 + c_1 \sum_{n=0}^{\infty} x^{2n+1}/[(2n+1)3^n]$

6. $c_{n+2} = [(n-3)(n-4)/(n+1)(n+2)]c_n$; $y = c_0(1 + 6x^2 + x^4) + c_1(1+x)$

7. $c_{n+2} = -[(n-4)^2/3(n+1)(n+3)]c_n$; $y = c_0(1 - \frac{8}{3}x^2 + \frac{8}{27}x^4) + c_1(x - \frac{1}{2}x^3 + \frac{1}{120}x^5 + 9 \sum_{n=3}^{\infty} \{[(2n-5)!!]^2(-1)^n/(2n+1)!3^n\}x^{2n+1})$

8. $c_{n+2} = [(n-4)(n+4)/2(n+1)(n+2)]c_n$; $y = c_0(1 - 4x^2 + 2x^4) + c_1(x - \frac{5}{4}x^3 + \frac{7}{32}x^5 +$
$\sum_{n=3}^{\infty} [(2n-5)!!(2n+3)!!/(2n+1)!2^n]x^{2n+1})$

9. $c_{n+2} = [(n+3)(n+4)/(n+1)(n+2)]c_n$; $y = c_0 \sum_{n=0}^{\infty} (n+1)(2n+1)x^{2n} + \frac{1}{3}c_1 \sum_{n=0}^{\infty} (n+1)(2n+3)x^{2n+1}$

10. $c_{n+2} = -[(n-4)/3(n+1)(n+2)]c_n$; $y = c_0(1 + \frac{2}{3}x^2 + \frac{1}{27}x^4) + c_1(x + \frac{1}{6}x^3 + \frac{1}{360}x^5 + 3\sum_{n=3}^{\infty} [(-1)^n(2n-5)!!/(2n+1)!3^n]x^{2n+1})$

11. $c_{n+2} = [2(n-5)/5(n+1)(n+2)]c_n$; $y = c_1(x - \frac{4}{15}x^3 + \frac{4}{375}x^5) + c_0(1 - x^2 + \frac{1}{10}x^4 + \frac{1}{750}x^6 - 15\sum_{n=4}^{\infty} [(2n-7)!!2^n/(2n)!5^n]x^{2n})$

12. $c_2 = 0$; $c_{n+3} = c_n/(n+2)$; $y = c_0\left(1 + \sum_{n=1}^{\infty} x^{3n}/[2 \cdot 5 \cdots (3n-1)]\right) + c_1 \sum_{n=0}^{\infty} x^{3n+1}/(n!3^n)$

13. $c_2 = 0$; $c_{n+3} = -c_n/(n+3)$; $y = c_0 \sum_{n=0}^{\infty} (-1)^n x^{3n}/(n!3^n) + c_1 \sum_{n=0}^{\infty} (-1)^n x^{3n+1}/[1 \cdot 4 \cdots (3n+1)]$

14. $c_2 = 0$; $c_{n+3} = -c_n/[(n+2)(n+3)]$; $y = c_0\left(1 + \sum_{n=1}^{\infty} (-1)^n x^{3n}/[3^n \cdot n! \cdot 2 \cdot 5 \cdots (3n-1)]\right) +$
$c_1 \sum_{n=0}^{\infty} (-1)^n x^{3n+1}/[3^n \cdot n! \cdot 1 \cdot 4 \cdots (3n+1)]$

15. $c_2 = c_3 = 0$, $c_{n+4} = -c_n/[(n+3)(n+4)]$; $y = c_0\left(1 + \sum_{n=1}^{\infty} (-1)^n x^{4n}/[4^n \cdot n! \cdot 3 \cdot 7 \cdots (4n-1)]\right) +$
$c_1 \sum_{n=0}^{\infty} (-1)^n x^{4n+1}/[4^n \cdot n! \cdot 5 \cdot 9 \cdots (4n+1)]$

16. $y(x) = x$ **17.** $y(x) = 1 + x^2$ **18.** $c_{n+2} = -c_n/(n+2)$ **19.** $y = \frac{1}{3} \sum_{n=0}^{\infty} (2n+3)(x-1)^{2n+1}$; converges if $0 < x < 2$

20. $y = 2 - 6(x-3)^2$; converges for all x **21.** $y = 1 + 4(x+2)^2$; converges for all x

23. $2c_2 + c_0 = 0$; $(n+1)(n+2)c_{n+2} + c_n + c_{n-1} = 0$ for $n \geq 1$; $y_1 = 1 - \frac{1}{2}x^2 - \frac{1}{6}x^3 + \cdots$; $y_2 = x - \frac{1}{6}x^3 - \frac{1}{12}x^4 + \cdots$

25. $c_2 = c_3 = 0$; $(n+3)(n+4)c_{n+4} + (n+1)c_{n+1} + c_n = 0$ for $n \geq 0$; $y_1 = 1 - \frac{1}{12}x^4 + \frac{1}{126}x^7 + \cdots$; $y_2 = x - \frac{1}{12}x^4 - \frac{1}{20}x^5 + \cdots$

26. $y = c_0(1 - x^6/30 + x^9/72 + \cdots) + c_1(x - x^7/42 + x^{10}/90 + \cdots)$

27. $y = 1 - x - \frac{1}{2}x^2 + \frac{1}{3}x^3 - \frac{1}{24}x^4 + \frac{1}{30}x^5 + \frac{29}{720}x^6 - \frac{13}{630}x^7 - \frac{143}{40,320}x^8 + \cdots$; $y(0.5) \approx 0.4156$

28. $y = c_0(1 - x^2/2 + x^3/6 + \cdots) + (x - x^3/6 + x^4/12 + \cdots)$ **29.** $y_1 = 1 - \frac{1}{2}x^2 + \frac{1}{720}x^6 + \cdots$; $y_2 = x - \frac{1}{6}x^3 - \frac{1}{60}x^5 + \cdots$

30. $y = c_0(1 - x^2 + 13x^4/72 + \cdots) + c_1(x - x^3/3 + x^5/24 + \cdots)$

SECTION 3.3

1. Ordinary point **2.** Ordinary point **3.** Irregular singular point **4.** Irregular singular point
5. Regular singular point; $r_1 = 0, r = -1$ **6.** Regular singular point; $r_1 = 1, r_2 = -2$ **7.** Regular singular point; $r = -2, -3$
8. Regular singular point; $r = 1/2, -3$ **9.** Regular singular point $x = 1$ **10.** Regular singular point $x = 1$
11. Regular singular points $x = 1, -1$ **12.** Irregular singular point $x = 2$ **13.** Regular singular points $x = 2, -2$
14. Irregular singular points $x = 3, -3$ **15.** Regular singular point $x = 2$
16. Irregular singular point $x = 0$, regular singular point $x = 1$ **17.** $y_1 = \cos \sqrt{x}$, $y_2 = \sin \sqrt{x}$

18. $y_1 = \sum_{n=0}^{\infty} x^n/[n!(2n+1)!!]$, $y_2 = x^{-1/2} \sum_{n=0}^{\infty} x^n/[n!(2n-1)!!]$

19. $y_1 = x^{3/2}\left(1 + 3\sum_{n=1}^{\infty} x^n/[n!(2n+3)!!]\right)$, $y_2 = 1 - x - \sum_{n=2}^{\infty} x^n/[n!(2n-3)!!]$

20. $y_1 = x^{1/3} \sum_{n=0}^{\infty} (-1)^n 2^n x^n / [n! \cdot 4 \cdot 7 \cdots (3n+1)]$, $y_2 = \sum_{n=0}^{\infty} (-1)^n 2^n x^n / [n! \cdot 2 \cdot 5 \cdots (3n-1)]$

21. $y_1 = x \left(1 + \sum_{n=1}^{\infty} x^{2n} / [n! \cdot 7 \cdot 11 \cdots (4n+3)] \right)$, $y_2 = x^{-1/2} \sum_{n=0}^{\infty} x^{2n} / [n! \cdot 1 \cdot 5 \cdots (4n+1)]$

22. $y_1 = x^{3/2} \left(1 + \sum_{n=1}^{\infty} (-1)^n x^{2n} / [n! \cdot 9 \cdot 13 \cdots (4n+5)] \right)$, $y_2 = x^{-1} \left(1 + \sum_{n=1}^{\infty} (-1)^{n-1} x^{2n} / [n! \cdot 3 \cdot 7 \cdots (4n-1)] \right)$

23. $y_1 = x^{1/2} \left(1 + \sum_{n=1}^{\infty} x^{2n} / [2^n \cdot n! \cdot 19 \cdot 31 \cdots (12n+7)] \right)$, $y_2 = x^{-2/3} \left(1 + \sum_{n=1}^{\infty} x^{2n} / [2^n \cdot n! \cdot 5 \cdot 17 \cdots (12n-7)] \right)$

24. $y_1 = x^{1/3} \left(1 + \sum_{n=1}^{\infty} (-1)^n x^{2n} / [2^n \cdot n! \cdot 7 \cdot 13 \cdots (6n+1)] \right)$, $y_2 = 1 + \sum_{n=1}^{\infty} (-1)^n x^{2n} / [2^n \cdot n! \cdot 5 \cdot 11 \cdots (6n-1)]$

25. $y_1 = x^{1/2} \sum_{n=0}^{\infty} (-1)^n x^n / (n! 2^n) = x^{1/2} e^{-x/2}$, $y_2 = 1 + \sum_{n=1}^{\infty} (-1)^n x^n / (2n-1)!!$

26. $y_1 = x^{1/2} \sum_{n=0}^{\infty} x^{2n} / (n! 2^n) = x^{1/2} \exp(x^2/2)$, $y_2 = 1 + \sum_{n=1}^{\infty} 2^n x^{2n} / [3 \cdot 7 \cdots (4n-1)]$

27. $y_1 = (\cos 3x)/x$, $y_2 = (\sin 3x)/x$ **28.** $y_1 = (\cosh 2x)/x$, $y_2 = (\sinh 2x)/x$ **29.** $y_1 = (1/x)\cos(x/2)$, $y_2 = (1/x)\sin(x/2)$

30. $y_1 = \cos x^2$, $y_2 = \sin x^2$ **31.** $y_1 = x^{1/2} \cosh x$, $y_2 = x^{1/2} \sinh x$

32. $y_1 = x + 3x^2/5 + 3x^3/35 + x^4/315$ (terminates), $y_2 = x^{-1/2}(1 - 3x/2 - 3x^2/8 + x^3/48 + \cdots)$

33. $y_1 = x^{-1}(1 + 10x + 5x^2 + 10x^3/9 + \cdots)$, $y_2 = x^{1/2}(1 + 11x/20 - 11x^2/224 + 671x^3/24{,}192 + \cdots)$

34. $y_1 = x(1 - x^2/42 + x^4/1320 + \cdots)$, $y_2 = x^{-1/2}(1 - 7x^2/24 + 19x^4/3200 + \cdots)$

SECTION 3.4

1. $y_1 = x^{-2}(1 + x)$, $y_2 = 1 + 2 \sum_{n=1}^{\infty} x^n/(n+2)!$ **2.** $y_1 = x^{-4}(1 + x + \frac{1}{2}x^2 + \frac{1}{6}x^3)$, $y_2 = 1 + 24 \sum_{n=1}^{\infty} x^n/(n+4)!$

3. $y_1 = x^{-4}(1 - 3x + \frac{9}{2}x^2 - \frac{9}{2}x^3)$, $y_2 = 1 + 24 \sum_{n=1}^{\infty} (-1)^n 3^n x^n/(n+4)!$

4. $y_1 = x^{-5}(1 - \frac{3}{5}x + \frac{9}{50}x^2 - \frac{27}{250}x^3 + \frac{27}{1000}x^4)$, $y_2 = 1 + 120 \sum_{n=1}^{\infty} (-1)^n 3^n x^n/[(n+5)!5^n]$

5. $y_1 = 1 + \frac{3}{4}x + \frac{1}{4}x^2 + \frac{1}{24}x^3$, $y_2 = x^5 \left(1 + 120 \sum_{n=1}^{\infty} (n+1)x^n/(n+5)! \right)$ **6.** $y_1 = x^4 \left(1 + \frac{8}{5} \sum_{n=1}^{\infty} \frac{(2n+5)!! x^n}{n!(n+4)!2^n} \right)$

7. $y_1 = x^{-2}(2 - 6x + 9x^2)$, $y_2 = \sum (-1)^{n-1} 3^n x^n/(n+2)!$ **8.** $y_1 = 3 + 2x + x^2$, $y_2 = x^4/(1-x)^2$

9. $y_1 = 1 + x^2/2^2 + x^4/2^2 4^2 + x^6/2^2 4^2 6^2 + \cdots$, $y_2 = y_1(\ln x - x^2/4 + 5x^4/128 - 23x^6/3456 + \cdots)$

10. $y_1 = x(1 - x^2/2^2 + x^4/(2^2 4^2) - x^6/(2^2 4^2 6^2) + \cdots)$, $y_2 = y_1(\ln x + x^2/4 + 5x^4/128 + 23x^6/3456 + \cdots)$

11. $y_1 = x^2(1 - 2x + \frac{3}{2}x^2 - \frac{2}{3}x^3 + \cdots)$, $y_2 = y_1(\ln x + 3x + \frac{11}{4}x^2 + \frac{49}{18}x^3 + \cdots)$

12. $y_1 = x^2(1 - x/2 + 3x^2/20 - x^3/30 + x^4/168 - \cdots)$, $y_2 = y_1(-1/(3x^3) + 1/(20x) + 3x/100 + \cdots)$

13. $y_1 = x^3(1 - 2x + 2x^2 - \frac{4}{3}x^3 + \cdots)$, $y_2 = y_1(2\ln x - 1/2x^2 - 2/x + \frac{4}{3}x + \cdots)$

14. $y_1 = x^2(1 - 2x/5 + x^2/10 - 2x^3/105 + x^4/336 - \cdots)$, $y_2 = y_1(-1/(4x^4) + 1/(15x^3) + 1/(100x^2) - 13/(1750x) + \cdots)$; y_2 contains no logarithmic term

16. $y_1 = x^{3/2} \left(1 + \sum_{n=1}^{\infty} (-1)^n x^{2n} / [2^n \cdot n! \cdot 5 \cdot 7 \cdots (2n+3)] \right)$, $y_2 = x^{-3/2} \left(1 + \sum_{n=1}^{\infty} (-1)^n x^{2n} / [2^n \cdot n! \cdot (-1) \cdot 1 \cdot 3 \cdots (2n-3)] \right)$

SECTION 3.5

5. $J_4(x) = (1/x^2)(24 - x^2)J_0(x) + (8/x^3)(6 - x^2)J_1(x)$ **12.** 3   **13.** $x^2 J_1(x) + xJ_0(x) - \int J_0(x)\,dx + C$

14. $(x^3 - 4x)J_1(x) + 2x^2 J_0(x) + C$ **15.** $(x^4 - 9x^2)J_1(x) + (3x^3 - 9x)J_0(x) + 9\int J_0(x)\,dx + C$ **16.** $-xJ_0(x) + \int J_0(x)\,dx + C$

17. $2xJ_1(x) - x^2 J_0(x) + C$ **18.** $3x^2 J_1(x) + (3x - x^3)J_0(x) - 3\int J_0(x)\,dx + C$ **19.** $(4x^3 - 16x)J_1(x) + (8x^2 - x^4)J_0(x) + C$

20. $-2J_1(x) + \int J_0(x)\,dx + C$ **21.** $J_0(x) - (4/x)J_1(x) + C$

SECTION 3.6

1. $y = x[c_1 J_0(x) + c_2 Y_0(x)]$ **2.** $y = x^{-1}[c_1 J_1(x) + c_2 Y_1(x)]$ **3.** $y = x[c_1 J_{1/2}(3x^2) + c_2 J_{-1/2}(3x^2)]$
4. $y = x^3[c_1 J_2(2x^{1/2}) + c_2 Y_2(2x^{1/2})]$ **5.** $y = x^{-1/3}[c_1 J_{1/3}(\frac{1}{3}x^{3/2}) + c_2 J_{-1/3}(\frac{1}{3}x^{3/2})]$
6. $y = x^{-1/4}[c_1 J_0(2x^{3/2}) + c_2 Y_0(2x^{3/2})]$ **7.** $y = x^{-1}[c_1 J_0(x) + c_2 Y_0(x)]$ **8.** $y = x^2[c_1 J_1(4x^{1/2}) + c_2 Y_1(4x^{1/2})]$
9. $y = x^{1/2}[c_1 J_{1/2}(2x^{3/2}) + c_2 J_{-1/2}(2x^{3/2})]$ **10.** $y = x^{-1/4}[c_1 J_{3/2}(\frac{2}{5}x^{5/2}) + c_2 J_{-3/2}(\frac{2}{5}x^{5/2})]$
11. $y = x^{1/2}[c_1 J_{1/6}(\frac{1}{3}x^3) + c_2 J_{-1/6}(\frac{1}{3}x^3)]$ **12.** $y = x^{1/2}[c_1 J_{1/5}(\frac{4}{5}x^{5/2}) + c_2 J_{-1/5}(\frac{4}{5}x^{5/2})]$

SECTION 4.1

1. $1/s^2$, $s > 0$ **2.** $2/s^3$, $s > 0$ **3.** $e/(s - 3)$, $s > 3$ **4.** $s/(s^2 + 1)$, $s > 0$ **5.** $1/(s^2 - 1)$, $s > 1$ **7.** $(1 - e^{-s})/s$, $s > 0$
8. $(e^{-s} - e^{-2s})/s$, $s > 0$ **9.** $(1 - e^{-s} - se^{-s})/s^2$, $s > 0$ **11.** $\frac{1}{2}\sqrt{\pi}s^{-3/2} + 3s^{-2}$, $s > 0$ **13.** $s^{-2} - 2/(s - 3)$, $s > 3$
15. $s^{-1} + s/(s^2 - 25)$, $s > 5$ **16.** $(s + 2)/(s^2 + 4)$, $s > 0$ **17.** $\cos^2 2t = \frac{1}{2}(1 + \cos 4t)$; $\frac{1}{2}(s^{-1} + s/(s^2 + 16))$, $s > 0$
18. $3/(s^2 + 36)$, $s > 0$ **19.** $s^{-1} + 3s^{-2} + 6s^{-3} + 6s^{-4}$, $s > 0$ **20.** $1/(s - 1)^2$, $s > 1$ **21.** $(s^2 - 4)/(s^2 + 4)^2$, $s > 0$
22. $\frac{1}{2}(s/(s^2 - 36) - s^{-1})$ **23.** $\frac{1}{2}t^3$ **24.** $2\sqrt{t/\pi}$ **25.** $1 - 8t^{3/2}/(3\sqrt{\pi})$ **26.** e^{-5t} **27.** $3e^{4t}$ **28.** $3\cos 2t + \frac{1}{2}\sin 2t$
29. $\frac{5}{3}\sin 3t - 3\cos 3t$ **30.** $-\cosh 2t - \frac{9}{2}\sinh 2t$ **31.** $\frac{3}{5}\sinh 5t - 10\cosh 5t$ **32.** $2u(t - 3)$ **37.** $f(t) = 1 - u_a(t) = 1 - u(t - a)$

SECTION 4.2

1. $x(t) = 5\cos 2t$ **2.** $x(t) = 3\cos 3t + \frac{4}{3}\sin 3t$ **3.** $x(t) = \frac{2}{3}(e^{2t} - e^{-t})$ **4.** $x(t) = \frac{1}{2}(7e^{-3t} - 3e^{-5t})$ **5.** $x(t) = \frac{1}{3}(2\sin t - \sin 2t)$
6. $x(t) = \frac{1}{3}(\cos t - \cos 2t)$ **7.** $x(t) = \frac{1}{8}(9\cos t - \cos 3t)$ **8.** $x(t) = \frac{1}{9}(1 - \cos 3t)$ **9.** $x(t) = \frac{1}{6}(2 - 3e^{-t} + e^{-3t})$
10. $x(t) = \frac{1}{4}(2t - 3 + 12e^{-t} - 9e^{-2t})$ **11.** $f(t) = \frac{1}{3}(e^{3t} - 1)$ **12.** $f(t) = \frac{3}{5}(1 - e^{-5t})$ **13.** $f(t) = \frac{1}{4}(1 - \cos 2t) = \frac{1}{2}\sin^2 t$
14. $f(t) = \frac{1}{9}(6\sin 3t - \cos 3t + 1)$ **15.** $f(t) = t - \sin t$ **16.** $f(t) = \frac{1}{9}(-1 + \cosh 3t)$ **17.** $f(t) = -t + \sinh t$
18. $f(t) = \frac{1}{2}(e^{-2t} - 2e^{-t} + 1)$

SECTION 4.3

1. $24/(s - \pi)^5$ **2.** $\frac{3}{4}\sqrt{\pi}(s + 4)^{-5/2}$ **3.** $3\pi/((s + 2)^2 + 9\pi^2)$ **4.** $\sqrt{2}(2s + 5)/(4s^2 + 4s + 17)$ **5.** $\frac{3}{2}e^{2t}$ **6.** $e^{-t}(t - t^2)$ **7.** te^{-2t}
8. $e^{-2t}\cos t$ **9.** $e^{3t}(3\cos 4t + \frac{7}{2}\sin 4t)$ **10.** $\frac{1}{36}e^{2t/3}(8\cos \frac{4}{3}t - 5\sin \frac{4}{3}t)$ **11.** $\frac{1}{2}\sinh 2t$ **12.** $2 + 3e^{3t}$ **13.** $3e^{-2t} - 5e^{-5t}$
14. $2 + e^{2t} - 3e^{-t}$ **15.** $\frac{1}{25}(e^{5t} - 1 - 5t)$ **16.** $\frac{1}{125}(e^{2t}(-2 + 5t) + e^{-3t}(2 + 5t))$ **17.** $\frac{1}{16}(\sinh 2t - \sin 2t)$
18. $e^{4t}(6 + 12t + 24t^2 + \frac{32}{3}t^3)$ **19.** $\frac{1}{3}(2\cos 2t + 2\sin 2t - 2\cos t - \sin t)$ **20.** $\frac{1}{32}(e^{2t}(2t - 1) + e^{-2t}(2t + 1))$
21. $\frac{1}{2}e^{-t}(5\sin t - 3t\cos t - 2t\sin t)$ **22.** $\frac{1}{64}e^{t/2}((4t + 8)\cos t + (4 - 3t)\sin t)$ **27.** $\frac{1}{4}e^{-3t}(8\cos 4t + 9\sin 4t)$ **28.** $\frac{1}{4}(1 - 2e^{2t} + e^{4t})$
29. $\frac{1}{8}(-6t + 3\sinh 2t)$ **30.** $\frac{1}{10}(2e^{-t} - e^{-2t}(2\cos 2t + \sin 2t))$ **31.** $\frac{1}{15}(6e^{2t} - 5 - e^{-3t})$ **32.** $\frac{1}{2}(\cosh t + \cos t)$
33. $x(t) = r(\cosh rt \sin rt - \sinh rt \cos rt)$ where $r = 1/\sqrt{2}$ **34.** $\frac{1}{2}\sin 2t + \frac{1}{3}\sin 3t$ **35.** $\frac{1}{16}(\sin 2t - 2t\cos 2t)$
36. $\frac{1}{50}(2e^{2t} + (10t - 2)\cos t - (5t + 14)\sin t)$ **37.** $\frac{1}{50}((5t - 1)e^{-t} + e^{-2t}(\cos 3t + 32\sin 3t))$
38. $\frac{1}{510}e^{-3t}(489\cos 3t + 307\sin 3t) + \frac{1}{170}(7\cos 2t + 6\sin 2t)$

SECTION 4.4

1. $\frac{1}{2}t^2$ **2.** $(e^{at} - at - 1)/a^2$ **3.** $\frac{1}{2}(\sin t - t\cos t)$ **4.** $2(t - \sin t)$ **5.** te^{at} **6.** $(e^{at} - e^{bt})/(a - b)$ **7.** $\frac{1}{3}(e^{3t} - 1)$ **8.** $\frac{1}{4}(1 - \cos 2t)$
9. $\frac{1}{54}(\sin 3t - 3t\cos 3t)$ **10.** $(kt - \sin kt)/k^3$ **11.** $\frac{1}{4}(\sin 2t + 2t\cos 2t)$ **12.** $\frac{1}{5}(1 - e^{-2t}(\cos t + 2\sin t))$
13. $\frac{1}{10}(3e^{3t} - 3\cos t + \sin t)$ **14.** $\frac{1}{3}(\cos t - \cos 2t)$ **15.** $6s/(s^2 + 9)^2$, $s > 0$ **16.** $(2s^3 - 24s)/(s^2 + 4)^3$, $s > 0$
17. $(s^2 - 4s - 5)/(s^2 - 4s + 13)^2$, $s > 0$ **18.** $2(3s^2 + 6s + 7)/((s + 1)^2(s^2 + 2s + 5)^2)$, $s > 0$
19. $(\pi/2) - \arctan s = \arctan (1/s)$, $s > 0$ **20.** $\frac{1}{2}\ln (s^2 + 4) - \ln s$, $s > 0$ **21.** $\ln s - \ln (s - 3)$, $s > 3$
22. $\ln (s + 1) - \ln (s - 1)$, $s > 1$ **23.** $-2t^{-1}\sinh 2t$ **24.** $2t^{-1}(\cos 2t - \cos t)$ **25.** $t^{-1}(e^{-2t} + e^{3t} - 2\cos t)$ **26.** $t^{-1}e^{-2t}\sin 3t$
27. $2t^{-1}(1 - \cos t)$ **28.** $\frac{1}{8}(t\sin t - t^2\cos t)$ **29.** $(s + 1)X'(s) + 4X(s) = 0$; $x(t) = Ct^3e^{-t}$, $C \neq 0$
30. $X(s) = A/(s + 3)^3$; $x(t) = Ct^2e^{-3t}$, $C \neq 0$ **31.** $(s - 2)X'(s) + 3X(s) = 0$; $x(t) = Ct^2e^{-2t}$, $C \neq 0$
32. $(s^2 + 2s)X'(s) + (4s + 4)X(s) = 0$; $x(t) = C(1 - t - e^{-2t} - te^{-2t})$, $C \neq 0$
33. $(s^2 + 1)X'(s) + 4sX(s) = 0$; $x(t) = C(\sin t - t\cos t)$, $C \neq 0$ **34.** $x(t) = Ce^{-2t}(\sin 3t - 3t\cos 3t)$

1. $f(t) = (t - 3)u_3(t)$

3. $f(t) = e^{-2(t-1)}u_1(t)$

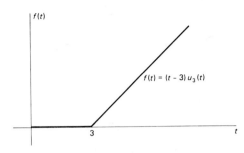

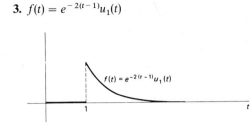

5. $f(t) = u_\pi(t) \sin (t - \pi) = -u_\pi(t) \sin t$

7. $f(t) = (1 - u_{2\pi}(t)) \sin t$

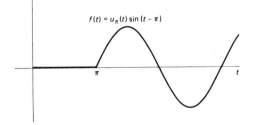

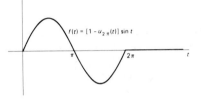

9. $f(t) = (1 - u_3(t)) \cos \pi t$ **11.** $f(t) = 2(1 - u_3(t))$; $F(s) = 2(1 - e^{-3s})/s$ **12.** $F(s) = 3(e^{-s} - e^{-4s})/s$ **13.** $F(s) = (1 - e^{-2\pi s})/(s^2 + 1)$
14. $F(s) = s(1 - e^{-2s})/(s^2 + \pi^2)$ **15.** $F(s) = (1 + e^{-3\pi s})/(s^2 + 1)$ **16.** $F(s) = 2(e^{-\pi s} - e^{-2\pi s})/(s^2 + 4)$
17. $F(s) = \pi(e^{-2s} + e^{-3s})/(s^2 + \pi^2)$ **18.** $F(s) = 2\pi(e^{-3s} + e^{-5s})/(4s^2 + \pi^2)$ **19.** $F(s) = e^{-s}(s^{-1} + s^{-2})$ **20.** $F(s) = (1 - e^{-s})/s^2$
21. $F(s) = (1 - 2e^{-s} + e^{-2s})/s^2$ **28.** $F(s) = (1 - e^{-as} - ase^{-as})/(s^2(1 - e^{-2as}))$ **31.** $x(t) = \frac{1}{2}(1 - u_\pi(t)) \sin^2 t$
32. $x(t) = g(t)$ if $t < 2$; $x(t) = g(t) - g(t - 2)$ if $t \geq 2$ **33.** $x(t) = \frac{1}{8}(1 - u_{2\pi}(t))(\sin t - \frac{1}{3} \sin 3t)$
34. $x(t) = t - \sin t$ if $t < 1$, $x(t) = -\sin t + \sin (t - 1) + \cos (t - 1)$ if $t > 1$
35. $x(t) = \frac{1}{4}\{t - 1 + (t + 1)e^{-2t} + u_2(t)[1 - t + (3t - 5)e^{-2(t-2)}]\}$ **36.** $i(t) = e^{-10t} - u_1(t)e^{-10(t-1)}$ **37.** $i(t) = (1 - u_{2\pi}(t)) \sin 100t$
38. $i(t) = g(t)$ if $t < \pi$, $i(t) = 0$ if $t > \pi$ **39.** $i(t) = \frac{1}{50}(1 - e^{-50t})^2 - \frac{1}{50}u_1(t)\{1 + 98e^{-50(t-1)} - 99e^{-100(t-1)}\}$
40. $i(t) = g(t) - u_1(t)g(t - 1) - u_1(t)h(t - 1)$
41. The complete solution $x(t) = 2|\sin t| \sin t$ is periodic, so the transient solution is zero.
42. $x(t) = g(t) + 2 \sum (-1)^n u_{n\pi}(t)g(t - n\pi)$

1. $x(t) = \frac{1}{2} \sin 2t$ **2.** $x(t) = \frac{1}{2} \sin 2t$ if $t < \pi$, $x(t) = \sin 2t$ if $t > \pi$ **3.** $x(t) = \frac{1}{4}(1 - e^{-2t}) - \frac{1}{2}te^{-2t} + u_2(t)(t - 2)e^{-2(t-2)}$
4. $x(t) = t - 2 + 2e^{-t} + 3te^{-t}$ **5.** $x(t) = 0$ if $0 \leq t < \pi$, $x(t) = -2e^{-(t-\pi)} \sin t$ if $t \geq \pi$ **6.** $x(t) = -\frac{1}{3}u_{3\pi}(t) \sin 3t + \frac{1}{6}t \sin t$
7. $x(t) = (2 - e^{2\pi}u_\pi(t) + e^{4\pi}u_{2\pi}(t))e^{-2t} \sin t$ **8.** $x(t) = (5t + 2)e^{-t} - u_2(t)(t - 2)e^{-(t-2)}$ **9.** $x(t) = \int_0^t \frac{1}{2}(\sin 2\tau)f(t - \tau) \, d\tau$
10. $x(t) = \int_0^t \tau e^{-3\tau}f(t - \tau) \, d\tau$ **11.** $x(t) = \int_0^t (e^{-3\tau} \sinh \tau)f(t - \tau) \, d\tau$ **12.** $x(t) = \frac{1}{2} \int_0^t (e^{-2\tau} \sin 2\tau)f(t - \tau) \, d\tau$
13. (a) $mx_\epsilon(t) = p(t^2 - u_\epsilon(t)(t - \epsilon)^2)/(2\epsilon)$; (b) If $t > \epsilon$, then $mx_\epsilon(t) = p(2\epsilon t - \epsilon^2)/(2\epsilon)$, and hence $mx_\epsilon(t) \to pt$ as $\epsilon \to 0$;
(c) $mv = (mx)' = (pt)' = p$.
15. The transform of each of the two given initial value problems is $(ms^2 + k)X(s) = mv_0 = p_0$.
17. (b) $i(t) = e^{-100(t-1)}u_1(t) - e^{-100(t-2)}u_2(t)$. If $t > 2$, then $i(t) = -(e^{100} - 1)e^{100(1-t)} < 0$.

1. $x_1' = x_2$, $x_2' = -7x_1 - 3x_2 + t^2$ **2.** $x_1' = x_2$, $x_2' = x_3$, $x_3' = x_4$, $x_4' = -x_1 + 3x_2 - 6x_3 + \cos 3t$
3. $x_1' = x_2$, $t^2 x_2' = (1 - t^2)x_1 - tx_2$ **4.** $x_1' = x_2$, $x_2' = x_3$, $t^3 x_3' = -5x_1 - 3tx_2 + 2t^2 x_3 + \ln t$
5. $x_1' = x_2$, $x_2' = x_3$, $x_3' = x_2^2 + \cos x_1$ **6.** $x_1' = x_2$, $y_1' = y_2$, $x_2' = 5x_1 - 4y_1$, $y_2' = -4x_1 + 5y_1$
7. $x_1' = x_2$, $y_1' = y_2$, $x_2' = -kx_1(x_1^2 + y_1^2)^{-3/2}$, $y_2' = -ky_1(x_1^2 + y_1^2)^{-3/2}$
8. $x_1' = x_2$, $y_1' = y_2$, $x_2' = -4x_1 + 2y_1 - 3x_2$, $y_2' = 3x_1 - y_1 - 2y_2 + \cos t$
9. $x_1' = x_2$, $y_1' = y_2$, $z_1' = z_2$, $x_2' = 3x_1 - y_1 + 2z_1$, $y_2' = x_1 + y_1 - 4z_1$, $z_2' = 5x_1 - y_1 - z_1$
10. $x_1' = x_2$, $y_1' = y_2$, $x_2' = x_1(1 - y_1)$, $y_2' = y_1(1 - x_1)$ **11.** $x = -\frac{1}{8}\sin 2t$, $y = -\frac{3}{8}\sin 2t$ **13.** $x = -\frac{3}{8}\cos 3t$, $y = \frac{1}{4}\cos 3t$
15. $x(t) = 1$, $y(t) = -2$ **17.** $x(t) = -(2/r)\sinh(t/r)$, $y(t) = \cosh(t/r) + (1/r)\sinh(t/r)$ where $r = \sqrt{3}$
19. $x(t) = \frac{1}{3}(2 + e^{-3t/2}\{\cos(rt/2) + r\sin(rt/2)\})$, $y(t) = \frac{1}{21}(28 - 9e^t + 2e^{-3t/2}\{\cos(rt/2) + 4r\sin(rt/2)\})$ where $r = \sqrt{3}$
20. $x(t) = \cos t + \sin t$, $y(t) = e^t - \cos t$, $z(t) = -2\sin t$
21. $x(t) = 25(1 + 1/\sqrt{33})e^{at} + 25(1 - 1/\sqrt{33})e^{bt}$, $y(t) = 25(1 - 15/\sqrt{33})e^{at} + 25(1 + 15/\sqrt{33})e^{bt}$ where $a = (-9 - \sqrt{33})/40$ and $b = (-9 + \sqrt{33})/40$
22. $I_1(t) = 2 + e^{-5t}\{4r\sin(5rt/3) - 2\cos(5rt/3)\}$, $I_2(t) = (20/r)e^{-5t}\sin(5rt/3)$ where $r = \sqrt{6}$
29. $2(I_1' - I_2') + 50I_1 = 100\sin 60t$, $2(I_2' - I_1') + 25I_2 = 0$ **30.** $I_1' = -20(I_1 - I_2)$, $I_2' = 40(I_1 - I_2)$

1. $x = a_1 e^{-t} + a_2 e^{2t}$, $y = a_2 e^{2t}$ **2.** $x = (c_1 + c_2 t)e^{-t}$, $y = (c_1 - \frac{1}{2}c_2 + c_2 t)e^{-t}$ **3.** $x = \frac{4}{5}(e^{3t} - e^{-2t})$, $y = \frac{2}{5}(6e^{3t} - e^{-2t})$
4. $x = \frac{1}{2}(3e^{2t} - e^{-2t})$, $y = \frac{1}{2}(3e^{2t} - 5e^{-2t})$ **5.** $x = e^{-t}(a_1 \cos 2t + a_2 \sin 2t)$, $y = -\frac{1}{2}e^{-t}\{(a_1 + a_2)\cos 2t + (a_2 - a_1)\sin 2t\}$
6. $x = e^{-2t}(3\cos 3t + 9\sin 3t)$, $y = e^{-2t}(2\cos 3t - 4\sin 3t)$ **7.** $x = a_1 e^{2t} + a_2 e^{3t} - \frac{1}{3}t + \frac{1}{18}$, $y = -2a_1 e^{2t} - a_2 e^{3t} - \frac{2}{3}t - \frac{5}{9}$
8. $x = c_1 e^t + c_2 e^{3t}$, $y = -c_1 e^t + c_2 e^{3t}$ **9.** $x = 3a_1 e^t + a_2 e^{-t} - \frac{1}{5}(7\cos 2t + 4\sin 2t)$, $y = a_1 e^t + a_2 e^{-t} - \frac{1}{5}(2\cos 2t + 4\sin 2t)$
10. $x = e^t$, $y = -e^t$ **11.** $x = a_1 \cos 3t + a_2 \sin 3t - \frac{11}{20}e^t - \frac{1}{4}e^{-t}$, $y = \frac{1}{3}\{(a_1 - a_2)\cos 3t + (a_1 + a_2)\sin 3t\} + \frac{1}{10}e^t$
12. $x = c_1 e^{2t} + c_2 e^{-2t} + c_3 e^{3t} + c_4 e^{-3t}$, $y = -c_1 e^{2t} - c_2 e^{-2t} + \frac{3}{2}c_3 e^{3t} + \frac{3}{2}c_4 e^{-3t}$
13. $x = a_1 \cos 2t + a_2 \sin 2t + b_1 \cos 3t + b_2 \sin 3t$, $y = \frac{1}{2}(a_1 \cos 2t + a_2 \sin 2t) - 2(b_1 \cos 3t + b_2 \sin 3t)$
14. $x = c_1 \cos 2t + c_2 \sin 2t + \frac{1}{3}\sin t$, $y = c_1 \cos 2t + c_2 \sin 2t + c_3 \cos 2t\sqrt{2} + c_4 \sin 2t\sqrt{2} + \frac{4}{21}\sin t$
15. $x = a_1 \cos t + a_2 \sin t + b_1 \cos 2t + b_2 \sin 2t$, $y = a_2 \cos t - a_1 \sin t + b_2 \cos 2t - b_1 \sin 2t$,
17. $x = a_1 \cos t + a_2 \sin t + b_1 e^{2t} + b_2 e^{-2t}$, $y = 3a_2 \cos t - 3a_1 \sin t + b_1 e^{2t} - b_2 e^{-2t}$
18. $x = \frac{1}{6}(4c_1 e^{3t} - 3c_2 e^{-4t})$, $y = c_1 e^{3t} + c_2 e^{-4t}$, $z = \frac{1}{6}(-4c_1 e^{3t} + 3c_2 e^{-4t})$
19. $x = a_1 + a_2 e^{4t} + a_3 e^{8t}$, $y = 2a_1 - 2a_3 e^{8t}$, $z = 2a_1 - 2a_2 e^{4t} + 2a_3 e^{8t}$
20. $x = a_1 e^{2t} + a_2 e^{-t} + \frac{2}{3}te^{-t}$, $y = a_1 e^{2t} + b_2 e^{-t} - \frac{1}{3}te^{-t}$, $z = a_1 e^{2t} - (a_2 + b_2 + \frac{1}{3})e^{-t} - \frac{1}{3}te^{-t}$ **23.** Infinitely many solutions
24. No solution   **25.** Infinitely many solutions   **26.** Two arbitrary constants   **27.** No arbitrary constants   **28.** No solution
29. Four arbitrary constants   **31.** $I_1(t) = 2 + e^{-5t}\{-2\cos(10t/r) + 4r\sin(10t/r)\}$, $I_2(t) = (20/r)e^{-5t}\sin(10t/r)$ where $r = \sqrt{6}$
32. $I_1(t) = \frac{1}{1321}(120e^{-25t/3} - 120\cos 60t + 1778\sin 60t)$, $I_2(t) = \frac{1}{1321}(-240e^{-25t/3} + 240\cos 60t + 1728\sin 60t)$
33. $I_1(t) = \frac{2}{3}(2 + e^{-60t})$, $I_2(t) = \frac{4}{3}(1 - e^{-60t})$
37. (a) $x = a_1 \cos 5t + a_2 \sin 5t + b_1 \cos 5t\sqrt{3} + b_2 \sin 5t\sqrt{3}$, $y = 2a_1 \cos 5t + 2a_2 \sin 5t - 2b_1 \cos 5t\sqrt{3} - 2b_2 \sin 5t\sqrt{3}$; (b) In the natural mode with frequency $\omega_1 = 5$ the masses move in the same direction, while in the natural mode with frequency $\omega_2 = 5\sqrt{3}$ they move in opposite directions. In each case the amplitude of the motion of m_2 is twice that of m_1.
39. $x = a_1 \cos t + a_2 \sin t + b_1 \cos 3t + b_2 \sin 3t$, $y = a_1 \cos t + a_2 \sin t - b_1 \cos 3t - b_2 \sin 3t$
40. In the natural mode with frequency $\omega_1 = 1$ the two masses move in the same direction with equal amplitudes of oscillation. In the natural mode with frequency $\omega_2 = 2$ the two masses move in opposite directions with the amplitude of motion of m_2 half that of m_1.
41. $x = a_1 \cos t + a_2 \sin t + b_1 \cos t\sqrt{5} + b_2 \sin t\sqrt{5}$, $y = a_1 \cos t + a_2 \sin t - b_1 \cos t\sqrt{5} - b_2 \sin t\sqrt{5}$
42. In the natural mode with frequency $\omega_1 = \sqrt{2}$ the two masses move in the same direction; in the natural mode with frequency $\omega_2 = 2$ they move in opposite directions. In each natural mode their amplitudes of oscillation are equal.
43. $x = a_1 \cos t\sqrt{2} + a_2 \sin t\sqrt{2} + b_1 \cos 2t\sqrt{2} + b_2 \sin 2t\sqrt{2}$, $y = a_1 \cos t\sqrt{2} + a_2 \sin t\sqrt{2} - \frac{1}{2}b_1 \cos 2t\sqrt{2} - \frac{1}{2}b_2 \sin 2t\sqrt{2}$

1. (a) $\begin{bmatrix} 13 & -18 \\ 23 & 17 \end{bmatrix}$; (b) $\begin{bmatrix} 0 & -1 \\ 2 & 19 \end{bmatrix}$; (c) $\begin{bmatrix} -9 & -11 \\ 47 & -9 \end{bmatrix}$; (d) $\begin{bmatrix} -10 & -37 \\ 14 & -8 \end{bmatrix}$

2. $(AB)C = A(BC) = \begin{bmatrix} -33 & -7 \\ -27 & 103 \end{bmatrix}$; $A(B + C) = AB + AC = \begin{bmatrix} -18 & -4 \\ 68 & -8 \end{bmatrix}$

3. $AB = \begin{bmatrix} -1 & 8 \\ 46 & -1 \end{bmatrix}$; $BA = \begin{bmatrix} 11 & -12 & 14 \\ -14 & 0 & 7 \\ 0 & 8 & -13 \end{bmatrix}$ **4.** $Ay = \begin{bmatrix} 2t^2 - \cos t \\ 3t^2 - 4\sin t + 5\cos t \end{bmatrix}$, $Bx = \begin{bmatrix} 2t + 3e^{-t} \\ -14t \\ 6t + 2e^{-t} \end{bmatrix}$

5. (a) $\begin{bmatrix} 21 & 2 & 1 \\ 4 & 44 & 9 \\ -27 & 34 & 45 \end{bmatrix}$; (b) $\begin{bmatrix} 9 & 21 & -13 \\ -5 & -8 & 24 \\ -25 & -19 & 26 \end{bmatrix}$; (c) $\begin{bmatrix} 0 & -6 & 1 \\ 10 & 31 & -15 \\ 16 & 58 & -23 \end{bmatrix}$; (d) $\begin{bmatrix} -10 & -8 & 5 \\ 18 & 12 & -10 \\ 11 & 22 & 6 \end{bmatrix}$; (e) $\begin{bmatrix} 3-t & 2 & -1 \\ 0 & 4-t & 3 \\ -5 & 2 & 7-t \end{bmatrix}$

7. $\det(A) = \det(B) = 0$ **8.** $\det(AB) = \det(BA) = 144$ **9.** $(AB)' = \begin{bmatrix} 1 - 8t + 18t^2 & 1 + 2t - 12t^2 + 32t^3 \\ 3 - 3t^2 - 4t^3 & 8t + 3t^2 + 4t^3 \end{bmatrix}$

11. $x = \begin{bmatrix} x \\ y \end{bmatrix}$, $P(t) = \begin{bmatrix} 3 & -2 \\ 2 & 1 \end{bmatrix}$, $f(t) = \begin{bmatrix} 0 \\ 0 \end{bmatrix}$ **13.** $x = \begin{bmatrix} x \\ y \end{bmatrix}$, $P(t) = \begin{bmatrix} t & -e^t \\ e^{-t} & t^2 \end{bmatrix}$, $f(t) = \begin{bmatrix} \cos t \\ -\sin t \end{bmatrix}$

15. $x = \begin{bmatrix} x \\ y \\ z \end{bmatrix}$, $P(t) = \begin{bmatrix} t & -1 & e^t \\ 2 & t^2 & -1 \\ e^{-t} & 3t & t^3 \end{bmatrix}$, $f(t) = \begin{bmatrix} 0 \\ 0 \\ 0 \end{bmatrix}$ **16.** $x(t) = c_1 \begin{bmatrix} 1 \\ 3 \end{bmatrix} e^{3t} + c_2 \begin{bmatrix} 2 \\ 1 \end{bmatrix} e^{-2t}$ **17.** $W(t) = 4$, $x(t) = \begin{bmatrix} c_1 e^{2t} + c_2 e^{-2t} \\ c_1 e^{2t} + 5c_2 e^{-2t} \end{bmatrix}$

19. $W(t) = 7e^{-3t}$, $x(t) = \begin{bmatrix} 3c_1 e^{2t} + c_2 e^{-5t} \\ 2c_1 e^{2t} + 3c_2 e^{-5t} \end{bmatrix}$ **20.** $x(t) = c_1 \begin{bmatrix} 2 \\ 2 \\ 1 \end{bmatrix} e^t + c_2 \begin{bmatrix} -2 \\ 0 \\ 1 \end{bmatrix} e^{3t} + c_3 \begin{bmatrix} 2 \\ -2 \\ 1 \end{bmatrix} e^{5t}$

21. $W(t) = 3$, $x(t) = \begin{bmatrix} c_1 e^{2t} + c_2 e^{-t} \\ c_1 e^{2t} + c_3 e^{-t} \\ c_1 e^{2t} - (c_2 + c_3)e^{-t} \end{bmatrix}$ **23.** $x = 2x_1 - x_2$ **24.** $x = 7x_1 - 2x_2$ **25.** $x = 3x_1 + 4x_2$

26. $x = 3x_1 - 2x_2$ **27.** $x = x_1 + 2x_2 + x_3$ **28.** $x = 7x_1 + 3x_2 + 5x_3$

29. (a) $x_2 = tx_1$, so neither is a constant multiple of the other. (b) $W(x_1, x_2)$ is identically zero, whereas Theorem 2 implies that $W \neq 0$ if x_1 and x_2 were independent solutions of a system of the indicated form.

34. $x(t) = c_1 \cos t + c_2 \sin t + c_3 \cos 3t + c_4 \sin 3t$, $y(t) = c_1 \cos t + c_2 \sin t - c_3 \cos 3t - c_4 \sin 3t$

35. $x_1 = a_1 \cos t + a_2 \sin t + b_1 \cos 2t + b_2 \sin 2t - \frac{7}{40} \cos 3t$, $x_2 = a_1 \cos t + a_2 \sin t - \frac{1}{2}(b_1 \cos 2t + b_2 \sin 2t) + \frac{1}{40} \cos 3t$

36. $x_1 = c_1 \cos t + c_2 \sin t + c_3 \cos t\sqrt{5} + c_4 \sin t\sqrt{5} + \frac{2}{3} \sin 2t$, $x_2 = c_1 \cos t + c_2 \sin t - c_3 \cos t\sqrt{5} - c_4 \sin t\sqrt{5} - \frac{4}{3} \sin 2t$

37. $x = a_1 \cos t\sqrt{2} + a_2 \sin t\sqrt{2} + b_1 \cos 2t + b_2 \sin 2t + \frac{4}{3} \cos t$, $y = a_1 \cos t\sqrt{2} + a_2 \sin t\sqrt{2} - b_1 \cos 2t - b_2 \sin 2t + \frac{5}{3} \cos t$

38. $x_1 = c_1 \cos rt + c_2 \sin rt + c_3 \cos 2rt + c_4 \sin 2rt$, $x_2 = c_1 \cos rt + c_2 \sin rt - \frac{1}{2}(c_3 \cos 2rt + c_4 \sin 2rt)$ where $r = \sqrt{2}$

SECTION 5.4

1. $x_1 = \text{Re}[\frac{2}{3}e^{10ti}] = \frac{2}{3}\cos 10t$, $x_2 = \text{Re}[-\frac{14}{3}e^{10ti}] = -\frac{14}{3}\cos 10t$

3. $x_1 = \text{Re}[-\frac{1}{15}(6 - 8i)e^{10ti}] = -\frac{1}{15}(6\cos 10t + 8\sin 10t)$, $x_2 = -x_1$

5. $x_1 = (3450\cos 10t + 8720\sin 10t)/9881$, $x_2 = (-19610\cos 10t - 40400\sin 10t)/9881$

6. Natural frequencies: $\omega_1 = 1$, $\omega_2 = 3$. In the first natural mode, the two masses move in the same direction with equal amplitudes of oscillation. In the second natural mode, they move in opposite directions with equal amplitudes.

7. Natural frequencies: $\omega_1 = 1$, $\omega_2 = 2$ (rad/s). In the natural mode with frequency $\omega_1 = 1$ the two masses move in the same direction with equal amplitudes of motion. In the natural mode with frequency $\omega_2 = 2$ the two masses move in opposite directions, with the amplitude of motion of m_1 twice that of m_2.

9. Natural frequencies: $\omega_1 = \sqrt{2}$, $\omega_2 = 2$ (rad/s). In each natural mode the two masses have equal amplitudes of oscillation. They move in the same direction at frequency $\omega_1 = \sqrt{2}$, in opposite directions at frequency $\omega_2 = 2$.

11. $x_1(t) = x_2(t) = e^{-t}(\cos 2t + \frac{1}{2}\sin 2t)$ **13.** $m_2 = 0.1$ (slug)

15. Natural frequencies: $\omega_1 = \sqrt{20}$, $\omega_2 = \sqrt{30 - 10\sqrt{7}}$, and $\omega_3 = \sqrt{30 + 10\sqrt{7}}$. The ratios of amplitudes $A_i : B_i : C_i$ for the natural mode with frequency ω_i are given by $A_1 : B_1 : C_1 = 1 : 0 : -1$, $A_2 : B_2 : C_2 = 1 : (-1 + \sqrt{7}) : 2$, and $A_3 : B_3 : C_3 = 1 : (-1 - \sqrt{7}) : 2$.

18. $\omega_1 = 2\sqrt{10}$, $v_1 \approx 40.26$ ft/s ≈ 27 mi/h, $\omega_2 = 5\sqrt{5}$, $v_2 \approx 71.18$ ft/s ≈ 49 mi/h

19. $\omega_1 \approx 6.1311$, $v_1 \approx 39.03$ ft/s ≈ 27 mi/h, $\omega_2 \approx 10.3155$, $v_2 \approx 65.67$ ft/s ≈ 45 mi/h

20. $\omega_1 \approx 5.0424$, $v_1 \approx 32.10$ ft/s ≈ 22 mi/h, $\omega_2 \approx 9.9158$. $v_2 \approx 63.13$ ft/s ≈ 43 mi/h

SECTION 5.5

1. Eigenvalues: $\lambda_1 = -1$, $\lambda_2 = 3$; eigenvectors: $\mathbf{v}_1 = \begin{bmatrix} 1 & -1 \end{bmatrix}^T$, $\mathbf{v}_2 = \begin{bmatrix} 1 & 1 \end{bmatrix}^T$; solution: $x_1 = c_1 e^{-t} + c_2 e^{3t}$, $x_2 = -c_1 e^{-t} + c_2 e^{3t}$

2. Eigenvalues: -1 and 4; solution: $x_1 = c_1 e^{-t} + 3c_2 e^{4t}$, $x_2 = -c_1 e^{-t} + 2c_2 e^{4t}$. **3.** $x_1 = \frac{1}{7}(-e^{-t} + 8e^{6t})$, $x_2 = \frac{1}{7}(e^{-t} + 6e^{6t})$

4. $x_1 = c_1 e^{-2t} + c_2 e^{5t}$, $x_2 = -6c_1 e^{-2t} + c_2 e^{5t}$ **5.** $x_1 = c_1 e^{-t} + 7c_2 e^{5t}$, $x_2 = c_1 e^{-t} + c_2 e^{5t}$

6. Eigenvalues: 3 and 4 **7.** $x_1 = c_1 e^{t} + 2c_2 e^{-9t}$, $x_2 = c_1 e^{t} - 3c_2 e^{-9t}$

9. Eigenvalue: $4i$; eigenvector: $\begin{bmatrix} 5 & 2 - 4i \end{bmatrix}^T$; solution: $x_1 = 2 \cos 4t - \frac{11}{4} \sin 4t$, $x_2 = 3 \cos 4t + \frac{1}{2} \sin 4t$

11. $x_1 = -4e^{t} \sin 2t$, $x_2 = 4e^{t} \cos 2t$ **13.** $x_1 = 3e^{2t}(c_1 \cos 3t - c_2 \sin 3t)$, $x_2 = e^{2t}\{(c_1 + c_2) \cos 3t + (c_1 - c_2) \sin 3t\}$

14. Eigenvalue: $3 + 4i$ **15.** $x_1 = 5e^{5t}(c_1 \cos 4t - c_2 \sin 4t)$, $x_2 = e^{5t}\{(2c_1 + 4c_2) \cos 4t + (4c_1 - 2c_2) \sin 4t\}$

16. $\lambda = 2$ is a reported eigenvalue. **17.** $x_1 = e^{3t}(c_1 + c_2 t)$, $x_2 = -\frac{1}{2}e^{3t}(2c_1 + c_2 + 2c_2 t)$ **18.** $\lambda = 4$ is a repeated eigenvalue.

19. $\lambda = 5$ is a repeated eigenvalue. Solution: $x_1 = e^{5t}(c_1 + c_2 t)$, $x_2 = e^{5t}(-2c_1 + c_2 - 2c_2 t)$ **20.** $\lambda = 5$ is a repeated eigenvalue.

21. $x_1 = c_1 e^{9t} + c_2 e^{6t} + c_3$, $x_2 = c_1 e^{9t} - 2c_2 e^{6t}$, $x_3 = c_1 e^{9t} + c_2 e^{6t} - c_3$ **22.** Eigenvalues: 9, 6, 0

23. $x_1 = c_1 e^{6t} + c_2 e^{3t} + c_3 e^{3t}$, $x_2 = c_1 e^{6t} - 2c_2 e^{3t}$, $x_3 = c_1 e^{6t} + c_2 e^{3t} - c_3 e^{3t}$ **24.** Eigenvalues: 9, 6, 2

25. $x_1 = 6c_1 + 3c_2 e^{t} + 2c_3 e^{-t}$, $x_2 = 2c_1 + c_2 e^{t} + c_3 e^{-t}$, $x_3 = 5c_1 + 2c_2 e^{t} + 2c_3 e^{-t}$

27. $x_1 = c_1 e^{2t} + c_3 e^{3t}$, $x_2 = -c_1 e^{2t} + c_2 e^{-2t} - c_3 e^{3t}$, $x_3 = -c_2 e^{-2t} + c_3 e^{3t}$

28. $x_1 = c_1 e^{t} + c_2(2 \cos 2t - \sin 2t) + c_3(\cos 2t + 2 \sin 2t)$, $x_2 = -c_1 e^{t} - c_2(3 \cos 2t + \sin 2t) + c_3(\cos 2t - 3 \sin 2t)$, $x_3 = c_2(3 \cos 2t + \sin 2t) + c_3(3 \sin 2t - \cos 2t)$

29. Eigenvalues: 0, $2 + 3i$, $2 - 3i$; solution: $x_1 = c_1 + e^{2t}\{(-c_2 + c_3) \cos 3t + (c_2 + c_3) \sin 3t\}$, $x_2 = -c_1 + 2e^{2t}(c_2 \cos 3t - c_3 \sin 3t)$, $x_3 = 2e^{2t}(-c_2 \cos 3t + c_3 \sin 3t)$

31. $x = c_1 \cos 5t + c_2 \sin 5t + c_3 \cos 5t\sqrt{3} + c_4 \sin 5t\sqrt{3}$, $y = 2c_1 \cos 5t + 2c_2 \sin 5t - 2c_3 \cos 5t\sqrt{3} + 2c_4 \sin 5t\sqrt{3}$

33. $x_1 = c_1 e^{t} + e^{2t}(c_2 + c_3 t)$, $x_2 = -c_1 e^{t} + e^{2t}(-c_2 + c_3 - c_3 t)$, $x_3 = e^{2t}(c_2 - c_3 + c_3 t)$

34. A general solution: $x_1 = c_1 e^{2t} + c_2(te^{2t} + e^{2t}) + c_3(t^2 e^{2t} + 2e^{2t})$, $x_2 = -c_1 e^{2t} - c_2 te^{2t} + c_3(2e^{2t} - t^2 e^{2t})$, $x_3 = c_1 e^{2t} + c_2 te^{2t} + c_3(t^2 e^{2t} - 4e^{2t})$

36. $x_1 = c_1 e^{-t} + c_3 e^{t} + c_4 te^{t}$, $x_2 = c_2 e^{-t}$, $x_3 = c_1 e^{-t} - c_4 e^{t}$, $x_4 = c_2 e^{-t} - c_3 e^{t} - c_4 te^{t}$

37. $x_1 = x_4 = 2e^{10t} + e^{15t}$, $x_2 = x_3 = -e^{10t} + 2e^{15t}$

SECTION 5.6

1. $x = \frac{7}{3}$, $y = -\frac{8}{3}$ **2.** $x = (1 + 12t)/8$, $y = -(5 + 4t)/4$

3. $x = (864e^{-t} + 4e^{6t} - 504t^2 + 840t - 868)/756$, $y = (-864e^{-t} + 3e^{6t} + 378t^2 - 882t + 861)/756$

4. $x = (99e^{5t} - 8e^{-2t} - 7e^{t})/84$, $y = (99e^{5t} + 48e^{-2t} - 63e^{t})/84$ **5.** $x = -(12 + e^{-t} + 7te^{-t})/3$, $y = -(6 + 7te^{-t})/3$

7. $x = (369e^{t} + 166e^{-9t} - 125 \cos t - 105 \sin t)/410$, $y = (369e^{t} - 249e^{-9t} - 120 \cos t - 150 \sin t)/410$

9. $x = (\sin 2t + 2t \cos 2t + t \sin 2t)/4$, $y = (t \sin 2t)/4$ **11.** $x = (1 - 4t + e^{4t})/2$, $y = (-5 + 4t + e^{4t})/4$

13. $x = \frac{1}{2}(1 + 5t)e^{t}$, $y = -\frac{5}{2}te^{t}$ **15.** $\Phi(t) = \begin{bmatrix} e^{t} & e^{3t} \\ -e^{t} & e^{3t} \end{bmatrix}$, $\mathbf{x}(t) = \frac{1}{2}\begin{bmatrix} 5e^{t} + e^{3t} \\ -5e^{t} + e^{3t} \end{bmatrix}$

17. $\Phi(t) = \begin{bmatrix} 5 \cos 4t & -5 \sin 4t \\ 2 \cos 4t + 4 \sin 4t & 4 \cos 4t - 2 \sin 4t \end{bmatrix}$, $\mathbf{x}(t) = \frac{1}{4}\begin{bmatrix} -5 \sin 4t \\ 4 \cos 4t - 2 \sin 4t \end{bmatrix}$

19. $\Phi(t) = \begin{bmatrix} 2 \cos 3t & -2 \sin 3t \\ -3 \cos 3t + 3 \sin 3t & 3 \cos 3t + 3 \sin 3t \end{bmatrix}$, $\mathbf{x}(t) = \frac{1}{3}\begin{bmatrix} 3 \cos 3t - \sin 3t \\ -3 \cos 3t + 6 \sin 3t \end{bmatrix}$

21. $\Phi(t) = \begin{bmatrix} 6 & 3e^{t} & 2e^{-t} \\ 2 & e^{t} & e^{-t} \\ 5 & 2e^{t} & 2e^{-t} \end{bmatrix}$, $\mathbf{x}(t) = \begin{bmatrix} -12 + 12e^{t} + 2e^{-t} \\ -4 + 4e^{t} + e^{-t} \\ -10 + 8e^{t} + 2e^{-t} \end{bmatrix}$ **23.** $\mathbf{x}(t) = \frac{1}{150}\begin{bmatrix} 666 - 120t - 575e^{-t} - 91e^{5t} \\ 588 - 60t - 575e^{-t} - 13e^{5t} \end{bmatrix}$

25. $\mathbf{x}(t) = \frac{1}{16}\begin{bmatrix} 3e^{3t} - 3e^{-t} + 4te^{-t} \\ 3e^{3t} - 3e^{-t} + 20te^{-t} \end{bmatrix}$ **27.** $\mathbf{x}(t) = \frac{1}{28}\begin{bmatrix} -6 \cos 4t + 28 \sin 4t + 8 \cos 3t - 28 \sin 3t \\ -31 \cos 4t + 8 \sin 4t + 32 \cos 3t - 8 \sin 3t \end{bmatrix}$

29. $\mathbf{x}(t) = \Phi(t)\begin{bmatrix} t - \sin t + \ln|\cos t + \sin t \cos t| \\ 1 - 2t - \cos t + \ln|\cos t| \end{bmatrix}$ where $\Phi(t) = \begin{bmatrix} \cos t & -\sin t \\ 2 \cos t + \sin t & \cos t - 2 \sin t \end{bmatrix}$

31. $\mathbf{x}(t) = \frac{1}{4}e^{3t}\begin{bmatrix} -2t\sin 2t \\ \sin 2t + 2t\cos 2t \end{bmatrix}$

SECTION 5.7

In the first six answers we give the eigenvector matrix $\mathbf{P}$, the diagonal eigenvalue matrix $\mathbf{D}$, and the fundamental matrix $\boldsymbol{\Phi}$.

1. $\mathbf{P} = \begin{bmatrix} 2 & 1 \\ 1 & 1 \end{bmatrix}$, $\mathbf{D} = \begin{bmatrix} 3 & 0 \\ 0 & 1 \end{bmatrix}$, $\boldsymbol{\Phi}(t) = \begin{bmatrix} 2e^{3t} - e^t & -2e^{3t} + 2e^t \\ e^{3t} - e^t & -e^{3t} + 2e^t \end{bmatrix}$ **3.** $\mathbf{P} = \begin{bmatrix} 3 & 1 \\ 2 & 1 \end{bmatrix}$, $\mathbf{D} = \begin{bmatrix} 3 & 0 \\ 0 & 2 \end{bmatrix}$, $\boldsymbol{\Phi}(t) = \begin{bmatrix} 3e^{3t} - 2e^{2t} & -3e^{3t} + 3e^{2t} \\ 2e^{3t} - 2e^{2t} & -2e^{3t} + 3e^{2t} \end{bmatrix}$

5. $\mathbf{P} = \begin{bmatrix} 4 & 1 \\ 3 & 1 \end{bmatrix}$, $\mathbf{D} = \begin{bmatrix} 3 & 0 \\ 0 & 1 \end{bmatrix}$, $\boldsymbol{\Phi}(t) = \begin{bmatrix} 4e^{3t} - 3e^t & -4e^{3t} + 4e^t \\ 3e^{3t} - 3e^t & -3e^{3t} + 4e^t \end{bmatrix}$

7. $\mathbf{P} = \begin{bmatrix} 5 & 2 \\ 2 & 1 \end{bmatrix}$, $\mathbf{D} = \begin{bmatrix} 2 & 0 \\ 0 & 1 \end{bmatrix}$, $\boldsymbol{\Phi}(t) = \begin{bmatrix} 5e^{2t} - 4e^t & -10e^{2t} + 10e^t \\ 2e^{2t} - 2e^t & -4e^{2t} + 5e^t \end{bmatrix}$

9. $\mathbf{P} = \begin{bmatrix} 1 & 1 \\ 1 & -1 \end{bmatrix}$, $\mathbf{D} = \begin{bmatrix} 4 & 0 \\ 0 & 2 \end{bmatrix}$, $\boldsymbol{\Phi}(t) = \frac{1}{2}\begin{bmatrix} e^{4t} + e^{2t} & e^{4t} - e^{2t} \\ e^{4t} - e^{2t} & e^{4t} + e^{2t} \end{bmatrix}$

11. $\mathbf{P} = \begin{bmatrix} 2 & -1 \\ 1 & 2 \end{bmatrix}$, $\mathbf{D} = \begin{bmatrix} 10 & 0 \\ 0 & 5 \end{bmatrix}$, $\boldsymbol{\Phi}(t) = \frac{1}{5}\begin{bmatrix} 4e^{10t} + e^{5t} & 2e^{10t} - 2e^{5t} \\ 2e^{10t} - 2e^{5t} & e^{10t} + 4e^{5t} \end{bmatrix}$ **13.** $e^t\begin{bmatrix} 1 & 2t \\ 0 & 1 \end{bmatrix}$ **14.** $e^{3t}\begin{bmatrix} 1 & 5t \\ 0 & 1 \end{bmatrix}$ **15.** $e^t\begin{bmatrix} 1 & t & \frac{1}{2}t^2 \\ 0 & 1 & t \\ 0 & 0 & 1 \end{bmatrix}$

17. $\mathbf{x}(t) = e^t\begin{bmatrix} 12t + 6t^2 \\ 6t \end{bmatrix}$ **19.** $\mathbf{x}(t) = e^t\begin{bmatrix} 3t^2 + 2t^3 + t^4 \\ 6t^2 + 4t^3 \\ 12t^2 \end{bmatrix}$ **23.** $\mathbf{x}(t) = \begin{bmatrix} c_1\cosh t + c_2\sinh t \\ c_1\sinh t + c_2\cosh t \end{bmatrix}$

25. $\mathbf{x}(t) = e^t\begin{bmatrix} c_1 + 2c_2 t + 2c_3 t^2 \\ c_2 + 2c_3 t \\ c_3 \end{bmatrix}$

SECTION 6.1

Note: Most of the problems in Chapter 6 call for a table of values. Space limitations prohibit reproduction of the complete tables in this book; we have chosen in most cases to give only data from the last line of the table. Your answers may differ slightly because of variations in hardware and software.

1. $y_{n+1} = y_n + h(-y_n)$; $y(x) = 2e^{-x}$. Results:

x	$h = 0.1$ y	$h = 0.05$ y	Actual y
0.0	2.0000	2.0000	2.0000
0.1	1.8000	1.8050	1.8097
0.2	1.6200	1.6290	1.6375
0.3	1.4580	1.4702	1.4816
0.4	1.3122	1.3268	1.3406
0.5	1.1810	1.1975	1.2131

2. $y_{n+1} = y_n + h(2y_n)$; $y(x) = \frac{1}{2}e^{2x}$. At $x = 0.5$, we should see $y = 1.2442$ ($h = 0.1$), 1.2969 ($h = 0.05$), and 1.3591 (true value).
3. At $x = 0.5$ we should see $y = 2.2210$ ($h = 0.1$) and $y = 2.2578$ ($h = 0.05$); the true value is 2.2974.
4. At $x = 0.5$ we should see $y = 0.6810$ ($h = 0.1$) and $y = 0.6975$ ($h = 0.05$); the true value is 0.7131.
5. At $x = 0.5$ we should see $y = 0.8895$ ($h = 0.1$) and $y = 0.8711$ ($h = 0.05$); the true value is 0.8513.
6. At $x = 0.5$ we should see $y = 1.6272$ ($h = 0.1$) and $y = 1.5912$ ($h = 0.05$); the true value is 1.5576.
7. At $x = 0.5$ we should see $y = 2.7373$ ($h = 0.1$) and $y = 2.6930$ ($h = 0.05$); the true value is 2.6475.
8. At $x = 0.5$ we should see $y = 0.4198$ ($h = 0.1$) and $y = 0.4124$ ($h = 0.05$); the true value is 0.4055.
9. At $x = 0.5$ we should see $y = 1.2785$ ($h = 0.1$) and $y = 1.2828$ ($h = 0.05$); the true value is 1.2874.
10. At $x = 0.5$ we should see $y = 1.2313$ ($h = 0.1$) and $y = 1.2776$ ($h = 0.05$); the true value is 1.3333.
11. At $x = 1.0$ we should see $y = -0.7048$ ($h = 0.01$) and $y = -0.7115$ ($h = 0.005$); the true value is -0.7183.

12. At $x = 1.0$ we should see $y = 2.9864$ ($h = 0.01$) and $y = 2.9931$ ($h = 0.005$); the true value is 3.0000.
13. At $x = 2.0$ we should see $y = 4.8890$ ($h = 0.01$) and $y = 4.8940$ ($h = 0.005$); the true value is 4.8990.
14. At $x = 2.0$ we should see $y = 3.2031$ ($h = 0.01$) and $y = 3.2304$ ($h = 0.005$); the true value is 3.2589.
15. At $x = 3.0$ we should see $y = 3.4422$ ($h = 0.01$) and $y = 3.4433$ ($h = 0.005$); the true value is 3.4444.
16. At $x = 3.0$ we should see $y = 8.8440$ ($h = 0.01$) and $y = 8.8445$ ($h = 0.005$); the true value is 8.8451.
17. At $x = 1.0$ we should see $y = 0.2925$ ($h = 0.1$), $y = 0.3379$ ($h = 0.02$), and $y = 0.3477$ ($h = 0.004$); the true value is 0.3497.
18. At $x = 2.0$ we should see $y = 1.6680$ ($h = 0.1$), $y = 1.6771$ ($h = 0.02$), and $y = 1.6790$ ($h = 0.004$); the true value is 1.6794.
19. At $x = 2.0$ we should see $y = 6.1831$ ($h = 0.1$), $y = 6.3653$ ($h = 0.02$), and $y = 6.4022$ ($h = 0.004$); the true value is 6.4096.
20. At $x = 2.0$ we should see $y = -1.3792$ ($h = 0.1$), $y = -1.2843$ ($h = 0.02$) and $y = -1.2649$ ($h = 0.004$); the true value is -1.2610.
21. At $x = 2.0$ we should see $y = 2.8508$ ($h = 0.1$), $y = 2.8681$ ($h = 0.02$), and $y = 2.8716$ ($h = 0.004$); the true value is 2.8723.
22. At $x = 2.0$ we should see $y = 6.9879$ ($h = 0.1$), $y = 7.2601$ ($h = 0.02$), and $y = 7.3154$ ($h = 0.004$); the true value is 7.3264.
23. At $x = 1.0$ we should see $y = 1.2262$ ($h = 0.1$), $y = 1.2300$ ($h = 0.02$), and $y = 1.2306$ ($h = 0.004$); the true value is 1.2307.
24. At $x = 1.0$ we should see $y = 0.9585$ ($h = 0.1$), $y = 0.9918$ ($h = 0.02$), and $y = 0.9984$ ($h = 0.004$); the true value is 0.9997.

25.

x	$h = 0.15$ y	$h = 0.03$ y	$h = 0.006$ y
-1.0	1.0000	1.0000	1.0000
-0.7	1.0472	1.0512	1.0521
-0.4	1.1213	1.1358	1.1390
-0.1	1.2826	1.3612	1.3835
0.2	0.8900	1.4711	0.8210
0.5	0.7460	1.2808	0.7192

26.

x	$h = 0.1$ y	$h = 0.01$ y
1.8	2.8200	4.3308
1.9	3.9393	7.9425
2.0	5.8521	28.3926

27.

x	$h = 0.1$ y	$h = 0.01$ y
0.7	4.3460	6.4643
0.8	5.8670	11.8425
0.9	8.3349	39.5010

SECTION 6.2

1.

x	Improved Euler y	Three-term Taylor y	Actual y
0.1	1.8100	1.8100	1.8097
0.2	1.6381	1.6381	1.6375
0.3	1.4824	1.4824	1.4816
0.4	1.3416	1.3416	1.3406
0.5	1.2142	1.2142	1.2131

2. At $x = 0.5$ the improved Euler value of y is 1.3514, the three-term Taylor value is 1.3514, and the actual value is 1.3591.
3. At $x = 0.5$ the improved Euler value of y is 2.2949, the three-term Taylor value is 2.2949, and the actual value is 2.2974.
4. At $x = 0.5$ the improved Euler value of y is 0.7142, the three-term Taylor value is 0.7142, and the actual value is 0.7131.
5. At $x = 0.5$ the improved Euler value of y is 0.8526, the three-term Taylor value is 0.8526, and the actual value is 0.8513.
6. At $x = 0.5$ the improved Euler value of y is 1.5575, the three-term Taylor value is 1.5542, and the actual value is 1.5576.
7. At $x = 0.5$ the improved Euler value of y is 2.6405, the three-term Taylor value is 2.6580, and the actual value is 2.6475.
8. At $x = 0.5$ the improved Euler value of y is 0.4053, the three-term Taylor value is 0.4046, and the actual value is 0.4055.
9. At $x = 0.5$ the improved Euler value of y is 1.2873, the three-term Taylor value is 1.2871, and the actual value is 1.2874.
10. At $x = 0.5$ the improved Euler value of y is 1.3309, the three-term Taylor value is 1.3233, and the actual value is 1.3333.
11. At $x = 1.0$, we obtain $y = -0.71824$ ($h = 0.01$) and $y = -0.71827$ ($h = 0.005$); the true value is -0.71828.
12. At $x = 1.0$, we obtain $y = 2.99995$ ($h = 0.01$) and $y = 2.99999$ ($h = 0.005$); the true value is 3.00000.
13. At $x = 2.0$, we obtain $y = 4.89901$ ($h = 0.01$) and $y = 4.89899$ ($h = 0.005$); the true value is 4.89898.
14. At $x = 2.0$, we obtain $y = 3.25847$ ($h = 0.01$) and $y = 3.25878$ ($h = 0.005$); the true value is 3.25889.
15. At $x = 3.0$, we obtain $y = 3.44445$ ($h = 0.01$) and $y = 3.44445$ ($h = 0.005$); the true value is 3.44444.
16. At $x = 3.0$, we obtain $y = 8.84511$ ($h = 0.01$) and $y = 8.84509$ ($h = 0.005$); the true value is 8.84509.
17. At $x = 1.0$ we obtain these values of y: 0.35183, 0.35030, 0.35023, and 0.35023.
18. At $x = 2.0$ we obtain these values of y: 1.68043, 1.67949, 1.67946, and 1.67946.
19. At $x = 2.0$ we obtain these values of y: 6.40834, 6.41134, 6.41147, and 6.41147.

20. At $x = 2.0$ we obtain these values of y: -1.26092, -1.26003, -1.25999, and -1.25999.

21. At $x = 2.0$ we obtain these values of y: 2.87204, 2.87245, 2.87248, and 2.87247.

22. At $x = 2.0$ we obtain these values of y: 7.31578, 7.32841, 7.32916, and 7.32920.

23. At $x = 1.0$ we obtain these values of y: 1.22967, 1.23069, 1.23073, and 1.23073.

24. At $x = 1.0$ we obtain these values of y: 1.00006, 1.00000, 1.00000, and 1.00000.

25. $y_{n+1} = y_n + (1 + y_n^2)(h + h^2 y_n + \frac{1}{3}h^3(1 + 3y_n^2))$

27. $y_{n+1} = y_n + h(x_n + y_n)^2 + h^2((x_n + y_n) + (x_n + y_n)^3) + \frac{1}{3}h^3(1 + 4(x_n + y_n)^2 + 3(x_n + y_n)^4)$

SECTION 6.3

1. $y(0.25)$ is approximately 1.55762; the true value is 1.55760. $y(0.5)$ is approximately 1.21309; the true value is 1.21306. The solution is $y = 2e^{-x}$.

2. $y(0.5)$ is approximately 1.35867; the true value is 1.35914; the solution is $y = (0.5)e^{2x}$.

3. $y(0.5)$ is approximately 2.29740; the true value is 2.29744; the solution is $y = 2e^x - 1$.

4. $y(0.5)$ is approximately 0.71309; the true value is 0.71306; the solution is $y = 2e^{-x} + x - 1$.

5. $y(0.5)$ is approximately 0.85130; the true value is 0.85128; the solution is $y = -e^x + x + 2$.

6. $y(0.5)$ is approximately 1.55759; the true value is 1.55760; the solution is $y = 2\exp(-x^2)$.

7. $y(0.5)$ is approximately 2.64745; the true value is 2.64749; the solution is $y = 3\exp(-x^3)$.

8. $y(0.5)$ is approximately 0.40547; the true value is 0.40547; the solution is $y = \ln(x + 1)$.

9. $y(0.5)$ is approximately 1.28743; the true value is 1.28743; the solution is $y = \tan\frac{1}{4}(x + \pi)$.

10. $y(0.5)$ is approximately 1.33337; the true value is 1.33333; the solution is $y = (1 - x^2)^{-1}$.

11. Actual solution: $y(x) = 2 - e^x$.

x	$h = 0.2$ y	$h = 0.1$ y	**Exact** y
0.0	1.00000	1.00000	1.00000
0.2	0.77860	0.77860	0.77860
0.4	0.50818	0.50818	0.50818
0.6	0.17789	0.17788	0.17788
0.8	-0.22552	-0.22554	-0.22554
1.0	-0.71825	-0.71828	-0.71828

12. At $x = 1.0$, we obtain $y = 2.99996$ ($h = 0.2$) and $y = 3.00000$ ($h = 0.1$); the exact value of y is 3.00000 and the actual solution is $y = 1 + 2/(2 - x)$.

13. At $x = 2.0$, we obtain $y = 4.89900$ ($h = 0.2$) and $y = 4.89898$ ($h = 0.1$); the exact value of y is 4.89898 and the actual solution is $y = (8 + x^4)^{1/2}$.

14. At $x = 2.0$, we obtain $y = 3.25795$ ($h = 0.2$) and $y = 3.25882$ ($h = 0.1$); the exact value of y is 3.25889 and the actual solution is $y = 1/(1 - \ln x)$.

15. At $x = 3.0$, we obtain $y = 3.44445$ ($h = 0.2$) and $y = 3.44444$ ($h = 0.1$); the exact value of y is 3.44444 and the actual solution is $y = x + 4x^{-2}$.

16. At $x = 3.0$, we obtain $y = 8.84515$ ($h = 0.2$) and $y = 8.84509$ ($h = 0.1$); the exact value of y is 8.84509 and the actual solution is $y = (x^6 - 37)^{1/3}$.

17. At $x = 1.0$ we obtain these values for y: 0.350258, 0.350234, 0.350232, and 0.350232.

18. At $x = 2.0$ we obtain these values for y: 1.679513, 1.679461, 1.679459, and 1.679459.

19. At $x = 2.0$ we obtain these values for y: 6.411464, 6.411474, 6.411474, and 6.411474.

20. At $x = 2.0$ we obtain these values for y: -1.259990, -1.259992, -1.259993, and -1.259993.

21. At $x = 2.0$ we obtain these values for y: 2.872467, 2.872468, 2.872468, and 2.872468.

22. At $x = 2.0$ we obtain these values for y: 7.326761, 7.328452, 7.328971, and 7.329134.

23. At $x = 1.0$ we obtain these values for y: 1.230735, 1.230731, 1.230731, and 1.230731.

24. At $x = 1.0$ we obtain these values for y: 1.000000, 1.000000, 1.000000, and 1.000000.

25. $v(10) \approx 104.98$, $s(10) \approx 839.56$, $v(20) \approx 105.05$; $s(20) \approx 1889.95$; $s(t) = 0$ when $t \approx 97.2$

26. (b) $s(20) \approx 1461.43$; true value: about 1461.37; (c) terminal velocity: 80 ft/s; $s(t) = 0$ when $t \approx 127$

27. (a) 100 ft/s; (b) about 368 ft, about 63.21 ft/s; (c) approximately 21 s

28. (a) 100 ft/s; (b) About 434 ft, about 76.16 ft/s; (c) approximately 14 s.
29. (a) About 302.5 ft, about 4.05 s; (b) about 4.7 s, about 120 ft/s.
30. (a) About 294 ft, about 4.08 s; (b) about 4.5 s, about 119 ft/s.

SECTION 6.4

The format for the first eight answers is this: $(x(t), y(t))$ at $t = 0.2$ by the Euler method, the improved Euler method, the Runge-Kutta method, and finally the actual values.

1. (0.8800, 2.5000), (0.9600, 2.6000), (1.0027, 2.6401), (1.0034, 2.6408)
2. (0.8100, −0.8100), (0.8200, −0.8200), (0.8187, −0.8187), (0.8187, −0.8187)
3. (2.8100, 2.3100), (3.2200, 2.6200), (3.6481, 2.9407), (3.6775, 2.9628)
4. (3.3100, −1.6200), (3.8200, −2.0400), (4.2274, −2.4060), (4.2427, −2.4205)
5. (−0.5200, 2.9200), (−0.8400, 2.4400), (−0.5712, 2.4485), (−0.5793, 2.4488)
6. (−1.7600, 4.6800), (−1.9200, 4.5600), (−1.9029, 4.4995), (−1.9025, 4.4999)
7. (3.1200, 1.6800), (3.2400, 1.7600), (3.2816, 1.7899), (3.2820, 1.7902)
8. (2.1600, −0.6300), (2.5200, −0.4600), (2.5320, −0.3867), (2.5270, −0.3889)
9. At $t = 0.1$ we obtain $(x, y) = (3.99261, 6.21770)$ $(h = 0.1)$ and $(3.99234, 6.21768)$ $(h = 0.05)$; the actual value is $(3.99232, 6.21768)$.
10. At $t = 0.1$ we obtain $(x, y) = (1.31498, 1.02537)$ $(h = .01)$ and $(1.31501, 1.02538)$ $(h = 0.05)$; the actual value is $(1.31501, 1.02538)$.
11. At $t = 0.1$ we obtain $(x, y) = (-0.05832, 0.56664)$ $(h = 0.1)$ and $(-0.05832, 0.56665)$ $(h = 0.05)$; the actual value is $(-0.05832, 0.56665)$.
12. We solved $x' = y$, $y' = -x + \sin t$, $x(0) = y(0) = 0$. With $h = 0.1$ and again with $h = 0.05$ we obtain the actual value $x(1.0) \approx 0.15058$.
13. Runge-Kutta, $h = 0.1$: about 1050 ft in about 7.7 s
14. Runge-Kutta, $h = 0.1$; about 1044 ft in about 7.8 s
15. Runge-Kutta, $h = 1.0$: about 83.83 mi in about 168 s
16. At 40°: 5.0 s, 352.9 ft; at 45°: 5.4 s, 347.2 ft; at 50°: 5.8 s, 334.2 ft (all values approximate, of course)
17. At 39.0° the range is about 352.7 ft. At 39.5° it is 352.8; at 40.0°, 352.9; at 40.5°, 352.6; at 41.0°, 352.1.
18. Just under 57.5°   **19.** Approximately 253 ft/s
20. Maximum height: about 1005 ft, attained in about 5.6 s; range: about 1880 ft; time aloft: about 11.6 s
21. Runge-Kutta with $h = 0.1$ yields these results: (a) 21,400 ft, 46 s, 518 ft/s; (b) 8970 ft, 17.5 s; (c) 368 ft/s (at $t \approx 23$).

SECTION 7.1

1. $x(t) = x_0 e^{-t}$; $x = 0$ is a stable critical point.

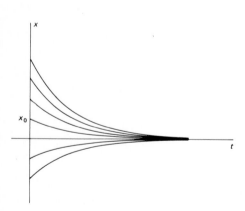

2. $x(t) = 2 + (x_0 - 2)e^t$; $x = 2$ is an unstable critical point.

3. $x(t) = x_0/\{x_0 + (1 - x_0)e^{-t}\}$; $x = 1$ is a stable critical point, whereas $x = 0$ is an unstable critical point.

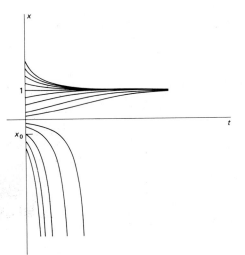

4. $x(t) = 2x_0/\{x_0 + (2 - x_0)e^{2t}\}$; $x = 0$ is a stable critical point, whereas $x = 2$ is an unstable critical point.

5. $x(t) = x_0/\{(x_0 + 1)e^{-t} - x_0\}$; $x = 0$ is an unstable critical point, whereas $x = -1$ is a stable critical point.

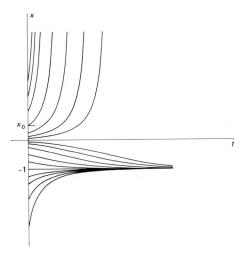

6. $x(t) = \{x_0 - t(x_0 - 1)\}/\{1 - t(x_0 - 1)\}$; $x = 1$ is an unstable critical point.

7. $x(t) = \{1 + x_0 + (x_0 - 1)e^{-2t}\}/\{1 + x_0 - (x_0 - 1)e^{-2t}\}$; $x = 1$ is a stable critical point, whereas $x = -1$ is an unstable critical point.

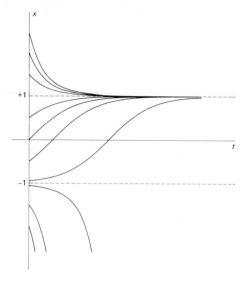

8. $x(t) = 2\{(x_0 + 2) + (x_0 - 2)e^{4t}\}/\{(x_0 + 2) - (x_0 - 2)e^{4t}\}$; $x = -2$ is a stable critical point, whereas $x = 2$ is an unstable critical point.

9. $x(t) = \{x_0 + 2 + 2(x_0 - 1)e^{-3t}\}/\{x_0 + 2 - (x_0 - 1)e^{-3t}\}$; $x = 1$ is a stable critical point, whereas $x = -2$ is an unstable critical point.

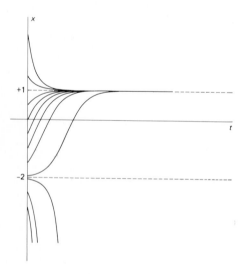

10. $x^2 = (x_0)^2/\{(x_0)^2 + (1 - (x_0)^2)e^{-2t}\}$; $x = 0$ is an unstable critical point, whereas $x = -1$ and $x = 1$ are stable critical points.
11. $x(t) = \{2(x_0 - 3) - 3(x_0 - 2)e^{-t}\}/\{(x_0 - 3) - (x_0 - 2)e^{-t}\}$; $x = 2$ is a stable critical point, whereas $x = 3$ is an unstable critical point.
12. $x(t) = 2 + (x_0 - 2)/\{1 + 2t(2 - x_0)^2\}^{1/2}$; $x = 2$ is a stable critical point.
14. $x(t) = ax_0/\{bx_0 + (a - bx_0)e^{-at}\}$ **15.** $x(t) = ax_0e^{-at}/\{(a - bx_0) + bx_0e^{-at}\}$ **16.** (a) $M = (a - k)/b$

SECTION 7.2

1. $(0, 0)$ **2.** All points on the line $x = 2y$ **3.** $(-2, -3)$ **4.** $(0, 0)$ **5.** $(0, 0)$ and $(3, 2)$ **6.** $(0, 0)$ and $(1, 2)$
7. All points of the form $(n\pi, 0)$ where n is an integer   **8.** $(0, 0)$ **9.** $x(t) = 0, 2, -2$ **10.** $x(t) = 0$
11. $x(t) = n\pi$ where n is an integer   **12.** $x(t) = 0$
13. The origin $(0, 0)$ is a stable proper node, similar to the one shown in Fig. 7.4.
14. Here, $(0, 0)$ is an unstable saddle point; reverse all arrows in Fig. 7.6 to see the trajectories.
15. The origin is a stable improper node, similar to the one shown in Fig. 7.5, except that the trajectories consist of the x-axis and parabolas of the form $x = ky^2$.
16. The origin is an unstable improper node.   **17.** The origin is a stable center.   **18.** The origin is a stable center.
19. The origin is a stable center.   **20.** The origin is an asymptotically stable spiral point.
23. The origin and the circles $x^2 + y^2 = C > 0$ **24.** The origin and the hyperbolas $y^2 - x^2 = C$
25. The origin and the ellipses $x^2 + 4y^2 = C > 0$ **26.** The origin and the ovals of the form $x^4 + y^4 = C > 0$

SECTION 7.3

1. Asymptotically stable node   **2.** Unstable improper node   **3.** Unstable saddle point   **4.** Unstable saddle point
5. Asymptotically stable node   **6.** Unstable node   **7.** Unstable spiral point   **8.** Asymptotically stable spiral point
9. Stable, but not asymptotically stable, center   **10.** Stable, but not asymptotically stable, center
11. Asymptotically stable node: $(2, 1)$ **12.** Unstable improper node: $(2, -3)$ **13.** Unstable saddle point: $(2, 2)$
14. Unstable saddle point: $(3, 4)$ **15.** Asymptotically stable spiral point: $(1, 1)$ **16.** Unstable spiral point: $(3, 2)$
17. Stable center: $(\frac{5}{2}, -\frac{1}{2})$ **18.** Stable (but not asymptotically stable) center: $(-2, -1)$ **19.** Asymptotically stable node
20. Unstable improper node   **21.** Unstable saddle point
22. Here, $(0, 0)$ is either a center or a spiral point, but its stability is not determined by Theorem 2.
23. Asymptotically stable spiral point   **24.** Unstable spiral point   **25.** Asymptotically stable node or spiral point
26. Either an unstable node or an unstable spiral point
27. Here, the origin is either a center or a spiral point, but its stability is not determined by Theorem 2.
28. The origin is either a center or a spiral point, but its stability is not determined by Theorem 2.
29. The origin is an unstable saddle point. The point $(1, 1)$ is either a center or a spiral point, but its stability is indeterminate.
30. There is an unstable saddle point at $(1, 1)$ and an asymptotically stable spiral point at $(-1, 1)$.
31. There is an unstable saddle point at $(1, 1)$ and an asymptotically stable spiral point at $(-1, -1)$.
32. There is an unstable saddle point at $(2, 1)$ and an asymptotically stable spiral point at $(-2, -1)$.
36. Note that the differential equation is homogeneous.

SECTION 7.4

1. Eigenvalues: $-2, -3$; stable node   **2.** Eigenvalues: $1, 3$; unstable node   **3.** Eigenvalues: $-3, 5$; unstable saddle point
4. Eigenvalues: $-1 + 2i, -1 - 2i$; stable spiral point
5. Critical points: $(0, n\pi)$, n an integer; an unstable saddle point if n is even, a stable spiral point if n is odd
6. Critical points: $(n, 0)$, n an integer; an unstable saddle point if n is even, a stable spiral point if n is odd
7. Critical points: $(n\pi, n\pi)$, n an integer; an unstable saddle point if n is even, a stable spiral point if n is odd
8. Critical points: $(n\pi, 0)$, n an integer; an unstable node if n is even, an unstable saddle point if n is odd
10. Unstable saddle points at $(2, 0)$ and $(-2, 0)$, a stable center at $(0, 0)$
11. Unstable saddle points at $(2, 0)$ and $(-2, 0)$, a stable spiral point at $(0, 0)$
12. Stable centers at $(2, 0)$ and $(-2, 0)$, an unstable saddle point at $(0, 0)$
13. A stable center at $(0, 0)$, an unstable saddle point at $(-4, 0)$
14. Stable centers at $(2, 0)$, $(0, 0)$, and $(-2, 0)$, unstable saddle points at $(1, 0)$ and $(-1, 0)$

SECTION 7.5

3. At $(0, 0)$ the linearized system is $x' = 60x$, $y' = 42y$, with general solution $x(t) = Ae^{60t}$, $y(t) = Be^{42t}$.
4. At $(0, 14)$ the linearized system has eigenvalues $\lambda_1 = 4$ and $\lambda_2 = -42$.
5. At $(20, 0)$ the linearized system has eigenvalues $\lambda_1 = 2$ and $\lambda_2 = -60$.

6. At (12, 6) the linearized system has the two negative eigenvalues $-27 + 3\sqrt{73}$ and $-27 - 3\sqrt{73}$.

8. At (0, 0) the linearized system is $x' = 5x$, $y' = -2y$, with general solution $x(t) = Ae^{5t}$, $y(t) = Be^{-2t}$.

9. Substitute $u = x - 5$, $v = y$, then show that the linearized uv-system has eigenvalues -5 and 3.

10. Substitute $u = x - 2$, $v = y - 3$, then show that the linearized uv-system has eigenvalues $-1 + i\sqrt{5}$ and $-1 - i\sqrt{5}$.

13. At (0, 0) the linearized system has general solution $x(t) = Ae^{-2t}$, $y(t) = Be^{-4t} = kx^2$.

15. Substitute $u = x - 2$, $v = y$. **16.** Substitute $u = x - 3$, $v = y - 1$. **20.** Approximately 200.43 days

SECTION 7.6

1. $y_0 = 3$, $y_1 = 3 + 3x$, $y_2 = 3 + 3x + \frac{3}{2}x^2$, $y_3 = 3 + 3x + \frac{3}{2}x^2 + \frac{1}{2}x^3$, $y_4 = 3 + 3x + \frac{3}{2}x^2 + \frac{1}{2}x^3 + \frac{1}{8}x^4$; $y(x) = 3e^x$

3. $y_0 = 1$, $y_1 = 1 - x^2$, $y_2 = 1 - x^2 + \frac{1}{2}x^4$, $y_3 = 1 - x^2 + \frac{1}{2}x^4 - \frac{1}{6}x^6$, $y_4 = 1 - x^2 + \frac{1}{2}x^4 - \frac{1}{6}x^6 + \frac{1}{24}x^8$; $y = \exp(-x^2)$

5. $y_0 = 0$, $y_1 = 2x$, $y_2 = 2x + 2x^2$, $y_3 = 2x + 2x^2 + \frac{4}{3}x^3$, $y_4 = 2x + 2x^2 + \frac{4}{3}x^3 + \frac{2}{3}x^4$; $y(x) = e^{2x} - 1$

7. $y_0 = 0$, $y_1 = x^2$, $y_2 = x^2 + \frac{1}{2}x^4$, $y_3 = x^2 + \frac{1}{2}x^4 + \frac{1}{6}x^6$, $y_4 = x^2 + \frac{1}{2}x^4 + \frac{1}{6}x^6 + \frac{1}{24}x^8$; $y(x) = \exp(x^2) - 1$

9. $y_0 = 1$, $y_1 = (1 + x) + \frac{1}{2}x^2$, $y_2 = (1 + x + x^2) + \frac{1}{6}x^3$, $y_3 = (1 + x + x^2 + \frac{1}{3}x^3) + \frac{1}{24}x^4$; $y(x) = 2e^x - 1 - x = 1 + x + x^2 + \frac{1}{3}x^3 + \cdots$

11. $y_0 = 1$, $y_1 = 1 + x$, $y_2 = (1 + x + x^2) + \frac{1}{3}x^3$, $y_3 = (1 + x + x^2 + x^3) + \frac{2}{3}x^4 + \frac{1}{3}x^5 + \frac{1}{9}x^6 + \frac{1}{63}x^7$; $y(x) = 1/(1 - x) = 1 + x + x^2 + x^3 + \cdots$

12. $y_0 = x$, $y_1 = 1 + \frac{1}{2}x$, $y_2 = 1 + \frac{1}{2}x + \frac{3}{8}x^2 + \frac{1}{64}x^4$, $y_3 = 1 + \frac{1}{2}x + \frac{3}{8}x^2 + \frac{5}{16}x^3 + \frac{13}{64}x^4 + \cdots$; $y(x) = (1 - x)^{-1/2}$

13. $\begin{bmatrix} x_0 \\ y_0 \end{bmatrix} = \begin{bmatrix} 1 \\ -1 \end{bmatrix}$, $\begin{bmatrix} x_1 \\ y_1 \end{bmatrix} = \begin{bmatrix} 1 + 3t \\ -1 + 5t \end{bmatrix}$, $\begin{bmatrix} x_2 \\ y_2 \end{bmatrix} = \begin{bmatrix} 1 + 3t + \frac{1}{2}t^2 \\ -1 + 5t - \frac{1}{2}t^2 \end{bmatrix}$, $\begin{bmatrix} x_3 \\ y_3 \end{bmatrix} = \begin{bmatrix} 1 + 3t + \frac{1}{2}t^2 + \frac{1}{3}t^3 \\ -1 + 5t - \frac{1}{2}t^2 + \frac{5}{6}t^3 \end{bmatrix}$

14. $\mathbf{x}(t) = \begin{bmatrix} e^t + te^t \\ e^t \end{bmatrix}$ **16.** $y_3(1) \approx 0.350185$

INDEX

A

Abel's formula, 116
Abramowitz, Milton, 56, 179, 269, 534
Acceleration, 12
    gravitational, 14
Addition of matrices, 367
Advanced Calculus, 525
*A First Course in Partial Differential
        Equations*, 535
Age of universe, 51
Air resistance, 89
Airy equation, 272
Almost linear system, 483, 490
American Mathematical Monthly, 248
Amperes, 194
Amplification factor, 185, 189
Amplitude, 140
    time-varying, 144
Analytic function, 213
*An Introduction to Ordinary Differential
        Equations*, 226, 240, 534
Annihilator, 158
Applied Linear Algebra, 535
Argument of complex number, 134
Artin, Emil, 263
Asymptotically stable critical point, 468,
        477
Asymptotic approximations, 270
*A Treatise on the Theory of Bessel
        Functions*, 261
Autonomous equation, 466

Autonomous system, 472
Auxiliary equation, 127

B

Balmer series, 277
Baseball example, 459
Beam:
    deflection, 15
Beats, 184, 199
Bernoulli, Daniel (1700–1782), 261
Bernoulli equation, 66
Bessel, F. W. (1784–1846), 261
Bessel function identities, 266
Bessel function of order p:
    first kind, 264
    second kind, 266
Bessel function of order zero:
    first kind, 244
    second kind, 258
Bessel's equation, 171, 212, 243, 261
    order zero, 244, 257, 317
    order l, 258
    parametric, 268
Bifurcation point, 471
Binomial series, 213
Birkhoff, G., 436, 528, 534
Birth rate, 81
Black hole, 101
Bohr, Niels (1885–1962), 278
Borrelli, R. L., 277

Boundary and Eigenvalue Problems in Mathematical Physics, 535
Boundary conditions, 17
Boundary value problem, 201
Brauer, F., 534
Braun, M., 534
Broughton Bridge, 186
Brown, J. W., 534
Buckled column or rod, 208, 272
Buoy, 147

C

Cable (hanging), 167
Calculus and Analytic Geometry, 92, 463
Calculus of Variations, 535
Cantilever, 17
Capacitor, 193
Cart-with-flywheel system, 181
Cascade of tanks, 62
Catenary, 169
Cat safe in tree, 171
Cauchy-Schwarz inequality, 533
Center, 477, 487
   of series, 213
Characteristic equation, 127, 395, 483
Characteristic value, 202, 395
Characteristic vector, 395
Churchill, R. V., 293, 319, 320, 321, 534
Circular drumhead, 261
Circular frequency, 140, 144
Clairaut equation, 71
Clepsydra, 51
Closed trajectory, 479
Coddington, E. A., 226, 240, 534
Coleman, C. S., 277
Column of matrix, 367
Column vector, 368
Competing species, 509
Competition equations, 510
Complementary function, 124, 379
Complete eigenvalue, 403
Complex exponential function, 132
Complex roots, 131, 133
Complex-valued function, 132
Compound interest, 44
Computation of exp (At), 422
Conte, S. D., 534
Continuous dependence of solutions, 531
Convolution of functions, 314
Cooling and heating, 47

Cooling law, 3, 47
Coulomb potential, 279
Coulombs, 194
Crescent of glittering steel, 277
Criterion for exactness, 74
Critical buckling force, 209
Critical damping, 142, 143
Critical point, 466, 473
   asymptotically stable, 468, 477
   isolated, 482
   stable, 467, 476
   unstable, 468, 476
Critical speed, 208
Cumulative error, 430
Current:
   steady periodic, 195
   transient, 195
Curve:
   curvature, 172
   deflection, 16
   solution, 17, 57

D

Damped motion, 138, 188
   nonlinear, 497, 503
Damping constant, 138
Dashpot, 138
Death rate, 81
deBoor, C., 534
Deflection of beam, 15
Deflection curve, 16
Degenerate system, 360
Delta function, 335
Density of force, 16
Derivative:
   complex-valued function, 132
   matrix-valued function, 372
Determinant, 371
   Vandermonde, 126
Determinism, 148
Differential equation, 2
Differential Equations, 535
Differential Equations and Their Applications, 534
Differential Equations, Dynamical Systems, and Linear Algebra, 535
Differential form, 72
Differential operator (linear), 358
Differentiation of power series, 215
Differentiation of transform, 316

Dirac, P. A. M. (1902–1982), 335
Dirac delta function, 335
Discrete Variable Methods in Ordinary Differential Equations, 535
Doomsday, 86
Drag coefficient, 91
Drug elimination, 45
Duhamel's principle, 341

E

Easy out, 462
Eigenfunction, 203
Eigenvalue, 202, 394
   complete, 403
   incomplete, 403
   multiplicity, 402
Eigenvalue matrix, 421
Eigenvalue method for systems, 394
Eigenvalue problem, 202
Eigenvector, 394
Eigenvector matrix, 421
Electrical resonance, 198
Element of matrix, 367
Elementary Linear Algebra, 396, 403, 421
Elementary Numerical Analysis, 534
Elimination constant, 45
Elimination method for systems, 357
Elliptic integral of first kind, 503
Endpoint conditions, 17
Endpoint value problem, 17
Energy:
   kinetic, 186
   potential, 186
Equality of matrices, 367
Equation:
   autonomous, 466
   auxiliary, 127
   Bernoulli, 66
   characteristic, 127, 483
   Clairaut, 71
   competition, 510
   differential, 2
   equidimensional, 162
   Euler-Cauchy, 162
   exact, 73
   general population, 82
   homogeneous, 64, 107
   Legendre, 171, 212, 232
   linear, 28, 107
   logistic, 82, 467

predator-prey, 507
reducible second-order, 165
Riccati, 71, 172
rocket propulsion, 98
second order, 109
separable, 25, 39
Equidimensional equation, 162
Equilibrium solution, 466, 473
Error in Euler method, 436
Euler, Leonhard (1707–1783), 261, 294
Euler buckling force, 209
Euler-Cauchy equation, 162
Euler's formula, 116, 132
Euler's method, 427
   improved, 437
   for systems, 454
Exact equation, 73
   criterion, 74
Existence of solution, 27, 56, 110, 117, 524, 527
Exponential growth equation, 45
Exponential matrix, 418
Exponential matrix solutions, 421
Exponential order, 292
Exponents of differential equation, 240
Extinction of population, 86

F

Farads, 193
Fermentation tanks, 356
First order equation, 52
First order system, 349
   homogeneous, 351
   nonhomogeneous, 351
   solution, 351
Flagpoles, 275
Flight trajectories, 68
Floating buoy, 147
Folia of Descartes, 493
Forced motion, 138, 188
Forced vibrations, 108
Formal multiplication of series, 214
Formula:
   Abel's, 116
   Euler's, 116
Fourier, Joseph (1768–1830), 261
Fourier Series, 535
Fourier Series and Boundary Value Problems, 534
Fractional mass of rocket, 99

Free motion, 138, 362
   damped, 142, 147
   undamped, 140
Free vibrations, 108
Frequency, 141, 144
   natural, 182
Fresnel integrals, 179
Frobenius, Georg (1848–1917), 239
Frobenius method, 238
Frobenius series, 239
Frobenius series solutions, 241
Fundamental matrix, 411
Fundamental matrix solutions, 412

G

Gamma function, 262
Gauss's hypergeometric equation, 248
Generalized function, 342
General population equation, 82
General solution, 11, 114, 123
   of homogeneous system, 377
Geometric series, 213
Glittering steel crescent, 277
Grandfather clock, 147
Gravitational acceleration, 14
Green's functions, 176, 181
Grouping method, 77
Gzyx, 20

H

Hailstone problem, 62
Half-life, 46
Handbook of Mathematical Functions, 56,
   179, 269, 534
Hanging cable, 167
Hard spring, 495
Harmonic motion (simple), 140
Heap of rope, 79
Heaviside, Oliver (1850–1925), 294
Henrici, P., 535
Henries, 193
Hermite equation, 234
Hermite polynomial, 234
Hirsch, M. W., 535
Hobson, D. D., 277
Hole through the earth, 147
Home run, 462
Homogeneous equation, 64, 107
Hooke's law, 137, 494

Hydrogen spectrum, 277
Hypergeometric equation and series, 248
Hypocycloid, 366
Hypozeuxis, 403

I

Identity matrix, 371
Identity principle (for power series), 216
Imaginary part, 132
Impedance, 196
Improper integral, 286
Improper node, 475, 484, 486
Improved Euler method, 437
Impulse, 333
Ince, E. L., 360, 535
Incomplete eigenvalue, 403
Indicial equation, 239
Inductor, 193
Initial condition, 4, 8
Initial position, 12
Initial value problem, 8, 110, 117
Initial velocity, 12
Insecticide, 516
Integrating factor, 52, 75, 76
Integration of transform, 318
Integrodifferential equation of RLC circuit,
   326
Intermediate Differential Equations, 535
Intermediate frequency, 199
Inverse Laplace transform, 289
Inverse of matrix, 371
Irregular singular point, 236
Isocline, 22
Isolated critical point, 482

J

Jump, 291
Jump rope, 206

K

Kansas City, 186
Keen as a razor, 277
Kepler's laws, 463
Kinetic energy, 186
Kirchhoff's laws, 194
Kutta, Wilhelm (1867–1944), 445

L

Laguerre polynomial, 283
Laguerre's equation, 280
Lake Erie, 59
Laplace, Pierre Simon de (1749–1827), 294
Laplace transform, 286
  existence, 292
  linearity, 288
  notation, 290
  of system, 352
  table, 290, 344
Laplacian, 278
Law of cooling, 3
Lebedev, N., 535
Legendre polynomial, 233
Legendre's equation, 171, 212, 232
Levinson, N., 534
Limited populations, 82
Linear Algebra and Its Applications, 535
Linear differential operator, 358
Linear equation, 28
  first order, 52
  nth order, 107
  second order, 109
Linearized system, 483
Linearly dependent functions, 112, 119, 375
Linearly independent functions, 112, 119, 375
Linear systems, 483
  and matrix exponentials, 418
Lipschitz continuous function, 523
Local error, 430
Local existence of solutions, 529
Logarithmic decrement, 148
Logistic equation, 82, 467
Lunar lander, 13, 96, 457

M

Macabre classic, 277
McLachlan, N. W., 535
MACSYMA, 248
Mann, W. R., 525
Many-term recurrence relation, 230
Mass-spring-dashpot system, 108
Mathematical modeling, 5
Mathematics Magazine, 277
Matrix, 367
  addition, 367
  determinant, 371
  equality, 367
  exponentials, 418
  fundamental, 411
  identity, 371
  inverse, 371
  multiplication, 369
  nonsingular, 371
  order, 371
  singular, 371
  square, 368
  subtraction, 368
  transpose, 368
  zero, 367
Matrix-valued function, 372
  derivative, 372
Mechanical-electrical analogy, 194
Method of elimination for systems, 357
Method of Frobenius, 238
Method of grouping, 77
Method of reduction of order, 159
Method of successive approximations, 519
Method of undetermined coefficients, 149
  for systems, 409
Method of variation of parameters, 173
  for systems, 414
Mexico City, 186
Mixture problems, 58
Modulus of complex number, 134
Motion:
  simple harmonic, 140
Multiplication:
  matrices, 369
  series, 214
Multiplicity of eigenvalue, 402

N

Natural frequency, 182, 363
Natural growth and decay, 43
Natural growth equation, 45
Natural mode of oscillation, 363
Nether extremity, 277
Newton:
  law of cooling, 3, 47
  law of gravitation, 94
Noble, B., 535
Node, 475
Nohel, J., 534
Nonelementary function, 427
Nonhomogeneous equation, 107, 123
Nonlinear pendulum, 499

Nonsingular matrix, 371
*n*th order equation, 7, 107
Numerical Methods, Software, and Analysis: IMSL Reference Edition, 535

O

Ohms, 193
One-parameter family of solutions, 6
Operational determinant, 359
Operational Mathematics, 293, 319, 320, 321, 534
Operator, 123
  linear differential, 358
Orbital, 278
Order (reduction), 159
Order of equation, 7
Order of matrix, 371
Ordinary Differential Equations, 360, 436, 528, 534
Ordinary equation, 8
Ordinary Non-Linear Differential Equations in Engineering and Physical Sciences, 535
Ordinary point, 225
Oscillations (undamped), 182
Overdamped motion, 143
Over the top, 502, 504

P

Parameters (variation of), 173
Parametric Bessel equation of order *n*, 268
Partial differential equation, 8
Partial fraction decomposition, 306
Particular solution, 11
Pendulum, 139, 147
  nonlinear, 499
  steadily descending, 277
Penney, Dr. Carol W., 79
Period, 141, 327
Periodic function, 327
  transform, 327
Periodic solution, 189
Phase angle, 140
Phase plane, 472
Phase portrait, 469, 474
Phenylthiourea, 51
Physical units, 14

Picard, Emile (1856–1941), 519
Piecewise continuous function, 290
Piecewise smooth function, 296
Planck's constant, 278
Poe, Edgar Allen (1809–1849), 277
Polar form, 134
Polynomial operator, 129
Population growth, 43, 81
  competition situation, 85
  doomsday, 86
  extinction, 86
  joint proportion situation, 85
  limited environment, 85
  U.S., 83
Position function, 12
Potential energy, 186
Power series, 212
  convergence, 212
Power series method, 214
Power series representation, 212
Predator-prey situation, 506
Predictor-corrector method, 438
Principal diagonal of matrix, 371
Principle of superposition, 109, 117, 155, 375
Probability amplitude function, 278
Proper node, 475, 486
Pseudoperiod, 144
Pursuit problem, 169

Q

Qualitative Theory of Ordinary Differential Equations, 534

R

Radio, 199
Radioactive decay, 44
Radiocarbon dating, 44
Radius of convergence, 218
Rainville, E., 535
Reactance, 196
Real part, 132
Recurrence relation, 217, 230
Reducible second-order equations, 165
Reduction of order, 159
Regular singular point, 236
Repeated roots, 129
  complex, 134

Resistor, 193
Resonance, 185
  electrical, 198
Resonance frequency, 198
Riccati equation, 71, 172, 276
Rice, J. R., 535
RLC circuit, 193
  integrodifferential equation, 326
Rocket propulsion, 97
Rodrigue's formula, 234
Rope heap, 79
Rota, G.-C., 436, 528, 534
Roundoff error, 432
Row of matrix, 367
Row vector, 368
Runge, Carl (1856–1927), 445
Runge-Kutta method, 445
  for systems, 455
Ruth, George Herman ("Babe," 1895–1948), 463
Rutherford, Ernest (1871–1937), 278
Rydberg constant, 277, 282

S

Saddle point, 475, 485
Sagan, H., 535
Scalar product of vectors, 369
Schrödinger equation, 278
Schwarz, Laurent, 342
Second order linear equations, 109
Separable equation, 25, 39
Separatrix, 495
Shift of index of summation, 217
Simmons, G. F., 535
Simple harmonic motion, 140
Simple pendulum, 139
Sine integral function, 55
Singular matrix, 371
Singular point, 225, 236
Singular solution, 42
Skalskaya, I. P., 535
Skydiver, 449, 453
Skywalk, 186
Slope field, 17
Smale, S., 535
Snowplow problem, 51, 52
Soft spring, 495
Solution curve, 17, 57, 472
Solution of equation, 7, 21
  equilibrium, 466

general, 11, 41, 114, 123
  implicit, 41
  nonhomogeneous case, 124
  particular, 11, 41
  singular, 42
  steady periodic, 189
  transient, 189
Solutions near an ordinary point, 226
Solutions of nonhomogeneous systems, 379
Special Functions and Their Applications, 535
Speed:
  critical, 208
  terminal, 90
Spiral point, 479, 487
Spring constant, 137
Square matrix, 368
Stability:
  of almost linear system, 490
  of linear system, 488
Stable critical point, 467, 476
Stacked polynomial, 443
Static equilibrium position, 139
Steady period current, 195
Steady periodic solution, 189, 385
Stegun, Irene A., 56, 179, 269, 534
Stirling's approximation, 62
Strang, W. G., 535
String (whirling), 206
Stokes' drag law, 148
Stonehenge, 47
Subtraction of matrices, 368
Successive approximation method, 1519
Superposition principle, 109, 117, 155, 375
Swimming dog, 72
System:
  first order linear, 373
  general solution, 377
  nonhomogeneous, 379
  solution, 374
Systems analysis, 341

T

Table of Laplace transforms, 290, 344
Taylor, A. E., 525
Taylor series, 213
Taylor series method, 441 order k, 443
Terminal speed, 90

Termwise addition of series, 214
Termwise inverse transformation of series, 321
The Gamma Function, 263
"The Pit and the Pendulum," 277
Theorem:
  complex roots, 133
  computation of exp($\mathbf{A}t$), 422
  continuous dependence of solutions, 531
  convolution property, 315
  criterion for exactness, 74
  differentiation on transforms, 316
  distinct real roots, 128
  eigenvalue solutions of systems, 395
  error in Euler method, 436
  exceptional cases (Frobenius method), 255
  existence and uniqueness, 27, 110, 117, 351
  existence for linear system, 527
  existence of Laplace transform, 292
  exponential matrix solutions, 421
  Frobenius series solutions, 241
  $F(s)$ for s large, 293
  fundamental matrix solutions, 412
  general solutions, 114, 123, 377
  global existence of solution, 524
  identity principle, 216
  integration of transforms, 318
  linear first order equation, 56
  linearity of Laplace transform, 288
  local existence of solutions, 529
  principle of superposition, 109, 117, 375
  radius of convergence, 218
  reduction of order, 160
  repeated roots, 131
  solution in Bessel functions, 271
  solution near an ordinary point, 226
  solutions of nonhomogeneous equations, 124
  stability of almost linear systems, 490
  stability of linear systems, 488
  termwise differentiation of power series, 215
  transforms of derivatives, 296, 298
  transforms of integrals, 301
  transforms of periodic functions, 327
  translation on the s-axis, 306
  translation on the t-axis, 322
  uniqueness of inverse Laplace transform, 294
  uniqueness of solutions, 529
  variation of parameters, 176, 415
  Wronskians of solutions, 113, 122, 376
Theory of Ordinary Differential Equations, 534
Thermal diffusivity, 8
Three-term Taylor series method, 442
Time-independent Schrödinger equation, 278
Time lag, 196
Time-varying amplitude, 144
Tolstov, G. P., 535
Torricelli's law, 3, 48
Trajectory:
  aircraft, 68
  of autonomous system, 472
  closed, 479
  projectile, 356
Transfer function, 341
Transform of periodic functions, 327
Transient current, 195
Transient solution, 189, 385
Translation on the s-axis, 306
Translation on the t-axis, 322
Transpose of matrix, 368
Two-axle automobile, 389
Two-term recurrence relation, 230
Two very long equations, 242

U

Ulfyand, Y. S., 535
Undamped motion, 138, 182
Underdamped motion, 144
Undetermined coefficients, 149
  for systems, 409
Unicycle model of car, 187
Uniqueness of inverse Laplace transform, 294
Uniqueness of solution, 27, 56, 110, 117, 529
Unit impulse response, 342
Unit step function, 291, 322
Unit step response, 342
Unstable critical point, 468, 476
U.S. population data, 83

V

V-2 rocket, 102
Vandermonde determinant, 126

Variable gravitational acceleration, 94
Variation of parameters, 173
  for systems, 414
Vector, 368
  scalar product, 369
Velocity, 12
Verhulst, Pierre-François (1804–1849), 83
Vertically suspended chain, 261
Very long equation, 250
Viscosity measurement, 148
Voltage drops, 194
Volterra, Vito (1860–1940), 506
Volts, 194

W

Washboard road, 187
Water clock, 51

Watson, G. N. (1886–1965), 261
Weight, 14
Weight function, 341
Weinberger, H. F., 535
Weinstock, R., 535
Whirling string, 206
Why flagpoles are hollow, 275
Worked Problems in Applied Mathematics,
  535
World population, 46
Wronski, J. M. H. (1778–1853), 120
Wronskian, 113, 120, 122, 376

Z

Zero matrix, 367